ADVANCES IN MEASUREMENT TECHNOLOGY, DISASTER PREVENTION
AND MITIGATION

Advances in Measurement Technology, Disaster Prevention and Mitigation collects papers resulting from the conference on Measurement Technology, Disaster Prevention and Mitigation (MTDPM 2022), Zhengzhou, China, 27–29 May, 2022. The primary goal is to promote research and developmental activities in measurement, disaster prevention and mitigation, and another goal is to promote scientific information interchange between scholars from the top universities, business associations, research centers and high-tech enterprises working all around the world.

The conference conducts in-depth exchanges and discussions on relevant topics such as measurement, disaster prevention and mitigation, aiming to provide an academic and technical communication platform for scholars and engineers engaged in scientific research and engineering practice in the field of measurement application, measurement in civil engineering and disaster reduction. By sharing the research status of scientific research achievements and cutting-edge technologies, it helps scholars and engineers all over the world comprehend the academic development trend and broaden research ideas. So as to strengthen international academic research, academic topics exchange and discussion, and promote the industrialization cooperation of academic achievements.

PROCEEDINGS OF THE 3RD INTERNATIONAL CONFERENCE ON MEASUREMENT TECHNOLOGY, DISASTER PREVENTION AND MITIGATION (MTDPM 2022), ZHENGZHOU, CHINA, 27–29 MAY 2022

Advances in Measurement Technology, Disaster Prevention and Mitigation

Edited by

Zongming Li
North China University of Water Resources and Electric Power, China

Mohd Johari Mohd Yusof
Faculty of Design and Architecture, Universiti Putra Malaysia, Malaysia

CRC Press
Taylor & Francis Group
Boca Raton   London   New York   Leiden

CRC Press is an imprint of the
Taylor & Francis Group, an **informa** business

A BALKEMA BOOK

First published 2023
by CRC Press/Balkema
4 Park Square, Milton Park / Abingdon, Oxon OX14 4RN / UK
e-mail: enquiries@taylorandfrancis.com
www.routledge.com – www.taylorandfrancis.com

CRC Press/Balkema is an imprint of the Taylor & Francis Group, an informa business

Library of Congress Cataloging-in-Publication Data
A catalog record has been requested for this book

ISBN: 978-1-032-36087-4 (hbk)
ISBN: 978-1-032-36088-1 (pbk)
ISBN: 978-1-003-33017-2 (ebk)

DOI: 10.1201/9781003330172

Typeset in Times New Roman
by MPS Limited, Chennai, India

Advances in Measurement Technology, Disaster Prevention and
Mitigation – Li & Mohd Yusof (Eds)
© 2023 The Editor(s), ISBN: 978-1-032-36087-4

Table of contents

Disaster prevention and intelligent disaster reduction project

Research on civil and hydraulic structure and geological characteristics

Advances in Measurement Technology, Disaster Prevention and
Mitigation – Li & Mohd Yusof (Eds)
© 2023 The Editor(s), ISBN: 978-1-032-36087-4

Preface

The 2022 3rd International Conference on Measurement Technology, Disaster Prevention and Mitigation (MTDPM 2022) was planned to be held in Zhengzhou, China from May 27–29, 2022. However, it is uncertain when the COVID-19 will end, so it remains unclear for postponement time, while many scholars and researchers wanted to attend this long-waited conference and have academic exchanges with their peers. Therefore, in order to actively respond the call of the government, and meet author's request, MTDPM 2022 was organized in a virtual mode. This approach not only avoids people gathering, but also meets their communication needs.

This conference is mainly for experts, scholars and engineers from universities, research institutes, enterprises and institutions at home and abroad, providing an academic platform to share scientific research results, explore cutting-edge engineering issues, discuss opportunities and challenges, promote cooperation and exchange, and promote the industrialization of scientific research results.

This scientific event brings together more than 120 national and international researchers in Measurement Technology, Disaster Prevention and Mitigation. On top of the local participants coming from different national universities, international participants are also registered from different countries, namely Algeria, Turkey, Australia, Malaysia and India. During the conference, the conference model was divided into three sessions, including oral presentations, keynote speeches, and online Q&A discussion. In the first part, some scholars, whose submissions were selected as the excellent papers, were given about 5–10 minutes to perform their oral presentations one by one. Then in the second part, keynote speakers were each allocated 30–45 minutes to hold their speeches.

We were greatly honor to have invited two distinguished experts as our keynote speakers. Professorial Senior Engineer Zongming Li, North China University of Water Resources and Electric Power was the first one to perform his thought-provoking speech. And then we had Senior Engineer Feiyan Meng, Nanjing University of Science and Technology. Their insightful speeches had triggered heated discussion in the third session of the conference. The online discussion lasted for about 30 minutes. Every participant praised this conference for disseminating useful and insightful knowledge.

We are glad to share with you that we received lots of submissions from the conference and we selected a bunch of high-quality papers and compiled them into the proceedings after rigorously reviewed them. These papers feature following topics but are not limited to: Measurement Technology, Application and Modern Detection Technology, Measurement Technology in Civil Engineering, Measurement Applications in Construction Engineering, Hydraulic Engineering and Surveying Applications, Intelligent Disaster Reduction and Protection Engineering, etc. All the papers have been through rigorous review and process to meet the requirements of international publication standard.

Lastly, we would like to warmly thank all the authors who, with their presentations and papers, generously contributed to the lively exchange of scientific information that is so vital to the endurance of scientific conferences of this kind.

The Committee of MTDPM 2022

*Advances in Measurement Technology, Disaster Prevention and
Mitigation – Li & Mohd Yusof (Eds)
© 2023 The Editor(s), ISBN: 978-1-032-36087-4*

Committee members

Conference Chair
Professorial Senior Engineer Zongming Li, *North China University of Water Resources and Electric Power, China*

Program (Academic) Committee Chair
Assoc. Prof. Mohammad Russel, *School of Ocean Science and Technology, Dalian University of Technology, Bangladesh*

Program (Academic) Committee Member
Prof. Bachir Achour, *University of Biskra, Algeria*
Prof. Behnam, *University of Tabriz, Iran*
Prof. Haluk Akgün, *Middle East Technical University, Turkey*
Prof. Fauziah Ahmad, *Universiti Sains Malaysia, Malaysia*
Senior Engineer Feiyan Meng, *National Quality Supervision and Inspection Centre for Building Decoration and Decoration Materials, China*
Dr. Xingbo Han, *University of New South Wales, Australia*
Dr. R. S. AJIN, *Idukki District Disaster Management Authority (DDMA) Collectorate, India*

Organizing Committee Chair
Assoc. Jianjun Ma, *Henan University Of Science And Technology, China*

Organizing Committee Member
Prof. Keren Dai, *Chengdu University of Technology, China*
Prof. Tetsuya HIRAISHI, *Kyoto University, Japan*
Assoc. Prof. Mohammadreza Vafaei, *Universiti Teknologi Malaysia, Iranian*
Dr. Zhe Yang, *Hainan Tropical Ocean University, China*
Dr. Sadegh Rezaei, *Shargh-e Golestan Institute of Higher Education, Iran*

Publication Chair
Prof. Sudip Basack, *MAKA University of Technology, Kolkata, India*
Professorial Senior Engineer Zongming Li, *North China University of Water Resources and Electric Power, China*
Assoc. Prof. Ts. Gs. Dr. Mohd Johari Mohd Yusof, *Universiti Putra Malaysia, Malaysia*
Assoc. Prof. Ziyan Zhang, *Hainan Tropical Ocean University, China*

Keynote lectures

Advances in Measurement Technology, Disaster Prevention and
Mitigation – Li & Mohd Yusof (Eds)
© 2023 The Author(s), ISBN: 978-1-032-36087-4

Construction and whole-process management for urban lifeline projects

Zongming Li

North China University of Water Resources and Electric Power

ABSTRACT: Urban lifeline projects mainly include urban pipeline networks, road traffic, electric power, communications, gas, heat, water supply and drainage. Due to rapid economic development, most modern urban infrastructures are overloaded, and the performance of materials and structures is fatigued or degraded due to aging, corrosion, and repeated loading. Besides, the scarcity of primary data could also lead to vulnerability of these infrastructures under disaster conditions.

Advances in Measurement Technology, Disaster Prevention and Mitigation – Li & Mohd Yusof (Eds)
© 2023 The Author(s), ISBN: 978-1-032-36087-4

Research on WPC-based key technologies

Feiyan Meng
Nanjing University of Science and Technology

ABSTRACT: WPCs have beautiful appearance of wood products, with advantageous features such as antibacterial effect, anti-corrosion, and recyclability, making them an optimal choice in manufacturing. We believe that through studying key technologies of advanced processing and manufacturing processes, our project will achieve a breakthrough in high performance and cost performance of WPC products. This could help us establish a technology platform to support the development of WPC industry, as well as its related industries, including building materials, automobile manufacturing, logistics and transportation, and furniture manufacturing. This project could also generate substantial economic benefits for WPC manufacturers. It can, for instance, effectively utilize natural wood fiber and improve reasonable utilization of wood resources. In addition, it can also reduce environmental pollution, contributing to the coordination of man and nature, and to the national economy as well as to sustainable social development.

DOI 10.1201/9781003330172-2

*Measurement technology and
data predictive risk assessment*

Measurement technology and intercomparability assessment

Advances in Measurement Technology, Disaster Prevention and Mitigation – Li & Mohd Yusof (Eds)
© 2023 The Author(s), ISBN: 978-1-032-36087-4

Temporal and spatial characteristics of satellite remote sensing precipitation data in Naqu, Tibet

Jiarui Hong, Jing Zhang* & Yongyu Song
Beijing Laboratory of Water Resources Security, Beijing, China
Key Laboratory of 3D Information Acquisition and Application of Ministry of Education, Capital Normal University, Beijing, China

ABSTRACT: Due to the special geographical and topographic conditions of mountainous alpine areas, the hydro-meteorological monitoring stations are scarce, which makes the measured data less representative. In order to study the hydrological patterns in the plateau region where data are severely scarce, this paper compares and evaluates the consistency, detection capability and spatial and temporal characteristics of the measured precipitation and the multi-source precipitation datasets using the Naqu watershed on the Qinghai-Tibet Plateau, China as the study area. The results show that Multi-Source Weather (MSWX) and the China Meteorological Assimilation Driving Datasets (CMADS) are better than Global Precipitation Climatology (GPCP) and the Precipitation-PERSIANN Climate Data Record (PERSIANN-CDR), among which MSWX is the best. On the whole, MSWX shows better precipitation inversion accuracy in the Naqu Basin and has certain applicability. The distribution of precipitation in the Naqu Basin is relatively uneven among months, and the spatial distribution of annual precipitation in the multi-source precipitation datasets varies greatly.

1 INTRODUCTION

The watershed area of Naqu is 16822 km^2, which is the source area of the Nujiang River. Runoff is mainly derived from precipitation, and the inter-annual variation and regional distribution of runoff are quite closely related to the variation of precipitation (Gong 2019). However, there is only one meteorological station in the basin, and there is a lack of long-series precipitation information. The local livestock industry is the mainstay of development, and the economic level is backward, so water management needs to improve pasture production to promote economic development. The scarcity of hydrological information alone to measure precipitation in the area is inadequate to support water resource management.

Since the accuracy of multi-source precipitation datasets can vary significantly in different regions due to multiple factors such as topography, climate, estimation algorithms, and the number of actual measurement stations (Ma 2021). Luo et al. (2011) compared and analyzed the applicability of TRMM-3B42 with station information in the region of China and concluded that the applicability is higher in the eastern region. Deng et al. (2018) evaluated the accuracy and reliability of MSWE precipitation products in China and found that there is an overall overestimation phenomenon, and the agreement between MSWEP and the measured data is the lowest in the southwest region due to the influence of local factors such as topography. Sheng et al. (2021) used GPM-IMERG to build and calibrate the down-scaling model, which can better reflect the precipitation characteristics in the humid region. Chen et al. (2022) selected five precipitation products to evaluate the applicability to the Yuanjiang-Hongjiang River basin, and the study found that the accuracy of most precipitation products was related to the elevation.

---

*Corresponding Author: 5607@cnu.edu.cn

DOI 10.1201/9781003330172-3

Therefore, based on the Naqu Basin with few stations, this paper will evaluate the multi-source precipitation data set, avoid the calibration instability caused by the scarcity of stations and the large random error of a single station, and further improve the precipitation estimation accuracy in the area with few stations on the Qinghai Tibet Plateau. It is hoped that it can provide scientific reference for selecting appropriate precipitation data in areas lacking data and enhance the understanding of the temporal and spatial distribution characteristics of precipitation in the basin.

2 STUDY AREA AND DATA COLLECTION

2.1 *Overview of the study area*

In this study, the Naqu watershed in the upper reaches of the Nujiang River was selected as the study area, which is located between $91° \sim 92°$E, $30° \sim 32°$, with an average elevation above 4500 m. The climate type in the basin mainly belongs to the highland sub-cool monsoonal semi-humid climate zone. The annual precipitation distribution in the Naqu Basin is extremely uneven, concentrated from June to September, accounting for 80% of the year, and its multi-year average annual precipitation is about 550 mm (Liu 2017). The geographical location of the Naqu River Basin is shown in Figure 1. It is a typical area with a serious lack of data, and the precipitation observation data of the stations are extremely scarce, and there is only one Dasa-Naqu station in the basin.

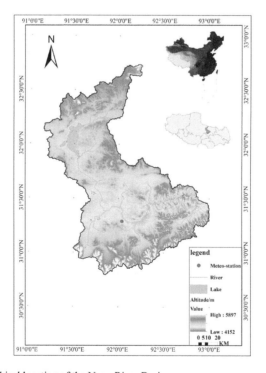

Figure 1. The geographical location of the Naqu River Basin.

2.2 *Data sources*

In this paper, the data from the national conventional meteorological stations will be taken as the benchmark for assessing and analyzing the quality of other precipitation numbers. There is only one meteorological station in the Naqu Basin, although it can only represent the precipitation values in a small area and is the most direct and reliable data source at present (Meng & Wang 2017;

Meng 2008). Four widely used multi-source precipitation datasets with high accuracy worldwide are collected, their applicability in the Naqu Basin is compared and analyzed, and the characteristics of each data set are shown in Table 1.

Table 1. Characteristics of multi-source precipitation datasets.

No	Short Name	Coverage Area	spatial resolution	Time-series
1	CMADS	East Asia	1/3°	2008–2019
2	GPCP	Global	1°	1998–2018
3	MSWX	Global	0.1°	1984–2021
4	PERSIANN-CDR	Global	0.25°	1990–2021

2.3 *Methodology*

Considering the differences between the time scales, spatial resolutions, sequences and accesses of the four multi-source precipitation datasets, the study will preprocess the data and interpolate the spatial resolutions using the inverse distance weighting (IDW) method to unify them to $0.1° \times 0.1°$, while selecting the period 2008–2017 as the study period. IDW is the most popular geometric method, and due to its simple nature, it has been applied to different spatial and temporal scales (Kurtzman et al. 2009).

$$W_i = \frac{h_i^{-p}}{\sum_{i=1}^{n} h_i^{-p}} \tag{1}$$

$$h_i = \sqrt{(x - x_i)^2 + (y - y_i)^2} \tag{2}$$

$$Wi = \frac{(\frac{R-h}{Rh})^2}{\sum_{j=1}^{n} (\frac{R-h}{Rh})^2} \tag{3}$$

p is the weight index, usually $p = 2$; h_i is the distance from the discrete point to the interpolation point; (x, y) is the coordinates of the interpolation point; (x_i, y_i) is the coordinates of the discrete point; R is the distance from the interpolation point to the farthest discrete point; n is the number of discrete points.

Four basic statistical indexes, root mean square error (RMSE), correlation coefficient (CC), relative bias (bias) and mean error (me), are used to quantitatively evaluate the performance and applicability of four multi-source precipitation data sets in the Naqu Basin. Three classified statistical indexes, probability of detection (POD), false alarm rate (far) and critical success index (CSI), are used to measure the detection ability of precipitation data set for precipitation events of moderate rain and above in Naqu Basin (Zeng & Yong 2021).

Table 2. Calculation index calculation formula.

Indicators	Formula	Optimum value
Basic statistical indicators	$RMSE = \sqrt{\frac{1}{n}\sum_{i=1}^{n}(S_i - G_i)^2}$	0

(continued)

Table 2. Continued.

Indicators	Formula	Optimum value
	$CC = \dfrac{\sum_{i=1}^{n}(G_i-\overline{G})(S_i-\overline{S})}{\sqrt{\sum_{i=1}^{n}(G_i-\overline{G})^2} * \sqrt{\sum_{i=1}^{n}(S_i-\overline{S})^2}}$	1
	$BIAS = \dfrac{\sum_{i=1}^{n}(S_i-G_i)}{\sum_{i=1}^{n}G_i}$	0
	$ME = \dfrac{\sum_{i=1}^{n}(S_i-G_i)}{N}$	0
Classification statistics index	$POD = \dfrac{H}{H+M}$	1
	$FAR = \dfrac{F}{F+H}$	0
	$CSI = \dfrac{H}{F+H+M}$	1

S_i is the precipitation of the multi-source data set; G_i is the actual precipitation measured at the weather station; n is the number of samples.

H indicates the number of days when both weather stations and precipitation datasets monitored moderate rain and above; F indicates the number of days when only precipitation datasets monitored moderate rain and above; M indicates the number of days when only weather stations monitored moderate rain and above.

3 RESULTS AND DISCUSSION

3.1 *Consistency analysis of multi-source precipitation datasets*

In this paper, the four multi-source precipitation datasets are compared with the meteorological stations within the Naqu Basin as a benchmark for one-dimensional linear analysis. According to Table 3, it can be found that the CC values of the four precipitation datasets are low, and there is no obvious linear relationship. Comparing the RMSE, only GPCP has a value greater than 5 mm/d in the four precipitation datasets, indicating its worst accuracy, and MSWX has a minimum RMSE of 2.897 mm/d, indicating its better accuracy. The BIAS of MSWX and CMADS datasets are smaller at 4.5% and –5.1%, respectively, of which only CMADS has a negative BIAS value, indicating that the daily-scale CMADS dataset is underestimated. ME is consistent with the indicator trend and positive and negative cases of BIAS.

Table 3. Comparison of basic statistical indexes of multi-source precipitation datasets.

Index	GPCP	PERSIANN-CDR	CMADS	MSWX
RMSE	5.149	4.033	3.186	2.897
CC	0.265	0.364	0.495	0.504
BIAS	0.541	0.464	−0.051	0.045
ME	0.712	0.610	−0.067	0.059

In summary, it can be seen that MSWX reflects the best indicator situation and has the highest accuracy, followed by CMADS, which also has higher accuracy. The consistency of MSWX in the Naqu region, where the stations are sparse, is higher than in the rest of the datasets, proving a better representation. This is consistent with Peng Zhenhua's study statistics found that MSWX in the Tibetan Plateau region performed significantly better than other multi-source precipitation datasets (Peng et al. 2019).

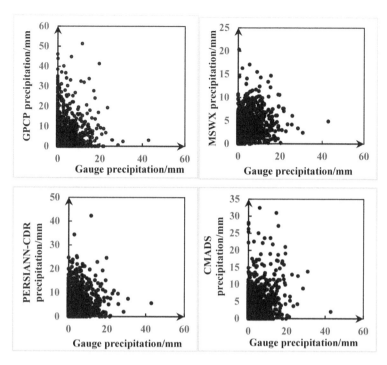

Figure 2.   Scatter comparison between multi-source precipitation data set and measured precipitation of meteorological stations.

3.2   *Comparison of precipitation level detection performance*

According to the classification standard of precipitation intensity level specified by the National Meteorological Administration, the daily precipitation is divided into five levels, as shown in Table 4 below. Statistical multi-source precipitation data sets in different precipitation intensity occurrence frequency and frequency error are compared and analyzed to show the detection capability.

Table 4.   Classification standard of precipitation intensity level.

Precipitation intensity (mm/d)	No rain	Light rain	Moderate rain	Heavy rain	Rainstorm
Precipitation (mm)	<0.1	0.1~9.9	9.9~24.9	24.9~49.9	49.9~100

Figure 3 shows the frequency distribution of occurrence of actual precipitation data from meteorological stations and four multi-source precipitation data sets at different precipitation intensities. According to Figure 3, it can be seen that all four precipitation data sets have different degrees of underestimation for no rain prediction and different degrees of overestimation for light rain prediction. For the occurrence frequency of moderate rainfall, GPCP and PERSIANN-CDR are more frequent than the measured ones, and CMADS and MSWX are lower than the measured ones. The frequency of GPCP is over-represented in the frequency of occurrence of heavy rainfall and heavy rainfall, while the frequency of MSWX is zero in both, indicating that it underestimates the frequency of occurrence of precipitation intensity at heavy rainfall and above.

Combined with the statistical indicators of the detection ability of medium rainfall and above in Table 5, it can be seen that the difference of FAR of the four multi-source precipitation datasets is not large, among which the FAR of CMADS is relatively small, which indicates that the probability of the datasets being incorrectly predicted for the measured precipitation is small. The probability of precipitation being correctly detected for medium rainfall and above is small, and the POD of CMADS is the highest but only 0.33. CSI can reflect the comprehensive index of correct detection ability, and similar to POD, CMADS has the best effect of 0.162.

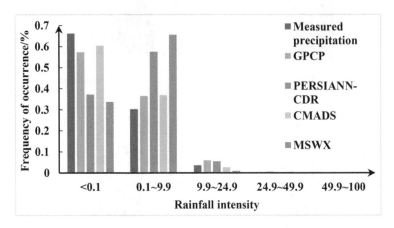

Figure 3.   Frequency distribution of multi-source precipitation data set under different precipitation intensity.

Table 5.   Comparison of statistical indicators of detection ability of multi-source rainfall data set to detect moderate and above precipitation.

	FAR	POD	CSI
GPCP	0.880	0.120	0.083
PERSIANN-CDR	0.870	0.130	0.086
CMADS	0.670	0.330	0.162
MSWX	0.781	0.219	0.045
Optimum value		1	1

In summary, the combination of the three statistical indicators shows that the multi-source datasets are not effective in detecting medium rainfall and above, and the daily precipitation in the Naqu River basin has a significant mismatch with the daily precipitation in the four precipitation datasets, and the CMADS is relatively effective in comparison. This may be because the precipitation datasets reflect regional averages, resulting in a regional "normalization" of heavy precipitation, which is confirmed by the TRMM satellite (Li et al. 2012).

3.3   *Spatial-temporal characteristics*

The inter-annual variation curves of the actual and multi-source precipitation datasets at the meteorological stations are shown in Figure 4. According to Figure 4, it can be seen that the trends of the five curves are generally consistent, but there are some differences in the magnitude of the values. The variability of the measured precipitation curves at the meteorological stations is greater than that of the four multi-source precipitation datasets, with a slight jump in inter-annual variability in 2015. As seen in Figure 5, the distribution of inter-annual precipitation in the Naqu Basin is more uneven each month, mainly concentrated in May-September, and its rainy season begins with the shift of atmospheric circulation from winter to summer (Zou et al. 1995). The precipitation from November to February is extremely low.

The spatial distribution of multi-year average precipitation in the Naqu Basin is shown in Figure 6. As a whole, there are large differences in the spatial distribution of annual precipitation in the Naqu Basin for each data set, with GPCP and PERSIANN-CDR gradually increasing from north to south, and the southeast direction is the maximum precipitation area with annual precipitation of 900~1000 mm. the difference between the MSWX maximum value is smaller than that of GPCP.

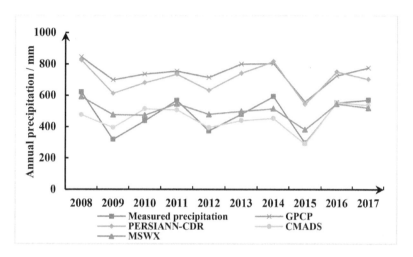

Figure 4.   Inter-annual variation of measured precipitation and multi-source precipitation datasets.

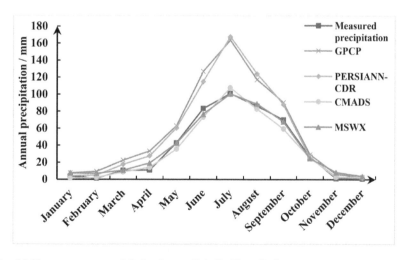

Figure 5.   Multi-year average precipitation by month in the Naqu Basin.

The CMADS precipitation distribution is more consistent with the glacier distribution (Liu et al. 2017). In addition, topography and local circulation also become important factors affecting precipitation (Li & Li 1992; Zhang & Liu 2018).

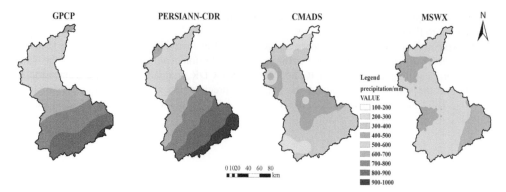

Figure 6. Spatial distribution of multi-year average precipitation in the Naqu Basin from multi-source precipitation datasets.

4 CONCLUSION

In this paper, four multi-source precipitation datasets of the Naqu River basin from 1998-2017 are compared with the corresponding measured precipitation at meteorological stations. The statistical results show that MSWX and CMADS outperform GPCP and PERSIANN-CDR in each precipitation class, and MSWX is more consistent than the remaining three multi-source datasets. The detection ability to effectively identify real precipitation events is evaluated based on the three statistical indicators. The multi-source datasets are not effective in detecting medium rainfall and above. There is an obvious mismatch between the daily precipitation in the Naqu River basin and the daily precipitation in the four precipitation datasets. CMADS is relatively better in comparison. According to the spatial and temporal distribution maps of the four multi-source precipitation datasets, it can be seen that the inter-annual precipitation distribution is consistent with the perception of flood and non-flood periods, but the distribution of precipitation in each month is more uneven, mainly concentrated in May-September. From the overall perspective of the basin, the spatial distribution of annual precipitation in the Naqu Basin from multi-source precipitation datasets varies greatly due to the effects of topography, local circulation and glacier distribution.

In summary, based on the precipitation characteristics and accuracy performance of the datasets, it is concluded that MSWX is more consistent with the remaining three multi-source datasets and shows better inverse precipitation accuracy and applicability. In future studies, the precipitation characteristics and precipitation phase identification in highland areas can be further analyzed to calibrate the multi-source datasets. Coupled with the driven distributed hydrological model coupled with the snow-melt module to simulate and predict the runoff in the Naqu Basin, it is beneficial to understand the hydrological cycle in the Naqu Basin to support water resources management and economic development.

ACKNOWLEDGMENTS

This research was funded by the National Key R&D Program of China (2017YFC0406004) and the NSFC (41271004).

REFERENCES

Chen YD, Wang J, Ma XC, et al. (2022) Artemisia Meng,Wu Sushu,Hu Xiaodong. Adaptability assessment of five precipitation data products in the Yuanjiang-Hongjiang River Basin[J]. *Hydropower Energy Science*, 40(03): 9–12+16.

Deng Y, Jiang WG, Wang XY, et al. (2018) Accuracy assessment of MSWEP precipitation products in mainland China region[J]. *Advances in Water Science*, 29(04): 455–464.

Gong BY. (2019) Analysis of water sources in the Naqu watershed based on natural geographical features[D]. *China Institute of Water Resources and Hydropower Research*.

Kurtzman D, Navon S, Morin E. (2009) Improving interpolation of daily precipitation for hydrologic modeling: Spatial patterns of preferred interpolators [J]. *Hydrological Processes: An International Journal*, 23(23): 3281–3291.

Li SJ, Li SD. (1992) Quaternary glaciation and environmental evolution in the Cocosili region of Qinghai[J]. *Glacial Permafrost*, 1992(04): 316–324.

Li XH, Zhang Q, Xu CY. (2012) Suitability of the TRMM satellite rainfalls in driving a distributed hydrological model for water balance computations in Xinjiang catchment, Poyang lake basin[J]. *Journal of Hydrology*, 426: 28–38.

Liu J, Liu SY, Shangguan DH, et al. (2017) Evaluation of applicability of three precipitation datasets CMADS, ITPCAS and TRMM 3B42 in Yulongkashi River Basin[J]. *Journal of North China University of Water Resources and Hydropower* (Natural Science Edition), 38(05): 28–37.

Liu SH. (2017) Evolution of the water cycle in the upper Nujiang River basin and its response to climate change[D]. *China Institute of Water Resources and Hydropower Research*.

Luo S, Miao JF, Niu T, et al. (2011) Comparative analysis of TRMM rainfall measurement product 3B42 and station data in the Chinese region[J]. *Meteorology*, 37(09): 1081–1090.

Ma X. (2021) Accuracy evaluation of different types of precipitation products in different subdivisions of China and applicability study of drought monitoring[D]. *Northwest Agriculture and Forestry University of Science and Technology*.

Meng XY, Wang H. (2017) Significance of the China Meteorological Assimilation Driving Datasets for the SWAT Model (CMADS) of East Asia. *Water*. 9(10), 765.

Meng XY. (2008) Establishment and Evaluation of the China Meteorological Assimilation Driving Datasets for the SWAT Model (CMADS)[J]. *Water*, 10(11) : 1555–1555.

Peng ZH, Li YZ, Yu WJ, et al. (2019) Study on the applicability of remote sensing precipitation products in different climate zones in China[J]. *Journal of Geoinformation Science*, 23(07): 1296–1311.

Sheng X, Shi YL, Ding HY. (2021) Spatial downscaling of GPM precipitation data on the Qinghai-Tibet Plateau[J]. *Remote Sensing Technology and Applications*, 36(03): 571–580.

Zeng XK, Yong B. (2021) Accuracy assessment of Global Precipitation Program IMERG and GSMaP inversion precipitation in Sichuan region[J]. *Journal of Geography*, 74(07): 1305–1318.

Zhang Z, Liu SY. (2018) Remote sensing monitoring study of glacier changes and material balance in Gangzari, Qinghai-Tibet Plateau, 1970–2016[J]. *Journal of Geoinformation Science*, 20(09): 1338–1349.

Zou JS, Zhang DQ, Wang L. (1995) Precipitation characteristics and long-term climate change in Tibet[J]. *Journal of Nanjing University* (Natural Science Edition), (04): 691–695.

Advances in Measurement Technology, Disaster Prevention and Mitigation – Li & Mohd Yusof (Eds)
ISBN: 978-1-032-36087-4

Design of a tension measuring instrument for cable structure

Yue Hua*, Ting-Ting Liu*, Tian Shi* & Guo-Hui Chen*
Xi'an Institute of Space Radio Technology, Xi'an, China

ABSTRACT: In order to accurately obtain the tension of flexible cable net structure on Spaceborne mesh antenna with high precision, a tension measuring instrument is designed. The instrument eliminates the influence of the three-point bending of the rope on the measurement results through two actions and realizes the determination of the tension value of the fixed pretension rope at both ends. By building a laboratory prototype, the tension measurement system is designed, and the application test is carried out, which has a certain application value.

1 INTRODUCTION

Spaceborne mesh antenna reflector has the characteristics of lightweight, easy folding, high storage ratio and large aperture, which provides an important solution for the demand for spaceborne large aperture antenna.

According to the different support forms and deployment driving modes of the flexible reflection network, it can be divided into the annular reflector, umbrella reflector, frame reflector and so on. Its core is to form a stable support structure through the deployment and locking of the deployment mechanism with high stiffness, and the rope connected to its connection point forms a rope grid. By designing the rope grid, the reflective surface grid meeting the use accuracy is realized. The reflective network is fixed to the node of the rope grid or the middle position of the rope system, and finally, the surface accuracy is guaranteed through the accuracy of the rope grid.

The accuracy of the rope grid is determined by the material characteristics of the rope and the tension characteristics after installation. The core of reflector profile accuracy lies in the design and realization of rope tension. In order to realize the preset tension in engineering, Chen Lu et al. studied the principle and method of cable tension measurement in cable structure (Chen et al. 2006) and mainly determined the tension of cable through vibration method, three-point bending method and elastic magnetism method. Li Zhao et al. studied the static tension measurement method of anchor lifting and dropping streamer applied to the marine operation and adopted the working principle of maintaining the balance between brake force and cable tension (Li & Chen 2017). Xie Xin et al. studied the analysis process and measurement method of the tension of precision flexible cable transmission. The measurement principle is the improved three-point bending method. Its core is to invert the real-time tension on the flexible cable through the measurement of the bending stiffness of the flexible cable, which greatly improves the measurement accuracy of the tension in the cable (Xie et al. 2018).

In this paper, a tension measuring instrument is proposed. The equipment adopts the three-point bending method. Through the operation of two measuring straight lines, the problem that it is difficult to eliminate the additional deformation caused by the substitution of the instrument in the one-time three-point bending method is eliminated, and the equipment has high measurement accuracy.

---

*Corresponding Authors: hysl2002@163.com; 1316518575@qq.com; 1044533303@qq.com and 409020581@qq.com

        DOI 10.1201/9781003330172-4

2 PRINCIPLE OF CABLE TENSION MEASUREMENT WITH QUASI RIGID CONSTRAINTS AT BOTH ENDS

The test principles are shown in Figures 1 and 2.

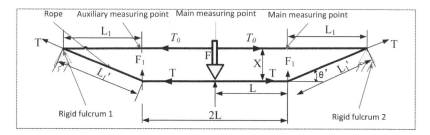

Figure 1. Test principle (step 1).

In the figure, T_0 is the initial tension of tension wire rope. F is the load exerted by the main measuring point on the tension mesh rope. F_1 is the vertical component of the load applied by the auxiliary measuring point to the tension net rope. T is the tension of the tension net rope when the instrument is operating in the established state. L is the projection distance between the main measuring point and the auxiliary measuring point. L_1 is the projection distance between the rope restraint point and auxiliary measuring equipment. L_1' is the distance between the rope restraint point and auxiliary measuring equipment.

Mechanical conditions:

$$F_1 = \sin \theta' \cdot T = \frac{X}{L_1'} \cdot T \tag{1}$$

F_1 is the vertical component of the load exerted by the auxiliary measuring point on the tension net rope (N); θ' is the included angle between the position after the tension net rope is offset and the position before it is not offset during measurement (°). T is the tension of the tension net rope when the measuring instrument is in the established state of operation (N).

Physical conditions:

$$T = \frac{EA \cdot \Delta L'}{(2L_1 + 2L - \Delta L_0)} \tag{2}$$

In the formula, E is the elastic modulus of tension net rope (N/m^2), A is the cross-sectional area of the rope (m^2). $\Delta L'$ is the total elongation of the rope when the test instrument is established. ΔL_0 is the pre-elongation of the rope before measurement.

The elongation of the rope factor test instrument when it is in the established state is:

$$\Delta L' = \Delta L_0 + \Delta L_C \tag{3}$$

Geometric deformation conditions:

$$\Delta L_C = 2 \left(L_1' - L_1 \right) \tag{4}$$

Geometric conditions:

$$L_1^2 + X^2 = \left(L_1' \right)^2 \tag{5}$$

In the formula, ΔL_c is the additional elongation generated by the instrument when the test instrument is in the established state.

The first step of the measurement process belongs to the construction state of the tension measuring instrument. The measuring instrument is in contact with the measured object, resulting in the rope offset x, but the x value is difficult to determine. At this time, we fix the measuring instrument, start the measuring operation, and move the main measuring point forward DX.

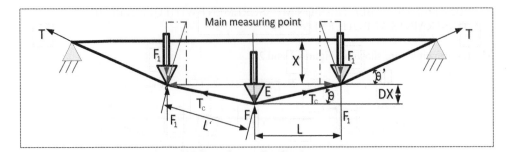

Figure 2.   Test principle (step 2).

In the figure, T_c is the tension of the tension net rope when the main tension measuring point is in the working state; L' is the distance between the main measuring point and auxiliary measuring point; θ' is the included angle between the position of the tension net rope after offset and before offset during measurement; θ is the angle between the line formed by the rope in the working state of the main tension measuring point and the contact point between the auxiliary measuring point and the tension net rope and the two auxiliary measuring points when the test instrument is established after the measurement operation is started.

Step 2 of the measurement method is the formal measurement operation. $\Delta L''$ refers to the total elongation of the measured object after starting the measurement operation; $\Delta L'_C$ is the additional elongation during the measurement operation of the test instrument; X is the offset distance of the rope caused by the measuring device when the influence of the established state operation of the measuring instrument is caused; DX refers to the distance that the main measurement point moves forward after the measurement operation is started.

According to the mechanical conditions:

$$F = \sin\theta \cdot T_c = \frac{DX}{L} \cdot T_c \tag{6}$$

Physical conditions:

$$T_C = EA \cdot \Delta L'' \tag{7}$$

The elongation of the rope due to the measurement operation is:

$$\Delta L'' = \Delta L' + \Delta L'_C \tag{8}$$

Geometric deformation conditions:

$$\Delta L'_C = 2\left(L' - L\right) \tag{9}$$

Geometric conditions:

$$L^2 + (DX)^2 = \left(L'\right)^2 \tag{10}$$

Initial physical condition:

$$T_0 = \frac{EA \cdot \Delta L_0}{2L_1 + 2L_2 - \Delta L_0} \tag{11}$$

Combining formula (1), formula (2) and formula (5), we have:

$$X = \frac{F_1 \cdot (2L_1 + 2L - \Delta L_0) \cdot \sqrt{L_1^2 + X^2}}{EA \cdot \Delta L'} \tag{12}$$

Combining formula (3), formula (4) and formula (5), we have:

$$\Delta L_0 = \Delta L' - 2\left(\sqrt{L_1^2 + X^2} - L_1\right) \tag{13}$$

18

From formula (6) to formula (10), we have:

$$\Delta L' = \frac{L}{DX \cdot EA} \cdot F - 2\left(\sqrt{(L)^2 + (DX)^2} - L\right) \tag{14}$$

Substitute formula (14) into formula (12) to obtain:

$$X = \frac{F_1 \cdot (2L_1 + 2L - \Delta L_0) \cdot \sqrt{L_1^2 + X^2}}{EA\left[\frac{L}{DX \cdot EA} \cdot F - 2\left(\sqrt{L^2 + (DX)^2}\right) - L\right]} \tag{15}$$

X can be obtained from formula (15). We bring formula (14) and x value into formula (13) to obtain ΔL_0, and substitute ΔL_0 into formula (11) to obtain the initial tension T_0.

Where L, L_1, DX, F, F_1 can be measured by the testing instrument.

3   DESIGN PRINCIPLE OF MEASURING EQUIPMENT

On the basis of theoretical research, a corresponding instrument for measuring the pretension of two-end restraint rope is developed. This design is different from the traditional three-point bending rope measuring instrument. It will not lead to excessive changes in the initial shape of the rope and excessive additional stress in the test process.

The hardware design is divided into two parts: test instrument design and data processing instrument. The structure of the measuring equipment is shown in Figure 3. Three measuring contacts of equal length are designed. Among them, the auxiliary measuring contacts on both sides have high axial and radial stiffness. The radial deformation generated in the test process is small, which has little impact on the calculation of the geometric deformation of the rope and can be ignored. A force sensor along the axial direction is arranged inside the contacts at both ends, which can directly measure the axial load.

The middle main measuring point is a combined measurement, which can measure the linear displacement and axial load at the same time.

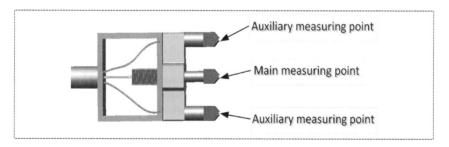

Figure 3.   Structure diagram of test equipment.

During the test, the test equipment must have high stiffness support to reduce its own deformation in the test process, so as to reduce the error caused by the overall deformation or translation of the test system.

The hardware system design block diagram of the data acquisition and processing end of the measuring equipment is shown in Figure 4 (Zhang et al. 2015).

When the test system works, the power to drive the main and auxiliary measuring heads to make a linear motion is provided by the stepping motor, which is driven by the control system to complete the forward or backward action as required. In order to reduce the measurement error caused by the gap between the actuating link mechanism of the main and auxiliary measuring points, the

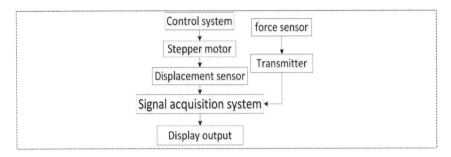

Figure 4.   Hardware system design block diagram.

linear motion value here is given by the displacement sensor, and the linear displacement value converted by the stepping motor only plays the role of controlling the motion. The force sensor at the front end of the stepping motor obtains the force value generated by the measuring point on the tension rope of the measured object and realizes its digital display and communication through the transmitter. The measured signal of the displacement sensor and the force value signal of the force sensor is summarized, and the tension value is calculated and displayed by using the program in combination with equation (11), equations (13), (14) and (15).

4   MEASUREMENT TEST

Spaceborne high-precision mesh antenna is a prestressed system composed of a cable net and support structure with high rope tension requirements. During installation, the rope tension shall be controlled by repeated measurement. In order to verify the accuracy of the measuring equipment, a verification test is designed, as shown in Figure 5. By connecting the springs at both ends of the tension rope in series, the spring shape variable is measured synchronously in the tension measurement process. The two groups of test data are compared to verify the accuracy of the measurement results of the measuring equipment.

At the beginning of the measurement, the tensiometer needs to be checked first. The probe length of the three groups of main and auxiliary measuring points is consistent with the theoretical require-ments. During measurement, the main measuring point will move axially. Before measurement, zero and calibrate the measuring instrument, and adjust the three probes to the same plane. We adjust the position and support of the measuring equipment so that there is a relatively good posi-tion relationship between the measuring instrument and the measured object. The support needs to have high overall stiffness, which can reduce the measurement error caused by stiffness. Then we adjust the measuring equipment close to the measured part, place the measuring equipment within 2 mm from the measured rope and fix it so that the head of the measuring equipment has high support stiffness. Generally speaking, the maximum deformation of the test equipment caused by the reverse force applied by the tested object to the test equipment shall not exceed 1/10 of its own deformation; otherwise, the measurement result error is large.

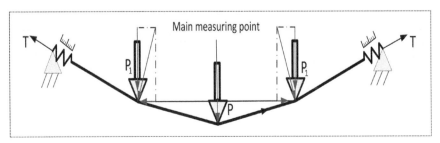

Figure 5.   Verification of accuracy of Pre-tension cable net measurement system.

It turns on the test equipment and controls the computer. The purpose of clearing the sensing quantity of the sensor is to eliminate the initial deviation. The difference between the left and right measuring heads shall be less than 0.2 g. If the error exceeds this value, it is necessary to manually adjust the placement angle between the measuring instrument and the measured object. Finally, we input the constants of the test system, the distance between the auxiliary measuring heads, the maximum value of the linear travel of the main measuring head, etc.

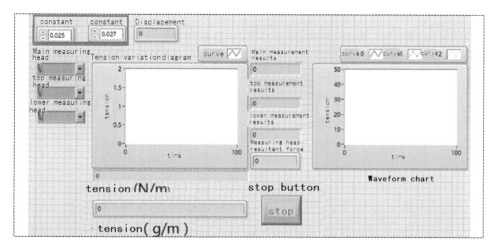

Figure 6. Terminal interface of measurement system.

Step1: Select the leveling of the auxiliary measuring head, and the leveling speed is ≤ 1mm / s. stop when the force value difference after the leveling of the auxiliary measuring head is ≤ 0.05 N and the force value is ≤ 0.5 N.

Step2: Select the leveling of the main measuring head, the leveling speed is ≤ 1 mm/s, the middle measuring head presses the measured tension rope steadily forward, observe the measurement reading of the main measuring head, and stop when the force value of the auxiliary measuring head is ≤ 1 n. After the measurement is completed, the tension value of the measured object displayed on the measurement interface is the actual tension value.

Step3: Record the data, select the "reverse" option, start the motor and return the main measuring head to the starting position. Repeat the measurement according to the above steps, take the average value, save the data and exit the program.

5 CONCLUSION

Aiming at the difficult problem of tension measurement of pretension cable net structure of space-borne high-precision mesh antenna, a tension measurement instrument for pretension cable is developed in this paper. The measurement equipment is developed by establishing the measurement principle of the measurement equipment, and the test verification of the test equipment is carried out. The following conclusions can be drawn:

1) In the process of tension measurement of pretension rope, the smaller the additional tension introduced by the measurement system, the higher the accuracy of tension measurement value;
2) Under the framework of the existing measurement principle, the measurement accuracy can be improved by increasing the spacing between the sub-measuring heads;
3) The construction of a contact measurement system is complex, which is unsuitable for many rapid measurement conditions. It can be used as an inspection measurement link of high-precision and convenient measurement equipment in engineering.

REFERENCES

Chen Lu, Zhang Qi-lin, Wu Ming-er, *Industrial Construction*[J]. Principle and of Measuring Cable Tesion in a Cable Structures, 2006(1): 368–371.

Li Zhao, Chen Qiang. *Ship&Ocean Engineering*[J]. Measuring Method of Static Tension Mooring Line for Large-tonnage Anchor Handling/Towing Winch. 2017, 36(1): 138–140.

Xie Xin, Qi Chao, Jiang Xian-liang et al. *Optics and Precision Engineering*[J].Tension analysis and measurement of precision cable drive. 2018, 26(10): 2423–2429.

Zhang Qiao, Ma Xiao-fei, Yan Xiang-cheng. *Computer Measurement & Control*[J]. Cable Tensility Testing Method Based on Non-contact Measurement. 2015, 23(12): 3987–3989.

Advances in Measurement Technology, Disaster Prevention and Mitigation – Li & Mohd Yusof (Eds)

Practical application of radar wave testing technology in soil cavity damage detection

Liyang Wan*

First Water Conservancy Engineering Bureau of Henan Province, Zhengzhou, Henan, China

ABSTRACT: The existing buildings will be affected by rainstorms, floods and other disasters during use. A large amount of rainwater scouring leads to the change in the current situation of the structural soil, which also leads to the change in the service load of the buildings and the overpressure failure of the waterproof structural layer of the soil adjacent structural surface. It will bring irreversible damage to the safety performance and service performance of buildings in use. After the flood, it is of great significance to find out the void of the surrounding soil to restore the safety performance and normal use function of the building. The conventional soil core drilling method is not only time-consuming and laborious, but also needs huge economic support and can only detect the damage to soil point by point. At this time, the advantages of radar wave non-destructive testing technology, which is mostly used for the detection of tunnel lining, highway pavement, water conservancy anti-seepage walls and other projects, are prominent, and it has become the main technical means to detect the damage of soil mass of flood building structure. This paper introduces the relevant principles and achievement analysis methods of radar detection in detail, and introduces the radar detection technology into the field of industrial and civil buildings, which has guiding significance for the post-disaster maintenance and reinforcement of similar projects.

1 INTRODUCTION

The torrential rain disaster in Zhengzhou on July 20 has affected tens of millions of people and countless buildings and structures, which has significantly impacted people's life safety, economic and social development and national security. After the disaster, many buildings suffered structural damage, their reliability was seriously reduced, and some even lost their use function. The foundation of roads, bridges, and industrial and civil buildings is largely damaged, and the soil is washed and soaked, resulting in internal looseness, cavity, ground subsidence, house inclination and other phenomena from time to time. In order to restore production and life as soon as possible, the damaged buildings must quickly find out the damage to formulate the recovery and reinforcement plan in time. The timeliness of traditional detection methods is difficult to ensure, and the advantages of radar wave non-destructive detection technology are prominent. It plays a great role in the detection of foundation damage and saves a lot of time and cost for society in the evaluation of the safety and use of the function of building structures.

2 RADAR WAVE TEST TECHNOLOGY

The radar wave test is commonly referred to as ground-penetrating radar. Ground-penetrating radar uses the reflection of pulsed electromagnetic waves of different frequencies to detect target objects and various geological phenomena (Zeng 2005). It emits electromagnetic waves down from the surface of target objects to achieve the purpose of detection, so it is also called geological radar. The ground-penetrating radar uses the characteristics of different frequencies of high-frequency

---

*Corresponding Author: 158781617@qq.com

electromagnetic waves with different penetration and resolution to send them underground in the form of broadband short-pulses from the ground through the transmitting antenna T, return to the ground after being reflected by the stratum medium or target object, and be received by the receiving antenna R (as shown in Figure 1) and the pulse wave travel time $t = \sqrt{4z^2 + x^2}/v$. When the wave velocity V in the underground medium is known, it can be calculated according to the measured time t. The depth z of the reflector can be calculated from the above formula. In the formula, the value of X is fixed in profile detection, and the value of velocity V can be measured directly by the wide angle method. When the conductivity of the medium is very low, it can also be calculated according to the $v \approx c/\sqrt{\varepsilon}$ approximation, where c is the speed of light (c = 0.3m / ns), and ε is the relative permittivity of the electromagnetic wave propagation medium (Shen 2017). The latter can be obtained using the empirical data of similar media or direct measurement. According to this principle, the position and depth of the change of the object interface of the measured target can be detected.

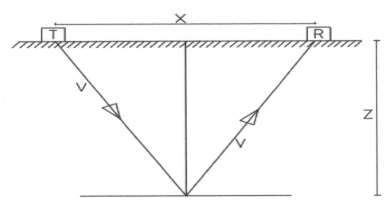

Figure 1.   Principle of reflection detection.

In the process of electromagnetic wave propagation in the medium, due to the dispersive and uneven nature of the medium, the electromagnetic wave will be absorbed or attenuated to varying degrees. So, when the pulse electromagnetic wave is received by the receiving antenna, the electromagnetic wave energy is reduced, the wave amplitude is reduced, and the waveform changes greatly. In addition, the interference of various circuit pipelines or vibration at the detection site will also be received by the receiving antenna, resulting in the distortion of the measured waveform. Therefore, the received signal must be properly processed to improve data quality. Radar data is usually recorded in the form of a waveform when a pulse reflected wave is received. Reflected waves are represented by white, gray or gray scales, respectively. In this way, the in-phase axis or equal gray and equal color lines can intuitively show the position characteristics of the measured medium interface (Zhong et al. 2010), as shown in Figure 2.

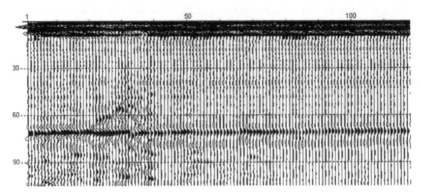

Figure 2.   Typical radar profile.

The pulse radar wave propagates in the measured medium. As long as there is an electrical differ-ence in the measured medium, the radar reflected wave corresponding to the electrical difference interface could be found in the radar image section accumulated by the reflected wave group. The curve formed by connecting the same phase of adjacent reflected waves in the reflected wave group is the in-phase axis of the radar profile. In the process of radar measurement, the point distance used for data acquisition is very small. Under the condition of no sudden change in the measured medium, the characteristics of the two adjacent reflected wave groups will remain unchanged, and the reflected waveforms will be similar. Therefore, the waveform, amplitude, period and envelope of the reflected wave from the dielectric layer with similar electrical characteristics will form certain characteristics.

The data of radar wave detection results are mainly reflected in the form of radar profiles. The time profile reflected by the radar profile is the main data for analyzing the change in the dielectric layer. During the detection process, in case of breakage, crack development and large change in water content, the electrical characteristics of the medium will change, which is manifested in the obvious dislocation of the phase axis of the radar reflected wave on the time profile. The greater the change in electrical characteristics, the more obvious the in-phase axis dislocation is (Wang et al. 2017), as shown in Figure 3.

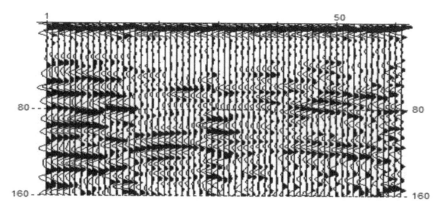

Figure 3.   Typical radar profile of in-phase axis staggering.

The development of cracks and the change of electrical characteristics in the measured soil medium are often unbalanced, so the degree of absorption and attenuation of radar waves and reflected waves is also different. The local absence of an in-phase axis in the radar profile is often caused by the inconsistent effect of the change of electrical characteristics in different parts on the absorption and attenuation of radar reflected waves, as shown in Figure 4.

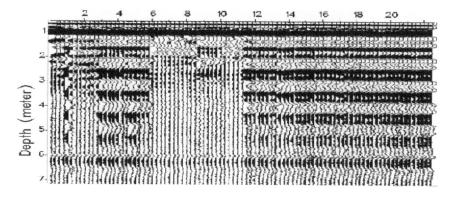

Figure 4.   Typical radar profile with partial loss of in-phase axis.

In the process of soil detection, the development of underground cracks and fissures, the content of soil composition and salinity will change the electrical characteristics of the soil. Due to its relaxation effect, absorption and attenuation of electromagnetic waves, it will sometimes cause waveform distortion or frequency reduction of radar reflected waves. At this time, it is necessary to comprehensively evaluate the reasons for the change in soil electrical characteristics in combination with soil drilling data.

3   EXAMPLES OF RADAR WAVE DETECTION

3.1   *Project overview*

The project is an underground civil air defense basement on the second floor, with a floor height of 5m. The foundation form is an integral raft foundation, the main structure form is a frame structure, and the soil covering thickness above the structural roof is 4m. It is an underground shopping mall in peacetime and a wartime shelter for civil air defense. Under the influence of a rainstorm, a large amount of soil sediment around the structure washes out. In order to detect the cavity of underground soil in time, with the consent of the project construction party, it is decided to use a radar wave detection method to detect the cavity of soil around the building.

3.2   *Detection equipment, detection technology and analysis of measured data*

SIR-20 radar produced by the American laurel company is used for this detection. The center frequency of the radar antenna is 400 MHz and 80 MHz. According to the propagation speed of electromagnetic waves in different media, we set the dielectric constant and propagation speed after calibrating the parameters on site. The 400MHz antenna is measured by manual dragging. The scanning rate is 100 scans/m, the time window length is 50 ns, the gain points are 5, the sampling points are 512, and the detection depth is 2.5m ~ 3m. The 80MHz antenna is measured by point measurement. The distance between measuring points is 0.2m, the length of the time window is 250ns, the number of gain points is 5, the number of sampling points is 512, and the detection depth is 11m ~ 12m.

GPR data processing is divided into pre-processing and processing analysis, such as marking stake numbers and adding title and identification. Its purpose is to suppress rules and eliminate random interference, display reflected waves on the GPR image section with as high resolution as possible, and highlight useful abnormal reflected signals to help interpretation. The ground-penetrating radar receives reflected waves from different underground electrical interfaces, and its correct analysis depends on factors such as reasonable selection of detection parameters, proper data processing, simulation experiment analogy and map reading experience (Ma et al. 2011).

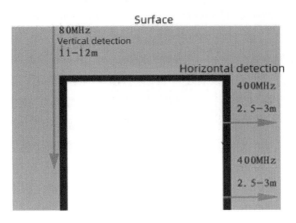

Figure 5.   Detection mode.

According to the opinions of the construction parties and the actual situation of the site, a 400 MHz high-frequency antenna is used to detect horizontally from the outer wall of the basement, and the detection depth is 2.5m–3.0 m, as shown in Figures 5 to Figure 8. An 80 MHz low-frequency antenna is used to detect vertically from the ground downward, and the detection depth is 11.0m–12.0m, as shown in Figures 9 and 10:

Typical radar profile analysis in the detection result data: the detection position is 4-7 / D of the outer wall of zone B on the second floor, 1.5m away from the bottom plate. The radar profile is shown in Figure 6.

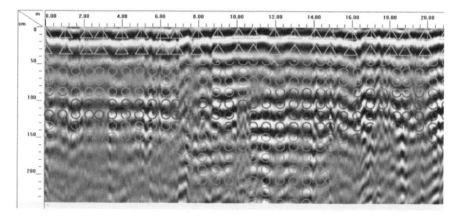

Figure 6.   Radar profile at the height of 1.5m from the bottom plate of the outer wall 4-7 / D of zone B on the second basement.

According to the measured radar profile, the grayscale of the reflected wave within the range of 0mm–500mm is obvious, clear and stable, indicating that the medium has good continuity and basically no electrical change. The field data show the grayscale display of reflected waves is obvious in the range of 50mm–100mm, and the dielectric electrical property in this area has changed. Compared with the red area, the reflected wave grayscale is obvious in the depth of 100mm–140mm, proving that a large medium change has occurred at this interface. After comprehensive study and judgment, the red circled area in the figure is a non-dense and void reflected signal. The green circle area in the figure is the loose and water-rich reflection signal. Therefore, the depth of non-dense and void parts in this area is 1.0m–1.3m. The depth of loose and water-rich parts is 0.5m-1.0m, and the depth is 1.3m–2.5m. There are local loose and water-rich phenomena.

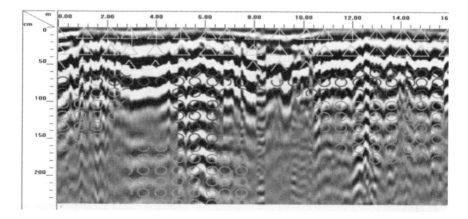

Figure 7.   Exterior wall 33–36 / D of zone B on the second basement, 1.5m from the bottom plate.

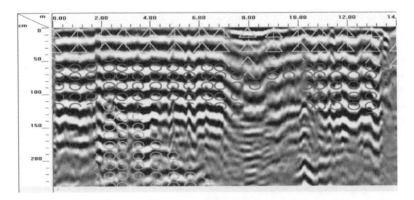

Figure 8.   Exterior wall 10-12 / A of zone B on the second basement, 1.5m from the bottom plate.

Based on this principle, the data analysis results in Figure 7 are as follows: the yellow triangle area is a 500-mm reinforced concrete outer wall. The red circled area is the non-dense and void reflection signal. The green circle area is a loose and water-rich reflection signal. The depth of non-dense and vacant parts is 0.5m–0.8m. The depth of locally loose and water-rich parts is 0.8m–2.5m. The data analysis results in Figure 8 are as follows: the yellow triangle area is a 500-mm reinforced concrete outer wall. The red circled area is the non-dense and void reflection signal. The green circle area is a loose and non-dense reflected signal. The position depth of non-dense and void parts is 0.5m–1.2m. The depth of partially loose and non-dense parts is 1.2m–2.5m.

The typical profile of radar detection from the surface down is:

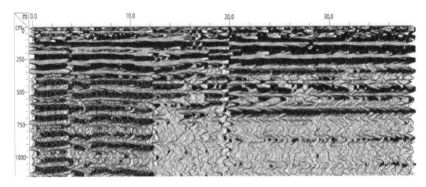

Figure 9.   13 / A-35 / A external soil mass of the exterior wall.

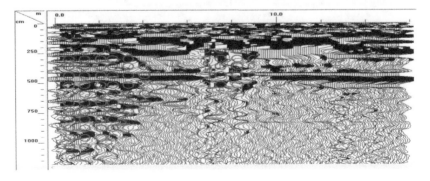

Figure 10.   38 / A-43 / A external soil mass of the exterior wall.

28

Comprehensive analysis and judgment on Figures 9 and 10: the red area 13 / A + 29560 on the left side of Figure 9 is 4.0 m-11.5 m deep, hollow and not dense. Figure 10 red area 43 / A-9750 on the left, with a depth of 2.5 m-11.0 m, empty and not dense. The middle red area 41 / A-975-41 / A-5850, with a depth of 2.5 m-5.5 m, is empty and not dense.

Through the comprehensive analysis of indoor detection data, the soil cavity results outside the exterior wall of the basement are analyzed and judged. The soil within the range of 0.5m–1.5m outside the exterior wall is loose and rich in water, and there are hollow and non-dense parts within the range of 1.5 m–2.5 m outside the exterior wall. The water-rich part will form too high water pressure on the outer wall of the basement, resulting in structural water leakage. During the field investigation, it has been confirmed that there is a large amount of water leakage and local water gushing at the construction joints of the outer wall. Through the analysis of the ground detection results, there is water-rich, void and non-dense soil in the range of 3.0 m–12.0 m below the ground outside the eastern exterior wall of the basement. Combined with the indoor detection results, it can be determined that the range of 0.5 m–1.5 m outside the outer wall of the basement and the range of 3.0 m–12.0 m below the surface are greatly affected by rainstorm scouring and there is soil loss. Its development is very easy to cause pavement collapse and affect people's life safety and social and economic development. Special reinforcement measures should be taken for the soil within this range to restore the prosperity and development of nearby businesses as soon as possible.

4 CONCLUSION

In recent years, the frequent occurrence of natural disasters has had a great adverse impact on people's production and life, and the building structure is also deeply affected. The rapid recovery of the safety and use function of buildings after disasters has become the top priority of disaster reduction. Compared with the traditional drilling technology, it will become the backbone of the advanced, convenient and non-destructive detection technology.

REFERENCES

Ma Yan, Yang Yongguo, Wang Hu, et al. Application of geological radar in the detection of the cut-off wall [J] *Urban Survey*, 2011 (3): 3

Shen Yaowei. Application of geological radar combined with seismic imaging in railway concealed karst exploration [J] *Chinese Sci-tech Journal Database (Abstract Edition) Engineering Technology*: 00298–00298, 2017

Wang Jiahui, Zhou Jianchun, XiongXuexiang. Application of comprehensive geophysical method in tunnel geological prediction [J] *Low-temperature Building Technology*, 2017, 39 (2): 4

Zeng Yuzhen. Progress of FHWA GPR bridge deck non-destructive evaluation technology [J] *Sino Foreign Highway*, 2005

Zhong Jinning, Hua Anzhong, Yang Yali, et al. Research on the application of geological radar to detect underground cavities [C] *Member Congress of Underground Pipeline of Professional Committee of China Urban Planning Association*, 2010

Advances in Measurement Technology, Disaster Prevention and Mitigation – Li & Mohd Yusof (Eds)
© 2023 The Author(s), ISBN: 978-1-032-36087-4

Analysis of passive flood inundation and assessment of ecosystem service value loss in Chittagong

Panpan Zhang
National Ocean Technology Center, Tianjin, China

Zhikun Zhang*
School of Civil and Architectural Engineering, Shandong University of Technology, Shandong, China

Xin Teng
National Ocean Technology Center, Tianjin, China

Mohammad Saydul Islam Sarkar
Department of Oceanography, University of Chittagong, Bangladesh

ABSTRACT: Chittagong city of Bangladesh has been seriously eroded by the flood disaster. The evaluation of flood loss can provide a reference for Chittagong to guide life and production and carry out post-disaster assessments. Taking Chittagong city as the research object, this paper, based on DEM data and land-use dataset, extracts the scope of flood inundation by the passive inundation analysis method and assesses the losses caused by floods to the city's ecological environment. The results show that: (1) The lands in Chittagong city were mainly used as farmlands and woodlands with a quite high land use degree; (2) there existed a quadratic polynomial growth relationship between the inundation area and the height of water level, and when the water level rose to 4m, farmlands became the main land use type in each inundated area; (3) from the perspective of ecological environment, the flood caused the greatest loss to wetlands. The relevant research results can provide effective assistance for government departments in disaster prevention.

1 INTRODUCTION

With the rapid development of society and economy and the constant increase of population, climate change has intensified the occurrence of natural disasters and brought great challenges to the national economic development, so the rapid and accurate simulation and prediction of disasters have become urgent demands for the current social development. Particularly, flood inundation analysis and prediction have been emphasized by many scholars and have become important research topics (Ansori et al. 2021; Azizian & Brocca 2020; Balogun et al. 2020). Many scholars combine GIS (Geographic Information Science) spatial analysis technology with hydrological analysis model to achieve rapid, accurate and scientific simulation and prediction of flood inundation scope (Ding et al. 2013, 2004; Karamouz & Mahani 2021; Try et al. 2020). Liu et al. (2015) classified flood inundation into "active inundation" and "passive inundation" according to the causes, which has been widely used in flood analysis since then. On this basis, Cheng (2015), in view of such defects as inefficiency and instability of active flood calculation, proposed a seed spread algorithm for active flood inundation analysis based on the buffer stack in the inundated area and then conducted relevant practice and verification. With the increasing maturity of GIS technology, some scholars have designed a more comprehensive flood inundation model system for comprehensive simulation and analysis of the whole inundated area. Luo et al. (2018) proposed a

---

*Corresponding Author: 1870506231@qq.com

DOI 10.1201/9781003330172-6

calibrated flood inundation model to evaluate the impact of extreme rainfall events on the inundation area in the center of Central Hanoi. Ao et al. (2018) achieved the macro dynamic monitoring of inundation scope and depth in river basins without referential data by virtue of multi-source remote sensing data. Sophal et al. (Try et al. 2018) proposed the rainfall-run-off inundation (RRI) model, which can simultaneously simulate rainfall, runoff and flood inundation in the Mekong River Basin. However, in reality, due to the influence of such factors as complex river geometry, large scale of river basins and coarse resolution of topographic data, the flood inundation model may produce various errors in the simulation of flood inundation scope. Borah et al. (2018), by synthetic aperture radar (SAR) data, conducted flood monitoring in Kaziranga National Park during the monsoon season in 2017, proving that the accuracy of the model adopted is improved through such data. The studies above show that in flood prevention and control, GIS technology can simulate and predict flood inundation and conduct loss assessments rapidly, accurately and scientifically.

According to the Global Risks Report 2015, Bangladesh has been identified as the sixth most disaster-prone country in the world, with the number of losses caused by floods accounting for 23.23% of the total amount of disaster losses in the country (Bangladesh Bureau of Statistics. Bangladesh Disaster-related Statistics 2015). Flood has become the most frequent and most serious natural disaster in Bangladesh.

In view of the phenomenon above, this paper, taking Chittagong city as the research object, adopts the passive inundation analysis method to rapidly and accurately analyze and predict the scope of flood inundation in Chittagong City and, in combination with relevant land use data, assesses the loss caused by flood to the ecological environment in the inundated area. This research aims to provide data support for disaster loss reduction and assessment as well as production and life guidance in Chittagong City by applying GIS-related technologies and models and providing data guarantee for promoting regional economic development under such strategic opportunities as the "Belt & Road Initiative," "Maritime Silk Road," and "Bangladesh-China-India-Myanmar Economic Corridor."

2 OVERVIEW OF THE RESEARCH OBJECT

Chittagong City is located in the hilly region of Southeastern Bangladesh, adjacent to the Bay of Bengal, with latitude and longitude of 22°18′N and 91°48′E. The city is an important junction of the "Silk Road Economic Belt" and the "21st Century Maritime Silk Road" that has the geographical advantages of connecting the east and west, linking the south and north, and bridging the land and sea. Chittagong city has a developed water system and a dense river network, with the Karnaphuli River flowing from north to south into the Bay of Bengal from the east side of the city. As to climate, Chittagong City is featured by warm and humid tropical monsoon climate with an average annual temperature of roughly 25.9°C. As to industrial distribution, the city's commercial and port areas are distributed in the southern and western parts, industrial areas in the southeastern, northeastern and northwestern parts, and residential areas in the northern part. Chittagong City contributes 80% of the volume of Bangladesh's international trade and 40% of the country's industrial output.

3 DATA SOURCES AND RESEARCH METHODS

3.1 *Data sources and processing*

In this research, Landsat 5 TM images of Chittagong city in 2019 were selected as the data source, and the data came from THE United States Geological Survey (http://glovis.usgs.gov/), with a spatial resolution of 30m. On this basis, the method of artificial visual interpretation was adopted to classify the land use types within the researched area into grassland, farmland, construction land, woodland, unused land, wetland and waters in combination with the classification standards for remote sensing interpretation and the land use characteristics of Chittagong City, and its overall classification accuracy was verified to be above 90%, which can meet the study requirements (Liu et al. 2014). The DEM data used in this research were obtained from the NASA Data Center with a resolution of 10m.

3.2 Land use degree analysis

The index of land use degree can quantitatively reveal the comprehensive level and changing trend of land use in the researched area, reflecting not only the breadth and depth of land use but also the comprehensive effects of human factors and the natural environment. Liu et al. (1992) classified land use degree into four grades according to the natural balance state of the land natural complex under the influence of social factors and assigned indices for different levels, respectively, as shown in Table 1, thus providing the quantitative expression of land use degree. The formula for the comprehensive index of land use degree is as follows:

$$L_a = 100 \times \sum_{i=1}^{n} B_i \times C_i \qquad (1)$$

wherein: L_a indicates the comprehensive index of land use degree; B_i indicates the grading index of grad i land use degree; C_i indicates the percentage of the grading area of grad i land use degree.

Table 1. Value assignment table for land use degree.

	Grade 1	Grade 2	Grade 3	Grade 4
Land-use type	Unused land	Woodland, grassland and wetland	Farmland	Construction land
Graded index	1	2	3	4

3.3 Flood inundation analysis

According to the causes of flood inundation, flood inundation can be divided into two categories: passive inundation and active inundation (Guo & Long 2002). Passive inundation means that only water level rise caused by precipitation is considered, that is, ensuring uniform precipitation within the researched area without considering the inflow of surface runoff water. Active inundation refers to flood inundation caused by river flow or reservoir burst. In the case of active inundation, such complex problems as regional connectivity, depression merging and surface runoff should be considered based on passive inundation. The two kinds of inundation analyses are both of great significance in reality. Passive inundation is mostly applicable to areas where the terrain is relatively flat, and the flood source is difficult to determine, while active inundation is mostly applicable to mountainous and hilly areas (Jin & Xiao 2014). The complexity of the active inundation model entails a relatively long calculation time, which is not conducive to rapid flood simulation. Therefore, only passive inundation is selected for analysis in this paper.

When widespread precipitation occurs in the researched area, and the distribution of precipitation amount is relatively uniform, it can be deemed that the researched area experiences passive inundation. At this time, water logging disasters will occur in areas with low elevation points within the researched area. Based on the DEM data and through the spatial analysis function of ArcGIS software, this paper extracts inundated areas with elevation points lower than water level and, after a series of processing, conducts overlay analysis in combination with the land-use dataset, finally obtaining the information about various land-use types in each inundated area.

3.4 Ecosystem service value assessment

Based on the table of ecological service equivalent per unit area improved by Xie (Xie et al. 2008), this paper draws on his research results—the economic value of each ecological service value equivalent factor is equal to 1/7 of the economic value of food production service function provided by farmland per unit area. According to the statistics released by the Bangladesh Bureau of Statistics, the average grain yield in Bangladesh in 2018–2019 was 2,571 kg/ha, and the average comprehensive grain price was $0.24 /kg. Therefore, it can be concluded that the economic value of each ecosystem service value equivalent factor in Chittagong City was $88.15 /ha/yr. After substituting this data into the improved table of ecological service equivalent per unit area, we finally obtained the table of ecosystem service value coefficients for Chittagong City, as is shown in Table 2.

Table 2. Table of ecosystem service value coefficients per unit area for each land use types in Chittagong. City

Primary classification	Secondary classification	Unit value ($/ha/yr)					
		Grassland	Farmland	Construction land	Woodland	Unused land	Wetland
Supply service	Food production	37.90	88.15	0.00	29.09	1.76	31.73
	Raw material production	31.73	34.38	0.00	262.69	3.53	21.16
Regulation service	Gas regulation	132.23	63.47	5.29	380.81	5.29	212.44
	Climate regulation	137.51	85.51	11.46	358.77	11.46	1194.43
	Water regulation	133.99	67.88	0.00	360.53	6.17	1184.74
	Waste treatment	116.36	122.53	0.00	151.62	22.92	1269.36
Support services	Soil conservation	197.46	129.58	0.00	354.36	14.99	175.42
	Biological diversity	164.84	89.91	0.00	397.56	35.26	325.27
Cultural services	Recreation	76.69	14.99	183.35	183.35	21.16	413.42
Total		1028.71	696.39	200.10	2478.78	122.53	4827.98

The total ecosystem service value can be calculated in combination with the table of ecosystem service value coefficients and the area of each land-use type, and the calculation formula is as follows:

$$ESV = \sum (A_p \times V_c) \tag{2}$$

wherein: ESV indicates the total ecosystem service value; A_p indicates the area of land use type p; V_c indicates the coefficient of ecosystem service value.

4 RESULTS AND ANALYSIS

4.1 Analysis of land use distribution characteristics and land use degree in Chittagong city

This paper conducted statistical analysis on the land use dataset of Chittagong City, and finally obtained the area and proportion of each land-use type in the research area, as shown in Table 3 and Figure 1. The chart shows that farmland was the largest land-use type in Chittagong City, accounting for 39.30%, followed by woodland and construction land, accounting for 36.20% and 16.14%, respectively. In addition, based on the area statistics, the comprehensive land use index of Chittagong City was calculated in combination with Formula (1), which was 271.43, indicating that Chittagong City was highly developed and utilized by human beings.

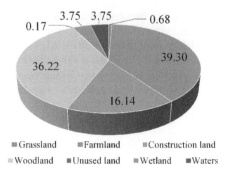

Figure 1.   Proportions of areas of various land-use types in Chittagong city.

To visualize the spatial distribution of various land-use types in Chittagong City, this paper formulated the land use distribution map (Figure 2(a)) of Chittagong City. As shown in Figure 2, land use types in Chittagong City are mainly farmland and woodland. Farmlands are mainly distributed in low-altitude plain and hilly areas in an extensive way, making the land use index of farmland the highest among all land use types in the city. The distribution of woodlands is relatively concentrated, mainly in high-altitude mountainous areas in the eastern, northern and southern parts of the city. The land use index of woodland is 72.44. Construction land is mainly concentrated in the estuary of Karnaphuli River, where human activities are the most frequent and land use degree is the highest. Wetlands are distributed in coastal areas in the form of marsh, and waters are dominated by the Karnaphuli River and its tributaries. Grasslands and unused lands are relatively small in areas without obvious distribution laws.

Table 3.   Statistics of land use areas and the comprehensive index of land use degree in Chittagong city.

	Grassland	Farmland	Construction land	Woodland	Unused land	Wetland	Waters	Total
Area (km²)	29.79	1731.02	710.99	1595.38	7.65	165.15	165.12	4405.09
Proportion (%)	0.68	39.30	16.14	36.22	0.17	3.75	3.75	100.00
Land use index	1.36	117.90	64.56	72.44	0.17	7.50	7.50	271.43

4.2   *Analysis of flood inundation in Chittagong city*

According to the application scope of inundation analysis methods and the characteristics of the research object, the passive inundation analysis method based on the water level was adopted to analyze the flood inundation in Chittagong City in combination with the DEM data. Based on the analysis of contour lines extracted from the DEM data and other relevant materials, it was concluded that the water level in Chittagong City was roughly 0.5m at low tide and 4m at high tide. Therefore, in this paper, the water level was set as 3m, 3.5m, 4m, 4.5m, 5m and 5.5m, respectively. When the water level reaches the set value, all the pixels with elevation points below the water level in the DEM grid will be extracted. In Figure 2, (a)–(f) respectively show the scope of flood inundation when the water level rises from low to high.

By virtue of software tools, this paper had the inundation scope extracted and transformed into vector data for overlay analysis in combination with the land-use dataset, based on which obtained the relationship between the area of each land use type (excluding waters) in an inundated area and the height of water level, as shown in Table 4. As can be seen from the table, when the water level is lower than 4m, the wetland is the largest land-use type in each inundated area, while when the water level exceeds 4m, farmland becomes the largest land-use type. Therefore, it can be concluded that when the water level exceeds 4m, the continuous rise in flood level will cause greater damage to the social economy. Figure 3 more intuitively shows the changing trend of the land-use area in each inundated area with the height of the water level. It can be seen from the figure that the wetland area

34

shows slow linear growth with the rise of water level, and the proportion thereof keeps decreasing. With the rise of water level, farmland area and its proportion both show increasingly high growth, similar to construction land. The inundation area of woodland is not large because of the high-altitude terrain. Grassland and unused land are less inundated because of small areas. Overall, the growth trend of the total inundation area is similar to that of farmland. It can be seen from the fitting function obtained through software calculation of the original data that the inundation area and the height of the water level show a quadratic polynomial growth relationship. Using the fitting function as a prediction function, the inundation area was calculated to be 869.05 km² when the water level reached 6m, with an error of 5.43% compared with the true value. Therefore, it can be concluded that this fitting function can well predict the changing trend of inundation areas with the height of the water level.

Table 4.   Statistics of land use area in each inundated area.

Water level (m)		Grassland	Farmland	Construction land	Woodland	Unused land	Wetland	Total
3	Area (km²)	0.38	22.53	4.57	0.67	0.91	87.18	58.13
	Proportion (%)	0.44	25.84	5.24	0.77	1.04	100	66.68
3.5	Area (km²)	0.87	50.42	9.49	1.72	1.08	134.72	71.14
	Proportion (%)	0.65	37.42	7.04	1.28	0.80	100	52.81
4	Area (km²)	1.67	104.46	18.89	4.12	1.32	217.03	86.56
	Proportion (%)	0.77	48.13	8.70	1.90	0.61	100	39.89
4.5	Area (km²)	2.62	185.71	34.67	7.75	1.71	332.99	100.53
	Proportion (%)	0.79	55.77	10.41	2.33	0.51	100	30.19
5	Area (km²)	3.58	293.94	58.42	13.93	2.19	485.70	113.64
	Proportion (%)	0.74	60.52	12.03	2.87	0.45	100	23.40
5.5	Area (km²)	5.03	413.74	88.69	23.60	2.68	655.72	121.98
	Proportion (%)	0.77	63.10	13.53	3.60	0.41	100	18.60

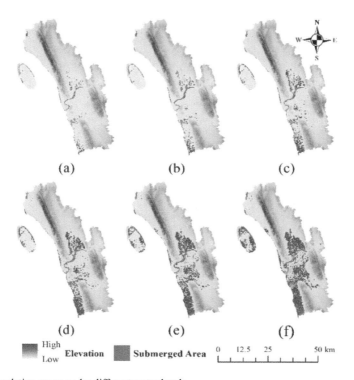

(a) (b) (c)

(d) (e) (f)

High
Low Elevation   Submerged Area   0 12.5 25   50 km

Figure 2.   Inundation scope under different water levels.

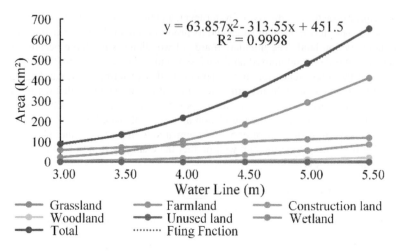

Figure 3. Changing trend of the land-use area in each inundated area with the height of water level.

4.3 Assessment of ecosystem service value loss in each inundated area

This paper assessed the losses caused by flood inundation in Chittagong City from the perspective of the ecological environment and, in combination with Formula (2), calculated the total ecosystem service value in each inundated area at different water levels, as shown in Table 5. Figure 4 shows the changing trend of ecosystem service value in each inundated area with the rise in water level. As can be seen from the table and figure below, the land use type with the highest ecosystem service value in each inundated area is wetland, although the proportion thereof has been declining. The ecosystem service value of farmland is seconded to that of wetland, which increases with the rise of water level at an increasingly fast speed. On the whole, ESV in each inundated area shows a quadratic polynomial growth relationship with the height of the water level. According to the results of prediction by the fitting function, when the water level reaches 6m, ESV in each inundated area was 1.15×108, with an error of 4.68% compared with the true value (1.10×108), indicating that the fitting function can well predict the changing trend of ecosystem service value in each inundated area with the height of water level.

Table 5. Ecosystem service value in each inundated area in Chittagong city.

Water level (m)		Grassland	Farmland	Construction land	Woodland	Unused land	Wetland	ESV
3	Value (10⁴$/ha)	3.91	156.90	9.14	16.61	1.12	2806.50	2994.18
	Proportion (%)	0.13	5.24	0.31	0.55	0.04	93.73	100
3.5	Value (10⁴$/ha)	8.95	351.12	18.99	42.64	1.32	3434.62	3857.64
	Proportion (%)	0.23	9.10	0.49	1.11	0.03	89.03	100
4	Value (10⁴$/ha)	17.18	727.45	37.80	102.13	1.62	4179.10	5065.27
	Proportion (%)	0.34	14.36	0.75	2.02	0.03	82.50	100
4.5	Value (10⁴$/ha)	26.95	1293.27	69.37	192.11	2.10	4853.57	6437.36
	Proportion (%)	0.42	20.09	1.08	2.98	0.03	75.40	100
5	Value (10⁴$/ha)	36.83	2046.97	116.90	345.29	2.68	5486.52	8035.19
	Proportion (%)	0.46	25.48	1.45	4.30	0.03	68.28	100
5.5	Value (10⁴$/ha)	51.74	2881.24	177.47	584.99	3.28	5889.17	9587.90
	Proportion (%)	0.54	30.05	1.85	6.10	0.03	61.42	100

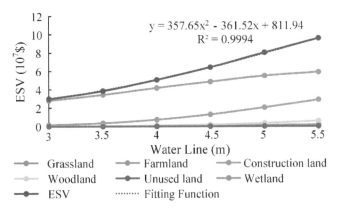

Figure 4. Changing trend of ecosystem service value in each inundated area with the height of water level.

5 CONCLUSION

Taking Chittagong city as the research object, this paper, based on the DEM data and land-use dataset, extracted the scope of inundated area in Chittagong City by the passive inundation analysis method, assessed the losses caused by floods to the city's ecological environment by the equivalent factor method, and further revealed the changing trend of land use, inundation area and ecosystem service value loss with the flood level. The main conclusions obtained are as follows: (1) The land-use types in Chittagong City were mainly farmland and woodland, accounting for 39.30% and 36.22%, respectively, and the distribution characteristics thereof were consistent with local topography. Specifically, the farmlands were mainly distributed in the plains and hilly areas with low terrain, woodlands were distributed in the mountainous areas with high terrain, and construction lands were concentrated in the estuary of Karnaphuli River. The land use degree of Chittagong City was relatively high overall. (2) There was a quadratic polynomial growth relationship between the inundation area and the height of the water level; when the water level was lower than 4m, the wetland was the main land-use type in the inundated area, while when the water level exceeded 4m, farmland became the main land use type with the area thereof increasing at an increasingly fast speed with the rise of water level. In view of this, it is suggested that Chittagong city take 3.5–4m as the early warning water level and adopt countermeasures to prevent and mitigate disaster by planting crops with different growth cycles to avoid the flood season, thus reducing property losses. (3) As can be seen from the assessment of flood losses from the perspective of the ecological environment, wetland became the land use type with the highest loss value for reasons of low topography and coastal distribution. In view of this, it is suggested that Chittagong City further protect the ecological environment of mangrove wetlands and other wetlands and, especially considering the rapid development of tourism, establish and improve relevant laws and regulations to strengthen the protection of ecological resources. As Chittagong City is the second-largest city and the largest port city in Bangladesh, the rational development and utilization of land resources and the improvement of disaster prevention and mitigation measures in the city will surely play an important role in achieving the development goal of the blue economy in the Bay of Bengal, which is conducive to giving play to the city's geographical advantage of connectivity in the "Bangladesh-China-Burma-India Economic Corridor" and promoting the healthy and sustainable development of the regional economy.

Flood inundation is typically complex and changeable in reality. In future studies, the connectivity between terrains, surface runoff and other factors may be considered to establish a more realistic flood model. Meanwhile, more social and economic data may be referred to and introduced to make assessment results more in-depth and reliable.

ACKNOWLEDGMENT

This work was also supported by the Asian Cooperation Fund "The Building of China-ASEAN Blue Partnership" Project, "Promoting Marine Spatial Planning Advancing Blue Economy Development" Maritime Silk Road Project, the National Natural Science Foundation of China (Grant No. 42171413), and the Shandong Provincial Natural Science Foundation (Grant No. ZR2020MD015).

REFERENCES

A. Azizian and L. Brocca, "Determining the best remotely sensed DEM for flood inundation mapping in data-sparse regions," *International Journal of Remote Sensing*, vol. 41, no. 5, pp. 1884–1906, 2020.

Bangladesh Bureau of Statistics. Bangladesh Disaster-related Statistics 2015 [EO/OL]. http://www.bbs.gov.bd/site/page/76c9d52f-0a19-4563-99aa-9f5737bbd0d7/-, 2022-01-31.

G. D. Xie, L. Zhen, C. X. Lu, Y. Xiao, and C. Chen, "Expert knowledge-based valuation method of ecosystem services in China," *Journal of Natural Resources*, vol. 23, no. 5, pp. 911–919, 2008

J. Cheng, "DEM-based algorithm for active flood inundation analysis," *Science & Technology Economy Market*, no. 12, p. 2, 2015.

J. Y. Liu et al., "Spatiotemporal characteristics, patterns, and causes of land-use changes in China since the late 1980s," *Journal of Geographical Sciences*, vol. 24, no. 2, pp. 195–210, 2014.

J. Y. Liu, Land use in Tibet Autonomous Region, China Science Publishing, Beijing, 1992.

L. H. Guo, and Y. Long, "Analysis of flood submerging based on DEM," *Bulletin of Surveying and Mapping*, no. 11, pp. 25–27, 2002.

M. B. Ansori, A. A. N. S. Damarnegara, N. F. Margini, and D. A. D. Nusantara, "Flood inundation and Dam break analysis for disaster risk mitigation (a case study of way apu dam)," *Geomate Journal*, vol. 21, no. 84, pp. 85–92, 2021.

M. Karamouz and F. F. Mahani, "DEM uncertainty based coastal flood inundation modeling considering water quality impacts," *Water Resources Management*, vol. 35, no. 10, pp. 3083–3103, 2021.

O. Balogun, A. Ibrahim, S. Boyi, A. A. Okewu, and I. T. Philip, "Flood Inundation Analysis of Lower Usuma River in Gwagwalada Town Abuja, Nigeria," *World Wide Journal of Multidisciplinary Research and Development*, vol. 6, no. 7, pp. 19–27, 2020.

P. Luo et al., "Flood inundation assessment for the Hanoi Central Area, Vietnam under historical and extreme rainfall conditions," *Scientific reports*, vol. 8, no. 1, pp. 1–11, 2018.

R. Y. Liu and N. Liu, "A GIS-based model for calculating of flood area," *Acta Geogrphica Sinica*, vol. 56, no. 1, p. 6, 2001.

S. B. Borah, T. Sivasankar, M. Ramya, and P. Raju, "Flood inundation mapping and monitoring in Kaziranga National Park, Assam using Sentinel-1 SAR data," *Environmental monitoring and assessment*, vol. 190, no. 9, pp. 1–11, 2018.

S. Try et al., "Projection of extreme flood inundation in the Mekong River basin under 4K increasing scenario using large ensemble climate data," *Hydrological Processes*, vol. 34, no. 22, pp. 4350–4364, 2020.

S. Try, G. Lee, W. Yu, C. Oeurng, and C. Jang, "Large-scale flood-inundation modeling in the Mekong River basin," *Journal of Hydrologic Engineering*, vol. 23, no. 7, p. 05018011, 2018.

W. Gao, Q. Shen, Y. K. Zhou, X. Li, "Analysis of flood inundation in ungauged basins based on multi-source remote sensing data," *Environmental monitoring and assessment*, vol. 190, no. 3, pp. 1–13, 2018.

Y. L. Ding, Z. Q. Du, Q. Zhu, and Y. T. Zhang, "Adaptive water level correction algorithm for flooding analysis," *Acta Geodaetica et Cartographica Sinica*, vol. 42, no. 4, p. 0, 2013.

Z. Jin, and N. N. Xiao, "Analysis of the Area of Flood based on GIS," *Jilin Water Resources*, vol. 030, no. 003, pp. 30–32, 2014.

Z. X. Ding, J. R. Li, and L. Li, "Method for flood submergence analysis based on GIS grid model," *Journal of Hydraulic Engineering*, vol. 6, pp. 56–60, 2004.

Advances in Measurement Technology, Disaster Prevention and Mitigation – Li & Mohd Yusof (Eds)
© *2023 The Author(s), ISBN: 978-1-032-36087-4*

Application of single hole acoustic testing technology in blasting excavation of reservoir dam foundation

Sen Ma*

First Water Conservancy Engineering Bureau of Henan Province, Zhengzhou, Henan, China

ABSTRACT: The quality of dam foundation rock mass is a key factor related to the life cycle and operation safety of the whole reservoir project. Blasting excavation is mostly used for the dam foundation of reservoir dams. In view of the uncontrollable divergence of blasting energy, the rock mass below the foundation surface will inevitably be affected by blasting, resulting in rock mass diseases such as breakage and cracks that are detrimental to structural safety. Therefore, after the blasting excavation of the reservoir dam foundation, it is particularly important to find out the degree and depth of the rock mass below the foundation surface affected by the blasting. The transverse discontinuity of damaged rock mass medium limits the conventional cross-hole ultrasonic nondestructive testing method. At this time, the advantages of the single-hole acoustic testing method and segmented testing technology are prominent. Using the single hole acoustic testing technology, we drill holes within 5 m below the base elevation and carry out the rock mass acoustic testing in sections; the acoustic parameter data of the rock mass in the longitudinal direction along the hole depth are obtained. The data results can reflect the rock mass characteristics in the depth direction of the dam foundation, and the results are reliable, providing a scientific and effective basis for whether to take reinforcement measures for the rock mass of the dam foundation affected by blasting in the next step.

1 INTRODUCTION

The rock mass is an aggregate of rocks with uneven medium, internal discontinuity and different mechanical properties of each structural plane formed by gradual condensation, cooling and crystallization under high temperature and pressure caused by the long geological transformation process and crustal movement. In the formation process of rock mass, it will not only bear the effects of structural change and external weathering and erosion, but also experience the destruction or transformation of unloading effects such as fracture and fracture. Therefore, there are joints, faults and other structural planes in the rock mass, resulting in the rock mass becoming an anisotropic, multi fissure and discontinuous medium. Generally, the rock mass is relatively stable without natural disturbance. However, in the actual engineering construction process, the excavation and blasting of rock mass will inevitably cause interference and damage to the natural rock mass. If the degree and depth of interference and damage cannot be clearly known, it is bound to cause infinite potential safety hazards to the construction and operation of the project. The single hole acoustic testing technology uses the law that the acoustic parameters of elastic waves, such as wave velocity, wave amplitude and frequency, will change with the changes in the structure, fracture development and density of the medium rock mass. The damage degree of the medium rock mass is evaluated through the detection of acoustic parameters (Fan et al. 2021), so as to provide accurate and reliable data support for the acceptance of dam foundation excavation quality and the optimization of design and construction scheme.

---

*Corresponding Author: 847620032@qq.com

DOI 10.1201/9781003330172-7

2 SINGLE-HOLE ACOUSTIC TESTING TECHNOLOGY

The principle of this method is based on the wave theory of elastic waves. When an elastic wave propagates in various media and media, the change in the physical state of the media will lead to the change in wave parameters of elastic waves. To transmit elastic waves to the medium, first, physical excitation is needed to generate a vibration source. The medium around the particle of the vibration source will be driven to vibrate with the vibration source, and the elastic waves in the medium will propagate with the particle vibration. The receiving equipment is used to receive the propagation speed, vibration amplitude, vibration frequency and other fluctuation parameters of the elastic wave in the transmission carrier at a certain spatial distance. Through the analysis and processing of the numerical changes of the fluctuation parameters, The wave parameters are connected with the engineering status of the medium carrier to solve the relevant engineering problems.

In general, in order to master the vertical engineering quality of rock mass in engineering, a vertical hole is drilled from top to bottom at the designated part of the rock mass. During detection, the integrated transceiver transducer is placed in the vertical borehole. The transmitting transducer is generally made of special materials that can convert mechanical energy and electric energy to each other, which is used as the excitation facility of elastic waves. In order to excite the acoustic wave beam we need, we need to use the refraction principle of the wave. When the elastic wave enters another medium from one medium, the propagation direction of the transmitted wave will change. As long as the angle of the incident wave entering the medium is appropriate, the transmitted wave propagating along the hole wall will be generated. The transmitted wave propagating along the hole wall will become a new wave source and refracted into the borehole with the same principle (Feng 2016). The refracted acoustic signal is received by the receiving transducers S_1 and S_2, as shown in Figure 1.

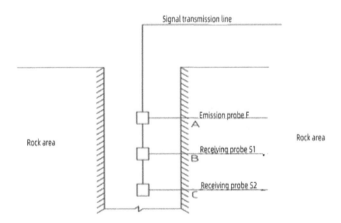

Figure 1.   Schematic diagram of single hole acoustic detection.

Therefore, the incident angle of the wave source emitted by the excitation facility should be appropriate, and the spacing between the acoustic transmitting transducer and the receiving transducer should also be appropriate. In this way, the incident wave propagating along the borehole wall is received by the receiving transducer before the wave propagates without the borehole. Since the connector between the transmitting transducer and the receiving transducer may also become the medium of sound wave propagation, sound insulation materials or grid design should be adopted to eliminate the influence of the transducer itself. In this way, the sound wave is transmitted from the transmitting transducer. The sound wave that first reaches the receiving end is the target beam that we need to propagate along the hole wall. The propagation time is defined as t_1 and t_2, and the sound wave velocity in section BC of the hole wall can be calculated by:

$$\Delta t = t_2 - t_1 \quad V = \Delta L / \Delta t$$

Where V is the wave velocity of the measured medium in the test section. ΔL is the length of the test section, that is, the distance between two receiving transducers. t_1 and t_2 are the propagation time of the sound wave received by the two receiving transducers.

In order to obtain the accurate sound wave propagation time, it is necessary to accurately identify the arrival time of the first wave of rock mass medium in the waveform diagram. There is a coupling agent (usually clear water) in the borehole. How to distinguish whether the arrival first wave is the sound wave propagating in the rock mass medium or the sound wave propagating in the coupling agent is particularly important. The sound wave propagates along the relatively complete rock mass, and the take-off point of the waveform received by the receiving transducer is the arrival time of the first wave of the sound wave in the rock mass medium, as shown in Figure 2. If the rock mass medium is relatively broken, the sound wave propagating in the coupling agent will arrive at the receiving transducer and be received first, then the sound wave propagating in the rock mass medium. At this time, the take-off point in the waveform cannot be used as the arrival time of the first wave of the rock mass medium sound wave. Instead, the waveform take-off point after excluding the acoustic interference of the coupling agent is the arrival time of the first wave of the acoustic wave in a rock mass medium (Yuan et al. 2011), as shown in Figure 3. If the rock mass medium is seriously broken, or even if there will be no receiving waveform or serious distortion of receiving waveform, it is necessary to comprehensively study and judge the integrity of the drilling column core.

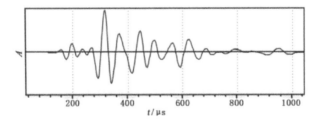

Figure 2. Sound waves propagating in rock mass media are first received.

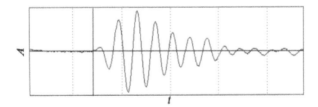

Figure 3. Sound wave propagating in the coupling agent is first received.

The first wave determination is accurate. At this time, the position of the test section is the rock mass medium on the hole wall parallel to the direction of transducers S_1 and S_2, and the height is the distance from S_1 to S_2. Then, we take the distance between two receiving transducers as the detection sampling step and continuously pull and measure to obtain the continuous fluctuation parameter information in the vertical rock around the borehole, so as to establish a relationship between the depth of rock around the borehole and the fluctuation parameters. Thus, it can be judged whether the rock mass around the borehole is uniform and continuous and whether the rock mass quality is complete and reliable (Zhang 2011).

3 DETECTION EXAMPLES OF SINGLE HOLE ACOUSTIC TECHNOLOGY

3.1 *Project overview and detection reasons*

Chushandian reservoir in Henan Province is a class I project, which is mainly composed of the main dam, auxiliary dam, irrigation tunnel, power plant, etc. It focuses on flood control, combined with water supply, irrigation, power generation and other comprehensive utilization. After completion, it can effectively regulate and store the flood in the upstream mountainous area and reduce the peak flow of the mainstream, which plays an important role in reducing the flood control pressure in the middle reaches of the Huaihe River. During the construction of the concrete dam, the engineering rock mass above the foundation surface is excavated by blasting. When it is excavated to the foundation surface, it is found that the blasting process has also caused certain interference and damage to the engineering rock mass below the foundation surface. In order to find out the influence depth and damage degree of blasting on the rock mass below the foundation surface, it is decided to drill down at the elevation of the foundation surface for detection and provide scientific and reasonable data support for whether the rock mass is reinforced or optimized in the later stage.

3.2 *Inspection hole layout*

In order to find out the damage degree of the rock mass below the foundation surface caused by the rock mass blasting of the dam foundation, combined with the suggestions of the project construction parties, a total of six test holes are arranged in this test, which are located in 3#, 4# dam sections respectively. Each dam section is arranged with three single holes, the hole depth is about 5 m, and the number of test holes is as follows. Test hole 3-1: the elevation of the hole top is 64.554 m, and the hole location coordinates are 356 940 8.53 1496 061. 687. Test hole 3-2: the hole top elevation is 64.706m, and the hole position coordinates are 356 940 7.33 6496 060.945. Test hole 3-3: the hole top elevation is 64.711m, and the hole position coordinates are 356 940 8.58 0496 060.324. Test hole 4-1: hole top elevation 64.591m, hole location coordinates are 356 938 7.28 2496 070.877. Test hole 4-2: hole top elevation 64.571m, hole location coordinates are 356 938 8.35 0496 071.745. Test hole 4-3: the hole top elevation is 64.609m, and the hole location coordinates are 356 938 7.05 9496 072.199. The location of the test hole is shown in Figure 4:

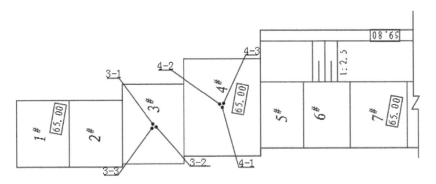

Figure 4.   Layout of measuring holes.

3.3 *Test process*

According to the relevant provisions of the code for rock test of water resources and Hydropower Engineering SL 264-2001 and the standard for test methods of engineering rock mass GB / T 50266-2013, the test holes are numbered one by one, and the holes are washed repeatedly with clean water to reduce or eliminate the interference of impurities in the clean water coupling agent on sound waves. After cleaning, the holes are filled with clean water. The RSM series non-metallic ultrasonic detector of Zhongyan technology is used for the test, which is equipped with an integrated probe of one transmitting and two receiving. According to the drilling profile and working conditions of the test section, we pull the transducer cable from bottom to top and record the waveform at an interval of 20cm. For the rock mass greatly affected by blasting, the change of rock mass structure such as crack and cavity caused by blasting will aggravate the attenuation of elastic wave in the propagation process, increase the propagation path, reduce the wave velocity, and reduce the amplitude of the first wave received by the receiving transducer, which is reflected in the great attenuation of wave energy. On the contrary, the energy attenuation of rock mass with uniform, dense and reliable quality in the process of sound wave propagation is very small, the sound wave propagation time is relatively short, and the wave propagation speed is much higher than that of incomplete rock mass (He, 2020). The influence degree and depth of the blasting on the foundation rock can be obtained through the comparative analysis of the wave speed of the whole hole section.

The wave velocity test results of each hole are shown in Figure 5 (the upper row is 3-1, 3-2 and 3-3 test holes from left to right; the lower row is 4-1, 4-2 and 4-3 test holes from left to right).

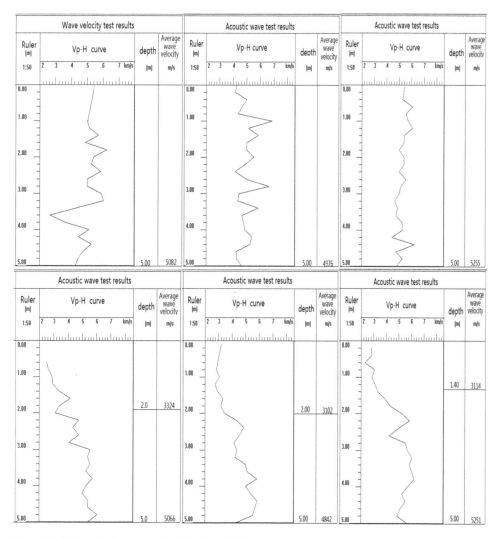

Figure 5.   Wave velocity test results of each test hole.

According to the research and judgment of the acoustic test results of each test hole, the sound velocity division characteristics of each test hole are summarized and analyzed along the hole depth direction. The results are shown in Table 1.

Table 1. Statistical analysis of acoustic wave test results of rock mass.

Dam section	Number	Lowwave velocity Depth (m)	Average value of sound velocity in low wave velocity zone (m/s)	Average value of sound velocity in stable section (m/s)	Result
3#	3-1	3.6	307 7	508 2	Except for the 3.6m test point, the sound velocity of the sound wave test curve in other depths in the hole has not decreased continuously, and the wave velocity is stable
	3-2	—	—	497 6	There is no continuous decrease in the sound velocity of the sound wave test curve, and the wave velocity is stable
	3-3	—	—	525 5	There is no continuous decrease in the sound velocity of the sound wave test curve, and the wave velocity is stable
4#	4-1	0.00-2.00	332 4	506 6	The sound wave test curve changes at 2.00m. The sound velocity in the test hole is not stable, and the sound velocity in the other test holes is not stable.
	4-2	0.00-2.00	310 2	484 2	The sound wave test curve changes at 2.00m. The sound velocity in the test hole is not stable, and the sound velocity in the other test holes is not stable
	4-3	0.00-1.40	311 4	525 1	The sound wave test curve changes at 1.40M. The sound velocity in the test hole is not stable, and the sound velocity in the other test holes is not stable

It can be seen from the test data in Figure 5 and Table 1 that the wave velocity of the upper and lower sections of the 3# dam section changes little, and the sudden change point is suspected to be the normal fracture development of the rock mass. The 4# dam section has a continuous reduction of sound velocity within 2m from the foundation surface. The vibration damage, fracture and crack caused by blasting are one of the main reasons for the reduction of sound wave velocity. Therefore, the blasting impact has little impact on the 3# dam section of the project, and has a great impact on the rock mass within 2m below the foundation surface of the 4# dam section. This part has formed a potential safety hazard that significantly impacts the project. It is suggested that consolidation grouting and other reinforcement measures be taken for the rock mass within 2m below the foundation surface of the 4# dam section to ensure high safety and stability of the dam foundation.

4 CONCLUSION

During the blasting excavation of Chushandian Reservoir 3# and 4# dam foundations, the 3# dam section is complete, and the sound velocity within 2m below the foundation surface of the 4# dam section is continuously low. The rock mass within this range is evaluated as the damaged part affected by blasting, and the rock mass is relatively broken. Before subsequent construction, consolidation grouting and reinforcement shall be carried out for the rock mass within the damaged parts affected by blasting in time to ensure the safety and stability of the dam construction and operation period.

Single-hole acoustic testing technology positively impacts the integrity testing of rigid materials such as rock mass and foundation in hydraulic engineering. Its testing results can provide accurate and reliable data support for the construction organization design and reinforcement of the project in the process of subsequent project construction and play an important role in project construction.

REFERENCES

Fan Liqun, Song Wenyan, Cheng Gang, et al. Cross hole acoustic detection and analysis of rock mass quality of dam foundation excavation [J] *Water Conservancy Technical Supervision*, 2021 (8): 4

Feng Yin. Discussion on the application of single hole acoustic method in the test of rock mass loose circle [J] *Science and Technology Shangpin*, 2016 (8): 3

He Bangjing. Single-hole acoustic testing and analysis of dam foundation quality of Xianrendong reservoir [J] *China Hydropower and Electrification*, 2020 (11): 37–40 + 55

Yuan Bo, He Haohui, Yao Liming Principle and application of single hole "one generator and two receivers" ultrasonic testing [J] *Geotechnical Foundation*, 2011, 25 (4): 2

Zhang Zaiyun. Propagation and stability analysis of elastic waves in different media [D] *Central South University*, 2011

Advances in Measurement Technology, Disaster Prevention and Mitigation – Li & Mohd Yusof (Eds)
© 2023 The Author(s), ISBN: 978-1-032-36087-4

Displacement prediction method of cofferdam under condition of exceeding warning value based on adaptive Kalman filter

Mingjie Chen* & Pingjie Li*
CCCC Fourth Harbor Engineering Institute Co., Ltd., Guangzhou, Guangdong, China
CCCC Key Lab of Environmental Protection & Safety in Foundation Engineering of Transportation,
Guangzhou, Guangdong, China

ABSTRACT: At present, China has increased investment in inland waterway locks. More and more locks have occurred in China. However, the safety monitoring of cofferdams in locks remains in the traditional way, especially the displacement prediction of cofferdam in the lock. When the cofferdam displacement exceeds the warning value, the safety of the cofferdam would be seriously threatened. But the cofferdam displacement is affected by various factors simultaneously so that the system state and data change trend cannot be accurately predicted. The adaptive Kalman filter method is used to effectively fuse the surface horizontal displacement and the deep horizontal displacement closest to the surface at the same position whose displacement changed fastest. Besides, the grey GM (1, 1) model is used to predict the displacement of the cofferdam, and the relevant indicators of the model are tested to evaluate the prediction accuracy of the model. Then, the prediction results are appropriately corrected according to the state transition probability in the Markov chain. The results show that after the data processing of the combined model, the prediction data are better estimation for the displacement of cofferdam compared with the data of a single measurement point. At the same time, the prediction accuracy is significantly improved compared with linear prediction, indicating that this combined data processing model can be effectively used for cofferdam displacement prediction.

1 INTRODUCTION

With the rise of a large number of inland waterway locks, the construction safety of cofferdams used for ship lock construction has attracted more and more attention. The role of cofferdam deformation monitoring for construction safety has become increasingly prominent. Once the cofferdam failed, the consequences were extremely serious. Cofferdam deformation monitoring data often face the problem of large monitoring data errors and inaccurate monitoring results. The traditional monitoring means mainly relies on optical instruments for horizontal displacement measurement. Weather conditions and human factors have a great influence on the monitoring results, especially for monitoring in harsh weather conditions. Not only is monitoring frequency required to be increased, but also critical data quality needs to be improved. In addition, the inclinometer is also widely used. The horizontal offset of each measuring point is calculated according to the angle between the axis of the instrument and the pendulum vertical line and the distance between the vertical measuring points so that the deformation curve of the measuring section is formed by accumulation. Temperature effect, random vibration and human factors greatly affect the inclinometer measurement results. How to integrate the two data is an important link to further improve the accuracy of monitoring data. Although they belong to multi-source heterogeneous data and it is relatively difficult to fuse them, considering that their essence is horizontal displacement and the description object is generally the same plane position as the cofferdam, it is feasible to select deep horizontal displacement and surface horizontal displacement for data fusion.

---

*Corresponding Authors: 365836145@qq.com and 278393325@qq.com

 DOI 10.1201/9781003330172-8

There are many factors affecting the deformation of the cofferdam. It is difficult to define the specific role of a factor in it by traditional methods, and it is also difficult to evaluate the magnitude of the action. Therefore, the deformation of the cofferdam is very consistent with the definition of the grey system. The grey system theory was established by Professor Julong Deng, a famous scholar, in 1982 and has been widely used, improved and continuously deepened. This system is based on small sample data aiming at the characteristics of partial known information and partial unknown information, which takes uncertainty to predict the next state of the system (Deng 1990; Ding et al. et al. 2013; Lin 2013). At present, there are many improved algorithms for the grey model, such as the research in reference (Wu et al. 2007), which enriches the grey theory from the scope of use. The research in reference (Hu 2006) improved the prediction accuracy. However, cofferdam deformation monitoring often contains a variety of random noise, but the traditional grey model cannot eliminate random noise. If the data are directly predicted, the data sample is small, and the information is poor, the prediction results will seriously deviate from the actual situation, so as to misjudge the safety state of the cofferdam. Markov chain is often used to determine the random state emphasizing the transition probability of the state. The current state is related to the occurrence probability and has nothing to do with the change characteristics of the previous data. When applied to the data processing of the system with large volatility, it has high stability and prediction accuracy, which is complementary to the grey prediction model (Cheng 2020).

Currently, the research on cofferdam monitoring data processing mainly focuses on the single prediction model and the combination prediction model. For example, Zhu (2017) et al. used the Kalman filter to eliminate the disturbance error in the measured data and established the Grey Markov prediction analysis combination model using the processed data. The research results have been significantly improved. The core ideas of these studies mostly focus on the improvement of prediction accuracy with a combination of traditional prediction models. But there is hardly any study to improve the accuracy of prediction data based on data fusion. In this paper, according to the surface horizontal displacement and deep horizontal displacement commonly used in cofferdam monitoring, the data fusion was carried out by an adaptive Kalman filter. Then, the grey GM (1, 1) model was established, and the predicted value was corrected by Markov chain. The results showed that the deviation value gradually converged after the fusion of the adaptive Kalman filter. Compared with the single measuring point data, the fused data is a better estimation for the cofferdam state. Using the grey model prediction combined with the Markov chain to modify the prediction results, the variation range of data residual and relative residual after processing is greatly reduced, and the accuracy is greatly improved. Therefore, the combined model algorithm in this paper has great guiding significance for cofferdam displacement prediction and has an important reference value for data processing in other scenes.

2 BASIC ALGORITHM OF MODEL

2.1 *Adaptive Kalman model*

Adaptive Kalman filter theory introduces the state space theory into the physical system, uses a recursive algorithm to linear filter the data, takes the minimum mean square error as the optimization criterion, establishes the state-space equation, and calculates the optimal estimation of the current system. Generally, it includes two main steps. First, we estimate the state variables by the least square method. Second, the observation value is introduced to update the state variables. The core idea of Kalman filter theory is the Kalman gain. The existence of the gain factor is equivalent to assigning weights between the prior estimation and the measured value. When the estimation is more accurate and credible, the gain factor tends to be 0. When the measured value is more accurate and credible, the gain factor tends to be 1. The solving process is an optimization process. The optimal gain factor is obtained based on the minimum mean square deviation condition. At the same time, the gain factor is updated iteratively by the covariance matrix, and the optimal estimation of the system is formed in the iterative process.

$$Z(k) = HX(k) + V(k) \tag{1}$$

$$X(k|k - 1) = AX(k - 1|k - 1) + BU(k) \tag{2}$$
$$P(k|k - 1) = AP(k - 1|k - 1)A^T + Q \tag{3}$$
$$X(k|k) = X(k|k - 1) + K_g(k)(Z(k) - HX(k|k - 1)) \tag{4}$$
$$K_g(k) = P(k|k - 1)H^T / (HP(k|k - 1)H^T + R) \tag{5}$$
$$P(k|k) = (I - K_g(k)H)P(k|k - 1) \tag{6}$$

$X(k)$ represents the system state at the time k. $X(k|k - 1)$ represents the current estimated value due to the previous state. $X(k - 1|k - 1)$ represents the estimate of the previous state. H represents the parameters of the measurement system. $V(k)$ and $U(k)$ denote Gaussian white noise, and obey normal distribution. The corresponding mean square error matrices are R and Q respectively. $Z(k)$ represents the measured value of the system. $K_g(k)$ represents the gain factor. A and B represent the state transition matrix of the system from $k - 1$ to k. $P(k|k - 1)$ represents the current time mean square error matrix. $P(k - 1|k - 1)$ represents the mean square error matrix of the previous moment. $P(k|k)$ represents the mean square error matrix for the next moment. I is the unit matrix.

2.2 Grey GM (1,1) model

Suppose the original data sequence is $x^{(0)} = \{x^0(1), x^0(2), \cdots , x^0(n)\}$ (7)

The cumulative sequence is $x^{(1)} = \{x^1(1), x^1(2), \cdots , x^1(n)\}$ (8)

Among them $x^{(1)}(k) = \sum_{i=1}^{k} x^0(i)(k = 1, 2, \cdots , n)$ (9)

So next to the mean generating sequence will be $z^{(1)} = \{z^1(1), z^1(2), \cdots , z^1(n)\}$ (10)

Among them $z^{(1)}(k) = 0.5x^1(k) + 0.5x^1(k - 1)(k = 2, 3, \cdots , n)$ (11)

Establish differential equations $x^0(k) + az^1(k) = b(k = 2, 3, \cdots , n)$ (12)

So that whitening differential equation is $\dfrac{dx^{(1)}(t)}{dt} + ax^{(1)}(t) = b$ (13)

Suppose $\hat{a}$ is the parameter vector to be estimated, so $\hat{a} = \begin{bmatrix} a \\ b \end{bmatrix}$ (14)

Solving by the least square method, the formula is $J(\hat{a}) = (Y - B\hat{a})^T(Y - B\hat{a})$ (15)

So that $\hat{a} = (B^T B)^{-1} B^T Y$ (16)

Among them $Y = \{x^0(2), x^0(3), \cdots , x^0(n)\}^T$ (17)

$$B = \begin{pmatrix} -z^1(2) & 1 \\ -z^1(3) & 1 \\ \vdots & \vdots \\ -z^1(n) & 1 \end{pmatrix} \tag{18}$$

Further, calculate $\hat{x}^{(1)}(k) = (x^0(1) - \dfrac{b}{a})e^{-a(k-1)} + \dfrac{b}{a} (k = 2, 3, \cdots , n)$ (19)

The reduction value is $\hat{x}^{(0)}(k) = \hat{x}^{(1)}(k) - \hat{x}^{(1)}(k - 1)$ (20)

Generally, there are two important preconditions to establish the sequence of models. First, the quasi-smooth condition must be satisfied, that is

$$\rho(t) = \frac{x^0(k)}{\sum_{i=1}^{k-1} x^0(i)} < 0.5 \tag{21}$$

Second, it's necessary to meet the test conditions of quasi-exponential law, that is

$$\delta^{(1)}(t) = \frac{x^1(k)}{x^1(k-1)} \in [1, 1.5) \tag{22}$$

The Grey GM (1, 1) model can be established only when the change speed of the time series is controlled in a certain range to ensure that the grey model is suitable for data sequence.

After the Grey GM (1, 1) model is established and completes the prediction of relevant data, in order to strictly evaluate the accuracy of the prediction results of the model, the prediction results and related evaluation indexes need to be tested again. This paper mainly adopts the posterior difference test method. According to the residual definition, the expression is

$$\varepsilon(k) = x^0(k) - \hat{x}^0(k) \tag{23}$$

By definition of relative residual, it can be expressed as $\Delta(k) = \varepsilon(k)/x^0(k)$ (24)

The original data sequence variance and residual variance can be further expressed as:

$$S_1^2 = \frac{1}{n} \sum_{k=1}^{n} (x^0(k) - \frac{1}{n} \sum_{k=1}^{n} x^0(k))^2 \tag{25}$$

$$S_2^2 = \frac{1}{n} \sum_{k=1}^{n} (\varepsilon(k) - \frac{1}{n} \sum_{k=1}^{n} \varepsilon(k))^2 \tag{26}$$

According to the accuracy test level reference table, the variance ratio is $C = \dfrac{S_2}{S_1}$ (27)

The small error probability is $P = P\{|\varepsilon(k) - \bar{\varepsilon}| < 0.6745 \cdot S_1\}$ (28)

2.3 Markov model

According to the Markov model theory, the change process of the system is closely related to the state that the current state of the system is close to but unrelated to the state that the current state is far away from. This property is called the after-effectlessness of the Markov model (Zhang et al. 2014). Markov correction mainly includes three steps. Step one is the state division; generally, three to five states are appropriate. Second, the state transition probability matrix is calculated in detail according to the number of occurrences of states and the number of state transitions. Third, the predicted value is modified.

The detailed calculation steps are as follows: according to the range of relative residual variation obtained from the grey model results, the model is divided into k states, each state contains the upper and lower limits of the range, and its mathematical expression is

$$E_i = [a_i, b_i) \quad (i = 1, 2, \cdots, k), \tag{29}$$

$$P_{ij} = \frac{M_{ij}}{M_i} \tag{30}$$

$$P_h = \left[P_{ij}^{(h)} \right] = \begin{bmatrix} P_{11}^h & \cdots & P_{1n}^h \\ \vdots & \vdots & \vdots \\ P_{n1}^h & \cdots & P_{nn}^h \end{bmatrix} \tag{31}$$

In the formula, $P_{ij}^{(h)}$ is the h step state transition probability of Markov chains. M_{ij} is the number of times from state E_i to state E_j. M_i is the number of occurrences of state E_i. P_h is the state transition probability matrix of the Markov process in h step.

Assuming $S_{(0)}$ is the initial state vector, the state vector transformed by P steps will be

$$S_{(p)} = S_{(0)} \times (P)^p \tag{32}$$

The probability of the grey prediction value turning to each state interval is predicted based on the initial state vector and the state transition probability matrix. According to the state interval with the maximum probability, the grey prediction value is further corrected.

$$\xi_k = \frac{\hat{x}^{(0)}(k)}{1 - 0.5(a_i + b_i)} \tag{33}$$

3   PROJECT PROFILE

In a lock cofferdam project, the earth-rock dam is used as a lateral pressure force structure, and a plain bite waterproof pile is used as a flexible waterproof curtain. Due to poor crack resistance of plain bite waterproof pile, it is necessary to strictly control cofferdam deformation by cofferdam monitoring. In the monitoring process, the surface horizontal displacement monitoring points, the surface vertical displacement monitoring points, the deep horizontal displacement monitoring points, the water level monitoring points and other different types of monitoring points are arranged in key parts of the top of the cofferdam, the top of the bite pile and the internal part of the bite pile. Among them, the surface horizontal displacement monitoring point W2 and the deep horizontal displacement monitoring point CX2 are at the same position. The change rate of them is the fastest in the construction process and reaches or exceeds the alarm value from time to time which is 5 mm/d. In order to determine whether the cumulative displacement will exceed the allowable value of the cofferdam, it is necessary to accurately predict the next state. Considering that W2 and CX2 are multi-source heterogeneous data, the measurement points closest to the surface of CX2 and W2 are selected for data fusion. There are 13 days of data from 2019-7-10 to 2019-7-22. The data for the first ten days are used to establish the prediction model, and the data of the last three days are used to verify the accuracy of the model prediction results.

4   ANALYSIS OF PREDICTION RESULTS

4.1   *Analysis of adaptive Kalman model results*

It can be seen in Table 1 and Figures 1 ~ 2 that the variation trends of surface horizontal displacement, deep horizontal displacement and adaptive Kalman filter fusion value are basically the same, and the values of them are very close. The deviation value of surface horizontal displacement is -1.729 ~ 5.273mm, and the deviation value of deep horizontal displacement is -4.671 ~ 2.270mm. The deviation value is large in the early stage but gradually tends to converge in a later stage. The horizontal displacement at the same position is independent of test means, test method and test instrument without considering the error in terms of theory, but the existence of error leads to the inconsistency of the indication. It can be seen from Table 1 and Figure 1 that after Kalman data fusion, the Gaussian noise and the one-sidedness of the data itself are eliminated, and the data after fusion are better estimated for the displacement of the cofferdam at this position compared with the data before fusion.

4.2   *Grey GM (1, 1) model results analysis*

The quasi-smoothness test and quasi-exponential rule test are carried out before the grey model is established according to the fusion value of the Kalman filter. It can be seen from Figure 3 that the smoothness ratio is less than 0.5, and the level ratio is basically ranged from 1.0 to 1.5, indicating that the test conditions for establishing the grey model are satisfied. Based on the process of grey model establishment, the model parameters $a = -0.1090$ and $b = 38.7778$ are obtained. After model establishment variance ratio is $C = 0.15 < 0.35$. Residuals ranged from -4.188 mm to 6.069 mm, with an average of -0.205 mm. The formula $|\varepsilon(k) - \bar{\varepsilon}| < 0.6745 \cdot S_1 = 15.239$ is always true, indicating that from 2019-7-10 to 2019-7-19 formula $P = 1$ is always true with residual relative values ranging from -0.079 to -0.080. Compared with the dividing limit of Table 2, the prediction accuracy is grade one, and the prediction accuracy is good. The data from 2019-7-20 to 2019-7-22 is predicted based on the model. The residual relative value ranged from -0.062 to -0.025, and the prediction accuracy is high. But the residual ranged from -8.76 to -3.309 mm, indicating that the fluctuation range is greater than the rate of alarm value, which still needs to be further optimized.

Table 1.   Kalman data fusion results and grey model prediction results.

Date	Surface horizontal displacement /mm	Deep horizontal displacement (mm)	Kalman filter fusion (mm)	Surface horizontal displacement deviation (mm)	Deep horizontal displacement deviation (mm)
2019-7-10	41.3	31.9	36.027	5.273	−4.127
2019-7-11	46.6	37.8	42.421	4.179	−4.671
2019-7-12	48.8	44.0	46.608	2.192	−2.608
2019-7-13	53.6	51.7	52.975	0.625	−1.325
2019-7-14	61.8	63.1	62.806	−0.986	0.244
2019-7-15	77.8	74.2	75.832	1.968	−1.632
2019-7-16	83.1	82.9	82.998	0.102	−0.098
2019-7-17	86.6	90.3	88.030	−1.430	2.270
2019-7-18	94.1	97.3	95.601	−1.501	1.699
2019-7-19	105.7	101.4	103.684	2.016	−2.284
2019-7-20	117.7	112.2	114.770	2.930	−2.570
2019-7-21	129.1	132.0	130.829	−1.729	1.121
2019-7-22	139.5	142.0	140.820	−1.320	1.180

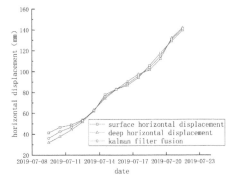

Figure 1. Adaptive Kalman filter fusion curve.

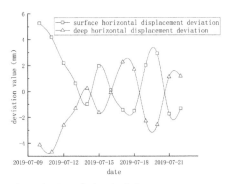

Figure 2. Data fusion deviation curve.

4.3   Analysis of Markov model results

According to the formula (29) ~ (33), the prediction results of the grey model are modified by means of the Markov chain. Based on the range of relative residual error -0.079 ~ 0.08 from 2019-7-10 to 2019-7-19, the grey model is divided into four states: E1 ~ E4 by roughly uniform and the occurrence times are counted. Such as E_1 is [-0.079,-0.039), E_2 is [-0.039,0.001),E_3 is [0.001,0.041), E_4 is [0.041,0.081). E_1 appears four times, E_2 appears two times, E_3 appears two times, and E_4 appears two times. According to the times of state transitions, the corresponding state transition probability matrix is calculated and finally, the results are $S_{(1)} = [0.4\ 0.1\ 0.2\ 0.2]$, $S_{(2)} = [0.3\ 0.1\ 0.2\ 0.2]$, $S_{(3)} = [0.25\ 0.1\ 0.175\ 0.2]$. It can be seen that the probability of state transition to E_1 is the largest, and it is reasonable to believe that the system states during 2019-7-20~2019-7-22 are transferred to E_1 consecutively for three days. Therefore, the prediction value of the grey model is corrected according to the upper and lower limits of E_1 state. The specific calculation results are shown in Table 3.

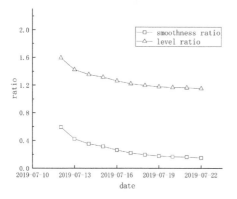

Figure 3. Grey model smoothness and level ratio curve.

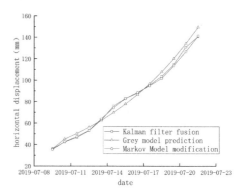

Figure 4. Horizontal displacement-time curve.

Table 2. Classification table of prediction accuracy.

Grade of prediction precision	Probability of small error	Variance ratio
Good (Grade one)	$P \geq 0.95$	$C \leq 0.35$
Qualify (Grade two)	$0.8 \leq P < 0.95$	$0.35 < C \leq 0.5$
Marginal (Grade three)	$0.7 \leq P < 0.8$	$0.5 < C \leq 0.65$
Unqualify (Grade four)	$P < 0.7$	$C > 0.65$

Table 3. Comparison of grey model prediction results and Markov correction results.

Date	Prediction from the grey model (mm)	Residual of grey model prediction	Relative residual of grey model prediction	State	Prediction after Markov correction (mm)	Residual after correction	Relative residual after correction
2019-7-10	36.027	0.000	0.000	E2	35.360	0.667	0.019
2019-7-11	45.117	−2.696	−0.064	E1	42.600	−0.179	−0.004
2019-7-12	50.311	−3.703	−0.079	E1	47.510	−0.902	−0.019
2019-7-13	56.102	−3.127	−0.059	E1	52.980	−0.005	0.000
2019-7-14	62.561	0.245	0.004	E3	63.900	−1.094	−0.017
2019-7-15	69.763	6.069	0.080	E4	74.290	1.542	0.020
2019-7-16	77.794	5.204	0.063	E4	82.850	0.148	0.002
2019-7-17	86.749	1.281	0.015	E3	88.610	−0.580	−0.007
2019-7-18	96.736	−1.135	−0.012	E2	94.930	0.671	0.007
2019-7-19	107.872	−4.188	−0.040	E1	101.860	1.824	0.018
2019-7-20	120.290	−5.520	−0.048		113.590	1.180	0.010
2019-7-21	134.138	−3.309	−0.025		126.660	4.169	0.032
2019-7-22	149.580	−8.760	−0.062		141.250	−0.430	−0.003

It can be seen from Table 3 and Figures 4 ∼ 6 that after Markov chain correction, the overall range of residual error decreases from -8.76 ∼ 6.069 mm to -1.094 ∼ 4.169 mm, and the average residual error decreases from - 1.511 mm to 0.539 mm. The range of relative residual error decreases from -0.079 ∼ 0.08 to -0.019 ∼ 0.032. The data accuracy is greatly improved.

The range of residual predicted for the three-day decreases from-8.76 ∼ -3.309 mm to -0.43 ∼ 4.169 mm. The average residual decreases from -5.863 mm to 1.640 mm. The range of relative residual decreases from -0.062 ∼ -0.025 to -0.003 ∼ 0.032.

The prediction accuracy is greatly improved. Compared with the results of the grey model predicting only, the horizontal displacement-time curve corrected by the Markov chain is closer to the fusion value of the Kalman filter. The fluctuation ranges of the residual and relative residual curves are significantly reduced, indicating that Markov chain correction significantly improves the accuracy of grey prediction.

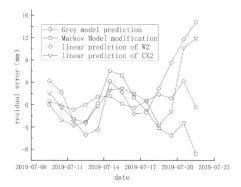

Figure 5. Comparison of residual under different prediction methods.

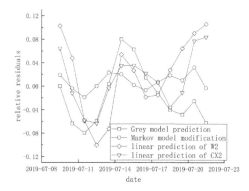

Figure 6. Comparison of relative residuals under different prediction methods.

4.4 Comparison between combination model prediction and linear prediction

In order to further illustrate the superiority of the combined model prediction accuracy, the surface horizontal displacement W2 and the deep horizontal displacement CX2 are predicted separately by linear fitting. Also, the data of a total of 13 days from 2019-7-10 to 2019-7-22 are selected. The data of the first ten days are used to establish a prediction model. The data of the last three days are used to check the accuracy of the model prediction results. The details are shown in Table 4 and Figures 5∼8.

It can be seen in Table 4 and Figures 5∼8 that compared with the linear prediction method of a single measurement point, the overall range of residual error of the combined prediction model based on data fusion decreases from -5.382 ∼ 14.759 mm of W2 and -3.647 ∼ 11.885 mm of CX2 to -1.094 ∼ 4.169 mm. The average residual error decreases from 2.615 mm of W2 and 1.606 mm of CX2 to 0.539 mm. The range of relative residual error decreases from -0.100 ∼ 0.106 of W2 and −0.063 ∼ 0.084 of CX2 to -0.019 ∼ 0.032, so the data accuracy is greatly improved.

The range of residuals predicted for the three-day decreased from 7.572 ∼ 14.759 mm of W2 and -1.203 ∼ 11.885 mm of CX2 to -0.43 ∼ 4.169 mm. The average residuals decreased from 11.332 mm of W2 and 6.957 mm of CX2 to 1.640 mm. The range of relative residuals decreased from 0.064 ∼ 0.106 of W2 and -0.011 ∼ 0.084 of CX2 to -0.003 ∼ 0.032. The prediction accuracy was good.

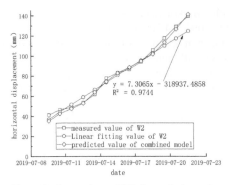

Figure 7. Comparison of W2 linear fitting value with the method of the combined model.

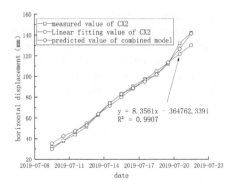

Figure 8. Comparison of CX2 linear fitting Values with the method of the combined model.

Table 4. Linear fitting prediction results in surface and deep horizontal displacement.

Date	Linear prediction result of W2 (mm)	Residual of W2 (mm)	Relative residual of W2	Linear prediction result of CX2 (mm)	Residual of CX2 (mm)	Relative residual of CX2
2019-7-10	37.1	4.237	0.103	29.8	2.057	0.064
2019-7-11	44.4	2.231	0.048	38.2	−0.449	−0.012
2019-7-12	51.7	−2.876	−0.059	46.6	−2.555	−0.058
2019-7-13	59.0	−5.382	−0.100	54.9	−3.261	−0.063
2019-7-14	66.3	−4.469	−0.072	63.3	−0.217	−0.003
2019-7-15	73.6	4.205	0.054	71.6	2.577	0.035
2019-7-16	80.9	2.198	0.026	80.0	2.921	0.035
2019-7-17	88.2	−1.608	−0.019	88.3	1.965	0.022
2019-7-18	95.5	−1.415	−0.015	96.7	0.609	0.006
2019-7-19	102.8	2.879	0.027	105.0	−3.647	−0.036
2019-7-20	110.1	7.572	0.064	113.4	−1.203	−0.011
2019-7-21	117.4	11.665	0.090	121.8	10.191	0.077
2019-7-22	124.7	14.759	0.106	130.1	11.885	0.084

5 SUMMARY AND RECOMMENDATIONS

The properties described by surface horizontal displacement and deep horizontal displacement of cofferdam are consistent. The monitoring point data at the same position, such as the monitoring point data of deep horizontal displacement closest to the surface and the monitoring point data of surface horizontal displacement, can be selected to integrate effectively. Compared with the data before fusion, the one-sidedness and Gaussian noise of the single measuring point data after Kalman filtering fusion are eliminated, indicating that Kalman filtering fusion is a better estimation for the system. The data accuracy can be effectively guaranteed before grey prediction.

The quasi-smoothness and quasi-exponential law are tested before establishing the grey model, and the variance ratio and small error probability of the precision grade evaluation index are tested after establishing the grey model. The results show that the deformation monitoring of the cofferdam meets the grey system. When the grey model GM (1, 1) is used to predict the displacement of the cofferdam, the prediction accuracy is high and basically meets the engineering needs.

The next state of the predicted value of the grey model is calculated by the state transition probability matrix. The predicted value corrected by the Markov chain is closer to the real value.

Markov modification can greatly improve the accuracy of the prediction. For cofferdam monitoring data, multi-model combination prediction can achieve a better prediction effect than single model prediction.

In general, the data is fused by an adaptive Kalman filter, then predicted by the grey model and modified by the Markov chain. Each step improves the prediction accuracy, reduces random error and optimizes the estimation value. After three steps of data processing, the prediction results tend to the real state of the cofferdam. Compared with the linear prediction of a single measuring point, the maximum residual of the combined prediction model based on data fusion is reduced from 14.759 mm to -0.43 mm, and the maximum relative residual is reduced from 10.6 % to 0.3 %.

Although prediction accuracy has improved greatly, previous data from instruments is vital after all. Improving measurement methods and using advanced measurement equipment remains the focus of future research.

ACKNOWLEDGMENTS

Sponsored by the Guangzhou Pearl River Science and Technology Star Special Fund (201806010162) and the South China Marine Science and Engineering Guangdong Laboratory (Zhuhai) estuary coast and reef engineering innovation team construction project (311020009).

REFERENCES

Ge Ding, Xianghong Hua, Mao Tian, et al. Application of Improved Wavelet Neural Network in Settlement Prediction[J]. *Surveying and Mapping Geographic Information*, 2013(6): 27–29.

Guji Lin. Data Processing of Cofferdam Deformation Monitoring Based on Grey Model GM (1,1) [J]. *Surveying and Mapping of Geology and Mineral Resources*, 2013, 29(4): 41–42

Julong Deng. *Grey System Theory Course*[M]. Wuhan: Huazhong University of Technology Press, 1990.

Juntao Zhu, Dongxu Xiong, Yawei Li. Application of grey Markov combination model based on Kalman filter in foundation pit deformation monitoring [J]. *Journal of Guilin University of Technology*, 2017, 37(4): 653–657.

Ruifang Zhang, Xiao-hui Cheng, Zihang Song, et al. Copy-paste tamper detection algorithm for binary wavelet image based on grey Markov theory[J]. *Journal of Guilin University of Technology*, 2014, 34 (4): 775–781.

Wusheng Hu. *Neural Network Theory and Its Engineering Application*[M]. Beijing: Surveying and Mapping Press, 2006.

Yiping Wu, Weifu Teng, Yawei Li. Application of Grey: Neural Network Model in Landslide Deformation Prediction[J]. *Chinese Journal of Rock Mechanics and Engineering*, 2007, 26(3): 632–636.

Yong Cheng. *Application of Grey Markov Model in Settlement Prediction of Ningbang Super High-rise Building*[D]. Xiangtan University, 2020.

Advances in Measurement Technology, Disaster Prevention and Mitigation – Li & Mohd Yusof (Eds)
© 2023 The Author(s), ISBN: 978-1-032-36087-4

Glacierized and snow-cover area variation in O'Higgins lake area from 2000 to 2020

Jiantan Chen*

School of Geographical & Earth Sciences, University of Glasgow, Glasgow, UK

ABSTRACT: There are many glaciers in the O'Higgins Lake area with perennial snow. It is of great scientific and practical significance to understand glaciers' current situation and changes here. Using Landsat remote sensing images, climate, and glacier inventory data, this paper studies the glacierized and snow-cover area changes in the O'Higgins Lake area from 2000 to 2020. The results show that the glacierized and snow-cover area in the O'Higgins Lake area has an obvious downward trend in 2020. Compared with July 7, 2000, the area decreased by 9.34% on July 14, 2020. This downward trend is a climate response to the average temperature rise rate of 0.0724 °C / A in the O'Higgins Lake area from 2000 to 2020.

1 INTRODUCTION

The areas covered by snow and ice on the earth are collectively referred to as the cryosphere. As an important part of the cryosphere, mountain glacier is a sensitive indicator of climate change and known as 'solid reservoir' (Wang et al. 2021). As important water resources, glaciers, perennial snow resources, and snow melt water's response to climate change can significantly change the coastal rivers, soil, and local climate (Yan et al. 2005). From 1880 to 2012, the global average temperature has increased by 0.85°C (Shen & Wang 2013), which leads to the 11.3% shrinkage of 16 first-order glacier regions (excluding the Antarctic and Greenland ice sheet) in the world over the past half century (Mu et al. 2018). The indicative function of glacierized and snow-cover area variation on global climate change is becoming increasingly obvious in this century. It also has an obvious impact on water resources. Therefore, the research on glacier change has attracted much attention. The glacier thickness and area changes are directly or indirectly related to climate change. Although glaciers are mostly developed in alpine areas with poor climatic conditions and inconvenient transportation, high-resolution remote sensing data make it possible to acquire plane change of glaciers in large-scale areas. Existing research focuses primarily on a few glaciers around Patagonia, while much less is known about the glacierized and snow-cover area in the whole O'Higgins Lake area (Chen & Liu 2015).

In order to explore the regularity of glacier and snow-cover area around the O'Higgins Lake, data such as Landsat images and SRTM1 are used as the basis for research on glacierized and snow-cover area variation in the O'Higgins Lake area from 2000 to 2020. This paper collects the coverage data of the glacierized and snow-cover area each year, extracts the glacierized and snow-cover area variation information, and analyzes the information combined with relevant thematic data.

2 REGIONAL BACKGROUND

The lake known as O'Higgins in Chile or San Martín in Argentina is located around 48°50'S, 72°36'W in Patagonia, between the Aysén del General Carlos Ibáñez del Campo Region and the Santa

---

*Corresponding Author: 2700341C@student.gla.ac.uk.

DOI 10.1201/9781003330172-9

Cruz Province (Figure 1). The lake consists of a series of finger-shaped flooded valleys, of which 554 square kilometers are in Chile and 459 square kilometers in Argentina. The lake is the deepest in the Americas, with a maximum depth of 836 meters (2,743 ft) near O'Higgins Glacier, which is also the fifth deepest lake in the world. To the west of O'Higgins Lake are the tall Andes mountains, which make a large number of glaciers exist around O'Higgins Lake. The latitude and longitude range of the study area in this paper is 48.30° ~ 49.24° S, 72.10° ~ 73.33° W, with the highest altitude of about 2991m and an average altitude of about 956.03m.

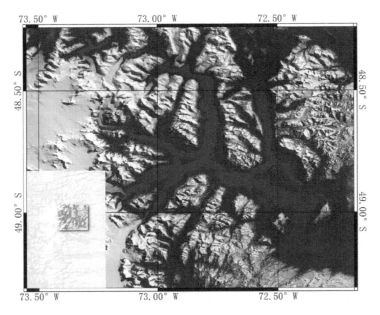

Figure 1.   Location of the study area.

The study area covers O'Higgins Lake and some surrounding areas. Mountain glaciers are widely distributed in this area, most of which exist in Chile. The study of the glacierized and snow-cover area variation in this region has implications for the climate change research of the O'Higgins Lake area.

3   DATA AND METHODS

3.1   *Data*

The main data utilized in the study include Landsat images, SRTM1, glacier inventory datasets, and related meteorological data. Depending on cloud cover and the date of acquisition, images with little cloud or no cloud from July to October from 2000 to 2020 were uniformly selected in this study. Because the images with the low cloud-cover area around 2010 are too few in the study region, and most of them are of poor quality during the expected period, we utilized the satellite image in December 2009. We carefully selected ten scenes from two satellite sensors in different years: Landsat 7 Enhanced Thematic Mapper (ETM+) and Landsat8-OLI. Each year, we acquired one scene from World Reference System (WRS-2) path 231 and row 94 to ensure each scene can cover the whole study area. Landsat scenes are downloaded from the United States Geological Survey (USGS; https://www.usgs.gov/) (United States Geological Survey 2020). Because of the scan-line corrector (SLC) failure of the Landsat 7 Enhanced Thematic Mapper Plus (ETM+) sensor, the Band-specific gap mask files are used to fill the gaps in certain scenes before pre-processing. After that, radiometric calibration and FLAASH atmospheric correction were performed to obtain the surface reflectance data of each scene in the study area. The data of Glacier Inventories is downloaded from Global Land Ice Measurements from Space (GLIMS; http://www.glims.org/) (Global Land Ice Measurements from Space glacier database 2018). SRTM1 is used as the digital elevation data, which is downloaded from EarthExplorer (https://earthexplorer.usgs.gov/) (The Shuttle Radar Topography Mission 2000). The meteorological data is downloaded from National Oceanic and Atmospheric Administration's National Centers for Environmental Information (NOAA-NCEI, https://www.ncei.noaa.gov/) (National Oceanic and Atmospheric Administration's National Centers for Environmental Information 2020).

Table 1. Resolution and dates acquisition for each satellite scene used in this study.

Sensor	Bands used	Spatial Resolution (m)	Date of acquisition (mm/dd/yyyy)
Landsat ETM+	5,4,3	30	07/07/2000,07/05/2005, 09/13/2007,12/07/2009, 09/29/2013,07/14/2020, 08/31/2020,10/02/2020
Landsat OLI	6,5,4	30	09/29/2016,09/06/2019

3.2 Methods

Remote sensing imagery classification uses the computer to analyze the spectral and spatial information of all kinds of surface features in the imagery, assign pixels into different classes, and then make the classification based on the corresponding relation between pixels and ground features (Wang et al. 2021; Wang & Wang 2016; Yan et al. 2011; Yu 2013). In this study, supervised classification is used to classify the satellite imagery. In supervised classification, the user guides the image processing software to specify the land cover classes of interest. Some areas are defined as "training sites," representing particular land cover classes. Before using the classified classes and pixels as feature information, it establishes prior knowledge to classify other unknown pixels (Adili 2021). Each pixel in the image is assigned to the most closely matching class based on its spectral signature. The training site is evaluated based on the separability of the ROI (Region of Interest). When the separability value is greater than 1.9, the separation is good; when it is less than 1.8, the ROI needs to be selected again (Zu 2018). This study used a support vector machine to perform the supervised classification. The support vector machine method is simple and rapid and can better perform the small sample, nonlinear and high-dimensional data classification (Li 2007; Yan 2016).

Bands 3, 4, 5 of Landsat ETM+ and bands 4, 5, and 6 of Landsat OLI are used to make false-color images using the Environment for Visualization of Imagery (ENVI) v5.3. Ground features such as rocks, vegetation, water, and glacierized area can be detected readily in combined false-color images before performing the supervised classification. The training areas are defined and assigned into four training classes: vegetation, water, rock, glacierized, and snow-cover. More than ten training areas are selected for each training class except the water class with visual interpretation. Due to the obvious characteristics of water areas, we selected a relatively small number of water training areas that are representative (Yin et al. 2021). The training areas are uniformly distributed in the study area, and each training class's ROI separability values are more than 1.9.

3.3 Post classification

Post classification is the necessary step in the process of supervised classification. The classification results will inevitably include some minor spots, which must be eliminated from the perspective of both the thematic mapping and practical application. Post classification can reduce noise and remove excessively refined minor spots (Liao et al. 2019; Yuan et al. 2021). In this paper, majority/minority analysis in ENVI software is primarily used for post-classification to improve the quality of remote sensing imageries and the accuracy of classification results. Figure 2. Shows the classification results after post-classification.

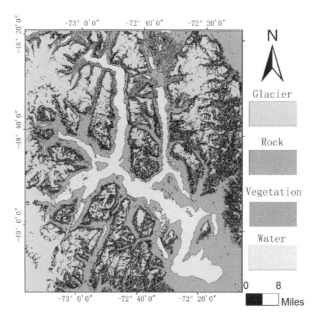

Figure 2.   Results of supervised classification.

Confusion Matrix Using Ground Truth ROIs in ENVI is used to validate the classification accuracy of the results. The main parameters are overall accuracy and the Kappa coefficient. When the overall accuracy and Kappa coefficient values are greater than or equal to 0.8, the classification results meet the accuracy requirements. Based on the confusion matrix, the following formulas are used to calculate the overall accuracy and Kappa coefficient (Ning et al. 2021; Wang & Lei 2021).
Overall accuracy:

$$P_c = \sum_{k=1}^{m} P_{kk}/N \tag{1}$$

Kappa coefficient:

$$K = \frac{N\sum_{i=1}^{m} P_{ii} - \sum_{i=1}^{m}(G_i * C_i)}{N^2 - \sum_{i=1}^{m}(G_i * C_i)} \tag{2}$$

where P_c is the overall accuracy; m is the number of training classes; N is the total number of values; P_{kk} is the number of correctly classified values for class k; K is Kappa coefficient; P_{ii} is the number of values belonging to the truth class i that have also been classified as class i; G_i is the total number of truth values for class i; C_i is the total number of predicted values for class i. Table 2 lists the results of accuracy validation.

Table 2.   Results of accuracy validation.

Date of Acquisition	Kappa Coefficient	Overall Accuracy
July 7, 2000	0.9630	98.0411%
July 5, 2005	0.9860	99.1344%
September 13, 2007	0.9668	97.8675%
December 7, 2009	0.9762	98.4751%
September 29, 2013	0.9571	97.2672%

(continued)

59

Table 2.   Continued.

Date of Acquisition	Kappa Coefficient	Overall Accuracy
September 29, 2016	0.9413	96.8936%
September 6, 2019	0.9768	98.7540%
July 14, 2020	0.9506	98.0418%
August 31, 2020	0.9315	96.6237%
October 2, 2020	0.9574	97.2011%

The classification has high reliability and meets the requirements of subsequent data processing.

4   RESULTS AND ANALYSIS

This study analyzed the average size and temperatures of the glacierized and snow-cover area each year from 2000 to 2020. Temperature data were obtained from the two meteorological stations closest to the study area, Cochrane in Chile and El Calafate Aero in Argentina. Except for 2009, all the area and temperature data were obtained from July to October. For example, the data for 2020 was obtained on August 31.

Figure 3 shows that the size of the glacierized and snow-cover area was (6025.09 ± 118.03) km^2 in 2000. In 2005, the size was the largest, which was (6102.68 ± 52.82) km^2. Since the warming of temperature in December, the size of the glacierized and snow-cover area decreased significantly in 2009. The changing trend of the size of the glacierized and snow-cover area has an obvious opposite relationship with the changing trend of temperature.

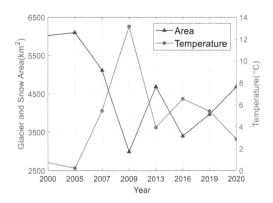

Figure 3.   Trends in the glacierized and snow-cover area and temperature of corresponding months were analyzed from study data.

On August 31, 2020, the size of the glacierized and snow-cover area was (4685.12 ± 158.18) km^2. It is seen that the size of the glacierized and snow-cover area in the study region changes with the temperature change. Landsat 7 ETM+ images of the study area on June 28, 2020, were compared with images taken on July 14, August 31, and October 2, 2020. Through visual interpretation of the glacier, whose glacier ID is G287406E48735S, it is observed that there are significant changes in the snow line, and the snow line reached its lowest altitude on July 14 (Figure 4). Combined with SRTM1 digital elevation data, it can be concluded that the altitude of the eastern snow line increased by about 350m continuously from July 14 to October 2. The size of the glacierized and snow-cover area in the study region decreased from (5462.15 ± 106.96) km^2 on July 14 to (4486.87 ± 125.58) km^2 on October 2 in 2020. Combined with the monthly average temperature data of meteorological stations in 2020, it is found that the lowest temperature occurred in July, reaching $-1.61°C$, and it is also the period when the altitude of the snow line is the lowest. In this period, the size of the glacierized and snow-cover area reaches the highest value in a year (Figure 5).

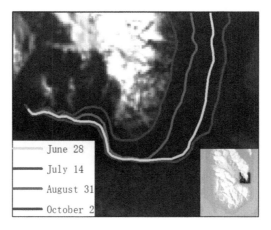

Figure 4.   Change of the snow line.

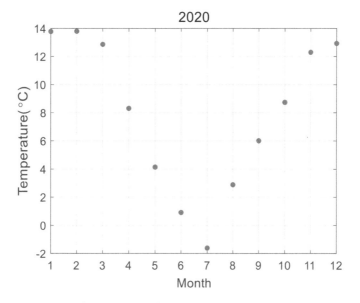

Figure 5.   Temperature change in the study area in 2020.

To better compare the variation trends of the size of the glacierized and snow-cover area, the remaining nine remote sensing images except December 2009 were divided into three groups according to the principle of similar acquisition time (Table 3).

Table 3.   Groups of images.

Group	Date of acquisition(mm/dd/yyyy)
July	07/07/2000, 07/05/2005, 07/14/2020
September	09/13/2007, 09/06/2019, 08/31/2020
October	09/29/2013, 09/29/2016, 10/02/2020

By comparing and fitting the variation trends of the size of the glacierized and snow-cover area in each group with the least square method, it is seen that the size of the glacierized and snow-cover area have a downward trend by 2020 compared with ones in previous years (Figure 6). Compared with July 7, 2000, the size of the glacierized and snow-cover area on July 14, 2020, is (5462.15±106.96) km², decreasing 9.34%. On September 13, 2007, the size of the glacierized and snow-cover area was (5115.08±109.08) km², which decreased by 8.41% on August 31, 2020. The size of the glacierized and snow-cover area on October 2, 2020, decreased by 4.32% compared with the one on September 29, 2013 (4689.3114±128.15) km².

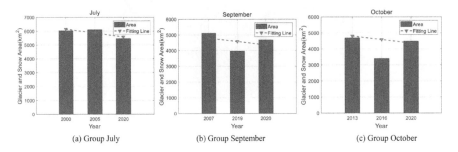

Figure 6.   Comparison of the glacierized and snow-cover area in 2020 with the same period in previous years.

The annual average temperature data of meteorological stations from 2000 to 2020 were analyzed. After fitting it with the least square method, it is found that the average increase rate of local temperature is 0.0724°C/a from 2000 to 2020 (Figure 7). Temperature and climate data cannot fully reflect the detailed characteristics of glacier and snow distribution areas. Still, from the relationship between glacierized and snow-cover area and temperature data in the past 20 years, it is seen that in July and August of each year, the size of the glacierized and snow-cover area peaked decreased since melting. Furthermore, with the continuous rise of temperature in recent 20 years, the size of the glacierized and snow-cover area has declined compared with the ones in previous years.

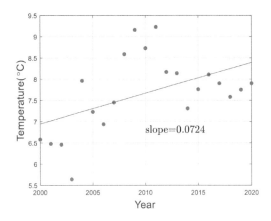

Figure 7.   Scatter and fitting trend of temperature in the study region from 2000 to 2020.

5 CONCLUSION

This study mainly analyzed the variation of the size of the glacierized and snow-cover area around O'Higgins Lake in the past 20 years and the causes for the distribution characteristics of the glacierized and snow-cover area.

The size of the glacierized and snow-cover area around O'Higgins Lake accumulates to the maximum in July. On July 14, 2020, the glacierized and snow-cover area size in the study area (5462.15 ± 106.96) km^2 decreased by 9.34% compared with the one on July 7, 2000 (6025.09 ± 118.03) km^2. In 2020, the size of the glacierized and snow-cover area in September and October decreased by 8.41% and 4.32% compared with the ones in 2007 and 2013. The average temperature increase rate around the O'Higgins Lake area from 2000 to 2020 is 0.0724°C/a, and the temperature change is inversely correlated with the trend of the size of the glacierized and snow-cover area.

There are still some improvements for this study. The variation of snow lines in some other regions could be studied, and some subtle differences could be summarized. More investigations could be conducted on the factors influencing the glacierized and snow-cover area, such as the influence of greenhouse gases, industrial emissions, and other factors. In the future, satellite imageries in some other periods could be used to analyze more objective laws.

REFERENCES

Adili Keremu. (2021). *Extraction of Cotton Planting Area Information in Alar City Based on Landsat 8 Satellite Remote Sensing*. Tarim University. MA thesis.

Chen Jian, & Liu Hanhu. Remote Sensing Monitoring of Glacier Change and Mass Balance (2000-2015) in Geladandong. *Henan Science*. 39, 02(2021): 282–289.

EarthExplorer (https://earthexplorer.usgs.gov/). (2000). SRTM 1 Arc-Second Global (Digital Object Identifier (DOI) number: /10.5066/F7PR7TFT).

United States Geological Survey (USGS; https://www.usgs.gov/). (2020). Landsat-7, Landsat-8 image courtesy of the U.S. Geological Survey.

GLIMS and NSIDC (2005, updated 2018): Global Land Ice Measurements from Space glacier database. Compiled and made available by the international GLIMS community and the National Snow and Ice Data Center, Boulder, CO, U.S.A. DOI:10.7265/N5V98602

Li Miao. (2007). *Land Cover Classification with SVM Applied to MODIS Imagery*. Liaoning Technical University. MA thesis.

Liao Dong, Dai Hongbao, & Xu Jiying. Comparison of Land Use Classification Methods Based on Landsat-8 Supervised Classification and Unsupervised Classification. *Journal of Henan Science and Technology*. 08(2019): 14–16.

Mu Jianxin, Li Zhongqin, Zhang Hui, & Liang Pengbin. The Global Glacierized Area: Current Situation and Recent Change, Based on the Randolph Glacier Inventory (RGI 6.0) Published in 2017. *Journal of Glaciology and Geocryology*. 40, 02(2018): 238–248.

National Oceanic and Atmospheric Administration's National Centers for Environmental Information (NOAA-NCEI, https://www.ncei.noaa.gov/). (2020).

Ning Quanke, Xie Shicheng, Zhong Chen, Yu Dan, & Yu Xiaoman. Remote Sensing Analysis of Land Use/Cover Change. *Beijing Surveying and Mapping*. 35, 07(2021): 921–925.

Shen Yongping, & Wang Guoya. Key Findings and Assessment Results of IPCC WGI Fifth Assessment Report. *Journal of Glaciology and Geocryology*. 35, 05(2013): 1068–1076.

Wang Hongbin, Jia Bowen, Wang Yetang, & Hou Shugui. Glacier Changes in Shulenan Mountain from 1973 to 2018. *Journal of Arid Land Resources and Environment*. 35, 06(2021): 60–65.

Wang Shengnan, & Wang Xiyuan. Research on Land Use Classification Method of High-Resolution Remote Sensing Image Based on ENVI. *Digital Technology and Application*. 10(2016): 105–106.

Wang Xiaoyu, & Lei Jun. Extraction of Cotton Planting Area Information in Xinjiang Based on GF-1 Satellite Data. *Xinjiang Agricultural Science and Technology*. 01(2021): 23–26.

Wang Zhuoxin, Sun Ziying, Zhou Mei, Li Mengzhen, Shu Yang, Zhao Pengwu, et al. Research on Land Use Change in Saihanwula National Nature Reserve Based on Landsat Image. *Journal of Inner Mongolia Agricultural University* (Natural Science Edition). 2021, 42(5): 26–31.

Yan Xiaotian. (2016). *Study on Evolution of Urban Spatial Pattern of Nanchang Base on Support Vector Machine and MODIS Data*. East China University of Technology. MA thesis.

Yan Xueying, Zhang Qinqin, Zhang Sicong, & Gu fang. Monitoring Research on Glaciers and Perennial Snow Change in Xinjiang from 2005 to 2015. *Geospatial Information*. 17, 08(2019): 36–39.

Yan Yan, Dong Xiulan, & Li Yan, The Comparative Study of Remote Sensing Image Supervised Classification Methods Based on ENVI. *Beijing Surveying and Mapping*. 03(2011): 14–16.

Yin Menghan, Ai Dong, & Ye Jing. Climate Change and Land Use Response of Metropolis of the Yangtze River Economic Belt: A case study of Wuhan. *Journal of China Agricultural University*. 26, 06(2021): 126–140.

Yu Guobin. (2013). *The Research on Intelligent Division Technology of Forest Resources Based on The Remote Sensing Image*. Central South University of Forestry and Technology. MA thesis.

Yuan Xinyue, Gan Shu, & Yuan Xiping. Discussion on Supervised Classification Method Based on GF-2 Satellite Data ——The Valley of XiaoJiang River in Chuandong District. *Journal of Geological Hazards and Environment Preservation*. 32. 02(2021): 78–81.

Zu Jinyan. Shoreline Extraction of Zhuhai City Based on Supervised Classification and Multi-source Remote Sensing Data. *China Science & Technology Panorama Magazine*. 2018(22): 182–183.

Advances in Measurement Technology, Disaster Prevention and
Mitigation – Li & Mohd Yusof (Eds)

Experimental research on performance of a novel differential pressure flowmeter: Olive-shaped flowmeter (OSF)

Chen Gu*, Guozeng Feng* & Yuejiao Guo*
Department of Energy and Power, Jiangsu University of Science and Technology, Zhenjiang, China

ABSTRACT: To study the flow performance of the olive-shaped flowmeter (OSF), a prototype model of the flowmeter was made, and an airflow measurement experiment was carried out. The inner diameter of the pipe used in the experiment is 50 mm, the flow medium is air, and the flow range is 50-400 m³/h. Experimental results show that the OSF has stable performance under test flow conditions with an average outflow coefficient of 0.9743 and an average pressure loss rate of 0.32, which proves it has a good energy-saving effect. Besides, the fitting model of the outflow coefficient of the OSF is obtained by analyzing the experimental data. The fitting outflow coefficient and experimental value have a high degree of agreement with a relative error of less than 1%.

1 INTRODUCTION

Flowmeter is widely used in daily life, and industrial production for the measurement of various fluid flows (Thornton & Abt 2013). There are many kinds of flowmeters, among which the differential pressure (DP) flowmeters have been widely concerned because of their versatility and low installation cost. Although the traditional DP flowmeters are convenient to use, it has the disadvantages of low measurement accuracy, poor repeatability, large pressure loss, and so on.

Researchers have made an innovative design for the geometric structure and proposed a new DP flowmeter. DiNardo et al. proposed a conical flowmeter with new geometry, which can obtain better performance than the original flowmeter. In particular, it provides a higher discharge coefficient (C_d) for a wider range of operating conditions (Dinardo et al. 2019). Sun et al. introduced a new retractable portable multifunctional flowmeter for measuring sewage flow rate (Sun et al. 2020). Fang et al. designed a new internal and external tube differential pressure flowmeter with good sensitivity and accuracy (Tlf et al. 2021).

To study the influence of different pressure loads on the accuracy of the orifice flowmeter, the mathematical model is established by simulation software, and the different pressure loads of the orifice flowmeter are analyzed under the temperature load of 20°C (Tong et al. 2019). He et al. provided insights into the effect of cavitation on the measurement of the V-Cone flowmeter and opened a new avenue for metering the cryogenic fluid (He et al. 2019). Nasiruddin et al. evaluated the performance and the flow characteristics of the Wafer cone flowmeter using Particle Image Velocimetry (PIV) (Nasiruddin et al. 2020). Tomaszewska-Wach et al. proposed a method of gas mass flow correction (Tomaszewska-Wach & Rzasa 2021).

Most of the experiments on the flow characteristics of DP flowmeters use water as the measuring fluid, and there are also a few experimental studies taking air as the measured fluid. The relationship between flow rate and pressure difference needs to be calibrated separately for different DP flowmeters and measured fluids. Previous experiments using oil as the measured fluid showed that

---

*Corresponding Authors: 199210015@stu.just.edu.cn, Frank7792@sina.com and
189210030@stu.just.edu.cn

the pressure loss of the olive flowmeter was smaller than that of the orifice flowmeter under the same flow condition, which proved that OSF has a certain energy-saving effect (Feng et al. 2020). In this paper, the outflow coefficient of the OSF is calibrated, and the pressure loss curve of air flowing through OSF is obtained. The fitted equation of the outflow coefficient is derived and verified by the experimental value.

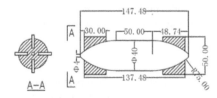

(a) The actual model of the olive cone

(b) Processing drawing of the olive cone

Figure 1.   Model and drawing of the olive cone.

2   AIRFLOW CALIBRATION EXPERIMENT

The olive cone model is shown in Figure 1. The olive throttle element is made of 304 stainless steel and supported by fins with a length of 30 mm. The experimental model of the OSF is shown in Figure 2, where the measuring pipe section is made of plexiglass, the high-pressure pressure-taking hole P_A is located 1D upstream from the cone (D is the inner pipe diameter), and the pressure-taking hole P_B is located at the waist center of the cone. The low-pressure taking hole P_C is located 1D downstream from the cone of the pipe section.

This experiment is conducted at Zhenjiang Metrological Verification and Testing Center. The flow calibration experimental platform is shown in Figure 3. The air enters the measuring pipe through the muffler and passes through the measured flowmeter, standard meter, valve, sonic nozzle metering chamber, pressure regulator tank, and vacuum pump in turn. The technical basis of calibration is JJG 640-2016 "Verification Regulation of Differential Pressure Flow Meter." The standard table is the critical flow nozzle gas flow standard device with a grade of 0.25. The measuring range of the device is 0.2~5000 m³/h, the uncertainty of the experimental system is Urel=0.3% (k=2), and the pressure stabilizing method is container pressure stabilizing. In addition, the front straight pipe length of the calibration section of the OSF is 1 m, and the back straight pipe length is 0.5 m.

Figure 2. Experimental model of the OSF.   Figure 3. Gas-phase flow calibration experimental platform.

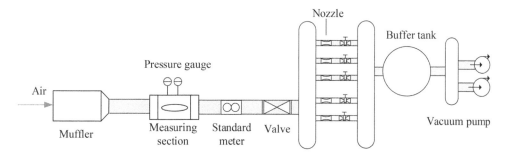

Figure 4. Airflow calibration flow chart.

The airflow experiment was carried out at eight operating points in the flow range of 50-400 m³/h. The operating points were 50 m³/h, 100 m³/h, 150 m³/h, 200 m³/h, 250 m³/h, 300 m³/h, 350 m³/h, and 400 m³/h, respectively. Each measuring point was measured three times. When the number of differential pressure sensors did not change, the data was recorded. Each measurement time lasted for 30 s, and the data within 30 s were averaged and output.

3 EXPERIMENTAL RESULTS

The airflow calibration experiment of OSF was carried out under an atmospheric pressure of 102.6 kPa, temperature of 13.2°C, and relative humidity of 59.4 %. The setting of the flow value, the measured pressure difference, and air density are shown in Table 1. According to the data measured in the experiment, the flow characteristics of the OSF in airflow are analyzed.

The volume flow of fluid was calculated by XU and Ying in 2008:

$$q_v = \frac{C_d \cdot \varepsilon}{\sqrt{1 - \beta^4}} \times \frac{\pi}{4}\beta^2 D^2 \times \sqrt{\frac{2\Delta p}{\rho}} \qquad (1)$$

where, q_v is the volume flow of the fluid, m³/s; ε is the expansibility coefficient of the fluid, for incompressible fluids, $\varepsilon=1$; for compressible fluids such as gases and vapors, $\varepsilon<1$; D is the diameter of the measuring tube, m; β is the equivalent diameter ratio, dimensionless; Δp is the pressure difference, Pa; ρ is the density of the fluid, kg/m³; C_d is the outflow coefficient, dimensionless.

For the compressible fluid, its expansion coefficient is calculated by:

$$\varepsilon = \sqrt{\left(\frac{\kappa \tau^{2/\kappa}}{\kappa - 1}\right)\left(\frac{1 - \beta^4}{1 - \beta^4 \tau^{2/\kappa}}\right)\left(\frac{1 - \tau^{(\kappa-1)/\kappa}}{1 - \tau}\right)} \qquad (2)$$

where, ε is the expansion coefficient of fluid; β is the equivalent diameter ratio, where $\beta=0.6$; κ is isentropic index, for air $\kappa = 1.4$; τ is pressure ratio, $\tau = (P - \Delta p + 101.325) / (P + 101.325)$, P is gas pressure, $P = 102.6$ kPa; δp is the differential pressure.

The expansion coefficient under different flow conditions is calculated as shown in Table 1.

Table 1.   Airflow expansion coefficient of the OSF.

	Volume flow (m³/h)	Pressure difference (kPa)	Density (kg/m³)	τ	ε
First group	50	0.22	1.2433	0.998921	0.999282
	100	0.85	1.2400	0.995832	0.997298
	150	1.84	1.2347	0.990977	0.994178
	200	3.28	1.2276	0.983916	0.989636
	250	5.21	1.2188	0.974451	0.983540
	300	7.71	1.2081	0.962192	0.975629
	350	10.79	1.1962	0.947088	0.965860
	400	14.67	1.1825	0.928062	0.953514
Second group	50	0.22	1.2428	0.998916	0.999279
	100	0.85	1.2392	0.995832	0.997298
	150	1.85	1.2339	0.990928	0.994146
	200	3.28	1.2268	0.983916	0.989636
	250	5.21	1.2176	0.974451	0.983540
	300	7.70	1.2072	0.962241	0.975661
	350	10.81	1.1952	0.946990	0.965796
	400	14.69	1.1817	0.927964	0.953450
Third group	50	0.22	1.2419	0.998921	0.999282
	100	0.85	1.2385	0.995832	0.997298
	150	1.85	1.2334	0.990928	0.994146
	200	3.28	1.2256	0.983916	0.989636
	250	5.21	1.2160	0.974451	0.983540
	300	7.69	1.2062	0.962290	0.975692
	350	10.79	1.1946	0.947088	0.965860
	400	14.67	1.1814	0.928062	0.953514

According to equation (1), the outflow coefficient of compressible fluid is shown in equation (3).

$$C_d = \frac{4q_v\sqrt{1-\beta^4}}{\pi\varepsilon\beta^2 D^2 \sqrt{2\Delta p/\rho}} \tag{3}$$

where, q_v is the volume flow of the fluid, m³/s; D is the diameter of the measuring tube, the diameter of the experimental pipe is 50mm; β is the equivalent diameter ratio, dimensionless, and the equivalent diameter ratio of the experimental prototype is 0.6; Δp is the pressure difference, Pa; ρ is the density of the fluid, kg/m³, ε is the expansion coefficient of fluid.

The outflow coefficient under different flow rates is calculated as shown in Figure 5. The average outflow coefficient of the OSF under gas-phase conditions is 0.9743. The outflow coefficient increases first and then decreases gradually with the increase of flow rate and gradually tends to be stable. Figure 6 shows the change of permanent pressure loss of the OSF under different flow flows. The permanent pressure loss increases exponentially with the increase of flow, and its index is 2.34. Because the permanent pressure loss is related to speed, the greater the speed, the greater the resistance along the path; that is, the greater the permanent pressure loss.

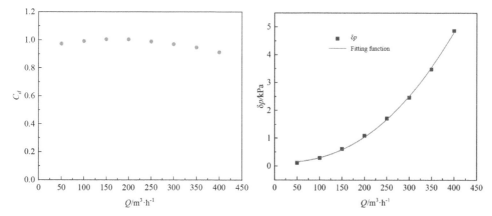

Figure 5. C_d of the OSF.　　　　　　　　　　　Figure 6. $Q - \delta p$ curve of the OSF.

Since the permanent pressure loss is proportional to the differential pressure, the pressure loss rate $\delta p/\Delta p$ (the ratio of the permanent pressure loss to the differential pressure between P_A and P_C) is introduced to measure the energy-saving characteristics of the flowmeter. In the flow range of 50-400 m³/h, the pressure loss rate is 0.6 as the initial flow rate is 50 m³/h. With the increase in flow rate, the pressure loss rate decreases and tends to be constant at 0.32.

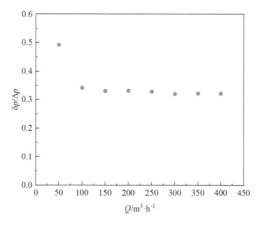

Figure 7.　$Q - \delta p/\Delta p$ of the OSF.

To obtain the airflow measurement model of the OSF, the fitting analysis of outflow coefficient C_d was carried out. The first set of data was used to perform the polynomial fitting for the outflow coefficient. The fitting diagram of the differential pressure measured in the experiment and the outflow coefficient C_d is shown in Figure 8. The fitting model coefficient of R^2 is equal to 0.99, indicating a significant fitting effect. The fitting equation is shown as follows.

$$C_d = 0.969 + 0.0322\Delta p - 0.0086\Delta p^2 + 7.48 \times 10^{-4} \times \Delta p^3 - 2.22 \times 10^{-5} \times \Delta p^4 \quad (4)$$

where, C_d is the outflow coefficient, and Δp is the pressure difference, kPa.

To verify the accuracy of the fitting model, the experimental data of the first, second, and third groups are brought into equation (4) to obtain the fitting outflow coefficient. The measured outflow coefficient and the fitting outflow coefficient are brought into equation (5) to calculate the relative error of the fitting model. The distribution of the relative error is shown in Figure 9. The relative error Δ_C of the fitting outflow coefficient is within ±1%. The fitting outflow coefficient has a high degree of coincidence with the experimental value.

$$\Delta_C = \frac{C_n - C_s}{C_s} \times 100\% \tag{5}$$

where, Δ_C is the relative error of outflow coefficient, %; C_n is the fitting value of outflow coefficient; C_s is the experimental value of outflow coefficient.

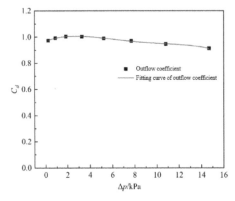

Figure 8. $\Delta p - C_d$ curve of the OSF.   Figure 9. Relative error of fitting flow coefficient.

4 CONCLUSION

The flow characteristics of the OSF in airflow are analyzed through the airflow calibration experiment. According to the experimental results, the following conclusions are obtained:

(1) The OSF has stable performance under airflow conditions, and the flow rate has little influence on the outflow coefficient. With the increase in flow rate, the outflow coefficient has a stable trend.
(2) The permanent pressure loss of the OSF increases with the increase in flow rate, and the average pressure loss rate tends to be constant at 0.32 with the increase in flow rate, which proves it has a certain energy-saving effect.
(3) The obtained fitting outflow coefficient has a high degree of coincidence with the experimental value with the relative error of less than 1%, which provides a reference for the establishment of the mathematical model of gas-phase flow measurement of the flow meter.

REFERENCES

Denghui, He, Bofeng, Bai, Senlin, & Chen, et al. (2019). Performance of v-cone flowmeter applied to cryogenic fluid measurement considering cavitation. IOP Conference Series: *Earth and Environmental Science*, 240(6), 62052–62052.
Dinardo, G., Fabbiano, L., & Vacca, G. (2019). Analysis and optimization of a cone flowmeter performance by means of a numerical and experimental approach. *Journal of Sensors and Sensor Systems*, 8(2), 269–283.
Feng, G., Guo, Y., Shi, D., Gu, C., & Sun, S. (2020). Experimental and numerical study of the flow characteristics of a novel olive-shaped flowmeter. *Flow Measurement and Instrumentation*, 76(101832), 1–14.
Nasiruddin, S., Singh, S. N., Veeravalli, S. V., & Hegde, S. (2020). Flow characteristics of back supported v-cone flowmeter (wafer cone) using PIV. *Flow Measurement and Instrumentation*, 73, 101750.

Sun, B., Chen, S., Liu, Q., Lu, Y., & Fang, H. (2020). Review of sewage flow measuring instruments. *Ain Shams Engineering Journal*.

Thornton, C. I., & Abt, S. R. (2013). Supercritical flow measurement using a large Parshall flume. *Journal of Irrigation and Drainage Engineering*.

Tlf, A., Bh, A., Xl, A., Yf, B., Zw, A., & Xl, A. (2021). Structural optimization and characteristics of a novel single-ended pressure vessel with inside and outside tubes. *Measurement*.

Tomaszewska-Wach, B., & Rzasa, M. (2021). A correction method for wet gas flow metering using a standard orifice and slotted orifices. *Sensors*, 21(7), 2291.

Tong, L., Zheng, S., Chen, X., Fan, J., & Li, J. (2019). Finite element analysis of the effect of internal pressure on orifice flowmeter. *American Journal of Mechanical and Industrial Engineering*, 4(4), 52.

XU, & Ying. (2008). Influence of key factors on discharge coefficient in v-cone flowmeter. *Chinese Journal of Mechanical Engineering*.

Advances in Measurement Technology, Disaster Prevention and Mitigation – Li & Mohd Yusof (Eds)
© 2023 The Author(s), ISBN: 978-1-032-36087-4

Study on site selection evaluation of post-earthquake reconstruction in active fault area—Taking Qingshuihe plain in Tongxin county of Ningxia as an example

Yunlin Liu & Chunfeng Li*

Shaanxi Seismological Bureau, Xi'an, China

ABSTRACT: To fully consider the destructive effect of rare earthquakes, the post-earthquake reconstruction address was selected in the Qingshuihe plain, Tongxin County, Ningxia, in the active fault area without rain. Firstly, based on the delimitation of seismic risk areas and surface fracture zones, the concept of an "effective slope length ratio" is used to evaluate the risk of geological disasters that may be induced by earthquakes in the future. Then, the paper uses the information entropy method to further narrow the site selection scope by evaluating construction and environmental factors. Finally, it selects three relatively suitable addresses and summarizes their advantages and disadvantages to provide a reference for the emergency and recovery planning of post-disaster reconstruction.

1 INTRODUCTION

Ningxia Hui Autonomous Region locates in Mainland China, at the junction of Ordos, Alexa and Qinghai Tibet. It has strong tectonic activities, fragile ecological resources and environment, and frequent and frequent natural disasters. Among them, earthquake disasters take the lead because of their strong destructive ability, large destructive area and the ability to instantly damage lifeline projects. The distribution and activity of active faults are the main causes of earthquakes (Zhang 2018). In Ningxia Hui Autonomous Region, there were two earthquakes with M 8, three with M 7 and 33 with M 5 ∼ 6 in modern times (Ma et al. 2016). The earthquake caused serious casualties and property losses in Ningxia Hui Autonomous Region. Therefore, in the urban construction and earthquake prevention and disaster reduction planning of Ningxia Hui Autonomous Region, the role of rare earthquakes should be fully considered, and it is more important to select a scientific and appropriate post-earthquake reconstruction address without rain.

The study area of this work is located in the Qingshuihe plain in the west of Tongxin County, Wuzhong City, Ningxia Hui Autonomous Region. It is adjacent to Haiyuan County, Zhongwei City, in the South and Shapotou District, Zhongwei City, in the West. The terrain is relatively flat. It is the core area of the arid zone in central Ningxia and the main belt of the "Yellow River irrigation area" (Figure 1). This area is an important transportation hub in Ningxia. There are G109, G70, Baozhong railway, S101, S304 and other roads passing through one after another. It is also distributed along Hexi Town, Tongxin county and Wangtuan Town, with concentrated infrastructure and great disaster vulnerability. This paper discusses the effective conditions of safety and convenience from the perspective of site selection, discusses the scientific principle of site selection from the perspective of safety and convenience, and seeks scientific site selection and effective disaster prevention ideas suitable for the earthquake disaster background in the study area,

---

*Corresponding Author: 514263941@qq.com

DOI 10.1201/9781003330172-11

which also provides a reference for the project construction site selection of earthquake disaster development areas in other regions.

2 BASIC INFORMATION AND RESEARCH METHODS

2.1 Basic information

2.1.1 Ground motion parameters in the study area
According to the seismic ground motion parameter zoning map of China (GB18306-2015), the basic seismic peak acceleration of a class II site is 0.2 g (i.e., VIII degree), and the characteristic period of basic seismic acceleration response spectrum of a class II site is 0.45 s, the rare seismic peak acceleration coefficient of class II site is 1.9, and the extremely rare seismic peak acceleration coefficient of class II site is 3.0.

2.1.2 Main active faults in the study area
Many earthquake disaster examples show that the active fault is not only the source of earthquakes, but also the area along the line with the most serious building damage and casualties. Rare earthquakes often cause the dislocation of several meters on the surface (Xu et al. 2002). Therefore, it is necessary to determine the width of the "avoidance zone" according to the nature of the fault to classify the risk level.

According to the detailed investigation and comprehensive study of 1:50000 geological hazards in Ningxia, the basic characteristics of the main active faults in the study area (Table 1)(Zhong et al. 2017). The following two faults are Holocene active fault zones with complex structures. According to the age analysis of the latest sediments covered on the fault, the latest fracture event occurred between 1100 and 1200 years ago.

Table 1. Characteristics of major active faults in the research area.

| Name | Length/km | Occurrence | | | Nature | Geomorphic significance |
		Trend	Inclination	Dip angle		
Yandongshan -Yaoshan Xiaoguan - east piedmont fault	>150	Arc (North East bend)	SW	60°	Shear of left-handed torsion Compression fracture	Northeast structural boundary of arc mountain
Tianjingshan fault	>220	Arc (North East bend)	SW	60 – 70°	Compression fracture	Tectonic boundary on the northeast side of middle row arc mountain

According to the research of Zhang Yongshuang et al. (Zhang et al. 2010), the width of the influence zone of earthquake surface rupture can be approximately expressed as follows:

$$D = (1 + c)L \qquad (1)$$

Where, D is the width of the influence zone of surface rupture (m), L is the inclined length of the ground with obvious topographic change (m), and is the ground shortening rate (%). According to the investigation and statistics, the researcher infers the expression between the vertical displacement of the surface fracture and the width of the influence zone as follows:

$$D = 10.11H + 16.0 \tag{2}$$

Where, H is the vertical displacement of the surface fracture.

According to the code for seismic design of buildings (GB50011-2001), when the seismic fortification intensity of class C buildings (general industrial and civil buildings) is VIII, the minimum avoidance distance of seismic fracture is 200m. When the fortification intensity is IX, the minimum avoidance distance of the seismogenic fault is 300 m (Bai et al. 2018). Based on the epicenter of the earth in the Chinese Mainland - long-term risk study (M7 Special Working Group 2012), the fault zone, historical earthquakes and other data are sorted out, and the distribution of the seismic fracture area and the judgment map of the large earthquake gap in the study area are obtained by using the ArcGIS software (Figure 1).

2.2 Research principles and methods

On the basis of conforming to and respecting the overall national policy principles, earthquake site selection should maximize the consideration of scientific, developmental, ecological conditions and people's livelihood. Among them, security and people's livelihood are the top priority of reconstruction site selection planning.

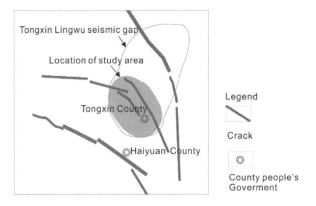

Figure 1.   Simple diagram of fault distribution in the study area.

Disaster risk assessment is the primary task of formulating reconstruction strategies, which involves the judgment of risk acceptability and risk severity. Information entropy is the basis of algorithm technology. With the help of algorithm technology, the uncertainty of people and things can be eliminated (Lin & Chen 2022; He 1995; Sun et al. 2007). Different from the common analytic hierarchy process, expert scoring and other methods, in this planning and site selection, the areas with strong restrictive factors and areas with unacceptable risks will be permanently excluded. The areas with general restrictive factors and areas with acceptable risks will use information entropy statistics to determine the weight and then comprehensively evaluate and select the final location. The evaluation process of information entropy is as follows:

$$I\,(f_i, E) = \ln \frac{N_i/N}{A_i/A} \tag{3}$$

Where: N_i is the number of f_i units distributed in existing towns within a specific category within the factor; N is the total number of units distributed in cities and towns in the study area; A_i is the number of units with evaluation factors in the study area; A is the total number of units in the study area. If $I(f_i, E) < 0$, it proves that the general restrictive factor has little relevance in the choice of urban construction; $I(f_i, E) >, 1$, the stronger the relevance.

3 EVALUATION SYSTEM

Due to the complexity of the site selection work, it is not suitable to select too many factors to participate in the evaluation. We should try to avoid the confusion and distortion caused by nature, form repetition, contradiction and other problems of the evaluation factors, and grasp the principles of objectivity, effectiveness and conciseness as much as possible. Referring to the principles, requirements and procedures for the preparation of post-disaster recovery and reconstruction planning specified in the regulations on post-Wenchuan earthquake recovery and reconstruction (Decree of the State Council of the people's Republic of China 2008), this paper selects the aspects of safety and people's livelihood to form an evaluation system:

3.1 Earthquake disaster risk

One of the principles for the selection of new sites is to firmly avoid the areas directly affected by active faults and the seismic risk areas as far as possible. The reason is that the existing building seismic technology can not compete with the surface rupture caused by sudden dislocation in an earthquake, and this risk must be avoided. Therefore, according to the introduction of fault characteristics in this area in 1.2, the Tianjingshan fault and Yandongshan Yaoshan Xiaoguan east piedmont fault are set as strongly restricted areas and unsuitable areas within 200m around; Other seismic risk areas are generally restricted areas and more suitable areas. Non-hazardous areas and areas far from other surrounding active faults are suitable.

3.2 Geological hazard risk

The risk of geological disasters shall take into account the regional vulnerability characteristics and the sections easily induced by earthquakes and causing more serious secondary disasters. Generally speaking, the risk is the combination of natural attributes and socio-economic attributes of disasters, which is expressed as the product of risk and vulnerability (Xue 2007). In the case of planning, vulnerability can be understood as equivalent, while the evaluation of risk in the future for a long time represents the overall characteristics of risk to some extent. To sum up, the concept of "effective slope length ratio" is introduced in this evaluation (Wischmeier & Smith 1978), and the calculation formula is:

$$L = (\frac{L_H}{22.13})^{\frac{\beta}{1+\beta}} \tag{4}$$

$$\beta = \frac{sina}{0.0896}/(3sina^{0.8} + 0.56) \tag{5}$$

Where, L is the effective slope length ratio, is the horizontal slope length, β Is the rill erosion rate, and a is the natural slope.

The reason why geological disaster-prone zoning is not used as the evaluation of geological disaster risk here is that in the previous evaluation of geological disaster-prone areas, the division of high, medium, low and non is made use of discrete and occurred geological disaster points to form regional classification results through various artificially constructed evaluation systems, which is highly subjective and macro (Huang 2012). If the prone zoning is selected as the evaluation standard, there is the possibility of "erasing" or "omitting" suitable areas. Since it is planning and site selection, it is based on all objective elements. Through the calculation of the effective slope length ratio, we can judge the change of surface movement in a geological environment in the future. If the activity is intense, it is easy to cause various geological disasters. Using formulas (4) and (5) to calculate 154 samples of disaster points in the study area and surrounding in the effective slope length ratio model and make sample distribution Table 2, it can be seen that 126 are located in the high-value area of effective slope length ratio (Table 2 and Figure 2), which proves that the figure and table are connected to some extent.

Table 2. Sample distribution of geohazard spots in and around the research area in the effective slope length model.

Number	Horizontal slope length/m	Slope/°	Rill erosion rate	Effective slope length ratio	Geological hazard points greater than this value range
1	42	36	2.6021	1.5886	85
2	30	29	2.4144	1.2401	41
3	22	17	1.9404	0.9961	26
4	16	7	1.6411	0.8175	1
5	9	5	0.9866	0.6397	0

According to the calculation results of the effective slope length ratio model (on the premise that there is no significant change in geomorphic space after the earthquake), the natural breakpoint method (Su 2018; Bai et al. 2018; Wang & Pan 2017; Zhu et al. 2019) is divided into the following three categories:

The effective wavelength ratio L > 1.21 is the strongly restricted area and unsuitable area, 0.65 < L < 1.21 is the general restricted area and more suitable area, and L < 0.65 is the suitable area. Then, the spatial similarity is counted according to the information entropy of the existing disaster point distribution. Among them, the area with L < 0.65 also represents greater development potential and better site engineering geological conditions, which is similar to the spatial ductility evaluation.

3.3 Evaluation of construction and environmental factors

Based on the above risk assessment and referring to similar literature (Huang 2012; Tang et al. 2015), the scope of site selection continues to be reduced by using construction elements.

3.3.1 Land type
Permanent basic farmland, lifeline project and other land are directly excluded as strong restrictive factors and unsuitable areas.

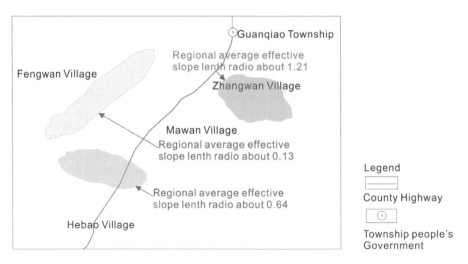

Figure 2. Mean statistics of effective slope length of 3 random areas in Guanqiao.

76

3.3.2 *Traffic*

Traffic conditions are the most critical factors in the reconstruction process. The study takes the trunk lines with good infrastructure conditions of surrounding expressways, national roads and provincial roads, relatively stable surrounding environment and strong earthquake tolerance as the main traffic network, and takes the development law of Hexi Town, Tongxin county and Wangtuan town as samples. Through statistics, it can be seen that the 1000 m area near the main traffic trunk lines is the most suitable place for reconstruction. The more adjacent roads, the higher the information entropy. The stronger the economic radiation capacity of the existing infrastructure. To sum up, the evaluation system has been basically constituted, and the summary is shown in Table 3.

Table 3. Classification of evaluating factors.

Regional classification	Type of evaluation factor
Strongly restrictive factors or unsuitable areas	Main active faults within 200m Area with an effective slope length ratio greater than 1.21 Permanent basic farmland, lifeline projects and other non-misappropriated lands
Strong restrictive factors or less suitable areas	Seismic risk area 2500 m away from the main traffic trunk line Areas with poor surface and groundwater resources
General restrictive factors or more suitable areas	Areas with an effective slope length ratio between 0.65 and 1.21 1000 ~ 2500 m away from the main traffic trunk line More than 800m away from surface water resources There are no towns and districts within 5km around
Suitable area	Areas with an effective slope length ratio less than 0.65 Area within 1000m from the main traffic trunk line Area within 800m from surface water resources There are towns and districts within 5km

3.3.3 *Water resources*

Water is a necessary condition for living. Taking the surrounding natural rivers with excellent water quality as the optimal condition, the suitability classification is defined according to the distribution of surface and groundwater.

3.4 *Comprehensive information entropy evaluation*

Based on the above layers, taking the existing urban distribution map spots in our region as the sample, the importance of each map spot classification is counted using information entropy to balance the unscientificity brought by subjective cognition in the process. The statistical results show that the five most important influencing factors for site selection are as follows (Table 4).

It is found that the calculation process of effective slope length ratio can basically cover spatial expansibility and geomorphic factors. A considerable number of sections with a low effective slope length ratio and large-area continuity have relatively open terrain, weak erosion and relatively high foundation bearing capacity, which are suitable areas for construction sites. According to the above table, we take the information entropy as the final score and delete the broken and small discontinuous areas. The final evaluation diagram is as follows (Figure 3).

Table 4. Entropy of information ranking of evaluating factors.

Importance ranking	Type of evaluation factor	Information entropy
1	Area within 150m from the main traffic trunk line	3.9813
2	There are towns and districts within 5km	2.5714
3	Areas with an effective slope length ratio less than 0.65	1.8035
4	Area within 300m from surface water resources	1.3636
5	Area 150 ~ 1500m away from main traffic trunk line	1.1302

4 CONCLUSIONS

According to the analysis results, Figure 3 shows that the most suitable area for construction is still largely located in the Qingshuihe plain, which can fully illustrate the great advantage of the original site. It is suggested to take local reconstruction measures after the earthquake when the risk is acceptable or reduced. The remaining three relatively suitable sites for reconstruction are located in Machun village, Haiyuan County–Xi'an town (site1), Magaozhuang Township, Tongxin County–Zhangjiayuan township (site2), and Xinzhuangji Township, Hongsibao district–Hongsibao town (site3), which are briefly described as follows:

(1) Machun village, Haiyuan County–Xi'an town
    It covers an area of 10.87 km² and is 26.37 km away from the designated dangerous area. It is connected by the Zhongwei Jingning highway around, and the Machunbao river passes through the center. It is 7.10 km and 12.59 km away from Xi'an town and Haiyuan county, respectively, with fair development potential and fewer geological disasters.
(2) Magaozhuang Township, Tongxin County Zhangjiayuan Township
    It covers an area of 23.93 km², 17.12 km away from the danger zone designated this time. Huinong Pingliang highway and Tongxin Qianwang highway are connected around it. It is rich in water resources, but the infrastructure conditions are relatively poor. Most of the surrounding areas are loess plateau areas, which greatly restricts economic development.
(3) Xinzhuangji Township, Hongsibao district–Hongsibao town
    It covers an area of 14.15 km² and is 5.14 km away from the danger zone designated this time. It is close to Hongsibao Town, Dahe Township and Xinzhuangji township. The surrounding infrastructure is perfect, water resources are rich, agricultural development is good, the geological environment is excellent, and transportation is very convenient. To sum up, the article concludes as follows:

4.1. According to the evaluation of information entropy, "short road" (the area within 150m from the main traffic trunk line), "near the city" (the area with towns and districts within 5km) and "spatial leveling" (the area with effective slope length ratio < 0.65) are the three most priority factors in reconstruction.

4.2. When the risk is acceptable or reduced, it is recommended to rebuild in situ after the earthquake.

4.3. If some people have to move, the recommended order is: Xinzhuangji Township, Hongsibao District–Hongsibao Town > Machun Village, Haiyuan County–Xi'an Town > Magaozhuang Township, Tongxin County–Zhangjiayuan Township.

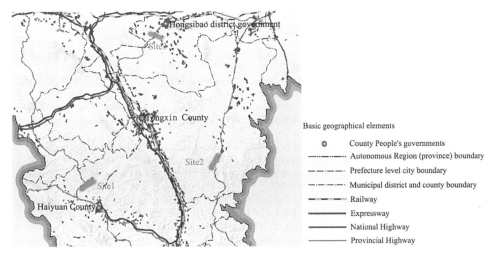

Basic geographical elements

◎	County People's governments
—··——··——··	Autonomous Region (province) boundary
—— —·—— —·	Prefecture level city boundary
—·——·—·——··	Municipal district and county boundary
————————	Railway
————————	Expressway
————————	National Highway
————————	Provincial Highway

Figure 3.   Reverse planning result of site selection of post-disaster reconstruction in the research area.

5   EXISTING PROBLEMS

5.1. This paper uses a simplified evaluation model of effective slope length ratio, which only provides researchers with an effective method, which must be revised in combination with field investigation in practical work.

5.2. The spatial evaluation of land resources is a very complex problem. The problems such as the mismatch between population and resources and the imbalance between economy and industry are always implemented. Based on anti-planning research, we should reasonably guide population flow, reasonably allocate living land and effectively gather and integrate industrial land (Li et al. 2010), so as to truly implement the spatial optimization of land resources after disasters.

REFERENCES

Bai Song, Jing Liandong, Feng Wenlan, et al. Study on the demarcation method of ecological protection red line of soil and water conservation in Sichuan Province. *Journal of Ecological Environment*, 2018, 27(4): 699–705.

Decree of the State Council of the people's Republic of China No. 526 regulations on post-Wenchuan earthquake recovery and reconstruction. 2008.

He Zongyi. Determination of map classification by information theory. *Sichuan Surveying and Mapping*, 1995, 18(1): 18–22.

Huang Siyi. Seismic risk prediction based research on disaster prevention space planning [D]. Guangxi: Guilin University of Technology, 2012.

Li Ting, Huang Zhengmin, Zhang Ximeng, et al. *Study on the carrying capacity of urban resources and environment* [M]. Shenzhen: Haitian Press, 2010

Lin Aijun, Chen Yixin. Information entropy, media algorithm and value guidance [J]. *Journal of Social Sciences of Hunan Normal University*, 2022, 51(2): 125–131.

M7 Special Working Group. *Study on long-term risk in large earthquakes in mainland China* [M]. Beijing: Seismological Press, 2012.

Ma Heqing, Xie Jiping, Huang Wei, et al. *Preliminary study on the planning of post-disaster reconstruction after major natural disasters in Ningxia* [R]. Seismological Bureau of the Ningxia Hui Autonomous Region, 2016.

Su Hongyu. *Study on ecological service function, ecological sensitivity evaluation and ecological red line delimitation in Qilian County* [D]. Xi'an: Xi'an University of Architecture and Technology, 2018.

Sun Juanjuan, Lin Zanwu, Zhang Yuling, et al. Statistical data hierarchical comprehensive evaluation model based on multi-attribute decision-making. *Surveying and Mapping and Spatial Geographic Information*, 2007, 30(5): 46–48.

Tang Qing, Xu Yong, Dong Xiaohui, et al. Evaluation of land resources security in rebuilt area after earthquake in Lushan [J]. *Journal of Geography*, 2015, 70(4): 650–663.

Wang C Y, Pan D L Zoning of Hangzhou Bay ecological red line using GIS-based multi-criteria decision analysis. *Ocean & Coastal Management* 2017, 139: 42–50.

Wischmeier W H, Smith D D *Predicting rainfall erosion losses: a guide to conservation planning* [M]. Washington District of Columbia USA: US Department of Agriculture Science and Education Administration 1978, (537): 285–291

Xu Xiwei, Yu Guihua, Ma Wentao, et al. The basis and method for determining the width of the 'avoiding belt' of the surface rupture of an active fault earthquake [J]. *Seismological Geology*, 2002, 24(4): 470–483.

Xue Qiang. Discussion on vulnerability and risk of susceptibility to geological hazards [J]. *Journal of engineering geology*, 2007, 15(S1): 124–128.

Zhang Yongshuang Characteristics, monitoring and prevention progress of seismic geological disasters in China [J]. *City and Disaster Reduction*, 2018, (3): 10.

Zhang Yongshuang, Sun Ping, Shi Jvsong, et al. Investigation of the impact zone of the surface rupture of Wenchuan earthquake and the discussion of the width of the building site [J]. *Journal of Engineering Geology*, 2010, 18(3): 312–319.

Zhong Jiaxin, Huang Wei, Yu Dongmei, et al. Comprehensive research on geohazards of 1:50000 scale of geological survey in Ningxia [R]. *Ningxia Institute of Survey and Monitoring of Land and Resources*, 2017.

Zhu Kangwen, Lei Bo, He Jun, et al. Study on the demarcation method of county scale ecological protection red line. *Three Gorges Ecological Environment Monitoring*, 2019, 4(1): 31–39.

Advances in Measurement Technology, Disaster Prevention and
Mitigation – Li & Mohd Yusof (Eds)
© 2023 The Author(s), ISBN: 978-1-032-36087-4

Study on the site selection method of highway dumping field in Jiangxi mountainous area based on GIS

Wukun Zou
Ji'an Highway Survey and Design Institute, Jian, China

Chunsheng Hu
Ji'an Highway Construction and Maintenance Center Ji Moisture Center, Jiangxi, China

Sangen Deng & Xuan Ding
Ji'an Highway Survey and Design Institute, Jian, China

Guangqing Yang*
China Merchants Chongqing Communications Technology Research & Design Institute Co., Ltd., Chongqing, China

Rongbin Zhang
Chongqing Jiaotong University, Chongqing, China

Feng Xu
China Merchants Chongqing Communications Technology Research & Design Institute Co., Ltd., Chongqing, China

ABSTRACT: In the construction of mountain roads, the dumping field is a more common off-line ancillary project, and the dumping field will have a certain impact on the surrounding environment. In view of the site selection method of the abandoned land for green highway construction, this paper studies the site selection of the abandoned land site. Firstly, the influencing factors of the abandoned land are analyzed, and three major types of influencing factors are proposed, namely ecological environment, economic performance, topography and geological structure. We need to consider the environmental impact, the economics of the cost, and the safety of site selection. For the influencing factor of site selection, we use the analytic hierarchy method to construct the structural diagram and then construct the judgment matrix of site selection. Quantified relative importance is used to indicate the importance of the impact factor to achieve the purpose of site selection. Secondly, based on the GIS system, a spatial analysis site selection model is built to analyze different factors. According to DEM, the characteristics of landforms such as soil dumps can be calculated. The data is standardized, and according to the weighted superposition analysis, the influence of multiple factors on the site selection of the abandoned land can be determined, so that the location of the abandoned land can be determined.

1   INTRODUCTION

In the research on the site selection of the abandoned land [1] for the construction of mountain roads in Jiangxi Province, in view of the lack of research methods and according to the characteristics of the nature of the dump site, three major types of influencing factors are proposed. It is necessary to protect the ecological environment while considering reasonable economic conditions, and the most critical factors are topography, landform and geological structure. Using the analytic hierarchy method and the GIS system, the collected data are standardized, and the superposition analysis

---

*Corresponding Author: 770293205@qq.com

81

method can finally obtain a reasonable dump site location. Studying the reasonable location of the abandoned land site is not only beneficial to the residents, but also beneficial to the environment. If the site selection location is improper, it will also harm the surrounding environment and bring about the impact of geological disasters. Therefore, we should pay attention to the location of the mountain road dump site and plan and study the reasonable location of the abandoned land [2].

2 INFLUENCING FACTORS OF GREEN SITE SELECTION OF WASTE DUMP

Aiming at the shortcomings of the research on the site selection method of the abandoned land based on the construction of green highways, according to the characteristics of the nature of the abandoned land engineering of the mountainous highway, three categories of the ecological environment, economic performance, topography and geological structure of the abandoned land site are proposed as the main influencing factors.

2.1 *Ecological environment*

The ecological protection redline of Jiangxi Province accounts for 28.06% of the national land area. It focuses on the protection of ecological function areas, ecologically sensitive and fragile areas, as well as other important ecological areas. The "Jiangxi Provincial Green Highway Construction Promotion Work Plan" proposes to implement ecological environmental protection design and ecological protection technology, focusing on strengthening the protection of natural landforms, native vegetation, topsoil resources, wetland ecology, and wild animals, etc. In the construction of environmentally sensitive areas, special plans for ecological environmental protection construction should be formulated, and environmental protection measures should be strictly implemented to reduce the impact of construction on the environment. In order to be able to study in detail the influence of ecological factors on the site selection of abandoned land, this project initially divides environmental factors into soil erosion, residential areas, secondary environmental disasters and other factors.

Soil erosion: The abandoned soil field has a large difference in the physical and mechanical properties of the material. The structure is loose, the pores are many, the water level is high, and there are many rainstorms in summer and autumn in Jiangxi Province. A large amount of sewage and slag are inevitably generated after the construction and completion of the abandoned soil field. These pollutants flow into the groundwater or flow to the surrounding area due to the osmosis effect, which eventually causes serious pollution to the surrounding water body, so it is necessary to consider the soil erosion of the abandoned soil field when choosing the site.

Location of residential areas: During the construction of the abandoned soil yard, the construction machinery and transport vehicles must cause serious dust, noise and other pollution, causing inconvenience to the surrounding residents in life and work, and at the same time, due to soil erosion, it will also pollute the residents' lives and irrigation water resources.

Secondary disasters: Under the influence of rainfall, earthquake or artificial engineering, the construction of abandoned land is very easy to form disasters such as mudslides and landslides in the abandoned land, which pose a threat to roads, villagers and environmental safety.

2.2 *Economic performance*

Economic factors in the highway dump site selection planning decision-making department of the primary and main control factors, according to the highway along the dumping party, we select the appropriate dump site location, achieve the sum of transportation costs and land costs to achieve the optimal value, so this category to transport costs, land costs and construction costs as the focus of research.

Transportation costs: Dumps in different locations will directly affect the distance of abandoned soil transportation and the corresponding transportation volume and eventually produce different transportation costs, so transportation must be considered an important economic factor.

Land cost: The construction of the abandoned land involves the land cost costs incurred in the process of land acquisition, which is divided into temporary land and permanent land.

Construction cost: The cost of the construction process of the abandoned land and the restoration of vegetation.

2.3 *Topography and geological structure.*

Topography and geological structure are the key factors in the safe site selection of the abandoned soil field, and the determination of the accumulation capacity, catchment area and bottom slope of the abandoned soil field directly affects the stability of the abandoned soil field in the later stage and is of great significance to the safe construction of the abandoned soil field.

Accumulation capacity: To maximize the drainage field capacity and occupy less arable land or farmland, the mountain abandoned land usually chooses a valley with steep terrain and a large height difference.

Catchment area: Surface catchment and atmospheric precipitation have a greater impact on the stability of the abandoned land and often play a decisive role. According to the survey, our dump site disasters are mostly related to water, so the catchment area is an important factor in the dumping site selection.

Geological structure: Poor geological structure will bring different geological disasters, directly threatening the safety of abandoned land. The undesirable geological phenomena that are dangerous to the project include joint fracture development, faults, rock formation non-integration, etc.

3 HIERARCHICAL ANALYSIS METHOD FOR SITE SELECTION OF ABANDONED SOIL

The site selection of the abandoned land involves three major influencing factors: ecological environment, economic performance, topography and geological structure, and dozens of influencing factors such as soil erosion, transportation costs, catchment area and accumulation capacity, etc. The target selection of multiple factors is a complex problem, and most of them are qualitatively selected by experienced designers, bringing many uncertainties.

The analytic hierarchy method refers to the decision-making method of decomposing the elements related to decision-making into levels such as goals, guidelines, and programs, and conducting qualitative and quantitative analysis on this basis. The analysis method organically combines qualitative methods and quantitative methods, so that the complex system decomposition can make people's thinking process mathematical, systematic, and easy for people to accept. Moreover, the multi-objective, multi-criteria and difficult-to-quantify decision-making problems can be transformed into multi-level single-objective problems, and the quantitative relationship between the same level of elements relative to the previous level of elements can be determined by comparison; finally, simple mathematical operations can be carried out.

3.1 *Analytic hierarchy process structure diagram*

The objectives of decision-making, the factors considered (decision-making criteria) and the decision-making objects will be selected as the most appropriate abandoned land site as the target layer according to their interrelationship, the three categories of ecological environment, economic performance, topography and geological structure are defined as the criterion layer, and dozens of site selection influencing factors such as soil erosion, transportation costs, catchment area and accumulation capacity are defined as the indicator layer. Finally, the several alternative abandoned site addresses in the previous period are defined as the scheme layer, as shown in Figure 1.

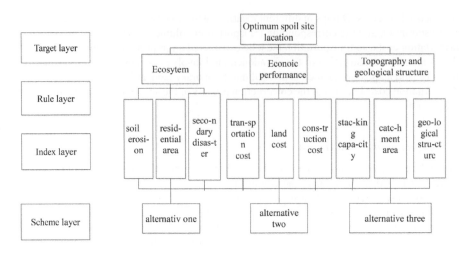

Figure 1.   Structure diagram of analytic hierarchy process.

3.2   *Constructing a site selection judgment matrix*

After the hierarchical model is constructed, for soil erosion, transportation costs, catchment area and accumulation capacity, etc., the expert scoring method is used to compare the importance of the ith element and the jth influencing factor relative to a factor in the previous layer. The quantitative relative importance is used to express it. Assuming that there are n elements participating in the comparison, the matrix is used.

$$A = \begin{pmatrix} a_{11} & \cdots & a_{1n} \\ \vdots & \ddots & \vdots \\ a_{n1} & \cdots & a_{nn} \end{pmatrix} = \left(a_{ij}\right)_{n \times n} \tag{1}$$

According to the research of many scholars, it is recommended to use 1 to 9 and its reciprocal as the scale to determine the value. This is shown in Table 4–1 below:

Table 1.   Index importance coefficient.

I is more important than j.	Be equal	Slightly strong	Strong	Very strong	Absolute strength
a_{ij}	1	3	5	7	9

Among them, values such as 2, 4, 6, and 8 can be taken to indicate the importance corresponding to 1, 3, 5, 7, and 9, respectively.

$$rule\, a_{ij} = 1/a_{ji} \tag{2}$$

4   SITE SELECTION METHOD OF MOUNTAIN WASTE SITE BASED ON GIS

The purpose of establishing a spatial analysis site selection model is to conduct spatial analysis on different influencing factors, accelerate the speed of spatial analysis, and improve the efficiency of time utilization. The spatial analysis model analyzes the influencing factors such as distance analysis, slope analysis, reclassification and weighted superposition, and finally obtains the optimal dump site location.

4.1 Digital elevation model

Digital Elevation Model (Digital Elevation Model), referred to as DEM, is to achieve digital simulation of ground terrain (that is, digital expression of terrain surface morphology) through limited terrain elevation data, which is a solid ground model that expresses the ground elevation in the form of a set of ordered numerical arrays and can calculate and obtain landform characteristics such as waste soil volume, catchment area, slope, and aspect according to DEM.

Currently, there are several ways to build a DEM, among which the data source and collection methods are as follows: (1) From the ground measurement, the instruments involved are horizontal guides, stylus, stylus holders and relative elevation measurement plates and other components, can also use GPS, total station, field measurement and other high-end instruments. (2) Another way is according to aviation or aerospace imagery, through photogrammetry access, such as stereoscopic coordinate instrument observation and air three encryption method, analytical mapping, digital photogrammetry, etc. (3) The last way is from the existing topographic map, such as grid reading point method, digitizer hand-in-hand tracking and scanner semi-automatic acquisition and then generate DEM by interpolation. It is planned to obtain DEM data around the highway selection from the geospatial data cloud or design unit.

4.2 Data standardization processing

Data standardization is mainly to unify the original values into a unified dimension through a variety of methods to facilitate comparative analysis. For example, the transport distance is divided into 1-10 levels according to the length of the transport distance, the volume of the abandoned soil accumulation according to the volume, and the catchment area according to the area. Among them, level 1 represents the best (shortest transportation distance, largest accumulation volume, smallest catchment area, etc.), level 10 represents the worst conditions (longest transport distance, smallest accumulation volume, largest catchment area, etc.), and the intermediate value can be calculated according to interpolation.

Classification principle

Primary data

4	8	10
11	4	13
2	15	18

Primary data	Standardized data
0-2	1
3-4	2
5-6	3
7-8	4
9-10	5
11-12	6
13-14	7
15-16	8
17-18	9
19-20	10

Standardized data

2	4	5
6	2	7
1	8	9

Figure 2.   Schematic diagram of data standardization processing.

4.3 *Weighted superposition analysis*

Weighted overlay analysis is one of the most commonly used overlay analysis methods in GIS, which is mostly used to solve decision-making problems with multiple influencing factors, including site selection problems. In this paper, the optimal site selection problem of the abandoned land-based on multiple influencing factors is determined by weighted superposition analysis, and the more scientific and reasonable location of the abandoned site is determined after the weighted superposition analysis based on the weighted superposition analysis.

5 CONCLUSION

This paper takes the selection and planning of abandoned land for highway construction in mountainous areas of Jiangxi Province as an example. The selection, planning and design of the site are the basic work of the construction of abandoned land. This paper studies the three major influencing factors of green site selection, ecological environment, economic performance, topography and geomorphology and geological structure. Since the choice of multiple factors is a complex problem, bringing many uncertainties, a hierarchical analysis of the location of the abandoned land is carried out. A structure diagram and a judgment matrix for site selection are constructed. The GIS platform is used to build a model, and different factors are analyzed. Through the analysis of the spatial model, we can finally determine the optimal location of the dump site. The location of the abandoned land is an important issue. If the design is not properly affected people's lives and the surrounding environment, the rational use of the abandoned land will bring many benefits. Since site selection and design issues are the basic work, we must study them and propose a reasonable site selection plan for the abandoned land. The future site selection study of abandoned land is still an important issue; hence, we should do a good job of designing from the impact factor to develop a reasonable location plan for the abandoned land.

ACKNOWLEDGMENT

The authors would like to acknowledge the financial support of the Science and Technology Project of the Jiangxi Provincial Department of Transportation (2020H0026, 2020H0052).

REFERENCES

[1] Han Xin, Zhang Minjing. Discussion on location selection and design principles and methods of spoil ground [J] *Highway*, September 2008, 282–285.
[2] Jiang Erzhong.Environmental Impact Assessment and Countermeasures of Expressway in Mountainous Areas. *Scientific and Technological Information Development and Economy*. 2004, 14(4): 207–208.
[3] Li Zhilin, Zhu Qing. *Digital Elevation Model* [M]. Wuhan: Wuhan University Press, 2003: 18–98.

Advances in Measurement Technology, Disaster Prevention and Mitigation – Li & Mohd Yusof (Eds)
© 2023 The Author(s), ISBN: 978-1-032-36087-4

Study on long-term performance monitoring technology of loess tunnel based on distributed optical fiber

Guang Yao*

Shanxi Transportation Technology Research & Development Co., Ltd., Taiyuan, Shanxi, China

ABSTRACT: A large number of tunnels in China cross loess strata, and the deterioration of surrounding rocks during operation leads to large stress on supporting structures. Long-term performance monitoring is of great significance for the formulation of maintenance plans and the reduction of tunnel operation risks. This paper uses numerical simulation, field tests and other methods to analyze the key monitoring positions of the lining of the loess tunnel during operation. Relying on the Qiaoyuan tunnel, a long-term performance monitoring system based on distributed optical fiber is constructed. The results show that distributed optical fiber can accurately monitor lining cracks in the loess tunnel operation stage, which is of great significance to the development of maintenance plans and the reduction of tunnel operation risk.

1 INTRODUCTION

About 75% of China's land area is mountainous or hilly. The tunnel scheme has the advantages of shortening the driving distance, improving the linear standard, ensuring the safety of operation and protecting the ecological environment, which has been widely used in the construction of mountain expressways. By the end of 2020, China had 21,316 road tunnels in operation, with a total length of more than 21,000 kilometers. However, due to China's vast territory, significant regional differences and complex geological conditions, a large number of tunnels inevitably pass through the loess stratum. The physical properties of loess are loose, porous, vertical joint development, water seepage, and a lot of soluble substances, easy to be eroded by running water. In addition, loess has significant structure and strong water sensitivity, resulting in serious deterioration of surrounding rock of loess tunnel, which is easy to produce geological disasters such as caves, water holes, ground fissures and local subsidences, so that the health status of loess highway tunnel faces severe challenges in the construction and operation process[1,2]. Liang Siming[3] used a z-shaped optical cable to monitor the differential settlement of segments in the shield tunnel. Hou Gongyu[4] used distributed optical fiber to monitor the roof of the coal mine roadway, and the results show that the optical fiber strain change can accurately reflect the deformation of the roof in real-time. Wu Jinghong[5] applied the fiber Bragg grating monitoring technology to the structural health monitoring of the large-section tunnel of the Jingxiong High-speed Railway. Shen Sheng[6] explored the convergence deformation monitoring method of shield tunnel cross-section based on distributed fiber optic strain sensing technology.

In the process of loess tunnel operation, the deterioration of surrounding rock leads to large stress on the supporting structure, which leads to collapse, block falling, lining cracking, water leakage, inverted arch cracking, basement uplift and other diseases. Due to the special operating environment of tunnels, it is relatively difficult to monitor the structural health, and the traditional monitoring method requires a large number of sensors, heavy workload and high cost. In recent years, with the optical fiber sensing technology progress, it has the advantages of distribution, long distance, strong real-time performance, high precision and durability, etc. It can perceive and remotely monitor every part of large-scale infrastructure projects like the nervous system, accurately locate

---

*Corresponding Author: typeeasy2009@sina.com

tunnel diseases, and provide strong technical support for loess tunnel monitoring. In this paper, numerical simulation, field test and other methods are used to analyze the key monitoring positions of loess tunnel lining during the operation period, and a long-term performance monitoring system based on distributed optical fiber is built based on the Qiaoyuan tunnel to monitor the stress of tunnel lining for a long time.

2 ANALYSIS OF KEY MONITORING POSITIONS OF LOESS TUNNEL LINING

2.1 *Tunnel numerical model*

Taking Qiaoyuan Tunnel in Jixian to Hejin highway as the supporting project, this paper uses numerical analysis software to establish a simulation model to simulate the changes of internal forces and other indicators of the tunnel during the operation period and analyzes the key monitoring positions of the lining of the loess tunnel during the operation period. The simulation model is established, as shown in Figure 1. The buried depth of the highway tunnel is 67m, the longitudinal length of the tunnel is 60m, and the distance between the side wall and both sides' boundaries is 36m. The boundary conditions all meet the requirements of three times hole diameter. According to the design document, the Qiaoyuan tunnel is mainly surrounded by Upper Pleistocene (Q_3^{eol}) and a drifted deposit layer of Middle Pleistocene (Q_2^{al+pl}), and the grade of surrounding rock is V. The three-bench-seven-step excavation method is adopted to simulate tunnel excavation, as shown in Figure 2.

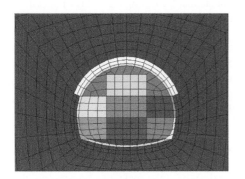

Figure 1.   Tunnel numerical model.

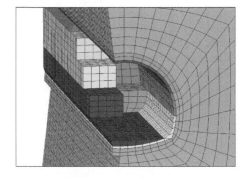

Figure 2.   Tunnel excavation method.

The main material parameters are selected as follows:

Table 1.   Physical and mechanical parameters of surrounding rock and supporting structure.

Material	E (GPa)	υ	ρ (kg/m^3)	C(kPa)	φ (°)
Surrounding rock	0.6	0.3	1800	40	25
Shotcrete	28.5	0.22	2400	–	–
Secondary lining	30	0.2	2600	–	–
Bolt	210	0.3	8000	–	–
Steel frame	210	0.3	8000	–	–

2.2 Result analysis

The stress characteristics of the second lining of the tunnel are analyzed. The internal force of the structure during tunnel construction is shown in Figure 3, and the stress on the arch waist and arch foot is large, which should be regarded as the focus of tunnel monitoring in the operation period.

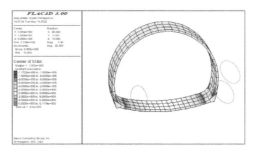

Figure 3.   Secondary lining maximum principal stress.

3   LONG-TERM PERFORMANCE MONITORING SYSTEM OF LOESS TUNNEL BASED ON DISTRIBUTED OPTICAL FIBER

In this project, the technology of grooving and gluing was adopted to lay the embedded long-distance optical fiber sensor in the concrete of the secondary lining of the loess tunnel, thus ensuring the accuracy and stability of the distributed optical fiber sensor in the long-term monitoring process. The specific steps are shown in Figure 4. The optical fiber layout includes annular sensing fibers

(a) Grooving on the surface   (b) Embedding optical fiber

(c) Fixed optical fiber by grouting   (d) Plastered carbon fiber cloth

Figure 4.   Procedure for laying optical fibers.

along the tunnel arch ring and longitudinal sensing fibers at the arch waist positions on both sides of the tunnel, as shown in Figure 5.

Figure 5.   Optical fiber global distribution.

At the same time as embedding distributed optical fiber, strain sensors are installed at the arch waist and arch foot to verify the monitoring results of distributed optical fiber. The internal strain gauge of concrete is shown in Figure 6.

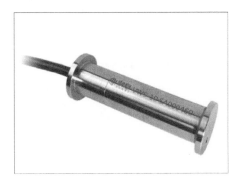   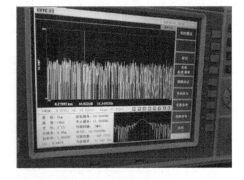

Figure 6.   Concrete internal strain gauge.   Figure 7.   Equipment debugging.

In the field monitoring of this project, the optical fiber modulation and demodulation instrument model AV6419 was adopted, with strain testing accuracy of $\pm20\mu\varepsilon$ and spatial resolution of 0.25m, as shown in Figure 7. The optical fibers in this section are configured with 43 meters in total, 0–15 meters in a horizontal layout, 15–32 meters in a circular layout, and 32–43 meters in a horizontal layout. In this monitoring period, the total time is seven months. The representative concrete strain values of tunnel lining five times are shown in Figure 8.

It can be seen that the strain values of most monitoring points in the whole monitoring area vary from 0 to $120\mu\varepsilon$, and the corresponding crack width is 0.120mm, which is smaller than the reinforced concrete crack width limit of 0.20mm. The lining of this part of the tunnel is in a stable and safe state. As can be seen from the figure, when the length of optical fiber is 15m, 23m and 31m, the strain values of these three places increase significantly, corresponding to the left arch waist, the vault and the right arch waist, respectively. It can be seen that the stress distribution obtained by using distributed optical fiber to monitor the secondary lining concrete is similar to that of the primary support, and the stress on the arch waist and corner of the loess tunnel is large, so the anchor bolt support should be strengthened in the construction.

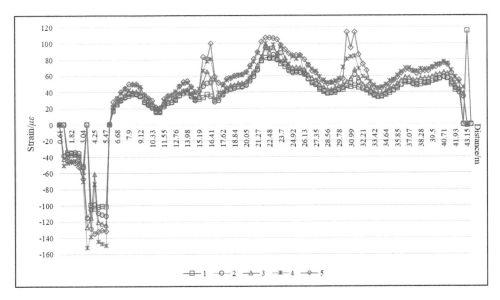

Figure 8. Strain distribution diagram of tunnel lining structure.

The distributed fiber optic monitoring of secondary lining stress is compared with the monitoring data of the concrete strain gauge. The stress changes in main positions are shown in Figure 9.

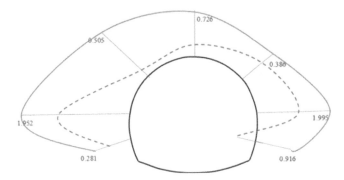

Figure 9. Stress distribution diagram of tunnel lining structure (MPa).

There is a certain difference between the strain value obtained by using distributed optical fiber monitoring and the vibrating string sensor. The possible reason is that the laying process needs to be further optimized, and the optical fiber is too sensitive, which causes a certain influence on the optical fiber when cleaning the lining.

Based on the analysis of monitoring results above, distributed optical fiber can accurately monitor lining cracks in the loess tunnel operation stage, which is of great significance to the formulation of the maintenance plan and the reduction of tunnel operation risk.

4   CONCLUSION

A large number of tunnels in China cross loess strata, and the deterioration of surrounding rocks during operation leads to large stress on supporting structures. Long-term performance monitoring is of great significance for the formulation of maintenance plans and the reduction of tunnel operation risks. This paper used numerical simulation, field test and other methods to analyze the key monitoring positions of the lining of the loess tunnel during operation. Relying on the Qiaoyuan tunnel, a long-term performance monitoring system based on distributed optical fiber was constructed to monitor the stress of tunnel lining for a long time, and the following conclusions were obtained.

1. Numerical simulation method was used to analyze the stress characteristics of the second lining of the tunnel. Arch waist and arch foot should be taken as the strengthening parts during construction and the key monitoring points of the operation tunnel.
2. The groove glue injection technology will be embedded in long-distance optical fiber sensor layout in the loess tunnel secondary lining concrete, can guarantee the distributed optical fiber sensors in the process of long-term monitoring accuracy and stability, the use of type AV6419 BOTDR distributed optical fiber strain tester is on-site readings, realize the operation period of loess tunnel lining structure health monitoring for a long time.
3. There is a certain difference between the strain value obtained using distributed optical fiber monitoring and the vibrating string sensor. The possible reason is that the laying process needs to be further optimized, and the optical fiber is too sensitive, which causes a certain influence on the optical fiber when cleaning the lining.

ACKNOWLEDGMENTS

This work was financially supported by the Research Project of Shanxi Transportation Holdings Group Co., Ltd. [20-JKKJ-18] and the Innovative Development Plan Project [21-JKCF-04].

REFERENCES

Hou GY, Hu T, Xv GCh, et al. (2020) Coal Mine Roadway Roof Monitoring System Based On Distributed Optical Fiber Technology. *Industry and Mine Automation*, 46(01):1–6.

Liang SM, Xie ChL, Jiang E, et al. (2020) Application of Distributed Optical Fiber Technology in Tunnel Deformation Monitoring. *Tunnel Construction*, 40(S1):436–443.

Shen Sh, Wu ZhSh, Yang CQ, et al. (2013) Convergence deformation monitoring of shield tunnels based on distributed optical fiber strain sensing technique. *China Civil Engineering Journal*, 46(09):104–116.

Wu JH, Ye ShM, Zhang JQ, et al. (2020) Structural Health Monitoring of Large-section Tunnel of Jingxiong High-speed Railway Based on Fiber. *Laser & Optoelectronics Progress*, 57(21):113–121.

Wu RB. (2021)Improvement and Upgrade Scheme Design of Highway Tunnel. *Shanxi Science & Technology of Communications*, 3:76–79.

Yao G. (2020). Loess Tunnel Lining Cracking Analysis and Monitoring System Research, In: ICTIM 2016. Xi'an. 760–766.

Advances in Measurement Technology, Disaster Prevention and Mitigation – Li & Mohd Yusof (Eds)
© 2023 The Author(s), ISBN: 978-1-032-36087-4

Research on permeability coefficient of fractured rock mass

Zhaowei Yu*
XAUAT UniSA An De College, Xi'an University of Architecture and Technology, Xi'an, China

Xiang Wang*
Guiyang Urban Rail Transit Group Co., LTD., Guiyang, China

ABSTRACT: Since the 21st century, the focus of China's transportation infrastructure has gradually shifted to the mountainous and karst areas in the central and western regions. The construction of tunnels and underground engineering is faced with severe tests of complex structure, changeable geological environment and frequent disasters. For this reason, this paper mainly conducts seepage research on fractured rock masses by constructing a dual-media seepage model. It is assumed that the fault fracture zone is a continuous medium composed of a pore system and a fracture system. The two systems overlap and are evenly distributed throughout the media system. The fractured rock mass is considered to be composed of two media: fractured media for water conduction and porous rock media for water storage. The seepage model of fractured media and the seepage model of porous media were established for the two media, respectively. The pores and pore rock mass alternate formulas are used to connect to form a coupled equation to solve. At the same time, for the dual-medium seepage model, two Richards equations combined with water exchange function are used to form coupled partial differential equations to express saturated and unsaturated seepage. Finally, the calculation formula of the total permeability coefficient K of the fault fracture zone is obtained.

1   INTRODUCTION

China is the country with the largest scale, quantity and difficulty of the tunnel and underground engineering construction in the world. The construction of transportation infrastructure plays an important role in the national economy. Since the beginning of the 21st century, a large number of major infrastructures, such as railway and highway projects in China, have been put on the construction schedule one after another. At the same time, the focus of China's transportation infrastructure has gradually shifted to the mountainous and karst areas in the central and western regions. The construction of tunnels and underground engineering is faced with severe tests of complex structure, changeable geological environment and frequent disasters. According to statistics, nearly 80% of major safety accidents in the construction of underground projects such as transportation and hydropower are caused by karst geological water inrush (mud outburst) disasters and improper disposal. The accident caused heavy casualties, economic losses, and delays in construction schedules and even forced tunnel construction to be suspended or rerouted. Therefore, the study of karst phenomena in these areas is of great significance.

This paper mainly studies the seepage of fractured rock masses by constructing a dual-media seepage model. The dual-medium seepage model was first proposed by Soviet scholar Barenblatt (Kong et al. 2007) in 1960. Wang Enzhi (Kovama & Jing 2007), Wu Yanqing (2010) and others conducted research. The dual-media seepage model regards the fractured rock mass as two media consisting of fractured media for water conduction and porous rock media for water storage. The fractured media seepage model and the porous media seepage model are established, respectively, and the pore and pore rock mass alternate formulas are used to connect to form a coupled equation solution. Different from the seepage properties of rock blocks, the dual media seepage model is

---

*Corresponding Authors: 1293500750@qq.com and 601498431@qq.com

divided into narrow dual media and general dual media. In a narrow sense, dual media is defined as "a water-containing medium with hydraulic connection formed by the coexistence of porous media and fractured media in a rock mass system." The generalized dual-medium is defined as "a continuous (or equivalently continuous) medium and a discontinuous network medium coexisting in a rock mass system to form a hydraulically connected water-containing medium."

2 MECHANISM OF WATER INRUSH AND MUD OUTBURST IN FAULT FRACTURE ZONE

When deep-buried tunnels pass through the water-rich fracture zone, water inrush disasters are prone to occur during the construction process. The influencing factors can be attributed to the special hydrogeological conditions of the fault fracture zone and the construction disturbance during the tunnel excavation and support process.

The special hydrogeological conditions in the fracture zone of the fault are the root cause of major water hazards during the construction of near-fault tunnels. Affected by multi-phase tectonic movement, fault geological structure is very complicated (Liu et al. 2011). The joints and fissures of the rock mass are abnormally developed, and the self-stabilization ability is extremely weak. In addition, the water-bearing structure and water-conducting structure in the fault fracture zone form a potential water inrush channel around the tunnel. It has a strong ability to cause disasters and determines the risk of water inrush to a large extent.

The construction disturbance during the excavation and support process is the direct cause of the tunnel water inrush accident. Past engineering practice has proved that the seepage field and the in-situ stress field of the original geological conditions are in a certain equilibrium state without being disturbed by the construction (Sherard et al. 1984). In other words, the direct cause of the tunnel water inrush accident is not the water control advantage of the structural junction. It is the engineering disturbance that changes the seepage state of groundwater and causes the formation of structural belts to be deformed and activated (Zhang & Xu 2010).

According to the research, it can be considered that the water inrush from the fault is essentially the mechanical balance of the broken surrounding rock and the balance of the groundwater seepage change drastically due to the disturbance of the tunnel construction. At the same time, it causes the redistribution of stress and the instantaneous release of groundwater energy, a kind of dynamic damage phenomenon that quickly rushes to the free surface of the project in a fluid state.

After the tunnel construction revealed the fault, the seepage field, stress field and displacement field of surrounding rock constantly change under the drag of groundwater penetration. The hydro-dynamic pressure on the rock particles near the water inrush exceeds the cohesion between the particles. The limit equilibrium state of the rock particles is broken. The fluid in the pore space migrates and forms a new percolation channel as the seepage action time prolongs. The seepage channel quickly becomes larger, making the exposed location a drainage channel for groundwater. For deep-buried tunnels, the upper rock and soil body can effectively prevent groundwater recharge by surface water. Therefore, the rapid drainage of groundwater will cause the groundwater level to drop sharply, causing a significant increase in hydraulic gradient. At this time, the rock and soil near the bedrock surface generate instantaneous negative pressure. The permeability of the rock mass also increased rapidly. Locally weak rock masses will gush out along with the water flow and eventually cause qualitative changes. Water and mud inrush accidents occurred in the tunnel. With the development of water inrush and mud, more and more groundwater and rock and soil are lost. The upper broken rock mass is less and less supported by the surrounding complete rock mass. Under the action of vacuum absorption force, the rock and soil in the weak and broken zone will collapse.

3 RELATED THEORIES OF EXISTING FAULT FRACTURE ZONE ROCK MASS SEEPAGE MODELS

The rock mass has its properties and complex geological environment, especially the fault zone rock mass, which has the characteristics of water richness and anisotropy. The rock mass in the fault zone weakens the strength of the rock mass and makes the seepage-stress coupling extremely complicated. The research on seepage of broken rock masses is mostly concentrated in water conservancy projects. The representative formulas include the Hazen formula (Zhang 2015), Cosen formula, Uyebrie formula and so on.

Hazen formula:

$$k = cd_{10}^2 \tag{1}$$

Cosson formula:

$$k_{18} = 780 \frac{n^3}{(1-n)^2} d_9^2 \tag{2}$$

Uyebrie formula:

$$k_{18} = c \frac{n^2}{(1-n)^2} d_{18}^2 \tag{3}$$

Kondrachov formula:

$$k_{18} = 105n(\eta D_{50})^2, \quad \eta = \frac{D_n}{D_{100-n}} \tag{4}$$

The formula of Water Resources and Hydropower Research Institute:

$$k_{10} = 234n^3 d_{20}^2 \tag{5}$$

In the above formula, d is the effective particle size; C is the coefficient, the median value of different formulas is 100–350; k is the rock mass permeability coefficient, and the subscript is temperature; n is the porosity; D is the particle size.

4 THEORETICAL DERIVATION OF DUAL-MEDIUM SEEPAGE MODEL

It is assumed that the fault fracture zone is a continuous medium composed of a pore system and a fracture system. The two systems overlap and are evenly distributed in the entire medium system. Select the model for permeability research. First, the actual distribution and development characteristics of geological fractures are generalized into the volume fraction of the fracture domain in the fractured zone (Zhang 2015). We determine the volume fraction of the fracture domain based on the development of the crack in the actual project. The seepage flow in the fracture zone is considered the priority seepage flow, represented by the subscript f. The seepage flow in the pore domain is considered matrix flow, denoted by the subscript m. The coupling between the fracture domain and the pore domain is realized by water exchange terms. The macroscopic model of the unit body seepage is shown in the figure. At the same time, for the dual-medium seepage model, two Richards equations combined with water exchange function are used to form coupled partial differential equations to express saturated and unsaturated seepage. The partial differential equations are:

$$(C_f + S_e S_s) \frac{\partial h_{pf}}{\partial t} - \nabla[K_f(\nabla h_{pf} + 1)] = -\frac{\Gamma_w}{w_f} \tag{6}$$

$$(C_m + S_e S_s) \frac{\partial h_{pm}}{\partial t} - \nabla[K_m(\nabla h_{pm} + 1)] = \frac{\Gamma_w}{1-w_f} \tag{7}$$

In the formula, C is the specific water capacity ($d\theta/dh$); S_e is the effective saturation; S_s is the storage coefficient; h_{pf} and h_{pm} are the pressure heads of the fracture domain and the pore domain, respectively; K_f and K_m are the unsaturated permeability coefficients of the fracture domain and the pore domain, respectively; Γ_w is the water exchange term; w_f is the volume fraction of the fissure domain.

The water exchange term is expressed as Γ_w in the coupled equations (1) and (2). When $\Gamma_w > 0$, the water exchange is from the fracture domain to the pore domain. On the contrary, when $\Gamma_w < 0$,

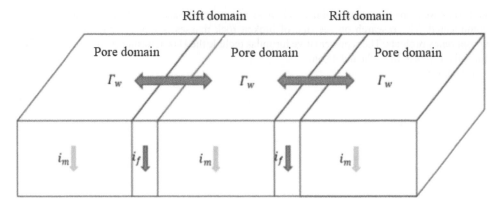

Figure 1.   Macroscopic model of unit infiltration in dual media.

it means that the water exchange flows from the pore domain into the fracture domain. At the same time, assuming that the water exchange term is proportional to the head pressure of the pore domain and the fracture domain, the water exchange term equation can be expressed as:

$$\Gamma_w = \alpha_w K_a(h_{pf} - h_{pm}) \tag{8}$$

In the formula, α_w is the effective water exchange coefficient; K_a is the effective permeability coefficient. The effective permeability coefficient K_a is the average value of the unsaturated permeability coefficients of the fracture domain and the pore domain, and the calculation formula is:

$$K_a = (K_f + K_m)/2 \tag{9}$$

At the same time, the unsaturated permeability coefficient of the fracture domain and pore domain can be calculated by the following formula:

$$K_f = K_{sf} k_{rf} \tag{10}$$

$$K_m = K_{sm} k_{rm} \tag{11}$$

In the formula, K_{sf} and K_{sm} are the saturated permeability coefficients of the fracture domain and the pore domain, respectively; k_{rf} and k_{rm} are the relative permeability coefficients of the fracture domain and the pore domain, respectively.

Finally, the calculation formula for the total permeability coefficient of the fault fracture zone is:

$$K = w_f K_f + (1 - w_f)K_m \tag{12}$$

The unsaturated hydraulic parameters in the above equations C, S_e, k_r, θ can be calculated by the van Genuchten model or the Brooks-Corey model.

5   CONCLUSION

This article assumes that the fault fracture zone is a continuum body composed of a pore system and a fracture system. Two Richards equations combined with the water exchange function are used to form a coupled partial differential equation system to express saturated and unsaturated seepage flow. The coupling between the fracture domain and the pore domain is realized by water exchange terms. It is further assumed that the water exchange term is proportional to the head pressure of the pore domain and the fracture domain.

Based on the derivation of the appeal formula, we can get this conclusion:

(1) Let $\Gamma_w > 0$ denotes the water exchange flows from the fracture domain into the pore domain and $\Gamma_w < 0$ denotes the water exchange flows from the pore domain into the fracture domain. The equation for the water exchange term can be obtained as:

$$\Gamma_w = \alpha_w K_a (h_{pf} - h_{pm}) \tag{13}$$

(2) The final calculation formula for the total permeability coefficient of the fault fracture zone is:

$$K = w_f K_f + (1 - w_f) K_m \tag{14}$$

Obtaining the total permeability coefficient K of the fault fracture zone is beneficial to the related permeability calculation. At the same time, in the permeability calculation process for other specific projects, the accuracy of the permeability coefficient can be further tested. The formula can be further modified and perfected according to the inspection in the follow-up.

REFERENCES

Kong Liang, Wang Yuan, Xia Junmin.(2007) Unsaturated fluid-solid coupling dual-porosity media model control method Cheng[J]. *Journal of Xi'an Shiyou University* (Natural Science Edition), 22(2): 163–165, 183.

Kovama T, Jing L.(2007) Effects of Model Scale and Particle Size on Micro Mechanical Properties and Failure Processes of Rocks - A Particle Mechanics Approach[J]. *Engineering Analysis with Boundary Elements*, 31(3): 458–472.

Liu Yang, Li Shihai, Liu Xiaoyu.(2011) Dual-medium seepage response based on discrete element of continuum Force coupling model[J]. *Chinese Journal of Rock Mechanics and Engineering*, 30(5): 951–959.

Sherard JL, Dunnigan LP and Talbot JR.(1984) Basic properties of sand and gravel filters[J]. *Geotechnical Engineering*, 110(6).

Zhang Guike, Xu Jia.(2010) Research on Orthotropic Equivalent Deformation Parameters of Jointed Rock Mass[J]. *Chinese Journal of Geotechnical Engineering*, 32(6): 908–915.

Zhang Linlin.(2015) *Numerical simulation study on the migration law of water and gas two-phase flow in fractured rock mass*[D]. Liaoning: Liaoning Technical University.

Zhang Linlin.(2015) *Unsaturated soil mechanics*[M]. Beijing: Higher Education Press.

Advances in Measurement Technology, Disaster Prevention and
Mitigation – Li & Mohd Yusof (Eds)
© 2023 The Author(s), ISBN: 978-1-032-36087-4

Comprehensive safety risk assessment of foundation pit construction based on FAHP-SPA coupling

Jian-feng Liu*
CCCC Fourth Harbor Engineering Co., Ltd., Guangzhou, China
Key Laboratory of Environment Protection & Safely of Transportation Foundation Engineering of CCCC, Guangzhou, China
Southern Marin Science and Engineering Guangdong Laboratory (Zhuhai), Zhuhai, China

Lu Yang*
China Harbour Engineering Co., Ltd, Beijing, China

Kun-peng Wu*
CCCC Fourth Harbor Engineering Co., Ltd., Guangzhou, China
Key Laboratory of Environment Protection & Safely of Transportation Foundation Engineering of CCCC, Guangzhou, China
Southern Marin Science and Engineering Guangdong Laboratory (Zhuhai), Zhuhai, China

Jun-xing Luo*, Hong-xing Zhou* & Qing-chang Qiu*
CCCC Fourth Harbor Engineering Co., Ltd., Guangzhou, China
China Harbour Engineering Co., Ltd, Beijing, China
Key Laboratory of Environment Protection & Safely of Transportation Foundation Engineering of CCCC, Guangzhou, China

ABSTRACT: To scientifically and effectively carry out a comprehensive safety risk assessment in the construction process of foundation pit engineering, a complete safety risk assessment model of foundation pit based on construction monitoring data coupled with fuzzy analytic hierarchy process and set pair analysis method (FAHP-SPA) is established. First, a comprehensive safety risk evaluation system for foundation pits is established with monitoring items as evaluation indicators; then, the traditional FAHP evaluation index empowerment process is improved, the principle of expert group decision-making is adopted, and the set pair analysis method is introduced to consider the identity and consistency of opinions among experts. Finally, the safety risk classification of the foundation pit is carried out with the monitoring and alarm value as a reference, and the on-site monitoring data is used as the analysis data.

1 INTRODUCTION

Foundation pit engineering is a multi-factor complex system with many uncertainties such as randomness, grayness, fuzziness, and so on. (Wang et al. 2020; Xia et al. 2016; Zhang et al. 2016). However, most evaluation methods only analyze some of the uncertain characteristics and cannot fully excavate the uncertainty information for security risk systems of foundation pits. In addition, due to the insufficient statistics of historical data on construction safety risks and their obvious regional characteristics, the safety evaluation process still needs to rely on the experience and wisdom of the expert group.

---
*Corresponding Authors: ljianfeng@ cccc4. com, 85722757@ qq. com, wkunpeng@ cccc4. com, ljunxing@ cccc4. com, zhongxin@ cccc4. com and qqingchang@ cccc4. com

DOI 10.1201/9781003330172-15

Therefore, in order to ensure the safety of the foundation pit engineering system, this paper attempts to introduce the set-pair analysis theory into the safety evaluation of the foundation pit engineering construction to fully describe the relationship between the certainty and uncertainty of the foundation pit construction risk factors. With the advantages of group decision-making and fuzzy theory and analytic hierarchy process, a fuzzy hierarchy and set pair analysis (FAHP-SPA) coupling model for comprehensive safety risk assessment of foundation pit construction was established based on on-site monitoring data. The evaluation results will be more scientific and reasonable to provide a new idea for foundation pit engineering safety evaluation.

2 BASIC THEORY

2.1 Fuzzy analytic hierarchy process

FAHP considers the fuzziness of decision-making information and has the advantages of a clear hierarchy, clear logic, and strong systematicness(Xu 1988). The implementation process is as follows:

(1) The triangular fuzzy judgment matrix is constructed based on the 1–9 scale criterion (refer to literature 4), but in order to expand the selection range of the importance scale value, the 1–9 scale criterion is improved. When the importance degree is between the aforementioned importance degree levels, we use the real number between adjacent importance level scales for quantization.

$$X = \left(x_{ij}\right)_{n \times n} = \begin{bmatrix} x_{11} & x_{12} & \cdots & x_{1n} \\ x_{21} & x_{22} & \cdots & x_{2n} \\ \vdots & \vdots & \vdots & \vdots \\ x_{n1} & x_{n2} & \cdots & x_{nn} \end{bmatrix} \tag{1}$$

Where x_{ij} is the i-th row and j-th column element of the triangular fuzzy judgment matrix X; m_{ij}, l_{ij}, u_{ij} are the possible evaluation value, the lowest evaluation value and the highest evaluation value of the relative importance of index i to index j considered by experts respectively; they have the following properties: 1) $x_{ij} = (l_{ij}, m_{ij}, u_{ij})$, 2) $u_{ij} \geq l_{ij} \geq m_{ij} > 1$, 3) $x_{ii} = (1,1,1)$, 4) $x_{ji} = (1/u_{ij}, 1/m_{ij}, 1/l_{ij})$; n is the order of the matrix.

(2) Consistency test of conversion judgment matrix Y(Pan et al. 2010), we use formula (2) to convert matrix X to obtain conversion judgment matrix $Y = (y_{ij})_{n \times n}$.

$$y_{ij} = \frac{l_{ij} + 4m_{ij} + u_{ij}}{6} \tag{2}$$

(3) To ensure the global logical consistency of expert judgment information, the method in the literature 4 is used to test the consistency of the transformation judgment matrix Y.

2.2 Set pair analysis theory

SPA theory (Zhao 2000) is a system theoretical method to analyze and determine uncertainty problems. This method first forms a set pair system between two sets with a certain relationship, such as set A and set B and sets pair $H = (A, B)$. Then in the context of a given problem, it analyzes the identification, difference, and opposition of the characteristics of the two sets and finally uses the connection degree expression to classify and quantitatively describe.

$$\mu = a + bi + cj \tag{3}$$

where μ is the main value of the connection number; a, b, and c are generally called the connection number components, which are the degree of identity, difference and oppositeness of the set pair analysis, where a and c represent certainty, b represents uncertainty; i is the difference coefficient, the value range is $[-1,1]$; j is the opposite coefficient, the value is -1.

3 SAFETY RISK ASSESSMENT METHOD FOR DEEP FOUNDATION PIT CONSTRUCTION

3.1 *Establishment of a safety risk evaluation index system based on monitoring*

A hierarchical structure system of three-level construction safety evaluation indicators for deep foundation pits is established based on the construction monitoring projects. The first level is the foundation pit support structure and surrounding environment affected by the foundation pit construction. The second level is the monitoring items corresponding to the first level. The third level is the accumulated value and change rate of the monitoring data of the second level monitoring item.

3.2 *The method to determine the weight of the evaluation index*

(1) First, we invite experts related to the project to negotiate and determine the importance order of the evaluation indicators and build a scoring table by ordering the evaluation indicators in descending order of importance. It is worth noting that only the upper triangle of the judgment matrix is scored and converted. The lower triangle is the reciprocal of the upper triangle, so the conversion judgment matrix $Y^{(1)}$, $Y^{(2)}$, ..., $Y^{(r)}$. When the transformation judgment matrix of each expert passes the consistency test, the comprehensive judgment matrix $Z=(z_{ij})_{n \times n}$ is constructed by the geometric mean method, and the consistency test of the comprehensive judgment matrix Z is carried out. If the expert individual conversion judgment matrix or the expert group comprehensive judgment matrix cannot pass the consistency test, the method recommended in literature 7 can be used.

$$z_{ij} = \left(\prod_{k=1}^{r} y_{ij}^{(k)} \right)^{\frac{1}{r}} \tag{4}$$

(2) The similarity and different model of the SPA method is introduced to consider the differences of opinions within the expert group (Wu et a. 2010), and the set pair judgment matrix U is constructed.

$$U = A + \alpha B = \begin{bmatrix} a_{11} & a_{12} & \cdots & a_{1n} \\ a_{21} & a_{22} & \cdots & a_{2n} \\ \vdots & \vdots & \vdots & \vdots \\ a_{n1} & a_{n2} & \cdots & a_{nn} \end{bmatrix} + \alpha \begin{bmatrix} b_{11} & b_{12} & \cdots & b_{1n} \\ b_{21} & b_{22} & \cdots & b_{2n} \\ \vdots & \vdots & \vdots & \vdots \\ b_{n1} & b_{n2} & \cdots & b_{nn} \end{bmatrix} \tag{5}$$

$$a_{ij} = \begin{cases} \min_{r} \{y_{ij}\} & y_{ij} \geq 1 \\ \max_{r} \{y_{ij}\} & y_{ij} \leq 1 \end{cases} \tag{6}$$

$$b_{ij} = \begin{cases} \left| \max_{r} \{y_{ij}\} - \min_{r} \{y_{ij}\} \right| & y_{ij} \geq 1 \\ -\left| \max_{r} \{y_{ij}\} - \min_{r} \{y_{ij}\} \right| & y_{ij} \leq 1 \end{cases} \tag{7}$$

Where α is the difference coefficient, which measures the difference in the internal understanding of the expert group; A is the identity matrix, indicating the evaluation value that the expert group can accept for the relative importance of two indicators; B is the difference matrix, indicating the difference between the evaluation values of the relative importance of two indicators by the expert group.

(3) Since the set pair judgment matrix U is generally not consistent, the compatibility matrix method (Ye et al. 2006) is used to obtain the compatibility judgment matrix $D=(d_{ij})_{n \times n}$.

$$d_{ij} = \left(\prod_{p=1}^{n} u_{ip} u_{pj} \right)^{\frac{1}{n}} \tag{8}$$

(4) Then, we calculate the index weight by using the square root method for the compatibility judgment matrix w_s.

$$w_s = \frac{\left(\prod\limits_{j=1}^{n} d_{sj}\right)^{\frac{1}{n}}}{\sum\limits_{i=1}^{n}\left(\prod\limits_{j=1}^{n} d_{ij}\right)^{\frac{1}{n}}} \qquad (9)$$

3.3 Safety risk assessment criteria

3.3.1 Criteria for classification of foundation pit safety risk levels based on monitoring

The commonly used five-level classification method (Zhang et al. 2016) establishes the classification standard of the safety risk level. The safety risk assessment, acceptance criteria, and countermeasures of foundation pit construction are proposed, as shown in Table 1.

Table 1.   Classification criteria and evaluation of safety risk levels for monitoring projects.

Risk level	Risk status	Risk assessment and acceptance criteria	Countermeasures	Monitoring value alarmvalue
I	Safety	Very low risk; negligible	Risk treatment measures, routine management, and review are not required.	<0.6
II	track	low risk; tolerable	There is no need to take risky treatment measures, attract attention, and routinely monitor and manage.	0.6~0.7
III	warning	medium risk; acceptable	Attention should be paid to formulate detailed risk prevention, treatment, and monitoring measures.	0.7~0.8
IV	alarm	High risk; partially acceptable	Take risk treatment measures to reduce the risk to three levels as much as possible, and strengthen monitoring and warning.	0.8~1.0
V	danger	Very high risk; refuse to accept	Pay great attention and take measures to avoid and transfer; otherwise, reduce the risk to at least level 4 at any cost.	≥1.0

3.3.2 Calculation of contact number and determination of security risk level

The evaluation level and the evaluation index are formed into a set pair. The 5-element connection coefficient expression of the bottom evaluation index is established by using the same-difference-inverse hierarchy method (Pan et al. 2010). The comprehensive evaluation connection number of each subsystem and the total system is obtained by using the method. Finally, the safety risk evaluation level is determined by referring to literature 11.

(1) Calculation of the underlying comprehensive evaluation connection number:

$$\mu_{mk} = \begin{cases} 1 + 0i_1 + 0i_2 + 0i_3 + 0j & c_{mk} \in [0, s_{(1,2)k}) \\[2ex] \dfrac{1}{2}\dfrac{c_{mk} - s_{(2,3)k}}{s_{(1,2)k} - s_{(2,3)k}} + \dfrac{1}{2}i_1 + \dfrac{1}{2}\dfrac{s_{(1,2)k} - c_{mk}}{s_{(1,2)k} - s_{(2,3)k}}i_2 + 0i_3 + 0j & c_{mk} \in [s_{(1,2)k}, s_{(2,3)k}) \\[2ex] 0 + \dfrac{1}{2}\dfrac{c_{mk} - s_{(3,4)k}}{s_{(2,3)k} - s_{(3,4)k}}i_1 + \dfrac{1}{2}i_2 + \dfrac{1}{2}\dfrac{s_{(2,3)k} - c_{mk}}{s_{(2,3)k} - s_{(3,4)k}}i_3 + 0j & c_{mk} \in [s_{(2,3)k}, s_{(3,4)k}) \\[2ex] 0 + 0i_1 + \dfrac{1}{2}\dfrac{c_{mk} - s_{(4,5)k}}{s_{(3,4)k} - s_{(4,5)k}}i_2 + \dfrac{1}{2}i_3 + \dfrac{1}{2}\dfrac{s_{(3,4)k} - c_{mk}}{s_{(3,4)k} - s_{(4,5)k}}j & c_{mk} \in [s_{(3,4)k}, s_{(4,5)k}) \\[2ex] 0 + 0i_1 + 0i_2 + 0 + 0i_1 + 0i_3 + 1j & c_{mk} \in [s_{(4,5)k}, +\infty) \end{cases} \quad (10)$$

In the formula, m is the m-th evaluation unit; k is the kth evaluation index; c_{mk} is the sample value of the evaluation index. This paper refers to the cumulative value and change rate of the monitoring data; $s_{(x,y)k}$ is the value of the kth evaluation index The threshold value of the x_{th}, y_{th} grade standard , this parameter is determined according to the engineering design document; i_1, i_2, i_3 are the difference coefficients. Let i_1, i_2, i_3 be 0.5, 0, and -0.5, respectively.

The safety risk level increases as the monitoring value increases, and the evaluation indicators determined based on the monitoring items are all positive indicators. In addition, the safety risk level of the foundation pit is divided into five levels, and the associated number calculation formula is determined. The calculation formula of the contact number is as formula (10), Simplified to Equation 11:

$$\mu_{mk} = r_{mk1} + r_{mk2}i_1 + r_{mk3}i_2 + r_{mk4}i_3 + r_{mk5}j \quad (11)$$

$r_{mk1} \sim r_{mk5}$ are the contact number components corresponding to formula (13).

(2) Subsystem comprehensive evaluation connection number calculation:

$$\mu_m = \sum_{p=1}^{k} w_{mp}r_{mp1} + \sum_{p=1}^{k} w_{mp}r_{mp2}i_1 + \sum_{p=1}^{k} w_{mp}r_{mp3}i_2 + \sum_{p=1}^{k} w_{mp}r_{mp4}i_3 + \sum_{p=1}^{k} w_{mp}r_{mp5}j \quad (12)$$

$r_{m1} \sim r_{m5}$ are the connection degree components of the subsystem index relative to the security risk level; $w_{mp1} \sim w_{mp5}$ are evaluation index weights f the secondary subsystem obtained in Section 3.2.

(3) Calculation of contact numbers for comprehensive evaluation of the total system of the foundation pit:

$$\mu = \sum_{m=1}^{k} w_m r_{m1} + \sum_{m=1}^{k} w_m r_{m2}i_1 + \sum_{m=1}^{k} w_m r_{m3}i_2 + \sum_{m=1}^{k} w_m r_{m4}i_3 + \sum_{m=1}^{k} w_m r_{m5}j \quad (13)$$

$r_1 \sim r_2$ are connection degree components of the system relative to the security risk level; $w_{m1} \sim w_{m5}$ are the evaluation index weights of the first-level subsystem obtained in Section 3.2.

(4) Safety risk level evaluation

According to the "principle of equal division," the $[-1,1]$ interval is divided into five sub-intervals, namely $(0.6,1)$, $(0.2,0.6)$, $(-0.2,0.2)$, $(-0.6,-0.2)$, $(-1,-0.6)$, corresponding to the safety risk levels I, II, III, IV, and V. The larger the main value of the connection number, the lower the safety risk level, which means that the foundation pit system is safer. By comparing the main value of the obtained connection number at different levels with the sub-intervals, the safety risk level of the foundation pit at each level can be obtained, respectively. Then the safety risk assessment of the foundation pit can be carried out at different levels.

102

4 CONCLUSION

Combining Fuzzy Analytic Hierarchy Process (FAHP) and Set Pair Analysis (SPA) organically, based on on-site monitoring data and adopting a group decision-making method, a FAHP-SPA evaluation model for comprehensive safety risk of foundation pit construction is proposed. At the same time, the traditional 1–9 scale importance judgment criterion is improved. The qualitative judgment and scoring quantitative judgment process are separated, which can reduce the occurrence probability of the judgment matrix consistency requirement to a certain extent, and then introduce the triangle. The fuzzy number constructs a judgment matrix, which makes the judgment and scoring process more in line with the way of thinking of the human brain.

REFERENCES

Li Lingjuan, Dou Kun. Research on the Consistency of Judgment Matrix in AHP [J]. *Computer Technology and Development*, 2009, 19(10):131–133.

Pan Zhengwei, Jin Juliang, Wu Kaiya, et al. System Comprehensive Evaluation Model Based on Link Function and Its Application [J]. *Practice and Understanding of Mathematics*, 2010, 40(023):40–47.

Qin Zhihai, Qin Peng. Evaluation coupling model for high slope stability based on fuzzy analytical hierarchy process- set pair analysis method [J]. *Chinese Journal of Geotechnical Engineering*, 2010, 32(005):706–711.

Wang Weidong, Ding Wenqi, Yang Xiuren, et al. Foundation pit engineering, and underground engineering: high efficiency and energy saving, low environmental impact and sustainable development new technology [J]. *Chinese Journal of Civil Engineering*, 2020, (7):78–98.

Wu Jianjun, Cai Yao, Liu Zhengjiang. Set pair analysis of index weights in comprehensive safety evaluation [J]. *China Navigation*, 2010, 33(003):60–63.

Xia Yuanyou, Chen Chunshu, Chen Jinpei, et al. Dynamic risk assessment of deep foundation pit construction based on on-site monitoring [J]. *Chinese Journal of Underground Space and Engineering*, 2016, (12):1378–1384.

Xu Shubai. *Practical Decision-Making Methods: The Principles of Analytic Hierarchy Process* [M]. Tianjin: Tianjin University Press, 1988.

Ye Yicheng, Ke Lihua, Huang Deyu. *System Comprehensive Evaluation Technology and Its Application* [M]. Beijing: Metallurgical Industry Press, 2006.

Zhang Shengxi, Chen Weigong, Wang Huihui, et al. Risk assessment of deep foundation pit construction based on G-FAHP [J]. *Journal of Civil Engineering and Management*, 2016, 33(005): 104–109.

Zhao Keqin. *Set Pair Analysis and its Preliminary Application* [M]. Zhejiang: Zhejiang Science and Technology Press, 2000.

Advances in Measurement Technology, Disaster Prevention and Mitigation – Li & Mohd Yusof (Eds)
© 2023 The Author(s), ISBN: 978-1-032-36087-4

Urban flood risk analysis based on GIS and SWMM: A case study in Dujiangyan

Mingxia Lu*, Ting Ni* & Xuejin Ying*
College of Environment and Civil Engineering, Chengdu University of Technology, Sichuan, China

ABSTRACT: With the gradual increase in urbanization rate and people's activity area, the change of land use type leads to the increase of imperviousness and thus the increasing problem of urban flooding. In this paper, a hierarchical index system combining SWMM and GIS simulation with hierarchical analysis (AHP) is proposed to assess urban flooding risk. Taking Dujiangyan as an example, it is divided into 154 sub-catchment areas in GIS using Tyson polygons for urban flood risk assessment. Finally, the risk values obtained were entered into the GIS, the 154 areas were divided into 5 risk levels, and different colors were used to represent the study area to show the magnitude of risk in the study area so as to visualize the risk.

1 INTRODUCTION

In recent years, urban flooding has gradually become a common hazard in many cities due to the increase in extreme rainfall caused by climate change and the change in land use patterns caused by increased urbanization rates (Chen et al. 2021; Zhou et al. 2021). Urban flooding is a phenomenon in which heavy or short-duration rainfall exceeds the capacity of the urban drainage system, and waterlogging occurs (Jin 2015; Xue et al. 2016), which has threatened the safety of the society and harmed the economic development of the city to some extent (Feng 2021; Wu et al. 2019).

The Ministry of Water Resources released the "China Water and Drought Disaster Defense Bulletin" statistics on 28 provinces in 2020 (autonomous regions and municipalities directly under the Central Government) for flooding disasters of varying degrees. Among them, flooding disasters in Anhui, Sichuan, Jiangxi, and Hubei Provinces are heavy, and direct economic losses accounted for more than 61% of the country. Among them, Sichuan affected 8,511,000 people, 45 people died, 23 people missing, and direct economic losses of 42.52 billion yuan. But this hazard is inevitable, so our focus should be on the response to the post-disaster situation (Tia & Daniel et al. 2020; Wang et al. 2021). It held that the risk evaluation work is very important (Shi et al. 2017).

There are three general approaches to assessing storm flood hazard vulnerability: vulnerability assessment based on historical hazard data (Boudou et al. 2016), assessment based on indicator systems (Hai-Min et al. 2018), and scenario simulation based on hydrologic models (Goncalves et al. 2018). In this study, the method of establishing an assessment model based on a system of indicators was chosen.

2 PROBLEM STATEMENT

2.1 *Study area*

The total area of Dujiangyan is 1,208 square kilometers, the city's road mileage is 1,584km, the road network density is 1.31km/km2, the elevation is between 574m and 4,436m, and the terrain is undulating. Situated in the western Chengdu Plain on the Min River about 50 km from downtown

---

*Corresponding Authors: lumingxia@stu.cdut.edu.cn, niting17@cdut.edu.cn and yingxuejin0911@163.com

     DOI 10.1201/9781003330172-16

Chengdu, between 30°44′-31°22′ N latitude, 103°25′-103°47′ E longitude. This city is bordered by Pengzhou City, Pidu District, and Wenjiang District in the east, connected with Wenchuan County in the west and north, and attached to Chongzhou City in the south; the topography of the city is high in the northwest and low in the southeast. The urbanization rate of Dujiangyan city in 2018 is 60.2%, the resident population of the city is 697,000, the population density is 577 inhab/km², the greening coverage rate is 43.01%, and the drainage network density is 14.72 km/km². the maximum average precipitation is usually in July and August, and the average rainfall is 257.6 mm and 254.5 mm. In this paper, the region is selected as the research object and the method of creating Tyson polygons in GIS is chosen to divide the study area into 154 sub-catchment areas (Wang 2020), as shown in Figure 1.

Figure 1.   Study area sub-catchment.

2.2   *Data preparation*

The index data in this paper mainly comes from the GIS processing of original basemaps such as DEM maps and land use maps. Some come from SWMM model simulations, but most of the SWMM model data also comes from GIS processing of the original basemap. In this simulation, the rainfall data input to SWMM was calculated using the revised rainstorm intensity formula in the central urban area of Chengdu in 2015, and the return period of the rainstorm was once every ten years.

3   RISK ASSESSMENT OF WATERLOGGING DISASTERS

3.1   *Evaluation model construction*

The indicators in this paper are selected according to the structural system of the regional hazard system of Peijun Shi, which consists of three factors, namely, disaster-pregnant environment (E), hazard factor (H), and disaster-bearing body (B)(Shi 2005). According to some basic conditions of the study area, river network density and greening rate were selected as indicators of the disaster-pregnant environment, the indicators of the disaster-bearing body were population density and imperviousness, and peak runoff and nodal overload hours were used as indicators of the disaster-causing factors. The index system was divided into target, criterion, and indicator layers according to the above indicators as the evaluation index system of the study area, as shown in Figure 2.

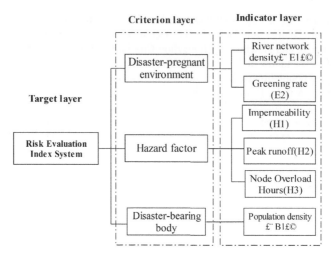

Figure 2.   Risk evaluation index system.

Since the units of each indicator are not consistent and the indicators include both positive and negative indicators, it is necessary to standardize and homogenize the indicators using Equation (1) in order to facilitate the calculation.

$$\overline{X} = \frac{X - X_{min}}{X_{max} - X_{min}} \qquad (1)$$

Where X is the value of the indicator, X_{min} is the minimum value of the indicator, and X_{max} is the maximum value of the indicator.

3.2   GIS and SWMM-based metrics calculation

Among the six selected metrics, peak runoff and nodal overload hours were simulated in SWMM to obtain data based on Tyson polygons created in GIS to build the corresponding models.

River network density, greenness, and imperviousness can be derived through GIS processing. For example, when calculating the river network density, the river network density of each small area can be obtained by processing the water system map in GIS through Spatial Analyst Tools, as shown in Figure 3.

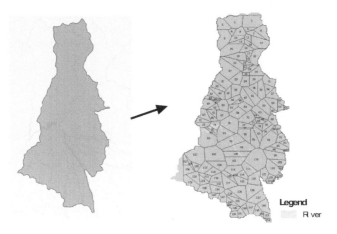

Figure 3.   River network density.

3.3 *The weighting of each index based on hierarchical analysis*

Hierarchical analysis (Deng et al. 2012) is a multi-option or multi-objective decision evaluation method proposed by a professor (T L Saaty) at the University of Pittsburgh in the mid-1970s. The basic idea is that by analyzing the various factors and their interrelationships contained in a complex problem, all the elements of the problem studied are classified according to different levels, and the connections between the elements of the upper and lower levels are marked so as to form a multi-level structure. The relative importance of the elements in each level is judged to determine the weights of the elements, and finally, the combined weights of the elements in each level to the total objective are calculated.

The importance judgments of the indicators in this study were evaluated by three experts in the professional field according to the nine importance levels given by Saaty in the hierarchical analysis method, and the final results were taken as the average of the three experts, and the weights of the different levels were calculated as shown in Tables 1~3.

Table 1. Results of the calculation of the weight of the criterion layer.

A	Importance discriminant matrix			Weights
	E	H	B	
E	1	1/5	1/2	0.122
H	5	1	3	0.648
B	2	1/3	1	0.230
Inspection parameters	$\lambda_{max} = 3.003$, CI=0.0015; CR=0.0026<0.1, passing the consistency test			

Table 2. Calculation results of pregnancy and disaster-pregnant environment weights.

E	Importance discriminant matrix		Weights
	E1	E2	
E1	1	1/5	0.167
E2	5	1	0.833

Table 3. Calculation results of hazard factor weights.

A	Importance discriminant matrix			Weights
	H1	H2	H3	
H1	1	1/7	3	0.277
H2	7	1	1/2	0.423
H3	1/3	2	1	0.300
Inspection parameters	$\lambda_{max} = 3.026$, CI=0.013; CR=0.022<0.1, passing the consistency test			

3.4 Regional Risk Map

According to Section 3.2.1, the risk value of the study area can be calculated based on the results of the weight calculation according to Equation (2).

$$R = f(E, H, B) = \omega_E \sum_{i=1}^{k} E_i \cdot \omega_{Ei} + \omega_H \sum_{i=1}^{k} H_i \cdot \omega_{Hi} + \omega_B \sum_{i=1}^{k} B_i \cdot \omega_{Bi} \qquad (2)$$

Where R is the flood risk index; ω_E, ω_H, ω_B are the weights of the pregnant environment, disaster-causing factor, and disaster-bearing body; E_i, H_i, B_i are the values of the ith pregnant environment index, disaster-causing factor index, disaster-bearing body index; ω_{Ei}, ω_{Hi}, ω_{Bi} are the weights of the ith pregnant environment index, disaster-causing factor index, and disaster-bearing body index.
For example, the risk value for sub-catchment area 1 is:

$$
\begin{aligned}
R &= \omega_E(E_1 \cdot \omega_{E1} + E_2 \cdot \omega_{E2}) + \omega_H(\cdot \omega_{H1} + H_2 \cdot \omega_{H2} + H_3 \cdot \omega_{H3}) + \omega_B \cdot B_1 \cdot \omega_{B1} \\
&= 0.122 \times (0.03 \times 0.167 + 1 \times 0.833) + 0.648 \times (0 \times 0.277 + 0.37 \times 0.423 + 0.12 \times 0.3) \\
&\quad + 0.23 \times 0 \\
&= 0.227
\end{aligned}
$$

The calculated risk values are between 0.153 and 0.693, which are divided into five risk levels in this paper, $0.153 \leq X < 0.28$ for the very low-risk area, $0.28 \leq X < 0.363$ for the low-risk area, $0.363 \leq X < 0.439$ for the medium-risk area, $0.439 \leq X < 0.544$ for the high-risk area, and $0.544 \leq X \leq 0.0693$ for the very high-risk area. With this, the risk distribution of the study area was drawn, as shown in Figure 4.

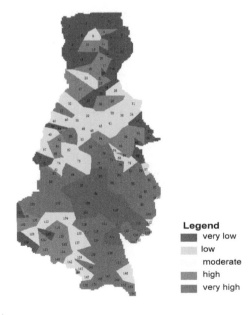

Legend
- ▮ very low
- ▯ low
- moderate
- ▮ high
- ▮ very high

Figure 4.   Risk distribution map.

4 CONCLUSION

In this study, reasonable evaluation indexes were selected from three aspects of the disaster-inducing environment, disaster-causing factors, and disaster-bearing bodies, and a disaster risk evaluation system was constructed, analyzed, and simulated for the study area by combining GIS and SWMM software. The weights of each index were calculated by using hierarchical analysis, and finally, the risk distribution map of the study area was drawn in GIS, in which there were 27 very low-risk areas, 35 low-risk areas, 40 medium-risk areas, 43 high-risk areas, and 9 very high-risk areas. Comparing the risk distribution map with satellite images of the study area and on-site conditions, it is found that the areas with higher risk are areas with more roads and higher impermeability, while areas with a high greening rate have lower risks. Therefore, increasing the greening rate of the city can reduce water impermeability to a certain extent, thereby reducing flood risk.

REFERENCES

Boudou M, Danière B, Lang M . (2016) Assessing changes in urban flood vulnerability through mapping land use from historical information. *J. Hydrology and Earth System Sciences*, 20, 1(2016-01-18), 12(1):161–173.

Chen Z, Li K, Du J, et al. (2021) Three-dimensional simulation of regional urban waterlogging based on high-precision DEM model. *J. Natural Hazards*, 1–25.

Deng Xue, Li Jiaming, Zeng Haojian, et al. (2012) Analysis and Application of AHP Weight Calculation Method. *J. Practice and Understanding of Mathematics*, 24(7):8.

Feng P . (2021) Risk Assessment of Urban Floods Based on a SWMM-MIKE21-Coupled Model Using GF-2 Data. *J. Remote Sensing*, 13.

Goncalves M, Zischg J, Rau S, et al. (2018) Modeling the Effects of Introducing Low Impact Development in a Tropical City: A Case Study from Joinville, *Brazil. J. Sustainability*, 10(3):728.

Hai-Min L, Jack S, Arul A . (2018) Assessment of Geohazards and Preventative Countermeasures Using AHP Incorporated with GIS in Lanzhou, China. *J. Sustainability*, 10(2):304.

Jin Zhisen. (2015) Research on urban waterlogging risk assessment and zoning method based on AHP: Taking Quanzhou Donghai Group as an example *.J. Urban Housing*, 2015(12):5.

Shi Peijun. (2005) Four Discussions on the Theory and Practice of Disaster System Research. *J. Journal of Natural Disasters*, 2005(6):7.

Shi Xiaojing, Zha Xiaochun, Liu Jiahui, et al. (2017) Flood disaster risk assessment in Ankang City in the upper reaches of the Han River based on cloud model. *J. Progress in Water Resources and Hydropower Science and Technology*, 37(3):7.

Tia D, Daniel H, et al. (2020) Managing urban flood risk: An expert assessment of economic policy instruments. *Journal of Urban Affairs*.

Wang Huimin, Cao Weiwei, Huang Jing, et al. (2021) Assessment of public flood risk cognition level based on SEM: Taking Jingdezhen City as an example. *J. Water Conservancy Economy*, 39(3):11

Wang Sirun. (2020) *Rainfall-runoff simulation based on SWMM and analysis of LID combination scheme*. D. Xi'an University of Technology.

Wu, Shen, Wang. (2019) Assessing Urban Areas' Vulnerability to Flood Disaster Based on Text Data: A Case Study in Zhengzhou City. *J. Sustainability*, 11(17):4548.

Xue F, Huang M, Wang W, et al. (2016) Numerical Simulation of Urban Waterlogging Based on FloodArea Model.*J. Advances in Meteorology*,2016:1–9.

Zhou M, Feng X, Liu K, et al.(2021) An Alternative Risk Assessment Model of Urban Waterlogging: A Case Study of Ningbo City. *J. Sustainability*, 13.

Advances in Measurement Technology, Disaster Prevention and Mitigation – Li & Mohd Yusof (Eds)
© 2023 The Author(s), ISBN: 978-1-032-36087-4

Analysis of monitoring results of spillway slope of Wacun hydropower station

Xiaochuang Ding*

NARI Group Corporation, Nanjing, China

ABSTRACT: As a discontinuous medium, the deformation of rock slopes is affected by many factors. Based on the actual measured data from the slope multi-point displacement meter, through the process, the holes' depths, changing values of measured displacement, and the deformation characteristics of the slope during the construction period are analyzed from the perspective of time and space, and abnormal conditions of the slope are analyzed and warned.

1 INTRODUCTION

With the development of large hydropower projects, some high and steep slopes take form after excavation. Due to the effect of stress relief and release or unloading of tectonic stress, the mountain slope changes its original equilibrium state, and the slope will adjust the stress distribution through deformation to seek a new equilibrium state. When the deformation surpasses a particular threshold, the slope will finally produce unstable failure. Numerous factors influence the stability of slopes, and it is difficult to understand the actual deformation law of a slope solely by a theoretical and model study (Chen et al. 2006; Gong et al. 2013; Liu et al. 2010). Multi-point rod extensometers have been commonly used in engineering slope monitoring (Dai et al. 2015; Huang et al. 2009; Zhu et al. 2004), and their monitoring data can reveal deformation characteristics of the slope, which is useful for determining the slope's stability during construction and operation.

Using the spillway slope of Wacun Hydropower Station in Guangxi as an example, slope deformation characteristics are analyzed using measured data results from multi-point rod extensometers. It serves as a reference for similar rock-high slope projects.

2 SLOPE GEOLOGICAL PROFILE AND MONITORING FACILITY LAYOUT

2.1 *Profile of wacun hydropower station spillway slope*

Wacun Hydropower Station is the second phase of the overall usage plan for Yujiang River. The planned dam is in the upper sections of the Yujiang River in Tianlin County. The open spillway on the right bank is arranged about 900m downstream of the dam site, using the saddle-shaped mountain terrain between the right bank of the Tuaniang River and the Ba Nang Gou to arrange the spillway of the riverbank, and the axis is roughly perpendicular to the Tuoniang River. The spillway is composed of an inlet channel section, control chamber section, discharge trough section, and outlet flow section and has a total length of 458.14 m. The spillway layout is seen in Figure 1.

Rock mass of the spillway slope is generally a lithological slope, and the lithology is mainly gray medium-thick-layer to thin-layer (microcrystalline, siliceous, argillaceous) limestone mixed with silty mudstone, some lithic feldspar sandstone, silty sandstone mudstone and so on. The surface layer of the slope consists of gravel loam, and hard rocks such as lithic feldspar sandstone and (microcrystalline, siliceous, argillaceous) limestone in the bedrock that are the main rocks in the area, accounting for approximately 90%. Except for the concentrated distribution of mudstone

---

*Corresponding Author: dingxiaochuang@sgepri.sgcc.com.cn

DOI 10.1201/9781003330172-17

and other soft rocks above 360 m elevation on the right bank, they are primarily distributed in interlayers or interlayers, accounting for about 10%. Except for weathering and unloading of the superficial layer, the slope rock mass is generally a micro-new rock mass, and its overall stability is controlled by the structural plane properties and their combination relationships. The overall structure is monoclinic, the rock formation is N60~70°E/SE∠30~40°, and the stratum fractures are relatively developed. Apart from small deflections and randomly-distributed small faults in local sections of the rock, no large-scale distribution of controlling weak structural surfaces is found, and the slope rock mainly develops a system of jointed fractures. Apart from layer fissures, there are mainly ①, ②, and ③ group fissures developed in the weathering unloading zone. The left side of the spillway outlet generally constitutes a forward slope, and under boundary conditions of f2 faults, layer fissures, and other structural planes, there is a possibility of plane sliding.

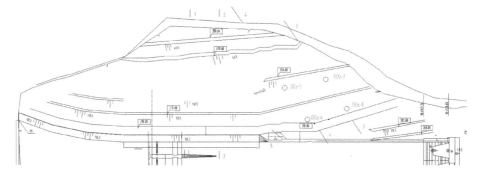

Figure 1.   Spillway floor plan.

2.2   *Spillway slope monitoring layout*

The left bank slope of the spillway outlet of Wacun Hydropower Station adopts the NVD series multi-point rod extensometers of NARI Group, which are vibrating wire multi-point rod extensometers. The extensometers installed on the slope are all four-point extensometers.

Sections 3-3 and 4-4 on the slope system are selected, and a group of multi-point rod extensome-ters is set up at 343 elevations and 313 elevations, respectively. The measuring point depths are 12 m, 11 m, 14 m, and 13 m for 343 m elevation multi-point rod extensometers. The measuring point depths are 10 m, 10 m, 140 m, and 10 m for 313 m elevation multi-point rod extensometers, and the layout of the monitoring apparatus section is seen in Figure 2.

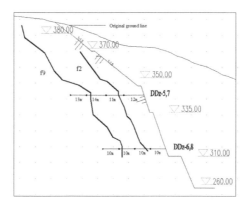

Figure 2.   Layout of high slope multi-point rod extensometers section.

3 ANALYSIS OF SLOPE DEFORMATION AND MONITORING RESULTS

3.1 *Slope deformation process*

During the excavation and support period, three large-scale slump failures occurred on the left exit slope of the spillway as a result of variables including torrential rainfall, support lag, construction blasting control, etc.

1) On August 24, 2017, the slope collapsed with a square measuring site of about 30,000 m³. The mechanism of landslide is seen below: (1) The layer fissure Lc1 begins to track and extend along the layer, establishing the primary sliding surface; the downstream side slip is confined and expands, creeps to the empty surface beyond the slope, and shears and stretches along f2 to produce a lateral slip. The slope's trailing edge is gradually fractured, forming a discontinuous arc-shaped steeply dipping trailing edge cutting surface. The aforesaid structural surfaces together constitute the wedge-shaped failure boundary condition of the slope with Lc1 as the main sliding surface and f2 as the lateral sliding surface. Deep anchor cable support measures for the slope in this section are yet to be implemented, resulting in the failure of effective containment of the deformation of the rock mass in this section before leading to final failure and instability and pulling the rock mass above EL360 m to collapse, which is a typical traction landslide; ② The physical and mechanical qualities of the slope rock mass and structural plane have deteriorated due to constant heavy rain prior to the landslide, lowering the slope's stability.

2) The slope slipped or dropped blocks with a cumulative volume of about 8,000 m³ on 5 January 2018 and subsequent time due to lagging support work, essentially in the same area, with the same deformation mechanism, and eventually formed a cavity rock body in the ultimate stable state at the slump area, see Figure 3. The slope of the exit section of the spillway is exposed to the air on two sides in comparison to the Tuoniang River and spillway. As the slope of this section intersects the rock direction at a small angle and they have the same inclination, it is a downward slope, so the rock body is prone to plane sliding failure along the cut-foot air face in the excavation process. Upstream protection adopts the system mortar bolt and hanging net shotcrete support. The collapsed section and downstream slope deploy the combined support of the C25 frame beam, anchor cable, mortar bolt, and hanging net shotcrete. However, during the construction phase, continuous excavation was not fully supported in time, and site inspection still exhibited evidence of cracking and deformation in partial places. There were indicators of pulling fractures in the slope range of outflow of the left bank of the spillway, which may be broadly split into Lf1, Lf2, and Lf3 in the transverse slope and Lf4 and Lf5 in the longitudinal slope, and their distribution sites are distributed in Figure 4. After local cracks are detected on the site, the site strictly restricts excavation operation of the construction unit and requires excavation operation after completing support based on design requirements. Also, multi-point rod extensometers are added in sections 3-3 and 4-4 to strengthen slope deformation monitoring.

Figure 3.   Second landslide of spillway slope.   Figure 4.   Fracture distribution of spillway slope.

3) From April 25th to May 1st, 2018, four groups of multi-point rod extensometers in sections 3-3 and 4-4 were installed and monitored.

4) During the flood season, the Wacun Hydropower Station spillway went through a temporary overflow from August 8 to August 15, 2018. During the construction period, the flood was not effectively controlled due to the incomplete completion of the left chute, overhang, and slope support on the left bank. As a result, the toe of the slope on the left was heavily brushed by floods, and the spillway slope produced a bottom-to-up traction landslide with an elevation range of EL403-230 m. As seen in Figure 5, the collapse of the engineering slope and traction drive the slope 15 to 50 meters beyond the range's opening line to produce deformations such as tensile cracks and staggered platforms, forming the slope deformation area for the environment.

Figure 5.   The third landslide of the left bank slope.

3.2   *Analysis of multi-point rod extensometers monitoring results*

3.2.1   *Analysis of multi-point rod extensometers measured process*
At the end of April 2018, four sets of multi-point rod extensometers began to be observed. Figures 6–9 show that deformation laws of the three sets of multi-point rod extensometers of DDz-5 and DDz-7 are essentially the same, and slope displacement changes of the spillway's left bank can be divided into three stages: 1) acceleration deformation; 2) slope-supported deformation stabilization stage; and 3) unstable failure stage. The DDz-8 measured process is basically unchanged. The major reason is that during the installation process, PVC protection pipe joints were not properly sealed, and the transmission rod was sealed by slurry leakage, resulting in no response to the construction process by monitoring data at the 10 m, 20 m, and 30 m measurement sites. The anchor cable tension demonstrates noticeable compression deformation in the rock wall, which is related to the tension of the anchor cable. For EL343 elevation multi-point rod extensometers DDz-5 and DDz-7, after the excavation is suspended and the slope anchor cable is tensioned, the measured value changes slowly, indicating that the slope deformation slows down. Spillway overflowed in early August 2018, and the water flow washed the slope toe. The measured value of multi-point rod extensometers changed significantly, and the spillway slope underwent unstable failure.

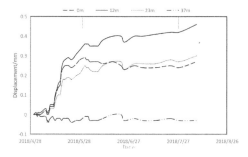

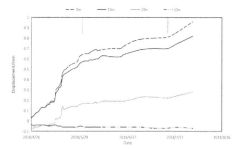

Figure 6.   Displacement-time curves of DDz-5.   Figure 7.   Displacement-time curves of DDz-6.

EL313 elevation multi-point rod extensometers DDz-6 anchor cable is not tensioned yet. After construction of the lower excavation is suspended, slope deformation slows down. It enters the stable creep stage. After spillway overflows, slope deformation intensifies, followed by unstable failure.

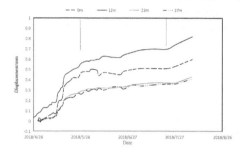

Figure 8. Displacement-time curves of DDz-7.   Figure 9. Displacement-time curves of DDz-8.

3.2.2 Multi-point rod extensometers hole depth displacement analysis

With the observation point with maximum burial depth as the relatively fixed point (i.e., the base point), it is needed to convert the deep observation value of the observing sites into the displacement value $u_1 \sim u_4$ (rock wall $\sim$ deepest observation point) relative to the base point, and illustrate displacement of the multiple measurement points. Figures 11–13 depict the distribution of hole depth, and Table 1 details statistics of cumulative maximum displacement of multi-point rod extensometers.

Deformation of the slope rock mass is primarily toward the air surface, and its deformation magnitude is not large, with the highest value being 0.96 mm. The main reason is that three sets of multi-point rod extensometers were later added to the monitoring facilities, and during normal observation, this part of the slope and the lower slope have been largely excavated, and the main deformation has occurred. So, we failed to observe the main deformation. Under normal circumstances, the deformation law of slope rock and soil mass should gradually decrease from the outside to the inside, alternatively put, it shows that $u_1 \geq u_2 \geq u_3 \geq u_4$ (Chen et al. 2020), but since mid-May 2018, the EL343 elevation multi- Point rod extensometers DDz-5 and DDz-7 fail to meet the deformation law of slope rock and soil mass under normal conditions. And in particular, displacement of different depths in DDz-5 is evident. The concentrated tension of the anchor cable is the main reason for this trend. From the anchoring mechanism of the anchor cable, it is evident that compression deformation will be produced after the anchor cable is tensioned, and the same deformation will be produced at the same time as the rock mass anchored with it is driven, which can be expressed as that the slope surface generates displacement towards the slope. The effect of this factor gradually weakens with depth, so displacement generated by the slope is also reduced from the outside to the inside. After superimposing with early-stage displacement, the deformation amount on the surface is smaller than that of the deep measuring points, as shown in Figures 10~12.

Table 1. Statistics of slope multi-point rod extensometers.

Code	Observation displacement value/mm			
	$u1$	$u2$	$u3$	$u4$
DDz-5	0.27	0.46	0.3	−0.03
DDz-6	0.96	0.82	0.28	−0.07
DDz-7	0.6	0.61	0.43	0.41

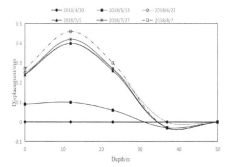

Figure 10.   Displacement-depth curves of DDz-5.

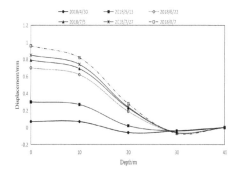

Figure 11.   Displacement-depth curves of DDz-6.

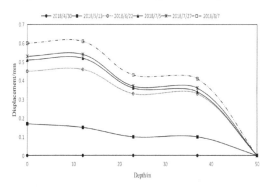

Figure 12.   Displacement-depth curves of DDz-7.

3.2.3 *Slope deformation monitoring alarm*

The slope deformation began to increase again in late July 2018. Even though the increase in deformation when compared to the stable period is minor (in comparison to the elastic deformation amount in the excavation phase), it bypasses unstable deformation during the elastic stage and should cause concern. Figures 13 to 15 show displacement changes of three groups of multi-point rod extensometers displacement at different hole depths, indicating that from late July to August, in which the deformation amount grew both on the slope surface and in-depth. On August 8, shortly after the instability precursor emerged, a landslide occurred. Though the deformation amount recorded by the device was extremely tiny prior to the slope instability, it bypasses unstable deformation during the elastic stage and should cause concern.

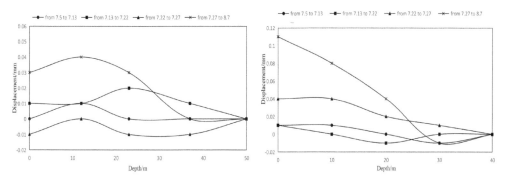

Figure 13.   Variation-depth of DDz-5.

Figure 14.   Variation-depth of DDz-6.

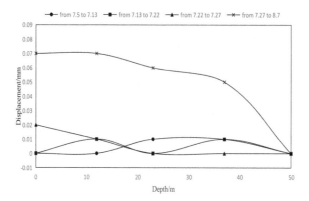

Figure 15. Variation-depth of DDz-7.

4 CONCLUSION

The slope deformation characteristics of the left bank of the spillway at Wacun Hydropower Station are evaluated using multi-point rod extensometer data, and the following conclusions can be drawn:

(1) Outlet slope of the Wacun spillway is mainly divided into the excavation acceleration deformation stage, deformation stabilization after support stage, and unstable deformation stage.
(2) Overall deformation amount of the slope at the exit of the left bank of the spillway is minimal. The main reason for this is that the monitoring facility's installation is lagging, and support that has been put in place effectively restricts slope deformation. During the construction period, attention should be paid to the timeliness of slope deformation monitoring and support measures.
(3) After slope elastic deformation, abnormal deformation increment should draw enough attention, which is likely to cause instability.

REFERENCES

Chen Fei, Deng Jianhui, Wei Jinbing, Gao Chunyu. Analysis of the reliability of monitoring data of rod-type multi-point rod extensometers[J]. *Chinese Journal of Underground Space and Engineering*, 2020, 16(03):897–902.
Chen Qiang, Zou Zhengming, Wang Jialin. Case analysis of slope monitoring and dynamic design and construction of expressway in mountainous areas [J]. *Chinese Journal of Disaster Prevention and Mitigation Engineering*, 2006, 26(3):332–336.
Dai Feng, Li Biao, Xu Nuwen, et al. Analysis of features of excavation damage area of deeply -buried underground powerhouse at the Monkey Rock Hydropower Station [J]. *Rock Mechanics and Engineering*, 2015, 34(4): 735–746.
Gong Jing, Yuan Youcang, Dong Yufei. Analysis of deformation depth of high slope excavation period based on multi-point rod extensometers and external deformation observation pier monitoring results [J] *Sichuan Hydropower*, 2013, 32(3):164–168.
Huang Qiuxiang, Wang Jialin, Deng Jianhui. Analysis of slope deformation characteristics based on multi-point rod extensometers monitoring results[J]. *Journal of Rock Mechanics and Engineering*, 2009, 28(S1): 2667–2673.
Liu Wenqing, Zhao Fei, Dong Jianhui, et al. Applying of multi-point rod extensometers in high slope stability monitoring of expressways[J]. *Geological Hazards and Environmental Protection*, 2010, 21(4):104–107.
Zhu Quanping, Xu Bo, Zhang Dexiang. Analysis on deformation of high slope multi-point rod extensometers of ship lock of the Three Gorges Project[J]. *China Three Gorges Construction*, 2004(6):36–38.

Advances in Measurement Technology, Disaster Prevention and
Mitigation – Li & Mohd Yusof (Eds)
© 2023 The Author(s), ISBN: 978-1-032-36087-4

Study of land surface temperature in Beijing based on remote sensing technology

Shizheng Guan & Jing Zhang*
Beijing Laboratory of Water Resources Security, Capital Normal University, Beijing, China
The Key Lab of Resource Environment and GIS of Beijing, Capital Normal University, Beijing, China

ABSTRACT: Land Surface Temperature is an important parameter in the field of climatological and meteorological research. Changes in land surface temperature are directly related to the balance of material and energy exchange between the earth and the atmosphere, and have a direct impact on changes in crops and local soil quality. This paper investigates the spatial and temporal evolution of land surface temperature in Beijing from 2010 to 2015, using the land surface temperature monthly synthetic product for land surface temperature inversion and analysis. The results show that the land surface temperature showed an overall increasing trend from 2010 to 2015, and the spatial distribution was relatively uneven, with a pattern of "southern than northern". The expansion of built-up areas has a significant warming effect on the surface.

1 INTRODUCTION

Land Surface Temperature is one of the most important parameters in the study of regional and global land flow processes and climate evolution in China, and the degree of energy exchange between land and air is extremely relevant. Changes in land surface temperature can directly affect the material and energy balance between the ground and the atmosphere, causing changes in many natural environmental factors such as temperature, precipitation, flora and fauna, which in turn have a significant impact on the maintenance and evolution of the ecological environment within the region over time (An et al. 2021; Liu et al. 2020; Zhang et al. 2018; Zhao 2020). Land surface temperature is influenced by a combination of environmental conditions such as solar radiation, vegetation, topography, humidity, temperature, and precipitation. Interannual and seasonal variations have a direct impact on crop and local soil quality changes.

The traditional way of obtaining land surface temperature is done through meteorological station data or ground measurements. Usually, the land surface temperature data can be obtained from a single point on the ground. However, as the land surface temperature is influenced by various factors, especially elevation, elevation, latitude and longitude, and subsurface conditions, which have a significant impact on the variability of land surface temperature distribution, there may be a problem with the large variability in the distribution in space. Therefore, the spatially land surface temperature data obtained by the traditional method of interpolation cannot fully meet the requirements of the study in terms of accuracy and reliability, mainly because the density of ground-based meteorological stations or measurement sites is far from meeting the required accuracy of the study.

In addition to the traditional measurement methods, satellite thermal infrared technology based inversion of land surface temperature values is widely used in China. Compared to the current traditional methods of ground-based fixed-point measurements, satellite thermal infrared technology inversion of land surface temperature values can still directly obtain large-scale and spatially

---

*Corresponding Author: 5607@cnu.edu.cn

continuous land surface temperature values. The accuracy is much higher than the traditional fixed-point measurement methods, for which Landsat infrared band and Moderate Resolution Moderate Resolution Imaging Spectrometer (MODIS) have become the main source of data information for the study of temperature change in the inversion process of the earth's surface. Infrared technology allows real-time access to the data and offers a great advantage in terms of convenience over traditional methods.

To date, many scholars have used thermal infrared techniques to invert land surface temperature to study the spatial and temporal evolution of land surface temperature and the mechanisms of influencing factors at small and medium scale levels, but due to the huge data processing workload, there are relatively few macroscopic studies at large scales, large scales and long time series. In this paper, the improved single-window algorithm (IWM) proposed by Wang et al (Chen et al. 2020; Du & Zhang 2020; Ren et al. 2020; Zhao et al. 2020). The Landsat-8 TIRS10 band is used to maintain an inversion accuracy of about 1.4K for real temperature data by adjusting the atmospheric averaging parameters, which greatly meets the accuracy requirements of various studies.

However, due to the limitation of satellite transit time, it is difficult to maintain synchronization with observations from ground-based observation points, and the insufficient density of ground-based observation points, the time and space scales of the measured near-land surface temperature are seldom considered in most land surface temperature inversion studies, thus there is a certain impact on the accuracy of the land surface temperature field (Chen et al. 2020; Du & Zhang 2020; Li et al. 2020; Ren et al. 2020; Zhao et al. 2020; Zhao et al. 2020; 2021). Current methods for land surface temperature remote sensing acquisition include single channel algorithms for a single band, multi-band algorithms, split-window algorithms, and the two-channel correction algorithm proposed by Gerace et al. to reduce the effect of errors caused by stray light. As land surface temperature inversion research progresses, MODIS-based products are becoming more widely used for the geographical analysis of land surface temperature at large scales, mainly for the analysis of land surface temperature under the influence of large-scale factors such as elevation, slope, slope direction, and latitude. However, there is a lack of precision in this method for small-scale influencing factors, mainly neglecting the influence of microtopography, anthropogenic influences, water distribution, and vegetation cover (Ma 2021; Zhang & ji 2019; Zhao et al. 2021). It is important to integrate the constraints of multiple influencing factors in the study and try to take into account the need for precision in all aspects of the study.

This study focuses on the processing of data using ArcGIS software. The mask extraction tool was used to crop the spatial distribution map of land surface temperature in Beijing among various raster data. The spatial distribution of annual mean land surface temperature from 2010 to 2015 was later calculated by the map algebra method. To further analyze the spatial evolution of the land surface temperature data, the resulting raster data of the annual mean land surface temperature distribution in Beijing for each year were reclassified, and the land surface temperature was divided into equal intervals, and each temperature interval after the division was color-coded to provide a more intuitive analysis of the land surface temperature distribution in Beijing. The normalized and reclassified annual mean land surface temperature data from 2010 to 2015 were then used to extract the image elements, classify the image elements in each interval and analyze the spatial and temporal evolution of land surface temperature in the Xining area.

2 MATERIALS AND METHOD

2.1 *Overview of the study area*

Beijing (39°26′N-41°03′N, 115°25′E-117°30′E) has a total area of about 1.06 million km^2, of which about 38% is plain and 62% is mountainous. Topographically, it is high in the northwest and low in the southeast, with an average altitude of 43.5m, of which 20–60m in the plains and 1000–1500m in the mountains. Beijing has a temperate continental climate, with four distinct seasons, hot and rainy in summer and cold and dry in winter, with a multi-year average temperature of 12.3°C. The annual

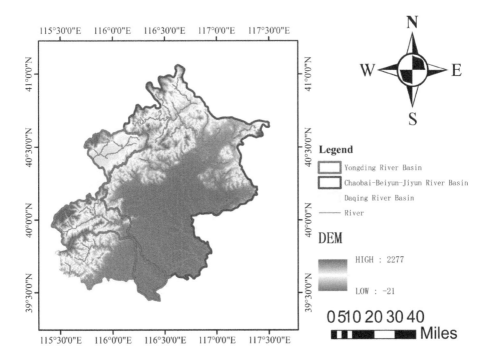

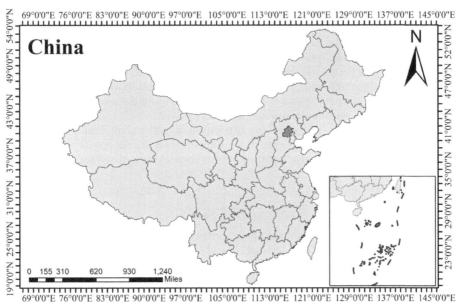

Figure 1.　Digital elevation distribution map of Beijing.

precipitation is about 600mm, with the majority of rainfall concentrated in summer. The overall economic level of Beijing is at a leading position in the country, and the Beijing-Tianjin-Hebei city cluster, with Beijing at its core, is striving to develop into a world-class city cluster.

2.2 Methodology

2.2.1 Phase land surface temperature inversion and single window algorithm

In this study, the multi-window area matching algorithm (IMW) with continuity constraint and brightness gradient constraint was chosen as the land surface temperature inversion method. The structure of the IMW algorithm is mainly based on the surface thermal radiation transmission equation, which is derived to invert the land surface temperature using thermal infrared band data (Ji 2018; Wang et al. 2021; 2020), using the 10th band of land-sat-8 as an example, and its calculation formula is given in (1).

$$TS = [a10(1 - C10 - D10) + (b10(1 - C10 - D10) + C10 - D10)T10 - D10Ta]/c10 \quad (1)$$

where Ts is the land surface temperature based on the inversion of the 10th band of Landsat-8; Ta is the mean interaction temperature in the atmosphere; $T10$ is the bright temperature on the surface stars that can be observed in the 10th band of Landsat-8; $a10$ and $b10$ are a correlation coefficient (where $a10$=-67.35535, $b10$=0.458608); $C10$ and $D10$ are internal parameters of the algorithm.

2.2.2 Normalized analysis

The different indicators of different evaluation methods (i.e., different indicators with different definitions in a different vector space) often need to have different scales and different units of definition, which may directly affect the results of the analysis of the corresponding indicator data. In order to effectively reduce the impact of different definitions of the indicators, it is necessary to standardize the analysis of the indicators in order to effectively solve the problem of comparability between different evaluation methods and different data evaluation indicators. After the raw data has been standardized, all quantitative indicators should be of the same order of magnitude, allowing for an overall and comprehensive quantitative comparison and evaluation. One of the most typical solutions is to normalize the data for all data models. Normalization is a dimensionless method of processing, whereby the absolute value of a physical system becomes a relative relationship, effectively simplifying the calculation of the system and reducing the relative magnitude. In short, the main purpose of normalization is to completely limit all sample data to a certain interval (e.g. [0,1] or [−1,1]) after sample pre-processing, thus completely eliminating the negative effects of data singularities that make samples difficult to process.

This paper uses Min-Max Normalization, also known as outlier normalization, which aims to make the resultant values map to between [0, 1]. This study divides the results within [0, 1] into five intervals to facilitate the analysis of the time evolution of the land surface temperature. The conversion functions used are.

$$x' = \frac{x - \min(x)}{\max(x) - \min(x)} \quad (2)$$

It is important to note that the normalization method of maximum-minimum normalization is more suitable when the data is concentrated. In this study, the data is generally dense, and the normalization method of maximum-minimum normalization is more compatible, so this method is chosen. If the distribution of the maximum and minimum values of the data is very unstable, the normalized results will be very unstable and the subsequent application will be very unstable. If the maximum and minimum values are unstable, they can be replaced by empirical constants for the maximum and minimum values, and the normalized results can be made more stable.

3 RESULTS AND DISCUSSION

This paper takes Beijing as the study area and uses Landsat 8 remote sensing image data to perform remote sensing inversion of land surface temperature in Beijing from 2010 to 2015. The inversion results are analyzed to explore the feasibility of applying remote sensing images to land surface temperature inversion.

3.1 *Preliminary research results*

The monthly synthetic land surface temperature products (1KM) obtained from the Land-sat-8 satellite were used to extract the monthly synthetic land surface temperature products for Beijing, and the extracted monthly synthetic products were further processed using a raster calculator to obtain the spatial distribution of the annual mean land surface temperature from 2010 to 2015, and the obtained annual mean land surface temperature distribution was divided into temperature intervals of 4°C. This allows for a more intuitive study of the evolution of land surface temperature in the Beijing area between 2010 and 2015.

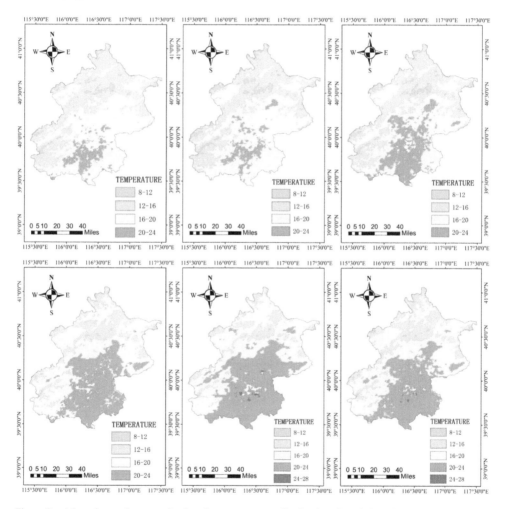

Figure 2.   Map of annual average land surface temperature distribution (from left to right, 2010 to 2015).

It is clear from the graph that the 16–24°C range is widely distributed in Beijing and that there is a spatial pattern of high temperatures in the south and low temperatures in the north. As the southern plain is the main area of population concentration and the city is gradually developing towards the north, it is clear that between 2010 and 2015 there is a tendency for the southern high-temperature zone to spread towards the north. In time, the land surface temperature in the Beijing area tends to increase year by year, especially in the densely populated southeastern region, where the proportion of the 20–24°C range increases year by year.

Based on the results of the previous study, the data will be normalized and the normalized land surface temperature data will be mapped to the interval [0,1], dividing [0,1] into five intervals [0,0.2], [0.2,0.4], [0.4,0.6], [0.6,0.8], [0.8,1], which are sub-tabulated to represent the low temperature zone, the sub-low temperature zone, the medium temperature zone, sub-high temperature zone, and high temperature zone. After the data were reclassified and organized, the number of image elements in each temperature zone was statistically analyzed as a percentage, as shown in the table.

Table 1.   Share of image dollars by the year.

| | Percentage | | | | | |
Zone	2010	2011	2012	2013	2014	2015
Low	0.843%	0.862%	0.644%	0.559%	0.788%	0.732%
Sub-low	13.100%	10.264%	11.132%	10.415%	11.863%	13.504%
Medium	34.669%	30.179%	33.541%	34.642%	31.656%	35.167%
Sub-high	36.984%	32.709%	31.017%	33.815%	35.004%	35.303%
High	14.404%	25.987%	23.667%	20.569%	20.688%	15.294%

This table was processed into a discounted graph as shown below.

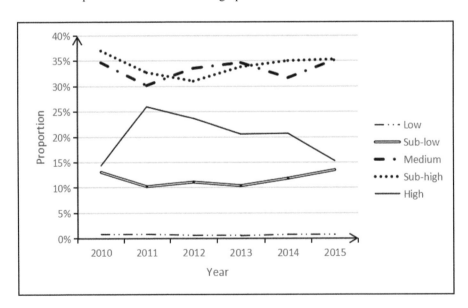

Figure 3.   Folded graph of the percentage of image elements.

As can be seen from the table above, the medium and medium-high temperature zones have the highest share, the low-temperature zone has the lowest share, and the medium and high-temperature zones are at intermediate levels. There is a more pronounced change around 2011, with the exception of the low-temperature zone share, before levelling off again and returning to the share level of around 2010.

122

3.2 *Discussion*

The results show that from 2010 to 2015, the land surface temperature showed an upward trend, the spatial distribution was relatively uneven, and there was a law of "south is greater than north". The surface temperature high-temperature area was highly consistent with the urban and rural residents' gathering area, the impact of different surface cover types on the surface temperature was large, and the expansion of urban built-up areas had obvious warming effects on the surface.

The land surface temperature high-temperature areas in Beijing are highly coincident with the areas where urban and rural residents congregate. The influence of different ground cover types on land surface temperature varies greatly, and the expansion of urban built-up areas has a significant warming effect on the surface, so the high-temperature areas in Beijing are distributed in the southeastern urban areas.

4 CONCLUSION

Land surface temperature is mainly affected by the combined action of many environmental conditions, such as solar radiation, vegetation, terrain, humidity, temperature, and precipitation. Interannual and seasonal variations in surface temperature have a direct impact on both crop and local soil quality changes. In order to better study the production and development of the Beijing region, this paper focuses on the spatial and temporal evolution of land surface temperature in the Beijing region.

From 2010 to 2015, the land surface temperature showed an overall increasing trend and a more uneven spatial distribution, with a general pattern of "south over north", with the proportion of low, medium, and high-temperature zones increasing by 0.404%, 0.498%, and 0.89%, respectively, from 2010 to 2015. The proportion of low and sub-high temperature zones decreased by 0.111% and 1.681% respectively. The change in land surface temperature is mainly concentrated in the urban and rural settlements in the southeast. The main reason for this may be the change in vegetation cover, greenhouse gas emissions, and subsurface conditions caused by urban development. The impact of the opening of the South-North Water Transfer Line to Beijing at the end of 2014 cannot be excluded as well.

The land surface temperature data inferred from remotely sensed thermal infrared data are strongly influenced by cloud cover, which can cause a lack of land surface temperature data information in cloud-covered areas. In future research ground stations, data should be applied to the process of reconstructing land surface temperature data under cloud cover, which is more representative of the land surface temperature conditions in cloud contaminated areas, and further research is needed in improving the accuracy of the product. However, the accuracy of the reconstructed results can also be affected by the number of ground stations, as well as the topography and surface type. In future studies, these two aspects of information can be considered and the number of ground stations can be further increased to further improve the accuracy of the reconstructed land surface temperature data.

The data selected for this study are from 2010 to 2015, and only the general trends between 2010 and 2015 have been analyzed in terms of the amount of data processing. At the same time, more relevant influencing factors, such as DEM, NDVI, temperature, precipitation, and land use type, can be selected to further analyze the influencing factors in the spatio-temporal evolution of land surface temperature. Further detailed studies of daily and quarterly spatial and temporal evolution can provide a theoretical basis for exploring more influencing factors and more complex impression mechanisms, and further data processing can be carried out in the future to obtain more detailed spatial and temporal evolution patterns.

REFERENCES

An B, Xiao W, Zhang S, Zhu N, Zhang J D. (2021) Spatial and temporal variability of land surface temperature on the Loess Plateau from 1960–2017[J]. *Arid Zone Geography*, 44(03):778–785.

Chen Qiuji,Hang Mengru,Guo Zhao. (2020) Spatial and temporal evolution of land surface temperature in Xi'an in summer[J]. *Mapping Science.*, 45(11):139–146.

Du Yanting, Zhang Dongyou. (2020) Analysis of the spatial and temporal distribution of land surface temperature and influencing factors in the Daxinganling region from 2001–2019[J]. *Forest Engineering*, 36(06):9–18.

Ji Ran. (2018) *Analysis of spatial and temporal variation characteristics of land surface temperature and influencing factors in Liaoning Province from 1960–2016*[D]. Liaoning Normal University.

Li Kexiang, Xiao Jiujun, Xie Yuangui, Chen Yuan, Chen Yang, Zhang Langyue. (2020) Spatial and temporal characteristics of land surface temperature on the Yunnan-Guizhou Plateau based on MODIS data[J]. *Guizhou Science*, 38(06):43–48.

Liu Dandan,Liu Jiang,Jiang Hongbo. (2020) Spatial and temporal variation of heat island effect in Harbin City based on Landsat imagery[J]. *Mapping and spatial geographic information*, 43(12):5–7+13.

Ma Songchao. (2021) Research on the spatial and temporal characteristics and evolution of urban heat islands based on remote sensing: an example of Yangzi River city cluster[J/OL]. *Mapping and Geographic Information*:1–4. https://doi.org/10.14188/j.2095-6045.2020039.

Pan M X, Zhang L J, Qu C J, Pan T, Zhang F. (2021) Characteristics of spatial and temporal variation of soil moisture in spring in Heilongjiang Province from 1983 to 2019 and factors influencing it[J]. *Geographical Research*, 40(04):1111–1124.

Ren Jingquan, Liu Yuxi, Wang Liwei, Wang Liang, Wang Dongni, Guo Chunming, Cao Tiehua. (2020) Study on the spatial and temporal variation of land surface temperature and influencing factors in Jilin Province[J]. *China Agronomy Bulletin*, 36(05):103–109.

WANG Jingwen, ZHAO Wei, YE Jiangxia, ZHU Hongqin, ZHANG Mingsha. (2021) Application of Landsat-8 data to analyze land surface temperature patterns and influencing factors in mountainous areas[J]. *Journal of Northeast Forestry University*, 49(05):97–104.

Wang Yanqiang, Du Tingting, Ye Xichen, He Linqian, Xu Chang, Wang Decai, Yue Qingling. (2020) Analysis of heat island effect in Zhengzhou City based on land surface temperature and vegetation index[J]. *Journal of Henan University of Science and Technology* (Natural Science Edition), 48(04):56–61.

Zhang Hongqiang,Luo Chunyu,Cui Ling,Qu Yi,Li Haiyan,Zeng Xingyu,Gao Yuhui. (2018) Analysis of spatial and temporal variation of land surface temperature in the growing season of the Flexi river basin from 2011–2016[J]. *Land and Natural Resources* Research, (02):68–69.

ZHANG Wei, JI Ran. (2019) Spatial and temporal variation of land surface temperature and influencing factors in Liaoning Province[J]. *Journal of Ecology*, 39(18):6772–6784.

ZHAO Bing, MAO Kebiao, CAI Yulin, MENG Xiangjin. (2020) Study on the spatial and temporal evolution of land surface temperature in China[J]. *Remote sensing of land resources*, 32(02):233–240.

Zhao Bing. (2020) *MODIS-based time series reconstruction of land surface temperature and analysis of driving factors*[D]. Shandong University of Science and Technology.

ZHAO Meiliang, CAO Guangchao, CAO Shengkui, LIU Fugang, YUAN Jie, ZHANG Zhuo, DIAO Erlong, FU Jianxin. (2021) Spatial and temporal variability of land surface temperature in Qinghai Province from 1980-2017[J]. *Arid Zone Research*, 38(01):178–187.

ZHAO Qiuyang, LI Mingxi, DENG Liansheng. (2020) Study on the spatial and temporal variation of land surface temperature in Huangshi City based on remote sensing data[J]. *Journal of Hubei Institute of Technology*, 36(03):26–31.

ZHOU Siyan, TAN Yongbin, HOU Mengfei, AI Jinquan, ZHANG Zezhi, ZHU Shangjun. (2021) Correlation analysis of population density and land surface temperature in ChangZhuTan area combined with night-light remote sensing data[J]. *Jiangxi Science*, 39(01):105–110.

Advances in Measurement Technology, Disaster Prevention and
Mitigation – Li & Mohd Yusof (Eds)
© 2023 The Author(s), ISBN: 978-1-032-36087-4

Research on noise reduction method of slope monitoring data based on wavelet analysis

Dexin Liu*, Wenqing Wang & Peng Liu*
South China Engineering Co, LTD, Guangzhou, China

Feng Xu
China Merchants Chongqing Communications Research & Design Institute Co., Ltd, Chongqing, China

ABSTRACT: Slope monitoring data often contain various noises, which makes the actual defor-mation characteristics of the slope unable to be identified. Wavelet analysis can effectively separate the real deformation signal and noise signal of the slope with different time-frequency character-istics. Different methods and decomposition layers also affect the noise reduction effect. The experimental results show that the noise reduction effect is better when the sym4 wavelet is used for three-layer decomposition.

1 INTRODUCTION

In the processing of signals containing interference, wavelet analysis is a novel theory. It overcomes the shortcoming that traditional Fourier analysis can not do time-frequency analysis at the same time. It has good locality in both the time and frequency domains, and because of this property, the use of wavelet analysis can effectively reduce noise and extract the true deformation characteristics of the slope (Jiang 2019, Tang & Tang 2014). In recent years, wavelet analysis has developed rapidly and has become a new research direction (Wu & Zhu 2011).

At present, adaptive intelligent monitoring instruments such as laser range finders and deep inclinometer have been widely used in slope monitoring work. However, the actual engineering situation is complicated. Due to the influence of instrument installation, maintenance, weather, and construction, the monitoring data are usually mixed with noise. Noise in signal will have an adverse effect on signal transmission, reception, analysis, and processing, resulting in the reduction of useful signals in the signal (Liang et al. 2008, Qin & Lu 2006, Wei et al. 2011). Research on the signal denoising method to eliminate and reduce the noise in the signal is beneficial to obtaining more useful signals in the detection data. Selecting different parameters for different noise reduction methods will also affect the noise reduction effect. This paper mainly studies the influence of choosing different wavelet bases and decomposition layers on the effect of noise reduction in order to find a reasonable parameter to achieve a better effect of noise reduction.

2 WAVELET NOISE REDUCTION THEORY ANALYSIS

Wavelet analysis has the multi-resolution characteristics of adjustable time-frequency windows for non-stationary sequences such as slope monitoring sequences, so this paper uses the wavelet transform method to remove noise signals and restore the original signals. Signals and noise have different patterns. Mainly when the scale changes, its amplitude will produce different trends. Changes in the general scale do not cause changes in the amplitude of the effective signal, and the

---

*Corresponding Author: 294924266@qq.com and liupeng_luis@163.com

increases in the scale will cause the amplitude of the noise to attenuate to zero. Based on this prior knowledge, the wavelet denoising algorithm is constructed by mathematical methods to achieve the purpose of signal noise reduction and even noise reduction.

2.1 Noise reduction methods

Since the 1980s, wavelet analysis has developed rapidly, and a large number of analytical methods have emerged so far, of which wavelet noise reduction methods can be divided into three types: 1. Based on maximum denoising, due to the different characteristics of signal and noise in scale change, the maximum point generated by removing the noise and the corresponding modulo maximum point of the retaining signal make the remaining models' maximum point reconstruct the wavelet coefficient and restore the signal; 2. Based on correlation noise reduction, since the effective signal has a strong correlation at any scale, the noisy signal is directly reconstructed by calculating the correlation of the wavelet coefficients between adjacent scales after the wavelet transformation, distinguishing the effective signal and noise according to the correlation, and determining the type of wavelet coefficient; 3. Based on threshold noise reduction, because the signal corresponding to the wavelet coefficient contains important information, its amplitude is large but with a small quantity and the number of wavelet coefficients corresponding to the noise is small but numerous. It is needed to set the wavelet coefficients with smaller absolute values to zero, and retain or contract the coefficient with a larger absolute, and the signal is reconstructed by performing a wavelet inverse transformation of the processed wavelet coefficient. Threshold method wavelet de-noising is currently the most widely used method. It is simple to calculate, easy to implement, and has a good de-noising effect. By setting a threshold for the wavelet coefficients, the threshold is used to determine whether to retain the wavelet coefficients of each layer.

2.2 Wavelet threshold denoising

The noisy signal can be expressed as $s(n) = f(n) + \sigma e(n)$, where $e(n)$ is the noise, σ is the intensity of the noise. In this article, $f(n)$ is the real deformation signal and $s(n)$ is the monitoring ground displacement data. The Matlab toolbox has three types of noise: standard Gaussian white noise, white noise with unknown variance, and non-white noise. The noise reduction results are also affected by the noise model. In actual engineering, unless there is a certain noise source, it is generally assumed to be white noise with unknown variance. The essence of noise reduction is to suppress $e(n)$ and restore $f(n)$. Its process can be divided into three steps. Firstly, it is needed to determine the number of analysis layers N, and select the wavelet function for decomposition. Since more than 90% of the best noise reduction effect can be achieved when the number of wavelet decomposition layers is 5 combined with other engineering examples, this paper adopts 3-5 layers of wavelet analysis layers. Secondly, the decomposed wavelet will be divided into two parts: high frequency and low frequency. Let n be the number of high-frequency coefficients, and the threshold is $\lambda = \sigma \sqrt{2 \log(n)}$. Finally, the denoised wavelet is reconstructed through the decomposed low-frequency part and the threshold-processed high-frequency part. The noise reduction process is shown in Figure 1.

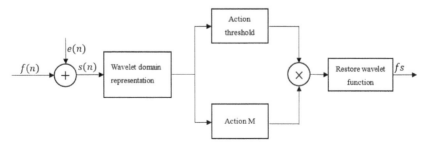

Figure 1.   Wavelet threshold denoising model.

After applying the threshold value, the wavelet coefficient whose absolute value is less than the threshold value is set to zero, and the wavelet coefficient whose absolute value is greater than the threshold value is retained. Function M refers to the wavelet coefficients that retain a particular value, and the others are set to zero.

3 EXPERIMENTAL RESEARCH ON WAVELET DENOISING

3.1 Experimental methods

This paper is based on the wavelet toolbox in MATLAB to realize the noise reduction of the monitoring point signals, and the original monitoring sequence of the surface displacement of the soil slope on the right side of K10 + 585 ~ K10 + 820 of a certain expressway in Guangzhou is analyzed. After analysis, because the number of decomposition layers of a wavelet has a greater impact on the signal noise reduction effect, and when the number of wavelet decomposition layers is 3–5 layers, there is a better noise reduction effect, so two types of db4 and sym4 are used for the original signal. Wavelet decomposes 3–5 layers for noise reduction experiments. The specific process is shown in Figure 2.

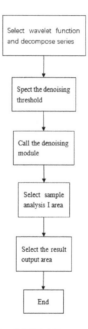

Figure 2. Wavelet denoising process based on MATLAB.

3.2 Experimental accuracy index

1. Root mean square error

The root mean square error (RMSE) refers to the root mean square error between the final reconstructed signal obtained by wavelet decomposition and threshold processing and the original signal collected by the instrument. Its formula is:

$$RMSE = \sqrt{\frac{1}{n}\sum_{i=1}^{n}[s(n) - fs]^2} \tag{1}$$

2. Signal-to-noise ratio

Signal-to-noise ratio (SNR) is one of the classical standards used to measure signal quality and is widely used in signal quality evaluation. It is defined as the ratio of useful signal to noise:

$$SNR = 10 \times \lg(p_s/p_n) \tag{2}$$

$$p_s = \frac{\sum\limits_n f^2(n)}{n} \tag{3}$$

$$p_n = \frac{\sum\limits_n [f(n) - fs]^2}{n} \tag{4}$$

Here p_s is the real signal power, p_n is the noise signal power, $f(n)$ is the deformation monitoring data, and fs is the signal after noise reduction. The larger the signal-to-noise ratio is, the better the noise reduction effect will be.

3. Peak signal-to-noise ratio

The peak signal-to-noise ratio (PSNR) represents the relationship between the maximum possible power of a signal and the destructive noise power that affects its accuracy. It is often used as an evaluation index for signal reconstruction quality. Under normal circumstances, the general benchmark for PSNR is 30dB. Its formula is:

$$PSNR = 10 \log_{10} \frac{MAX[s(n)]^2}{\frac{1}{n}\sum\limits_{i=1}^{n}[s(n) - fs]^2} \tag{5}$$

This article uses the above three indicators to evaluate the noise reduction effect. The smaller the RMSE after noise reduction is, the better the noise reduction effect will be. The larger the PSNR and SNR are, the better the noise reduction effect will be.

3.3 *Analysis of experimental results*

Monitoring sequence of this slope surface displacement based on the above research conclusions and using the most commonly used db4 and sym4 wavelets to perform 3–5 layers decomposition and noise reduction respectively based on the MATLAB program, they have achieved good noise reduction effects. The noise reduction results are shown in the figure.

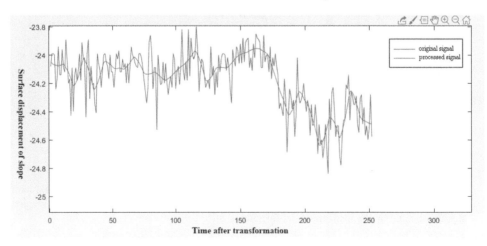

Figure 3.   Sym4 Wavelet three-layer noise reduction result.

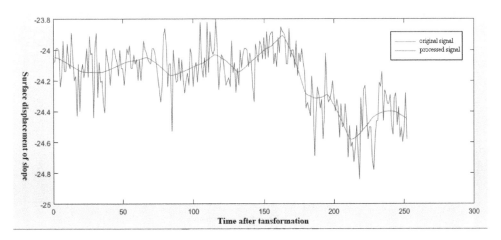

Figure 4.   Sym4 Wavelet four-layer noise reduction result.

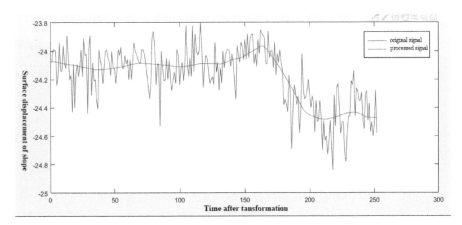

Figure 5.   Sym4 Wavelet five-layer noise reduction result.

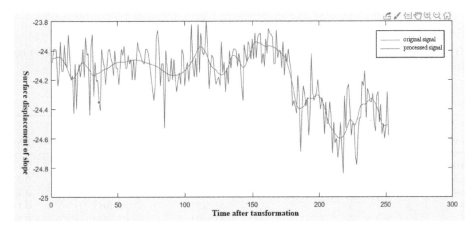

Figure 6.   Db4 Wavelet three-layer noise reduction result.

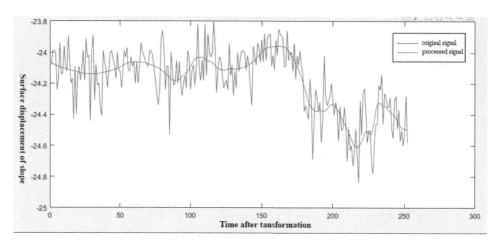

Figure 7.　Db4 Wavelet four-layer noise reduction result.

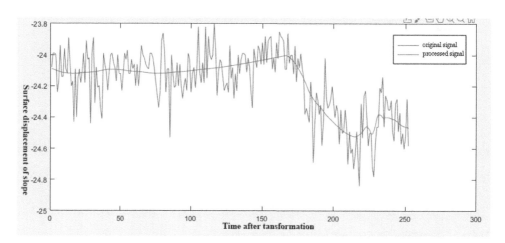

Figure 8.　Db4 Wavelet five-layer noise reduction result.

By analyzing the results of noise reduction, the accuracy evaluation indexes of different noise reduction algorithms are calculated as shown in the table.

Table 1.　Comparison of the effects of different noise reduction algorithms.

Wavelet name	Decomposition layer	SNR	PSNR	RMSE
sym4	3	46.4400	46.3015	0.1152
sym4	4	45.7325	45.5942	0.1250
sym4	5	45.1699	45.0314	0.1334
db4	3	46.2471	46.1085	0.1178
db4	4	45.9511	45.8124	0.1219
db4	5	45.4444	45.3058	0.1292

Through the analysis of six commonly used noise reduction methods, it can be seen that when the number of wavelet decomposition layers increases, the curve will become smoother, but if the number of wavelet decomposition layers increases, more effective signals will also be filtered out. By analyzing the signal-to-noise ratio (SNR), peak signal-to-noise ratio (PSNR), and root mean square error (RMSE) of the six algorithms, it is found that the two present commonly used wavelets, sym4, and db4, are decomposed in 3, 4, and 5. The number of layers has a good noise reduction effect. The SNR and PSNR are both greater than 40, and the error is small. However, when the sym4 wavelet is used for the 3-layer decomposition of the slope surface displacement monitoring sequence, the SNR and PSNR are both the largest, indicating the true signal can get the most retention. The smallest RSME indicates the smallest error and the best noise reduction effect, so this article uses this algorithm to analyze the monitoring sequence.

4 CONCLUSION

According to the original monitoring sequence of the surface displacement of the soil slope on the right side of K10+585 ∼ K10+820 in a certain expressway in Guangzhou, the wavelet de-noising experiment was carried out. After the performance of 6 wavelet de-noising methods was compared and analyzed, the sym4 wavelet 3-layer de-noising method was selected.

In this paper, two different wavelet bases and different decomposition layers are used to de-noise the original signal, and the de-noising effect of db4 and SYM4 wavelets is studied when different decomposition layers are selected, which provides some references for the selection of small wavelet bases and decomposition layers in wavelet de-noising. Experiments show that the effect of noise reduction is influenced by the threshold, wavelet base, and the number of noise reduction layers, and further research is needed on the influence of threshold on the effect of noise reduction. In actual engineering, appropriate noise reduction schemes should be selected according to the requirements of the above factors.

REFERENCES

Jiang Bo. (2019) Research on wavelet denoising based on MATLAB. *J. Electronics Practice*, 87–88.
Liang G, Xu W, Tan X, et al. (2008) Denoising processing of safety monitoring data for high rock slope based on wavelet transform. *J. Chinese Journal of Rock Mechanics and Engineering*, 27: 1837–1844.
Qin Hao, Lu Yang. (2006) Wavelet Analysis for Inclination Monitoring. *J. Highway Traffic Technology*, 23: 57–61.
Tang Liwen, Tang Dongfeng. (2014) Wavelet signal denoising technology based on Matlab. J. *Journal of Hunan University of Science and Technology* (Natural Science Edition), 29: 84–87
Wei G F, Su F, Jian T. (2011) Study on the robust wavelet threshold technique for heavy-tailed noises. *J. Journal of Computers*, 6: 1246–1253.
Wu Qilin, Zhu Mo. (2011) Wavelet de-noising of slope deformation monitoring based on cubic hermit interpolation. *J. Highway Engineering*, 36: 16–19.

Advances in Measurement Technology, Disaster Prevention and
Mitigation – Li & Mohd Yusof (Eds)
© 2023 The Author(s), ISBN: 978-1-032-36087-4

Research on remote sensing technology application in water quality parameter inversion

Xin Cao & Jing Zhang*

Beijing Laboratory of Water Resources Security, Capital Normal University, Beijing, China
The Key Lab of Resource Environment and GIS of Beijing, Capital Normal University, Beijing, China

ABSTRACT: As a basic natural resource, the safety of the water environment is a direct threat to human health. Therefore, the monitoring of water resources and the accurate and effective control of water pollution are the key to the treatment and protection of water resources. With the development of remote sensing technology in recent years, it has become a new way of water quality monitoring. Remote sensing technology is widely used by researchers for its wide range, low cost, and real-time dynamic monitoring. In this paper, the method of water quality monitoring based on remote sensing image data is investigated. The Yellow River Delta is taken as the study area, and the rationality and reliability of the simulation effect of coupling remote sensing technology are explored. Meanwhile, the applicability of remote sensing technology in water quality monitoring is verified. The combination of remote sensing data and actual water quality data can improve the temporal resolution of remote sensing monitoring, achieve complementary multi-platform observation information, and obtain more comprehensive and accurate water quality monitoring.

1 INTRODUCTION

As the source of life and the key to production and the foundation of ecology, water resources play an important role in human production and life. The World Comprehensive Water Resources Assessment Report, released by the United Nations in 1997, stated that water problems would seriously constrain global economic and social development in the 21st century and could lead to conflicts between countries (Zhang & Xia 2009). With rapid socioeconomic developments and a rapidly increasing population, the demand for water resources has increased and the discharge of industrial wastewater and domestic sewage is growing. Coupled with the rapid progress of urbanization construction, water pollution is a serious phenomenon. Water shortage and water pollution have become the two factors that seriously restrict China's sustainable development. Ensuring the safety of drinking water is the core content of water pollution prevention and control. Water quality monitoring is the basis for water quality assessment and pollution prevention, and is also a key aspect in water environment management. The ability to accurately and effectively grasp the water quality pollution situation and tailor it to local conditions is the key to treatment and protection (Zhao 2021).

The traditional method of water quality monitoring is to collect water samples on the spot, and then get various water quality parameters by physical and chemical analysis in the laboratory. At present, all kinds of testing instruments are constantly refined and developed, and almost all kinds of water quality parameters can be detected. This method produces accurate water quality data, but the process is complex and lengthy, time-consuming, and laborious (Zhao 2021). In addition, only local data can be obtained by this method, which can not meet the requirement of real-time dynamic monitoring of the whole water area.

---

*Corresponding Author: 5607@cnu.edu.cn

                    DOI 10.1201/9781003330172-20

The different spectral characteristics of different pollutants in water provide a theoretical basis for monitoring water quality by remote sensing technology. At present, multi-spectral remote sensing data are commonly used in water quality monitoring, including Landsat-TM ETM + data, SPOT-HR V data, weather satellite AVHRR data, MODIS data, and so on (Ma & Li 2002; Hu et al. 2017; Xu et al. 2007; Zhao 2021). Barrett et al. (Barrett & Amy 2016) extracted reflectivity measurements from Landsat TM/ETM+ data processing and found that the ratios of reflectivity to other bands were significantly correlated with chlorophyll and turbidity in a group of lakes in eastern Oklahoma. Using Landsat 8 remote sensing data, Markogianni et al. (Vassiliki et al. 2018) quantified ammonia in lake crispiness and found that Landsat 8 has a good ability to estimate water quality components in nutrient-poor freshwater bodies. Song et al. (Song et al. 2011) concluded that the visible and near-infrared bands of Landsat-TM data combined with neural network models were more effective in the inversion assessment of four water quality parameters, chlorophyll A, turbidity, total dissolved organic matter, and total phosphorus in surface water in Lake Chagan. Han et al. (Han et al. 2020) used Landsat 8 image data to invert dissolved organic matter (DOM) in Tai Lake Waters, and explored and analyzed the source and influence factors of DOM. With the increasing demand for monitoring, the diversified development of remote sensing data sources provides many options for water quality monitoring using remote sensing. Sentinel-2, GF-1 WFV, and other data sources are also widely used in research, even with the development of UAV technology, it is gradually combined with remote sensing technology and applied to water quality monitoring. Tan et al. (Tan et al. 2020) used GF-2 image data to analyze the water quality of the Wenjiang section of Jinma River in Chengdu, and the results showed that the water quality of this area was in a medium nutrition state in 2016 and 2018. In the water quality monitoring of the Xingyun Lake River and Maozhou River, Huang et al. (Huang et al. 2020) used an unmanned aerial vehicle (UAV) equipped with high-spectral instrument to obtain hyperspectral imaging data and establish monitoring models.

At present, the model analysis method, empirical analysis method, and semi-empirical analysis method are the three main methods for water quality monitoring and inversion based on remote sensing data (Zhao 2021). Among them, the model analysis method relies on the water radioactive transfer model. It uses remotely sensed images linked to the absorption and backscatter coefficients of the components in the water to invert and obtain the concentration content of the water quality parameters. The empirical analysis method is based on the statistical relationship between measured water quality data and remotely sensed band data, which in turn leads to water quality inversions. The semi-empirical analysis method is to use the statistical relationship between the best band or combination of band data and the actual measured water quality data, then select the appropriate mathematical method to construct a water quality inversion model, and finally further derive water quality parameter estimates. The semi-empirical analysis is one of the three methods widely used, but now many researchers will combine empirical analysis with semi-empirical analysis in research (Wang & Bai 2013; Xu et al. 2007; Zhao 2021). Ma et al. (Ma & Li 2002) combined the field water spectral measurement data with satellite remote sensing data (CBERS-1, TM) to monitor the water pollution in the Daliao estuary and quickly obtain the water pollution information.

In recent years, during the construction of a high-efficiency economic zone in the Yellow River Delta, the large-scale development of the region and the rapid growth of the economic development level have also resulted in the deterioration of the local ecological environment and serious water pollution. Water pollution has become one of the important concerns hindering development. Traditional water quality monitoring is mainly based on the field sampling method. Although this method can accurately obtain the accurate values of various water quality parameters in the monitoring area, it needs a lot of time, manpower, and cost. The collected data are point data in the monitoring area, which can not meet the needs of real-time and large-scale water quality monitoring. Because of its low cost and wide range, remote sensing has become a new technology for the dynamic monitoring of water quality. Therefore, it is particularly important for water quality monitoring to explore effective methods for the change of main water quality parameters and realize the rapid inversion of main water quality parameters (Xu et al. 2007; Zhao 2021). In addition, there are few studies on water quality monitoring using remote sensing technology in the Yellow River

Delta. Based on the above factors, remote sensing (RS) technology is used to retrieve the water quality of the Yellow River Delta, which provides a theoretical basis for the water quality and pollution prevention work of the Yellow River Delta and a scientific reference for the ecological environment protection of the Yellow River Delta (Xu et al. 2007; Zhao 2021). The key aspects of this paper are: taking the Yellow River Delta as the study area, using the Landsat 8 RS image data and the water quality data of the water quality monitoring station, the ammonia nitrogen as a water quality parameter is retrieved by RS, and the inversion results are analyzed, to explore the feasibility of RS image application in water quality inversion.

2 MATERIALS AND METHOD

2.1 Study area and data collection

The Yellow River Delta is an alluvial plain formed by the sediment carried by the Yellow River at its mouth into the sea and deposited in the Bohai Sea depression, which culminates in the Ninghai of Kenli and extends from the mouth of the Tuol River in the north to the mouth of the branch ditch in the south. It covers an area of about 5,400 sq. km, of which about 96% is within the city of Dongying in Shandong Province. With superior geographical location and rich natural resources, the Yellow River Delta has Shengli Oilfield, the second-largest oilfield in China (Zhang et al. 2000). Due to the demands of economic development, with the development of oil and gas exploitation and refining industry, the local industry has formed an industrial system based on petrochemical, textile, paper-making, and electromechanical industry. The sewage discharge is increasing. The rivers in the Yellow River Delta are polluted to varying degrees, and water pollution is becoming more and more serious (Sun et al. 2008; Zhang et al. 2012). The problem of water pollution directly affects the drinking water safety and quality of life of the residents in the area, threatens the local ecological environment, and restricts the economic development of the Yellow River Delta (Zhang et al. 2012). Therefore, it is of great significance to study the current situation of water quality in the Yellow River Delta and improve the ability of pollution control. Combined with the existing water quality research on the Yellow River Delta and the water quality measurement of various monitoring stations (Figure 1), it can be seen that the pollution in the Yellow River Delta is mainly organic pollution, in which COD, permanganate index, and ammonia nitrogen exceed the standard. Zhang et al. (Zhang et al. 2012) analyzed the water pollution characteristics of the Dongying Yellow River Delta and evaluated the pollution of main rivers. The results are shown in Table 1.

Table 1. Main river pollution assessment.

River name	Mean combined pollution index	Pollution level
Shenxian Ditch	1.67	Severe Pollution
Pick River	1.31	Severe Pollution
Overflow River	1.10	Severe Pollution
Guangli River	0.95	Moderate pollution
Branch Vein River	0.86	Moderate pollution

2.2 Research methods

(1) Correlation analysis method

To characterize the correlation between the surface reflectance and the measured concentration data of water quality, the beams with high correlation or the combination of bands for the inversion model were selected, and Pearson correlation coefficient method was adopted (Zhao 2021). The Pearson correlation coefficient is the quotient of the covariance, and standard deviation between two variables, which measures the degree of correlation between two variables. Its calculation formula is shown below.

$$r_{xy} = \frac{\sum_{i=1}^{n}(x_i - \bar{x})(y_i - \bar{y})}{\sqrt{\sum_{i=1}^{n}(x_i - \bar{x})^2}\sqrt{\sum_{i=1}^{n}(y_i - \bar{y})^2}} \tag{1}$$

In the formula: r_{xy} is the correlation coefficient between the surface reflectance and the water quality concentration parameters. x_i and y_i are the sample values of the surface reflectance and water quality concentration parameters, respectively. $\bar{x}$ and $\bar{y}$ are the means of the samples of the two variables, respectively. The closer $|r_{xy}|$ tends to 1, the stronger its correlation.

(2) Unary or multiple regression

Unary regression is the fitting of unary linear or nonlinear regression curves based on a single variable. Common models include monic linear functions, logarithmic functions, inverse power functions, exponential functions, polynomial functions, etc. Multivariate regression (MR) is a method of regression analysis of two or more variables based on the statistical principle to establish a quantitative relationship between multiple variables in a linear or non-linear mathematical model.

$$y = kx + m \tag{2}$$

$$y = a + b \ln x \tag{3}$$

In Formulas (2) and (3): k is the slope of the linear model. a and b are the pending parameter of the equation.

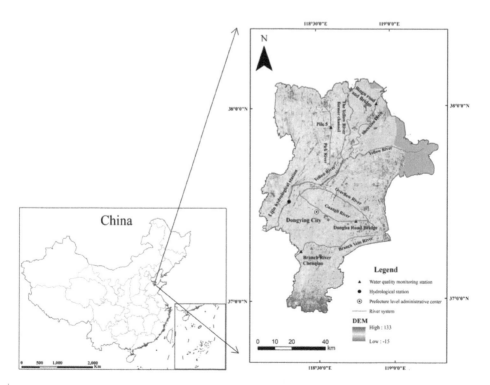

Figure 1.   Yellow river delta water system and water quality monitoring stations.

(3) Partial least squares regression

The partial least squares (PLS) method is a mathematical optimization technique developed on the basis of the traditional least-squares method. It combines the advantages of multiple linear regression analysis (MLR), canonical correlation analysis (CCA), and principal component analysis (PCA). This approach provides a more rational regression model, as well as similar research content to multiple linear regression analysis and canonical correlation analysis, and provides more depth and richness of information.

(4) Machine learning

Machine learning is currently a hot element in the field of data analysis. It is a special algorithm, not a particular algorithm. Machine learning is not a specific algorithm, but a collective term for many algorithms. Widely used algorithms in existing research include support vector machines (SVM) and random forests (RF). The support vector machine method has a simple structure and strong generalization ability, which is suitable for linear and nonlinear data and is not prone to over-fitting. The random forest method is widely used in nonlinear and high-dimensional data, which can improve computational efficiency.

(5) Accuracy evaluation

Model accuracy measures the comprehensive estimation ability of a model and evaluates the fit effect between the measured and predicted values. The coefficients of determination (R^2) and the root mean squared difference (RMSE) were used as the evaluation indicators of the model fit superiority.

$$R^2 = 1 - \frac{\sum_{i=1}^{n} (\hat{y}_i - y_i)^2}{\sum_{i=1}^{n} (y_i - \bar{y})^2} \tag{4}$$

$$RMSE = \sqrt{\frac{\sum_{i=1}^{n} (y_i - \hat{y}_i)^2}{n}} \tag{5}$$

In Formulas (4) and (5): y_i and $\hat{y}_i$ are measured and predicted values of the water quality parameter content, respectively. $\bar{y}$ is the average of the measured values of the water quality parameter content. n is the sample number.

3   RESULTS AND DISCUSSION

The RS inversion model of water quality parameters is mainly achieved by following steps: (1) Collecting and sorting out various water quality parameters of water samples in the study area; (2) combining the water quality parameters that need to be inverted, obtaining the corresponding RS images, and pre-processing the RS images; (3) analyzing the correlation between the reflectance data obtained by RS image processing and the measured concentration data of water quality, and selecting band combinations or high correlations for model inversion; (4) selecting unary or multivariate regression according to the actual situation of the research area. Partial least squares regression and machine learning methods are used to construct water quality inversion models; (5) comparative analysis and accuracy evaluation of various models are carried out, models with higher inversion effect are screened, and inversion results are analyzed. The main ideas are as follows (Figure 2).

The collected water quality data are rejected and classified as abnormal data, and the average value of water quality parameters of each monitoring station is calculated (Figure 3).

As can be seen from Figure 3, the concentration of total nitrogen of Pile 5, Binggu Road Bridge, and Dongba Road bridge is at a low value, and the average value of monitoring points has little difference. But the concentration of Chenqiao in the tributary river is sharply higher. Therefore, in terms of total nitrogen parameters, tributary river pollution is the most serious. The cause of this phenomenon needs to synthesize the development situation of this area and the inversion result of other water quality parameters to analyze the source and influence factors of pollutants.

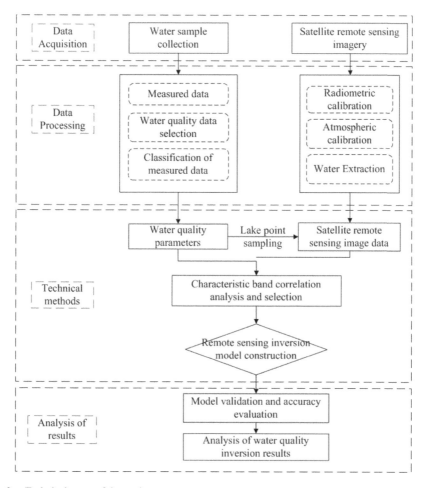

Figure 2.   Technical route of the study.

In the process of water quality monitoring based on multi-spectral RS image, the spectrum range of characteristic band or band combination is mainly between 400 mm and 900 mm. And it mainly appears in blue, green, red, and near-infrared bands (Ma et al. 2021). Through addition, subtraction, multiplication, and division, as well as molecular formula, we can arrange and combine many kinds of waveband combinations. In the preliminary study of water quality inversion in the Yellow River Delta, Pearson correlation analysis was carried out by calculating the measured value of total nitrogen and the corresponding band and band combination. The band combined with a larger correlation coefficient was obtained. The characteristic band combination of total nitrogen parameters is (b2+b5)/b4. The next research plan is to construct the model of RS inversion by various mathematical methods and choose the model with the best fitting effect.

The total water quality of the Yellow River Delta from November to October 2020 is below Category IV at Binggu Road Bridge station, Dongba Road Bridge station, and Zhimaihe Chenqiao station. The water quality of Pile 5 is Category III~IV. Analysis of the total nitrogen fraction of the measured water quality data yielded lower concentrations at Pile 5 monitoring station compared to

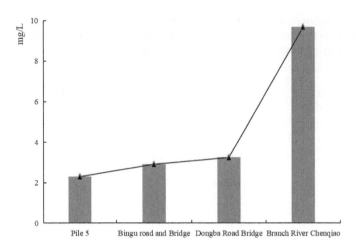

Figure 3. Average value of total nitrogen parameter at each monitoring station.

the other three stations. According to the current research results, the results based on RS image data are consistent with some previous studies in the Yellow River Delta. It shows that the reliability of the simulation effect and RS technology are effective means for water quality monitoring (Xu et al. 2022). The limited water quality monitoring stations in this paper lead to the shortage of water quality data. It is very valuable to inverse water quality parameters with RS image data.

However, the study of the Yellow River Delta in this paper is just beginning, and there are still a lot of points to pay attention to in the next step of the study:

(1) For single-parameter analysis, the result has a low reference value. Of course, in addition to total nitrogen, pollutants in the Yellow River Delta also include other water quality parameters. The retrieval of various water quality parameters is carried out by RS combined with the inversion results for spatial analysis, and the sources and trends of pollutants are explored. This is the next step.
(2) Looking for two or three data sources to carry out the inversion of various water quality parameters, constructing the water quality monitoring scheme based on different data sources, and comparing and analyzing the inversion results, the established model can provide the model basis for the water quality monitoring of the Yellow River Delta.

4 CONCLUSION

Water quality monitoring based on RS image data has been applied in many scholars' studies and combined with existing literature and preliminary studies in the Yellow River Delta. I think the current research should focus on the following trends and research limitations. Also, there are some of my suggestions for future research on remote sensing monitoring of water quality in the Yellow River Delta.

(1) At present, there are many satellite data sources for water quality RS monitoring, most of which are multi-spectral images. And many hyperspectral satellite data have not been studied and applied. So we hope to see more hyperspectral satellite data in future research (Wang 2019; Xu et al 2022). At the same time, it can also combine the satellite RS image and the measured spectral data on the water surface. The most direct performance is that the model inversion effect may be more prominent. At the same time, it can improve the time resolution of RS monitoring, a realization of complementary observation information of multi-platform.

(2) There are many water quality parameters in a water body, and most of the existing studies focus on the parameters of chlorophyll A, total nitrogen, total phosphorus, ammonia nitrogen, turbidity, and more. In the follow-up study of the Yellow River Delta, we can extend other parameters with a higher pollution index, to realize more comprehensive and accurate water quality monitoring by remote sensing (Wang 2019).

REFERENCES

D Barrett, Amy F. (2016) Automated Method for Monitoring Water Quality Using Landsat Imagery[J]. *Water*, 8(6) :257.

Han zenglei, Xiao Min, Wang Zhongliang. (2020) Application of Landsat data in analyzing the influence mechanism of dissolved organic carbon in Taihu Lake [J] *Journal of ecology*, 39 (07): 2446–2455.

Hu Hong, Hu Guangxin, Li Xinhui. (2017) Review on remote sensing monitoring of water quality [J] *Environment and development*, 29 (08): 158 + 160.

Huang Yu, Chen Xinghai, Liu Yelin, et al. (2020) Retrieval of river and lake water quality parameters based on UAV hyperspectral imaging technology [J] *People's Yangtze River*, 51 (03): 205–212.

Ma Fangkai, Gao Zhaobo, Ye Bangling. (2021) Remote sensing inversion of water quality of Tangxun Lake based on high-resolution remote sensing satellite image [J] *Water resources development and management*, (05): 69–75.

Ma Gang, Li Guoying. (2002) Demonstration study on satellite remote sensing monitoring of water pollution in Daliao estuary [J] *Liaoning urban and rural environmental science and technology*, (06): 26–29 + 36.

Song K, Wang Z, Blackwell J, et al. (2011) Water quality monitoring using Landsat Themate Mapper data with empirical algorithms in Chagan Lake, China[J]. *Journal of Applied Remote Sensing*, 5(1): 3506.

Sun Jianhui, Feng Jinglan, Chai Yan, et al. (2008) Research progress and Control Countermeasures of water pollution in the Yellow River [J] *Journal of Henan Normal University* (NATURAL SCIENCE EDITION), 36 (06): 82–87.

Tan Xiaoqin, Luo Yong, Zhao Zheng, et al. (2020) Study on river water quality inversion based on high-resolution remote sensing – Taking Wenjiang section of Jinma River as an example [J] *Environmental ecology*, 2 (07): 29–36.

Vassiliki M, Dionissios K, George P, et al. (2018) An Appraisal of the Potential of Landsat 8 in Estimating Chlorophyll-a, Ammonium Concentrations and Other Water Quality Indicators[J]. *Remote Sensing*, 10(7): 1018-.

Wang Lin, Bai Hongwei. (2013) Review on the inversion of lake water quality parameters based on remote sensing technology [J] *Global positioning system*, 38 (01): 57–61 + 72.

Wang Shan. (2019) *Monitoring and analysis of water quality changes in Panjiakou Daheiting reservoir based on multi-source remote sensing images* [D] Lanzhou Jiaotong University.

Xu Jinhong, Deng Mingjing, Liu Guodong. (2007) Application of remote sensing technology in water pollution monitoring [J] *Soil and water conservation research*, (05): 324–326 + 330.

Xu Xin, Zhang Yanjun, Dong Wenxun, et al. (2022) Water quality simulation of Xiangxi River Based on initial conditions of remote sensing inversion [J] *Rural water conservancy and hydropower in China*: 1–15.

Zhang Hongjin, Cao Guirong, Sun Tao. (2000) Investigation on the current situation of water environment in the Yellow River Delta [J] *Environmental monitoring management and technology*, (04): 24–25.

Zhang Jianwei, Guo Xiuyan, Yuan Xilong, et al. (2012) Water pollution characteristics and provenance analysis of Dongying Yellow River Delta [J] *Shandong land and resources*, 28 (03): 15–18.

Zhang Liping, Xia Jun. (2009) Hu Zhifang Analysis of water resources situation and water resources security in China [J] *Resources and environment in the Yangtze River Basin*,18 (02): 116–120.

Zhao song. (2021) *Retrieval of water quality parameters of Fuyang River in Handan City Based on multi-source remote sensing data* [D] Hebei University of engineering.

Advances in Measurement Technology, Disaster Prevention and Mitigation – Li & Mohd Yusof (Eds)
© 2023 The Author(s), ISBN: 978-1-032-36087-4

Application of air-coupled seismic waves to explosion yield estimation

Liangyong Zhang

College of Meteorology and Oceanography, National University of Defense Technology,
Changsha, Hunan, China
Northwest Institute of Nuclear Technology, Xi'an, Shaanxi, China

Weiguo Xiao, Xiaolin Hu, Xin Li, Yanjun Ma, Ao Li & Qiang Lu*

Northwest Institute of Nuclear Technology, Xi'an, Shaanxi, China

ABSTRACT: Explosion yield estimation is essential for explosion accident monitoring, weapon effectiveness testing, and verification monitoring. In this paper, the variation of the amplitude and the arrival time of the primary wave with distance are analyzed. Subsequently, based on data of the air-coupled seismic wave and the primary wave, the joint inversion of the explosion source parameters is carried out. Ultimately, the following conclusions are drawn: first, as the primary wave is affected by the underground reflecting interface, after a long-range propagation, the reflected wave will catch up with the direct wave, changing the relationships between the first peak displacement and the arrival time with distance; second, compared to the true values, the inverted results of explosion yield are larger, but the inverted values of 100-kg-magnitude experiments have small relative errors with high estimation accuracy, while the inverted value of the 8-kg-magnitude experiment has a large relative error; third, the overall error of inverted height-of-burst (HOB) is larger, but the symbol of inversion results is consistent with that of true HOB, that is, the inversion results can be used to judge whether the explosion is over or under the ground.

1 INTRODUCTION

Near-surface explosion yield estimation is significant to explosion accident monitoring (Jiang et al. 2019), weapon effectiveness testing (Driels 2012), and verification monitoring (Arrowsmith et al. 2010; Pasyanos & Myers 2018), but since near-surface explosion involves a complex process of seismo-acoustic energy partitioning which is closely related to factors such as geological composition, surface structure, and atmospheric environments, its inversion of yield has long been a difficulty (Arrowsmith & Bowman 2017; Bonner et al. 2013; Ford et al. 2014; Jiang et al. 2019; Pasyanos & Kim 2018; Templeton et al. 2018). The seismic method is one of the commonly used methods in explosion yield inversion, but due to the trade-off relationship between height-of-burst (HOB) and explosion yield, the seismic wave method can be used alone to revert explosion yield, which is of great uncertainty (Bonner et al. 2013; Ford et al. 2014; Pasyanos & Myers 2018; Pasyanos & Kim 2018). To solve the uncertainty of the single inversion method, the seismo-acoustic analysis method has been developed rapidly. It improves the estimation accuracy of source parameters (Arrowsmith et al. 2010; Bonner et al. 2013; Ford et al. 2014, 2021; Pasyanos & Myers 2018; Pasyanos & Kim 2018; Zhang et al. 2021) by fusing acoustic and seismic data to implement multiple constraints on source parameters. However, requiring both acoustic and seismic data, the above method adds a huge workload to the outfield measurement work, especially in

---

*Corresponding Author: luqiang@nint.ac.cn

DOI 10.1201/9781003330172-21

the harsh outdoor environment, as it is extremely difficult to install a large number of seismic and acoustic measurement equipment at the same time. In this paper, to estimate the explosion yield using the seismometer only, it is attempted to revert the parameters of the near-surface explosion source by combining data of air-coupled seismic wave and the primary seismic wave. The basic theory for inversion of explosion source parameters by fusing data of air-coupled seismic wave and the primary seismic wave is firstly introduced, and then the experimental data are analyzed. On this basis, the explosion source parameters are reverted by combining data of the air-coupled seismic wave and the primary seismic wave.

2 ESTIMATION THEORY OF EXPLOSION YIELD

The air-coupled seismic waves contain rich wave information about explosion waves coupled to the air. Using the precursor seismic wave to revert explosion yield by joining the air-coupled seismic wave can obtain wave information from air and ground medium simultaneously to apply multiple constraints on the source parameters so that the estimation accuracy of source parameters can be improved. The explosion yield inversion method of fusing data of air-coupled seismic wave and the primary wave includes three parts: air-coupled seismic wave model, seismic model of first peak displacement, and data fusion method. The air-coupled seismic wave model uses the coefficient of acoustic-to-seismic coupling to establish the relation between the characteristic quantity of air-coupled seismic wave and the acoustic quantity; the seismic model of first peak displacement establishes the relationship among the first peak displacement of the primary wave, explosion yield, propagation distance, and burial depth; the two models apply multiple constraints on the explosion source parameters by using a data fusion method.

2.1 The air-coupled seismic wave model

In the far-field, there is a linear relationship between the acoustic characteristic quantity (overpressure peak and positive acoustic impulse, etc.) and the characteristic quantity of air-coupled seismic waves.

$$q_{aco} = C_p q_{seis} \tag{1}$$

In the equation, q_{aco} is the acoustic characteristic quantity such as the overpressure peak p (Pa) and the positive acoustic impulse i(Pa·s); q_{seis} is the characteristic quantity of air-coupled seismic waves such as Zhalf, ENZhalf 1&2Cf, Z2, and ENZ2 0&1, etc. (Zhang et al. 2022); C_p is the coefficient of acoustic-to-seismic coupling and it is a constant when surface parameters are consistent. A study (Zhang et al. 2022).. shows that the relation between the overpressure peak and air-coupled seismic parameters Zhalf and ENZhalf 1&2 has the minimum dispersion in the sand medium. The coefficients of acoustic-to-seismic coupling are 0.0081 Pa·s/(um/s) and 0.0047 Pa·s/(um/s) respectively.

For the empirical acoustic model of acoustic impulse, the KG85 model of acoustic impulse has a high accuracy of explosion yield estimation (Zhang et al. 2022), and the KG85 model of acoustic impulse is (Kinney & Graham 1985) as follows:

$$i_s = \frac{6.7 \cdot \sqrt[2]{1 + (r_s/0.23)^4}}{r_s^2 \cdot \sqrt[3]{1 + (r_s/1.55)^3}} \tag{2}$$

where i_s (Pa·s/kg$^{1/3}$) is scaled acoustic impulse, h_s (m/kg$^{1/3}$) is scaled HOB, and r_s (m/kg$^{1/3}$) is scaled distance.

When the explosion source is close to the ground, the ground reflection and the coupling of the ground medium near the explosion source will apparently affect the overpressure waveform parameters generated by the explosion. For ground reflection, the equivalent yield factor (Zhang et al. 2022) is as below:

$$F = \begin{cases} 2, & h_s \leq 0 \\ 1.9513 - 0.10458 h_s + 4.05707 e^{-(h_s - 3.9047)^2/3.4280}, & 0 < h_s \leq 9.11 \\ 1, & h_s > 9.11 \end{cases} \quad (3)$$

where F is the equivalent yield factor and the relation between equivalent yield and actual yield is as follows:

$$W_{eq} = FW \quad (4)$$

where W is the actual explosion source yield (kg).

For the coupling of the ground medium near the explosion source, the coupling coefficient of acoustic impulse caused by seismo-acoustic energy partitioning (Ford et al. 2014; Pasyanos & Kim 2018) is:

$$C_F = \begin{cases} 10^{0.4343[\alpha_3 h_s - \ln(1 + e^{\alpha_3 h_s})]}, & \text{hard rock} \\ 10^{(\alpha_3 h_s - 0.1 \cdot \log_{10}(1 + 10^{10\alpha_3 h_s}))}, & \text{soft rock} \end{cases} \quad (5)$$

where α_3 is an undetermined coefficient related to geology characteristics and the value is 5.22 for hard rock and 2.15 for soft rock. The acoustic characteristic quantities before and after the ground coupling have the following relationship:

$$q_s = C_F \cdot q_0 \quad (6)$$

where q_0 is the acoustic characteristic quantity in the free sound field, and q_s is the acoustic characteristic quantity after seismo-acoustic energy partitioning.

Combining Equations (2) to (6), the relation between acoustic impulse, HOB, explosion yield, and propagation distance can be obtained. On this basis, by substituting Equation (1) into the above relation, the relation between the characteristic quantity of air-coupled seismic wave, explosion yield, distance, and HOB can be obtained, which can be used to revert explosion source parameters.

2.2 Seismic model of first peak displacement

For the seismic wave model, studies show that using the first peak displacement of the P wave to revert explosion source parameters is more robust than using waveform parameters such as velocity and acceleration. The relation between the first peak displacement of the P wave (primary wave), explosion yield, and distance is as follows (Ford et al. 2014; Templeton et al. 2018):

$$\log_{10}(d_s) = \beta_1 + \beta_2 \log_{10}(r_s) + \beta_3 \tanh(\beta_4 h_s + \beta_5) \quad (7)$$

where d_s is the scaled displacement (m/kg$^{1/3}$), β_1, β_2, β_3, β_4, and β_5 are undetermined coefficients closely related to geology characteristics, and the undetermined coefficients of soil medium are shown in Table 1.

Table 1. Undetermined coefficients of soil medium (Ford et al. 2014).

Undetermined coefficients	β_1	β_2	β_3	β_4	β_5
Soil	−3.39	−1.74	−0.22	4.84	1.23

2.3 Data fusion method

The relative error method is a common method of data fusion(Bonner et al. 2013). The relation of relative error is used to combine data of the primary wave and air-coupled seismic wave, and error weights are assigned to the data, then the grid searching method is used to solve them. The total error of the primary and air-coupled seismic data is as follows:

$$\text{Metric} = w_d \left(\left| \frac{d_s - d_0}{d_0} \right| \right) + w_i \left(\left| \frac{q_s - q_0}{q_0} \right| \right) \tag{8}$$

where w_d is the weight of the primary wave data, while w_i is the weight of the data of air-coupled seismic wave. This paper takes 0.5 for each weight, taking w_d as 0 and w_i as 0 respectively to obtain the trade-off curves for the primary wave and air-coupled seismic wave.

3 EXPERIMENTAL DATA

3.1 Experimental conditions

There are three experiments with explosion yields of 100 kg and 8 kg respectively. The explosion sources and the meteorological conditions are shown in Table 2, while the distribution of seismic measurement points is shown in Figure 1. Sand medium is almost in the whole site, but in a small area near the explosion sources, there is a thin layer of hard-rock medium covering the sand medium.

Table 2. Explosion sources and meteorological conditions.

Explosion source number	Yield/kg	HOB/m	Meteorological conditions		
			Atmospheric pressure/hPa	Temperature/°C	Wind speed
EX01	100	10	860	−0.7	Breeze
EX02	8	0.5	864	−1.45	Breeze
EX03	100	0.5	861	−2.65	Breeze

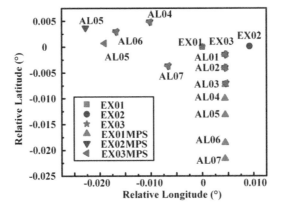

Figure 1. Distribution of measurement points. MPS mean measurement points.

143

3.2 Seismic data

Figure 2 is a typical waveform of the primary wave, which presents a good signal-to-noise ratio (SNR). All the measurement points of explosion sources EX01and EX03 have better SNR, but for EX02, because of its small explosion yield, only the first three measurement points have good SNR, while the remote measurement points have poor SNR.

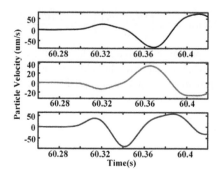

Figure 2. Typical waveform of the primary wave. From up to down are the east (E), north (N), and vertical (Z) components of the seismic wave.

Figures 3 and 4 indicate the changing relationship between the first peak displacement and arrival time with distance. For EX01, it can be seen from Figure 3(a) that the first displacement peak in logarithmic coordinate decays linearly with distance, and there is a small deviation at 1500 m with a significant deviation after 1600 m. It can also be seen from the arrival time curve (Figure 4(a)) that the farthest measurement point (nearby 2400m) obviously deviates from the solid line (fitted line of data). This is because there is an underground interface, and as the distance increases, the reflected wave gradually catches up with the direct wave and they overlay at 1500 m. Therefore, the decay curve of the first peak appears linear deviation in logarithmic coordinate. As the reflected wave further catches up with the direct wave, they overlay obviously at 2100 m, which makes the decay curve of the first peak deviate obviously, but the reflected wave hasn't exceeded the direct wave, as shown in the arrival time curve (Figure 4(a)) where the arrival time of the measurement

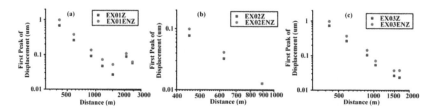

Figure 3. Variation of the first peak displacement of the primary wave with distance. (a) EX01, (b) EX02, (c) EX03.

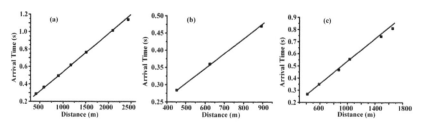

Figure 4. Arrival time of the primary wave. (a) EX01, (b) EX02, (c) EX03.

point at 2100 m is still on the arrival time curve. As the distance further increases, the reflected wave exceeds the direct wave at 2400 m and the arrival time of the primary wave begins to deviate from the linear curve of arrival time. For EX01, the data of the first four measurement points, are used for the inversion of explosion source parameters. For EX03 (Figure 3(c) and Figure 4(c)), similar to the above analysis, the first four measurement points are used for the inversion of explosion source parameters. But for EX02 (Figure 3(b) and Figure 4(b)), as only the first three measurement points of the primary wave have good SNR, only the data of the first three measurement points are used. The data of the air-coupled seismic wave can be found in reference (Zhang et al. 2022).

4   INVERSION RESULTS

Since the medium nearby the explosion sources are hard rock with the sand medium distributed in the other places where the energy coupled to the air induced by the explosion sources is reflected and coupled by the nearby hard interface and then acoustic waves are formed and propagated far away, while the energy coupled to the ground will form seismic waves in the sand medium and spread far. The undetermined coefficients of the hard or soft rock cannot simply be used to solve the explosion source parameters, and thus the acoustic coupling coefficient of the ground medium nearby the explosion source takes the hard rock coefficient, while the seismic coupling coefficient of the seismic wave model takes soil medium coefficient due to the site which is mainly covered with sand medium and where the seismic waves propagate mainly in the sand medium.

The inversion results of explosion source parameters are shown in Figure 5 and Table 3. In the figure, there is an intersection between the trade-off curves of EX01 and EX03, and the intersection

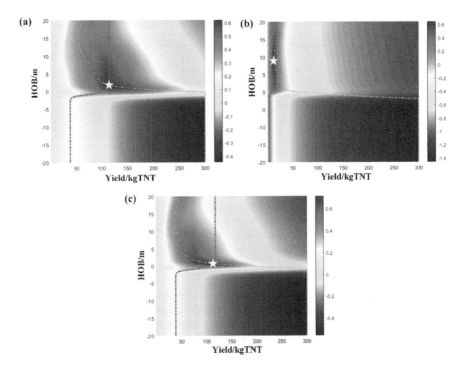

Figure 5.   Inversion results. (a) EX01, (b) EX02, (c) EX03. The larger the value of the color bar, the smaller the error. The green and red dotted lines are the trade-off curves of the air-coupled seismic wave model and the seismic model of first peak displacement respectively (trade-off curves composed of explosion yield and burial depth corresponding to minimum error). The five-pointed white star locates at the intersection and minimum error position.

position and minimum error position of the relative error method overlap. The trade-off curves of EX02 do not intersect, but the minimum error position of the relative error method is the middle of the nearest points of the two trade-off curves.

As shown in Table 3, the inversion results of the three experiments are larger than the actual values. The relative error of the explosion yield of a 100 kg explosion is smaller than that of an 8kg explosion, and the relative error of the 100 kg explosion yield estimation doesn't exceed 14% while the relative error of the low-yield 8 kg is more than half of the actual value. These results indicate that the above method has high accuracy in explosion yield estimation for 100 kg magnitude, but the error of low-yield estimation is relatively large. Moreover, the results of height-of-burst in the three experiments deviated from the actual value apparently, but are consistent with the symbol of the actual values, which indicates that using the above method is difficult to predict height-of-burst accurately, but it can predict whether the explosion source is above or under the ground.

The errors of the above inversion results mainly come from the modeling errors of the mixed site and the discreteness of the coupling coefficients of air-coupled seismic waves. The explosion energy is coupled to the sand medium through the hard rock to form seismic waves propagated far away, but the seismic wave model in this paper adopts the soil medium model, which is different from the actual situation. Besides, due to the uneven geological distribution, the coefficient of acoustic-to-seismic coupling of air-coupled seismic waves is discrete with distance, so its value is not constant, but fluctuates in a range.

Table 3. Inversion results of source parameters.

Explosion source Number	Explosion yield			Height-of-burst	
	Actual value/kg	Inversion value/kg	Relative error	Actual value/m	Inversion value/m
EX01	100	114	0.14	10	1.8
EX02	8	12.4	0.55	0.5	8.9
EX03	100	112	0.12	0.5	0.7

5 CONCLUSION

In this paper, the changing relationship of the amplitude and the arrival time of the primary wave with distance is firstly analyzed, and then, based on the data of the air-coupled seismic wave and the primary wave, the joint inversion of explosion source parameters is carried out. The following conclusions are drawn:

(1) As the primary wave is affected by the underground reflecting interface, after a long-range propagation, the reflected wave will catch up with the direct wave, changing the relationships of the first peak displacement and the arrival time with distance.
(2) Compared to the true values, the inverted results of explosion yield are larger, but the inverted values of 100-kg-magnitude experiments have small relative errors with high estimation accuracy, while the inverted value of the 8-kg-magnitude experiment has a large relative error.
(3) The overall error of inverted height-of-burst (HOB) is larger, but the symbol of inversion results is consistent with that of true HOB, that is, the inversion results can be used to judge whether the explosion is over or under the ground.

REFERENCES

Arrowsmith, S., Bowman, D. (2017) Explosion yield estimation from pressure wave template matching. *J. Acoust. Soc. Am.*, 141(6): L519–L525.

Arrowsmith, S.J., Johnson, J.B., Drob, D.P., et al. (2010) The seismoacoustic wavefield: a new paradigm in studying geophysical phenomena. *Reviews of Geophysics*, 48(RG4003): 1–23.

Bonner, J., Waxler, R., Gitterman, Y., et al. (2013) Seismo-acoustic energy partitioning at near-source and local distances from the 2011 Sayarim explosions in the Negev Desert, Israel. *Bulletin of the Seismological Society of America*, 103(2A): 741–758.

Driels, M.R. (2012) *Weaponeering: Conventional weapon system effectiveness*. 2 ed. Virginia: American Institute of Aeronautics Astronautics, Inc.

Ford, S.R., Bulaevskaya, V., Ramirez, A., et al. (2021) Joint Bayesian inference for near-surface explosion yield and height-of-burst. *Journal of Geophysical Research: Solid Earth*, 126(e2020JB020968): 1–20.

Ford, S.R., Rodgers, A.J., Xu, H., et al. (2014) Partitioning of seismoacoustic energy and estimation of yield and height-of-burst/depth-of-burial for near-surface explosions. *Bulletin of the Seismological Society of America*, 104(2): 608–623.

Jiang, W.B., Chen, Y., Peng, F. (2019) The yield estimation for the explosion at the Xiangshui, Jiangsu chemical plant in March 2019. *Chinese J. Geophys*. 63(2): 541–550.

Kinney, G.F., Graham, K.J. (1985) *Explosion shocks in air*. New York: Springer Science+Business Media.

Pasyanos, M.E., Kim, K. (2018) Seismoacoustic analysis of chemical explosions at the Nevada National Security Site. *Journal of Geophysical Research: Solid Earth*, 124: 908–924.

Pasyanos, M.E., Myers, S.C. (2018) The coupled location depth yield problem for North Korea's declared nuclear tests. *Seismological Research Letters*, 89(6): 2059–2067.

Templeton, D.C., Ford, S.R., Rodgers, A.J., et al. (2018) Seismic models for near-surface explosion yield estimation in alluvium and sedimentary rock. *Bulletin of the Seismological Society of America*, 108: 1384–1398.

Zhang, L., Li X., Xiao W., et al. (2022) Acoustic-to-seismic coupling characteristics of a specific poroelastic site. 2022 *International Conference on Materials Engineering and Applied Mechanics* (ICMEAAE 2022), Changsha, China.

Zhang, L., Li, X., Liang, X., et al. (2021) Improved seismoacoustic analysis model and its application to source parameter inversion of near-surface small-yield chemical explosions. *Applied Geophysics*, 18(1): 17–30.

Zhang, L., Lu, Q., Xiao, W., et al. (2022) Yield prediction method of surface explosions at soil site. *Acta Scientiarum Naturalium Universitatis Sunyatseni* (Accepted).

Advances in Measurement Technology, Disaster Prevention and Mitigation – Li & Mohd Yusof (Eds)
© 2023 The Author(s), ISBN: 978-1-032-36087-4

Rational evaluation of substation engineering cost based on fuzzy evaluation theory

Ye Ke, Minquan Ye, Huiying Wu, Wenqi Ou & Jiawei Lin
State Grid Economic and Technological Research Institute of Fujian Electric Power Co., Ltd., Fuzhou, Fujian, China

Cheng Xin
State Grid Economic and Technology Research Institute Co., Ltd., Beijing, China

Jinpeng Liu*
School of Economics and Management, North China Electric Power University, Beijing, China

ABSTRACT: The substation project is an important part of the power system and one of the key facilities for the investment and construction of power grid enterprises. Therefore, the issue of whether the substation project cost is reasonable is directly related to the scale of capital cost expenditure of power grid enterprises but also related to the scale of enterprise operating benefits. In this context, this paper proposes the rationality evaluation method of substation project cost: firstly, we construct the rationality of substation cost based on the main component analysis, then determine the index weight, and finally, construct the engineering cost evaluation model based on fuzzy theory.

1 INTRODUCTION

Carrying out the rationality evaluation of substation project cost is one of the important starting points to improve the level of project cost control and improve the efficiency and efficiency of project construction. In view of the importance of substation project cost control, relevant scholars have carried out a series of exploration and research on project cost management, and the specific analysis is as follows.

Kang, Li, Li, and Ge (2019) analyzed the new factors that affect the power grid engineering cost of new technology, and new policy, and then determined the background of the new technology situation affecting the power grid engineering cost evaluation index system. The AHP is established to influence the power grid engineering cost evaluation model, analyze and evaluate the corresponding power grid engineering examples quantitatively, and verify the effectiveness of the evaluation model. Ding and Peng (2021) proposed a T E transformation cost prediction method based on the MK-TESM method. Based on the historical data of the evaluation indicators of 110kV power transmission and transformation project cost from 2014 to 2018, the cost prediction model of power transmission and transformation project cost is established by using the Mann-Kendall trend inspection method and three-index smoothing method, so as to predict the trend and future data of the cost of power transmission and transformation project. The results show that the average absolute error rate of power transmission and transformation project cost data keeps the prediction rate within about 10% and the prediction accuracy is high. Zhang et al. (2018), from the perspective of technology and transformation, divided the construction stages of power transmission and

---

*Corresponding Author: juzen123@126.com

DOI 10.1201/9781003330172-22

transformation projects in detail, sorted out and identifies the technical and transformation risks of power transmission and transformation engineering from the perspective of the whole process, and evaluated the risk level of specific cases based on advanced index empowerment and comprehensive evaluation methods to provide theoretical reference for the development of specific work. Xu, Wen, Lu, and Wang (2018) proposed the determination method of cost deviation threshold by constructing the dynamic early warning model of cost deviation of power transmission and transformation projects. The target project cost information should be collected dynamically, and the trained neural network is used to predict the evaluated project cost, judge the deviation between the project cost forecast value and the target value at multiple future time points in real time, and carry out dynamic early warning. Qian (2020) combined the construction process of the power transmission and transformation project, and put forward the project cost control measures from the three stages of preliminary design estimate, development and contracting management, and construction.

To sum up, relevant scholars pay more attention to the research on the influencing factors of power grid engineering cost, cost prediction methods, and cost control measures, while the research on the rationality evaluation of engineering cost is relatively weak. Therefore, it is very necessary to carry out the research work on the rationality evaluation of the project cost.

2 CONSTRUCTION OF THE REASONABLE EVALUATION MODEL OF SUBSTATION PROJECT COST

2.1 Basic principles of the hierarchical analysis method

Analytic Hierarchy Process (AHP) decomposes the problem into different components according to the nature of the problem and the overall goal to be achieved, and aggregates and combines the factors at different levels according to the interrelated influence and affiliation between the factors to form a multi-level analysis structure model, so that the problem finally boils down to the determination of the relative importance of the lowest level (plans, measures, etc. for decision-making) relative to the highest level (total goal) or the relative order of superiority and inferiority (https://baike.so.com/doc/6012611-6225598.html).

The first step of the method is to construct the judgment matrix:

$$x = (x_{ij})_{n*n}(i,j = 1, 2, \ldots, n) \tag{1}$$

The second step of the method is to process the matrix data:

$$\bar{x}_{ij} = x_{ij} / \sum_{k=1}^{n} x_{kj}(i,j = 1, 2, \ldots, n) \tag{2}$$

The weights are calculated as follows:s

$$c_i = \bar{c}_i / \sum_{i=1}^{n} \bar{c}_i (i = 1, 2, \ldots, n) \tag{3}$$

The last step is to test the consistency of the judgment results: the process of testing A by using the consistency index and the numerical table of the consistency ratio <0.1 and the random consistency index.

2.2 Basic principles of fuzzy evaluation and analysis method

1) Establish the evaluation factor set $P = \{p_1, p_2, \ldots, p_n\}$
   It is set that P is composed of many influencing factors of the evaluation object, which constitutes the evaluation index system. Where $p_i(i = 1, 2, \ldots, n)$ is the evaluation index, that is, the influencing factors after screening.

2) Establish the judgment set $F = \{f_1, f_2, \ldots, f_n\}$

Evaluation set F consists of different evaluation levels, such as "excellent, good, qualified", "strong, strong and weak", which must be determined according to the characteristics of the evaluation object. m represents the number of evaluation grades and $f_j(j = 1, 2, \ldots, m)$ represents the comments of each index. Through the research on the applicable literature on fuzzy comprehensive evaluation method, it is found that the classification of evaluation grade in the evaluation process is too simple, so the evaluation grade standard can be established and properly expounded.

3) The fuzzy relation matrix H (membership matrix) from P to F is established

The fuzzy membership subset $H_i = \{h_{i1}, h_{i2}, \ldots, h_{im}\}(i = 1, 2, \ldots, n; j = 1, 2, \ldots, m)$ is established. Where H_i can be interpreted as a fuzzy vector relative to C obtained by single factor evaluation of the $i - th$ evaluation factor u_i, that is, the membership of the $i - th$ evaluation index to each evaluation standard f_j in the evaluation target set. The calculation formula is that h_{ij} is equal to the number of people who choose grade f_i in the evaluation index / the number of people participating in the evaluation. The fuzzy relation matrix R can be obtained by summarizing the fuzzy membership subsets of each index, that is:

$$H = \begin{bmatrix} H_1 \\ \ldots \\ H_n \end{bmatrix} = \begin{bmatrix} h_{11} & \cdots & h_{1m} \\ \vdots & \ddots & \vdots \\ h_{n1} & \cdots & h_{nm} \end{bmatrix} \tag{4}$$

The whole matrix H contains all the information obtained from the evaluation of the evaluation index set U by evaluation set F. Among them, h_{ij} is the factor P_i, with the degree of f_j $0 \leq r \leq 1$

4) Determine the weight vector of the evaluation factors $C = \{c_1, c_2, \ldots, c_n\}$

c_i represents the weight of factors, that is, the importance of evaluation indicators. Where $0 \leq c_i \leq 1$. The determination of weight is the key to a fuzzy comprehensive evaluation method, and the selection of an appropriate weight assignment method is related to the quality of weight.

5) Synthetic fuzzy comprehensive evaluation result matrix Q

According to the weight vector C and fuzzy matrix H obtained in the above steps, the evaluation result matrix is synthesized through fuzzy transformation. The multiplication and addition operator is used to perform the fuzzy operation between the weight vector C and the fuzzy matrix H. The calculation formula of multiplication additive fuzzy operator is as follows:

$$b_j = \sum_{j=1}^{n}(c_i h_{ij}) \tag{5}$$

To sum up, the final result of the fuzzy comprehensive evaluation (matrix Q) is the synthesis of weight vector W and membership matrix H as follows:

$$Q = C * H = [c_1, c_2, \ldots, c_n] * \begin{bmatrix} h_{11} & \cdots & h_{1m} \\ \vdots & \ddots & \vdots \\ h_{n1} & \cdots & h_{nm} \end{bmatrix} = [q_1, q_2, \ldots, q_m] \tag{6}$$

In the formula:

$$q_j = \sum_{i=1}^{n} c_i * h_{ij}, j = 1, 2, \ldots, m \tag{7}$$

3 THE CONSTRUCTION OF THE EVALUATION INDEX SYSTEM AND THE EMPIRICAL ANALYSIS

3.1 *Construction of the evaluation index system*

Combined with the characteristics of substation project construction of power grid enterprises, the reasonable evaluation index system of substation project cost constructed in this paper is as follows:

Table 1. Evaluation index system.

Serial number	Level 1 index	Secondary indicators
1	Design indicators	Reasonability of project site selection
2		reasonableness of design scheme
3		The rationality of the technical indicators
4		Design depth and quality
5		The reasonable value of ontology engineering cost
6		reasonableness of construction site and cleaning cost
7		reasonableness of equipment and materials fees
8		Reasonable cost of the production preparation fee
9	Cost index	Reasonable cost of technical service fee
10		reasonableness of auxiliary engineering cost
11		The rationality of the construction management fee
12		Reasonable cost of other expenses

3.2 *Weight calculation of evaluation indicators*

Combined with the characteristics of substation engineering construction of power grid enterprises, from the two aspects of design and cost, the rationality evaluation index system of substation engineering cost is constructed:

Table 2. Calculation results of the weight value of the evaluation indicators.

Order number	Level 1 index	Level 1 index weight	Secondary indicators	Secondary index weight
1	Design indicators	0.2	Reasonability of project site selection	0.0232
2			reasonableness of design scheme	0.0827
3			The rationality of the technical indicators	0.0108
4			Design depth and quality	0.0833
5			Reasonable value of ontology engineering cost	0.2849
6			reasonableness of construction site and cleaning cost	0.0977
7			reasonableness of equipment and materials fees	0.0757
8			Reasonable cost of production preparation fee	0.0255
9	Cost index	0.8	Reasonable cost of technical service fee	0.0198
10			reasonableness of auxiliary engineering cost	0.0344
11			The rationality of the construction management fee	0.0977
12			Reasonable cost of other expenses	0.1643

3.3 Build an evaluation result sett

The evaluation result set is similar to the comprehensive "scoring result". The difference is that the evaluation result set is not a collection of scores, but comments composed of fuzzy concepts, which need to be set as needed. This paper defines the evaluation result set: V={V1, V2, V3, V4, V5}, where V1, V2, V3, V4, and V5 represent very reasonable, relatively reasonable, moderately reasonable, generally reasonable, and unreasonable, respectively.

Table 3. Expert comments.

Level 3 indicators	Very reasonable	Relatively reasonable	Moderately reasonable	Generally reasonable	Unreasonable
Reasonabability of project site selection	0.52	0.25	0.13	0.1 0	0
reasonableness of design scheme	0.56	0.34	0.10		0
The rationality of the technical indicators	0.52	0.27	0.10	0.11	0
Design depth and quality	0.44	0.21	0.23	0.12	0
Reasonable value of ontology engineering cost	0.48	0.35	0.17		0
reasonableness of construction site and cleaning cost	0.41	0.36	0.12	0.11	0
reasonableness of equipment and materials fees	0.68	0.12	0.13	0.07	0
Reasonable cost of production preparation fee	0.63	0.22	0.15		0
Reasonable cost of technical service fee	0.52	0.25	0.13	0.1 0	0
reasonableness of auxiliary engineering cost	0.41	0.36	0.12	0.11	0
The rationality of the construction management fee	0.56	0.34	0.10		0
Reasonable cost of other expenses	0.44	0.21	0.23	0.12	0

3.4 The conclusion of rationality evaluation of power transmission and transformation project cost

According to the basic calculation process of the fuzzy comprehensive evaluation method, the membership degree set of the evaluation indicators is calculated as:

$$S = (0.496398 \ 0.288963 \ 0.159609 \ 0.05503 \ 0)$$

From the above calculation results, it can be judged that the cost rationality evaluation result of the power transmission and transformation project is: very reasonable.

4 CONCLUSION

Strengthening the control level of substation project investment and construction cost plays an important role in improving the project construction efficiency and improving the operating effi-ciency of enterprises. Based on the characteristics of substation construction, we construct the project cost rationality evaluation model based on fuzzy evaluation theory and verify the effectiveness by empirical analysis. However, due to the limitation of time and other factors, it still needs to further discuss the selection of evaluation indicators in the future, so as to continuously improve the applicability of the evaluation.

ACKNOWLEDGMENTS

The paper is supported by the Science and Technology Project of State Grid Corporation of China: Research on Optimal Selection of Transmission and Transformation Projects and Intelligent Evaluation of Costs Based on the Design of Wide-area Information Value Mining Links (No. 5200-202156080A-0-0-00).

REFERENCES

Ding Zhengzhong, Peng Luwei. Cost data prediction method of power transmission and transformation project based on MK-TESM method [J]. *Journal of the Shenyang University of Technology*, 2021,43 (02): 126–131.

Kang Yanfang, Li Dapeng, Li Xuyang, Ge Guowei. Impact analysis of the power transmission and transformation engineering cost based on AHP under the background of the new technical situation [J]. *China Management Informatization*, 2019,22 (20): 8–11.

Qian Wanxiang. Countermeasures and results of the whole process of the cost of power transmission and transformation project [J].*China Building Decoration*, 2020 (11): 88–89.

Xu Dan, Wen Weining, Lu Yanchao, Wang Xiaohui. Dynamic Warning of Power Transmission and Transformation Project Cost deviation Based on BP Neural Network [J]. *Management of China Electric Power Enterprises*, 2018 (12): 75–76.

Zhang Bo, Zhang Haoyu. Study on the process risk evaluation of power transmission and transformation Project [J].*China Electric Power Enterprise Management*, 2018 (27): 65–67.

Advances in Measurement Technology, Disaster Prevention and Mitigation – Li & Mohd Yusof (Eds)
© 2023 The Author(s), ISBN: 978-1-032-36087-4

Research on substation engineering cost prediction method based on index theory

Zheng Chenhong*

State Grid Economic and Technological Research Institute of Fujian Electric Power Co., Ltd., Fuzhou, Fujian, China

ABSTRACT: The substation project is one of the key investment points of power grid enterprises. Therefore, it is of great significance to strengthen the cost control level of the substation project. This paper presents a substation engineering prediction method based on engineering cost index theory and verifies the effectiveness combined with practical case analysis.

1 INTRODUCTION

At present, the project cost index is generally used in western developed countries to calculate and control the cost level of each stage of the project. However, China has implemented the traditional cost management mode of integrating quantity and price for a long time, and due to the limitation of the economic system, the research of the project cost index was not launched until the 1990s, which was relatively lagging behind that of western developed countries, and the lack of specific research on the index calculation and practical application.

Huang Xin (Huang 1994) expounded on the importance, specific methods, and precautions of calculating the cost index; Ge Weimin (Ge 1997) referred to the system framework of construction cost index according to the composition of construction and installation cost; Li Yuansheng (Li 2005) proposed the partial project cost index, making the project cost comparable in quantity and price; Peng Xiongwen et al (Peng & Du 2007) analyzed the calculation principle of project cost index and the project cost index model; Shen Weichun et al (Shen & Dong 2008) discussed the defects in calculating and releasing cost index, and put forward relevant improvement suggestions from the perspectives of information collection, system construction, and calculation method.

In general, the project cost index has not been comprehensively promoted and applied in China. Although some provinces and cities have already carried out the calculation and release of the project cost index, the development level is uneven, the index calculation basis is not sufficient, and there is still a certain gap between the requirements of the real index system.

2 THEORY AND METHOD OF COST INDEX ANALYSIS

2.1 *Project cost definition of the index*

The index is a ratio, it can compare the economic phenomena in time and space, and it is also one of the most commonly used economic analysis methods in society. Using the index to study the extent of quantitative changes and development trends of certain specific socio-economic phenomena can provide a more convincing basis for formulating relevant national policies to a certain extent.

---

*Corresponding Author: juzen123@163.com

DOI 10.1201/9781003330172-23

The definition of the index is reflected by the basic concept of the index and the understanding of the index analysis method. First, the basic concept of the index is the relative number change due to multiple influencing factors in different spaces and times; the second aspect is that the correlation analysis of the index is reflected by the change amount of its sub-index and its degree of change.

The project cost index is an index to explain the relative change trend and change range of individual prices and comprehensive prices in different periods, and it has a strong auxiliary role in studying the dynamics of project cost management.

Project cost index can be divided into single price index and comprehensive cost index. The main research object of the individual price index is the change degree and trend of individual price, while the comprehensive cost index is the main basis for studying the changing trend and degree of the total cost level.

2.2 *The role of the project cost index*

(1) The cost change trend and its reasons can be analyzed by using the project cost index.

   Since the index can reflect the change range and trend of social and economic phenomena, the price index of individual price changes in project cost provides reliable data to relevant departments, analyzes the causes of price changes in the construction industry, and provides a certain basis for formulating relevant regulation measures.

(2) The project cost index is an important basis for the project contractor to conduct the project valuation and price settlement.

   In engineering construction projects, there may be some projects with a long contract period. With the passage of time, these projects are often affected by various factors such as price fluctuations. In order to reasonably solve the risks borne by both parties due to market price fluctuations, the two parties usually sign adjustable contracts. The project cost index is a way to adjust the project cost. The project cost index is an index reflecting the range of market price change, which can provide necessary conditions for the realization of project valuation and dynamic settlement of project price, and make the signing of adjustable contracts more reasonable and scientific.

(3) The project cost index can provide a reference for the bid quotation.

   At present, the construction market, especially the private investment projects, has basically won the bid at a reasonable low price. In order to more accurately calculate the bidding of a construction enterprise as an acceptable cost price, the reference engineering cost index has a certain role in bidding pricing. Especially for the construction enterprises bidding in other places, when they do not understand the local price level, the project cost index and other information can provide some help for the calculation and pricing.

(4) Project cost index is an important tool to solve the static nature of the built project cost information.

   At the current stage of China, combining the representative project cost data and the project cost index as one of the bases for pricing can appropriately solve the problem of the static nature of the built project cost information, which has operational and important practical significance for the establishment of the project cost management mode suitable for China's construction market.

3 EMPIRICAL ANALYSIS

Project cost prediction refers to the prediction of the project cost based on the existing project cost data and historical data, through scientific calculation and synthesis, combined with the subjective experience and judgment ability of the prediction personnel. The accuracy of the project cost prediction is closely related to the choice of the prediction model. Therefore, the analysis of the commonly used economic prediction methods and the selection of prediction methods is an important link in the research of engineering cost prediction.

3.1 Grey GM (1,1) prediction model

(1) Construction of the gray GM (1,1) model

Step 1: A given observation number is the column $X^{(0)}$.

$$X^{(0)} = \{X^{(0)}(1), X^{(0)}(2), ..., X^{(0)}(N)\} \tag{1}$$

One accumulation generates the sequence $X^{(1)}$:

$$X^{(1)} = \{X^{(1)}(1), X^{(1)}(2), ..., X^{(1)}(N)\} \tag{2}$$

Among:

$$X^{(1)}(i) = \sum_{j=1}^{n} X^{(0)}(j)(i = 1, 2, ..., N) \tag{3}$$

For the ups and downs of the original columns themselves, after an accumulation, after the generation of the new columns, the ups and downs weaken, and the generated new columns appear in an incremental form.

Step 2: Supposing that $X^{(1)}$ meet the first-order differential equation:

$$\frac{dx^1}{dt} + ax^1 = u \tag{4}$$

Among them, a and u are constants.

When the equation satisfies the condition $t = t_0$, the solution of $X^{(1)} = X^{(1)}(t_0)$ is:

$$x^{(1)}(t) = \left[x^{(1)}(t_0) - \frac{u}{a}\right] e^{-a(t-t_0)} + \frac{u}{a} \tag{5}$$

The discrete values sampled at the same interval ($t_0 = 1$) are then:

$$x^{(1)}(k + 1) = \left[x^{(1)}(1) - \frac{u}{a}\right] e^{-ak} + \frac{u}{a}(k = 1, 2, ..., N - 1) \tag{6}$$

Step 3: The least squares method estimates the constants a and u:

$$\hat{U} = [a, u]^T = \left(B^T B\right)^{-1} B^T y \tag{7}$$

Among:

$$B = \begin{bmatrix} -\frac{1}{2}[x^{(1)}(2) + x^{(1)}(1)] & 1 \\ ... & 1 \\ -\frac{1}{2}[x^{(1)}(N) + x^{(1)}(N-1)] & 1 \end{bmatrix} \tag{8}$$

$$U = [a, u]^T \tag{9}$$

Step 4: The estimated values $\hat{a}$ and $\hat{u}$ are substituted respectively to obtain the response equation:

$$\hat{x}^{(1)}(k + 1) = \left[x^{(1)}(1) - \frac{\hat{U}}{\hat{a}}\right] e^{-\hat{a}k} + \frac{\hat{U}}{\hat{a}} \tag{10}$$

When $k = 1, 2, ..., N - 1$, $\hat{x}^{(1)}(k + 1)$ is the fitted value of $x^{(1)}$, and when $k = N$, $\hat{x}^{(1)}(k + 1)$ is the predicted value of $x^{(1)}$. Then the post-reduction and reduction operation is performed. When $k = 1, 2, ..., N - 1, \hat{x}^{(1)}(k + 1)$ is the fitted value of $x^{(0)}$, and when $k = N, \hat{x}^{(1)}(k + 1)$ is the predicted value of $x^{(0)}$.

3.2 Empirical analysis

Taking the comprehensive cost index of a power grid company as an example, the example analysis of the GM (1,1) prediction model was conducted. The following table is the comprehensive cost index from 2013 and 2014–2020.

Table 1. Comprehensive cost index of substation project in 2009 of 2016.

Date	2013	2014	2015	2016	2017	2018	2019	2020
index number	100	100.54	101.4	103.4	102.07	104.22	105.39	106.2

Step 1: Collect the original cost index and get the following columns:

$$X^{(0)} = \{100, 100.54, 101.4, 103.4, 102.07, 104.22, 105.39, 106.2\}$$

Add the column once and get the new column as follows:

$$X^{(1)} = \{100, 200.54, 301.94, 405.34, 507.41, 611.63, 717.02, 823.22\}$$

Step 2: $X^{(1)}$ Make it adjacent to the mean, and generate the following columns:

$$Z^{(1)} = \{150.27, 251.24, 353.64, 456.375, 559.52, 664, 325, 770.12\}$$

Step 3: To obtain the cumulative generation matrix B and the vector y are as follows:

$$B = \begin{bmatrix} -Z^{(1)}(2) & 1 \\ \dots & \dots \\ -Z^{(1)}(8) & 1 \end{bmatrix}$$

$$y = \begin{bmatrix} X^{(1)}(2) \\ \dots \\ X^{(1)}(8) \end{bmatrix} = \begin{bmatrix} 100.54 \\ \dots \\ 106.2 \end{bmatrix}$$

Get $\hat{U} = [a, u]^T = (B^T B)^{-1} B^T y = \begin{bmatrix} -0.0089 \\ 99.232 \end{bmatrix}$

Step 4: Find the GM (1.1) gray differential equation:

On the basis of $\hat{U} = [a, u]^T = (B^T B)^{-1} B^T y = \begin{bmatrix} -0.0089 \\ 99.232 \end{bmatrix}$ and $\frac{dx^1}{dt} + ax^1 = u$

Gray differential equations are available $\frac{dx^1}{dt} - 0.0089x^1 = 99.2329$.

Step 5: Solving the GM (1.1) gray differential equation:

Available by solving the differential equations:

$$x^{(1)}(k + 1) = \left[x^{(1)}(1) - \frac{u}{a}\right]e^{-ak} + \frac{u}{a}$$

Step 6: Solution to $X^{(1)}$. The simulation values are as follows:

$$X^{(1)} = \{100, 200.54, 301.94, 405.34, 507.41, 611.63, 717.02, 823.22\}$$

The restore gives a numbered column $X^{(0)}$:

$$X^{(0)} = \{100, 100.54, 101.4, 103.4, 102.07, 104.22, 105.39, 106.2\}$$

157

The comparison of the simulated value of the cost index with the actual value and the relative error rate is shown in Table 1, and the average relative error rate is 0.33%, indicating that the prediction accuracy is good.

Based on the prediction equation, the comprehensive cost index in 2021 is 106.9697. To predict the future cost index, under the premise of better prediction model selection and relatively high prediction accuracy, it can implement a reasonable evaluation of the project cost more quickly to a certain extent, which provides a basic reference for the cost management personnel to carry out the cost management work more timely and actively.

Table 2. Analysis of comprehensive cost index of a power grid company.

Date	2009	2010	2011	2012	2013	2014	2015	2016
The actual index	100	100.54	101.4	103.4	102.07	104.22	105.39	106.2
predictive index	100.00	100.57	101.47	102.38	103.30	104.22	105.16	106.10
fractional error	0%	0.03%	0.07%	0.98%	1.21%	0.01%	0.22%	0.09%

Table 3. Evaluation table of the model prediction accuracy level.

Prediction accuracy level	P	C
good	> 0.95	< 0.35
qualified	> 0.8	< 0.45
slightly qualified	> 0.7	< 0.5
unqualified	≤ 0.7	≥ 0.65

After calculation, the test calculation standard deviation C = 0.00218176, small error probability P = 1. Comparing it with Table 2, the accuracy level is good, so GM (1.1) can be used to predict the project cost index. The future substation project cost is mainly analyzed, which lays a good data foundation for cost management, so as to realize lean management and improve work efficiency.

4 CONCLUSION

Based on the application situation and application method of engineering cost index theory and gray prediction theory, the prediction model has good prediction accuracy, which can improve the quality of budgeting of power grid engineering projects, and then improve the level of engineering cost control.

REFERENCES

Ge Weimin. Project Cost Index and Dynamic Management of Project Cost [J]. *Journal of Taiyuan University of Technology*, 1997, (4): 5–8.
Huang Xin. Engineering Calculation of Engineering Cost Index [J]. *Anhui Architecture*, 1994, (1): 37–38.
Li Yuansheng. The Dynamic Management Project of the Cost Index [J].*Jiangxi Science*, 2005(06):793–795. DOI:10.13990/j.issn1001-3679.2005.06.030.
Peng Xiongwen, Du Hua. Calculation of engineering cost cost index [J]. *Shanxi Science and Technology*, 2007 (03): 71–72.
Shen Weichun, Dong Shibo. Research on Engineering Cost Index System and Computational Model [J]. *Technical Economy*, 2008 (10): 62–68.

Advances in Measurement Technology, Disaster Prevention and Mitigation – Li & Mohd Yusof (Eds)
© 2023 The Author(s), ISBN: 978-1-032-36087-4

Identification and assessment of safety risks for coastal tunnel construction

Bao Yang*

Road & Bridge International Co., Ltd., Beijing, China

ABSTRACT: A risk management scheme for the construction safety of coastal tunnel projects is presented. The WBS-RBS method was used to identify the risk sources affecting safe project construction. A fault tree analysis was performed in conjunction with the analytic hierarchy process for risk assessment, where objective and subjective analyses were combined to reduce the influence of human factors on the evaluation results. The risk assessment results were used to propose risk disposal measures. The results of this study can provide a reference for the successful realization of construction safety objectives for coastal tunnel projects.

1 INTRODUCTION

Continuously increasing construction for transportation infrastructure in China has been accompanied by a rise in the scale and complexity of construction projects, and therefore, the number of risk factors involved in engineering projects. Tunnel engineering projects in coastal areas are large in scale and have complex service environments. A large number of uncertain factors may arise during construction that can easily lead to safety accidents. There are three basic aspects of construction safety risk management: risk identification, risk assessment, and risk disposal (Ghosh & Jintanapakanont 2004). Risk identification is commonly performed using the work breakdown structure (WBS) and risk breakdown structure (RBS) (Siami-Irdemoosa et al. 2015; Yan et al. 2019), and risk assessment is commonly performed using the analytic hierarchy process (AHP) and fault tree analysis (FTA) (Ikwan et al. 2021; Szatmári 2021).

A coastal tunnel project in Zhejiang was used as a case study, where WBS-RBS was adopted to identify the risk sources affecting the safe construction of the project, and FTA was combined with AHP to evaluate risk. Risk disposal measures were then proposed to provide a reference scheme for the successful implementation of the project.

2 RISK IDENTIFICATION OF COASTAL TUNNEL CONSTRUCTION

The following steps are performed in risk identification: a group of comprehensive risk factors that may affect the respective project in one or more ways are identified; these factors are classified and ranked according to the nature and probability of risk occurrence and the degree of impact on the project; the entire project process is comprehensively analyzed from the perspective of systems theory.

A tunnel construction project in a coastal area in eastern China is evaluated in this study. The water quality analysis report shows that the seawater in the project area is mainly composed of chloride ions (Cl^-), potassium ions (K^+), and sodium ions (Na^+) that can corrode the concrete structure. In the beach area, the upper part of the formation lithology is mainly thin-layer silt and mucky soil, which are characterized by high water content, low strength, and poor properties. The

---

*Corresponding Author: 282684857@qq.com

middle and lower parts of the formation lithology are mainly silty clay, pebbly silty clay, cohesive soil, and gravel.

Considering the project characteristics, all the project tasks were organized into different layers as low-level work packages with suitable sizes for implementation, resulting in the tree-like work breakdown structure (WBS) shown in Figure 1. The possible risks in the project for each layer were then determined according to the types and attribution relationships of the risks involved to obtain the risk breakdown structure (RBS) shown in Figure 2. Finally, a coupling analysis was carried out on the WBS-RBS results. That is, a one-to-one coupling between the basic risk factors in RBS and the underlying work packages in WBS was used to generate a risk identification matrix, where the columns were the work packages and the rows were the basic risk factors. A risk breakdown matrix (RBM) was thus obtained.

3 RISK ASSESSMENT OF COASTAL TUNNEL CONSTRUCTION

The identified risk sources were used to perform a risk assessment to convert the probability and impact of risk events on the whole project into measurable variables and obtain the probability distribution for the project objectives under the overall action of the risk events. Currently, the universality of the risk assessment model is an urgent problem to be solved in the application. For example, the equal risk graph method is convenient, which can obtain the risk coefficient according to the probability of risk occurrence. However, it is difficult to obtain the two variables of risk occurrence probability and risk consequences. The dynamic decision tree has clear levels and obvious stages, which is suitable for solving complex decision-making problems such as multi-stage and multi-factor, but it has the disadvantages of heavy workload and easy omission. In this section, the quantitative analysis of the risk variables is presented, and the resulting fault tree for the project is shown in Figure 3. This fault tree was quantitatively analyzed using the descending method. The importance weights of the risk sources for safe construction of were determined by starting from the top event and extending down through the project layers.

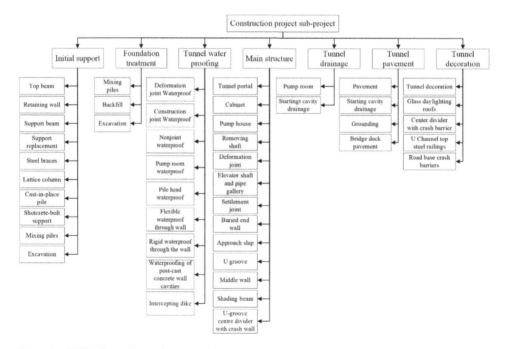

Figure 1.   WBS of coastal tunnel construction.

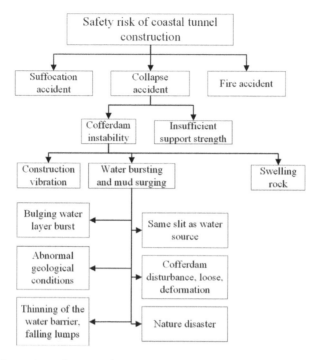

Figure 2.   RBS of coastal tunnel construction.

The purpose of FTA is to find the cause event under the top event and its causal relationship, that is, the minimum cut set which refers to a necessary and sufficient set of basic events in which all basic events contained in the set occur, and the top event will occur. The minimum cut set of fault tree in this project is {X1,X2}, {X13,X14,X15,X16,X17,X18}, {X3}, {X15}, {X16}, {X4}, {X5}, {X6}, {X7}, {X8}, {X9}, {X10}, {X11}, {X12}.

The AHP implementation steps include evaluating the relative importance of the indicators, calculating the indicator weights, and performing a consistency check. To calculate the index weights, an interactive judgment is carried out for a specific index layer to generate the judgment matrix shown in Equation (1):

$$A = (a_{ij})_{n \times n} = \begin{bmatrix} a_{11} & \cdots & a_{1n} \\ \vdots & \ddots & \vdots \\ a_{n1} & \cdots & a_{nn} \end{bmatrix} \tag{1}$$

where a_{ij} is a scale factor based on the importance of the respective object.
Equation (2) shows the eigenvectors of the judgment matrix.

$$\bar{w}_i = \sqrt[n]{\prod_{i=1}^{n} A_{ij}} \tag{2}$$

The maximum eigenvalue of the normalized judgment matrix is given below.

$$W_i = \frac{\bar{w}_i}{\sum_{i=1}^{n} \bar{w}_i} \tag{3}$$

$$\lambda_{max} = \sum_{i=1}^{n} \frac{(C \times W)_i}{nW_i} \qquad (4)$$

The consistency index (CI) is used as a measure of the deviation consistency of the judgment matrix and is given in Equation (5).

$$CI = \frac{\lambda_{max} - n}{n - 1} \qquad (5)$$

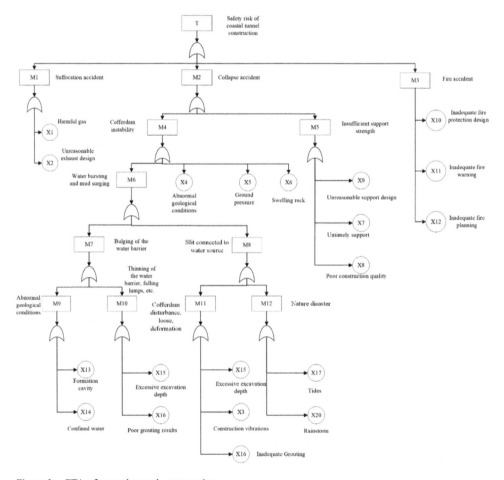

Figure 3.   FTA of coastal tunnel construction.

The analytic hierarchy process was carried out as follows: questionnaires were distributed to 5 senior experts in the field, the importance of the risk factors was scored, the arithmetic means of the results were used to generate a judgment matrix, and the importance weights of the risk sources were determined. The calculated consistency index shows that the analysis passed the consistency test. AHP and FTA were each used to obtain 50% of the results, and the final risk assessment results are shown in Table 1.

4 RISK DISPOSAL FOR COASTAL TUNNEL CONSTRUCTION

Risk disposal is used to analyze the benefits and costs of risk events for formulating and implementing risk disposal plans. Risk sources with low importance weights, such as confined water, are retained for disposal. Risk avoidance and risk transfer should be adopted for risk sources with higher importance weights, where disposal mainly involves foundation treatment, initial support, and tunnel drainage.

Table 1. Construction safety risk assessment for a coastal tunnel.

ID	Risk source	Weight
X1	Harmful gas	0.233
X2	Unreasonable exhaust design	0.238
X3	Construction vibration	0.0128
X4	Abnormal geological conditions	0.026
X5	Ground pressure	0.026
X6	Swelling rock	0.026
X7	Untimely support	0.054
X8	Poor construction quality	0.026
X9	Unreasonable support design	0.054
X10	Inadequate fire protection design	0.067
X11	Insufficient fire warning	0.065
X12	Inadequate fire planning	0.065
X13	Formation cavity	0.013
X14	Confined water	0.013
X15	Excessive excavation depth	0.026
X16	Inadequate grouting	0.026
X17	Tide	0.01155
X18	Rainstorm	0.013

Foundation treatment stage: Machinery and equipment should be used to determine the site geology in advance of excavation. In tunnel sections with high gas contents, grouting should be performed for gas sealing. During the excavation process, construction workers should intensify detection of surface settlement, surrounding rock shrinkage deformation, and support stress. Backfill grouting should be carried out as soon as possible after the tunnel lining is installed.

Initial support stage: Upon the completion of the excavation, construction workers should provide initial support and perform inverted arch construction of the tunnel in a timely manner and appropriately delay construction of a secondary lining. For areas with abnormal ground pressure, supporting pillars should be added to reduce stress on the surrounding rock of the excavation face. In case of water leakage, grouting should be performed for water plugging and seepage prevention and to reduce the soil porosity.

Tunnel drainage: The groundwater should be predrained before installing a lining, and the surrounding rock drainage channels should be excavated at the same time. Any groundwater that has infiltrated into the lining should be diverted.

5 CONCLUSIONS

A coastal tunnel engineering project was used as a case study: the WBS-RBS method was used to identify construction risk sources, and FTA was combined with AHP to evaluate the weights of the construction risks. The risk assessment results were used to propose risk disposal measures at three different stages of construction, i.e., foundation treatment, initial support, and tunnel drainage.

ACKNOWLEDGMENT

I would like to thank Mrs. Ping Wu, a lecturer at NingboTech University, for the technical discussions.

REFERENCES

Ghosh, S., Jintanapakanont, J. (2004) Identifying and assessing the critical risk factors in an underground rail project in Thailand: a factor analysis approach. *Int. J. Proj. Manag.*, 22: 633–643.

Ikwan, F., Sanders, D., Hassan, M. (2021) Safety evaluation of leak in a storage tank using fault tree analysis and risk matrix analysis. *J. Loss. Prevent. Proc.*, 73: 104597.

Siami-Irdemoosa, E., Dindarloo, S. R., Sharifzadeh, M. (2015) Work breakdown structure (WBS) development for underground construction. *Automat. Constr.*, 58: 85–94.

Szatmári, M. (2021) Proposal AHP method for Increasing the Security Level in the Railway Station. *Transp. Res. Procedia.*, 55: 1681–1688.

Yan, H. Y., Gao, C., Elzarka, H., Mostafa, K. (2019) Risk assessment for construction of urban rail transit projects. *Safety. Sci.*, 118: 583–594.

Advances in Measurement Technology, Disaster Prevention and
Mitigation – Li & Mohd Yusof (Eds)
© 2023 The Author(s), ISBN: 978-1-032-36087-4

Risk assessment of valley debris flow based on the residual structure—Taking Nujiang Lisu autonomous prefecture as an example

Fanshu Xu & Jun Han
School of Information and Technology, Yunnan Normal University, Yunnan Province, China

Baoyun Wang*
School of Mathematics, Yunnan Normal University, Yunnan Province, China

Kunxiang Liu
School of Information and Technology, Yunnan Normal University, Yunnan Province, China

ABSTRACT: In order to evaluate the hazard of the debris flow in the valleys of Nujiang Prefecture, a neural network model that simultaneously recognizes digital elevation data and remote sensing data is proposed based on the residual structure. After the model is trained on the historical disaster data, it scores the hazard risk of debris flow in the remaining valleys. The test results show that the model's hazard scores for typical valleys are basically in line with the results of the field investigation. The model can achieve an accuracy rate of 84% and a kappa coefficient of 0.81 on the valley classification task. The model can be used for risk assessment of debris flow valleys in the study area and has a certain reference value for the prevention and control of debris flow in the region.

1 INTRODUCTION

Debris flow is a sudden and harmful geological disaster. Nujiang Lisu Autonomous Prefecture (Nujiang Prefecture in short) is the hardest hit area by this kind of disaster (Kong et al. 2018). Debris flows generally occur in valleys. Therefore, how to evaluate the risk degree of the valley is of great significance for the follow-up prevention and control work. Many scholars have tried to use different methods to evaluate the risk of valleys or regions. Li Yimin(2019) et al. used the deterministic coefficient model and multi-factor superposition weight determination method to analyze the susceptibility of debris flow in Nujiang Prefecture. Li Kun (Li et al. 2022) and Hu Xiuyu (Hu et al 2019) used the random forest to evaluate the risk of debris flow. The grey correlation method is also a common method to analyze the susceptibility of debris flow (Hu & Xiong 2021; Niu et al 2021). Although the above methods have achieved certain results in the risk assessment of debris flow in a certain area, they all need to manually select the factors for risk assessment before using the corresponding model for risk assessment. Besides, the causes of debris flows are complex, and causes of debris flows are not exactly the same (Liu 2014), so these models do not have a good generalization effect. Therefore, there is an urgent need for a method that does not require manual selection of disaster factors and has a better generalization ability to evaluate the risk of debris flow.

A convolutional neural network is a special kind of model, which can adaptively acquire the features in the image without manual selection of disaster-causing factors, and has been used in

National Natural Science Foundation of China (61966040)
Identification of debris flow disaster-pregnant valleys in remote sensing images based on deep transfer learning-
Taking Yunnan Province as an example
*Corresponding Author: wspbmly@163.com

the identification of various disasters, such as landslides (Wang et al. 2021), earthquakes (Li et al 2020) and forest fire (Fu & Zhang 2020). We choose digital elevation (DEM) images and remote sensing images as the input data of the network. They are special image data that can provide a large amount of geological information and are used in the analysis of disasters including earthquakes (Ming et al. 2022) and landslides (Hu et al. 2022; Ma et al. 2022). We hope that the model can adaptively capture the geological information in the data and use it to evaluate the risk of debris flow disasters.

The problem with neural network is that it needs more data for training, but there is not so much debris flow disaster data. The residual network proposed by He Kaiming (He et al 2016) and others perform well in various problems with small sample sizes (Liu et al. 2022; Yuan et al. 2022). Therefore, this model is used as the basic model in this paper, and the model is designed according to the characteristics of the data.

2 STUDY AREA AND DATA

2.1 *Study area*

Nujiang Prefecture is located in the northwestern part of Yunnan Province, at the junction of the Eurasian and Indo-China plates, and has frequent geological activities. There are three rivers, the Nujiang River, the Lancang River, and the Dulong River in the territory. The mountains along both sides of the river are towering, and the height difference is large. In addition, the local rainfall in Nujiang Prefecture is frequent, and it is very easy to form debris flows. The study area is shown in Figure 1.

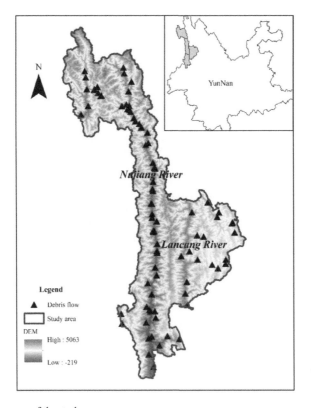

Figure 1.   Location map of the study area.

2.2 *Data*

The DEM data used in the experiments are from the USGS public dataset with a resolution of 30 meters. The multispectral images used were from the Gaofen-1 (GF-1) satellite.

By reviewing the "Yunnan Disaster Reduction Yearbook" from 2000 to 2018 and combining it with relevant news reports, it was confirmed that there were more than 100 debris flows in Nujiang Prefecture. Due to the inaccuracy of some recorded disaster locations, 50 debris flow valleys in Nujiang Prefecture were finally extracted as positive samples through ArcGis software. In order to keep the positive and negative sample sizes relatively balanced, 50 valleys without debris flow disaster records were selected as negative samples.

The extracted gully is divided into two categories based on whether there is debris flow, and then the length of the main gully and the area of the watershed are combined, and the data is divided into 6 categories. The specific division of data is shown in Table 1.

Table 1. Data classification.

Groove length(m)	Drainage area(km^2)	Happen	Yet to happen
(0,5000]	(0,15]	21	17
(5000,10000)	(15,35)	14	17
[10000,+∞]	[35,+∞)	15	16

Table 2. Schematic diagram of DEM.

		Drainage area		
		Small	Medium	Large
Happen/ Unhappen	Happen			
	Unhappen			

The DEM schematic diagrams of various valleys are shown in Table 2:

All DEM images were padded to 1080 × 1080 pixels after extraction. This is because the DEM image contains a lot of gully structural information, such as the length, shape, and height difference, which are very important for the generation of debris flow. Using simple stretching to unify the size of the image and then inputting it into the network will not only destroy these important features, but also make the model unable to distinguish the differences between the valleys, so that the geometric features of the debris flow valleys cannot be correctly identified. Multispectral images provide surface information including water bodies, vegetation, etc., which are also closely related to the occurrence of debris flows.

In order to comprehensively utilize the information provided by DEM and multispectral, the 4-channel multispectral image and DEM image of each valley are stacked into a 1080×1080×5 data block as the original input of the network. The construction method is shown in Figure 2.

3 MODEL CONSTRUCTION

DRNet is a network designed based on residual structure and has two residual structures. After the input data block of 1080 × 1080 × 5 is dimensionally reduced by the 1 × 1 × 5 convolution block, a 1080 × 1080 × 1 feature map is generated, which enters the residual structure 1, and passes through the 3 × 3 convolution and residual. The residual connection results in 64 feature maps of 360 × 360. After the feature map is subjected to 32 3 × 3 convolution blocks, 32 feature maps of 120 × 120 are generated and entered into the second residual structure. In this residual structure, the feature map undergoes 3 × 3 convolution. After residual concatenation, 32 feature maps of 40 × 40 are obtained. Finally, the similarity score of the corresponding category is output after convolution, average pooling, and full connection, and the risk of the valley is given according to the similarity score.

The model structure is shown in Figure 3.

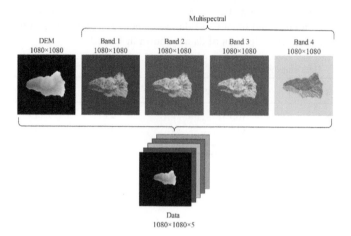

Figure 2. Pretreatment of data.

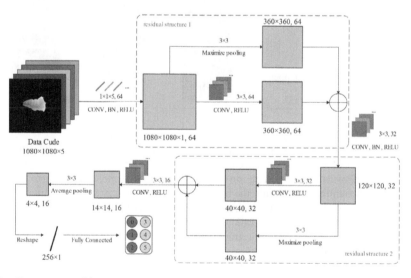

Figure 3. Pretreatment of data.

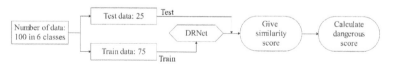

Figure 4. Experiment process.

4 EXPERIMENT AND RESULT

4.1 *Experiment platform and hyperparameters*

The environment of this experiment is as follows. Hardware environment: Intel Xeon(R) Gold 6271 CPU @ 2.60GHz, NVIDIA Tesla T4 GPU; software environment: Ubuntu 18.04, Python 3.8, CUDA 11.0, cuDNN 8.0, PyTorch 1.7.1, and image related third-party libraries.

GPU is used for model training, and the specific training settings are shown in Table 3.

4.2 *Experiment process*

First, the constructed neural network is trained using the data of the valleys where debris flows have occurred and the valleys that have not occurred. Then, the trained network is used to extract features from the valley data in the test set, and the risk of the valleys is calculated according to the similarity score. The dataset is divided as follows: categories 0, 1, and 2 are samples that confirm that debris flows have occurred, while categories 3, 4, and 5 are valleys where debris flows have not occurred. Finally, the similarity score of test images given by the neural network can be used to calculate the disaster risk of each valley. The specific experimental process is shown in Figure 4.

Let S represent the hazard risk score of the valley, y_i represents the similarity score between the valley to be evaluated and the i-th valley. For a unified metric, s_i is calculated from y_i, which is a normalized value between [0,1], specifically:

$$s_i = \frac{e^{y_i}}{\sum_{i=0}^{5} e^{y_i}} \tag{1}$$

$$S = \sum_{i=0}^{2} s_i - \sum_{i=3}^{5} s_i \tag{2}$$

Since $S \in [-1,1]$, the risk of debris flow in the valley can be divided according to the size of the value:

$$\begin{cases} -1 < S \leq -0.33 & \text{low risk} \\ -0.33 < S \leq 0.33 & \text{mid risk} \\ 0.33 \leq S \leq 1.00 & \text{high risk} \end{cases} \tag{3}$$

Table 3. Training settings.

Hyperparameter	Value
Max epoch	100
Batch size	2
Learning rate	0.0001
Optimization	Stochastic gradient descent (SGD)
Loss function	Cross entropy loss
Initialization	Kaiming normal (He et al. 2015)

4.3 *Experimental results and analysis*

4.3.1 *Case study*
Several examples of valley debris risk evaluations are given in Table 4.

Table 4. Example.

No.	DEM	Disaster situation	Risk
1		A total of 11,212 people were affected, and the direct economic loss exceeded 100 million RMB	High
2		There are mudslide disaster records, but no casualty records	Mid
3		No debris flow records	low

Sample No. 1 is the valley of Dongyue where the 8.18 huge debris flow occurred. The coordinates of the mouth of the valley are N27° 38′8.9″ and E 98° 43′50″. Figure 5 is a post-disaster aerial photograph.

Figure 5. Image after the disaster.

From the topographical point of view, the valley in Dongyue is the typical topography of alpine and canyon, and the main ditches are straight, which is conducive to the acceleration of debris flow. The drainage area of the valley is about 50 km^2, and the upper half of the catchment area is large, which is characterized by the occurrence of debris flow. In terms of provenance conditions, the rock masses where the upstream catchment surface is located are mostly quartz schist, marble, Yanshanian granite, etc. with poor weathering resistance. After the collapse, the crushed stone combined with the water flow accelerates down to form a debris flow. In addition, there are glaciers in the ditch heads of Dongyue, which can also constitute the source of debris flow under heavy rainfall conditions. The similarity of the model to each valley of the East Moon and the corresponding risk score are given below. It is calculated that the risk score here is $S = 1$, and the risk of disaster pregnancy is extremely high.

Table 5. Risk assessment of Dong Yuege.

Category	0	1	2	3	4	5
Output (y_i)	−123.2	−33.2	70.5	−91.2	−67.7	28.9
Score(s_i)	0	0	≈1	0	0	≈0

Sample No. 2 is the valley of Mujiajia Village, and the coordinates of the mouth of the groove are N27° 26′44″ and E 98° 48′59″. Although there is a large catchment area on both sides of the mountain, the main ditch is tortuous, so it is impossible to form a large destructive debris flow. The following table lists the model's risk score. The risk score after calculating is $S = -0.14$, medium risk.

Table 6. Risk assessment of Mu Jiajia village.

Category	0	1	2	3	4	5
Output (y_i)	9.5	−4.0	−11.3	9.8	−3.2	−9.7
Score(s_i)	0.43	0	0	0.57	0	0

Sample No. 3 is a valley of Shangchuda, and the coordinates of the mouth of the groove are N27° 28′18″ and E 99° 02′59″. The main ditch here is straight, and the gully on both sides is also straight, but both are short, which is not conducive to the collection of water flow. In addition, the weathering degree of the mountains on both sides is low, so mudslides are not easy to occur. According to the calculation, the risk score here is $S = -1$, low risk.

Table 7. Risk assessment of Mu Jiajia village.

Category	0	1	2	3	4	5
Output (y_i)	−13.1	−0.59	−9.4	−15.8	6.4	−5.8
Score (s_i)	0	0	0	0	1	0

4.3.2 *Comparison of different models*

In order to show the discriminative ability of the constructed DRNet for debris flow valleys, the most commonly used ResNet18 and ResNet34 in the residual family were selected for comparative experiments. The parameters such as the max epoch and learning rate set for each model training test are the same as those in Table 3. In order to make the DEM and remote sensing data with 5 channels enter the network correctly, the first convolution input parameter of ResNet18 and ResNet34 is changed to 5, and the rest of the structure is not changed. In the table below, "2 class accuracy" refers to the accuracy obtained when the sample is only regarded as occurrence and non-occurrence.

From the experimental results, DRNet can achieve the highest 2-class/6-class accuracy, the kappa coefficient is much higher than the original residual network, and the recall rate is slightly lower than ResNet18.

4.3.3 *The effect of residual structure*

The model constructed in this experiment uses two residual structures (residual structures in Figure 3). To illustrate the influence of the residual structure on the model effect, the feature map generated by maximizing pooling in the blue dashed box in Figure 4 is deleted. One structure, a model with no residual structure (NoRes) was obtained and compared with DRNet, and the training/test curve is shown in Figure 6.

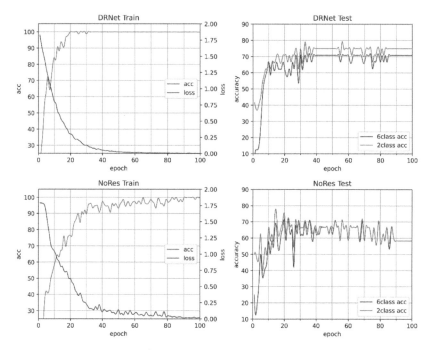

Figure 6.   With/without residual structure.

Table 8.   Test result.

Model	2 class accuracy/%	6 class accuracy /%	Precision/%	Recall rate/%	kappa
DRNet	**82.6**	**74.3**	**76.9**	71.4	**0.81**
ResNet18	69.9	70.8	68.8	**72.6**	0.61
ResNet34	75.0	71.6	69.2	60.1	0.67

Clearly, using the residual structure not only makes the loss curve more stable but also makes more use of the information in the feature maps generated by the previous convolution blocks, resulting in higher classification accuracy.

5 CONCLUSION

Taking the valley of Nujiang Prefecture as an example, this paper analyzes the risk of typical valleys in Nujiang Prefecture by constructing a neural network model based on residual structure. The analysis results are basically consistent with the conclusions drawn from the field investigation.

Through experiments and indicators such as accuracy rate and recall rate, the characterization ability of DRNet for DEM and multispectral data is verified. And through comparative experiments, the necessity of using the residual structure in the model is illustrated. The experimental results show that the DRNet based on the residual network design can simultaneously pick up features from DEM images and remote sensing data to evaluate the risk of debris flow disasters.

In order to further verify the generalization ability of the model, the next step will select debris flows in other areas as the evaluation object to improve the robustness of the model.

REFERENCES

Fu. Y.J., Zhang. L.H., "*Forest Fire Detection Method Based on Transfer Learning of Convolutional Neural Network*," Laser & Optoelectronics Progress. China, vol. 57, pp. 128–134, 2020

He K, Zhang X, Ren S, et al. "Deep residual learning for image recognition," *Proceedings of the IEEE conference on computer vision and pattern recognition*. 2016: 770–778.

He K, Zhang X, Ren S, et al. "Delving deep into rectifiers: Surpassing human-level performance on imagenet classification," *Proceedings of the IEEE international conference on computer vision*. 2015: 1026–1034.

Hu. F.H., Xiong. C.Z., "*Evaluation on the susceptibility of typhoon-triggered debris flow based on extension grey model*," Yangtze River. China, vol. 52, pp. 26–32, 2021

Hu. W.M., Yin. Z.Q., Yang. R.H., "Application of GF-2 satellite in the investigation of landslide disaster in Leijiashan, Hunan," *Satellite Application*. China, pp.50–54, 2022

Hu. X.Y., Qin. S.W., et al., "Susceptibility Analysis of Debris Flow Based on GIS and Random Forest—A Case Study of a Mountainous Area in Northern Taonan City, Jilin Province," *Bulletin of Soil and Water Conservation*. China, vol. 39, pp. 204–210+217+2, 2019

Kong. Y., Wang. B.Y., Yang. K., et al., "Analysis on the Spacial-temporal Distribution and the Typical Areas Disaster-pregnant Features of Debris Flows in Yunnan Province," *Journal of Yunnan Normal University*(Natural Sciences Edition). China, vol. 38, pp. 55–63, 2018

Li. J., Wang. X.M., Zhang. Y.H., et al, "Research on the seismic phase picking method based on the deep convolution neural network," *Chinese Journal of Geophysics*. China, vol. 63, pp. 1591–1601, 2020

Li. Kun, Zhao. J.S., Lin. Y.L., "Assessment of debris flow susceptibility in Dongchuan based on RF and SVM models," *Journal of Yunnan University*(Natural Sciences Edition). China, vol. 44, pp. 107–115, 2022

Li. Y.M., Yang. L., Wei. S.H., "Susceptibility Assessment of Debris Flow in Nujiang Prefecture Based on the Catchment," *Resources and Environment in the Yangtze Basin*. China, vol. 28, pp. 2419–2428, 2019

Liu. C.Z., "Genetic Types of Landslide and Debris Flow Disasters in China," *Geological Review*. China, vol. 60, pp. 858–868, 2014

Liu. F., Chen. R.W., Xing. K.L., "Fast fault diagnosis algorithm for rolling bearing based on transfer learning and deep residual network," *Journal of Vibration and Shock*. China, vol. 41, pp. 156–164, 2022

Ma. Y.L., Wu. H., Zhang. W.H., "Discussion on the application of remote sensing technology in landslide identification," *Technology Innovation and Application*. China, vol. 12, pp. 142–145, 2022

Ming. X.N., Yang. J.Q., Zhang. Y.S., "Preliminary assessment of the seismic capacity of buildings in Jianshui County, Yunnan Province based on remote-sensing images and the empirical-estimation method," *Journal of Seismological Research*. China, vol. 45, pp. 132–140, 2022

Niu. Q.F., Liu. M., Li. Y.F., et al, "Hazard assessment of debris flow in the Lanzhou City of Gansu Province based on methods of grey relation and rough dependence," *The Chinese Journal of Geological Hazard and Control*. China, vol. 30, pp. 48–56, 2019

Wang. Y., Fang. Z.C., Niu. R.Q., "Landslide Susceptibility Analysis based on Deep Learning," *Journal of Geo-information Science. China*, vol. 23, pp. 2244–2260, 2021

Yuan. P.S., Song. J., Xu. H.L., "Fish Image Recognition Based on Residual Network and Few-shot Learning," *Transactions of the Chinese Society for Agricultural Machinery*. China, vol. 53, pp. 282–290, 2022

Advances in Measurement Technology, Disaster Prevention and Mitigation – Li & Mohd Yusof (Eds)
© 2023 The Author(s), ISBN: 978-1-032-36087-4

Analysis of construction safety risk coupling behavior based on the EWM-DEMATEL model: A case study of the Xidian bay subsea tunnel project

Ping Wu*
NingboTech University, Ningbo, China

Guojun Lin
Ninghai Transportation Bureau, Ningbo, China

Hangcheng Zhang
NingboTech University, Ningbo, China

Yuanxiao Ma
Ninghai Transportation Bureau, Ningbo, China

ABSTRACT: Choosing the Ningbo Xidian Bay Subsea Tunnel project as a case study, nineteen risk factor coupling evaluation indices were obtained from five identified aspects: personnel, machinery and materials, management, technology, and environment. Then, via the construction of the EWM-DEMATEL coupling model, the weight of each safety risk factor and the coupling degree between all risk factors were obtained. The results indicated that priority should be given to the control of the coupling effect between personnel and management factors. The coupling effect between the factors of technology, machinery and materials, environment, and personnel should also be controlled to reduce the occurrence of construction safety risk-related accidents.

1 INTRODUCTION

Subsea tunnels provide the advantages of highway mileage shortening, time savings, and reduction in damage to the ecological environment. Due to the large scale, difficult construction technology, long construction period, and many risk factors of subsea tunnel construction projects, safety accidents easily occur during the construction process.

Research on safety risk mitigation and assessment can be divided into qualitative and quantitative methods. Qualitative research methods mainly include the safety checklist method, expert consultation method, and preliminary hazard analysis method (Sanni-Anibire et al. 2020; Hallowell & Gambatese 2010; Ameyaw et al. 2016). Quantitative research methods mainly include the analytic hierarchy process (AHP), fuzzy comprehensive evaluation (FCE), risk assessment of working conditions (LEC) and neural network and support vector machine (SVM) models (Kim & Lee 2020; Patel & Jha 2016; Zhang et al. 2020) Dai C (Dai 2012) used expert interviews and questionnaire surveys to obtain the tunnel construction risk list and construction risk factor database respectively, and took the Jiaozhou Bay Submarine Tunnel as an example to verify the fuzzy comprehensive assessment model of urban tunnel construction. Based on the decision-making trial and evaluation laboratory (DEMATEL) method and interpretative structural modeling (ISM) method, we took the Fuyang station of the Hanghuang high-speed railway as the subject and investigated 17 risk

---

*Corresponding Author: wuping@nit.zju.edu.cn

    DOI 10.1201/9781003330172-26

factors in four categories affecting construction safety, and the interaction relationship and degree of interaction among the factors were analyzed (Lin et al. 2021). However, existing construction risk assessment methods usually only focus on a single index but do not consider the coupling effect between various indices (Guo et al. 2019). Accidents are mainly caused by the interaction of multiple risk factors. The interaction among multiple uncertain factors in the process of submarine tunnel construction can greatly increase the risk of mutual interference. Therefore, it is of practical significance to analyze the safety risk coupling effect in subsea tunnel construction.

Choosing the Xidian Bay Subsea Tunnel as an example, this paper first identifies and classifies risk factors in combination with hydrogeological survey data and construction organization design documents. Second, based on the expert questionnaire method, the EWM-DEMATEL model is established for analysis. Finally, relevant measures are proposed, which provide a theoretical basis and methods for accident prevention.

2 EWM-DEMATEL COUPLING DEGREE MODEL

The entropy weight method (EWM) (Cheng 2010) is a method to calculate the weight of risk factors according to the decision matrix, which avoids the subjectivity of the AHP, fuzzy comprehensive evaluation, and other methods. However, the EWM cannot consider the coupling effect between various risk factors. The DEMATEL method (Zhang & Nie 2017) can assign values to the coupling influence relationship between risk factors, which constitutes an effective method to establish the causal relationship between risk factors and the degree of influence in a given system. By combining the advantages of the EWM and DEMATEL, the weight of risk factors can be obtained by establishing a coupling degree model, which can not only quantify the relationship between risk factors but can also ensure more scientific and reasonable established weights. The establishment process of the EWM-DEMATEL model is described below.

2.1 Construction of the judgment matrix

The proportion of the i-th evaluation sample value under the j-th evaluation index can be calculated as follows:

$$P_{ij} = \frac{X_{ij}}{\sum_{i=1}^{n} X_{ij}}, i = 1, 2, 3 \ldots, n; j = 1, 2, 3 \ldots, m \qquad (1)$$

The entropy value can be determined as follows:

$$e_j = -k \sum_{i=1}^{n} P_{ij} \ln \left(P_{ij} \right) \qquad (2)$$

The information entropy redundancy can be obtained as follows:

$$d_j = 1 = e, j = 1, \ldots, m \qquad (3)$$

The weight of each index can be calculated as follows:

$$w_j = \frac{d_j}{\sum_{j=1}^{m} d_j}, j = 1, \ldots, m \qquad (4)$$

2.2 Construction of the comprehensive impact matrix

After judgment matrix construction and weight calculation, the direct influence matrix A can be constructed. To assess the strength of the relationship between corresponding factors, matrix A is normalized to obtain matrix G.

$$G = \left(G_{ij} \right)_{n \times n} \qquad (5)$$

To analyze the indirect influence relationship between factors, the comprehensive influence matrix T is further constructed.

$$T = G(I - G)^{-1} \tag{6}$$

The influence degree f_i and the influenced degree e_i can be obtained as follows:

$$f_i = \sum_{j=1}^{n} t_{ij}; \quad e_i = \sum_{j=1}^{n} t_{ji}, i = 1, 2, 3 \ldots, n \tag{7}$$

The centrality can be obtained by adding the influence and influenced degrees of system factors m_i.

$$m_i = f_i + e_i \tag{8}$$

2.3 Determination of the efficacy function

According to the entropy weight obtained with the EWM and the centrality obtained via the DEMATEL, the combination weight S_j can be obtained through multiplication combination.

$$S_j = \frac{m_j \times w_j}{\sum_{j=1}^{m} m_j \times w_j}, j = 1, 2, 3 \ldots, m \tag{9}$$

Based on the efficacy coefficient, the linear weighting method is adopted to obtain the order parameter value U_i of each risk factor, as shown in Eq. (10).

$$U_i = \sum_{j=1}^{n} w_{ij} x_{ij} \tag{10}$$

2.4 Construction of the coupling function

Based on the coupling degree function measurement model and physical concepts, a coupling degree model of the subsea tunnel construction safety system can be further obtained.

$$C_m = \left\{ \frac{U_1 U_2 \ldots U_m}{[(U_1 + U_2 + \ldots + U_m)/m]^m} \right\}^{\frac{1}{m}} \tag{11}$$

According to the division of coupling in physics, for $C_m \in (0,0.3]$, low-intensity coupling occurs; for $C_m \in (0,0.7]$, the medium-intensity coupling can be determined, and for $C_m \in (0.7,1]$, high-intensity coupling occurs.

3 CASE STUDY

3.1 Identification of construction safety risk factors

The Xidian Bay Tunnel is located in Ninghai County, Ningbo City, Zhejiang Province, and the total length of the tunnel is 2.28 km. Safety risk coupling factors can be divided into five categories: personnel, machinery and materials, management, technology, and environment. The detailed contents are summarized in Table 1, and a diagram of the coupling behavior is shown in Figure 1.

Table 1. Safety risk factors in Xidian bay subsea tunnel construction.

Risk factor	Specific content
Personnel factors (F$_1$)	Poor safety awareness F$_{11}$ Poor physical or mental health F$_{12}$ Violation of safe operation procedures F$_{13}$ Construction not following the requirements of construction drawings F$_{14}$ Lack of professional ability F$_{15}$
Machinery and materials factors (F$_2$)	Unreasonable equipment selection, quantity, and layout F$_{21}$ Unqualified material quality F$_{22}$ Lack of safety protection devices F$_{23}$ The poor condition of equipment F$_{24}$
Management factors (F$_3$)	Insufficient safety education and training F$_{31}$ Inadequate safety technical disclosure F$_{32}$ Management organization confusion F$_{33}$
Technical factors (F$_4$)	Improper design of support and foundation treatments F$_{41}$ Improper dewatering and drainage scheme F$_{42}$ Poor effect of the off-site construction scheme (cofferdam) F$_{43}$
Environmental factors (F$_5$)	Adverse meteorological environmental conditions (rainstorm, flood, typhoon, etc.) F$_{51}$ Unfavorable geological conditions (muddy soil, weak soil layer, etc.) F$_{52}$ Poor hydrological conditions (coastal tidal action) F$_{53}$ The complex working environment at the construction site (such as on-site machinery cross operation) F$_{54}$

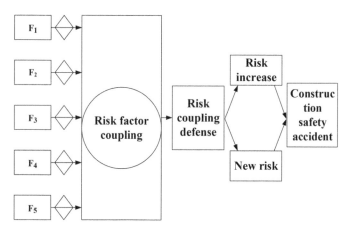

Figure 1. Schematic diagram of the safety risk coupling mechanism in subsea tunnel construction.

3.2 Weight calculation

According to the construction safety risk factors determined in Table 1, the on-site question-naire survey method was adopted, and questionnaires were distributed to relevant professionals with abundant project experience. A total of 10 questionnaires were distributed, each including a judgment matrix questionnaire and a direct impact matrix questionnaire. The entropy weight was calculated based on the judgment matrix questionnaire, as listed in Table 2.

Table 2. Entropy weight and the combined weight of the construction safety risk factors.

Primary index F_i	Secondary index F_{ij}	Entropy weight w_j	Centrality m_i	Combination weight s_j
Personnel factors (F_1)	F_{11}	0.0674	5.9915	0.0931
	F_{12}	0.0238	5.3857	0.0295
	F_{13}	0.0441	2.7487	0.0279
	F_{14}	0.0399	5.3517	0.0492
	F_{15}	0.0399	6.1372	0.0564
Machinery and materials factors (F_2)	F_{21}	0.0181	4.3327	0.0181
	F_{22}	0.1046	3.6476	0.0879
	F_{23}	0.1046	5.1804	0.1249
	F_{24}	0.0441	3.6801	0.0374
Management factors (F_3)	F_{31}	0.1613	5.1075	0.1899
	F_{32}	0.0230	5.6543	0.0300
	F_{33}	0.0181	5.3533	0.0223
Technical factors (F_4)	F_{41}	0.0181	4.0780	0.0170
	F_{42}	0.0674	4.4664	0.0694
	F_{43}	0.0399	3.7396	0.0344
Environmental factors (F_5)	F_{51}	0.0230	2.9309	0.0155
	F_{52}	0.0399	2.4889	0.0229
	F_{53}	0.1046	2.3839	0.0575
	F_{54}	0.0181	3.9680	0.0166

After the combined weight was determined, the order parameters between the various factors could be obtained according to Eq. (11). After calculation of the efficacy coefficient and order parameter of each safety risk factor, the coupling degrees among five factors, four factors, three factors, and two factors could be calculated and sorted. Details are listed in Table 3.

Table 3. Coupling degree among the construction safety risk factors.

Multifactor coupling	Risk coupling degree
Five-factor coupling	C(personnel-machinery-management-technology-environment) = 0.1497 C(personnel-machinery-management-environment) = 0.2352
Four-factor coupling	C(personnel-management-technology-environment) = 0.2032 C(personnel-machinery-technology-environment) = 0.1909 C(personnel-machinery-management-environment) = 0.1896 C(personnel-management-technology-environment) = 0.1876
Three-factor coupling	C(personnel-machinery-management) = 0.3272 C(personnel-management-environment) = 0.3246 C(machinery-management-environment) = 0.3143 C(personnel-machinery-environment) = 0.3142 C(personnel-technology-environment) = 0.2887

(continued)

Table 3. Continued.

Multifactor coupling	Risk coupling degree
	C(management-technology-environment) = 0.2765 C(personnel-management-technology) = 0.2714 C(machinery-technology-environment) = 0.2568 C(personnel-machinery-technology) = 0.2563 C(machinery-management-technology) = 0.2529
Two-factor coupling	C(personnel-management) = 0.4985 C(machinery-management) = 0.4968 C(personnel-environment) = 0.4943 C(personnel-machinery) = 0.4910 C(management-environment) = 0.4872 C(machinery-environment) = 0.4719 C(technology-environment) = 0.4635 C(personnel-technology) = 0.4329 C(management-technology) = 0.4149 C(machinery-technology) = 0.3863

3.3 Coupling analysis of the construction safety risk

The coupling degree among the various safety risk factors in the construction of the Xidian Bay Subsea Tunnel mainly indicated low- and medium-intensity coupling. In regard to five-factor coupling, the personnel-machinery-management technology-environment factor coupling degree indicated a low-intensity coupling state. In regard to four-factor coupling, the five cases indicated the low-intensity coupling state, among which the personnel-machinery-environment factor coupling degree was the highest. It could also be found that if the coupling degree with personnel factors could be reduced, the coupling degree could decrease. Regarding three-factor coupling, there were 10 cases of low- and medium-intensity coupling, among which the personnel-machinery-management, personnel-management-environment, machinery-management-environment, and personnel-machinery-environment factor coupling degrees indicated medium-intensity coupling. If the coupling degree between human and environmental factors and other factors were reduced, the coupling degree could be minimized. In terms of two-factor coupling, the 10 cases indicated medium-intensity coupling, among which the personnel-management factor coupling degree was the highest. Hence, the coupling of human and management factors could maximize the system risk. Therefore, much attention should be given to avoid the risk levels of human resources and management increasing simultaneously.

Based on the above analysis, it could be observed that the factors with the highest coupling degree were personnel and management factors, indicating that the simultaneous occurrence of these two factors could increase the risk of subsea tunnel construction safety systems. Via further analysis of the coupling proportion of each factor category, it could be concluded that the highest proportion was 20.58% associated with personnel factors, followed by 20.54% associated with environmental factors, 20.41% associated with management factors, 19.92% associated with equipment and material factors and 18.55% associated with technical factors. This demonstrates that the coupling effect with personnel factor subsystems is much greater than that with the other factor subsystems. This further indicates that the construction safety risk of the Xidian Bay Subsea Tunnel is closely related to personnel factors. Therefore, it is necessary to control the coupling between personnel factors and other subsystems to reduce the safety risk in subsea tunnel construction.

4 CONCLUSION

Choosing the Xidian Bay Subsea Tunnel as a case study, an analysis of construction safety risk coupling behavior based on the EWM-DEMATEL model was conducted. As indicated by the results, the combined weight of insufficient safety education and training F_{31} and the lack of safety protection devices F_{23} was the highest, which exerts a notable impact on the safety risk. The calculated ranking of the coupling degree among all factors reveals that the coupling degree between personnel and other factors was the highest, followed by management factors. Therefore, personnel factors readily interact with other factors. Priority should be given to controlling the coupling effect between personnel and management factors. The coupling effect between the factors of technology, equipment and materials, environment, and personnel should also be controlled to reduce the occurrence of construction safety risk-related accidents.

ACKNOWLEDGMENT

The authors are grateful for the financial support provided by the Science and Technology Project of Ningbo Transportation Bureau (Grant No. 202007).

REFERENCES

Ameyaw, E. E., Hu, Y., Shan, M., Chan, P. C., Le, Y. (2016) Application of Delphi method in construction engineering and management research: a quantitative perspective. *J. Civ. Eng. Manag.*, 22: 991–1000.

Cheng, Q. (2010) Structural entropy weight determination method for evaluation index weight. *Syst. Eng. Theor. Pract.*, 30: 1225–1228.

Dai, C. (2012) *Research on risk analysis and control technology of urban tunnel construction.* Tsinghua University Press, Beijing.

Guo, Z., Yin, K., Huang, F., Fu, S., Zhen, W. (2019) Evaluation of landslide susceptibility based on landslide classification and weighted frequency ratio model. *Chinese J. Rock. Mech. Eng.*, 38: 287–300.

Hallowell, M. R., Gambatese, J. A. (2010) Qualitative research: Application of the Delphi method to CEM research. *J. Constr. Eng. M.*, 136: 99-107.

Kim, Y., Lee, S. S. (2020) Application of artificial neural networks in assessing mining subsidence risk. *Appl. Sci.*, 10: 1302, 2020.

Lin, F., Wu, P., Xu, Y. D. (2021) Investigation of Factors Influencing the Construction Safety of High-Speed Railway Stations Based on DEMATEL and ISM. Adv. Civ. Eng., 2: 1–12.

Patel, D., Jha, K. (2016) Evaluation of construction projects based on the safe work behavior of co-employees through a neural network model. *Safety. Sci.* 89: 240–248.

Sanni-Anibire, M.O., Mahmoud, A.S., Hassanain, M.A., Salami, B.A. (2020) A risk assessment approach for enhancing construction safety performance. *Safety. Sci.* 121: 15–29.

Zhang, G., Wang, C., Jiao, Y., Wang, H., Zhong, G. (2020) Collapse risk analysis of deep foundation pits in metro stations using a fuzzy Bayesian network and a fuzzy AHP. *Math. Probl. Eng*, 2020: 4214379.

Zhang, S., Nie, L. (2017) Analysis of influencing factors of building construction safety management behavior based on DEMATEL method. *Safety. Environ. Eng.*, 24: 121–125.

Advances in Measurement Technology, Disaster Prevention and Mitigation – Li & Mohd Yusof (Eds)
© *2023 The Author(s), ISBN: 978-1-032-36087-4*

Evaluation of debris-flow risk in Xiaohongyan Gully of Baihetan hydropower station on the Jinsha River

Zhi-gang Shan & Wei-da Ni
POWERCHINA Huadong Engineering Corporation Limited, Zhejiang, China
Zhejiang Engineering Research Center of Marine Geotechnical Investigation Technology and Equipment, Hangzhou, China

Hao Wu* & Yi-huai Lou
Zhejiang Huadong Construction Engineering Corporation Limited, Zhejiang, China
Zhejiang Engineering Research Center of Marine Geotechnical Investigation Technology and Equipment, Hangzhou, China

ABSTRACT: Debris flows are common geological disasters that strike the southern mountainous areas in China frequently, posing harm to the construction and safe operation of cascade hydropower stations. In this study, we conduct a geological survey to analyze the topographic conditions for debris-flow triggering at the Xiaohongyan Gully of Baihetan Hydropower Station, Jinsha River. Then, we continue with a laboratory test and analyze the bulk density, speed, volume, and impact force of debris flows as well as the total volume of a debris flow and the total solid material content. Finally, we predict how a debris flow develops and how it affects engineering construction, thereby proposing suggestions on how to effectively control and prevent debris flows. The results revealed that debris flows in Xiaohongyan gully met the topographic, material source, and rainfall conditions. The total volume of loose solid material sources in this gully was 218.14×104 m^3, among which the volume of unstable material sources that might participate in debris flow activities was 26.46×104 m^3, indicating a gully rainfall-induced low-frequency large-scale diluted debris flow. The main parameters of Xiaohongyan debris flow gully with a 50-year return period were as follows: bulk density (1.52 t/m^3), speed (3.71 m/s), and peak volume (173.63 m^3/s) of debris flow, the total volume of a debris flow (55,000 m^3), solid material content (17,200 m^3) and overall impact force (2.83 kPa) of the fluid. The study results provide a beneficial reference for evaluating, preventing, and controlling debris flow disasters in major engineering construction areas.

1 INTRODUCTION

As common geological disasters in mountainous towns, debris flows usually occur at a large scale and lead to serious casualties and property losses, accompanied by enormous difficulties in disaster monitoring, early warning, prevention, and mitigation (Li et al. 2020a; Zhang et al. 2018). South-western China, one of the areas with the most complicated natural environment and engineering geological conditions throughout the world, is distributed with massive loose accumulations, thus being susceptible to debris flows under the action of earthquakes and rainfalls.

The Jinsha River is located on the upper reaches of the Yangtze River, which is a key area of hydroelectric development in China and is important to China's "West-to-East Power Transmission Project". The natural head of the Jinsha River accounts for 95% (5,100 m) of the total head of

---

This research was supported by the Zhejiang Provincial Natural Science Foundation (Grant No. LQ22E090008).
*Corresponding Author: wu_h5@hdec.com

the mainstream of the Yangtze River, containing enormous hydropower resources (theoretically, reaching 112,400,000 kilowatts). Besides, the technically developable hydropower resources reach 88,910,000 kilowatts, with an annual energy output of over 330 million kilowatts, and the enrichment degree of hydropower resources can be said to rank top across the world (Shan et al. 2021). Due to the development of deep-incised valleys in the Jinsha River, earthquakes and rainfalls occur frequently, thereby inducing frequent geological disasters, especially debris flows. Debris flows have been threatening the safety and stability of hydropower engineering in this area during the construction period and operation period, and mass deaths and casualties happen now and then (Cui et al. 2013; Hu et al. 2017). For example, a debris flow disaster burst out in the construction area of one hydropower station in Ganzi Prefecture in July 2009, giving rise to 54 deaths. In August 2012, a debris flow broke out outside and inside the construction area of a hydropower station in Liangshan Prefecture, leading to 24 deaths and missing persons. In June 2012, a debris flow occurred to nine gullies in the near field region at the dam site of Baihetan Hydropower Station in construction, among which Aizi Gully suffered from an extra-large debris flow, causing 41 deaths (He et al 2015). In August 2012, a debris flow disaster happened at Jinping Hydropower Station, during which 10 people died and went missing. In the Jinsha River basin, an area subjected to the most frequent disastrous debris flows in China, there are a total of 7,290 ascertained debris flow gullies, and over 50% of national debris flow disasters occur in this area (Hu et al. 2017; Zhong et al. 1994).

In the cascade hydropower station development area on the upper reaches of the Yangtze River, it is especially important to evaluate debris-flow risks to avoid and relieve the threats posed by debris flow disasters to hydropower engineering and to guarantee resident life and property safety in the hydropower station and the area of authority (Li et al. 2020a). Most gullies at the dam site of Baihetan Hydropower Station have experienced debris flows throughout geological history, and the debris flow activities are still very active in some gullies in the recent period. The structure of loose materials in debris flow gullies is looser than the original state, with a larger internal porosity, making the loose materials even more unstable under the action of external stresses, accompanied by lower critical hydraulic conditions for debris flows, i.e., debris flow activities are more violent (Fan et al. 2018; Li et al. 2020b; Wu et al. 2020). Lacking a detailed engineering geological survey on such frequent debris flow disasters, it is urgent to propose a risk evaluation method for typical debris flow gullies in this area (He et al 2013).

In this study, the topographic and material source conditions for the debris flows in Xiaohongyan Gully of Baihetan Hydropower Station in Baihetan Hydropower Station on the Jinsha River were systematically analyzed. Next, the accumulation characteristics of the debris flow in Xiaohongyan Gully were figured out through a laboratory test. Furthermore, the debris flow risk in Xiaohongyan Gully was evaluated from three aspects: speed, volume, and impact force of debris flow. This study aims to provide a reference for engineering construction and safe operation in this area.

2 REGIONAL ENGINEERING GEOLOGICAL BACKGROUND

2.1 *Engineering overview*

The Baihetan Hydropower Station is the second tier of four cascade hydropower stations (Wudongde, Baihetan, Xiluodu, and Xiangjiaba Hydropower Stations) on the lower reach (Panzhihua to Yibin) of the Jinsha River. It is 45 km away from upstream Qiaojia County, about 182 km from the dam site of upstream Wudongde Hydropower Station, and 195 km from downstream Xiluodu Hydropower Station. Liucheng Town, Ningnan County, and Sichuan Province are on the left bank of this dam, and Dazhai Town, Qiaojia County, and Yunnan Province are on the right bank. Hangudi stock ground is located on the upper reach of the dam site, at the right bank of the Jinsha River and on the mountain slope in the northeast of Qiaojia County, with a straight-line distance from the dam site of about 31 km and highway mileage of 55 km, serving as an aggregate quarry of Baihetan Hydropower Station dam project. As revealed by the engineering geological survey, a rainfall-induced low-frequency large-scale diluted debris flow fully—Xiaohongyan Gully—is developed at the north side of this quarry, which constitutes a potential safety threat during the engineering construction period and operation period.

2.2 Engineering geological conditions

• Topography and landform

The drainage area of Xiaohongyan Gully is about 16.32 km^2. Its highest altitude in the drainage basin reaches 3,250 m, and it flows into Bailong Gully at an elevation of 790 m. The height difference is about 2,460 m, and the main gully is about 10.37 km in length, with an overall longitudinal slope of 237‰. The whole drainage basin belongs to a tectonic erosional high-medium mountain landform, acting as the west slope of Jiaoding Mountain at the south foot of the Yaoshan Mountains. The main gully is generally spread in a nearly east-west (EW) direction, the plane of the drainage basin is of a long strip shape, and the gully is generally wide at the top and narrow at the bottom. Within the drainage basin, the overall vegetation coverage is about 50%, dominated by arbors and shrubs. Moreover, it is divided into three segments from top to bottom.

With an elevation of 3,250–2,700 m and a slope gradient of 50−70°, the upper segment is precipitous and developed with a dendritic drainage pattern in the gully. The elevation of the middle segment is 2,700–1,100 m, where the part above 2,200 m is a steep slope, and locally, there is a slope landform with an elevation of 2,200–1,100 and a slope gradient of about 18−25°. The elevation of the lower segment is 1,100–790 m, with a gentle platform and steep slope alternated landform, where a lot of residents settle on the Dapingzi gentle platform (slope gradient: 5−8°), and the gradient of the steep slope is about 30−40°. Qingmenkou Gully joins Xiaohongyan Gully nearby the elevation of 1,100 m and Bailong Gully at the elevation of 790 m. Collapse masses of Luopan Mountain are distributed on the north side nearby the elevation of 1,100 m.

• Formation lithology

Permian limestones and basalts are mainly exposed within the drainage basin. According to petrogenesis, the Quaternary system is divided into residual slope sediments, colluvial slope sediments, proluvial-alluvial sediments, and debris flow accumulations, as described in the following:

(1) Upper Permian Qixia-Makou Formation (P$_1$q+m): light grey-dark grey thick-ultrathick limestone and bioclastic limestone, which are brittle and weakly weathered with massive to sub-massive structures in the form of hard rock masses. Such rock masses are mainly distributed in Mujiaowan, Zhongcun, Pujialiangzi, and Hangudi stock ground with the elevation of below 3,000 m, being the main lithologies within the drainage basin.

Upper Permian Mount Emei Formation (P$_2\beta$) is a set of dark green and cinereous basalts with multiple eruption cycles, consisting of devonite, crypto-crystals, amygdaloid, tuff breccia, and tuff. The rocks are mainly weakly weathered with hard massive-mosaic structures and distributed above the elevation of 3,000 m and at the watershed.

(2) Quaternary accumulations (Q$_4$)

1) Residual slope sediments (Q$_4$edl): hard-plastic amaranth gravel-containing clay and clay with a thickness of about 0.5-2 m, and mainly distributed on high-altitude round hilltops, gentle slopes, planation surfaces and low concave areas. 2) Colluvial slope sediments (Q$_4$col+dl): grayish yellow-grey brown composite soil-crushed stones, crushed stone-composite soils, or crushed stones with uneven soil texture and the thickness of 3–5 m, and sporadically distributed at the foot of steep slopes (elevation: 2, 500–2, 700 m). 3) Riverbed sediments (Q$_4$apl): grey loose drifting pebbles and (crushed) sand-inclusion gravels with good sorting properties, and distributed in gulches. 4) Debris flow accumulations (Q$_4$sef): grey variegated crushed stones mixed with mud and sand, which are of different sizes in a medium compact-compact shape with a great accumulative thickness, and distributed on the Daping terrace below 1,100–1,000 m or at the gully port, the two sides of main gully and the gentle part of the gully bed.

3 ANALYSIS OF FORMATION CONDITIONS FOR XIAOHONGYAN GULLY DEBRIS FLOW

Debris flows are formed under the following conditions: topographic conditions, material source conditions, and rainfall-induced conditions. In this study, such conditions for the formation of debris flows in Xiaohongyan Gully were analyzed according to the practical reconnaissance and survey results.

3.1 *Topographic conditions*

The slope of the gully bed, an important condition for the conversion of potential energy into kinetic energy, plays a significant role in the formation and movement of debris flows. In general, the appropriate slope for riverbeds in debris flow source regions is 15−40°. Within this range, a greater slope of the gully bed is better for the formation of debris flows, and vice versa.

Xiaohongyan Gully drainage basin is high (highest point: 3,250 m) in the east and low in the west, so water flows from the east to the west. Besides, it flows into Bailong Gully at the elevation of 790 m and Shuinianhe Gully at 690 m, and the longitudinal slope of the gully bed is about 237‰. Above the elevation of 1, 400 m, the main gully is 6.03 km in length, with a drainage area of 8.54 km^2 and an overall longitudinal slope of 307‰ (Figure 1), accompanied by local scarps. In addition, the overall slope of the gully bed is about 17°, and the width and depth of the main gully are 20–30 m and 15–20 m, respectively, thus meeting the topographic conditions for debris flows.

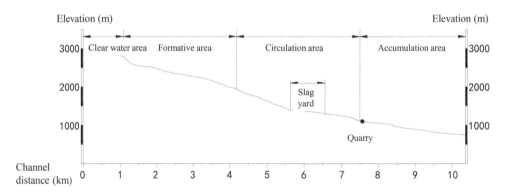

Figure 1.   Longitudinal Slope of Xiaohongyan Gully

Clearwater area: a denuded high and medium-mountainous area distributed at the elevation of 3,520–2,700 m, in which shallow etched tributary ditches are developed, with sparse vegetations. In this drainage basin, bedrocks (hard massive basalts) are exposed, along with a small number of loose material sources under weathering and unloading actions.

Formative area: an eroded medium-mountainous area distributed at the elevation of 2,700–1,940 m with a funnel-shaped landform and steep bank slopes. The whole gully presents a wide "V" shape. Moreover, a dendritic drainage system is developed in this area. Colluvial slope sediments are distributed on the rear edge of steep slopes and residual slope sediments are spread on gentle slopes. This area mainly exerts the function of water catchment and provides most of the material sources.

Circulation area: this area is distributed at 1,940–1,100 m, with a curved and narrow gully channel, and the gully is deeply incised. The gully is generally steep, accompanied by local waterfalls and scarps. The bank slope is steep (slope gradient: about 40−70°). Moreover, the main material sources in this area are sediments in the gully channel and the colluvial slope segments on the bank slope.

Accumulation area is an area distributed below the elevation of 1,100 m with a channel length of about 2.8 km. The gully is curved and the slope is relatively gentle. Old debris flow accumulations can be observed on the gentle terrace beneath the quarry. Following the elevation of 790 m, the kinetic energy of debris flows gradually disappears, and multiple gulches are gathered here, thus forming an alluvial terrace.

3.2 *Material source conditions*

The thickness of the earth source is one of the primary factors influencing the formation of debris flows. In this study, the critical earth thickness forming debris flows was about 15 cm according to the observation data and statistical analysis results. The earth thickness and its distribution data in Xiaohongyan Gully revealed that the thickness of loose earth in the gully bed accumulation of Xiaohongyan Gully met the critical thickness initiating debris flows.

Above 1,400 m of Xiaohongyan Gully, material sources are mainly distributed on the slope terraces at two sides of gulches, which are rich in loose solid material sources. In this investigation, a total of 14 material source points were discovered, including channel accumulations, collapse, landslide accumulations, and residual slope sediments on the slope surface. The total volume of solid material sources accumulated in the gully within the drainage basin was $11.20 \times 10^4 \, m^3$, usually under an erosion-siltation balance state. When a debris flow happens, it will scour channel beds and most alluvial-diluvial deposits replenish the material sources for debris flows, and materials were reaccumulated in the gully bed in the later phase. Besides loose alluvial-diluvial deposits in the gully, the waste slags produced by highway construction also constituted material sources (Figure 2). The volume of unstable material sources that might participate in debris flow activities was calculated by 30% of the total volume of material sources, thus being $3.36 \times 10^4 \, m^3$, and the average thickness of bed removal was roughly 60 cm.

Under the violent scouring of surface runoffs formed by heavy rainfalls, such gully accumulations will very probably be eroded to form a gully, and the slope sediments at two sides of the gully are continuously laterally eroded to form collapses at intervals. Subsequently, the gully will be both broadened and deepened, the collapsing loose accumulations are continuously gathered under the coercing action of floods, and meanwhile, a disastrous cascading effect will result from local clogging points inside the gully (Cui et al. 2013).

Figure 2.   Debris Flow Accumulation in Xiaohongyan Gully.

The total volume of collapse and landslide accumulations was $24.01 \times 104 \, m^3$, which were mainly distributed at 3 places above the elevation of 2,500 m with favorable overall stability, not going as far as to generate large-scale landslides and thus form concentrated material sources. There were collapse masses at two sides of some gulches, which were formed on a small scale that they were not calculated separately. The volume of collapse and landslide accumulation participating in debris flow activities as unstable material sources were $7.06 \times 10^4 \, m^3$.

The volume of eroding material sources was estimated according to Equation (1) (Chengdu Institute of Mountain Hazards and Environment, Institute of Soil and Water Conservation, CAS 2000):

$$V_s = M_r \cdot M_e \cdot A_b \tag{1}$$

Where V_s—the annual volume of eroding material sources (m^3);

M_r—rainfall coefficient, which is taken as 2 since the annual precipitation is 800-1,400 m in the Xiaohongyan Gully;

M_e—erosion modulus, which is set as 10,000 m^3/km^2 by considering the debris flow area in Xiaohongyan Gully as a general inferior land;

A_b—total area of loose accumulation sources within the drainage basin (taken as 2.116 km^2).

In Equation (1), the annual volume of eroding material sources provided by the Xiaohongyan Gully drainage basin was solved as 42,300 m^3, and the volume of eroding materials on the sloped surface that might take a part in debris flow activities as unstable material sources was calculated by 10% of total reserves, thus being 18.29×10^4 m^3. In conclusion, the total volume of loose solid material sources in the drainage basin was 218.14×10^4 m^3, among which the volume of unstable material sources participating in debris flow activities was 26.46×10^4 m^3.

3.3 *Characteristics of debris flow accumulations*

From the cross sections of two debris flow fans, the debris flow in this gully was mainly divided into two phases. In the early phase, the debris flow occurred on a large scale, some debris flow materials were washed into the river, particulate materials were carried away by water flow, and large block stones and huge stones were accumulated on the riverbed. In the later phase, the debris flow materials were of relatively small particle size, mainly being block gravelly soil locally mingled with a small number of huge stones, the content of block stones was about 60% with an average particle size of 30–50 cm or so, the content of gravelly soil was roughly 35% with an average particle size of 5–8 cm, and the remaining were huge stones with an average particle size of 1–2 cm.

The particles (< 200 mm) accumulated at the gully port were subjected to gradation and field bulk density experiments. As for the particle gradation experiment, the topsoil samples were collected from a trial trench and sieved on the field.

It could be known from the gradation curves that the uniformity coefficient and curvature coefficient of 1# soil sample were Cu = 5.93 and Cc = 1.09, respectively, and those of 2# soil sample were Cu=9.17 and Cc = 1.17, respectively. Both 1# and 2# soil samples met two conditions: uniformity coefficient Cu ≥ 5 and curvature coefficient Cc = 1-3, indicating good accumulation gradation at the port of Xiaohongyan Gully.

Overall, Xiaohongyan Gully drainage basin satisfied the three conditions inducing debris flows, i.e., topographic and landform conditions, material source conditions, and water source conditions. It was judged through an engineering survey that the debris flow in Xiaohongyan Gully belonged to rainfall-induced low-frequency large-scale diluted debris flow, being an area prone to debris flows.

4 QUANTITATIVE EVALUATION OF DEBRIS-FLOW RISKS

The debris-flow risks in Xiaohongyan Gully were quantitatively evaluated from three aspects: speed, volume, and impact force.

4.1 *Speed of debris flows*

The speed of debris flows in Xiaohongyan Gully was calculated using the improved formula recommended by the southwestern branch of the China Academy of Railway Sciences (Tang 2000):

$$V_c = \frac{1}{(\gamma_H \varphi + 1)^{1/2}} \cdot \frac{1}{n} \cdot R^{2/3} \cdot I^{1/2} \tag{2}$$

Where
V_c—speed of debris flows, m/s;
ϕ—sediment correction coefficient, which can be calculated through $\varphi = (\gamma_C - 1)/(\gamma_H - \gamma_C)$;
γ_H—bulk density of solid particles in debris flows (t/m^3), which is taken as 2.65 t/m^3 on basis of empirical value;
γ_C—bulk density of debris flows, t/m^3. The bulk density of debris flows with a 100-year return period is associated with the occurrence frequency, so this variable was calculated according to Tang (Tang 2000);
$1/n$—roughness coefficient of clear-water riverbed;
R—hydraulic radius (m), which is generally replaced by average water depth;
I—the hydraulic slope of debris flows, generally substituted by the longitudinal slope (‰) of gully bed, and it was 307‰ in Xiaohongyan Gully.

Above 1,400 m, Xiaohongyan Gully showed a narrow and steep landform, the most gully segments presented a narrow V-shaped valley with great cutting depth and vegetation development, and the vegetations at two sides of the gully channel were dominated by herbaceous plants and shrubs, accompanied by a small number of tall arbors in the gully channel. In this area, the gully channel was straight and smooth and slightly clogged with step-like head falls and scarps. The roughness coefficient was averaged as 8.8 according to the recommendation of Chengdu Institute of Mountain Hazards and Environment, Institute of Soil and Water Conservation, CAS (2000). The other parameter values and calculation results are listed in Table 1.

Table 1. Parameter values and calculation results.

Burst frequency of debris flows, P (%)	10	5	2	1
Bulk density of debris flows, γ_C (t/m^3)	1.32	1.40	1.52	1.60
Sediment correction coefficient, φ	0.24	0.32	0.45	0.57
Hydraulic radius, R (m)	0.8	1.0	1.2	1.5
Speed of debris flows, V_c (m/s)	3.29	3.58	3.71	4.03

4.2 The volume of debris flow

The volume of debris flows in Xiaohongyan Gully was calculated using the rain-flood Dongchuan formula, as shown in Equation (3) (Cui et al. 2013):

$$Q_C = Q_p(1 + \varphi)D_C \tag{3}$$

Where
Q_C—the volume of debris flow, m^3/s;
Q_p—flood volume in debris flow gully, m^3/s;
φ—sediment correction coefficient;
D_C—debris flow clogging coefficient.

All these parameters are determined through the attached Table 2 based on the *Specification of Geological Investigation for Debris Flow Stabilization* (T/CAGHP 006-2018) (issued by the Ministry of Land and Resources, PRC) according to the engineering geological conditions (1.5 for Xiaohongyan Gully).

Table 2. Parameter values and calculation results of volume of debris flow.

Burst frequency of debris flow, P (%)	10	5	2	1
Maximum volume, Q_p (m^3/s)	56.8	67.0	79.6	89.0
Volume of debris flows, Q_C (m^3/s)	105.63	133.05	173.64	209.79

4.3 The impact force of debris flow

The overall impact force of debris flows was calculated in Equation (4) (Department of Natural Resources of Sichuan Province 2006):

$$\delta = \lambda \frac{\gamma_c}{g} V_c^2 \sin \alpha \qquad (4)$$

Where
δ—impact force of debris flows, kPa;
λ—building shape coefficient, which is taken as $\lambda=1.33$ according to a rectangular building;
γ_C—the unit weight of debris flows (t/m^3);
V_c—the average speed of debris flows, m/s;
α—the included angle between the thrust face of a building and the impact direction of debris flows (°), which can be calculated according to the most unfavorable angle of 90°.

According to the field investigation, the calculation parameters and calculation results of the overall impact pressure of Xiaohongyan Gully debris flow are displayed in Table 3.

Table 3. Calculation results of overall impact pressure of debris flows in xiaohongyan gully.

Design frequency, P (%)	Building shape coefficient, λ	Unit weight of debris flows, γ_c (t/m^3)	The average speed of debris flows, V_c (m/s)	The included angle between thrust face and impact direction of debris flows, α (°)	Impact pressure of debris flows, δ (kPa)
10		1.32	3.29		1.93
5		1.40	3.28		2.44
2	1.33	1.52	3.71	90	2.83
1		1.60	4.03		3.53

5 CONCLUSIONS

Based on the engineering geological survey, the Xiaohongyan debris flow gully at the dam site of Baihetan Hydropower Station on the Jinsha River was comprehensively analyzed. Next, the debris flow risk in Xiaohongyan Gully was comprehensively evaluated from four aspects: speed, volume, burst scale, and impact force. Finally, we obtained the following conclusions:

(1) The debris flow occurring above the elevation of 1,400 m in Xiaohongyan Gully is gully rainfall-induced low-frequency large-scale diluted debris flow. Xiaohongyan Gully drainage basin, which meets the three major conditions for debris flows, is extremely susceptible to debris flows under heavy rainfalls.
(2) The drainage area of Xiaohongyan Gully is about 16.32 km^2, and the main gully is roughly 10.37 km in length and 237‰ in overall longitudinal slope. Above 1,400 m, the drainage area is 8.54 km^2, and the main gully is 6.03 km in length and 307‰ in overall longitudinal slope. The volume of loose solid material sources in the drainage basin totals 218.14×10^4 m^3, among which the volume of unstable material sources that may participate in debris flow activities is 26.46×10^4 m^3.
(3) As for the cross section at 1,400 m in Xiaohongyan Gully, the main parameters of debris flow with a 50-year return period are exhibited as follows: bulk density (1.52 t/m^3), speed (3.71 m/s), and peak volume (173.63 m^3/s) of debris flow, the total volume of a debris flow (55,000 m3), solid material content (17,200 m^3), and overall impact force of the fluid (2.83 kPa).

188

ACKNOWLEDGMENT

This research was supported by the Zhejiang Provincial Natural Science Foundation (Grant No. LQ22E090008). The authors are very grateful for the financial support.

REFERENCES

Cui P, Zhou G G D, Zhu X H, et al. 2013. Scale amplification of natural debris flows caused by cascading landslide dam failures[J]. *Geomorphology*, 182, 173–189.

Fan X M, Scaringi G, Xu Q, et al. 2018. Coseismic landslides triggered by the 8th August 2017 Ms 7.0 Jiuzhaigou earthquake (Sichuan, China): factors controlling their spatial distribution and implications for the seismogenic blind fault identification[J]. *Landslides*, 15, 967–983.

He N, Chen N S, Zeng M, et al. 2015. Study on debris flow rainfall threshold of Baihetan Hydropower station near-zone area[J]. *Shuili Xuebao*, 46(2), 9–18.

He N, Chen N S, Zhu Y H, et al. 2013. Research on influential factors and dynamic characteristics of a debris flow in Aizi gully[J]. *Jounal of Water Resources and Architechtural Engineering*, 11(1): 12–16.

Hu G S, Chen N S, Zhao C Y, et al. 2017. Effectiveness evaluation of debris flow control and mitigation strategies for cascade hydropower station in upper Yangtze river, China—A case study in Baihetan hydropower station of Jinsha river[J]. *Bulletin of Soil and Water Conservation*, 37(1): 241–247.

Li N, Tang C, Bu X, et al. 2020a. Characteristics and evolution of debris flows in Wenchuan county after" 5·12" earthquake[J]. *Journal of Engineering Geology*, 28, 1233–1245.

Li X, Zhao J, Kwan J S. 2020b. Assessing debris flow impact on flexible ring net barrier: A coupled CFD-DEM study[J]. *Computers Geotechnics*, 128, 103850.

Shan Z G, Ni W D, Hong W B, et al, 2021. Major engineering geological problems and coutermeasures of Baihetan Hydropower station[J]. *Journal of Engineering Geology*, 29(S1): 302–309.

Specification of geological investigation for debris flow stabilization. T/CAGHP 006-2018, Ministry of Land and Resources, PRC.

Tang B X. 2000. *Debris flow in China* [M]. Commercial Press.

Wu H, Nian T K, Chen, G Q, et al. 2020. Laboratory-scale investigation of the 3-D geometry of landslide dams in a U-shaped valley[J]. *Engineering Geology*, 105428.

Zhang N, Fang Z W, Han X, et al. 2018. The study on temporal and spatial distribution law and cause of debris flow disaster in China in recent years[J]. *Earth Science Frontiers*, 25(2): 299–308.

Zhong D L, Xie H, Wei F Q. 1994. Research on the regionalization of debris flow danger degree in the upper reaches of Changjiang river[J]. *Moutain Research*, 12(2): 65–70.

Advances in Measurement Technology, Disaster Prevention and Mitigation – Li & Mohd Yusof (Eds)
© 2023 The Author(s), ISBN: 978-1-032-36087-4

Technical analysis of the Guangli canal from the perspective of harmony between man and nature

Bo Zhou*
Institute of Water Conservancy History of China Institute of Water Resources and Hydropower Research, Beijing, China
Research Center on Flood Control, Drought Control and Disaster Mitigation of the Ministry of Water Resources, Beijing, China
Key Scientific Research Base of Water Heritage Protection and Research, State Administration of Cultural Heritage, Beijing, China

Tianliang Li
School of Public Administration of Hohai University, Nanjing, China

ABSTRACT: Today's society advocates the concept of water control that reveres nature and seeks harmony between man and nature. In 2021, we examined ancient eco-hydraulic projects in China through field research and literature searches, and discovered the Guangli Canal in Henan, a heritage irrigation project with a history of over 2,200 years but relatively poorly studied academically. Located in northwest Henan, the Guangli Canal is one of the longest-established irrigation projects in the lower reaches of the Yellow River in China, and one of the few tunnel-gate water diversion hubs in ancient China. From the perspective of harmony between man and nature, this paper, considering its natural environment and historical evolution, technically analyses the construction techniques of the Guangli Canal that are still in full use for irrigation and summarizes its technical essence and ideological core in the historical period, aiming to provide scientific ideas and ideals for the development of contemporary water conservancy projects.

1 INTRODUCTION

In 2021, we examined ancient eco-hydraulic projects in China through field research and literature search, and discovered the Guangli Canal in Henan, a heritage irrigation project with a history of over 2,200 years. The Guangli Canal is one of the few tunnel-gate water diversion hubs in ancient China that still serves an important irrigation purpose, but is relatively poorly studied academically, almost none from the perspective of harmony between man and nature. This paper analyses the construction concepts and technical essence of this ancient eco-hydraulic project from the perspectives of its natural environment and historical evolution: the construction of the Guangli Canal was tailored to local conditions, following the principles of hydraulics, and achieving the maximum hydraulic efficiency in the simplest form of engineering. This paper intends to inspire the development of contemporary water conservancy projects, i.e., developing new water conservancy projects that meet the needs of both human economic and social development and

---

*Corresponding Author: 369129958@qq.com

DOI 10.1201/9781003330172-28

healthy water ecosystems based on respecting nature and seeking harmony between man and nature.

2 OVERVIEW OF THE ENVIRONMENT

The Guangli Canal is located in the northwest of Henan Province, with the head of the canal located in Jiyuan City and the irrigation area in Jiaozuo City, belonging to the Qinhe River system in the Yellow River Basin. The project is located at $34°52' - 35°8'$ N and $112°40' - 113°15'$ E, adjacent to the Qinhe River in the North, and the Manghe River in the South, bordering Kongshan Wulongkou, Jiyuan City in the West, and Qinnan area, Wuzhi County in the East, extending 65 km from east to west, and 27 km from north to south, covering a total area of 509 km², involving 16 townships (towns) and 439 administrative villages in the four cities (counties) of Jiyuan, Qingyang, Wenxian and Wuzhi, and with an irrigated area of 510,000 mu.

The Guangli Canal irrigation area belongs to the north-tempered continental monsoon climate with four distinct seasons. There are many north-westerly winds and dry climates in winter, and many south-easterly winds in summer. The perennial mean temperature in the region is 14.5°C, with a maximum of 40.7°C (1999) and a minimum of -16.4°C (1990).

The multi-year average rainfall in the irrigation area is 593.52 mm, with the maximum registering at 721.6 mm (1954) and the minimum registering at 126.9 mm (1965) from June to September, accounting for 66.0% and 48.3% of the annual rainfall, respectively. The multi-year average evaporation is 1668.3 mm, 2.8 times that of rainfall. Large inter-annual variations in rainfall and highly uneven distribution within the year have led to frequent droughts in winter and spring, as well as frequent floods in summer and autumn. The mixture of droughts and floods makes irrigation projects particularly important for agricultural production.

3 HISTORICAL EVOLUTION

The Guangli Canal is over 2,200 years old. As early as the time of Emperor Qin Shi Huang, people built the earliest water diversion project at the head of the Qinhe River with "squared wood as the gate", which was generally called the "Qin Canal" in later times. As a result of water irrigation, Henei County, where the irrigation area was located, became an important economic area around the royal capital and became the base for the unification of the Eastern Han Dynasty around 5 BC. The Fangkou Weir in the Qin Dynasty, with squared wood as the inflow gate, was easily destroyed by floods. In the 3rd century AD, Sima Fu of Wei converted the inflow gate from a wooden one to a stone one, and the stable inflow gate gave life to the water diversion project, which had been used for generations ever since (Chen & Li 2012). In the 9th century AD, the water diversion project irrigated an area of more than 5,000 hectares, benefiting the five counties of Jiyuan, Henei, Wenxian, Wude, and Wuzhi in Henei County, making it one of the main grain-producing areas in the country. In the middle of the 13th century, the water diversion project was rebuilt and named the "Guangji Canal", which consisted of barrage dams, diverting gates, one flood diversion and drainage river, and four main canals (with a total length of 338.5 km), with an irrigated area of more than 3,000 hectares. The headworks of the Guangli Canal, which survive today, were mostly built in the 16th and 17th centuries. During this period, five diversion canals were built at the head of the Qinhe River in succession: Guangji, Yongli, Lifeng, Xingli, and Guanghui, with stone carvings of dragon heads in the shape of swallowing water standing above each diverting gate. Therefore, it is also known as the "Wulongkou" (Figure 1).

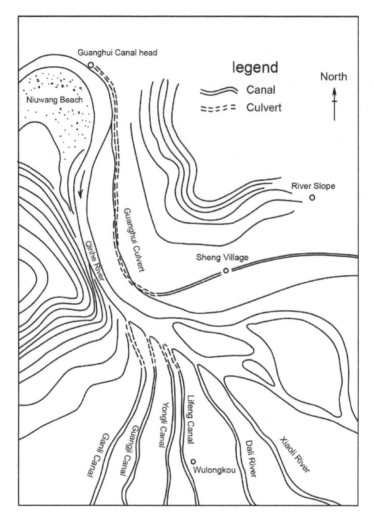

Figure 1.   Layout of the Guangli canal in ming and qing dynasties.

Figure 2 shows the site of the Guangji Canal, which was built in the early 17th century. Yuan Yingtai, the then governor of Henei, inspected the mountains and rivers in Henei on foot to build a canal to combat the droughts. It took five years to finally divert the water by overcoming mountains of difficulties, extending the irrigation area by 150 miles, and benefiting five counties and ten thousand households. This is the site of the Yongli Canal, which was started in the same year as the Guangji Canal. It was launched by Shi Jiyan, the then governor of Jiyuan, and was located downstream of the Guangji Canal, with a similar engineering structure to that of the Guangji Canal (Zhang et al. 2020). Figure 3 shows the basic condition of the diverting gates at the head of the Qinhe River in the 1950s, which still preserve the engineering structure of the 17th century. From left to right, they are the Lifeng Canal, the Guangji Canal, the Yongli Canal, and the Ganlin Canal. In the late 1950s, the water diversion project was integrated and expanded, and the three canals, Lifeng, Yongli, and Guangji, were combined into a single inflow gate called the "Guangli General Main Canal." In 1950, the Guangli Canal Authority was established, taking the most used words for each of these canals— "Guang" and "Li", and collectively calling them "Guangli Canal" (Tian 2020).

Figure 2.   The site of the Guangji canal.

Figure 3.   Overview of the diverting gates of the four main canals on the right bank of the Qinhe river in the 1950s.

The present Guangli Canal includes the Guangli General Main Canal and the Xinli Branch Canal on the right bank of the Qinhe River, and the Guanghui Main Canal on the left bank, with a total diversion flow of 20 m³/s. There are a total of 79 branch or above-canals, with a total length of 435.4 km; and there are more than 1,000 canal buildings of various types, with an irrigated area of 510,000 mu.

4 ENGINEERING FEATURES FROM THE PERSPECTIVE OF HARMONY BETWEEN MAN AND NATURE

The Guangli Canal is one of the few large irrigation areas in China where the main irrigation and water diversion engineering facilities in the Ming Dynasty have been preserved intact and is a typical model of dam-less tunnel-gate water diversion projects. The fact that it has been in use for more than 2,000 years is a result of its superb construction techniques and the balance between natural topography, mountains, water, and human use. Its engineering features are mainly reflected in the following aspects.

4.1 *The project siting is in line with the topographic conditions*

The head of the Qinhe River is geographically favorable, with relatively balanced alluvial siltation and a stable riverbed. Besides, it is located at the top of the alluvial fan of the Qinhe River, making for smooth water diversion. Most of the canal heads were located on the concave banks of the Qinhe River head and were cut into the stable cliff bedrock, making the construction extremely difficult but facilitating management and maintenance upon completion.

4.2 *The siting of the diverting gates conforms to the principles of circulating current in curved channels*

Historically, the diverting gates have developed from squared wooden gates in the Qin Dynasty through the stone gates, Shidou Gate and Lifeng Canal to the present head of the general main canal. These diverting gates are located at Wulongkou, except for the Guanghui Canal. Located at the head of the Qinhe River, Wulongkou has superior geographical conditions, with relatively balanced alluvial siltation and a stable riverbed. The riverbanks of the diverting gates boast favorable geological conditions as the diverting gates of the Guangji Canal, the Yongli Canal, the Lifeng Canal, and the Ganlin Canal are all located on steep limestone slopes cut by the Qinhe River. Wulongkou is located at the top of the alluvial fan of the Qinhe River, making for smooth water diversion. According to the principles of circulating current in curved channels, the water in a curved channel is subject to centrifugal forces, which cause the water level to rise on the concave bank and fall on the convex bank, creating a water level difference. The combined forces generated by the two create a lateral circulation of water. The surface water flows towards the concave bank, while the bottom water flows towards the convex bank. Water on the concave bank is clearer and prone to flushing, while water on the convex bank is more turbid and prone to siltation. The head of the Guangli Canal is mostly located slightly downstream on the solid concave bank, conforming to the principles of circulating current in curved channels (Zhang 1984).

4.3 *Combination of tunnel-chamber diverting gates and culvert-type diverting gates*

Ganlin Canal, Yongli Canal, Guangji Canal, and Lifeng Canal, all located on the southern bank of the Qinhe River, have tunnel-chamber diversion works as part of their diverting gates, while on the northern bank of the Qinhe River there are culvert-type diversion works along Guanghui Canal.

For tunnel-chamber diverting gates, the head section of the canal was cut into a stone cavern in the limestone cliff, with a gate inside the cavern, following which stone canals and earth canals were excavated downstream in order. This is the most representative interior of the Guangji Canal: the boulder with the carved dragon head at the top of the diverting gate, the cover stone at the top of the diverting gate, and the masonry under the balustrade are all about 1 m^3 in size and of very solid masonry. There are two stone carved water-swallowing beasts in the lower corners of the water-gazing gate, and inside the gate is a workshop of the chambers. The Guangji Canal has two diverting gates and chambers, which are for maintenance and water regulation. The two symmetrical diverting chambers are divided into three levels: the bottom level is the diverting hole; the middle level holds the lifting gate plate; and the upper level is the workshop, with steps on each level. The gate plate is a movable wooden board connected by two iron-carrying poles in tandem, with an iron cable in the middle. The forward and reverse rotation of the pulley above were leveraged to tighten or loosen the cable and raise or lower the gate plate. To divert the low water from the Qinhe River, the diverting gate of the Guangji Canal was set very low, with an absolute elevation of 138.3 m, just less than 1 m higher than the deepest part of the river bottom today. The gate will be closed during floods to prevent turbid flooding into the canal. Behind the Guangji Canal is a water transfer tunnel, which is 70 m long, 4 m wide at the bottom, and 3.3 m high. The 50-m section near the river bank is composed of solid rock with a narrower section, which gradually widens at the back. There is also a receding river to drain the floodwater and flush the sediment (Zhang 1993).

The Guanghui Canal is a culvert-type diversion project and the only one that diverts water on the north bank of the Qinhe River. As it was located on the convex bank, the Guanghui Canal was not conducive to water diversion and was repeatedly silted up after its construction. In the early 1800s, the head of the canal was built into a bunker-style diverting chamber, and a 1610-m stone culvert was built as a diversion canal. When the gate was closed during the flood season, the floodwater could flow over the top; and during the irrigation season, the water was diverted from the Qinhe River and transferred via the culvert. It took 239 years to complete the project, fully illustrating the hardships and twists and turns experienced. The unique head structure is also rare in China for dam-less water diversion projects.

4.4 Ingenious water diversion works and sand flushing channels

The Guangli Canal in the Ming and Qing Dynasties adopted the water diversion scheme between the distribution channels. The diversion stone was ingeniously used to divide the water, with stable diversion channels being built, fixed elevation of the diversion canal bottom, as well as reasonable diversion weir cross-sectional size and angle between the diversion branch channels, conforming to the principles of hydraulics. These simple water diversion methods help reduce conflicts over water. In addition, sand flushing channels were built. There was usually a water-reducing river at the head of the ancient Guangli Canal, intending to drain the overflowing flood water, flush the sediment in the channels and increase the flow rate of the canal water, making for the long-term stability of the irrigation channels for diverting turbid water.

4.5 A form of irrigation in which water is diverted through multiple gates and channels

Each canal in the Guangli Canal irrigates land at different elevations and in different ranges. During irrigation, silting can be prevented by avoiding using the control gate to raise the water level. In the meantime, diverting water through multiple gates and channels is also conducive to solving the conflicts over water between different administrative divisions. The whole canal system was adapted to the topography and hydrology of the Qinhe River, as well as to the ancient farming practices of smallholder production.

4.6 Tackling the technical problems of water diversion by mountain tunnels

Niu Cunxi, an outstanding hydraulician of the Ming Dynasty, tackled the key technical problems of measuring the elevation, positioning the hole, and moving the boulders when cutting the Guangji Canal. To accurately determine the elevation of the diverting gate and the outlet, he adopted the "method of measuring mountains and exploring water," which enabled him to "equalize the elevation of the two ends of the mountain." Concerning how to pierce a hole on both ends of a mountain slope that cannot be seen from each other and how to penetrate a smooth and straight canal line, he employed the "ground-piercing method" to anticipate deviations at both ends of the hole; when building the diverting chamber, he used the "mechanical method," similar to today's levers," to move and lift boulders weighing tons, without the need for large numbers of people or force. For their contemporaries, these advanced construction techniques were innovative.

5 CONCLUSIONS

In this paper, the perspective of harmony between man and nature is adopted to study the technical analysis of the Guangli Canal. The main conclusions can be summarized as follows:

(1) The Guangli Canal, built on convenient and favorable natural terrain, was tailored to local conditions, allowing for smooth water diversion as well as project management and maintenance.
(2) The water diversion of the Guangli Canal conformed to the hydraulic principles of circulating current in curved channels. In addition, different forms of diverting gates and multiple channels were used to divert water, which was suitable for irrigation in a variety of terrains and the hydrological characteristics of the Qinhe River, and helped tackle many technical problems in measurement for their contemporaries.
(3) The Guangli Canal is a model of ancient water conservancy projects that respect nature and achieve "harmony between man and water," as it achieves maximum irrigation efficiency in the simplest form without destroying nature, fully demonstrating the essence of the ancient concept of nature in water conservancy planning.

In terms of future work, this research perspective and concept of water conservancy heritage should be carried out to provide important enlightenment to contemporary water conservancy construction.

ACKNOWLEDGMENTS

This work was funded by the Key Scientific Research Base Project of the State Administration of Cultural Heritage (Grant No. 2020ZCK207), National Natural Science Foundation of China (Grant No. 42142029), and IWHR Scientific Research Projects (Grant Nos. JZ1208B022021, JZ0199B212019, JZ1003A112020, and WH0166B012021).

REFERENCES

Chen Junliang, Li Baohong. (2012) Investigation and Analysis of Jiyuan Wulongkou Water Conservancy Facility. *Journal of North China Institute of Water Conservancy and Hydroelectric Power* (Social Science), pp.24–27.

Tian Zhonghua. (2020) A Comparative Study of Caogong Ditch in Taiwan and the Guangji Canal on the Qinhe River in Northwest Henan, *Journal of Jiaozuo Teachers College*, pp.63–66.

Zhang Pengqi, Ye Wenqing, Zhu Xibin. (2020) Construction and Management of Water Diversion and Irrigation Project of Guangji Canal, Huaiqing Mansion, Ming Dynasty: A Survey Centered on Water Inscriptions. *Journal of Yunnan Agricultural University* (Social Science), pp.143–149.

Zhang Ruyi. (1984) A Preliminary Study of the Ancient Hydraulic Structures of the Guangli Canal on the Qinhe River. *Journal of Hydraulic Engineering*, pp. 65–71.

Zhang Ruyi. (1993) *A Brief History of the Guangli Canal Project on the Qinhe River* [M]. Nanjing: Hohai University Press.

Advances in Measurement Technology, Disaster Prevention and
Mitigation – Li & Mohd Yusof (Eds)
© 2023 The Author(s), ISBN: 978-1-032-36087-4

Simulation of damage mode of RC bridge pier under contact explosion

Pingming Huang* & Zerui Liu*

School of Highway, Chang'an University, Xi'an, China

ABSTRACT: In the past half a century, terrorist forces have been rising, and local wars and
conflicts have continued. As an important part of the transportation system, bridge structures have
suffered from an increasingly severe threat of explosion. As the key stress component of the bridge
structure, the pier is related to the overall safety of the bridge, so the damage to the pier under the
explosion is worth special attention. This paper establishes a high-precision 3D numerical model,
studies the destruction mode of RC piers under contact explosion, and divides the destruction mode
of RC piers into spalling destruction and penetrating destruction. At the same time, four damage
grades of RC piers under contact explosion are given, which can provide a reference for structural
anti-explosion design and structural damage assessment.

1   INTRODUCTION

In recent years, local wars and conflicts have continued, and explosion accidents have occurred
frequently. As an important part of the traffic lifeline project, the bridge structure is faced with an
increasingly severe threat of explosion (Hao 2018; Stewart & Mueller 2014; Williamson 2010).
The reinforced concrete piers, as the key stress components of the bridge structure, are easy to
produce local material failure under the action of contact and explosion, which seriously affects the
safety of the bridge structure. Therefore, the destruction of the RC pier is worth special attention.

In recent years, scholars at home and abroad have carried out tests and numerical simulation
studies on the destruction of RC components under the action of near-range explosions. The US
National Transportation Research Commission (TRB) (Williamson 2010) carried out a batch of RC
pier explosion tests earlier. Based on the test results, the damage mode of RC piers is summarized
as no flaking damage, peeling damage and penetrating damage. Zong Zhouhong (Zong et al. 2017)
tested the dynamic response and failure mechanism of RC pier under contact explosion. The results
showed that the axial pressure load could adversely affect the anti-explosive performance of RC
pier under contact explosion. Li Minghong (Li 2020) carried out a series of contact explosion tests
and put forward the formula for the residual bearing performance of a double-layer concrete-filled
steel tube column.

Based on the numerical model based on nonlinear explicit dynamic analysis software, the explosive resistance performance of RC pier can provide an important reference for the design of bridge
structure.

2   MODEL ESTABLISHMENT AND VALIDATION

2.1   *Establishment of the finite element model*

Three-dimensional numerical model of explosive-air-RC pier based on ANSYS / LS-DYNA.RC
piers adopt a scale ratio of 1:3, with a section diameter of 0.4 m, and a column height of 3.5m.
Reinforcement and concrete are established by the common node method, air and explosives
are coupled by ALE, and the keyword * CONSTRAINED_LAGRANGE_IN_SOLID is used

---

*Corresponding Authors: hpming@chd.edu.cn and 598362053@qq.com

DOI 10.1201/9781003330172-29

to realize the interaction of blast shock wave and structure. At the same time, the keywords * RIGIDWALL_PLANAR_FINITE and * BOUNDARY_NON_REFLECTING are used to realize the reflection of air in the test environment. Considering the solution accuracy and efficiency of numerical models, the grid of the air domain was determined to be 20 mm for explosives and 10 mm for explosives.

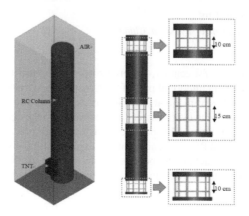

Figure 1.   Schematic representation of the 3 D FEM model.

2.2   Material constitutive model

2.2.1   Explosives and air

Air is constructed using the material model card * MAT_NULL combined with the equation of state card * EOS_LINEAR_POLYNOMIAL, where the equation of state is as follows:

$$P = C_0 + C_1\mu + C_2\mu^2 + C_3\mu^3 + (C_4 + C_5\mu + C_6\mu^2) E \qquad (1)$$

Formula: $C_0 \sim C_6$ are the polynomial equation coefficients; μ is the ratio of air density to initial density; E is the internal energy per unit volume.

Table 1.   Air model parameters.

ρ/ (kg·m^{-3})	C_0	C_1	C_2	C_3	C_4	C_5	C_6	E /(MJ·m^{-3})
1.292 9	0	0	0	0	0	0.4	0	0.25

The TNT is constructed using the material model * MAT_HIGH_EXPLOSIVE_BURN and combined with the equation of state * EOS_JWL, where the equation of state is as follows:

$$P = A\left(1 - \frac{\omega}{R_1 V}\right) e^{-R_1 V} + B\left(1 - \frac{\omega}{R_2 V}\right) e^{-R_2 V} + \frac{\omega E}{V} \qquad (2)$$

Formula: P is the pressure of the detonation; V is the relative volume; E is the internal energy per unit volume; A, B, R1, R2, and ω are material constants; the specific parameters are shown in Table 2.

Table 2.   Exploite model parameters.

ρ/ (kg·m^{-3})	v /(m·s^{-1})	pCJ/GPa	A /GPa	B /GPa	w	R1	R2	E /(MJ·m^{-3})
1630	6930	21	374	3.231	0.3	4.15	0.95	7000

198

2.2.2 Concrete and steel bars

Use the keywords * Mat_Concrete_Damage_Rel3 and * MAT_Plastic_Kinematic to establish concrete and reinforcement, C40 concrete, HRB400 for longitudinal bars and HPB300 for longitudinal bars.

2.3 Validation of the numerical model effectiveness

This paper uses Tang Biao's RC pier explosion test to verify the validity of the model, and the results are shown in Figure 2. Under the action of contact explosion of 1 kg TNT, the RC pier column only suffered partial spalling of the concrete protective layer and slight crushing of core concrete near the explosive height, but did not show overall deformation. The test results show that the peeling height of the concrete protective layer on the side of the column is 60 cm; many vertical and semi-ring cracks appear on the back burst surface, and the deformation of the longitudinal bar is inward; the maximum damage depth of the core concrete is 4 cm. The numerical simulation results show that the peeling height of the concrete protective layer is between 5 cm and 58 cm; the maximum damage depth of core concrete is about 4.5 cm, occurring near the explosive surface; the longitudinal bars and stirrups near the explosive surface, which is consistent with the test results.

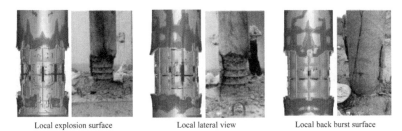

Local explosion surface   Local lateral view   Local back burst surface

Figure 2.   Comparison of numerical simulation and test phenomena.

By comparing the numerical simulation results and test results, the numerical model to simulate the destruction of concrete is accurate and reasonable. Although the numerical simulation results and test results are biased in some places, the difference is acceptable considering the many accidental factors of the explosion process.

3   RC PIER DAMAGE MODE UNDER CONTACT EXPLOSION

In the study on the destruction mode of RC piers under contact explosion, the circular pier with a column height of 3.5 m and 0.4 m diameter was taken as an example. The height of the explosive was kept at 0.3 m, and the numerical simulation of TNT loading was equivalent to 1 kg, 1.5 kg, 1.5 kg, 5 kg and 8 kg, respectively. The simulation results are shown in Figure 3.

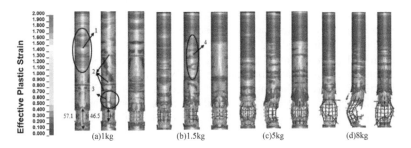

Figure 3.   Bridge pier damage results under different equivalents.

At the loading equivalent of 1 kg, The loss of concrete protective layer in the middle and lower pier within about 5 cm~57 cm, the crushing and damage height of the concrete protective layer on the explosive surface is 57.1 cm, and the shedding height of the side is 46.5 cm. The middle and upper part of the pier is not as seriously damaged as the middle and lower parts. The large area of plastic damage in the height range of 1.7 m to 2.9 m (mark 1), and appear a small number of annular cracks (mark 2) and mesh cracks (mark 3). The overall core concrete is relatively complete. A certain fragmentation and peeling occur on the explosive surface; its maximum depth is 4.2 cm, and the corresponding height is 0.31 m.

Figure 4.   Failure of the test column under contact and explosion.

With the increasing loading equivalent, the damage to the concrete in the middle and lower parts of RC piers is further intensified, the falling height of the concrete protective layer is increasing, and the crushing and peeling of the core concrete are more serious. The plastic damage to the concrete protective layer in the explosive surface of the piers (mark 4), and the previous ring and mesh cracks are further developed. When the loading equivalent is 1.5 kg, the maximum damage depth of the core concrete in the burst surface is about 8.6 cm, and the corresponding height is 0.29 m. The RC square pier column numbered E-300-500 completed by Dua et al. (2020) also observed the large shedding of the concrete protective layer in the middle and lower part of the pier column and the severe crushing and peeling of the core concrete. Figure 3(b) is similar to the damage shown in the test of Figure 4(a) (Dua et al. 2020).

When the loading equivalent is 5 kg, the RC pier runs through within the height of about 54.9 cm. At the larger loading equivalent, the penetration height of the core concrete reaches 61.8 cm. The RC square pier column numbered E-1000-300 is observed after (Dua et al. 2020), and the RC pier column corresponding to working condition case6 is also observed after the contact explosion test, as shown in Figure 4(b) and (c).

In conclusion, the damage to RC piers in the contact explosion is mainly concentrated in the area contacting explosives, mainly for local damage. With the increase of loading equivalent, the concrete is crushed in the explosion surface, stretched and peeled in the back blasting surface, the longitudinal bars and stirrups near the explosive contact surface are directly blown off, and the local plastic damage of the steel cage is more serious. Therefore, according to the destruction characteristics of RC piers under contact explosion, the destruction mode of RC piers is divided into peeling destruction and penetrating destruction, and the destruction classification of RC piers under contact explosion is proposed, as shown in Table 3.

Table 3. Declassification of reinforced concrete under contact explosion.

Damage grade	Destruction degree	Damage mode
Minor damage	The concrete protective layer falls off locally, a small amount of steel bar breaks, the steel cage is basically kept elastic, and the core concrete is slightly broken but basically complete	Spalling
Moderate damage	The concrete protective layer falls off locally, a small amount of steel bar fracture, the steel cage appears small plastic deformation, the core concrete is broken in the contact explosive area is more serious but no penetration damage	Spalling
Heavy damage	Concrete protective layer partial fall off, the right amount of steel bar fracture, steel cage appeared large plastic deformation but basically complete, the core concrete is penetrated, and the depth is small	Penetrating
Serious damage	The concrete protective layer falls off seriously, a large number of steel bars break, the steel cage is seriously damaged, and the integrity is lost. The core concrete is seriously penetrated, and the depth is very large	Penetrating

It is worth mentioning that, from the perspective of the loss of concrete, with the increase of loading equivalent, the damage degree of RC pier is significantly aggravated, in which the shedding range of concrete protective layer has been expanded to a certain extent, and the penetration height of core concrete is increasing. For the cylindrical test studied in this paper, 1.5 kg TNT explosive is a critical charge equivalent. If the charge equivalent is less than or equal to 1.5 kg, the RC pier falls off, and the core concrete of the pier is basically complete and still has certain functionality; if the charge equivalent is more than 1.5 kg, the RC pier will be damaged and completely lose functionality.

4 CONCLUSION

This paper establishes a high-precision 3D finite element model based on LS-DYNA, studies the damage mode of RC piers under contact explosion, and tentatively divides the damage level and damage mode of RC piers under contact explosion. The following conclusions are obtained, which can provide a reference for bridge anti-explosion design and damage assessment.

(1) According to the damage characteristics of RC piers under contact and explosion, the damage mode of RC piers is divided into spalling damage and penetrating damage.
(2) It qualitatively divides four damage grades of RC piers under contact explosion: minor, moderate, severe and serious damage.
(3) With the increase of loading equivalent, the damage degree of RC piers is significantly intensified, but the loss rate of the concrete protective layer and core concrete gradually slows down. For the specimens studied in this paper, 1.5 kg is a critical charge equivalent to determining what damage to the column will occur.

However, due to the complexity of the contact explosion, many problems remain in subsequent studies. For example, due to the strain rate effect of the structure, the test results of the scale model are quite different from the foot size structure, so more explosion tests of the scale model need to be carried out to provide more basis for numerical simulation analysis. In the future, the partition method of destruction mode should be deeply studied, and the corresponding rapid evaluation method should be established to realize the rapid evaluation of structural damage as soon as possible and greatly improve the economy and safety of the structural design.

REFERENCES

Dua Alok, Braimah Abass, Kumar Manish. Experimental and numerical investigation of rectangular reinforced concrete columns under contact explosion effects[J]. *Engineering* Structures, 2020, 205(C).

Hao Hong, Hao Yifei, *New progress in the research of engineering structure protection under the action of power multi-disaster* [M], Beijing: China State Construction Industry Press, 2018.

Hospital Sujing. *Study on collapse mechanism of concrete beam bridge under explosive load* [D]. Southeast University, 2019, Nanjing.

Li Minghong. *Study on the mechanism and evaluation method of explosion damage* [D]. Southeastern University, 2020.

Stewart M G, Mueller J. Terrorism risks for bridges in a multi-hazard environment[J]. *International Journal of Protective Structures*, 2014, 5(3): 275–289.

Tang Biao. *Study on explosive resistance of reinforced concrete pier column* [D]. Southeastern University, 2016.

Williamson E B. *Blast-resistant highway bridges: Design and detailing guidelines*[R]. Transportation Research Board, 2010.

Zong Zhou Hong, Tang Biao, Gao Gao, et al. Anti-performance test of RC pier column [J]. *Chinese Highway Journal*, 2017, 30(9): 51–60.

Advances in Measurement Technology, Disaster Prevention and Mitigation – Li & Mohd Yusof (Eds)
© 2023 The Author(s), ISBN: 978-1-032-36087-4

A numerical simulation study of curtain grouting boreholes for Gravity Dams

Yapeng Zhang*
Sinohydro Bureau12co., Ltd., Hangzhou, Hangzhou, Zhejiang, China

ABSTRACT: Two types of deformation can occur in rocks, i.e., brittle damage, and plastic damage, and a large number of irreversible deformations before damage occurs. The porosity of the rock is inversely proportional to the modulus of elasticity: the denser the rock, the greater the modulus of elasticity; the more developed the microfractures in the rock, the smaller the modulus of elasticity. The deformation of the rock varies greatly under different stress conditions: as the depth of the borehole becomes deeper, the pressure on the surrounding rock gradually increases, and the rock transitions from the brittle state to the plastic state; the deformation characteristics of the rock change with the nature of the load, as in the case of rocks under low-speed static load The deformation characteristics of the rock vary with the nature of the load, as the rock is plastic at low static loads and brittle at high dynamic loads.

1   INTRODUCTION

In the early stages of water conservancy project construction, the curtain grouting requirements of small-scale water conservancy projects could be achieved by adjusting the hole slope and drilling depth of the curtain grouting project and taking certain technical measures. However, in the face of increasing specifications and technical equipment requirements, the efficiency and quality of curtain grouting boreholes became increasingly important (Xiao et al. 2022). Research results are limited to traditional drilling methods, drilling considerations, etc. There is limited research on the causes of drilling offset, drilling offset conditions, and drilling offset rules for hydropower curtain works, and less research on curtain drilling methods, drilling protocol parameters, drilling process control, and special anti-slip drilling tools for special geological conditions (Hu 2022). Therefore, given the process characteristics and stratigraphic features of the Dagangshan Hydropower Station, the mechanical properties of drilling tools and pipe columns were specifically analyzed and studied in relation to the stratigraphic features of the work area.

2   ANALYSIS OF THE FORCES ON THE DRILL BIT WHEN BREAKING ROCK

2.1   *Flat-bottomed cylindrical drill bits*

Studies have shown that during the drilling of curtain grouting, when the cylindrical flat-bottomed indenter is pressed into the rock, the bit is subjected to a compressive force p provided by the drilling rig, resulting in elastic deformation within the rock, with the indenter coming into contact with the rock along the round face during the drilling process (Yang 2022). At the start of drilling, the pressure distribution on the contact surface of the bit is not uniform, resulting in stresses on the bit being mainly concentrated on the peripheral edges of the bit, which are mainly characterized by localized crushing or plastic deformation. During the subsequent press-in of the drill, the pressure on the drill bit tends to become more evenly distributed. During the drilling of a flat-bottomed

*Corresponding Author: 15632138258@163.com

DOI 10.1201/9781003330172-30

cylindrical bit, the pressure between the bit and the pressurized surface is not distributed as a constant, but as a function of the bit radius r (Figure 1).

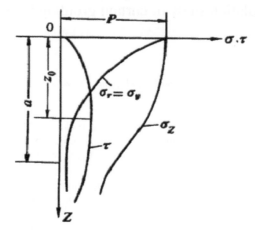

Figure 1. Stress distribution along the axis of symmetry in the compression of a flat bottomed cylindrical drill.

As shown above, when a flat bottom drill bit is pressed in, two extreme zones of stress are formed at the bottom of the borehole as follows: one is the edge of the indenter formed by the extreme zone of stress $z = 0$ and $r = a$; the other is the area of maximum shear stress where the extreme zone of stress is $z = z0$ and $r = 0$. During the drilling process, these two extreme stress zones are the source of rock breakage during drilling (Deng 2022). The maximum shear stress is about one-third of the uniform compression when the shear stress is on the z-axis when the shear stress reaches a maximum at two-thirds of the radius of the indenter. This leads many scholars who have studied the mechanism of rock crushing to conclude that the starting point of rock crushing is often located where the maximum shear stress is concentrated (Figure 2).

2.2 Spherical drill bits

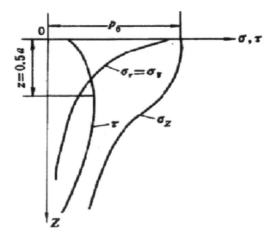

Figure 2. Stress distribution along the axis of symmetry of a spherical drill.

According to the Hertzian contact theory, the pressure distribution of a spherical drill bit can be calculated when it is on the surface of a semi-infinite body in Yaju. where P is the magnitude of the force acting on the spherical bit during drilling, R is the radius of the spherical bit, $1E$ and $2E$ are the moduli of elasticity of the indenter and the modulus of elasticity of the rock respectively, 1 and 2 are the Poisson's ratio of the bit and the Poisson's ratio of the rock during drilling, respectively.

As shown in Figure 2, the variation of each stress component with the z-axis: the shear stress τ starts to increase from small to large with z, reaching a maximum at the critical depth z0. It can be seen that the positive stress in the drill bit decreases as z increases as the hole is drilled deeper, but the stress reduction slows down at r and z.

2.3 Stress state of the rock below the indenter when axial and tangential forces act together

In rotary drilling, the rock crushing tool acts on the rock with both axial and tangential loads. Elastodynamic studies have shown that there is a significant difference in the distribution of stress contours within the elastic semi-infinite of the indenter under axial forces alone and under simultaneous axial and tangential forces (Ye 2022), with the former showing a symmetrical, uniform distribution and the latter a non-uniform, asymmetrical distribution. At the contact surface, the compressive stresses are located in front of the tangential force action, while the tensile stresses are located behind the tangential force action. Three stress zones are formed within the semi-infinite, namely the positive stress zone (I), the tensile stress zone (II), and the transition zone (III).

3 FINITE ELEMENT SIMULATION ANALYSIS OF ROCK STRESS STATE UNDER STATIC LOAD

3.1 Numerical analysis model

Generating geometric models directly in ANSYS software: When using ANSYS software to generate geometric models directly, the boundary conditions should first be set according to the actual situation of the construction area of the curtain grouting project of the hydropower station, and the shape and size of the model units in the curtain grouting construction project should be restricted, and the relevant commands of ANSYS software should be used to generate the relevant nodes and units.

Direct generation method: In the direct generation method, the nodes in the geometric model are first defined and then the units in the geometric model are defined after the nodes in the geometric model have been processed, resulting in a complete combined model (Ye 2022).

3.2 Isotropic rock finite element numerical simulation analysis

The isotropic rock and bottom drilling tool combination under static load was analyzed by numerical simulation: the Y-direction stress contours, shear stress contours, and equivalent force contours of the isotropic rock and rigid bottom drilling tool combination under static load were symmetrically and uniformly distributed (Deng 2022) the Y-direction displacement contours and combined displacement contours were also symmetrically and uniformly distributed (Figure 3–4).

Figure 3. Y direction stress contour map (left), Equivalent line map of shear stress (right).

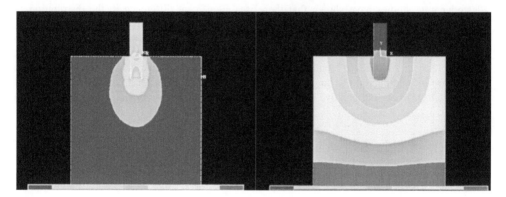

Figure 4. Equivalent stress contour map (left), Y direction displacement contour map (right).

3.3 *Anisotropic rock finite element numerical simulation analysis*

The stress analysis of the anisotropic rock and bottom drilling tool combination under static load was carried out by numerical simulation of finite elements (Sun & Li 2022). The Y-direction displacement contours and the combined displacement contours show that the displacement distribution is also elliptical and the direction of the long axis is perpendicular to the plane (Figure 5–6).

Figure 5. Y direction stress contour map (left), Contour of shear stress (right).

Figure 6.   Equivalent stress contour map (left), Y direction displacement contour map (right).

4   ANALYSIS OF ROCK CRUSHING PROCESSES

4.1   *Analysis of the rock crushing process under external load*

Drilling practice shows that the relationship between drilling speed Vm and drilling pressure P (shown in Figure 7) consists of three zones: in zone I, the rock surface is only surface grinding caused by friction, and because the drilling pressure is very small, the axial pressure on the cutting tool is far less than the rock hardness to invade the rock, the drilling speed is linearly related to the drilling pressure, which is called the surface crushing zone, and the drilling speed is very low; in zone II, the drilling pressure continues to increase so that the axial pressure on the cutting tool is equal to or greater than the rock hardness, and the rock reaches the condition of volume crushing (Xu & Mao 2021). The axial pressure is equal to or greater than the rock hardness, the rock reaches the condition of volume crushing, the cutting tool eats into the surface of the rock, large pieces of rock crumble away from the highest efficiency, and the drilling speed and drilling pressure is an exponential growth relationship called volume crushing zone; with the increase of drilling pressure, the rock from the surface crushing through a gradual process of fatigue crushing zone to volume crushing, the fatigue crushing zone drilling speed and drilling pressure is a curve relationship.

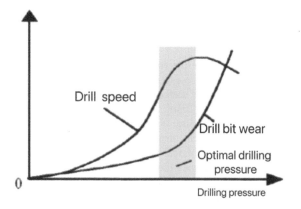

Figure 7.   Relationship between drilling speed and drilling pressure.

4.2   *Analysis of rock fragmentation under vertical load*

When a flat-bottomed indenter is applied to a rock under axial load, a first extreme zone of shear stress appears at the edge of the indenter with a conical ring-shaped crack extending towards the depth, which is cut off after a certain depth. The second dangerous shear stress extreme zone at a certain depth on the symmetry axis develops towards the edge, forming a limit zone similar to a sickle shape. The pressure and volume of the limit zone become larger as the axial load increases, and the forces acting on the outer side tend to drain and push out the rock at the edge also increase. As the compressive strength of the rock is much greater than the tensile and shear strength, the surrounding rock crumbles and forms a crushing cavity when the lateral pressure is equal to or greater than a certain limit value. The cone below the indenter is crushed after the surrounding rock has broken away, and the indenter then intrudes to a certain depth. The crushing mechanism of rocks under other shapes of indenter (spherical, square columnar) is similar to that described above (Figure 8).

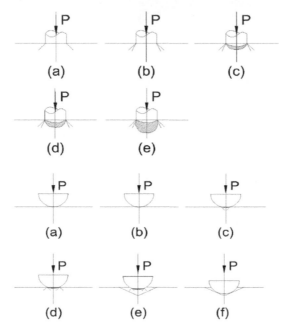

Figure 8.   Breaking process of rock under a static load of the flat bottom (left), Rock fracture processes under static load (right).

When the spherical indenter acts on the rock under axial load, the first contact between the cutting tool and the rock is a point, the external load increases, the cutting tool, and the rock produce elastic deformation, the contact surface begins to increase, the positive stress in the center of the contact surface is the largest, the first fissure appears, with the increase of external load, the contact surface increases, and a new fissure system parallel to the original fissure is produced, the external load increases again, the contact surface increases, the new fissure extends to the depth, the total elastic deformation of the spherical cutting The total elastic deformation of the tool is not proportional to the external load, when the external load continues to increase, the total value of the elastic deformation decreases, the stress increases, and the depth of the fissure extension increases; the subsequent crushing process is similar to that of the flat bottom compression mold.

5   CONCLUSION

This paper presents a numerical analysis of the stresses on flat-bottomed cylindrical drill bits and spherical drill bits in the curtain grouting drilling process from the perspective of rock mechanics mechanism analysis. By comparing the stress distribution states of the two types of rock, the analysis provides a theoretical basis for the analysis of the fracture mechanism at the bottom of small-diameter boreholes in deep holes by using ANSYS software for isotropic and anisotropic rocks.

REFERENCES

Deng G. (2022). Analysis of seepage control curtain grouting design of Juntiangou reservoir dam [J]. *Yunnan Hydropower*. 38(02): 71–74.

Hu Xiaoyun. (2022). Application of pressurized water test in the detection of initial seepage control effect of curtain grouting [J]. *Heilongjiang Water Conservancy Science and Technology*. 50(02): 169–171. DOI:10.14122/j.cnki.hskj.2022.02.046.

Liu Xinghua, Bai Chaowei, Qiao Bai. (2022). Design of curtain grouting for the dam base of the Lotus Terrace Hydropower Station Project[J]. *Groundwater*. 44(01):291–292. DOI:10.19807/j.cnki.DXS. 01-096.

Sun Jianwei, Li Yi. (2022). Study on the application of curtain grouting technology in the impermeability treatment of reservoir dams [J]. *Juye*. (01): 241–243.

Xiao Wei, Liu Shonan, Li Hongbin, Cai Hansheng. (2022). Design and construction of foundation curtain grouting for Jinsha hydropower station dam[J]. *Water Resources and Hydropower Express*. 43(03): 32–36. DOI:10.15974/j.cnki.slsdkb.2022.03.008.

Xu WF, Mao JX. (2021). Construction technology of curtain grouting for deep overburden layer under high head [C]//.2021 Technology innovation and development of water conservancy and hydropower foundation and foundation engineering. 76–79. DOI:10.26914/c.cnkihy.2021.048013.

Yang Yuru. (2022). Design and construction analysis of curtain grouting for dam removal and reinforcement of reservoirs [J]. *Northeast Water Conservancy and Hydropower*. 40(02): 25–27. doi:10.14124/j.cnki.dbslsd22-1097.2022.02.013.

Ye Zhi. (2022). Experimental analysis of curtain grouting of crushed concrete dams [J]. *Henan Water Conservancy and South-North Water Diversion*. 51(01): 73–74.

Disaster prevention and intelligent
disaster reduction project

Advances in Measurement Technology, Disaster Prevention and
Mitigation – Li & Mohd Yusof (Eds)
© 2023 The Author(s), ISBN: 978-1-032-36087-4

Construction and implementation of ontology model of dam break emergency plan

Dewei Yang*
Nanjing Hydraulic Research Institute, Nanjing, China
Dam Safety Management Center of the Ministry of Water Resources, Nanjing, China

Shuai Liu
Energy and Power Engineering School of Xihua University, Chengdu, Sichuan, China

Jinbao Sheng, Xuehui Peng & Huiwen Wang
Nanjing Hydraulic Research Institute, Nanjing, China
Dam Safety Management Center of the Ministry of Water Resources, Nanjing, China

ABSTRACT: The emergency plan for reservoir dam failure is one of the most important and effective non-engineering measures for reducing reservoir dam risks. The existing emergency plan for dam failure is primarily textual, which has some deficiencies and is inconvenient for emergency information management. In order to meet the needs of knowledge sharing, reuse, and interaction in the field of dam break emergency plan, a method for constructing the dam break emergency plan ontology model is proposed based on in-depth research on the existing ontology modeling methods and ontology construction criteria, which greatly improves emergency response and search efficiency of the plan.

1 INTRODUCTION

Ontology is a philosophical term and a philosophical problem investigating the essence of existence. In recent decades, this ontology has been applied to the computer field and has become increasingly significant in artificial intelligence, computer language, and database theory (Gruber 1993; Neches et al. 1991). In the computer field, ontology is a detailed description of conceptualization. It is often a formal vocabulary whose primary goal is to define professional terms in a certain field or domain and their relationships (Borst 1997; Studer et al. 1998). With the support of this series of professional words and their relationship, ontology can be applied to the research of the ontology model of the dam break emergency plan, and the efficiency of search, accumulation and sharing of the emergency plan will be greatly enhanced (Guarino 1998; 1997).

In general, ontology can not only be used to reason about the attributes of a domain, but also to define a domain. Over the past decade, ontology, which started from the subject of philosophy, has received extensive attention in knowledge engineering and related application fields. Ontology has been widely used in natural language understanding, software engineering, multi-problem solving, and multi-agent system, among other domains. Ontology has become a standard technology for knowledge representation, exchange, management, and reuse (Uschold & Gruninger 1996; Uschold et al. 1998).

2 PLANNING AND THINKING OF ONTOLOGY MODEL OF AN EMERGENCY PLAN FOR DAM BREAK

2.1 Purpose and hierarchy of ontology model

The purpose of developing the ontology model of the dam break emergency plan is to express the plan's concepts and their relationships using formal, standardized, and systematic computer

---

*Corresponding Author: dwyang@nhri.cn

DOI 10.1201/9781003330172-31

language and address requirements of information sharing, reuse, and interaction between different programs. The design of the ontology model (Gu 2004; Perez & Benjamins 1999) enables the intelligent administration and application of emergency plans for a dam break.

The ontology model of the dam break emergency plan constructed in this paper adopts a two-layer structure.

(1) The upper ontology model. The upper ontology is a universal basic concept to explain the knowledge of the dam break emergency plan, which can directly interact with other fields. In this paper, SUMO is selected as the upper ontology of the dam break emergency plan ontology model. SUMO epistatic ontology can classify the concepts to be described from abstract and concrete aspects, and SUMO epistatic ontology has rich expansibility.
(2) Domain ontology model. The domain ontology of the dam break emergency plan is based on the extension of the upper ontology of SUMO. Formal language is used to express the concepts and relationships between concepts in the field of the dam break emergency plan.

2.2 *Ontology modeling method process of dam break emergency plan*

The implementation of the dam break emergency plan ontology is a process of extracting and sorting out the knowledge in the dam break-related field. The objective is to develop an organic and intelligent concept, relationship, and example system in the field of the dam break emergency plan (Sanjurjo Rivo et al. 2015). The establishment of the ontology model for the dam break emergency plan mainly includes five steps listed below.

(1) Determine the discipline of ontology and define key concepts
   The important concepts involved in the ontology of the dam break emergency plan are extracted, and the extracted concepts are specified.
(2) Establish the conceptual structure of the ontology
   The concepts extracted in (1) are refined, and the relevant knowledge involved in the dam break emergency plan is expressed as accurately and concisely as possible. On this basis, a conceptual framework system of knowledge related to the dam break emergency plan is established (Doan et al. 2004).
(3) Design ontology model
   The design of the ontology model is mainly based on the conceptual framework system of the dam break emergency plan to define classes, class hierarchy, and class attributes. There are currently three primary methods for defining classes and class hierarchies: top-down, bottom-up, and hybrid. Since the hybrid method combines the benefits of the top-down method with the bottom-up method, the hybrid method is used to develop a comprehensive dam-break emergency plan model (Tutcher et al. 2015).
(4) Create an instance based on the ontology model
   Establishing an instance according to the ontology model, also known as semantic annotation, is the process of labeling an instance of a class based on the defined classes and attributes. There are two primary sorts of methods for semantic annotation: manual definition method and database information extraction method.
(5) Evaluation of Ontology
   The purpose of ontology evaluation is to assess the quality of the constructed ontology. The evaluation process for the ontology of the emergency plan for dam failure is to compare the ontology of the emergency plan for dam failure with reasonable standards through the index evaluation method and comprehensively evaluate various variables that affect the quality of the emergency plan for dam failure. The constructed ontology needs to be optimized.

3 ESTABLISHMENT OF ONTOLOGY MODEL OF DAM BREAK EMERGENCY PLAN

3.1 *The upper body of the dam break emergency plan*

SUMO is domain-independent, highly abstract, and encompasses the entirety of the objective world. In the class library of the SUMO upper ontology, the entity is the root class of all classes, and it is defined as all things in the objective world. It has two immediate subclasses: abstract and physical. The abstract class is used to convey the non-objectively existing concepts or information entities produced by the abstraction of human thinking, whereas the material class is used to express those entities that exist objectively in the real world and have space-time characteristics. This paper extracts relevant concepts and relationships from SUMO to construct an upper-level ontology for the field of the dam break emergency plan. On the one hand, it can cover all the concepts and relationships that may appear in the domain ontology; on the other hand, it ensures the simplicity and practicability of the upper-level ontology.

The abstract concept subtree of the SUMO ontology structure built in this paper is shown in Figure 1, and the physical substance concept subtree is shown in Figure 2. The primary concepts of SUMO are as follows:

(1) Event. Events are the process of things happening, developing, and altering. It is a subclass of both State-Transition and Process. An event can consist of multiple sub-events. Events are associated with the agent through hasPresence, and a single event may contain multiple sub-events.

(2) Roles. As a subclass of the abstract class, the role is used to express the role of the agent participating in the processing of an event. Participants are associated with their corresponding Roles through the attribute hasRole. Typically, the disposal of an event requires multiple agents, which may belong to different departments, and each agent performs a role through its role in the event. In the processing of complex events, a single agent may play multiple roles. Similarly, a role may also be shared by multiple agents. Therefore, the introduction of the concept of roles increases the adaptability of event management.

(3) Agent. The agent is a participant, a subclass of the object representing the action subject in the event handling process. An agent can be a person, an organization, or a system. The role of the Agent in event processing can be characterized by roles, and these roles are linked via the attribute hasRole.

(4) Organization. The organization represents the action subject in the event handling process, and it is a subclass of both the agent and the set. An organization can contain multiple departments and can also contain multiple sub-organizations or persons, which are linked through the attribute hasOrganizaiton.

(5) Intentional Process. It depicts the behavior of one or more agents driven by subjective consciousness. It is a subclass of the process.

Figure 1.   Abstract subtree of SUMO.

215

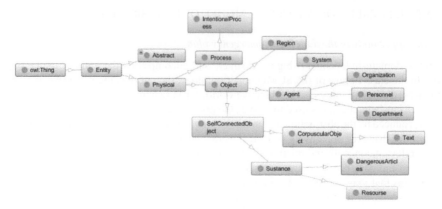

Figure 2.   Physical substance concept subtree of SUMO.

3.2   Ontology of dam break emergency plan field

3.2.1   Domain ontology modeling basis

The "Overall National Emergency Plan for Public Emergencies" promulgated by the State Council and the "Guidelines for the Preparation of Emergency Plans for Reservoir Dam Safety Management" promulgated by the Ministry of Water Resources serve as the main basis for the extraction of relevant concepts and the construction of the system framework in the field of emergency plans for dam failure. From the contents of the "National Overall Emergency Response Plan for Public Emergencies" and "Guidelines for the Preparation of Emergency Response Plans for Reservoir and Dam Safety Management," it is possible to summarize the concept of dam break emergencies, the classification and classification of dam break emergencies, and the dam break emergencies. The concepts include emergency plan, emergency organization system, and emergency support, among others. These contents serve as a foundation for constructing the ontology of emergency plans for dam failure.

3.2.2   Knowledge extraction and construction of domain ontology

The primary function of ontology knowledge extraction in the field of dam break emergency plan is to sort out and display the system of dam break emergency plan, the concepts it contains, and the relationships between concepts. Based on the SUMO upper ontology, this paper establishes the domain ontology of the dam break emergency plan.

The main concepts, attributes, and relationships of the dam break emergency plan domain ontology are described in detail as follows:

(1) Engineering Situation is a subclass of State inherited from the upper ontology. It comprises three subclasses: hydrology meteorological, hydraulic structures, and social and economic situation (reservoir downstream socio-economic profile).
(2) Dam Break Accident is a subclass that inherits events from the upper ontology. It is an enumerated class that replaces four subclasses: Flood Accident, Earthquake Accident, Engineering Accident, Terrorist Attacks, and War.
(3) Dam Break Accident contains an object property "hasLevel."
(4) Emergency Classification is a subclass that inherits Level from the upper ontology. It is an enumerated class with four subclasses: Level I, Level II, Level III, and Level IV.
(5) Dam Break Accident is associated with the class inherited in the upper ontology through the object attribute "hasLevel."

(6) Dam Break ERP is a man-made text that inherits a subclass of text from the upper ontology. It contains five data types of attributes, namely: ID (plan number), Title (plan name), Purpose (compiling Purpose), Principle (principle of preparation), and Date (published date).

(7) ER Organization is a subclass of Organization that inherits from the upper ontology. It contains five subclasses: Command Institution, Experts, Headquarters Office, Danger Elimination Team, and Rescue Team.

(8) ER Organization System is a subclass that inherits Set from the upper ontology and is linked to the five subclasses of ER Organization via the "has Organization" attribute, indicating that the ER Organization System is composed of Command Institution, Experts, Headquarters Office, Danger Elimination Team, Rescue Team.

(9) ER Role is a subclass that inherits the role from the upper body, and ER Personnel is a subclass that inherits Personnel from the upper body. ER Personnel is associated with ER Role via the "undertake" attribute, indicating that ER Personnel performs a certain ER Role within the emergency organization system.

(10) Operation Mechanism is a subclass that inherits Process from the upper body. It contains five subclasses: Forecast and Early Warning, Response, Disposal, Termination, and Notification.

(11) Early Warning Classification is a subclass that inherits Level from the upper ontology. It is an enumerated class with four subclasses, namely Warning LII (Level I is particularly severe red), Warning LII (level II is severe orange), Warning LIII (Grade III is more severe yellow), Warning LIV (grade IV is generally blue).

(12) Response Classification is a subclass that derives Level (classification) from the upper ontology. It is an enumerated class with four subclasses: Response LI, Response LII, Response LIII, and Response LIV.

(13) Guarantee is a subclass that inherits Process from the upper ontology. It contains eight subclasses: Funds Guarantee, Team Guarantee, Commodity Guarantee, Medical Guarantee, Living Guarantee, Traffic Guarantee, Communication Guarantee, Power Guarantee, and Security Guarantee.

(14) Funds, Rescue Materials and Equipment, Medical Supplies, Life Supplies, Transportation, Communication Equipment, and Power Equipment are subclasses that inherit the upper body resource.

(15) ERP Management Process is a subclass that inherits Process from the upper ontology. The ERP Management Process is associated with the Intentional Process through the attribute "has Stage."

(16) Draft, Check, Publish, Publicity, Train, and Manoeuvre are all subclasses that inherit the Intentional Process from the upper ontology. These intentional processes reflect several levels of the ERP Management Process, and they pass through the "has Draft Stage," "has Check Stage," "has Publish Stage," "has Publicity Stage," "has Train Stage," and "has Manoeuvre Stage," respectively, and are associated with the ERP Management Process.

4   EVALUATION OF ONTOLOGY MODEL OF DAM BREAK EMERGENCY PLAN

The purpose of ontology evaluation is to evaluate the quality of the constructed ontology. For the ontology of dam break emergency plan, the evaluation process compares the ontology of dam break emergency plan with reasonable standards using the index evaluation method, comprehensively evaluates various factors that affect the quality of the dam break emergency plan ontology, and draws an evaluation conclusion.

Ontology evaluation is divided into two categories: (1) Qualitative evaluation that tests the constructed ontology against reasonable standards and evaluates whether the ontology meets the standards. (2) Quantitative evaluation that calculates and determines whether the constructed ontology meets corresponding standards for relationship richness, attribute richness, and readability with the OntoQA method.

4.1 Qualitative evaluation

According to the ontology construction principle proposed by Gruber, the ontology of an emergency plan for dam failure can be evaluated qualitatively:

(1) Clarity and objectivity: The ontology of the dam break emergency plan objectively and abstractly describes the definitions of related concepts in the field of the dam break emergency plan. All the terms defined in the ontology are annotated with the language in the field of dam break emergency plan to guarantee that dam break emergency plan ontology may be applied without ambiguity.
(2) Completeness: All concepts in the dam break emergency plan ontology are compiled based on the "National Overall Emergency Plan for Public Emergencies" and "Guidelines for the Preparation of Emergency Plans for Reservoir Dam Safety Management." The legal and planning basis can express the meaning of all terms in the dam break emergency plan ontology.
(3) Consistency: The conclusion drawn from the relationship between the classes defined by the dam break emergency plan ontology is consistent with the meaning or principle expressed by the dam break emergency plan ontology, and there will be no contradiction.
(4) Maximum monotonicity: The classes defined in the dam break emergency plan ontology and the relationship between classes are the most basic concepts.
(5) Minimum Ontology Commitment: Adopt as few constraints as possible to ensure clarity and integrity of domain knowledge in the emergency plan for dam failure ontology description.

4.2 Quantitative evaluation

The OntoQA method is an ontology evaluation method presented by the LSDIS Laboratory of the University of Georgia (Pauwels et al. 2014; Samir & Arpinar 2007). It uses the calculation of a series of formulas to evaluate the quality of the ontology from different perspectives. This paper adopts three of the most general formulas. The richness and readability of the dam break emergency plan ontology are evaluated quantitatively from three perspectives:

(1) Richness of relationships

$$RR = \frac{P}{SC + P} \tag{1}$$

Where: RR represents the richness of relationships; P represents the number of non-parent-child relationships; SC represents the number of parent-child relationships.

In this paper, the number of non-parent-child relationships is 64, whereas the number of parent-child relationships is 26, and the calculated RR value is 0.71. The calculation results show that in the ontology of the emergency plan for dam failure, the proportion of non-parent-child relationship accounts for 0.71, exceeding 0.5. Therefore, the dam break emergency plan ontology can represent the relationship between classes in its domain and meet the richness of the ontology relationship.

(2) Richness of attributes

$$AR = \frac{att}{C} \tag{2}$$

Where: AR represents the richness of attributes; att represents the number of attributes; C represents the number of classes.

This paper contains 43 attributes and 62 classes, with an AR value of 0.69. The calculation results indicate that each class in the dam break emergency plan ontology contains an average of 0.69 attributes, exceeding the minimum requirement of 0.5 attributes. Therefore, the dam break emergency plan ontology has relatively rich domain knowledge, which satisfies the richness of ontology attributes.

(3) Readability

$$AR = |Q_{comment}| + |Q_{label}| \qquad (3)$$

Where: Rd represents readability; $Q_{comment}$ represents the number of comments; Q_{label} represents the number of labels. This article contains 62 annotations and 90 labels, with an estimated Rd value of 152. The calculation results show that the dam break emergency plan ontology contains 152 annotations and labels, which is more than five times the number of 28 annotations and labels in the SUMO upper ontology. It can be seen that the domain ontology is fully derived and extended based on the SUMO result. Therefore, the dam break emergency plan ontology meets the readability requirement of ontology construction.

At present, there is no unified international standard for ontology evaluation. The two evaluations employed in this paper can help guide the development of ontology and ensures that it meets the developer's goals to the fullest extent.

5 CONCLUSION

Construction of the ontology model of the dam break emergency plan is key to intelligentizing the dam break emergency plan. By analyzing the building and implementation method of the dam break emergency plan ontology model, emergency management in our country will evolve in a direction that is more modern, digital, and intelligent. The main contributions are as follows:

(1) Combining the ontology and construction purpose of the dam break emergency plan ontology model, the dam break emergency plan ontology model consists of two layers: the upper ontology model and the domain ontology model. From the perspective of ontology modeling semantics and knowledge level, an appropriate approach for developing the ontology model of the dam break emergency plan is proposed. In conjunction with this method, Protégé is used to model the ontology model of the dam break emergency plan.
(2) By interpreting SUMO's upper ontology concept and its characteristics and analyzing SUMO's core concepts, the SUMO ontology frame structure diagram is produced, and the SUMO upper ontology model is constructed. On the basis of the SUMO upper ontology model, appropriate concepts are selected and integrated with the knowledge extraction in the field of dam break emergency plan to develop an ontology model for the dam break emergency planning field.
(3) Based on comprehensive consideration of semantic expression ability, logical reasoning ability, and complexity of the relationship between dam break emergency plan process and field, in the ontology coding part of the dam break emergency plan, axioms, concepts/classes, relationships/attributes, notes, and explanations are implemented respectively.
(4) Based on ontology evaluation, the meanings of qualitative and quantitative evaluation of ontology are elucidated, and the constructed ontology model of the emergency plan for dam failure is evaluated. In conclusion, it is mentioned that ontology is repetitious and requires continuous maintenance and refinement.

ACKNOWLEDGMENTS

This study was funded by the National Natural Science Foundation of China (51909174) and NHRI, a national research institute for public services (Y722003 and Y722007).

REFERENCES

Doan B L, Bourda Y, Bennacer. *Using OWL to Describe Pedagogical Resources*[J]. Advanced Learning Technologies, 2004. *Proceedings IEEE International*, 2004:102–108.
Gu Fang. *Research on Ontology Design Method in Multidisciplinary Domain* [D]. Beijing: Chinese Academy of Sciences, 2004: 2–4.

Jonathan Tutcher, John M Easton, Clive Roberts. *Enabling Data Integration in the Rail Industry Using RDF and OWL: The RaCoOn Ontology*[J]. ASCE-ASME Journal of Risk and Uncertainty in Engineering Systems, 2015, 22: 2–12.

M. Uschold, M. Gruninger. *Ontologies: Principles, Methods, and Applications*[J]. The Knowledge Engineering Review, 1996 11(2): 93–155.

M. Uschold, M. King, S. Moralee, Y. Zorgios. *The Enterprise Ontology*[J]. The Knowledge Engineering Review 1998, 13(1): 31–89.

N. Guarino. *Formal Ontology and Information Systems*[J]. Proceedings of FOIS' 98-Formal Ontology in Information Systems, Trento, Italy, 1998:1–4.

N. Guarino. *Semantic Matching: Formal Ontological Distinctions for Information Organization* [M]. Springer Verlag, 1997: 139–170.

P Pauwels, E Corry, J O'Donnell *Representing Sim Model in the Web Ontology Language* [J]Computing in Civil and Building Engineering, 2014(22): 71–78.

Perez A Z, Benjamins V R. *Overview of Knowledge Sharing and Reuse Components: Ontologies and Problem Solving Methods*[J]. In: Proceedings of the IJCAI-99 Workshop on Ontologies and Problem-Solving Methods (KRRS). Stockholm, Sweden, 1999:1–15.

R. Neches, R. E. Fikes, T. Finin, T. R. Gruber, T. Senator, W. R. Swartout. *Enabling Technology for Knowledge Sharing*[J]. AI Magazine 1991, 12(3): 36–56.

R. Studer, V. R. Benjamins, and D. Fensel. *Knowledge Engineering, Principles Methods. Data and Knowledge Engineering* 1998, 25(1–2):161–197.

Sanjurjo Rivo M, Sánchez-Arriaga G, Peláez J *Efficient Computation of Current Collection in Bare Electrodynamic Tethers in and beyond OML Regime*[J]. Journal of Aerospace Engineering, 2015, 28(6): 1–7.

T. Gruber. *A Translation Approach to Portable Ontology Specifications*[J]. Knowledge Acquisition, 1993, 5(2):199–220.

Tartir Samir, Arpinar. I. *Ontology Evaluation and Ranking Using OntoQA* [C]. Semantic Computing (ICSC), 2007 1st IEEE International Conference on, California, 2007.

W. N. Borst. *Construction of Engineering Ontologies*[D]. University of Twenty, Enschede 1997:5–6.

Advances in Measurement Technology, Disaster Prevention and
Mitigation – Li & Mohd Yusof (Eds)
© 2023 The Author(s), ISBN: 978-1-032-36087-4

Research on the spatial distribution of natural disasters harmful to China's inland channel

Baoying Mu, Lixin Lu* & Yuchuan Wang
China Waterborne Transport Research Institute, Beijing, China

ABSTRACT: The channel is an important transportation infrastructure and the foundation of water transportation, especially the high-level channel, which plays an important role in the development of shipping. China suffers from various natural disasters every year. This article counts the natural disasters suffered by channels above the third level of inland rivers in China from 1978 to 2020. There are a total of about 868 natural disasters, which are mainly divided into seven types. Through the statistics of the distribution of various natural disasters, the study puts forward the conclusion that the overall distribution of disasters is "heavy in the south and light in the north, and light in the east and west," and the disasters that have suffered more are flood disasters and drought disasters.

1 INTRODUCTION

China has a vast territory and is severely affected by natural disasters (Shen et al. 2018). There are various types of disasters with a high incidence rate and a wide range of areas involved. China is one of the countries with the most developed shipping in the world. The Yangtze River and Pearl River systems have a large freight volume and a large number of ships. From the Beijing-Hangzhou Grand Canal in the Sui and Tang Dynasties to the Yangtze River Delta and the Pearl River Delta, the connectivity of water transportation has always been one of the important factors driving China's economic and social development. The channel is an important infrastructure for shipping, which is greatly affected by weather and water level. Once a natural disaster occurs, the channel will be damaged, ships will be interrupted, and freight will be interrupted, resulting in greater economic losses and serious casualties. The navigable mileage of China's inland channel is as high as 128,000 kilometers, covering the whole country. The channel is crisscrossed from Northeast China, North China to Central China, East China, and South China. Since the 14th Five-Year Plan period, the "Outline for Building a Strong Transportation Country" and "Outline for National Comprehensive Three-dimensional Transportation Network Planning" have put forward new requirements for water transportation and channel. This paper explores and analyzes the spatial distribution of various natural disasters suffered by the channel. The natural disasters suffered by the channel mainly include flood disasters, drought disasters, typhoon disasters, rainstorm disasters, landslide disasters, gale disasters, and collapse disasters. The occurrence frequency and location of various disasters are different, but the spatial distribution has certain rules to follow. This paper makes statistics of various natural disasters suffered by China's third-level and above channel from 1978 to 2020, and through analysis, studies the spatial distribution of various disasters and provides a reference for the development of the channel.

2 OVERALL SPATIAL DISTRIBUTION

The natural disasters suffered by China's inland channel have a certain spatial aggregation, and the overall situation is that "the south is heavy and the north is light, and the east is heavy, and the

---

*Corresponding Author: lulixin@wti.ac.cn

west is light." According to statistics, from 1978 to 2020, China's inland channel suffered a total of 868 natural disasters, including 486 flood disasters, 298 drought disasters, 48 typhoon disasters, 26 rainstorm disasters, seven landslide disasters, two collapse disasters, and one gale disaster.

The main inland channel that has suffered more natural disasters are Xiangjiang, Hanjiang, Ganjiang, Xijiang shipping trunk lines, Yuanshui River, Beijing-Hangzhou Canal, Songhua River, Minjiang Tonghai Channel, Minjiang Main Stream Taima Channel, Huaihe Main Stream Channel, Xinjiang, etc. The 65 channels mainly involve 14 provinces, including Hunan Province, Hubei Province, Jiangxi Province, Guangxi Zhuang Autonomous Region, Guangdong Province, Shandong Province, Fujian Province, Jiangsu Province, Chongqing City, Heilongjiang Province, Anhui Province, Liaoning Province, Sichuan Province, and Zhejiang Province. According to the number of natural disasters, the occurrence of natural disasters in each province is classified, as shown in Figures 1 and 2.

According to the statistics of the channel, the six channels of Xiangjiang, Hanjiang, Ganjiang, Xijiang, Yuanshui and Beijing-Hangzhou Canal suffered particularly serious natural disasters. A total of 618 natural disasters occurred, accounting for 71.20% of the total natural disasters in the country, resulting in direct economic losses. A total of 544.4366 million yuan, accounting for 48.67%.

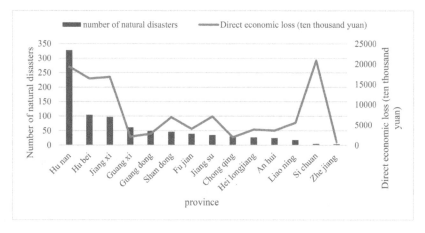

Figure 1. Distribution of the total number of disasters in the country.

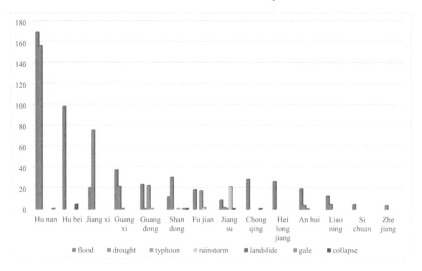

Figure 2. Number of various types of disasters.

3 SPATIAL DISTRIBUTION OF FLOOD DISASTERS

Flood disasters are distributed all over the country, from Heilongjiang Province in the north to Guangdong Province in the south, involving Hunan Province, Hubei Province, Guangxi Zhuang Autonomous Region, Chongqing City, Heilongjiang Province, Guangdong Province, Jiangxi Province, Anhui Province, Fujian Province, Liaoning Province, Shandong Province, Jiangsu Province, Sichuan Province and other 13 provinces. Among them, Hunan Province and Hubei Province are particularly serious, with 269 flood disasters, accounting for 55.35% of the total number of flood disasters in the country.

The channel with more flood disasters includes the Xiangjiang, Hanjiang, Xijiang shipping trunks, Yuanshui, Songhua River, Ganjiang, Huaihe trunks, and the Minjiang River to the sea and the Beijing-Hangzhou Canal. Among them, the Xiangjiang and Hanjiang floods were particularly serious, with a total of 220 incidents, accounting for 45.27% of the total number of floods in the country.

Floods and waterlogging disasters in the Xiangjiang River channel occurred in the entire channel section, with no specific disaster location. It involved five prefecture-level cities, including Yueyang, Zhuzhou, Changsha, Hengyang, and Xiangtan in Hunan Province. It generally occurred from June to July, with an average duration of 47. The average economic loss caused by each disaster is 422,500 yuan.

The locations of flood disasters on the Han River channel involve 13 channel segments, including Tianmen, Zhongxiang, Qianjiang, upper Hanjiang, Xiantao, and Shaxiang, and are located in Jingmen, Wuhan, Xiangyang, Xiaogan and directly under the provincial government in Hubei Province. The county-level administrative divisions are Tianmen City, Qianjiang City, and Xiantao City, which generally cause damage to remediation buildings and slope protection or loss of navigation marks, which mostly occur in September-October, and occasionally occur in February and June-July, with an average duration of for 47 days, the average economic loss caused by each disaster was 1.2413 million yuan.

This paper counts the losses caused by flood disasters in each province, as shown in Figure 3. It can be seen that the average direct economic loss caused by each disaster and the average daily direct economic loss in Sichuan Province are much higher than in other provinces. This is because there are many remediation buildings in Sichuan Province. After the flood disaster caused the damage to the remediation buildings, the repair workload was very large, which directly caused high economic losses.

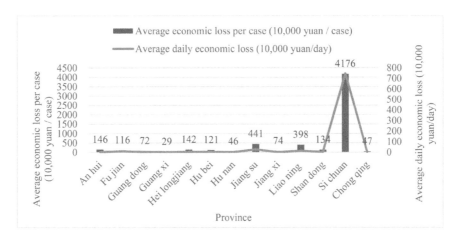

Figure 3.   Statistical chart of economic losses caused by flood disasters.

4   SPATIAL DISTRIBUTION OF DROUGHT DISASTERS

Drought disasters mostly occurred in central and southern China, including Hunan, Jiangxi, Shandong, Guangxi, Anhui, Jiangsu, and Guangdong. In addition, drought disasters occasionally occurred in the east port section of the Yalu River channel and the urban section of Dandong in Liaoning Province. Drought disasters were particularly severe in Hunan Province, with 157 cases accounting for 52.68% of the national drought disasters.

The channel with more drought disasters is mainly the Xiangjiang River, Ganjiang River, Yuanshui River, Beijing-Hangzhou Canal, and the shipping trunk line of the Xijiang River. Among them, the Xiangjiang River and Ganjiang River suffered more serious natural disasters, with a total of 185 disasters, accounting for 62.08% of the total number of drought disasters in the country.

The distribution of drought disasters in the Xiangjiang channel is relatively scattered. Among them, the Xiangyin Channel management section Haohekou-Xiangyin-Lulintan, Haohekou-Linzikou-Lulintan, Xiangtan channel management section, and Tiejiaozui channel management section Xiangyin County, Zhuzhou Channel Management Office Duan Tianyuan District, Changsha Channel Management Office Duan Tianyuan District, Lukou Channel Management Office Duan Tianyuan District, and Wangcheng Channel Management Office Duan Wangcheng District experienced relatively more drought disasters. Drought disasters in the Xiangjiang River Channel occur throughout the year, mainly in summer and autumn from June to November, with an average duration of 45 days, and the duration of drought in spring and winter is slightly shorter than that in summer and autumn.

The drought disasters in the Ganjiang channel are mainly distributed in the Fengcheng section, the new section, the Nanchang County section, the Yongxiu County section, the Zhangshu City section, the Xiajiang County section, and the Donghu section, etc., involving 16 channel sections. Most occurred in September-November, with an average duration of 62 days.

This paper analyzes the direct economic losses caused by drought disasters in various provinces, as shown in Figure 4. The average economic losses in Jiangsu, Jiangxi and Shandong provinces are relatively high, about 1–2 million yuan per case, and the average economic losses in Guangdong and Guangxi are relatively low. Overall, the direct economic losses caused by drought disasters are lower than those caused by flood disasters.

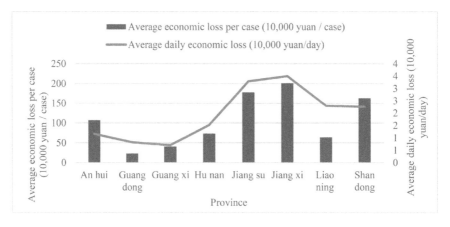

Figure 4.   Statistical chart of economic losses caused by drought disasters.

5   SPATIAL DISTRIBUTION OF OTHER DISASTERS

Typhoon disasters are mainly distributed in the eastern coastal areas, mainly in the southeast. Guangdong and Fujian provinces are the most severely affected by typhoon disasters. Zhejiang,

Jiangsu, Anhui, and Guangxi will also be affected by typhoons, with relatively few occurrences. The main channel affected by the typhoon disaster includes the Minjiang-to-sea channel, the Xijiang shipping trunk line, and the Taiwan-Malaysia channel, the mainstream of the Minjiang River.

The rainstorm disasters mainly involved nine channels, including the Beijing-Hangzhou Canal, the Yanhe River, and the Lianshen Line. Among them, the Beijing-Hangzhou Canal suffered eight torrential rain disasters, resulting in siltation of the channel, collapse of revetments or lack of topping fractures.

A total of seven landslide disasters occurred, involving the Badong Channel, Jibazi Channel, Qingtan Channel, Qinggan River Channel in Hubei Province, and Yanhe Channel in Jiangsu Province.

A total of two collapse disasters occurred, involving the entry and exit channel of the Binhu operation area in Shandong Province and the Lishui channel in Hunan Province.

There was only one gale disaster, which was located in the Zaozhuang section of the Beijing-Hangzhou Canal and downstream of Wannian Gate.

Among the losses caused by other disasters, the average economic loss of Hubei and Chongqing is higher than other provinces. This is due to the frequent landslide disaster in Hubei and Chongqing, which leads to higher average loss.

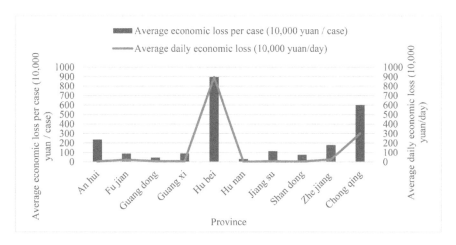

Figure 5.   Statistical chart of economic losses caused by other disasters.

6   CONCLUSION

This paper studies the spatial distribution law of natural disasters suffered by inland channels in China. There are many types of natural disasters suffered by channel, and their distribution is uneven. According to statistical data, it is mainly flood disasters and drought disasters; flood disasters are mainly concentrated in central and southern China, and drought disasters are mainly distributed in central and southern China. Multiple disasters occur simultaneously in some areas, causing serious losses. The occurrence of disasters is closely related to weather conditions. With global warming, the causes of natural disasters are more complicated, and extremely severe weather occasionally occurs.

REFERENCES

Cvetkoviæ, V., & Dragiceviæ, S. (2014). Spatial and temporal distribution of natural disasters. *Journal of the Geographical Institute Jovan Cvijic*, SASA, 64(3), 293–309.

De Juan, A., Pierskalla, J., & Schwarz, E. (2020). Natural disasters, aid distribution, and social conflict–Micro-level evidence from the 2015 earthquake in Nepal. *World Development*, 126, 104715.

El Morjani, Z. E. A., Ebener, S., Boos, J., Abdel Ghaffar, E., & Musani, A. (2007). Modeling the spatial distribution of five natural hazards in the context of the WHO/EMRO Atlas of Disaster Risk as a step towards the reduction of the health impact related to disasters. *International journal of health geographics*, 6(1), 1–28.

Liu, Y., Yang, Y., & Li, L. I. (2012). Major natural disasters and their spatio-temporal variation in the history of China. *Journal of Geographical Sciences*, 22(6), 963–976.

Pei, W., Guo, X., Ren, Y., & Liu, H. (2021). Study on the optimization of staple crops spatial distribution in China under the influence of natural disasters. *Journal of Cleaner Production*, 278, 123548.

Shen, S., Cheng, C., Song, C., Yang, J., Yang, S., Su, K., ... & Chen, X. (2018). Spatial distribution patterns of global natural disasters based on blustering. *Natural hazards*, 92(3), 1809–1820.

Wang, Q., Zhang, Q. P., Liu, Y. Y., Tong, L. J., Zhang, Y. Z., Li, X. Y., & Li, J. L. (2020). Characterizing the spatial distribution of typical natural disaster vulnerability in China from 2010 to 2017. *Natural Hazards*, 100(1), 3–15.

Advances in Measurement Technology, Disaster Prevention and Mitigation – Li & Mohd Yusof (Eds)
© 2023 The Author(s), ISBN: 978-1-032-36087-4

Application of super map technology to geological visualization in landslide monitoring

Jun Zhou
Beijing SuperMap Software Co., Ltd, Beijing, China

Xunsheng Bian & Qiong Liu*
Hainan Hydrogeological Engineering Investigation Institute, Haikou, Hainan, China

ABSTRACT: In this study, the integrated storage and management of the basic geological and geographic information in the forms of geological stratigraphic maps, images, tables, and word reports required by deformation monitoring were realized based on the new SuperMap GIS technology. Moreover, a 3D regional geological structural model was established using the data of boreholes and cross-sections. Then, the spatial distribution characteristics of regional geological tectonic units were intuitively and vividly expressed *via* the 3D visualization technology, realizing its integration and display. This technology serves the whole-process deformation monitoring, solves the deficiencies of traditional technological means, and substantially improves the informatization level during the monitoring of landslide-induced deformation.

1 INTRODUCTION

Landslide is a specifically formed natural phenomenon (China Building Industry Press 2014). The soil masses or rocks on a slope slide down as a whole or dispersedly along a certain soft plane or soft belt under the effect of gravity due to various factors like river erosion, groundwater activity, rainwater soaking, earthquake, and artificial slope cutting (China Building Industry Press 2013). Movable rocks (soil masses) are referred to as displaced bodies or sliding bodies, while unmovable underlying rocks (soil masses) as the slide bed (China Water Power Press 2006).

In the present landslide control, monitoring, analysis, and design work, 2D reconnaissance results are generated using relevant computational analysis software like Lizheng software to facilitate schematic design, construction, and monitoring (Modern Mining 2021, v.37; No.624(04):143–146). However, the traditional working modes are subjected to defects like nonintuitive geological findings, inaccurate design schemes, high field operation risks, and a low informatization degree, thus failing to pertinently guide the field dynamic design and construction and optimize the design parameters, not to mention to meet the timeliness demands of constructional and operational safety (Modern Mining 2017, v. 33; No.574(02):188–192). Hence, exploring geological visual modeling for landslide monitoring based on new technologies like GIS to realize digital multi-stage (landslide control design, construction, monitoring, operation, and use) seamless integration and the display will be of significant social value and economic value (Geological Publishing House 2019: 37–38).

In this study, the integrated storage and management of basic geological and geographical information in the forms of geological stratigraphic maps, images, tables, and word reports required by deformation monitoring were realized based on the new SuperMap GIS technology and combining the previous monitoring work of landslide-induced deformation(Electronic Technology & Software Engineering 2014 (16): 221). In addition, a 3D regional geological structural model was

*Corresponding Author: liuqiong2006@vip.qq.com

DOI 10.1201/9781003330172-33

constructed using the data of boreholes and cross-sections. Then, the spatial distribution characteristics of regional geological tectonic units were intuitively and vividly expressed by virtue of the 3D visualization technology, thus realizing its integration and display (China Energy and Environmental Protection 2020, v.42; No.289(01):21–24+29).

2 ADVANTAGES OF SUPERMAP IN GEOLOGICAL VISUALIZATION

SuperMap GIS 10i is a large-scale GIS basic software series developed by Beijing SuperMap Software Co., Ltd, and a 2D and 3D integrated tool software for acquiring, storing, managing, analyzing, processing, mapping, and visualizing spatial data (Shanxi Coking Coal Science & Technology 2015, v.39; No.246(12):52–55).

The traditional geological results related to deformation monitoring are limited to 2D maps and tables, which cannot vividly and intuitively describe the 3D spatial positions of strata and the deformation at different construction nodes(Building Technique Development 2018, 45 (13): 76–77). The SuperMap GIS 10i software is capable of integrated storage and management of basic geological and geographical information in the forms of geological stratigraphic maps, images, tables, and word reports required by deformation monitoring. Based on the data of boreholes and cross-sections, a 3D regional geological structural model can be established *via* the software, thus intuitively and vividly expressing the spatial distribution characteristics of regional geological tectonic units through 3D visualization technology (Marine Geology Frontiers 2015, 31 (5): 57–62). The specific advantages of SuperMap GIS 10i-based geological modeling are manifested as follows:

(1) Geological bodies are constructed backstage to support real-time sectioning and viewing inside geological bodies.
(2) Real-time analysis functions based on geological bodies are provided, *e.g.*, real-time clipping of geological bodies through a polygon to view the distribution of clipped regions. Besides, cylindrical and polygonal real-time excavated geological bodies are supplied to perform virtual hole drilling in such geological bodies.
(3) Given the thin rock strata in geological bodies, such geological bodies can be stretched along the direction Z to observe the distribution of rock strata in them clearly.
(4) SuperMap GIS 10i can also be used to edit and replace the textures of geological bodies and replace the rock strata display effect with real texture images to more intuitively express the internal structural and attribute information of rock strata.
(5) Moreover, different rock strata in geological bodies can be blown open to check the internal structure.

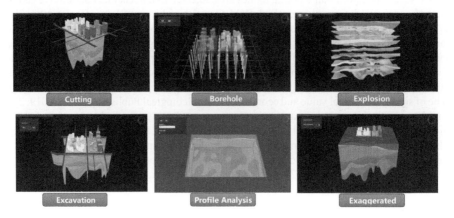

Figure 1. Functions related to visualization of geological bodies.

3 TECHNICAL ROUTE

The main technical route adopted for the geological modeling and application *via* SuperMap GIS 10i is described as follows: (1) The existing borehole data of each geological stratum are imported to generate stratified 3D geological observation point datasets. (2) Datasets are loaded using the function of "geological body construction," and geological models are constructed and saved. (3) A point dataset of "geological borehole analysis" is created, and the diameter and height of boreholes are self-defined through the function of "geological boreholes" to acquire the analysis results. (4) A line/plane dataset is created for "cross-sectional geological analysis," and the cross-sectional thickness and height are self-defined using the function of "geological cross-sections" to obtain the analysis results. The technical route is displayed in Figure 2.

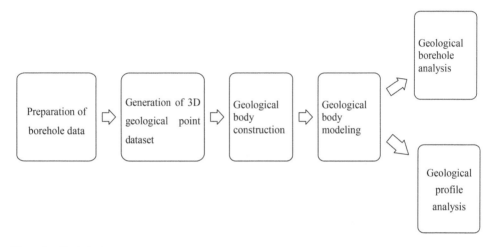

Figure 2.   Technical route.

The main specific steps are presented as follows:

(1) Create a new file-type data source in SuperMap iDesktop, click the right key to import the dataset, check "import as spatial data," choose the corresponding longitude, latitude, and elevation fields, and import the data of each geological stratum in batches.
(2) Click the pull-down button "geological body" in the "model" group of the "3D data" tab, choose "geological body construction" in the popup pull-down menu, click "+" at the top left corner to load the dataset for the to-be-established geological body, and check "color scheme" to generate a geological body model GeoBodyResult with clear and intuitive geological stratification.
(3) Click the pull-down button "geological body" in the "model" group of the "3D data" tab, choose "geological borehole analysis" in the popup pull-down menu, select "source data" and "borehole data," and set the borehole diameter and height to generate BoreholeResult analysis results.
(4) Click the pull-down button "geological body" in the "model" group of the "3D data" tab, choose "cross-sectional geological analysis" in the popup pull-down menu, select "source data" and "cross-sectional data," and set the cross-sectional thickness and height to generate Cross-SectionResult analysis results.

4 PROJECT EXAMPLE

4.1 *Project profile*

The project site is located on the slope belt in the north of a plaza in Chengmai County, with transitional geomorphic units from a volcanic weathering platform into an III-stage terrace in Nandu River. Sliding multiple times, the slide body generally presents the morphologies of a gentle slope with a slope gradient of 10°–20°, accompanied by local platform morphologies. In the reconnaissance area, above the trailing edge of the northern landslide is vacant land, which is a gentle slope belt with an elevation ranging from 93 m to 102 m, and tensile cracks are observed at the trailing edge. Due to multiple times of sliding, two landslide platforms and one landslide ridge, as well as multiple transverse cracks, are formed on the slide body. The first platform at the trailing edge is about 45 m in length, 15 m in width, and 87–88 m in elevation, and the landslide ridge is about 0.7–1.80 m in height. For the second platform, the length, width, and elevation are about 88 m, 22 m, and 81–82.5 m, respectively. Eucalyptuses on the right side of the main slide body lie on all sides like a "drunkard forest," and gentle slopes exist on other sides. Some accumulation bodies are manually cleared, the front edge of the main slide body is turned into a broken-line slope after artificial reforming, the surface is covered by grass, and the slope angle is about 30°–50°. The accumulation area at the front edge is manually cleared, a drainage ditch is constructed, the elevation is about 71.2–72.4 m, and no longitudinal cracks are observed. The field photos are as follows:

Figure 3.   Left side of the slide body.

Figure 4.   Middle part of the slide body.

In this project, the stratum mainly consists of Quaternary loose accumulation layer (Q^{ml}), basalt weathered soil (Q^{el}), tuff weathered soil (Q^{el}), and Quaternary lower Pleistocene alluvial-diluvial soil (Q_1^{al+pl}), and the strata closely related to the project, from new to old, are as follows:

(1) Artificial fill (Q_4^{ml}): Brown-red, loose and uncompacted artificial fill is distributed in all boreholes except zk2. The main component is plain fill. In the local cross-section above the platform of the slide body, construction wastes 1.0–2.5 m in thickness are contained, namely, concrete blocks and gravels. The landslide accumulation area is mainly filled with clayey soil and medium-fine sands for about three years. The exposed thickness is 0.50–9.40 m.

(2) Silty clay (Q^{el}): Silty clay is basalt-weathered soil, which is mainly brown-red and grey and only exposed in zk1 and zk2. Being plastic, the silty clay is mainly composed of powdery clay particles intermingled with weathered fragments, and the particle size of such fragments is 2–5 cm, along with edges and corners. In addition, the fragments are mainly distributed at the trailing edge of the slide body, with an exposed thickness of 3.60–4.30 m.

(3) Clay (Q^{el}): Clay is weathered tuff soil exposed in all boreholes except zk8, zk10, and zk14-zk17. It is yellow, light yellow, or brown-red. Being plastic and presenting local hard plastic characters, it is mainly composed of powdery clay particles, along with a small quantity of medium-fine quartz particles, which are locally enriched. Zk1 is not exposed. In the already exposed boreholes, the exposed thickness is 0.70–12.20 m. During exploratory boring, the clay at this stratum experiences hole shrinkage when encountering water, with medium expansion potential, as evidenced by the expansibility test results.

(4) Clay (Q_1^{al+pl}): The clay, which is only exposed in zk8, is mainly yellow and plastic with the main component of clay particles and the exposed thickness of 0.70 m.

(5) Coarse sands (Q_1^{al+pl}): Exposed in zk4, zk6-8, and zk14-15, coarse sands are mainly yellow, slight-to-medium dense, and slightly wet to wet, with quartz textures. Being sub-round, the coarse sands are mainly medium-coarse grains, followed by gravels, along with a small number of pebbles locally distributed. The exposed thickness is 0.70–3.50 m.

(6) Clay (Q_1^{al+pl}): Unexposed in zk1 and missing in zk16-17, the clay is light yellow and light green with certain plasticity. The main components are powdery particles and clay particles locally intermingled within thin-layered silts. The exposed thickness is 0.80–4.30 m.

(7) Medium sands (Q_1^{al+pl}): Medium sands are distributed in all cross-sections except the trailing edge of the slide body. It is mainly grey white, medium dense, and saturated with quartz textures. Being sub-round, medium sands are mainly medium-coarse grains, followed by gravels. The exposed thickness is 1.00–4.00 m.

(8) Silt (Q_1^{al+pl}): The silts are mainly distributed at the left side of the front edge of the slide body. They are mainly orange-yellow, loose and slightly dense, and saturated. The quartz textures are dominated by powdery particles, followed by fine particles. The exposed thickness is 1.60–2.50 m.

(9) Silt (Q_1^{al+pl}): The silts are distributed (yet unexposed) in all cross-sections except the trailing edge of the slide body. They are orange-yellow or light red, medium dense, and saturated. The quartz textures are mainly fine silty particles, followed by medium particles, and bulky clayey soil is locally distributed. The exposed thickness is 1.00–11.40 m.

(10) Silty clay (Q_1^{al+pl}): The silty clay, which is only exposed in zk7 and zk17, is yellow and plastic with the main component of silty clay particles. The exposed thickness is 1.00–1.40 m.

(11) Silt (Q_1^{al+pl}): Only exposed in zk7 and zk17, the silts are light red or purplish red, medium dense, saturated, and sub-round with quartz textures and main components of fine silty particles. The exposed thickness is 2.00–9.40 m.

Representative engineering geological cross-sections are displayed as follows (Figure 5).

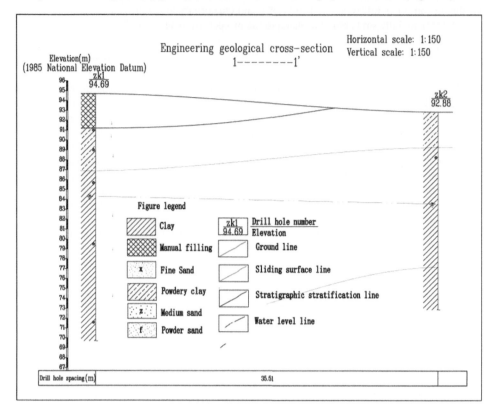

Figure 5.   Representative engineering geological cross-sectional diagram.

4.2   *Geological modeling*

(1) Layering of borehole data

The attribute table data of boreholes need to be layered, thus forming seven copies of layered table data, and a total of 17 sampling points exist at each layer.

(2) Generation of 3D geological point data

The table data are imported into SuperMap iDesktop data source to generate a 3D geological point dataset.

(3) Geological body construction

The "geological body" construction window is opened. The pull-down button "geological body" in the "model" group of the "3D data" tab is clicked, and the "geological body construction" in the popup pull-down menu is chosen. Then, the geological body results (GeoBodyResult) are loaded into a 3D scene for display.

4.3   *Visualized application of geological bodies*

(1) Geological borehole analysis

The "geological borehole analysis" function is implemented by setting borehole parameters according to the point dataset, which supports 2D and 3D points. With this function, the borehole diameter and height can be self-defined to acquire real-time analysis results of different borehole parameters.

232

The pull-down button "geological body" in the "model" group of the "3D data" tab is clicked, and the "geological borehole analysis" in the popup pull-down menu is chosen.

Source data: The geological body dataset requiring geological borehole analysis is selected.

Borehole data: The borehole dataset is chosen.

Parameter setting: The borehole diameter and elevation is set.

Result setting: The data source and dataset saved in result data are set. The dataset name is BoreholeResult by default or self-defined.

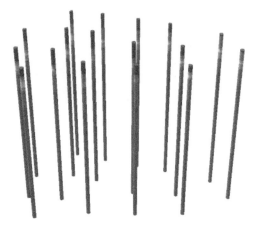

Figure 6.   Effect diagram of geological borehole analysis.

(2) Geological cross-sectional analysis

As for the "geological cross-sectional analysis" function, the geological model of a designated cross-section is extracted from the cross-sectional dataset. The cross-sectional dataset supports 2D and 3D lines and planes. The cross-sectional thickness and height can be self-defined through this function to acquire the real-time analysis results of different cross-sections.

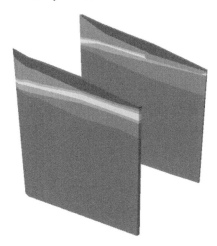

Figure 7.   Effect diagram of geological cross-sectional analysis.

Thanks to the integration and display of 3D geological visualization, the geological conditions, stratification line, groundwater line, etc. of this project become more intuitive, thus providing technical support for the follow-up design, construction, and security guarantee work of monitoring data results, analysis results, early warning and forecasting, and risk assessment.

Figure 8.   Effect diagram of geological body analysis.

5   CONCLUSIONS AND EXPECTATIONS

In this study, 3D visualized geological modeling of the landslide was conducted based on SuperMap GIS 10i and combined the 3D GIS technology and the monitoring needs of landslide-induced deformation. This 3D visualized model, which can effectively improve the informatization level while monitoring landslide-induced deformation, has been successfully integrated and displayed in some projects. In the future, the landslide monitoring stability analysis, 3D modeling, and BIM application can be more extensively investigated to produce higher-quality, multi-attribute, and good-looking models and further enhance the reliability, accuracy, and standardization level of deformation monitoring. In this way, updated methods and means can be provided for land-slide monitoring construction and operations management to accelerate the popularization and application of GIS technology in the monitoring field of landslide-induced deformation.

ACKNOWLEDGMENTS

This project is funded by the High-Level Talent Training Program of Hainan Natural Science Foundation (SQ2019MSXM0462) and the 2020 Provincial-Level State-Owned Capital Operation Project of the Department of Finance of Hainan Province (2020-01).

REFERENCES

Editorial Board of the Second Edition for Water Conservancy Encyclopedia China, Water Conservancy Encyclopedia China (the second edition) [M]. Beijing: China WaterPower Press, 2006: 576.
Jin Y B, Gong S Y, *et al. Automatic monitoring technique and practice of slope and foundation pit* [M]. Beijing: Geological Publishing House, 2019: 37–38.
Jin Y B, Gong S Y, *et al. Technical Code for Appraisal and Reinforcement of Building Slope* [S]. Beijing: China Building Industry Press, 2013.
Liu W J. Study on SuperMap-based dynamic mapping technique for mine gas geology [J]. *China Energy and Environmental Protection*, 2020, v.42; No.289(01):21–24+29.
Liu W, Niu P, Zhang P H, Yang T H. Establishment and application of WebGIS system in Wushan copper-molybdenum deposits [J]. *Modern Mining*, 2021, v.37; No.624(04):143–146.
Technical Code for Building Slope Engineering (GB 50330-2013) [S]. Beijing: China Building Industry Press, 2014.

Xiao F. Study of geological informatization based on IoT and cloud computing in the era of big data [J]. *Electronic Technology & Software Engineering*, 2014 (16): 221.

Yao Q H. Automatic monitoring system of geotechnical engineering and its application [J]. *Building Technique Development*, 2018, 45 (13): 76–77.

Yue J. Design and application of gas geology system to a working face of Shihao coal mine [J]. *Modern Mining*, 2017, v. 33; No.574(02):188–192.

Zhang X Q. Research on hydrogeological information system of Malan coal mine based on SuperMap object [J]. *Shanxi Coking Coal Science & Technology*, 2015, v.39; No.246(12):52–55.

Zhou J, Li C W, Song Y Q. Design and implementation of Quaternary marine geological borehole information management system based on component GIS [J]. *Marine Geology Frontiers*, 2015, 31 (5): 57–62.

Advances in Measurement Technology, Disaster Prevention and
Mitigation – Li & Mohd Yusof (Eds)
© 2023 The Author(s), ISBN: 978-1-032-36087-4

Comprehensive geological and geophysical exploration of loess landslides and high-precision 3D geological interpretation

Jie Li, Pingsong Zhang* & Yulin Xiao

School of Earth and Environment, Anhui University of Science and Technology, Huainan, Anhui, China

ABSTRACT: Loess landslides in China have large scale, large quantity and great harm, which seriously restrict the development of the regional economy and even threaten people's lives. The occurrence conditions and influencing factors of loess landslides are complex. In the research work of exploration and monitoring, transparent, comprehensive geological and geophysical methods have been more applied and popularized to avoid the limitations of inaccurate and intuitive information of a single method. This paper takes the Shuiwan landslide in Maiji District of Tianshui City as an example, uses the combination of shallow seismic reflection wave method and high-density electrical method of comprehensive geophysical techniques, and uses the method of drilling verification to carry out comprehensive detection on the site. The effectiveness of the fine detection method is studied and combined with drilling data, geospatial data, geological survey data, etc., to build a three-dimensional geological model of landslide mass. The structural characteristics of loess landslides are explained with high accuracy, and the underground geological information of landslide mass is analyzed intuitively. It can provide an effective technical reference for the exploration of loess landslides in China.

1 INTRODUCTION

Loess landslides is a kind of slope geological disaster that occurs frequently and is widely distributed in loess area of China. In recent years, with the gradual deepening of economic development in the loess region, the harm of loess landslides has become more serious. People are fully aware of the importance and urgency of accurately mastering geological conditions. Master the structure, geometric form and the sliding surface of a landslide could provide a favorable basis for exploring the formation mechanism, development law and stability of landslide, and then carry out reasonable planning and governance (Li et al. 2016; Liu et al. 2020). Therefore, it is particularly important to accurately determine the geological parameters of landslide mass.

Geophysical methods play a more and more important role in landslide geological hazard exploration. They have the characteristics of low cost, high efficiency, and high precision, can form continuous data profiles, and have been recognized and trusted by the majority of engineering workers (Yan et al. 2012; Xie et al. 2020). Wang Lei et al. used a high-density electrical method to detect a loess earthquake landslide in Xiji County, and its resistivity parameters effectively reflect the stratigraphic distribution characteristics of the study area (Wang et al. 2020). Long Jianhui et al. improved the traditional resistivity method to detect the loess landslides along the Jinghe river in Chengnan, Jingyang county, Shanxi province; the position of the sliding surface can be well determined according to the resistivity test curve (Long et al. 2007). In the process of research, due to the diversity of geophysical solutions, it is difficult to obtain satisfactory interpretation results by using a single geophysical method in some cases (Duan et al. 2013; He et al. 2013; Yan et al.

Corresponding Author: Jieli0909@163.com

  DOI 10.1201/9781003330172-34

2012). Therefore, the comprehensive geophysical method has been widely used because of its comprehensive and rich information, which is of great significance to the selection of landslide treatment scheme (Liu et al. 2013; Tiwari & Douglas 2012; Zheng et al. 2008). Guo Qiaoqiao et al. explored the ancient landslide space in Minjiang river valley of western Sichuan by means of the high-density resistivity method and conventional vertical symmetrical quadrupole resistivity sounding method, and the spatial structure characteristics of the landslide are effectively obtained (Guo et al. 2017). Lǚ Qingfeng et al. conducted a comprehensive geophysical exploration of the Gaijiayin mountain landslide of Lanzhou Chongqing railway with the help of the high-density electrical method and rayleigh surface wave method. The sliding surface, depth and active range of the landslide are further determined, and the development process and law of deformation and failure of landslide are recognized. It provides strong evidence for the determination of landslide treatment schemes (Lǚ et al. 2015). Tian Zhongying et al. studied the effectiveness of the fine detection method by using the high-density electrical method and ground-penetrating radar for the dominant channel of Heifangtai DH2 landslide mass, and its trailing edge, the fine characterization of dominant channels such as small cracks in loess is realized, and it proves the effectiveness of the application of fine geophysical prospecting (Tian et al. 2019).

This paper takes the Shuiwan landslide in Maiji District of Tianshui City as an example; combined with the geological survey data of landslide in this area and with the help of shallow seismic reflection wave method and high-density electrical method technology, to obtain parameters such as velocity and resistivity and conduct drilling verification. The stratum structure, buried depth of sliding surface, range of landslide mass and whether there is a deeper sliding surface is found. A three-dimensional geological model is constructed, and the landslide geological conditions are displayed transparently. It provides a basis for comprehensively understanding and evaluating the nature, scale, development characteristics and stability of landslides, and it can provide effective technical means for Landslide Exploration and treatment in loess area of China.

2 COMPLEXITY ANALYSIS OF GEOLOGICAL EXPLORATION OF SHUIWAN LANDSLIDE

There are many complex influencing factors in the field investigation and analysis of landslide areas. As shown in Figure 1, from the geological conditions of the landslide itself, the Shuiwan landslide is a typical historical earthquake landslide. The occurrence of landslide disaster is not only controlled by the slope structure, but also affected by the later rainfall, and the strata of the Shuiwan landslide are disordered. Quaternary loess, paleosol and Neogene mudstone are mixed and accumulated. At the same time, some areas are strongly transformed by rainfall and human activities, and the identification signs of landslide activities are not completely preserved.

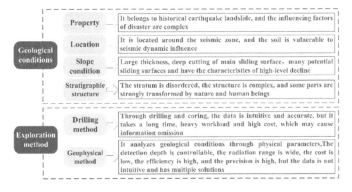

Figure 1.   Complexity analysis of Shuiwan landslide detection.

237

In addition, in terms of detection means, through drilling and coring, the data is intuitive and accurate. Still, it takes a long time, has a heavy workload and high cost, and a limited number of boreholes may result in the omission of important geological information. The geophysical prospecting method has a wide radiation range, low cost, high efficiency and high precision, but its data also has the characteristics of non-intuition and multi-solution. It often requires the comprehensive use of multiple geophysical means, combined with drilling verification, more accurate detection results are obtained.

3 COMPREHENSIVE GEOPHYSICAL EXPLORATION MODEL OF LOESS LANDSLIDES

Shuiwan landslide is mainly composed of clay, silty clay, paleosol and mudstone. The sliding bed comprises Q3 loess, paleosol, weathered zone soil and mudstone. The contact surface of the rock stratum is obvious, and the difference between electrical property and wave impedance is obvious in theory. It has the geophysical premise of developing a high-density electrical method and shallow reflected seismic wave method. As shown in Table 1, the advantages of each geophysical method and its sensitivity to engineering problems are analyzed based on the discussion of method theory. Considering its defects and influencing factors, geological purpose and accuracy requirements, the high-density electrical method and shallow reflected seismic wave method is selected to obtain underground geological information. On-site, the high-density electrical method is mainly used, supplemented by the shallow reflected wave method to comprehensively explore landslide mass.

Table 1. Characteristics of comprehensive geophysical research methods.

Method	Shallow seismic wave reflection method	High-density electrical method
Physical property basis	Wave impedance difference	Resistivity difference
Geological problems that can be solved	Quaternary stratigraphic division, acquisition of rock mass P-wave velocity, stratigraphic integrity division, fault and lithologic boundary survey	Stratigraphic boundary, geological structure, dissolution cave, stratigraphic integrity division and groundwater development
Advantage	It can directly obtain the P-wave velocity of underground rock mass; with strong stratification ability and accurate fault identification and location, it is indispensable to study structural characteristics	It has the advantages of flexible layout, high efficiency, wide measurement range, rich information and relatively low topographic requirements. It is the preferred method for exploration in mountainous areas
Disadvantage	The requirements for the terrain are high, the design of the slope observation system is troublesome, and the post-processing is complex	Limited detection depth; the detection accuracy decreases with depth, and the stratification ability is poor. There are multiple solutions under complex conditions
Disadvantages avoidance method	The working mode of small track spacing, multi-stack and high coverage is adopted, and a fine static correction in the later stage is carried out	Other methods are needed to reduce the multi-interpretation and improve the interpretation accuracy

According to the site requirements of the high-density resistivity method and the geological environment of the Shuiwan landslide, the survey line is arranged on the landslide along the direction of the main sliding surface. The section survey line is 617m long, and the rolling construction of the parallel electrical method is adopted. The field survey line of the shallow reflection seismic method coincides with that of the high-density resistivity method, the measuring points are arranged at the relatively flat part of the landslide mass, and the length of the measuring line is 544m. Figure 2 is the layout diagram of exploration lines and boreholes of the high-density electrical method and shallow reflection seismic method.

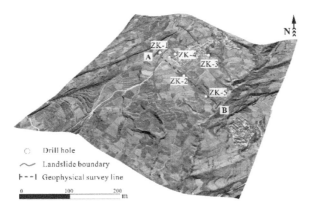

Figure 2.   Schematic diagram of survey line and borehole layout by high-density electrical method and shallow reflected wave method.

4   DETECTION EFFECT AND ANALYSIS

4.1   *High-density electrical method*

The electrical characteristics of the geotechnical medium in the resistivity profile of the Shuiwan exploration line are obvious, and it provides a basis for fine geological interpretation. The resistivity profile results are shown in Figure 3. Its specific characteristics are analyzed as follows: The surface layer of the survey line is a layer of 10–20m thick loess and weathered mudstone, there are tensile cracks with high resistance characteristics in some parts, and the medium resistivity reaches more than $50\Omega \cdot m$. Below it is a low resistivity dielectric layer with a resistivity value of $10-20\Omega \cdot m$, and has good continuity and remarkable characteristics in the tendency direction. It is inferred that it is the main sliding surface of landslide mass, and there are multiple secondary sliding surface shear outlets on the whole survey line. The lower part of the low resistance layer is an electrical layer with a resistance value of more than $50\Omega \cdot m$. There are $10-20\Omega \cdot m$ low resistivity areas locally, which are analyzed as anomalies such as fractures or small faults developed in the rock stratum.

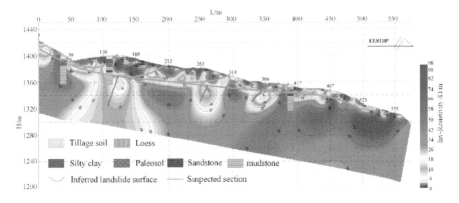

Figure 3.   Resistivity profile results of Shuiwan survey line.

4.2 Shallow reflection seismic wave method

The exploration conditions of the Shuiwan survey line are relatively good. Figure 4 shows the interpretation results of landslide mass along the seismic exploration line. From the seismic time profile, the wave group characteristics of the whole section are obvious, which is conducive to the corresponding phase tracking comparison and analysis. In the dip direction, the shallow wave component shows zoning, which is inferred to be the tensile crack caused by the creep deformation of the landslide. The wave group fault occurs in the transverse direction, which is analyzed as a fault or developed fracture in the sand mudstone medium layer, corresponding to the low resistivity area of the electrical profile. According to the continuous reflection phase and the characteristics of subsequent wave groups, the position of the landslide bottom interface is determined.

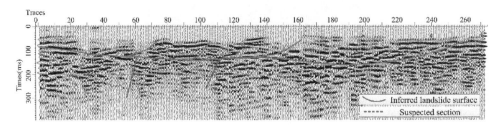

Figure 4.   Reflection seismic profile results of Shuiwan survey line.

4.3 3D geological modeling and comprehensive interpretation of landslide

After comprehensive geophysical exploration, drilling verification was carried out. A total of 5 geological boreholes were constructed, of which ZK1 is located on the sliding bed above the trailing edge of the landslide. The buried depth of the sliding surface is 21.5m for ZK4, 37.1m for ZK2 and 48.7m for ZK3, ZK5 is close to the front edge of the landslide, and the buried depth of the sliding surface is 14.5m. The borehole exposure results are basically consistent with the geophysical results, which verifies the effectiveness of the comprehensive geophysical fine detection method.

According to the comprehensive analysis of geophysical exploration and drilling verification, the upper part of the sliding bed of the Shuiwan landslide is mainly loess with several layers of ancient soil. The lower part is dominated by mudstone. The sliding surface is relatively straight, and the buried depth is 15~50m. The landslide mass is mainly composed of silt and paleosol, partially mixed with mudstone masses. The saturated area of the landslide is mainly distributed in the rear edge and middle, and the middle and trailing edge areas are thicker and gradually thinner as the slope goes down.

Constructing a three-dimensional geological model of landslide and displaying the geological conditions of landslide mass can show the spatial distribution characteristics of landslide mass, which is conducive to providing technical support for the study of landslide mechanism and the evaluation of landslide mass stability. And it is a new trend to assist in the research and analysis of landslide geological hazards. Combined with the field investigation, analysis of borehole lithology data, and comprehensive geophysical exploration results, the spatial distribution morphological characteristics of the stratigraphic structure and the sedimentary structure of sediments are determined. It provides basic data for constructing a 3D geological model of landslides. The overall three-dimensional geological model of landslide mass constructed using the limitations of topographic surface, sliding surface and bottom interface is shown in Figure 5 (a). In order to better understand the geological characteristics of landslide mass, after the part below the sliding bed is transparent, it forms the appearance of Figure 5 (b). The display direction is the rear edge of the landslide. At the same time, Figure 5 (c) shows the profile of landslide mass along the detection line of the complex.

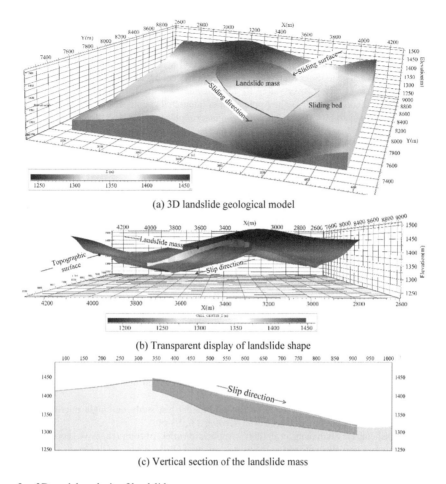

(a) 3D landslide geological model

(b) Transparent display of landslide shape

(c) Vertical section of the landslide mass

Figure 5.   3D model analysis of landslide mass.

5   CONCLUSION

This paper uses comprehensive geophysical exploration technology combined with shallow seismic reflection and high-density electrical and drilling verification methods to comprehensively detect the Shuiwan landslide in Maiji District, Tianshui city. The borehole exposure results are basically consistent with the geophysical results, which verifies the effectiveness of the comprehensive geophysical fine detection method. Based on geospatial, geological and geophysical data, the three-dimensional geological modeling of landslides is carried out. The visualization effect of the three-dimensional geological model is stronger than that of the two-dimensional section, which clearly expresses the spatial distribution and geological structure of the landslide. It lays a foundation for further developing the genetic mechanism and prevention of loess landslides and provides feasible ideas and scientific means for the research of landslide detection and treatment.

ACKNOWLEDGMENTS

This research was supported by the Anhui discipline and technology leader project (Program No. gxbjZD2016048).

REFERENCES

Duan Li, Jiang Wen-Jiang, Liu Zhao.(2013) Effect of Integrated Geophysical Methods on Dividing Unstable Areas of landslide. *J. Journal of Guizhou University* (Natural Sciences), 2: 33–36.

Guo Qiaoqiao, Guo Changbao, Shen Wei, et al. (2017) Geophysical interpretation of typical large-scale giant ancient landslides in Minjiang River Valley, *Western Sichuan. J.* 23(05): 788–797.

He Yu, Deng Zhuan, Zhou Lei. (2013) The Application of integrated geophysical methods to surface collapse exploration in Yuejiaqiao. *J. Chinese Journal of Engineering Geophysics*, 6: 814–821.

Li Hao, Yang Weimin, Huang Xiao, et al. (2016) Characteristics and deformation mechanism of Shuiwan seismic loess landslides in Maiji, Tianshui. *J. Journal of Geomechanics*, 22(1): 12–24.

Liu Chang, Zhang Pingsong, Yang Weiming, et al. (2020) Geotechnical Dynamic Characteristics and Stability Evaluation of Loess landslides in Shuiwan Earthquake, Tianshui, Gansu. *J Northwestern Geology*, 53(4): 176–185.

Liu Lei, Lei Wan, Jiang Fu-Peng, et al. (2013) The combined application of integrated geophysical methods in engineering survey. *J. Chinese Journal of Engineering Geophysics*, 10(1): 117–122.

Long Jianhui, Li Tonglu, Zhang Zhao. (2007) Experimental study on detecting sliding surface (zone) of loess landslides by resistivity method. *J. Journal of Engineering Geology*. 02: 268–272.

Lǚ· Qingfeng, Bu Simin, Wang Shengxin, et al. (2015) *Application of comprehensive geophysical prospecting method in stability evaluation of landslide J.* 37(S1): 142–147.

Tian Zhongying, Zhang Maosheng, Feng Li, et al. (2019) *Preferential Passage Detection of Loess Landside Based on Integrated Geophysical Exploration.* J. 52(02): 172–180.

Tiwari B, Douglas R. (2012) Application of GIS tools for three-dimensional slope stability analysis of pre-existing landslides. *J. Geo Congress*, 2012: 479–488.

Wang Lei, Cai Xiao-Guang, Li Xiao-bo, et al. (2020) Interpretation analysis of high-density electrical prospecting of typical seismic loess landslides in the southwestern mountainous area of Xiji county. *J. Progress in Geophysics* (in Chinese) , 35(1): 0351–0357.

Xie Xinglong, Ma Xuemei, Yang Qiang, et al. (2020) Application of comprehensive geophysical prospecting in complex landslides: a case study of Dujagou landslide in Wudu district. *J. Science Technology and Engineering*, 20(36): 14862–14868.

Yan Jinkai, Yin Yueping, Ma Juan. (2012) Large scale model test study on single micropile in landslide reinforcement. *J. Hydrogeology & Engineering Geology*, 4, 55–60.

Zheng Li, Ji Liansheng, He Zhanxiang, et al. (2008) Application effect of comprehensive geophysical prospecting technology in Loess Plateau area on the western edge of Ordos Basin. *J. Oil Geophysical Prospecting*. 02: 229–232+128.

Advances in Measurement Technology, Disaster Prevention and
Mitigation – Li & Mohd Yusof (Eds)
© 2023 The Author(s), ISBN: 978-1-032-36087-4

Establishment and evaluation of urban disaster prevention and mitigation model based on full-cycle monitoring

Yueyi Gao*
Jiangsu Academy of Safety Science and Technology, Nanjing, China

Haibo He
Emergency Management Department of Jiangsu Province, Nanjing, China

Gangfeng He
Jiangsu Academy of Safety Science and Technology, Nanjing, China

ABSTRACT: The construction of resilient cities has gradually become a practical choice and development trend for Chinese city managers to deal with disaster risks that are increasingly complex and uncertain. Based on the perspective of urban full-cycle governance, this paper uses bibliometrics and policy induction to analyze and compare relevant policy documents for urban resilience construction and build a theoretical model for urban resilience assessment, including six primary dimensions and 16 secondary dimensions. The evaluation results show that this new model can identify weak issues that affect urban safety and resilience and provides a reference for strengthening the construction of urban resilience.

1 INTRODUCTION

In the context of global climate change and rapid urbanization, cities face various threats such as natural disasters, accidents, pandemics, and terrorist attacks. Traditional models of urban governance have been unable to cope with the risks and challenges brought about by uncertain factors. The resilient city, a new concept and city type to deal with the urban crisis and uncertainty risk, continues to receive attention from academic and political circles (Sara & Joshua 2016; Zang & Wang 2019; Zhu 2021).

The construction of "resilient cities" has been included in "The Fourteenth Five-Year Plan for National Economic and Social Development of the People's Republic of China and the Outline of Long-Term Goals for 2035". Twenty-three provinces and autonomous regions have also made clear demands to strengthen the construction of "resilient cities" during the "14th Five-Year Plan" period, focusing on improving the quality of cities. It is clearly stated in the "14th Five-Year Plan for National Economic and Social Development of Jiangsu Province and the Outline of Long-term Goals for 2035" that advance planning of the urban lifeline system and emergency material reserve system to build resilient cities.

Since its inception, the concept of resilience has gone through three cognitive development stages: engineering resilience (Holling 1973), ecological resilience (Holling 2001) and evolutionary resilience (Davoudi 2012). At present, resilient cities are mainly applied in the field of disasters and climate change (Li 2020), urban and regional economic resilience, urban infrastructure resilience (Kim et al. 2021), urban terrorist attack resilience, spatial and urban planning (Brian et al. 2017; Liu 2018), etc. The research content mainly focuses on the evolution mechanism of resilient cities (Xu et al. 2017), the assessment of resilient cities (Li & Zhai 2017), and resilient cities planning.

Corresponding Author: ggs1@jssafety.ac.cn

DOI 10.1201/9781003330172-35

The research on the assessment system of resilient cities not only contributes to the enrichment of theories, but also can truly reflect the construction of resilient cities and guide their construction practices. It is an important link to promote the transformation of urban resilience research from Bohr's quadrant to Pasteur's quadrant (Wang et al. 2019).

2 RESILIENT CITY ASSESSMENT SYSTEM

2.1 *Development status*

The construction of the resilient city assessment system must follow the three major principles of comprehensiveness, accuracy and operability (Xu 2019). At present, there are three main construction ideas in the academic circle (Ni & Li 2021), including taking the basic constituent elements of the city as the core, taking the different characteristics of the resilient city as the core, taking the sequence of the stages of resilience as the core.

The process of the resilient city evaluation system is designed with the basic components of the city: firstly, we determine specific indicators through the Delphi method, analytic hierarchy process or other frameworks; secondly, we obtain indicator data from statistical data; finally, the calculation formula of the resilience index is expressed by the method of subjectively assigning the weight of each index; then we calculate the resilience coefficient of the specific system by the formula; thus the urban resilience evaluation result is obtained. ISO 37123 defines and establishes definitions and methodologies for a set of city resilience indicators (ISO 37123 Sustainable cities and communities indicators for resilient cities [S]. 2019). The indicators are classified into themes according to the different sectors and services provided by a city. The classification structure is used solely to denote the services and area of application of each type of indicator when reported on by a city. This classification has no hierarchical significance and is organized alphabetically according to themes.

The core of the resilient city assessment system, starting from the characteristics of urban resilience, lies in the capture of characteristics. With the deepening of research, the characteristics of urban resilience continue to expand, including self-organization, redundancy, diversity, immunity, synergy, intelligence, predictability, and so on. Supported by the Rockefeller Foundation, the City Resilience Index (CRI) is being developed by Arup. The CRI has been designed to enable cities to measure and monitor the multiple factors that contribute to their resilience. This provides a holistic articulation of city resilience, structured around four dimensions, 12 goals and 52 indicators that are critical for the resilience of our cities.

The resilient city assessment system is designed with the resilience stage process sequence as the core, and its specific indicators usually use a variety of basic urban components. Under different thinking paradigms, scholars have different perceptions of the resilience process. Engineering resilience generally includes three stages of resistance, absorption, and recovery; ecological resilience generally includes three stages of resistance, adjustment and adaptation; the division of evolutionary resilience stages has not yet reached a consensus, but more advanced stages such as learning and development have been added.

Based on policy documents, this research builds a resilient city assessment system with the whole process of city construction, management, and emergency as the main line and conducts empirical analysis. This research also carried out an empirical analysis using a county-level city as an example. It should be noted that the scope of the resilience of this study is limited to responding to emergencies such as accidents and natural disasters. Network security, social security, and public health are not considered.

2.2 *Index system framework*

The assessment system of policy-driven safety resilient cities (ASPSRC) is divided into four levels: the target level, the dimension level, the field level, and the index level. Figure 1 shows the framework of ASPSRC. This provides a holistic articulation of city resilience, structured around six dimensions, 16 fields and 52 indicators that are critical for the resilience of our cities.

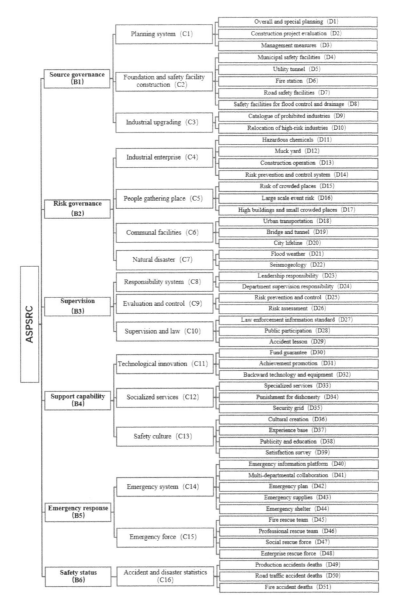

Figure 1. Assessment system of policy-driven safety resilient cities (ASPSRC).

2.3 *Index description*

The 51 specific indicators are mainly formulated based on 204 normative documents. As shown in Figure 2, local law accounted for 0.49%, local standard accounted for 0.98%, national regulation accounted for 2.94%, national law accounted for 3.43%, local regulation accounted for 14.71%, industry standard accounted for 23.53%, national standard accounted for 25.49%, and departmental rule accounted for 28.43%. For the specific content and assessment methods of the indicators, please refer to the "Guidelines for the Construction of a Model City for Safe Development in Jiangsu Province" compiled by the research group (Jiangsu Academy of Safety Science and Technology 2020).

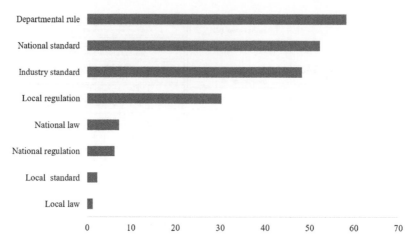

Figure 2. Normative documents and technical standards involved in ASPSRC.

3 ASSESSMENT MODEL

Taking a county-level city in Jiangsu Province as an example, the ASPSRC is used to assess urban safety resilience. The urban safety resilience index (P) adopts a 100-point system, and a score of 90 or more indicates good resilience. The weight relationship of each indicator is shown in table 1 through expert scoring and consistency check (consistency check<0.1) to ensure the scientificity and rationality of the weights. The weight relationship is directly integrated into the measurement of various indicators. For the scoring standards of each indicator, please refer to the [18].

$$P = \sum_{n=1}^{51} p_{cn} \div (100 - \sum_{n=1}^{51} q_{cn}) \times 100$$

In the formula, p_{Cn} is the score of various indicators, reflecting the resilience of cities in various fields, q_{Cn} is the index value that is not applicable to the city in each index. Because of the differences in the situation of each city, the applicability of the index is improved by the method of "not involving and not scoring."

Table 1. The weight of the dimension layer and the field layer.

Hierarchical framework	Index	Proportion	Consistency check
Dimension (6)	Source governance	0.18	0.009
	Risk governance	0.32	
	Supervision	0.11	
	Support capability	0.16	
	Emergency response	0.14	
	Safety status	0.09	
Field(16)	Planning system	0.05	
	Foundation and safety facility construction	0.09	0.003
	Industrial upgrading	0.04	

(continued)

Table 1. Continued.

Hierarchical framework	Index	Proportion	Consistency check
	Industrial enterprise	0.09	
	People gathering place	0.09	0.000
	Communal facilities	0.09	
	Natural disaster	0.05	
	Responsibility system	0.02	
	Evaluation and control	0.05	0.004
	Supervision and law enforcement	0.04	
	Technological innovation	0.04	
	Socialized services	0.03	0.004
	Safety culture	0.09	
	Emergency system	0.06	0.000
	Emergency force	0.08	
	Accident and disaster statistics	0.09	0.000

4 RESULT ANALYSIS

The expert group used the ASPSRC to conduct the assessment and obtained the scoring results of the city's resilience fields. As shown in Figure 3, the city has obvious shortcomings in terms of planning and emergency system construction. At the same time, infrastructure construction and urban risk assessment control need to be further improved. A detailed analysis of the scores of various indicators shows that the main problems in the city's resilience include:

(1) The construction of the urban safety risk management information platform did not realize the interconnection of information among various departments. Data barriers and information silos are still very significant. The information-based real-time perception, intelligent rapid warning, and automated, timely disposal of various risk elements in the city have not yet been realized.

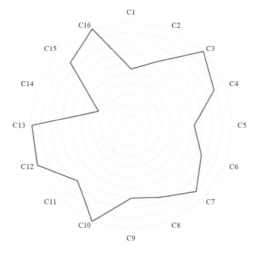

Figure 3. Distribution of urban resilience status.

(2) Existing urban planning does not take into account safety resilience. The overall plan does not fully reflect the requirements for comprehensive disaster prevention and public safety. There are no special plans such as comprehensive disaster prevention and mitigation plans, earthquake prevention and mitigation plans, and road traffic safety management plans in the existing special plans. There are still shortcomings in the formulation and revision of safety management measures for some urban facilities. The existing safety management methods do not solve the problems of what to manage and how to manage.

(3) There are still omissions in the management of municipal infrastructures, such as broken fire hydrants, damage to some water supply monitoring equipment, irregular road guardrails, and no safety signs on the roads around the school. Some production and business units have defects in management, such as incorrect installation of combustible gas alarm devices in catering establishments. Some old communities and streets have not been screened for hidden hazards, or the problems have not been rectified.

5 CONCLUSIONS

Safety resilient cities provide new ideas for urban development and are expected to realize the transformation of urban governance models from "safety defense" to "safe and worry-free." The resilient city assessment system established from the regional normative documents can effectively link the policy and the actual situation of the region and realize the linkage between the policy orientation and urban construction. It is envisaged that the ASPSRC will primarily be used by city governments in the best position to gather administrative data. It is hoped that the government will focus more on solving the problem rather than how much it scores because the ASPSRC aims to measure relative performance over time rather than a comparison between cities.

ACKNOWLEDGMENTS

This research was financially supported by the Jiangsu Natural Science Foundation (Grant No. BK20181516).

REFERENCES

Brian Deal et al. Urban Resilience and Planning Support Systems: The Need for Sentience[J]. *Journal of Urban Technology*, 2017, 24(1): 29–45.

Davoudi S. Resilience: A Bridging Concept or a Dead End?[J]. *Planning Theory& amp; Practice*, 2012, 13(2):299–333.

Holling C S. Resilience and Stability of Ecological Systems[J]. *Annual Review of Ecology and Systematics*, 1973:1–23.

Holling C S. Understanding the Complexity of Economic, Ecological, and Social Systems[J]. *Ecosystems*, 2001, 4(5): 390–405.

ISO 37123 *Sustainable cities and communities indicators for resilient cities* [S]. 2019.

Jiangsu Academy of Safety Science and Technology. *Guidelines for the Construction of a Model City for Safe Development in Jiangsu Province* [M]. Nanjing: Southeast University Press, 2020: 233–254.

Kim Yeowon et al. Capturing practitioner perspectives on infrastructure resilience using Q-methodology[J]. *Environmental Research: Infrastructure and Sustainability*, 2021, 1(2)

Li Tao. The construction and the development of international resilient cities based on the impact of climate disasters[J]. *Science & Technology Review*, 2020,38(08):30–39.

Li Ya, Zhai Guofang. Research on my country's Urban Disaster Resilience Assessment and Its Improvement Strategy[J]. *Planners*, 2017,33(08):5–11.

Liu Dan. A Review on Studies of Resilient Cities and Resilient Planning[J]. *Urban Planning*, 2018,42(05):114–122.

Ni Xiaolu, Li Xingqiang. Three types of resilient city evaluation systems and their new development directions[J]. *International Urban Planning*, 2021,36(03):76–82.

Sara Meerow and Joshua P. Newell and Melissa Stults. Defining urban resilience: A review[J]. *Landscape and Urban Planning*, 2016, 147: 38–49.

Wang Hui, Wang Tao, Xiang Weining. Analysis of Pasteur Paradigm in Urban Resilience Research[J]. *Chinese Garden*, 2019,35(07):51–55.

Xu Chan, Zhao Zhicong, Wen Tianzuo. Resilience: Conceptual Analysis and Reconstruction from a Multidisciplinary Perspective[J]. *Journal of Western Human Settlements and Environment*, 2017,32(05):59–70.

Xu Zhaofeng.Urban resilience evaluation system and optimization strategy from the perspective of disaster prevention[J]. *Chinese Safety Science Journal*, 2019,29(03):1–7.

Zang Xinyu, Wang Qiao. The conceptual evolution, research content and development trend of urban resilience[J]. *Science & Technology Review*, 2019,37(22):94–104.

Zhu Zhengwei. Resilient governance: the practice and exploration of China's resilient city construction[J]. *Public Management and Policy Review*, 2021,10(03):22–31.

Advances in Measurement Technology, Disaster Prevention and Mitigation – Li & Mohd Yusof (Eds)
© 2023 The Author(s), ISBN: 978-1-032-36087-4

Spatial pattern and interannual variation characteristics of rainstorms and flood disasters in Southern Xinjiang

Kan Chen & Xi Wang*

Faculty of Humanities and Social Sciences, Macao Polytechnic University, Macao, China

ABSTRACT: The ecological environment in southern Xinjiang is fragile. Rainstorms and flood disasters caused by short-duration heavy rainfall often lead to huge losses in local agriculture and animal husbandry. In this study, five disaster elements causing rainstorms and flood disasters in southern Xinjiang are used to construct a disaster exponent that can comprehensively represent the severity of the disasters by using the ratio method and dimensionless linear combination method. Based on the probability distribution function, the county and annual disaster loss indices are classified into four grades, mild (grade 1), moderate (grade 2), severe (grade 3), and extremely severe (grade 4). Further, the spatial and temporal distribution of disasters and their causes are analyzed. Results show that the five disaster elements exhibited significant geographical variability, and areas with the most severe (grade 4) rainstorms and flood disasters occurred in the Aksu and Kashgar regions. The disasters appeared from March to October and were concentrated from May to July. The interannual variation of the disaster exponent from 1986 to 2019 showed a significant linear decreasing trend, by 0.4 per 10 a. The 12-h rainfall threshold that triggered rainstorms and flood disasters in southern Xinjiang was 9 mm, and the spatial distribution of the number of occurrences of 12-h precipitation that exceeded the threshold determined the spatial distribution of disaster severity. There is a significant positive correlation between the annual occurrence of 12-h precipitation exceeding the threshold (N_{12}) and the annual disaster exponent (Z_K), and the decrease of N_{12} leads to a decrease in the intensity of rainstorms and flood disasters in southern Xinjiang.

1 INTRODUCTION

Rainstorms and flood disasters are natural disasters that cause damage to population, economy and property due to inundation of low-lying areas by large amounts of standing water and runoff from prolonged rainstorms or precipitation (Wang & Wang 2021). Rainstorms and floods are some of the most serious natural disasters causing huge economic losses and human casualties every year (Wan et al. 2016). Global warming and humidification have intensified the water cycle, increasing extreme precipitation, rainstorms and flood disasters. The increasing threat posed by these disasters (Cao et al. 2012; Zhang & Zhou 2020) has attracted wide attention among scholars.

Based on meteorological, hydrological and geographical information, various disaster risk assessment and prediction models have been established using principal component analysis (Tang & Wang 2015), fuzzy comprehensive evaluation method (Hu et al. 2014), information diffusion theory (Yao et al. 2012), grey correlation theory (Wen et al. 2017), and hierarchical analysis (Li et al. 2015). Since 1990, although the loss due to rainstorms and flood disasters has been decreasing in China overall, it has been increasing in the northern region, which should be the focus of disaster prevention in the future (Wan et al. 2016). Since most rainstorms and flood disasters occur south of the Yangtze River basin and lead to high cumulative losses, more research has been carried out

*Corresponding Author: xwang@ipm.edu.mo

DOI 10.1201/9781003330172-36

there, and less in the arid northwest. The ecological environment in arid northwest China is very fragile and more prone to disaster during rainstorms (Wang et al. 2017). Southern Xinjiang is a famous arid region in China, which is extremely susceptible to climate change. Damage caused by floods is severely restricting local economic development (Wang et al. 2020). In their analysis of rainstorms and floods in Xinjiang using the agricultural disaster area and annual precipitation during 1949-2014, Mansur et al. (Mansur & Wu 2012) pointed out that the increase in precipitation frequency and intensity led to an increase in rainstorm flood. Wang et al. (Wang et al. 2020) used linear regression analysis based on precipitation, GDP, and disaster losses during 1984-2016 to show a significant positive correlation between the annual frequency of rainstorms and flood disasters in Xinjiang with annual precipitation and GDP. Although the studies above have reached some conclusions, disaster evaluation based on a single disaster element is prone to more serious bias. So, an objective quantitative evaluation of disasters based on multiple disaster elements is needed. Therefore, in this study, multiple disaster elements are used to construct a disaster exponent, and the probability distribution function is used to objectively classify rainstorms and flood disasters. Then, the spatial distribution, monthly distribution and interannual variation of disasters are explored, and the climatic causes of the spatial and temporal distribution are discussed. This study aims to provide a scientific background for the early warning and prevention of rainstorms and flood disasters in southern Xinjiang.

2 EXPERIMENTAL

2.1 Overview of the study area

Southern Xinjiang, located to the south of the Tianshan Mountains and north of the Kunlun Mountains, has a temperate continental arid climate with long average sunshine hours and large diurnal temperature variation. Its unique climate provides a favorable environment for the growth of local crops such as cotton, wheat, melons, and fruits. Although there are fewer days per year with heavy precipitation in southern Xinjiang than in eastern China, heavy precipitation is prone to causing rainstorms and flooding due to the highly fragile ecological environment, which often leads to serious losses in local agriculture and animal husbandry. Southern Xinjiang includes seven prefectures and cities, Kizilsu Kirgiz Autonomous Prefecture (I, Kezhou for short), Kashgar Prefecture (II), Aksu Prefecture (III), Hotan Prefecture (IV), Bayangol Mongol Autonomous Prefecture (V, Bazhou for short), Turpan City (VI), and Kumul City (VII) (Figure 2a).

2.2 Data

This paper is based on the rainstorm and flood disaster information recorded by the Civil Affairs Department of Xinjiang Uygur Autonomous Region. It sorts out 48 counties and cities in southern Xinjiang during 1986-2019, including 1) the occurrence time (year, month and day), 2) the occurrence area (counties and cities), 3) and the number of deaths (persons); also, 4) the number of collapsed homes (houses), 5) the number of collapsed sheds (seats), 6) the number of livestock deaths (heads), and 7) the affected area of crops (hm^2). If there is a rainstorm and flood disaster in a county, the number of rainstorms and flood disasters in the county is recorded as 1. From 1986 to 2019, 1541 rainstorms and flood disasters occurred in southern Xinjiang.

This study uses day-by-day 12-h precipitation data from 44 meteorological observation stations in southern Xinjiang from 1986 to 2019 to analyze the impact of the number of occurrences (stations) of short-term heavy precipitation on the spatial distribution, monthly distribution, and interannual variation of disasters.

2.3 Methods

Since a disaster event consists of five disaster elements with different units, the intensity of different disasters can not be compared. In order to facilitate this comparison, it is necessary to construct

251

a disaster exponent (Z_i) that can represent the five disaster elements in an integrated manner. In the construction of Z_i, the weights of each disaster element are determined using the ratio method. Then, Z_i is obtained using the dimensionless linear combination method (Chen et al. 2022).

Assuming that each disaster element consists of n samples, a disaster element evaluation matrix $X_{n\times 5}$ can be obtained. Then the weights of the disaster elements a_j and the disaster exponent Z_i are calculated as follows

$$a_j = \frac{\sum_{i=1}^{n} \frac{X_{i,j}}{X_{ja}}}{\sum_{j=1}^{5} \sum_{i=1}^{n} \frac{X_{i,j}}{X_{ja}}} \tag{1}$$

$$Z_i = \sum_{j=1}^{5} a_j \frac{X_{i,j}}{\bar{X}_j} \tag{2}$$

Where $i = 1, 2, \ldots, n, j = 1, 2, \ldots, 5$. X_{ja} and $\bar{X}_j$ represent the maximum and average values of the j-th disaster element, respectively.

In order to analyze the spatial distribution and interannual variation of rainstorms and flood disasters from the perspective of counties and years, the per county and per year disaster elements were calculated respectively using Equation (2). When calculating the disaster elements by county, i equals 48, and j is the cumulative value of 34 years (1986 - 2019) for the j-th disaster element of a county. When calculating the annual disaster loss index, i equals 34, and j is the cumulative value of 48 counties for the j-th disaster element in a certain year.

After the calculation of Equation (2), the weight coefficient, the average and the maximum value of five disaster factors can be noted in Table 1.

Table 1. The weight, average value and maximum value of disaster elements.

		Deaths (persons)	Collapsed homes(houses)	Collapsed sheds(seats)	Livestock deaths(heads)	Crops affected area(hm^2)
County	Weight coefficient	0.19	0.28	0.10	0.29	0.14
	Average value	13.2	5513.1	2383.6	16406.0	47322.9
	Maximum value	62	17427	20611	49523	304120
Year	Weight coefficient	0.14	0.15	0.16	0.25	0.29
	Average value	18.6	7783.2	3365.0	23161.5	66808.9
	Maximum value	131	50571	20594	93444	225976.3

To objectively classify the hazard level of rainstorms and flood disasters by county and by year, first, the probability density functions of the county and year damage indices are determined using the histogram method and hypothesis testing. Then the thresholds for different grades of the county and year damage indices are determined based on the probability density. Using the graphical method, the histogram of the county and annual disaster exponents was obtained by dividing the sample of the county and annual disaster loss indices into ten groups according to the grouping rules (Chen et al. 1982) (Figure 1). The histogram showed that the disaster exponents approximately satisfied the gamma distribution. The mean $\mu = 1$ (1) and variance $\sigma^2 = 0.2807$ (0.8837) of the county and annual exponents were calculated, and the parameters of the county (annual) gamma distribution $\alpha = 3.5625$ (1.1316) and $\beta = 0.2807$ (0.8837) were calculated based on the relationships $\mu = \alpha\beta$, $\sigma^2 = \alpha\beta^2$.

Based on the test in the literature (Chen et al. 1982), the statistic of the county (annual) disaster exponent was calculated as $V = 4.13$ (3.55). Taking $\alpha = 0.01$ and checking the distribution table of

χ^2, the critical value λ=18.48 was obtained. As $V < \lambda$, the probability distribution of the county (annual) disaster exponent satisfied the gamma distribution. The specific expression is

$$f(Z) = \frac{1}{\beta^\alpha \Gamma(\alpha)} Z^{\alpha-1} e^{-Z/\beta} \tag{3}$$

Where Z represents the disaster exponent, Z_S denotes the county disaster exponent, and Z_K denotes the annual disaster exponent. α and β are the parameters of the gamma distribution.

The threshold Z_{PS} (Z_{PK}) of the county (annual) disaster exponent quantile was calculated from the probability of the gamma distribution function, and the county (annual) disaster exponent samples were classified into four grades according to the degree of disaster loss: mild, moderate, severe, and extremely severe, according to the different values of Z_{PS} (Z_{PK}) (Table 2).

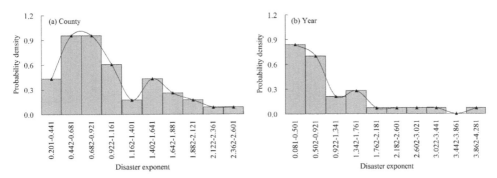

Figure 1.   Temporal (by year) and spatial (by country) probability density distribution of disaster exponent with rainstorm-flood.

Table 2.   Temporal (by year) and spatial (by country) classification criteria of rainstorm-flood losses levels.

Distribution function (F)	Spatial disaster exponent (Z_S)	Space disaster grade	Temporal disaster exponent (Z_K)	Annual disaster grade
$F \le 0.500$	$Z_S \le 0.9082$	Mild (Grade 1)	$Z_K \le 0.7256$	Mild (Grade 1)
$0.501 \le F \le 0.750$	$0.9083 \le Z_S \le 1.2892$	Moderate (Grade 2)	$0.7257 \le Z_K \le 1.3839$	Moderate (Grade 2)
$0.751 \le F \le 0.900$	$1.2893 \le Z_S \le 1.7104$	Severe (Grade 3)	$1.3840 \le Z_K \le 2.2334$	Severe (Grade 3)
$F \ge 0.901$	$Z_S \ge 1.7105$	Extremely severe (Grade 4)	$Z_K \ge 2.2335$	Extremely severe (Grade 4)

3   RESULTS

3.1   *Spatial distribution of rainstorm and flood disaster*

All 48 counties in southern Xinjiang have experienced rainstorms and floods, but the magnitude of the five disaster elements and the hazard level of the disasters are unevenly distributed (Figure 2). Regarding the annual average values of each disaster element, the regions with the highest number of deaths were Aksu and Kashgar, with an annual average of 1 to 2 deaths (Figure 2a). The number of collapsed houses was the highest in the Kashgar region, with an average of 390 to 513 (Figure 2b). The highest number of collapsed livestock sheds occurred in Luopu County, Hotan Region, with an average of 607 sheds collapsed (Figure 2c). Both the areas with high livestock deaths and the regions with the large affected area were concentrated in the Aksu region, with 1094–1457 dead livestock and 6745–8945 hm² affected area, respectively (Figure 2d, Figure 2e). In terms of disaster hazard level, the most severe areas were concentrated in the Aksu and Kashgar regions, followed by the Hotan region, with Turpan and Hami cities in third and Bazhou being the least severe (Figure 2f).

Using day-by-day 12-h precipitation data from 44 meteorological stations in southern Xinjiang, 1541 rainstorm and flooding events were statistically calculated, and the 12-h precipitation threshold that triggered rainstorms and floods in southern Xinjiang was 9 mm based on the percentile method with r = 20. The areas witnessing a high number of 12-h precipitation exceeding 9 mm from 1986 to 2019 were consistent with the areas where level 4 disasters.

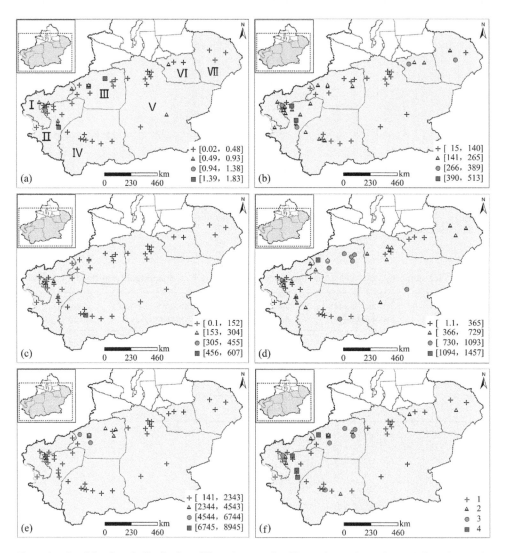

Figure 2.   Spatial and grade distribution of five rainstorm-flood losses factors in southern Xinjiang from 1986 to 2019. (a) average annual death toll, (b) average annual collapse of houses, (c) annual collapse shed, (d) average annual mortality of livestock, (e) average annual affected area, and (f) grade.

3.2 Monthly distribution of rainstorm and flood disaster

The monthly distribution of the five disaster elements was unimodal (figure omitted), with disasters occurring from March to October, fatalities concentrating from June to August, collapsed houses in April to July, collapsed livestock sheds from March to July, livestock death from May to July, and affected areas from April to July. In general, rainstorms and floods in southern Xinjiang were concentrated from May to July, with July being the most serious, which was similar to the monthly distribution of the number of occurrences where 12-h precipitation exceeded 9 mm at the 44 meteorological stations.

3.3 Interannual variation of rainstorm and flood disaster

The interannual variation of the disaster exponent (Z_K) of rainstorms and floods in southern Xinjiang from 1986 to 2019 showed a significant linear decreasing trend of 0.4 per 10 a (Figure 3). The univariate linear regression equation had a correlation coefficient $R = 0.40$ and a significant confidence level $P = 0.02$.

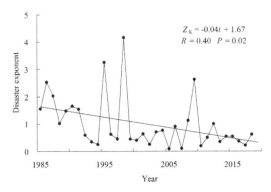

Figure 3.   Interannual variation of disaster exponent of rainstorm and flood in southern Xinjiang from 1985 to 2019.

The univariate linear regression equation between the annual occurrence of 12-h precipitation exceeding 9 mm at the 44 meteorological stations in southern Xinjiang (N_{12}) and the disaster exponent (Z_K) is $Z_K = 0.01N_{12} + 0.27$ ($R = 0.30$, $P = 0.09$), which indicates that the interannual variation of the loss index is determined by N_{12}, i.e., a decrease of N_{12} leads to the decrease in the intensity of rainstorms and floods in southern Xinjiang. It is also worth noting that severe grade 4 disasters occurred in 1987, 1996, 1999, and 2010. The one in 1999 was the most severe, with an annual disaster loss index of 4.17.

4 CONCLUSIONS

In terms of spatial distribution, all 48 counties and cities in southern Xinjiang have experienced rainstorms and flood disasters, and the five disaster elements showed diversity. The areas with the highest number of deaths were the Aksu and Kashgar regions, with an annual average of 1 to 2 deaths; the number of collapsed houses was the highest in the Kashgar region, with an average of 390 to 513; the highest number of collapsed livestock sheds occurred in Luopu county of the Hotan region, with an average of 607 livestock collapsed; Both the areas with high livestock deaths and the regions with the large affected area were concentrated in the Aksu region, with 1094-1457 dead livestock and 6745-8945 hm^2 affected area, respectively. The disasters occurred from March to October, with the number of deaths concentrated from June to August, collapsed houses from

April to July, collapsed shelters from March to July, livestock death from May to July, and affected areas from April to July.

The spatial distribution of hazard levels showed that the most severe (level 4) disasters were concentrated in the Aksu and Kashgar regions, followed by the Hotan region, with Turpan and Hami cities in third and Bazhou being the least severe. The 12-h precipitation threshold triggering rainstorms and flood disasters in southern Xinjiang was 9 mm. The areas witnessing a high number of 12-h precipitation exceeding 9 mm from 1986 to 2019 were consistent with the areas where level 4 disasters.

The interannual variation of the disaster exponent from 1986 to 2019 showed a significant linear decrease of 0.4 per 10 a. There was a significant positive correlation between the annual occurrence of 12-h precipitation exceeding 9 mm (N_{12}) with the annual disaster exponent (Z_K), and the decrease of N_{12} led to the weakening of the intensity of rainstorms and flood disasters in southern Xinjiang.

ACKNOWLEDGMENTS

This work was supported by the Foundation of Xinjiang Weather Modification (RYB202002).

REFERENCES

Cao J T, Li Y Y, Shen F X, *et al*. (2012) Drawing down our resources: estimating the total appropriation of water in China. *Water International*, 37 (5): 512–522.

Chen B X, Chen K, Wang X, et al. (2022) Spatial and temporal distribution characteristics of rainstorms and flood disasters around Tarim Basin. *Polish Journal of Environmental Studies* 31(3): 2029–2037.

Chen J D, Liu W R, Wang R G. (1982) Probability and statistics handout Beijing: Higher Education Press.

Hu B, Ding Y Y, He L D, *et al*. (2014) Risk division of rainstorm and water-logging disasters in Ningbo City based on fuzzy comprehensive evaluation. *Torrential Rain and Disasters*, 33(4): 380–385.

Li W J, Zuo J Q, Song Y L, *et al*. (2015) Changes in spatio-temporal distribution of drought/flood disaster in Southern China under global climate warming. *Meteorological Monthly*, 41(3): 261–271.

Mansur S, Wu M H. (2012) The spatio-temporal change characteristics of flood disaster in the southern Xinjiang in recent 60 years. *Scientia Geographica Sinica*, 32(3): 386–392.

Tang J F, Wang Q. (2015) The analysis of the geological disaster that occurred in the settlements space in 2013 for the Beichuan Qiang Autonomous county. *Journal of Catastrophology*, 30(1): 87–91.

Wan J H, Zhang B W, Liu J A, *et al*. (2016) The distribution of flood disaster loss during 1950–2013. *Journal of Catastrophology*, 31(2): 63–68.

Wang N, Cui C X, Liu Y. (2020) Temporal-spatial characteristics and the influencing factors of rainstorm-flood disasters in Xinjiang. *Arid Zone Research*, 37(2): 325–330.

Wang Q, Zhai P M, Qin D H. (2020) New perspectives on 'warming–wetting' trend in Xinjiang, China. *Advances in Climate Change Research*, 11 (3): 252–260.

Wang X Q, Wang X. (2021) Spatial distribution and temporal variation characteristics of rainstorm flood disasters with different intensities in southern Xinjiang from 1980 to 2019. *Journal of Glaciology and Geocryology*, 43(6): 1818–1828.

Wang Y J, Zhou B T, Qin D H, *et al*. (2017) Changes in Mean and Extreme Temperature and Precipitation over the Arid Region of Northwestern China: Observation and Projection. *Advances in Atmospheric Sciences*, 34(3): 287–305.

Wen Q P, Zhou Y H, Huo Z G, *et al*. (2017) Risk changes of storm flood disasters in southeast China under climatic warming. *Chinese Journal of Ecology*, 36(2): 483–490.

Yao J Y, Zhu H G, Nan J Y, *et al*. (2012) Analysis of flood and disaster forecast in Heilongjiang Province based on grey theory. *Journal of Catastrophology*, 27(1): 59–63.

Zhang W X, Zhou T J. (2020) Increasing impacts from extreme precipitation on population over China with global warming. *Science Bulletin*, 65(3): 243–252.

*Advances in Measurement Technology, Disaster Prevention and
Mitigation – Li & Mohd Yusof (Eds)*
© 2023 The Author(s), ISBN: 978-1-032-36087-4

Research on the time distribution of natural disasters harmful to inland channel in China

Baoying Mu, Yuchuan Wang* & Lixin Lu
China Waterborne Transport Research Institute, Beijing, China

ABSTRACT: China has a vast territory and suffers many natural disasters every year. Among
them, the natural disasters that are harmful to the inland channel are mainly flood disasters, drought
disasters, typhoon disasters, strong wind disasters, rainstorm disasters, landslide disasters, and
collapse disasters. This paper analyzes the time distribution of natural disasters that are harmful
to the inland channel. In recent years, there is no obvious trend in the number of natural disasters,
which is the result of the combined effect of natural disasters increasing damage to waterways
and preventing and strengthening maintenance in advance. Every year from June to October is
a period of the frequent occurrence of natural disasters, and inland channels are vulnerable to
damage. Among them, flood disasters from June to September are likely to cause greater harm to
the waterways. Drought disasters occur in stages throughout the year, mainly in autumn and winter,
including drought disasters, spring drought disasters, continuous drought disasters in winter and
spring, etc. Typhoon disasters generally occur from June to September, with an average duration of
6 days, rainstorm disasters may occur from January to September, with an average duration of 17
days. Landslide disasters, collapse disasters, and gale disasters have no obvious time distribution
pattern and are unpredictable natural disasters. By analyzing the time evolution law of natural
disasters that are harmful to the waterway, it is of great significance for early defence and timely
restoration of navigation.

1 INTRODUCTION

China has a vast territory, and frequent natural disasters have caused great damage to the transportation (Lu et al. 2020; Shuang et al. 2010), agriculture (Sivakumar et al. 2005), and the tourism
industry (Cvetkovi & Bokovi 2015; Rosselló et al. 2020). This paper mainly studies the impact
of natural disasters on the inland channel and provides a reference for China's inland channels to
resist natural disasters by exploring the time distribution of natural disasters that have an impact
on waterways.

The main natural disasters suffered by inland channels in my country include flood disasters,
drought disasters, typhoon disasters, gale disasters, rainstorm disasters, landslide disasters, and
collapse disasters. The types of waterway damage caused by various types of natural disasters are
quite different, and the degree of impact on the waterway is also different. Drought and flood
disasters generally have regularity and annual repetition. With the arrival of the flood season and
dry season, the waterway is often washed or shallow, and the buildings are damaged, leading to
displacement or loss; landslide disasters and collapse disasters are unpredictable to a certain extent,
which may cause blockage of the waterway and have a serious impact on navigation.

From 1978 to 2020, a total of 868 large-scale natural disasters occurred in my country's inland
channel and waterway facilities at level III and above, resulting in economic losses of 1,118,728,500
yuan, 48 deaths, 8 serious injuries, and 8 missing cases.

*Corresponding Author: wangyuchuan@wti.ac.cn

DOI 10.1201/9781003330172-37

The frequency of drought and flood disasters is the highest, and the impact on the waterway is the greatest. In recent years, the number of drought and flood disasters in my country accounted for 90.32% of the total natural disasters, and the direct economic losses accounted for 89.31% of the total, including 486 flood disasters, with an economic loss of 661.2963 million yuan, 298 drought disasters, and an economic loss of 337.8795 million yuan. Typhoons, rainstorms, landslides, collapses, and gale disasters also occur occasionally, causing varying degrees of damage to the waterway.

2 OVERVIEW OF TIME DISTRIBUTION

2.1 *Annual distribution law*

According to the frequency and distribution time of natural disasters, the period from June to October is a period of the frequent occurrence of natural disasters. Disasters such as floods and typhoons occur frequently, and inland channels are vulnerable to damage. According to statistics, in recent years, 660 natural disasters occurred from June to October, accounting for 76.04% of the whole year. The economic loss was 947.6352 million yuan, accounting for 84.71% of the whole year. On average, 132 natural disasters occurred every month, and the average monthly loss was 947.6352 million yuan. The number of natural disasters from January to May and November to December each year is relatively small, with an average of 30 per month and an average monthly loss of 24.4419 million yuan.

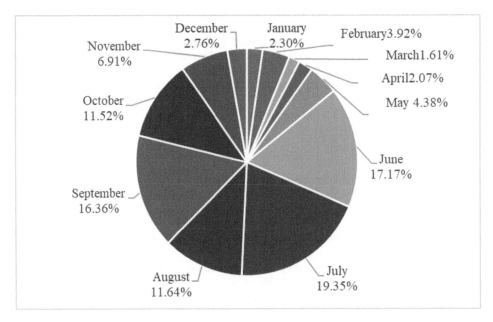

Figure 1.   The proportion of natural disasters occurring each month.

2.2 *The law of evolution over the years*

In order to analyze the annual change law of natural disasters, the number of natural disasters in recent years, and the direct economic loss, the average loss caused by each incident are counted. Considering that before 2008, the information on natural disasters suffered by inland channels in various places was not complete, since 2008, the central government has subsidized the waterways

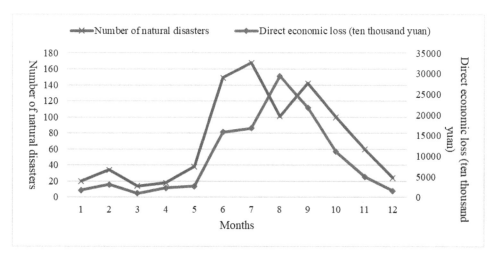

Figure 2. Monthly trend of natural disasters.

suffering from natural disasters in the form of emergency relief funds every year, and the ledger data is relatively complete. Therefore, the main analysis of the 2008–Change in the number of natural disasters in 2020 is shown in Figure 3.

It can be seen from the figure that in recent years, there is no obvious trend in the annual change in the number of natural disasters. Through comprehensive analysis, it is the result of the combined effect of natural disasters increasing the damage to the waterway and preventing and strengthening the maintenance in advance. On the one hand, with the increase in shipping demand, the channel conditions have been improved through channelization, dredging, and slope protection. The number increases accordingly. On the other hand, with the improvement of the waterway maintenance and management level, the waterway emergency response system is gradually improved. For regular natural disasters, the waterway maintenance and management departments at all levels are actively guarded against, and the number of natural disasters will be reduced accordingly. Considering the above factors, the number of natural disasters has not changed significantly over the years.

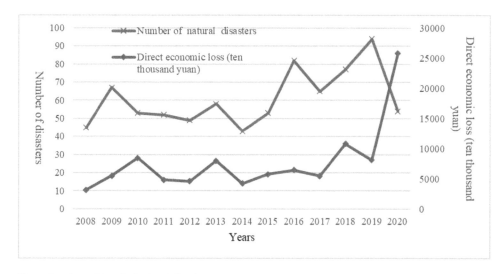

Figure 3. Annual trend of natural disasters.

3 TIME DISTRIBUTION OF VARIOUS TYPES OF DISASTERS

3.1 *Floods*

The time distribution of flood disasters is relatively concentrated, with obvious seasonality, mostly in summer. From June to September every year, flood disasters are prone to occur. A total of 418 flood disasters occurred, accounting for 86.01% of the year. Since then, the average monthly economic loss has been 153.0014 million yuan. During the rest of the month, floods and waterlogging disasters occur occasionally. Generally, the degree of disasters is relatively minor and the economic losses are relatively small.

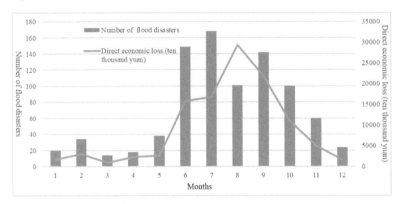

Figure 4.   Monthly distribution of flood disasters.

3.2 *Drought disaster*

Drought disasters occur in stages throughout the year, mainly manifested as consecutive drought disasters in autumn and winter, spring drought disasters, and consecutive drought disasters in winter and spring. Autumn and winter drought disasters are the most common, usually occurring from September to December, with an average of 54 disasters per month, with an average duration of 53 days. A total of 214 drought disasters occurred, accounting for about 71.81% of the whole year, and the economic loss was 236.4041 million yuan, accounting for 69.97% of the whole year. The second is the continuous drought in winter and spring, which usually occurs from December to February of the following year, with an average of about 19 cases per month, accounting for 19.46% of the year, and direct economic losses accounting for 15.23% of the year.

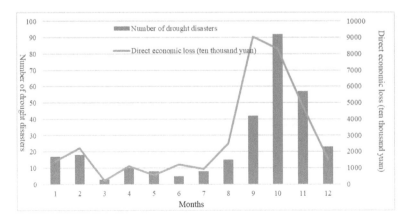

Figure 5.   Monthly distribution of drought disasters.

260

Drought disasters occasionally occur in summer drought and other months, slightly more in August, and the damage to the waterway is relatively minor.

3.3 *Other natural disasters*

Typhoon disasters generally occur in June-September, with a short duration, with an average duration of six days, causing great damage to the waterway, often accompanied by heavy rain, landslides, etc., resulting in sudden siltation of the waterway, loss or displacement of navigation marks, due to the weather conditions after the typhoon. Bad, it is difficult to implement emergency rush work.

Heavy rain disasters may occur from January to September, with an average duration of 17 days. Continuous rainstorms for many days can easily cause erosion of channel slopes, revetments, or damage to buildings.

Landslide disasters, collapse disasters, and gale disasters are rare, with a total of 10 occurrences in recent years. There is no obvious time distribution pattern, and they are unpredictable natural disasters.

4   CONCLUSIONS AND RECOMMENDATIONS

The main natural disasters suffered by inland channels in my country include flood disasters, drought disasters, typhoon disasters, gale disasters, rainstorm disasters, landslide disasters, and collapse disasters. The frequency of drought and flood disasters is the highest, and the impact on the waterway is the greatest. In recent years, there has been no obvious trend in the annual change in the number of natural disasters. Through comprehensive analysis, it is the result of the combined effect of natural disasters on the increase in waterway damage and early prevention and strengthening of maintenance. Every year, the period from June to October is a period of frequent natural disasters. Disasters such as floods and typhoons occur frequently, and inland channels are easily damaged. Among them, the time distribution of flood disasters is relatively concentrated, with obvious seasonality, mostly in summer. Drought disasters occur in stages throughout the year, mainly manifested as consecutive drought disasters in autumn and winter, spring drought disasters, and consecutive drought disasters in winter and spring. Typhoon disasters generally occur from June to September, with a short duration. Rainstorm disasters may occur from January to September, with an average duration of 17 days. Landslide disasters, collapse disasters, and gale disasters occur less frequently. In recent years, a total of ten disasters have occurred. It is an unpredictable natural disaster with no obvious time distribution pattern.

Therefore, it is recommended that the management department improve the emergency management system, formulate emergency plans, and prevent disasters in advance. After natural disasters cause damage to the waterway, it is necessary to repair the blocked waterway as soon as possible and restore navigation.

REFERENCES

Cvetkovi, V. M., & Bokovi, D. (2015). Analysis of the spatial and temporal distribution of drought as a natural disaster–analiza geoprostorne i vremenske distribucije sua kao prirodnih katastrofa. *Bezbednost Beograd*, 61(3), 148–164.

J Rosselló, Becken, S., & Santana-Gallego, M. (2020). The effects of natural disasters on international tourism: a global analysis. *Tourism Management*, 79, 104080.

Lu, H., Chen, M., & Kuang, W. (2020). The impacts of abnormal weather and natural disasters on transport and strategies for enhancing the ability for disaster prevention and mitigation. *Transport Policy*, 98.

Shuang, L., Zhu, X., & Lei, Q. (2010). Impact of unexpected natural disasters on transport demand management. *Road Traffic & Safety*, 80.

Sivakumar, M., Motha, R. P., & Das, H. P. (2005). *Impacts of natural disasters in agriculture, rangeland and forestry: an overview*. Springer Berlin Heidelberg, 10.1007/3-540-28307-2(Chapter 1), 1–22.

Advances in Measurement Technology, Disaster Prevention and Mitigation – Li & Mohd Yusof (Eds)
© 2023 The Author(s), ISBN: 978-1-032-36087-4

Research on resilience and disaster resilience evaluation system for older communities under hierarchical analysis and delphi method

CenSi Hu* & Xin Zhou*
Department of Environmental Art Design, Hubei University of Technology, Wuhan, China

ABSTRACT: In order to solve the problem of resilience and disaster resistance evaluation of old communities, this paper establishes a resilience and disaster resistance evaluation index system with 5 primary indicators, 10 secondary indicators, and 28 tertiary indicators, and then constructs a resilience and disaster resistance evaluation index model for old communities based on hierarchical analysis and Delphi method. The weights of each index are clarified and scored to effectively evaluate the resilience and disaster resistance of communities to cope with major disaster risks. This paper takes the Wuhan Huayuanshan community as a case study, and systematically evaluates the resilience and disaster-resilience evaluation index system of the old community by means of a field survey and questionnaire.

1   INTRODUCTION

In recent years, the fragility of the old urban community space environment has become increasingly prominent, which makes us strongly aware that the community, as the first catalyst for urban unit management, plays a vital role in urban development. However, due to the short history of resilience building in old communities in my country, the current theoretical research still mostly stays at the target level, mechanism level, and local functions and characteristics of community building. These studies have a high concept. It is instructive but lacks practical operability, so in the practice of community construction, the specific implementers generally feel that they have no idea where to start. Therefore, it is necessary and urgent to establish a set of evaluation index systems with both theoretical support and operability to guide the resilience construction of old communities.

This paper hopes to establish a set of disaster resilience evaluation systems for old communities through AHP and Delphi methods, so as to diagnose the disaster risk, vulnerability, and exposure of old communities in different aspects, stages, and modes. It is also helpful to target the problems, find appropriate solutions and optimization strategies, reduce the disaster vulnerability of old communities, improve the recoverability of old communities, and achieve the purpose of improving the overall resilience of old communities.

2   RESEARCH STATUS AND METHODOLOGY

2.1   *Current status of domestic and international research*

Under the urban resilience and disaster resistance index system, community resilience and disaster resistance are gradually being carried out. At present, the research on community construction based on resilience and disaster resistance at home and abroad aims to require the community to have a low vulnerability, that is, the occurrence of disasters is not easy to cause damage to the

---

*Corresponding Authors: 383033106@qq.com and 271035600@qq.com

DOI 10.1201/9781003330172-38

community; restoration denotes that the community is easy to recover or repair after a disaster occurs. Foreign research scholars Norris and others believe that community resilience is composed of different aspects of adaptive capacity, including economic development, social capital, information and communication, and community capacity. In 2008, foreign research scholars SL Cutter, L Barnes, etc. believed that community resilience is the ability of a social system to respond to and recover from disasters (Cutter 2010). In 2015, SA Alshehri et al. proposed a framework for community resilience against disaster risks using Saudi Arabia as an example and used the Delphi Method in their study. AHP and Analytic Hierarchy Process (AHP) extracted indicators related to community resilience through a large number of academic literature, and at the same time adopted the recommendations of the expert group to encode AHP weights to form a community resilience framework system, consisting of a total of 62 elements(Alshehri 2015). Domestic scholar Zhu Huagui defined community resilience as the resilience of a community and attributed its key indicators to four dimensions, including physical factors, institutional factors, demographic factors, economic factors, and 18 secondary evaluation indicators. In recent years, Yang Baoqing, Li Guicai, and others have applied the DPSRC model to the assessment of community resilience and selected 28 elements from the driving force layer, pressure layer, state layer, response layer, and control layer to form an international community social resilience evaluation index system.

2.2 *Research methodology*

(1) Hierarchical analysis method (AHP). It is a multi-criteria decision-making method proposed by American operations researcher T.L. Satty and others in the 1970s. It refers to taking a complex multi-objective decision-making problem as a system, decomposing the objective into multiple objectives or criteria, and then decomposing it. For several levels of multiple indicators (or criteria and constraints), the single-level ranking (weights) and total ranking are calculated by the fuzzy quantitative method of qualitative indicators, which are used as goals (multi-index) and multi-scheme optimization decision-making system method.
(2) Delphi method. Also known as the expert survey method, it was initiated and implemented by the RAND Corporation of the United States in 1946. It is essentially a feedback anonymous inquiry method. It is counted and then anonymously fed back to various experts, solicit opinions again, concentrate again, and then give feedback until a consensus is reached.

3 CONSTRUCTION OF RESILIENCE EVALUATION INDEX SYSTEM FOR OLDER COMMUNITIES

3.1 *Selection and establishment of index system*

Combining the practical experience of resilience and relevant literature studies at home and abroad, the evaluation indexes affecting the resilience of old urban communities can be broadly categorized into the ecological environment, community organizational measures, community demographic structure, community well-being index, community poverty level, economic status, social capital, building structure and infrastructure. Based on the selection principles of the resilience evaluation index system, the selection basis and the characteristics of the temporal and spatial dimensions of street resilience, the impact factors of street resilience are delineated, and the overall evaluation objectives are divided into the benchmark layer (first-level indicators), sub-benchmark layer (second-level indicators) and indicator layer (third-level indicators) layer by layer. Target layer A is the evaluation index system of resilience and disaster resistance in old communities; the benchmark level B_i (i = 1, 2, . . . , 5), the target layer is the community environmental resilience, community facility resilience, community economic resilience, community institutional resilience, and community organizational resilience; the sub-benchmark layer C_i (i = 1, 2, . . . , 10) is the second-level index in the evaluation index system; the indicator layer D_i (i = 1, 2, . . . , 28) is the three-level index in the evaluation index system. Together, they constitute the hierarchical structure model of the resilience evaluation index system of old communities (Table 1).

Table 1. Hierarchical structural model of the resilience and disaster resistance evaluation index system for old communities.

Target layer	First-level indicators (benchmark player)	Second-level indicators (sub-baseline layer)	Three-level indicators (indicator layer)
	Community Environmental Resilience B1	Ecological Environment C1	Geomorphology D1 Natural Disaster D2 Protected green space, street greening D3
		Space Organization C2	Site selection, floor area ratio D4 Traffic Organization D5 Disaster and risk avoidance capacity D6
	Community Facilities Resilience B2	Infrastructure C3	Commercial support scale D7 Medical support scale D8 Education Package Size D9 Public infrastructure services supporting scale D10
		Disaster prevention facilities C4	Emergency Shelter D11 Disaster prevention barrier D12 The effectiveness of various security facilities D13
Resilience Evaluation Index System for Older Communities	Community Economic Resilience B3	Economic stability C5	Proportion of resident-owned housing D14 Income level of residents D15 Resident employment rate D16
		Economic Diversity C6	Residential area industry type D17 Resident income channel D18
	Community System Resilience B4	System Assessment Power C7	Toughness evaluation system D19
		Management Mechanism C8	Disaster Prevention Management System D20 Emergency Command System D21 Resident management mechanism D22
	Community Organization Resilience B5	Organizational input C9	Number of sanitation facilities and equipment and personnel D23 Number of security facilities and equipment and personnel D24 Number of service facilities and equipment and personnel D25
		Organizational skills C10	Disaster prevention and emergency exercise organization frequency D26 Level of detail of emergency-related provisions D27 Resident self-organization capacity D28

3.2 Establishment of evaluation model

In order to realize the quantitative evaluation of resilience and disaster resistance of old communities, the importance of each index differs, and at the same time, the resilience level of each index of old communities is different, so the hierarchical analysis method is chosen for the determination of weights.

(1) Construction of judgment matrix. The judgment matrix can establish quantitative relationships among the evaluation indexes at each level. First of all, the "survey form of resilience impact factors of old communities" was distributed to experts in various fields such as professors and senior engineers of related professions. After obtaining the survey results, a judgment matrix is constructed based on the experts' scores, and the weights are calculated and tested for consistency, and the results will objectively reflect the weights and ranking of the impact factors on the resilience and disaster resistance of old communities through hierarchical analysis and impact factor weight evaluation. The "1–9 scale method" ($X = 3, 5, 7, 9$) is used, where 1 means equally important, 3 means slightly important, 5 means more important, 7 means very important, 9 means absolutely important, and 2, 4, 6, 8 means something in between. The importance of the two indicators is compared and the value is a_{ij}. Assuming that the factor layer exists, the judgment matrix is listed as $A = \{a_{ij}\}$, where $a_{ij} > 0$, $a_{ij} = 1/a_{ij}$, i , j, $i = 1, 2, 3, 4 \ldots, m$. Taking layer A and layer B as an example, the construction of the A − B judgment matrix is as below (Figure 1).

$$A = \begin{Bmatrix} 1 & 2 & 4 & 3 & 2 \\ 1/2 & 1 & 3 & 4 & 2 \\ 1/4 & 1/3 & 1 & 1/2 & 1/3 \\ 1/3 & 1/4 & 2 & 1 & 1/2 \\ 1/2 & 1/2 & 3 & 2 & 1 \end{Bmatrix}$$

Figure 1. A–B Judgment matrix.

(2) The characteristic root method is used to calculate the weight values. Firstly, this matrix row vector is multiplied, i.e. $M_i = a_{i1} \times a_{i2} \times a_{i3} \times \ldots \times a_{ij}$, followed by m times the square to obtain $\sqrt[m]{M_i}$. By taking this factor layer as m, all the indicators are calculated in this way. Then it is needed to sum up $H = \sum_{i=1}^{m} \sqrt[m]{M_i}$ to obtain the weight of the indicator. $Q_i = \sqrt[m]{M_i}/H$. Taking A layer and B layer as an example, the weight value is obtained (Table 2).

Table 2. AHP hierarchical analysis results.

Item	Eigenvector	Weight value Q	Maximum Eigenvalue	CI value
Environment B1	1.813	36.263%	5.144	0.036
Facility B2	1.393	27.861%		
Economy B3	0.360	7.202%		
System B4	0.525	10.501%		
Organization B5	0.909	18.173%		

From the above table, a 5th order judgment matrix was constructed for a total of 5 items: environment, facilities, economy, system, and organization. The eigenvectors were analyzed, and the corresponding weight values of the 5 items were obtained, which were 36.263%, 27.861%, 7.202%, 10.501%, and 18.173%. In addition, the maximum eigenroot of 5.144 can be calculated by combining the eigenvectors, and then the CI value of 0.036 CI is calculated by using the maximum eigenroot value for the following consistency test.

(3) Consistency test. Expert scoring is a subjective judgment, which is prone to produce undesirable results and requires consistent testing. The random consistency ratio is calculated, $CR = CI/RI$, where RI is the average random consistency index (Table 3). $CR < 0.1$ when the judgment matrix has satisfactory consistency. When $CR \leq 1$, the consistency is considered acceptable. If the consistency does not meet the $CR \leq 0.1$ requirements, it must be re-evaluated. In this paper, Spssau and Excel were used to calculate CI and RI. The calculation results show that the judgment matrix of

each CI is close to 0, and CR is less than 0.1, indicating that each judgment matrix has satisfactory consistency (Table 4).

Table 3. Random consistency RI table.

n	RI	n	RI	n	RI
3	0.52	9	1.46	15	1.59
4	0.89	10	1.49	16	1.593
5	1.12	11	1.52	17	1.606
6	1.26	12	1.54	18	1.613
7	1.36	13	1.56	19	1.620
8	1.41	14	1.58	20	1.629

Table 4. Consistency test results.

Maximum characteristic root	CI value	RI value	CR value	Consistency test results
5.144	0.036	1.120	0.032	Pass

(4) Determining the weight of resilience indicators of old communities: Selecting experts in the field of resilient cities, disaster prevention and mitigation, and comprehensive old community renewal and construction to score, the data is collected, and the geometric mean is calculated. The calculation steps are as follows: Assuming that n experts score a certain indicator as $W_1, W_2, ..., ...W_n$, then the geometric mean $W = \sqrt[n]{W_1 \times W_2 \times ... \times W_n}$, if $W \times (1 - 5\%) < W_i < W \times (1 + 5\%)$, then W is the score of the index, otherwise, the index is re-scored until the condition is satisfied, and finally, the score is multiplied by the corresponding weight to get the final value which is the toughness evaluation index R. The formula is as follows:

$$R_i = Q_i \sum \left[Q_{ij} \left(\sum Q_{ijk} \times W_{ijk} \right) \right]$$

$$R = \sum_{i=1}^{4} R_i$$

where: Q – Weights; W – evaluation value; i – The number of the reference layer; j – The number of the sub-basis layer; k – The number of the indicator layer.

4 EMPIRICAL CASES

4.1 *Basic overview of Wuhan Garden Hill Community*

Wuhan Grain Road Street Garden Hill Community is located in the ancient city historical area north of Wuchang Snake Hill, with a total community area of about 0.16 square kilometers. The existing residents are 3059 households and 7375 people. At present, the community is distributed into residential areas such as Shuangbai District, Han San Gong, Drum Rack Slope, and Shu Xiang Yuan, and also includes cultural, sports, and medical institutions such as Wuhan High School, Hubei University of Traditional Chinese Medicine and Hubei Provincial Hospital of Traditional Chinese Medicine. Garden Hill community belongs to the old urban community, the resilience index is not high, and the overall resilience and disaster-resistant construction still need to be improved (Figure 2).

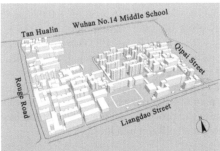

Figure 2. Analysis of the general situation of the Huayuanshan community.

4.2 *Evaluation results and analysis*

Through collating information, field research, and questionnaire collection, the proposed old community resilience evaluation index system was combined to evaluate the Garden Hill community, and the weights were multiplied with the corresponding scores to derive the scoring results and preliminary grading, as shown in Table 5.

Table 5. Weights and scores of resilience evaluation indicators in Huayuanshan old community.

First-level indicators (benchmarklayer)	Weights Q	Second-level indicators (sub-baseline layer)	Weights Q	Three-level indicators (indicator layer)	Weights Q	Score W
B1	0.36	C1	0.17	D1	0.038	70.26
				D2	0.042	75.34
				D3	0.07	62.54
		C2	0.19	D4	0.037	60.53
				D5	0.086	58.65
				D6	0.082	55.52
B2	0.27	C3	0.14	D7	0.026	75.28
				D8	0.069	78.69
				D9	0.025	76.45
				D10	0.037	74.82
		C4	0.15	D11	0.048	52.65
				D12	0.059	48.64
				D13	0.051	45.31
B3	0.07	C5	0.07	D14	0.013	68.29
				D15	0.014	72.47
				D16	0.014	65.34
		C6	0.08	D17	0.014	80.56
				D18	0.011	72.18
B4	0.1	C7	0.03	D19	0.029	65.52
		C8	0.03	D20	0.03	62.75
				D21	0.032	56.14
				D22	0.021	45.92
B5	0.18	C9	0.05	D23	0.023	58.68
				D24	0.02	52.35
				D25	0.02	54.47
		C10	0.06	D26	0.031	52.57
				D27	0.021	47.16
				D28	0.024	60.42

Taking R_1 as an example, $R_1 = 0.36 \times \{0.17 (0.038 \times 70.26 + 0.042 \times 75.34 + 0.07 \times 62.54) + 0.19 (0.037 \times 60.53 + 0.086 \times 58.65 + 0.082 \times 55.52)\} = 1.43$ (rounded to the last two digits). This gives the same reasoning as $R_2 = 0.77$; $R_3 = 0.07$; $R_4 = 0.02$; $R_5 = 0.08$. Then $R = R_1 + R_2 + R_3 + R_4 + R_5 = 2.37$. Therefore, we can obtain the evaluation index of resilience and disaster resistance of the old community of Garden Hill, which is 2.37. $R = 2.37$. From this analysis, we can get the conclusion that the old community has a poor overall resilience level, does not have strong stability and strong disaster adaptive capacity when dealing with various disasters, and has poor disaster resistance and mitigation response capacity, and the construction of this old community can provide a certain degree of the basis for the resilience and disaster resistance of other integrated old communities.

5   CONCLUSION

Based on the established community resilience index system, this paper constructs a comprehensive community resilience evaluation model based on the hierarchical analysis method, and finally validates it by applying it to the Garden Hill community in Wuhan, so as to draw the following conclusions:

(1) The examples show that community resilience can be studied quantitatively based on the resilience index system, thus achieving a comprehensive community resilience evaluation. It also shows that the indicators corresponding to environment, facilities, economy, system, and organization can better express the resilience of the community resilience to disasters.
(2) The hierarchical analysis method can achieve the quantitative evaluation of community resilience and disaster resistance, and the evaluation results are basically consistent with the reality, which can provide a reference for other community resilience and disaster resistance evaluation.

REFERENCES

Alshehri S A, Rezgui Y, Li H. Disaster community resilience assessment method: a consensus-based Delphi and AHP approach[J]. *Natural Hazards*, 2015, 78(1):395–416.
Cutter S L, Burton C G, Emrich C T. Disaster Resilience Indicators for Benchmarking Baseline Conditions[J]. *Journal of Homeland Security and Emergency Management*, 2010, 7(1).
Han L, Mei Q, Lu YM, et al. Analysis and research of AHP-fuzzy integrated evaluation method[J]. *Chinese Journal of Safety Science*, 2004, 14(7):86–89.
Sharifi A, Yamagata Y. Major principles and criteria for the development of an urban resilience assessment index[C]// International Conference & Utility Exhibition on Green Energy for Sustainable Development. IEEE, 2014.
Tan Yuejin. *Quantitative analysis methods* [M]. Beijing: People's University of China Press. 2012.
Tan, W.A., Wang, R.G. Research on the environmental quality evaluation system of intelligent community based on AHP method[J]. *Journal of Shanghai Second University of Technology*, 2019,36(03):188–194. DOI:10.19570/j.cnki.jsspu.2019.03.007.
Xu Xiaohan. *Research on Jinan Community Resilience Evaluation System and Planning Strategy under the Concept of Disaster Prevention* [D]. Shandong Jianzhu University, 2020.

Advances in Measurement Technology, Disaster Prevention and Mitigation – Li & Mohd Yusof (Eds)
© 2023 The Author(s), ISBN: 978-1-032-36087-4

Study on attenuation effect of buffer barrier on blast load in the soil

Zhao Zhang, Fei Liu*, Yonghong Gao, Kai Xin, Minhua Yan, Xu Huang, Yapeng Duan & Chaoyuan Huang
IDE, Academy of Military Science, Luoyang, China

ABSTRACT: Underground structures are vulnerable under blast loads, and it is difficult to protect them from being destroyed. This paper proposes a design method of buffer barrier based on the principle of wave interface transmission. A series of explosion tests of eight different material combinations were carried out to study the attenuation effect of buffer barrier on blast loads and explore the blast resistance against the close-in detonation effect in soil. The results show that the proposed design method can effectively reduce the peak value of blast loads, and the explosion wave passing through the buffer barrier attenuates into a compression wave with a lower peak value and longer duration. Different barrier materials can improve the peak load attenuation efficiency to varying degrees, in which barriers incorporating plastic pipes and steel pipes show a superior attenuation effect in the selected design scheme. The peak load attenuation efficiency decreases approximately linearly with the increase of the average density of the soft interlayer.

1 INTRODUCTION

Accidental explosion events (such as terrorist attacks) in the global world may make essential infrastructures susceptible to failure at home and abroad. Most of the underground infrastructures in cities are shallow-buried structures, such as underground tunnels, subways, utility tunnels, and basements, which are easily damaged under blast loads (Koneshwaran et al. 2015), and there are potential hazards in structural safety of blast resistance. Therefore, how to improve the blast-resistant performance of shallow-buried structures has become a focus in the research field of protective engineering (Mussa et al. 2017).

Due to many uncertain factors of explosion accidents outside shallow-buried structures, it is difficult to protect them from being destroyed. At present, a common protective measure is to set a barrier between the explosion source and the underground structure to buffer and isolate vibration (Kobielaka et al. 2007). Recently, researchers have studied the influence of different barrier materials, structural forms, and arrangement methods on the anti-explosion protection ability of underground structures. De and Zimmie (De & Zimmie 2007; Davies & Williams 1992; Davies 1994) initially reported the feasibility of the compressible porous barrier composed of polyurethane in underground tunnels to resist surface explosion and revealed its beneficial aspects in reducing the deformation of underground structures. De (De et al. 2013; 2016) et al. compared and evaluated the blast-resistant performance of polyurethane foam through centrifugal model test and numerical simulation. It was found that the increase of protective barrier thickness can reduce the stress, strain, and pressure at the top area of the tunnel under the same burial depth, thus reducing the dynamic response of the structure. However, when the thickness exceeds a particular critical value, the blast-resistant performance of the polyurethane barrier is no longer significant. Du (Du et al. 2006) et al. studied the anti-explosion technology of the soft backfill layer and selected foam concrete with low elastic modulus and density as the backfill material on the surface of the

*Corresponding Author: 13525944181@163.com

underground box structure. The numerical results show that the soft backfill layer with sufficient thickness can effectively reduce the peak acceleration of the structure, and the shock insulation effect mainly depends on the yield strength of the backfill material. Ma (2010) studied the blast resistance of foam concrete backfill layer to shallow-buried circular tunnel by numerical simulation method. It is found that the backfill layer can reduce the peak acceleration and displacement of the structure center, and ameliorate the stress and internal vibration condition of the structure. Baziar (2018) et al. carried out a model test of a rectangular section tunnel in soil on centrifuge and investigated the blast-resistant effect of the polystyrene insulation layer on underground structures. In the small-scale test, it is found that this material significantly reduced load impulse, peak strain, and main frequency of structural vibration. The protection effect was also compared among several foam layer positions in the blast test in which the explosive is set on top of the structure roof. The results show that the performance of the foam layer is better when arranged near the structure. Yang (2019) et al. studied the protection performance of polystyrene foam against an external explosion of the water culvert by using numerical methods. It is found that the polystyrene foam board laid on top of the structure can effectively absorb the energy of explosion shock wave and reduce dynamic loading on the roof, thus reducing the dynamic structural response and improving its anti-explosion ability.

Compared with related research, the design methods of buffer barriers can be divided into two categories according to the attenuation mechanism of explosive load in soil. In the first method, the energy absorption effect is generated by the deformation, dislocation, and compression of interlayer materials under blast loads to dissipate the explosive energy coupled into the soil, such as coarse sand and EPS geofoam (Kobielaka et al. 2007). In the second method, the interlayer is set to increase the difference in wave impedance between layers of blast load transmission medium to reduce the transmitted wave intensity. Among these design methods, the composite buffer barrier is an economical and applicable way of protection for underground engineering. The blast load intensity can be attenuated several times by setting a "hard and soft" interlayer in soil. The attenuation principle of the blast wave is shown in Figure 1. When the explosive compression wave propagates into the hard medium from the soft medium, the boosting time of the load is obviously increased due to the energy dissipation mechanism. When the wave propagates into the soft medium from the hard medium, the pressure value of the transmitted wave decreases according to the reflection and transmission law of shock waves at the interface, which significantly reduces the intensity of the explosive compression wave. After more than twice transmissions from different material interfaces, the explosive load finally attenuates into a compression wave with lower intensity.

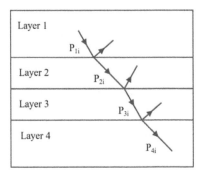

Figure 1.   Attenuate principle in composite buffer barrier.

The composite buffer barrier is designed with common building engineering materials, which possess the advantages of relatively low material cost, easy acquisition, and low construction difficulty. This study puts forward seven schemes of buffer barrier layout under blast conditions through the collocation and combination of materials to analyze the differences in the explosion energy dissipation process of typical engineering materials. The best collocation scheme was obtained through comparative analyses of the blast load attenuation effect in experimental research, which can provide a reference for its anti-explosion design and application.

2 TEST SETUP

In the blast test, the one-way reinforced concrete (RC) slab specimens were used to simulate the roof of the shallow-buried structure. The size details of RC specimens are shown in Figure 2. Figure 3 shows the arrangement of buffer barrier and sensor layout in soil. The depth of backfill soil above the RC slab is 2.5 m, and the buffer barrier is set at a depth of -1.0 m~-1.5 m in soil, with a hard layer of 0.3 m in thickness and a soft layer of 0.2 m in thickness. The hard layer material is gravel in all tests, and the materials of the soft layer are large coarse aggregate, coarse sand, wood, plastic pipe, iron pipe, cement brick and fly ash, respectively. The properties of each material are shown in Table 1. The laboratory test results of soil samples show that the backfill is silty clay with a density of 1.82 g/cm³ and water content of 15.45%. The influence of groundwater level is ignored in this study.

Figure 2.   RC slabs and dimensions.

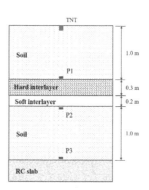

Figure 3.   Sensor layout.

Table 1.   Material properties.

Materials	Average density (g/cm³)	Specifications
Gravel	2.60	Building gravels meeting the requirements of the JGJ52-2006 standard of 5~25 mm continuous grading
Large coarse aggregate	2.20	Broken (egg) stone with a particle size of not less than 50 mm
Coarse sand	2.63	Building sand with a fineness modulus of 2.6 meeting the requirements of categorizing II in the JGJ52-2006 standard (medium sand).
Wood	0.53	Pine with section size of 2.5 cm×6 cm
Plastic pipe	–	PVC square pipes with section size of 10 cm×10 cm and thickness of 2 mm
Steel pipe	–	Q235 steel pipes with section size of Φ50mm and thickness of 2 mm
Cement brick	2.10	Light-weight porous cement brick with a single size of 20 cm×20 cm×40 cm
Fly ash	1.74	Fly ash with a fineness of 11.4% meeting the requirements of GB/T1596-2017 standard (Class I)

The laying size of the buffer barrier is designed to be 2.0 m× 2.0 m to reduce the diffraction of explosion waves in soil considering the span of the RC slab. The sensitive surface of pressure sensors is set 20 mm away from the boundaries of buffer layers, and fine sand is laid around the sensor to ensure the uniformity of surface pressure to prevent the error of sensor measurement caused by interlayer dislocation as much as possible. Sensor P1 is used to measure the incident pressure of the buffer barrier, and sensor P2 is used to measure the transmission pressure under the buffer barrier. Sensor P3 is set very close to the RC slab to measure the incident pressure of the structure. The laying process of partial buffer barriers is shown in Figure 4. Layered rolling was carried out when the covering soil was backfilled to make the backfill uniform in the property.

| (a) Pressure sensor | (b) Cement bricks | (c) Gravels |

Figure 4.   Layout process of buffer barrier and sensors.

The test conditions are shown in Table 2. A total of 8 materials are tested (H1~H8), in which H1 is the control group, i.e., the case without buffer barrier, which is used to compare the attenuation effect of composite buffer barriers on explosion waves in soil. TNT was used in the test, and the charge weight was 0.6 kg with a size of 7.5 cm×5 cm×10 cm under each condition. The explosive charge is set in the shallow burying mode (the upper surface of the explosive is flush with the ground surface) to ensure test operation safety and make explosive energy couple into the soil as much as possible. The electric detonator is used to detonate in the charge center. After each explosion test, the crater was backfilled and compacted in layers to restore the surface as before.

Table 2.   Test conditions.

Test	Charge weight/kg	Hard layer	Soft layer
H1		Soil	Soil
H2		Gravels	Large coarse aggregate
H3		Gravels	Coarse sand
H4		Gravels	Wood
H5	0.6	Gravels	Plastic pipe
H6		Gravels	Steel pipe
H7		Gravels	Cement brick
H8		Gravels	Fly ash

3   DISCUSSION OF TEST RESULTS

3.1   *Attenuation effect of buffer barrier*

Taking H1 (control working condition) and H5 working condition as examples, the comparison curve of pressure at measuring points P1~P3 recorded in the test is shown in Figure 5. It can be seen from the figure that the attenuation effect of the buffer barrier on blast load in the soil is mainly shown in the peak load and positive pressure time. From P1 to P3 in the H1 test condition, the peak load values decreased by 54.1% and 81.0%, respectively, while in the H5 test condition, the peak load values decreased by 72.7% and 87.1%, respectively. Both peak load values at P2 and P3 decrease, indicating that the buffer barrier can reduce the intensity of compressive waves in the soil below. The positive pressure time of measuring point P3 in the H1 test condition is 28.4 ms, and that in the H5 test condition is 33.5 ms, with a slight increase in time, which indicates that the explosion wave attenuates into a compressive wave with a lower peak value and longer duration time after propagating through the buffer barrier. Due to the shallow buried depth of P1 and P2 sensors, the explosive compression waves arrive at P1 and P2 measuring points successively within about 3 ms after explosive initiation. For the complete soil medium (H1), the arrival time of compression waves at P1 and P2 are almost the same, but the arrival time of P2 compression waves is about 5.0 ms later than that of P1 in the H2 test condition. Moreover, the arrival time of compression waves at P3 above the structure is 5.6 ms in the H1 test condition, whereas, in the H2 test condition, it is 10.8 ms, which indicates that the set of buffer barriers delays the arrival time of explosive compression waves.

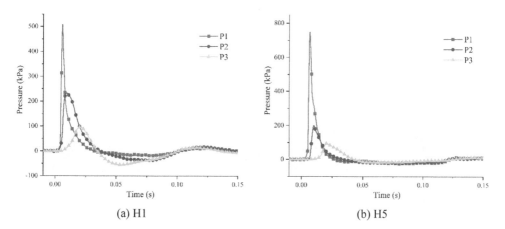

(a) H1   (b) H5

Figure 5.   Pressure-time curve.

3.2 *Influence of materials on attenuation effect of buffer barrier*

To evaluate the protective effect of buffer barrier on blast loads, the concept of "pressure attenuation efficiency" is put forward in this study for comparative analysis, which is defined as the ratio of pressure reduction of the far and near measuring points below the explosion center to the pressure of the near measuring point. Pressure attenuation efficiency indicates the reduction degree of the peak load when the blast load propagates through two points in soil. Given the characteristics of this test, the following simplified assumption is put forward for the close-in explosion to compare the attenuation effect: the explosion load has an attenuation effect in any medium, and because the thickness of the buffer layer is relatively small, it can be considered that the blast load intensity is constant when it propagates in the interlayer, and the attenuation effect inside the interlayer is ignored.

The pressure measurement results under different test conditions of buffer barrier settings are shown in Table 3. It can be found that there are differences in pressure P1 at several test conditions, which are mainly caused by the shallow buried depth of the sensor, and the error of explosive setting leads to apparent differences in pressure near the charge center. The load attenuation efficiency of different buffer barriers is shown in Figure 6. Compared with soil medium, all kinds of materials show different attenuation effects on the incident pressure. The attenuation efficiency of the buffer barrier containing plastic pipes and steel pipes is superior, which are 72.7% and 71.7%, respectively. This is mainly because the wave impedance difference between the upper and lower barrier layers is the largest by adding air medium, while the plastic pipes and steel pipes only react to support the barrier structure. The attenuation efficiency of other materials from high to low are large coarse

273

aggregate, wood, fly ash, coarse sand, and cement brick. The attenuation effect of cement brick is the least obvious. It shows little difference from that of the control group, which is mainly because the average density is close to silty clay density.

The attenuation efficiency of peak blast pressure on the structure surface is approximately the same as that of the buffer barrier, and the steel-tube buffer barrier shows the best performance, with an attenuation efficiency of 89.3%. The buffer barrier containing large coarse aggregate, cement brick and fly ash shows no noticeable attenuation effect of peak blast pressure on the structure surface. Their attenuation efficiency is only 2.2%, 4.0%, and 2.0% different from H1. Although the peak load attenuation efficiency under the sand and wood buffer barrier is low, the peak load attenuation efficiency on the structure surface is relatively high, which may be caused by the measurement error of the sensors for a small peak blast load.

Table 3. Peak pressure in soil (kPa).

Test	P1	P2	P3
H1	507.1	232.9	96.6
H2	305.1	108.2	63.6
H3	388.0	159.0	52.5
H4	674.8	241.6	84.2
H5	748.1	203.9	96.5
H6	1065.5	301.1	113.6
H7	296.5	130.7	65.7
H8	492.8	177.8	85.6

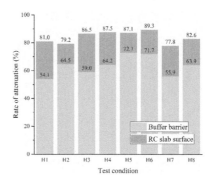

Figure 6. Pressure attenuation efficiency.

Figure 7 shows the variation law of peak pressure attenuation efficiency with the average density of soft interlayer materials. The average density of plastic and steel pipes is approximately taken as air density according to the energy dissipation mechanism, ignoring their weight. Figure 7 shows that the attenuation efficiency of the buffer barrier to peak blast pressure gradually decreases with the increase of the average density of the soft interlayer, which indicates that reducing the average density of the soft interlayer is beneficial to dissipation of blast energy propagated into the soil. The buffer barrier still shows the attenuation effect contrasting to the theoretical prediction, notwithstanding the soft interlayer density is higher than soil density. The empirical prediction value

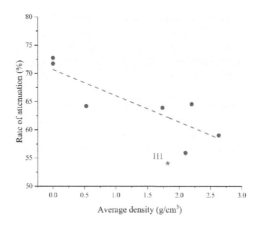

Figure 7. Variation of peak pressure attenuation efficiency.

of attenuation efficiency of the composite buffer barrier proposed in this paper can be approximately obtained by linear function fitting for different attenuation efficiencies. The fitting function is shown in the following formula, and it shows that the optimal attenuation efficiency is about 70.7%.

$$Y = -4.645X + 70.685$$

where X is the average density (g/cm^3) of the soft interlayer, and Y is the attenuation efficiency (%) of the peak pressure propagating through the barrier.

4 CONCLUSION

(1) Based on the principle of wave transmission at the interface, this paper puts forward a composite buffer barrier design scheme composed of hard and soft interlayers. The test results show that it can effectively attenuate the peak blast pressure under close-in detonation in soil and show good performance in decreasing blast loads on top of underground structures.

(2) Paving buffer barriers attenuates the blast load into a compression wave with a lower peak value and longer duration time and delays the arrival of blast loads at a specific position in the soil.

(3) Each soft interlayer material tested in this study can improve the attenuation efficiency of peak blast load to a certain extent, among which the buffer barrier incorporating plastic pipes and steel pipes show a superior attenuation effect, while cement brick shows the least apparent effect. Peak pressure attenuation efficiency decreases linearly with the increased average density of soft interlayer approximately, and the best optimal attenuation efficiency is about 70.7%.

REFERENCES

Baziar MH, Shahnazari H, Kazemi M, Mitigation of surface impact loading effects on the underground structures with geofoam barrier: Centrifuge modeling 2018 Tunnelling and Underground Space Technology, 80, 128–42.

Davies MCR. 1994 Dynamic soil-structure interaction resulting from blast loading. In: Proc of the International Conference on Centrifuge Modelling (Centrifuge 94) Singapore, pp. 319–24.

Davies MCR, Williams AJ 1992 Centrifuge Modelling the Protection of Buried Structures Subjected to Blast Loading. In: Proceedings of the International Conference on Structures Under Shock and Impact England.

De A, Morgante AN, Zimmie TF 2013 Mitigation of Blast Effects on Underground Structure Using Compressible Porous Foam Barriers. In: American Society of Civil Engineers Fifth Biot Conference on Poromechanics Vienna, Austria pp. 971–80.

De A, Morgante AN, Zimmie TF Numerical and physical modeling of geofoam barriers as protection against 2 effects of surface blast on underground tunnels 2016 Geotextiles and Geomembranes, 44(1), 1–12.

De A, Zimmie TF Centrifuge modeling of surface blast effects on underground structures 2007 *Geotechnical Testing Journal*, 30(5), 427–31.

Du X, Liao W, Tian Z, Li L *Dynamic response analysis of underground structures under explosion induced loads* 2006 Explosion and Shock Waves (in Chinese), 26(5), 474–8.

Kobielaka S, Krauthammerb T, Walczak, A Ground shock attenuation on a buried cylindrical structure by a barrier 2007 Shock and Vibration, 14(5), 305–20.

Koneshwaran S, Thambiratnam DP, Gallage C, *Blast Response of Segmented Bored Tunnel using*, Coupled SPH–FE Method 2015 Structures, 2, 58–71.

Ma L 2010, Centrifugal Modeling and Numerical Research For Urban Shallow-Buried Tunnel Under, *Blasting* (Beijing: Tsinghua University Press)

Mussa MH, Mutalib AA, Hamid R, Naidu SR, Radzi NAM, Abedini M, *Assessment of damage to an underground box tunnel by a surface explosion 2017, Tunnelling and Underground Space Technology*, 66, 64–76.

Yang G, Wang G, Lu W, Zhao X, Yan P, Chen M, Damage evaluation and protective effect analysis for a large-scale water transmission box culvert under the action of ground surface blast loads 2019 Journal of *Vibration and Shock* (in Chinese), 38(5), 29–37.

Advances in Measurement Technology, Disaster Prevention and Mitigation – Li & Mohd Yusof (Eds)
© 2023 The Author(s), ISBN: 978-1-032-36087-4

Supply and form of earthquake prevention and disaster reduction publicity and education products for the public

Yunlin Liu*
Shaanxi Provincial Earthquake Bureau, Xi'an, China

ABSTRACT: The publicity and education of earthquake prevention and disaster reduction serve the economic goal of minimizing disaster losses caused by earthquakes. At present, the insufficient supply of publicity and education on earthquake prevention and disaster reduction in China has become a short board to comprehensively improve the national capacity for earthquake prevention, disaster reduction, and disaster relief. As a kind of public goods, the mass publicity and education of earthquake prevention and disaster reduction is faced with the triple failure of market failure, government failure, and personal cognition failure, as well as the corresponding incentive problems. In order to establish a long-term supply of publicity and education products, based on the analysis of the current publicity and education products for earthquake prevention and disaster reduction from the five dimensions of form, platform, target, population, and cost, the article suggests integrating resources, perfecting the government investment mechanism and promoting diversified participation; this paper suggests that the supply of products and services should be expanded by means of government purchase of services; the product structure should be adjusted according to the communication situation, and the article suggests to improve the input-output ratio and put forward the list of products and services; referring to the new cooperative idea, the article suggests to focus on promoting the publicity and education of earthquake prevention and disaster reduction in schools.

1   INTRODUCTION

The basic national condition of our country is that earthquakes are numerous, intense, widely distributed, and disaster-prone. The practical experience of earthquake prevention and disaster reduction work proves that strengthening the scientific popularization of earthquake prevention and disaster reduction plays an important role in the public's correct understanding of earthquakes, enhances self-confidence, and overcomes panic. It is conducive to improving the public's awareness of earthquake prevention and disaster reduction, enhancing the public's ability to help and rescue each other, and improving the overall efficiency of earthquake emergency response (Liu & Xu 2021). It is one of the most important means to effectively reduce earthquake disaster losses and casualties. However, at present, the public's awareness of earthquake prevention and disaster reduction is generally low, and the mastery of skills is generally insufficient, which has become a short board for comprehensively improving the national earthquake prevention and disaster reduction and relief capabilities. The essence of the disaster problem is an economic problem, the ultimate goal of earthquake prevention and disaster reduction and relief is to minimize losses (Zheng 2021), and the development of earthquake prevention and disaster reduction publicity and education plays an important role in reducing disaster losses, and has its necessity and urgency.

---

*Corresponding Author: 514263941@qq.com

DOI 10.1201/9781003330172-40

2 THE NATURE OF EARTHQUAKE PREVENTION AND DISASTER REDUCTION PUBLICITY AND EDUCATION IS RELATED TO TRIPLE FAILURE

The concept of earthquake prevention and disaster reduction publicity and education includes two parts, earthquake prevention and disaster reduction point to specific areas and goals, and publicity and education refer to means, methods and carriers. Based on the two considerations of convention and policy research, this paper defines the concept of earthquake prevention and disaster reduction publicity and education as enabling the public to master the knowledge and skills of earthquake prevention and disaster reduction through publicity and education, enhance the awareness of earthquake prevention and disaster reduction, and form a cultural consciousness of earthquake prevention and disaster reduction. Its carrier includes a range of products and services. As a public good, publicity and education on earthquake prevention and disaster reduction face the triple failure of market failure, government failure, and personal cognitive failure, as well as the corresponding incentive problems.

2.1 *Public goods attributes and market failures*

According to the classification of goods in neoclassical economics (Paul Samuelson & William Nordhaus 2008), the publicity and education on earthquake prevention and disaster reduction are obviously non-competitive and non-exclusive, providing products and services to the public, and the use of one person does not affect the use of others, while potential users cannot be excluded, so it is a pure public good. Since it is a public good, the act of providing a public good is a kind of public management activity. Among them, the government is always the most important entity, although it is not necessarily the only provider of products and services, it should assume the role of the final funder. Although some enterprises provide publicity and education products, such as escape and self-help training, which are exclusive and competitive, making them private goods, to a certain extent, they are still differentiated upgraded products and services based on government products and services. Some foundational products and services, if delivered by the market, are inefficient and incentivized, and businesses will not survive due to short-sightedness, collective action, and free-riding problems.

2.2 *Government failure*

Government failure coexists with market failure. On the one hand, the publicity and education on earthquake prevention and disaster reduction are not mandatory, and almost all relevant laws and regulations require it to be carried out, but its provisions are too principled and there is no path, implementation plan, and means; on the other hand, the effect of publicity and education is difficult to measure objectively, and it is rarely included in performance appraisal. If there are few disasters, leading cadres do not pay attention to them, and there is insufficient public opinion, it is difficult for government departments to have sufficient incentives to provide public goods.

From the perspective of system tradition, in terms of earthquake prevention and disaster reduction methods, China has always been heavy in engineering disaster reduction and light on non-engineering disaster reduction, which is completely contrary to theoretical inference, from the cost-benefit cost performance and system complexity of the two dimensions of dividing the number of measures, formulating emergency plans, and carrying out earthquake prevention. Disaster reduction publicity and education should be easier to carry out (Kenny & Charles 2012), and from the perspective of our country, the tradition of engineering construction is an important reason. At the same time, the funds for earthquake prevention and disaster reduction publicity and education work are not included in the financial budget, and at the central level, such as the National Disaster Reduction Commission, there is no corresponding work funding, and funds can only be obtained through the form of projects. The investment of funds is limited and unstable, and the powers of the central and local governments are not clearly defined. At the same time as limited funds, there is also the problem of fragmentation of resources. Previously, due to the deep fragmentation of the earthquake prevention and disaster reduction and relief system and mechanism (Xu 2018), earthquake prevention and disaster reduction were prevented. Publicity and education resources are scattered in various disaster professional departments and science popularization departments. Activities have been carried out on "pepper noodles", the dispersion of funds makes many medium and high-cost forms of publicity and education unable to land, and low-cost print media products have become the main means, with the development of the times it is easier to flow in the form. With the formation of the Ministry of Emergency Management, this situation will hopefully be alleviated. Therefore, the publicity and education work of China's government departments on earthquake prevention and disaster reduction has shown an extremely weak situation, which is mainly manifested in three deficiencies: lack of long-term mechanisms, heavy sports-style activities, light daily publicity, limited coverage of the population and effectiveness; lack of systematic education, in-depth publicity and education for urban and rural communities, primary and secondary schools, and social forces involved in disaster relief need to be strengthened; lack of means and methods, multi-word indoctrination, less interactive experience, and insufficient publicity and education attractiveness. Information management and service capacity building are extremely backward.

2.3 *Personal cognitive failure*

The publicity and education on earthquake prevention and disaster reduction also face a third failure, that is, the failure of personal cognition. On the one hand, there is a gap between people's own feelings and actual emergency preparedness actions, which are far lower than the former (Kapucu 2008); on the other hand, psychological experiments show that people do not react to events that are less than a certain probability of occurrence, so it is difficult to take action to prevent such a small probability and large loss event as a disaster, and there is no personal need and the resulting social need. With the global climate change, the uncertainty of disasters increases, and the concept of uncertainty rather than risk is used here, thus, it is difficult for people to make their own judgments and form rational expectations of the probability of disasters according to past experience. With the process of urbanization, the "security illusion" of individuals themselves is becoming more and more serious. Coupled with the lack of accessibility of publicity and education products and services, the cost of acquisition is relatively high, and it is difficult for individuals to have the motivation to carry out earthquake prevention and disaster reduction publicity and education for the sake of the interests of ordinary individuals. When conducting the analysis of the dilemma of disaster reduction publicity and education, Feng Sijin (2015) pointed out that the reason for its dilemma is largely due to the regionality and non-normalization of natural disasters, the non-urgency, non-immediateness, and difficulty of evaluating the effects of publicity and education, resulting in a low correlation between disaster reduction publicity and education and relevant actors (Sikkim 2015).

In addition to the logic of interest, there is also the logic of common sense. Neumayer et al. (2013) constructed the concept of disaster propensity (frequency and density of experiencing specific disaster risks), using data from Munich Re to analyze and find that disaster preference has a great impact on incentives for investment in earthquake prevention and disaster reduction. The incentive to prefer a high preference is high, and some places that do not experience disasters often suffer greater losses in the event of sudden extreme disasters (Neumayer, Eric & Plümper, Thomas & Barthel, Fabian. (2014)). Unlike scientific knowledge, common sense is everyone's reliable hope and foresight in everyday life, a common property of a culture that draws a clear picture of arbitrariness and danger in action (Li 2008). Common sense corresponds to practical wisdom, and the application of phronesis in each specific context is "judgment." (Wang 2013) A common sense logic is that if they do not experience disasters frequently, local residents lack disaster knowledge, and in terms of personal preferences, they will rank earthquake prevention and disaster reduction in the lower position. Their utility evaluation is not high, the personal needs are low, and they hope that resources will be allocated more to other people's livelihood affairs, then the entire community will not be able to achieve local regional public affairs. There is a consensus on the importance of earthquake prevention and disaster reduction, and the consensus is the basis for autonomy and action.

3 ANALYSIS OF EARTHQUAKE PREVENTION AND DISASTER REDUCTION PUBLICITY AND EDUCATION PRODUCTS AND SERVICES

3.1 *Publicity and education objectives for earthquake prevention and disaster reduction*

The supply of earthquake prevention and disaster reduction publicity and education products and services must first serve the expected goals. The content of publicity and education can be divided into six aspects: disaster memory, principle, knowledge, skills, awareness, and action, that is, through the development of earthquake prevention and disaster reduction publicity and education, it is necessary to deepen disaster memory, understand the principle of disaster, master the knowledge of earthquake prevention and disaster reduction and escape self-help and mutual rescue, and master the skills of earthquake prevention and disaster reduction and escape self-help and mutual rescue awareness of earthquake prevention and disaster reduction, and the goal of carrying out daily earthquake prevention and disaster reduction actions. The goals to be achieved by different groups of people are not consistent (see Table 1), such as kindergarten children mastering simple escape self-help knowledge and skills, while primary and secondary school students need to understand the principle of disasters, "know why it is so", and master the knowledge and skills of earthquake prevention and disaster reduction and escape self-help and mutual rescue in detail, which can drive the whole family and improve earthquake prevention and disaster reduction awareness, carry out daily earthquake prevention and disaster reduction actions.

Table 1. Publicity and education targets for earthquake prevention and disaster reduction for different groups of people.

target / crowd	Disaster memory	Disaster principles	Earthquake prevention and disaster reduction and escape self-help and mutual rescue knowledge	Earthquake prevention and disaster reduction and escape self-help mutual rescue skills	Awareness of earthquake prevention and disaster reduction	Daily earthquake prevention and disaster reduction operations
Kindergarten children			✓	✓		
Large, Junior High, and Elementary School Students	✓	✓	✓	✓	✓	✓
worker	✓		✓	✓	✓	✓
senior citizen	✓		✓	✓	✓	✓

3.2 *Public earthquake prevention and disaster reduction publicity and education products and services and their platforms*

The products and services derived from the above objectives are various, mainly the following: (1) flat products and services, mainly including popular science books, folding manuals, picture books, comics, wall charts, understanding cards, display boards, print advertisements, mobile phone text messages, exhibitions, etc.; (2) video products and services, it mainly includes public welfare promotional films, documentaries, long and short animation series, animated films, micro-films, movies, etc.; (3) interactive products and services, mainly including games, live broadcasts, H5, artificial intelligence consulting responses and other network and mobile client interactive products; (4) curriculum products and services, including school courses, targeted training, various types of lectures, etc.; (5) experience products and services, including science education base experience projects, community games, emergency drills, etc. There are also activities such as knowledge contests.

The above products and services require corresponding platforms, mainly including the following: (1) new media with digital, integrated, interactive, and networked characteristics (Peng 2016), mainly including Weibo, WeChat, Kuaishou, Douyin, and other graphic and short video social platforms, vertical new media carriers such as today's headlines and a little information, and online live broadcasting platforms such as Yingke and Hua pepper Bilibili and other bullet screen websites, knowledge sharing platforms such as Zhihu, and Internet radio stations such as the Himalayas; (2) traditional media portals and news clients, industry websites; (3) publicity platforms for public places, such as mobile media, theaters, square electronic screens, etc.; (4) Schools; (5) Communities; (6) Earthquake prevention and disaster reduction science and education bases, including

The product service type	Crowd targets	cost
Form F Flat F_1 (popular science books, comics, wall charts, display boards, folding manuals, understanding cards, SMS, print advertisements, exhibitions). Video F_2 (public service films, documentaries, animated films, films, micro-films). interactionF_3(Games, live streams,H53. Artificial intelligence consultation Q&A Course F_4 (school curriculum, targeted training, various lectures). Experience F_5 (Science Education Base Experience Project, Community Games, Emergency Drills.)	Target T Memory T_1 Principle T_2 Knowledge T_3 Skill T_4 Consciousness T_5 Action T_6	Cost C Low cost C_1 (0 to 100,000 RMB). Medium cost C_2 (10 to 1million RMB). High cost C_3 (more than 1 million yuan).
Platform P New Media P_1 (graphic and short video social platform, webcast platform, network television, knowledge sharing platform, bullet screen website, vertical new media). Portal and news client P_2 Public places P_3 School P_4 Community P_5 Popular science education base P_6 Newspaper P_7 TV P_8	Crowd G Kindergarten children G_1 Large, Junior High and Elementary Students G_2 Worker G_3 Elderly G_4	

Figure 1. Analysis dimension of earthquake prevention and disaster reduction publicity and education products and services.

earthquake prevention and disaster reduction education experience halls, disaster site memorial halls, rescue training bases, disaster laboratories of scientific research institutes, etc. (Xu 2017); (7) Newspapers and periodicals; (8) TV.

The intersection of forms and platforms can form a variety of product services, and a product can be launched on different platforms. Different products and services have different costs, meet different goals, and have different effects depending on the specific presentation method and content and the public emotional connection, useful life, creative and interesting and other attraction points. At present, there are many kinds of earthquake prevention and disaster reduction publicity and education products and services in practice, and we can set the form (F), platform (P), goal (T), crowd (G), and cost (C). Five dimensions are analyzed, of which the form and platform definition constitute a specific product service, the target, crowd and cost are the basis for the selection, the analysis matrix is shown in Figure 1, taking the product launched by the National Disaster Reduction Center on the thirteenth anniversary of the Wenchuan earthquake as an example (see Table 2). As can be seen from the table, the traditional low-cost flat products can also achieve better results by using new forms such as animation to spread through new media. During the thirteenth anniversary of the Wenchuan earthquake, according to incomplete statistics, in addition to a large number of thirteenth-anniversary news reports that evoke disaster memories, there are relatively more products that improve knowledge through reading and viewing, and relatively few products that improve skills through experience interaction; there are more products that attract the public through emotional connection, and fewer products that are used to attract the public through creativity and fun and life. For the elderly, earthquake prevention and disaster reduction knowledge and skills products and services are relatively lacking; product costs are concentrated at the middle and low levels, and high-cost products are few.

Table 2. Products launched by the National Disaster Reduction Center Wenchuan on its 13th anniversary

products	form	platform	crowd	cost	effect
Earthquake Prevention and Disaster Reduction Science Popularization Manual (Earthquake Chapter).	F_1	$P_1/P_2/P_3/ P_4/P_5/P_7$	$G_2/G_3/G_4$	C_1	Households entered 1.1 million households, and the WeChat public account pushed a single article with a reading volume of 46,000
Are you willing to do one more thing for Wenchuan? H5	F_3	P_1/P_2	G_2/G_3	C_2	A direct push of 4 million people (excluding forwarding participation)
Earthquake prevention and disaster reduction science popularization series of animated short films (kindergarten, primary and secondary schools).	F_2	$P_1/P_2/ P_3/P_4$	$G_1/G_2/G_3$	C_2	It is broadcast on 37 children's channels such as CCTV Children's Channel, Beijing Kaku Children's Channel, Hunan Golden Eagle Cartoon, and so on
"5·12" National Earthquake Prevention and Disaster Reduction Day public welfare propaganda film (2021) and public welfare poster	F_2	P_3/P_8	$G_3/G_4/G_5$	C_2	The propaganda film was promoted on CCTV 1, 3, 4, 8, 9, 14, and other channels and in more than 10,000 cinemas across the country. The posters were displayed in 326 electronic display advertising booths of the Beijing Metro, with an audience of more than 5 million people per day

3.3 Input-output analysis of public disaster prevention and reduction publicity and education products

According to the attributes of public goods of disaster prevention and reduction publicity and education products, this paper attempts to construct an input-output analysis tool for public disaster prevention and reduction publicity and education products. The set of input elements is shown in Formula (1), where I is knowledge capital, K is capital and N is manpower; the output set is shown in Formula (2). These elements point to the mass level material (expressed in M), system (expressed in B), and culture (expressed in C). These three parts really promote the development of mass publicity and education on disaster prevention and reduction. The functional relationship with input-output is shown in Formula (3), in which socio-economic relations and external institutional guarantees are embedded in the function itself. Formulas (1), (2) and (3) constitute an input-output analysis of public disaster prevention and reduction of publicity and education products. This tool can be used to analyze the problems existing in current community disaster reduction.

$$F = \{K, N, I\}; \tag{1}$$

$$O = \{M, B, C\}; \tag{2}$$

$$f(F) = 0_\circ. \tag{3}$$

In recent years, China has gradually improved the fund investment mechanism for disaster prevention and reduction, further improved the systems and institutions for disaster prevention and reduction, and guaranteed personnel. However, the public's awareness and skills of emergency avoidance have not been well improved, that is, there are weaknesses at the cultural level. Compared with the implementation of materials and systems, there is also a "cultural gap" in the construction of community disaster reduction. Then, the community disaster reduction problems we analyzed can be summarized as follows:

$$\begin{cases} f(I, \overline{N}, \overline{K}) = C; \\ f(x) = I. \end{cases} \tag{4}$$

That is, the emergence of a "cultural gap" in the construction of public disaster prevention and reduction is directly related to the insufficient investment in public disaster prevention and reduction of knowledge capital, which originates from the insufficient formation of knowledge capital.

4 LONG-TERM SUPPLY OF PUBLICITY AND EDUCATION ON EARTHQUAKE PREVENTION AND DISASTER REDUCTION

4.1 Integrate resources, improve the government input mechanism, and promote diversified participation

Building a long-term publicity and education mechanism for earthquake prevention and disaster reduction requires stable government financial investment and sufficient incentives. Since earthquake prevention, disaster reduction publicity, and education are public goods. The government should be the last funder and resource provider, on the one hand, it is necessary to integrate the earthquake prevention and disaster reduction publicity and education resources scattered in various disaster-related departments, clarify the power and expenditure responsibilities of central departments and local governments at all levels, strengthen the linkage of resources at the upper and lower levels, and form a scale effect. Funds for publicity and education on earthquake prevention and disaster reduction have been included in the financial budget, changing the current situation of fighting "guerrilla warfare" through project applications. At the same time, in the context of China, the inclusion of "earthquake prevention and disaster reduction" into the performance appraisal system of governments at all levels will have the most effective incentive effect on government departments.

From the perspective of governance, the involvement of the third sector is an idea to solve the problem of public goods production and supply. In addition to the government's leadership, social forces, enterprises, and institutions should also join in, especially to make full use of social forces and the resources of disaster insurance enterprises, guide social forces to invest more resources in the pre-disaster stage, and give play to the publicity function of insurance enterprises in disaster prevention and loss reduction. When we examine the participation of social forces in publicity and education on earthquake prevention and disaster reduction, we have found some good trends. Previously, the social forces involved in the disaster were more enthusiastic about participating in disaster emergency rescue and paid less attention to the pre-disaster links. In recent years, many social organizations have turned their attention to publicity and education, such as the China Foundation for Poverty Alleviation to carry out the "Say NO to Disasters!"—5· 12 National Disaster Prevention Community Public Welfare Activities", promote the popularization of scientific knowledge on disaster reduction in schools, and compile the "Guide to Earthquake Prevention and Disaster Reduction Education for Primary School Students" (Liu 2018). With the development of disaster insurance, the publicity and education of insurance companies on earthquake prevention and disaster reduction can only gradually resume, for example, the Zunyi Civil Affairs Bureau of Guizhou Province and the Zunyi Branch of PICC Property insurance co-sponsored the earthquake prevention and disaster reduction publicity and education activities relying on policy-based farm house disaster insurance.

4.2 *Expand the supply of products and services by means of government procurement of services*

Changing trading venues, expanding transaction time, enriching transaction categories, speeding up transactions, and reducing intermediate links, the Internet has had a profound impact on the production and supply of products in all walks of life. Accompanied by the Internet, it is a new form of organization, and "intelligent organization" aims to achieve self-learning, self-adaptation, self-coordination, and self-evolution, to achieve flexible, elastic, light development, and external instability, future uncertainty, high complexity of the environment dynamic matching and integration of innovation, showing the characteristics of network ecology, global integration, platform operation, employee-based users, borderless development, self-organization management (Li et al. 2014). Using Internet thinking to create earthquake prevention and disaster reduction publicity and education products are needed by the times. Relevant government departments should transform from production departments to platform departments, integrate and provide resources, select appropriate market and social forces through the purchase of policy tools such as service and product outsourcing, expand diversified participation, create earthquake prevention and disaster reduction publicity and education products, improve the efficiency of capital use, and keep up with the communication situation at any time. It is mainly to establish a good product production cooperation relationship with high-quality cultural companies, establish a good cooperation relationship with the market-oriented head platform and its high-quality users for new media products, and establish a community-oriented publicity and cooperation relationship with social forces and earthquake prevention and disaster reduction science popularization and education bases. Among the earthquake prevention and disaster reduction science popularization and education bases, some rescue training bases are run by social organizations and enterprises, and the government can purchase services for the public to experience the field experience and master the skills of escape and self-help according to the number of people. In the community, the government can purchase social organization services and use its personalized services to carry out a number of earthquake prevention and disaster reduction game caravans in the community activities. This type of outsourced product also has the property of decreasing marginal costs, and the more activities are carried out, the lower the average cost.

4.3 *Adjust the product structure according to the communication situation, refine the delivery, and improve the input-output ratio*

In recent years, with the power of technological means of innovation, China's communication platforms and corresponding products are rapidly updating and iteration, from the decline of traditional media such as newspapers and television, the rise of new media, to the integration and development and rejuvenation of traditional media, as well as the replacement of new media itself from live broadcasting, Internet celebrities, self-media to short video social networking. The market share is changing rapidly, WeChat, Weibo, and Today's headlines and other head platforms have accumulated a dominant position so that the communication shows a trend of oligarchic competition, and the oligarchs themselves finely cover all user groups through market operations. At the same time, the industry clearly shows the tendency of supply to create demand, and people's reading and viewing needs and habits have gradually shown a trend of video, fragmentation, light reading, entertainment, and mobility. Worldwide, the results of a 2017 Pew Research Center survey showed that about 85 percent of Americans read news through mobile devices 13 percent more than in 2016 (NetEase News 2017). In China, as of December 2021, the number of Internet users reached 1.032 billion, the proportion of Internet users using mobile phones reached 99.7%, and users preferred to use targeted push information rather than active search in the past (China Internet Network Information Center 2022). Therefore, conducting annual research and understanding communication trends is a prerequisite for accurate delivery.

> ☑ *A national disaster prevention and mitigation manual that is updated annually*
> ☐ *Once a year, the community or enterprises and institutions will conduct drills*
> ☐ *A recommended list of household disaster relief supplies stockpiles, updated annually*
> ☐ *Multiple short graphic and short video products combining disaster events*
> ☐ *Once a year, the disaster prevention and mitigation science popularization and education base experience activities*
> ☐ *A number of community disaster prevention and mitigation caravans and game experience activities*

Figure 2.   List of recommendations for earthquake prevention and disaster reduction publicity and education products and services.

According to the current communication trend, under the constraint of funds, the relevant departments of earthquake prevention and disaster reduction publicity and education need to build appropriate product genealogies, make full use of mobile new media platforms and news clients of traditional media, focus on promoting disaster memory and earthquake prevention and disaster reduction and escape self-help knowledge graphics, disaster memory, knowledge and skills short videos, community skills, action games, and promote earthquake prevention and disaster reduction experience activities at the science popularization base (see Figure 2). The central departments focus on creating earthquake prevention and disaster reduction manuals, graphics, short videos, and family material proposal lists suitable for the whole people, and local governments focus on promoting the last three items in combination with local realities, and fully disseminating local disaster memories so that local residents can form disaster cognition, preferences, and expectations.

4.4   *With reference to the new cooperative thinking, focus on promoting publicity and education on earthquake prevention and disaster reduction in schools*

Corporatism is a very inclusive concept, Chinese so-called cooperativeism, corporatism, combinatorialism, unificationism, and corporatism, all refer to the same English word, and this article uses cooperativeism to express this concept. The study of co-operativeism belongs to the institutional economics branch of political economy and the field of modern political science, inseparable from the analysis of the politics of interests, mainly from the European experience, and introduced to China in the early 20th century before and after the May Fourth Movement (Zhao 2004). Zheng Bingwen (2002) argues that the cooperative welfare system is based on tripartite partnership (Zheng

2002). With reference to various definitions, there are roughly three definitions of co-operativeism: one is the theory of factors, which is a generalized analytical concept that quantitatively analyzes the co-operative factors contained in various countries; the second is an institutional theory, which exists as an institutional concept or practice; the third is the theory of co-operation, which emphasizes multi-party coordination and cooperation. German scholar Philippe C Schmitter (1974) proposed a representative ideal model (see Figure 3) (Schmitter 1974). In any case, the connotation of co-operativeism includes the integration, regulation, coordination, and cooperation of stakeholders, and their integration into some kind of institutional order.

Under the cooperative approach, multi-party cooperation between families, organizations, governments, and the social public sector should be carried out at the micro, meso, and macro levels, fully stimulating the orderly expression of various stakeholders, and fixing some of this cooperation through institutionalized forms. On many paths, the school system is very much in line with the requirements of cooperativeism, and the chain of scholars-school organizations (kindergartens, primary schools, secondary schools, universities) - public education departments - educational social organizations is quite complete, close to the mandatory education system interspersed with it. If mandatory earthquake prevention and disaster reduction are formulated in education policy, the system can be used to stimulate the interests of the appeal, which will greatly promote the development of this system, at the same time this system can be at the micro and meso levels can extrapolate the influence. Therefore, under the condition of various conditions and budget constraints, it is recommended to give priority to promoting the publicity and education of earthquake prevention and disaster reduction in schools, and put the promotion of compulsory school earthquake prevention and disaster reduction education in the first place. In reality, the education sector attaches the greatest importance to publicity and education on earthquake prevention, disaster reduction, and disaster relief. In 2007, the Ministry of Education formulated and forwarded the "Guiding Outline of Public Safety Education in Primary and Secondary Schools" by the General Office of the State Council, issued the "Guide to Emergency Evacuation Drills for

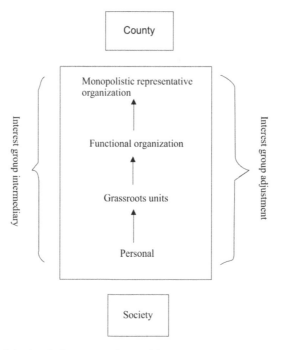

Figure 3.   Philippe C. Schmitter built a cooperative model.

Kindergartens in Primary and Secondary Schools" in 2017, and issued the "Guiding Outline of Comprehensive Practical Activities for Primary and Secondary Schools" in the same year. Social services, design and production, professional experience, and other ways are helpful to integrate and improve interactivity, experience, and practical skills (Zhang & Huang 2018).

5 CONCLUSION

Facing the triple failure of earthquake prevention and disaster reduction publicity, based on the analysis of the current earthquake prevention and disaster reduction publicity and education products from the five dimensions of setting form, platform, target, population, and cost, this paper suggests integrating resources and promote diversified participation; This paper suggests that the supply of products and services should be expanded by means of government purchase of services; the product structure should be adjusted according to the communication situation; the article suggests to improve the input-output ratio and put forward the list of products and services; referring to the new cooperative idea, the article suggests to focus on promoting the publicity and education of earthquake prevention and disaster reduction in schools.

China is located between the Pacific Rim seismic belt and the Eurasian seismic belt, and the real harm and potential threats to the security development of the economy and society caused by strong earthquakes and sudden earthquake disasters have existed for a long time. In order to prevent and resolve these risks, we must vigorously promote the modernization of earthquake prevention and disaster reduction and the modernization of science popularization (Zheng 2018). There is still a long way to go to optimize the supply of earthquake prevention and disaster reduction publicity products and promote the modernization of earthquake prevention and disaster reduction science.

REFERENCES

China Internet Network Information Center: *The 9th Statistical Report on the Development of China's Internet Network* [R].2022.
Kapucu, N. (2008). Culture of preparedness: Household disaster preparedness. *Disaster Prevention and Management: An International Journal*, 17(4), 526–535.
Kenny, Charles. (2012). *Disaster risk reduction in developing countries: Costs, benefits and institutions. Disasters*. 36. 559–588.
LI Haijian, TIAN Yuexin, LI Wenjie Internet Thinking and Traditional Enterprise Reengineering[J]. *China Industrial Economics*, 2014(10): 135–146
LI Xingmin. Knowledge, common sense and scientific knowledge [J]. *Northern Treatise*, 2008(1): 123–130
LIU Peng China Foundation for Poverty Alleviation innovates and carries out special earthquake prevention and disaster reduction projects[J]. *China Disaster Reduction*, 2018(7), 56–59
LIU Yunlin, XU Xianpeng. Emergency science popularization service strategy for destructive seismic events[J]. *Plateau Earthquake*, 2021, 33(01):146–147.
NetEase News Finished with print media? The proportion of mobile phones in the United States is as high as 85%. [2017-06-19] http://digi.163.com/17/0619/18/CNAHEQ22001680N8.html
Neumayer, Eric & Plümper, Thomas & Barthel, Fabian. (2014). The Political Economy of Natural Disaster Damage. *Global Environmental Change*. 24(2014):8–19.
Paul Samuelson, *William Nordhaus Economics* (18th Edition) [M]. Beijing: People's Post and Telecommunications Press, 2008: 32–33
Peng lan. Three threads for defining the concept of "new media". [J] *Journal of Journalism and Communication*, 2016, 23(3): 120–125
Schmitter, P. C. (1974), Still the Century of Corporatism? Review of Politics,36:85–131.
Sikkim Interest correlation analysis of disaster reduction publicity and education dilemma[J]. *China Disaster Reduction*, 2015(13): 56–59
Wang Ding Ding.Lecture Notes on The New Political Economy [M]. Shanghai: Shanghai People's Publishing House, 2013: 307–338
Xu Xuan Earthquake prevention and disaster reduction science popularization and education base "live" [J]. *China Disaster Reduction*, 2017(11): 10–13

Xu Xuan Reform of the Institutional Mechanism for Earthquake Prevention, Disaster Reduction and Relief: History, Process and Prospects [A]. Gong Weibin Report on the Reform of China's Social System (2018)[C]. Beijing: Social Sciences Academic Press, 2018: 275–284

Zhang Ying, Huang Qiong Comprehensive Practical Activity Course: A New Approach to Earthquake Prevention and Disaster Reduction Education[J]. *China Disaster Reduction*, 2018(11):44–47

ZHAO Quanmin.An Analysis of China's Cooperativeist Thought In the 1920s[J]. *Academic Monthly*, 2004(8): 87–94

ZHENG Bingwen Cooperativeism: The Restructuring of China's Welfare System Framework[J]. *Economic Research*, 2002(2):71–79

ZHENG Gongcheng From disaster management to disaster management[J]. China Disaster Reduction, 2021(13):41–45

ZHENG Guoguang. Vigorously carry out the popularization of earthquake prevention and disaster reduction science [N]. People's Daily, 2018-07-27(16).

Advances in Measurement Technology, Disaster Prevention and Mitigation – Li & Mohd Yusof (Eds)
© 2023 The Author(s), ISBN: 978-1-032-36087-4

Yuzhou mine section of the main canal of south-to-north water diversion mid-route phase i project deformation monitoring analysis

Xiaoying Liu*
South-to-North Water Diversion Central Line Project Construction Administration, Zhengzhou, Henan, China

ABSTRACT: The main canal of the South-to-North Water Diversion Mid-Route Phase I Project passes through a coal mined-out area in Yuzhou City, Henan Province. In order to ensure the safe operation of the main canal in the mining area, a unique deformation monitoring system was established to observe and analyze the deformation of the main canal during the construction period and the early operation period. The results show that the filling grouting and curtain grouting reinforcement treatment of the mined-out area canal section are effective, and the residual deformation of the site within the foundation treatment of the main canal section is small and stable, which is conducive to the safe operation of the main canal. This study provides a reliable technical basis for operating and managing the South-to-North Water Diversion Mid-Route Phase I Project in the mined-out area canal section and has reference value for the design, construction, and operation management of similar projects.

1   INTRODUCTION

The South-to-North Water Diversion Project is a super-large water transfer project across the river basin in China. It is an important strategic measure to achieve north-south transfer of water resources, optimize water resource allocation and solve the water shortage in northern China. Moreover, it is a project of vital and lasting importance for the sustainable economic and social development of the recipient area and the well-being of future generations.

The main canal of the South-to-North Water Diversion Mid-Route Phase I Project takes water from Taocha Canal in Nanyang, Henan Province, to Tuancheng Lake in Beijing. It is the "lifeline" of drinking water for nearly 80 million people along the route, so the operational safety of the project is paramount. The main canal crosses a mined-out coal mine area in Yuzhou City, Henan Province. It is the first large-scale water diversion project in China to be built in a mined-out area, lacking practical experience and technical standards of survey, design, and construction. In order to ensure the safety of the project, filling grouting and curtain grouting were carried out to reinforce the section of the canal in the mined-out area. The field tests were carried out in December 2010, the grouting construction of the entire mined-out area was completed in April 2012, the lining of the canal was completed in December 2013, the water was available in May 2014, and the water was officially available in December 2014.

In 2010, a unique deformation monitoring system was set up in the mine canal section to monitor the stability of the canal section in the mined-out area. This system is used to observe the deformation of the mined-out area before construction, during construction, and at the early stage of operation, to monitor the stability of the canal within the mined-out area, to evaluate the impact of the mined-out area on the canal and buildings in the main canal and to predict its development trend. It has great practical significance for the safe operation of the South-to-North Water Diversion Project, provides a reliable technical basis for the operation and management of the canal in the mined-out area, and has reference value for the design, construction, and operation management of similar projects.

_____
*Corresponding Author: 1147578993@qq.com

DOI 10.1201/9781003330172-41

2 OVERVIEW OF THE MINED-OUT AREA

The main canal passes through the mined-out area about 7km southwest of Yuzhou City, Henan Province. There are mainly four mined-out areas, including the former Xinfeng Mining Bureau No. 2 Mine, Guocun Coal Mine in Liangbei Town of Yuzhou City, Gongmao Coal Mine in Liangbei Town of Yuzhou City, and Fuli Coal Mine in Liangbei Town of Yuzhou City, with a cumulative length of 3.115km. The main distribution characteristics of the mined-out area are shown in Table 1. The four mined-out areas have both regular state-owned large mines and small collective coal mines. Significant differences exist in the closing time, mining method, degree of extraction, recovery rate, and buried depth of mined-out areas. There is no complete geological and coal mining information, so the deformation characteristics and stability are more complex.

Table 1. Characteristics of the distribution of mined-out areas in Yuzhou coal mines.

Name of mine	Area of mined-out (million m^2)	Length along the main canal (m)	The thickness of coal seam mined (m)	Coal seam floor elevation	Ground elevation	Buried depth of coal seam (m)	Shut downtime
Former Xinfeng Mining Bureau No. 2 Mine	138.7	747	0.9	−150 to +31	+116 to +138	107 to 266	1965
Guocun Coal Mine	52.5	1055	0.75 to 1.13	−150 to ±0	+126 to +140	126 to 290	1996
Gongmao Coal Mine	19.4	907	0.69 to 1.04	−100 to +20	+126 to +142	106 to 242	2005
Fuli Coal Mine	1.6	406	1	±0 to +50	+134 to +140	90 to 134	2003

3 MONITORING NETWORK DESIGN

The deformation monitoring network was divided into a benchmark network and a deformation monitoring network, arranged in two levels. The monitoring content included vertical displacement, horizontal displacement, and internal deformation of the geotechnical body (see Figure 1 for the distribution of the mined-out area and the arrangement of the monitoring network). A total of four horizontal displacement monitoring datum points, three vertical displacement monitoring datum points, five horizontal displacement deformation monitoring points, and 33 vertical displacement deformation monitoring points were buried, and the burial of signs was completed at the end of January 2010. The first observation of the Yuzhou Mining area started on 8 April 2010 after the markers were completely stabilized, and the observation lasted for 27 cycles and 83 months until November 2017.

3.1 *Benchmark network*

The benchmark points were located in the coal-free area outside the deformation-affected area, and their locations were stable and reliable. Horizontal displacement monitoring datum points YJP1 to YJP3 were located in the southwest outside the mined-out area of Fuli Coal Mine, Gongmao Coal Mine, and Guocun Coal Mine on the left bank of the main canal. YJP4 was located south of the Xinfeng Mining Bureau No.2 Mine, on the right bank of the main canal, as an auxiliary observation point. Vertical displacement monitoring datum points YJG1 to YJG3 were located in the middle

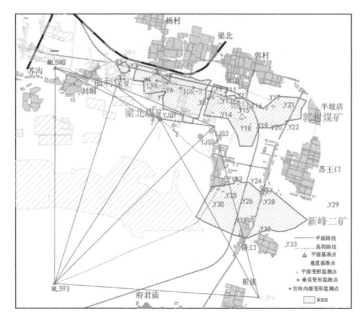

Figure 1. Diagram of the distribution of the Yuzhou mined-out area and the layout of the deformation monitoring network.

of the zone outside the mined-out area of the four coal mines, on the left bank of the main canal, facilitating the observation of the deformation monitoring points (Y1 to Y33).

3.1.1 *Horizontal displacement monitoring benchmark network*

The horizontal displacement monitoring benchmark network adopted a primary distribution network to jointly measure the C-class GPS points of the South-to-North Water Diversion Project. GPS was used to monitor the periodic horizontal displacement. Technical requirements for observation are shown in Table 2.

Table 2. Main technical requirements for GPS observations.

Effective observation Total number of satellites	Satellite altitude angle	Any satellite in the time slot Effective observation time	Length of the time period	Number of time slots	Sampling interval	PDOP
≥ 5	≥ 15°	≥ 15 min	≥ 60 min	≥ 2	10s to 30s	6

Double-difference fixed solution is adopted for the baseline solution (Li, 2010), and the rejection rate of observed data in the same period is less than 10%; the length difference of the retest baseline $Ds \leq 2\sqrt{2}\sigma$, which is calculated in Equation (1).

$$\sigma = \sqrt{a^2 \pm \left(b \times d \times 10^{-6}\right)^2} \qquad (1)$$

where: σ is the precision specified for the corresponding level according to the actual average edge length in mm; a is a fixed error in mm; b is the proportional error factor; d is the distance in km from adjacent points.

The adjustment treatment of the GPS network is firstly carried out with 3D unconstrained adjustment starting from the 3D coordinates of a single point in each network, and the gross error analysis and the internal coincidence accuracy of the GPS network are carried out. Then, two points are used as pegging points for 2D treatment among the points jointly measured in the network.

The closure of each coordinate component of the synchronous cycle and the closure of the full length of the cycle satisfies Equation (2).

$$Wx \leq \frac{\sqrt{n}}{5}, Wy \leq \frac{\sqrt{n}}{5}\sigma, Wz \leq \frac{\sqrt{n}}{5}\sigma, W = \sqrt{Wx^2 + Wy^2 + Wz^2}, W \leq \frac{\sqrt{3n}}{5}\sigma \qquad (2)$$

The closure of each coordinate component of the asynchronous cycle and the full-length closure of the cycle satisfies Equation (3).

$$Wx \leq 2\sqrt{n}\sigma, Wy \leq 2\sqrt{n}\sigma, Wz \leq 2\sqrt{n}\sigma, W = \sqrt{Wx^2 + Wy^2 + Wz^2}, W \leq 2\sqrt{3n}\sigma \qquad (3)$$

where: n is the number of synchronous (or asynchronous) cycle baseline edges; W is the closure of the full length of the cycle in mm.

The GPS control network carries out three-dimensional unconstrained adjustment under the WGS-84 coordinate system and two-dimensional constrained adjustment under the national coordinate system. The compulsory constraint is adopted for known coordinates, records or bearings, and the relative median error of side length between constrained points is less than 1/80000.

3.1.2 *Vertical displacement monitoring benchmark network*

For the first observation, a line conforming to the level shall be laid, the second level point of the South-to-North Water Diversion Project shall be measured jointly, and the elevation shall be directed to any marker of the benchmark network. Secondary leveling shall have a casual median error of less than 1.0mm per km of leveling. The main technical requirements are shown in Table 3.

Table 3. Main technical requirements for leveling measurement.

Main technical specifications	1) Line of sight length $\leq$ 50m and $\geq$ 3m, front-to-back sight distance difference 1.5m, front-to-back sight distance difference accumulated 6.0m at any station, line of sight height $\leq$ 2.80m and $\geq$ 0.55m.
	2) The difference between basic and auxiliary reading is 0.4mm, the difference between basic and auxiliary height measured is 0.6mm, the round-trip measurement height discrepancy is $4\sqrt{k}$, the route closure difference is $4\sqrt{L}$, the difference between the detection has been measured height is $6\sqrt{R}$. (Note: k, L, R for the length of the section or route, unit km, when the length of the section is less than 0.1km, according to 0.1km)
Decimal digits	Distance between survey sections 0.1km, the height difference between stations 0.01mm, the height difference between survey sections 0.1mm, level point elevation 0.1mm

3.2 *Deformation monitoring network*

The deformation monitoring points were selected in the mined-out area where the changing amplitude and deformation rate are large and representative. The monitoring lines were arranged in a straight line parallel to and perpendicular to the strike of the mine, taking into account the main canal channel, the ore bed strike, the mining method, and the overlying stratum production, with a length that exceeds the extent of surface movement and deformation. The observation line perpendicular to the strike of the ore bed was set on the inclined main section of the mobile basin (Zhu 2009). The observation points were spaced at approximately equal intervals of 100–200 m. While observing the surface deformation, the deformation of surface cracks and pits was observed (Zheng 2010).

The five horizontal displacement and deformation monitoring points were measured by three benchmark networks (YJP1 ~ YJP3) with a single distribution network, and GPS observation technical requirements were the same as the horizontal displacement monitoring benchmark network. Leveling observation in vertical displacement monitoring adopted an annexed leveling route to measure all monitoring points simultaneously, and the technical requirements were the same as the vertical displacement monitoring benchmark network.

3.3 *Geotechnical internal deformation monitoring*

The deformation monitoring instrument inside the geotechnical body was arranged in the representative channel section with a shallow burial depth and relatively short formation time in the mined-out area. Combined with the channel project, considering the direction of the mining layer and the distribution of the mined-out area, and taking into account factors such as not affecting the construction of the channel building, facilitating observation and protection, it was arranged near the red line on the left side of the channel in the mined-out area of Gongmao coal mine and Guocun coal mine (Zhang 2019). A total of three multi-point displacement meters and one inclinometer were embedded, and the monitoring depth was controlled within 60m below the bottom plate of the canal The anchorage points of the multi-point displacement meters were 15m, 30m, 40m, 50m, and 60m inside the hole, respectively. The inclinometer was equipped with 13 sensors at different depths. GPS observation technology requirements are the same as the horizontal displacement monitoring benchmark network, and the accuracy of side length can be relaxed to 1/10000.

4 ANALYSIS OF MONITORING RESULTS

4.1 *Stability analysis of benchmark networks and known reference points*

The horizontal benchmark network retest used relatively stable YJP1 and YJP4 as the starting points for the adjustment (YJP2 and YJP3 are in the middle of the mined-out area, YJP1 is at the edge of the mined-out area, and YJP4 is the farthest from the mined-out area). The displacement curve of the horizontal benchmark network is shown in Figure 2.

It can be seen from Figure 2 the displacement curves of YJP2 and YJP3 vary greatly because they are in the middle of the mined-out area, but within 10mm, YJP4 is farthest from the mined-out area, and its displacement curve does not change. When YJP1 and YJP4 are fixed, the adjacent periodic displacements of other points are less than two times the square error of the allowed accuracy, which proves that the benchmark point is relatively stable and can be used as a reference point for other monitoring points.

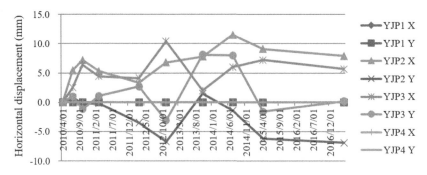

Figure 2.   Displacement curve of the horizontal benchmark network.

The vertical benchmark network retest used YJG1, which is relatively far away from the mined-out area, as the starting point to participate in the adjustment, and the vertical benchmark network displacement curve is shown in Figure 3.

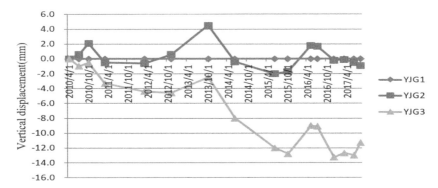

Figure 3. Displacement curve of the vertical benchmark network.

As can be seen from Figure 3, the monitoring data of YJG1 and YJG2 are close to each other, which proves that the benchmark point is relatively stable and can be used as a reference point for other monitoring points. YJG3, due to its proximity to the mine site and the fact that there are plants nearby that are susceptible to human activities, varies too much during the construction and opening period of the main trunk canal and tends to stabilize later.

The known reference point was the retest results of South-to-North Water Diversion in November 2009 (the starting reference value of the benchmark network), plane level C, elevation second class, the mark was buried at 0.5m underground. The results were recorded since 2005, retesting once every two years until October 2011. Three points near the mine area were taken for comparison, and their displacement variation is shown in Figure 4.

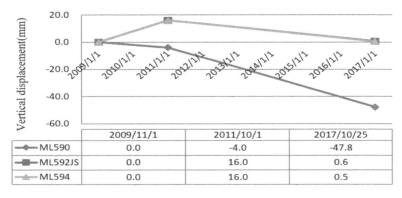

	2009/11/1	2011/10/1	2017/10/25
ML590	0.0	-4.0	-47.8
ML592JS	0.0	16.0	0.6
ML594	0.0	16.0	0.5

Figure 4. Displacement curve of known reference points.

As shown in Figure 4, the starting point ML590 shows little change within the period up to 2012, with significant settlement later in the year. Closing point ML594 conforms to the stabilizing trend of most points, rising by approximately 16mm within two years and stabilizing after the end of construction.

4.2 Horizontal displacement monitoring analysis

The horizontal displacement curves of the deformation monitoring points are shown in Figure 5. From Figure 5(a) to Figure 5(e), it can be seen that the difference between the coordinates of the adjacent cycles is generally less than the permissible range of the design accuracy (2 times the square error, i.e., ±8.5mm), indicating that there is no obvious deformation at each monitoring point.

From Figure 5(f) to Figure 5(h), it can be seen that the horizontal displacement curve of each monitoring point of the underground deformation has a large variation during the construction period but tends to be stable during its operation period, which indicates that with the increase of time, the variation of the displacement of the main canal in the mined-out area decreases, which is conducive to the safe operation of the main canal.

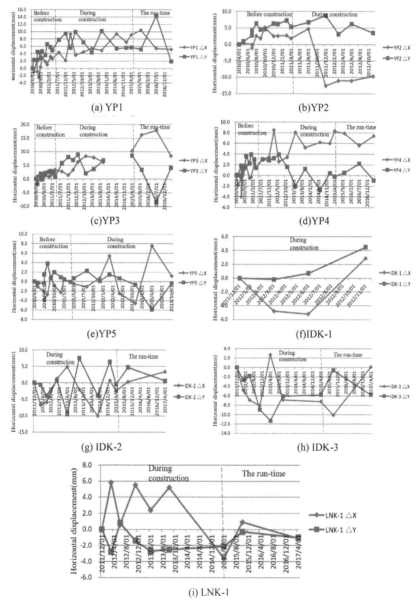

Figure 5.   Horizontal displacement curve of deformation monitoring points.

4.3 Vertical displacement monitoring analysis

4.3.1 Monitoring and analysis of vertical displacement in the mined-out area of the Fuli coal mine

There were three vertical displacement monitoring points in the area, and the vertical displacement curve is shown in Figure 6. Y1 and Y3 were within the excavation area of the main canal, and their service period ended in 2011, due to the canal construction. Before the grouting construction in 2011, the monitoring points showed no signs of deformation; during the grouting period, the oscillation amplitude of the vertical displacement of Y2 increased slightly, which is related to the stress adjustment of the rock body by grouting; after the construction, Y2 returns to normal. The accumulated vertical displacement of the monitoring point in 88 months is within the permissible error range, and the deformation is stable.

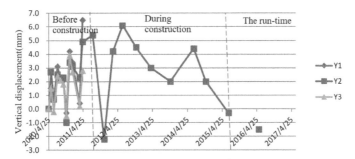

Figure 6. Vertical displacement curve of deformation monitoring points in the Fuli coal mine.

4.3.2 Monitoring and analysis of vertical displacement in the mined-out area of the Gongmao coal mine

There were five vertical displacement monitoring points in the area, and the vertical displacement curve is shown in Figure 7. Y5 was within the excavation area of the main canal, which ended its service period in 2011 Y4 and Y6 were within the grouting area, with stable deformation before construction and a slight increase in oscillation during grouting, of which the accumulated value of Y6 suddenly increased to 9mm in August 2011. According to the analysis, the ground on the south side of Y6 appeared the phenomenon of the ground bulge, crack uplift, and uplift during the grouting plugging period in July, and the rise of displacement is related to the high pressure of the grouting construction orifice and the adjustment of stress on the rock by grouting during the construction period. However, the relative displacement is not large, and the stability is restored after the grouting construction. Y7 and Y8 (the comparison points outside the engineering area) were farther away from the grouting area, and the accumulated settlement of 88 months is about 10mm, which is more stable.

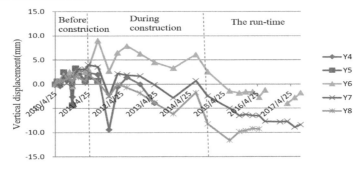

Figure 7. Vertical displacement curve of deformation monitoring points in the Gongmao coal mine.

4.3.3 Monitoring and analysis of vertical displacement in the mined-out area of the Guocun coal mine

14 monitoring points were set up in the mined-out area of the Guocun coal mine. Y11, Y12, Y16, Y18, and Y19 were located within the excavation area of the main canal, which ended its service period in 2011. The vertical displacement curves are shown in Figure 8

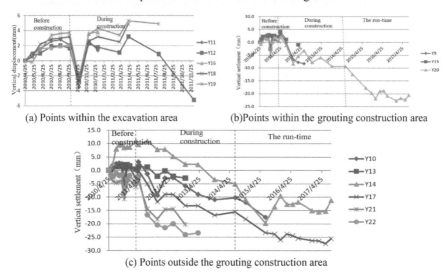

(a) Points within the excavation area   (b)Points within the grouting construction area

(c) Points outside the grouting construction area

Figure 8. Vertical displacement curve of monitoring points in the Guocun coal mine.

As shown from Figure 8(a) to Figure 8 (c), Y9, Y15, and Y20 within the influence range of the grouting area have stable deformation before construction, and the amplitude of oscillation increases slightly during grouting, but the relative displacement is not large. The stability is restored after the end of grouting (Liu et al. 2015). Since 2014, Y20 has a tendency for subsidence, with a cumulative settlement of 22.7mm, indicating that there is still a small amount of residual deformation at the extraction area of the main canal after grouting (see Figure 8 (b)). Y10, Y13, Y14, Y17, Y21, and Y22 (all are comparison points outside the engineering area) were far from the grouting area and were not affected by grouting, Y10, Y13, and Y14 are in a stable state, Y17, Y21, and Y22 have sunk since April 2011, and sunk significantly in December 2011, with a cumulative settlement of 26mm. After field investigation, new city water well was added near Y21 in May 2011 and new drainage well was added adjacent to Y22, which was affected by the construction and sunk (see Figure 8 (c)).

4.3.4 Monitoring and analysis of vertical displacement in the mined-out area of the former Xinfeng Mining Bureau No. 2 mine

A total of 11 monitoring points were located in the mined-out area of the former Xinfeng Mining Bureau No. 2 mine, and their vertical displacement curves are shown in Figure 9.

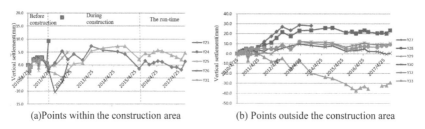

(a)Points within the construction area   (b) Points outside the construction area

Figure 9. Vertical displacement curve of monitoring points in the Xinfeng Mining Bureau No. 2 Mine.

As shown in Figure 9, Y24, Y26, and Y31 were located within the excavation area of the main canal, and their service period ended in 2011. Y23 and Y25 were within the influence range of the grouting area, and their deformation was stable Y27, Y28, Y29, Y30, Y32 and Y33 were comparison points outside the engineering area After 2011, the monitoring data of Y27, Y28, Y32 and Y33 show a clear rising trend, with a maximum cumulative rise of 28.7mm Especially Y27 and Y28 were in the village, which was vulnerable to human activities, Y29 has a sinking trend, with a maximum sinking value of 37.3mm.

4.4 *Monitoring and analysis of multi-point displacement and deformation within the geotechnical body*

The data from 15 sets of multi-point displacement meters in 3 holes were collected to calculate the displacement variation, and the change hydrograph was plotted in Figure 10. The displacement

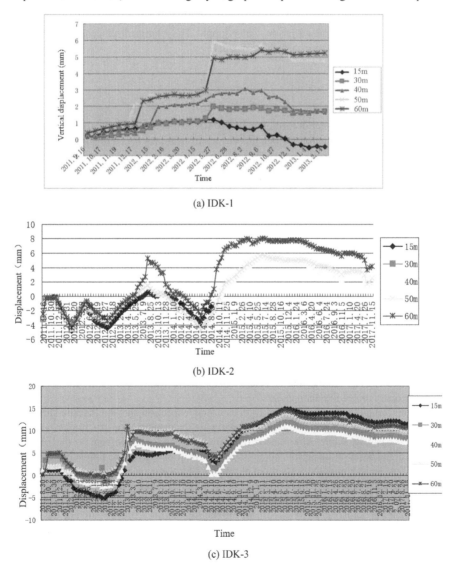

(a) IDK-1

(b) IDK-2

(c) IDK-3

Figure 10.   Multi-point displacement meter displacement hydrograph.

changes hydrograph of the 5-point multi-point displacement meters in hole IDK-1 of the Guocun coal mine shows that the shallower the burial depth is, the smaller the sinking displacement is. With the increase of burial depth, the subsidence value increases gradually, and the maximum subsidence at 50m depth is 5.92mm. It indicates that residual deformation still exists in the overlying rock and soil of the mined-out area, and the amount of deformation transmitted to the surface is much smaller.

The multi-point displacement meter hydrograph of IDK-2 between the mined-out area of the Guocun coal mine and the Gongmao coal mine shows a maximum cumulative change in rock mass displacement of 7.8mm. The duration curves of the displacement meter with different buried depths show narrow amplitude oscillation, the deformation rate is very low, and the deformation trend is basically stable.

The displacement hydrograph of hole IDK-3 in the mined-out area of the Gongmao coal mine shows that the rock mass displacement slightly raised during the grouting construction in 2011. After the opening of the main canal at the end of 2014, the accumulated displacement gradually increased to 14.6mm. It stabilized, indicating a small residual deformation within the overlying rock body of the mined-out area after the water supply. The relative displacement is small, which is conducive to the safe operation of the main canal.

5 CONCLUSION

Combining the results of surface deformation horizontal displacement, vertical displacement, internal geotechnical deformation monitoring, and surface inspections, the following main conclusions were drawn.

(1) The comparison points in the project area were all located outside the scope of foundation treatment in the mined-out canal section. During the observation period, some of the monitoring points had different degrees of residual deformation. The three points Y17, Y21 and Y22 in the Guocun coal mine mined-out area, which was far from the engineering area of the main canal, showed sinking. The four points Y27, Y28, Y32, and Y33 in the Xinfeng No.2 mine mined-out area, which was far from the engineering area of the main canal, showed an obvious rising trend with a maximum accumulated rise of 28.7mm, especially Y27 and Y28, which were significantly rising, related to the proximity to villages and human activities. Point Y29, which was 1.2km away from the engineering area, showed an accumulated sinking of 37.3mm. The above phenomenon indicates that there is still residual deformation or "activation" deformation in the site without grouting treatment.

(2) The multi-point displacement meter inside the geotechnical body gradually increased to 14.6mm after the main canal was opened to water in 2014 and then stabilized, indicating that there was a small residual deformation within the overlying rock body of the mined-out area after the water supply, and the relative displacement was small, which was conducive to the safe operation of the main canal.

(3) The monitoring results combined with the surface inspection showed that the surface deformation and foundation geotechnical deformation in the grouting construction area of the main canal was basically stable. After the main canal was opened to water from the end of 2014 until the end of the observation period, no obvious deformation was found in most of the monitoring points, but the accumulated settlement at point Y20 in Guocun mined-out area was 22.7mm, indicating that there was still residual settlement deformation locally in the mined-out area after grouting treatment, but the amount of residual deformation was not significant.

(4) The remaining deformation of the site within the foundation treatment area of the main canal was small and generally stable. The surface deformation of the mined-out area outside the project area was obvious. The analysis of the result showed that the remaining deformation of the geotechnical body could be effectively eliminated by grouting treatment of the mined-out area.

REFERENCES

Li YX. (2010) Exploration of deformation monitoring scheme in Yuzhou coal mine mining area of the South-North Water Diversion Central Trunk Canal[J]. *Resource Environment and Engineering*, 24(5):531–534.

Liu H D, Zhu H, Huang Y W. (2015) *Study on the stability of the Guocun mine mining section of the South-North Water Diversion Central Project* [C]. Wuhan: Geotechnics, pp. 519–524.

Zhang Jinyu. (2019) *Study on creep characteristics of the filling body and overlying rock in the extraction area and its influence on the long-term deformation of the channel* [D]. Yichang: Three Gorges University.

Zheng Baohua. (2010) Deformation monitoring method for the South-North Water Diversion Main Trunk Canal through coal mine mining areas [J]. *Henan Water Conservancy and South-North Water Diversion*, (8):22–23.

Zhu Dongming. Application of digital level in subsidence observation in Yuzhou mine area [J]. *Henan Science and Technology*, 2009(8 on): 66–67.

Advances in Measurement Technology, Disaster Prevention and Mitigation – Li & Mohd Yusof (Eds)
© 2023 The Author(s), ISBN: 978-1-032-36087-4

Installing air curtains for fire smoke control in long entrances and exits of metro stations without intermediate vertical evacuation staircases

Lu Zhou, Hongshan Dong & Miaocheng Weng*
School of Civil Engineering, Chongqing University, Chongqing, China

ABSTRACT: The vertical emergency staircase of a deep-buried subway station has many problems such as long distance, difficulty in getting out of the ground, and expensive costs, which makes against for people to escape a fire. In the paper, taking the Chongqing Railway Line 10 Red Land Station as a model, a new smoke control strategy by eliminating the vertical emergency staircase and installing an air curtain over the long entrance corridor has been proposed. The Fire Dynamics Simulator (FDS) was used to investigate the air curtain parameters and to model a scenario where people use an air curtain passageway to escape. The Pathfinder software was used to simulate the movement of the flow of people. The results show that the air curtain can be used in long entrances and exits of metro stations without intermediate vertical evacuation staircases. When the distance between the air curtain and fire source is 1m, the best combination for blocking the smoke is as follows: air curtain jet speed of 15m/s, jet angle of 15°, and two air curtains interval of 12 m. While the distance between the air curtain and fire source is 4 m, and the best and most economical condition is the jet speed of 11 m/s, jet angle of 15°, and two air curtains with an interval of 12 m. This project could provide theoretical support for the extension of air curtain application to metro inter-district tunnels.

1 INTRODUCTION

With the growth of the world's urban population and economic development, more and more metros are increasing rapidly. Although many subway stations are underground, the evacuation distance from the station to the ground may exceed 100 m in some subway stations such as Hong Kong and Chongqing. One example is the Chongqing Railway Line 10 Red Land Station, which is currently the deepest underground station in China at a depth of 95m, equivalent to a 31-story building. Typically, most subway stations are located underground with few exits and entrances. When a fire occurs in a subway station, the plume of heat and toxic products rises from the fire and spreads radially outward over the ceiling (Peng et al. 2021). Due to the dense population and underground space conditions, underground fires often result in death and injury on a large scale. Smoke is the most dangerous factor leading to death and injury in fires, rather than direct exposure to fire. A vast majority of fatalities connected with fire are triggered by smoke inhalation (Wang et al. 2018).

According to the requirements of the code for Subway Design GB 50157-2013, a safe evacuation staircase must be set up in the middle of the entrance passage. However, safety evacuation stairs are expensive and not normally used. Evacuees are not familiar with safe evacuation stairs, so it is easy to produce trampling events and cause secondary injuries. At present, air curtain techniques have been developed to control poisonous gases and remove heat generated by a fire in a compartment (Safarzadeh et al. 2021). The air curtain can send a curtain airflow with a certain velocity, temperature, and form two areas with different thermal environment characteristics. It allows people access to space while maintaining the temperature and/or chemical properties of the confined area (Ji et

---

*Corresponding Author: mcweng@126.com

DOI 10.1201/9781003330172-42

al. 2022). Huang Renwu of Chongqing University (Huang 2016) proposed to set the air curtain at the stairway entrance between the platform and concourse of a metro station. For a typical island metro station, the optimum jet velocity of 6-7m/s was used for a fire source power of 2-2.5MW. Sun Xiang (2018) recommended that when air curtains were applied to open staircases in metro stations, the air curtain parameters including the jet velocity of 7m/s and jet angle of 30° were guaranteed to be at least.

In this paper, a new smoke control approach for occupant evacuation along the corridor they normally walk along has been proposed. The Red Land Station in Chongqing was used as a model by removing the vertical emergency stairs and constructing an air curtain in the lengthy entry hallway. FDS (Fire Dynamics Simulator), a fire dynamics field simulation software developed by the National Institute of Standards and Technology (NIST), was applied to investigate the smoke exhaust when the air curtain devices were used for deeply buried metro stations without intermediate vertical evacuation staircases (Chi 2013). Pathfinder software was used to numerically simulate the evacuation time of personnel and to verify the effectiveness of the air curtain (Cheng 2014). These results demonstrate that the air curtain may be employed in lengthy metro station entrances and exits without the need for intermediate vertical evacuation staircases.

2 SOFTWARE AND CONCEPT

2.1 *ASET and RSET concept*

Available Safe Egress Time (ASET), refers to the time that it takes for a fire to develop into a dangerous state in which fire can cause direct harm to people. ASET includes the time when the fire is detected (t_d) and the time when the alarm is given until the fire poses a danger to people (t_h) (ASET=t_d+t_h). ASET is derived from the prediction of heat and smoke spread caused by a particular fire in a built environment. ASET is obtained according to the simulation of FDS.

Required Safe Evacuation Time (RSET) is the time taken to evacuate people to a safe area from the moment the fire starts. RSET includes fire alarm time (t_{alarm}), pre-action time (t_{pre}), and personnel evacuation movement time (t_{move}) (REST= t_{alarm}+ t_{pre}+ t_{move}). RSET is obtained by Pathfinder software.

RSET is the time required to exit the fire when warned, which should be less than the determined ASET (Schröder et al. 2020). The ASET and RSET are commonly accepted by criteria available to assess the safety risk of a fire scenario.

2.2 *Fire dynamics simulator*

The simulation model of a three-story continuous staircase for Chongqing Red Land Station hall is shown in Figure 1. Each staircase is 15 m long, 4.8 m wide and 5 m high. The middle platform is 3 m long and 4.8 m broad, with a 3 m difference in height between the ladder surface and the ceiling.

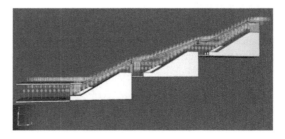

Figure 1. Simulation model of a three-story continuous staircase.

FDS (FDS version 2018) was adopted to investigate the fire spread to simulate pedestrian evacuation in a deeply buried subway under a fire emergency. The wall surface material is defined as concrete. The boundary of the model is the suspended ceiling and the handrails of the stairs. The opening boundary is established as the exit of the evacuation passage, which was in a state of natural ventilation. The ambient conditions of the evacuation passage are set to room temperature of 20°C, carbon monoxide (CO) concentration of zero, and visibility of 30 m at the start of the calculation.

Based on the FDS model, the effects of the air curtain jet velocity, jet Angle, and air curtain spacing on the smoke retaining effect are investigated, and the ASET under fire conditions is estimated. The air curtain's setting parameters are adjusted by comparing the ASET with RSET to achieve the goal of safe evacuation in case of a sudden fire in a deep underground station.

2.3 Pathfinder evacuation simulation

In order to obtain the required safe evacuation time (RSET), we used Pathfinder evacuation simulation software (pathfinder version 2019) to simulate personnel evacuation. In this study, an escalator outage was used as a normal staircase during an emergency evacuation of a subway station fire. The escalators are narrow, so the emergency evacuation can only allow a single flow unidirectional movement. The personnel flow condition in the escalator optimized personnel flow is similar to the stairs, and the transmission capacity is the same width as the stairs.

Normally, the passenger flow of subway stations during peak hours is mainly composed of commuter passenger flow and student passenger flow. According to the literature, the average shoulder width of personnel was set as 0.38 m, the clothing thickness was set as 0.02 m, and the buffer area was set as the minimum value of 0.48 m. Therefore, the platform pedestrian space was 1.63 person /m² (Cheng 2014).

In order to accommodate the flow of people, a bottom platform 13.5 m wide and 12.5 m long is set at the bottom of the stairs, with an area of 168.75 m².

2.4 The requirement of safety codes

In China, several building codes such as CFPDB (GB50016-2006) and Code for Design of Metro (GB50157-2013) can be referred to for the consideration of the untenable criteria in the fire design of subway stations. The period of the occupant evacuation under fire should be< 6 min. Thus, set up the simulation time to 300 s. The heat release rate of the platform fire is set as 3 MW (Wang et al. 2018). The NFPA130 (Standard for Fixed Guideway Transit and Passenger Rail System) of people stand on the temperature limit not be higher than 60°C.

Accordingly, the critical conditions for personnel safety under disaster conditions can be summarized as follows: (1) The temperature at a clear altitude is < 60°C. (2) The visibility at a clear altitude is >1.9 m. (3) The CO concentration at the clear height is < 500 ppm. (4) Smoke layer height is > 2 m.

2.5 Orthogonal experimental design

Orthogonal design is one of the most used experimental design methods, which makes statistical analysis by means of variance analysis to deduce the optimal level combination of multiple factors. In this project, smoke prevention efficiency (η) is used as an index and calculated following the Formula (1–2).

$$\eta = \frac{w_1}{w_0} \cdot \frac{b_1}{H} \left(1 + \varphi \sqrt{\frac{H}{b_1}}\right) \qquad (1)$$

$$\varphi = \frac{3}{2} \sqrt{\frac{a}{\cos\alpha}} \tanh \frac{\cos\alpha \sin\alpha}{a} \qquad (2)$$

Where η is smoke control efficiency, if η is 60%–100%, the solution is better. w_0 is the average rate of high-temperature flue gas invasion (m/s), which is assumed to be 3.5 m/s. w_1 is air jet rate (m/s). b_1 is the air curtain interval (m). a is the tested coefficient of air curtain tuyere, which assumed 0.22. α is the air jet angle (°); H is the height between the air curtain and floor (m), which is assumed to be 2.75m.

The distance between the air curtain and the fire source is set at 1 m for the whole orthogonal experiment. Four important factors including air jet velocity, air jet Angle, air curtain distance, and air curtain width were employed. When using an air curtain in open subway station stairwells, Wu Zhenkun (Wu et al. 2013) recommended that the air jet velocity is set to 5–7 m/s, the angle is 15–30°, the air curtain width is no more than 0.3 m, and air curtain interval is set not exceeding 15 m. Thus, an orthogonal experiment with four factors and three levels (L_9 (4^3)) was employed to find out the optimum smoke prevention conditions. The air jet velocity level was 5 m/s, 7 m/s, and 9 m/s. The air jet angle level was set at 0, 15°, and 30°. The air curtain width level was 0.2 m, 0.3 m, and 0.4 m, and the level of air curtain interval was 12 m, 15 m, and 18 m. The orthogonal experiment design is presented in Table 1. Range analysis of orthogonal designing software (II V3.1) is used for the evaluation of the statistical design.

3 RESULTS AND DISCUSSION

3.1 *Required safe evacuation time (RSET)*

Herein, we used Pathfinder evacuation simulation software to obtain the required safe evacuation time (RSET). According to the model, the passenger flow survey during the evening rush hour is carried out. The time period from one light rail of the station to the next light rail is about 3 min. The number of people on one side of the platform passing through the stairwell is 326.

In general, the density of people on the platform floor is about 1.5 people/m². Therefore, the average number of people on the platform is 326, of which 266 people are placed on the bottom platform, and the remaining 60 people are placed on the stair platform and stairs.

In order to accommodate the flow of people, a bottom platform 13.5 m wide and 12.5 m long is set at the bottom of the stairs. Thus, an area of 168.75 m² and a personnel density of 1.57 people/m² are obtained.

Figure 2 is the image of the flow of people during the process of evacuation. Evacuation started at 0 s, and everyone starts to evacuate to the exit. About 10 s, most people crowded onto the stairway; 82.6 s later, everyone passes the first air curtain. 112.5 s later, all the people passed through the

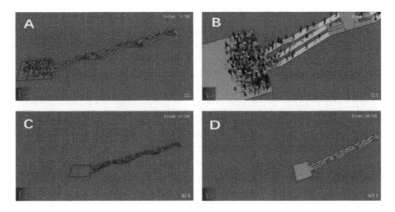

Figure 2. The images of the flow of people during the process of evacuation at different times (A: 0 s, B: 10s, C: 82.6 s, D:169.3 s).

second air curtain and came to a safer area. 169.3 s later, all 326 people were evacuated. Therefore, the RSET of this model under fire conditions is 169.3 s.

3.2 Orthogonal experimental analysis

The orthogonal experiment with four factors and three levels (L_9 (4^3)) and the smoke prevention efficiency (η) are used to discuss the influence of air curtains on the smoke prevention performance. The distance between the air curtain and the fire source is 1 m. Smoke prevention efficiency (η) is calculated following the Formulas (1-2) and the results were shown in Table 1.

Table 1.   The results of orthogonal experimental design.

Case Number\ Factors	Air jet angle(°)	Air jet velocity (m/s)	Air curtain width (m)	Air curtain interval (m)	Smoke prevention efficiency η (%)
Case1	0	5	0.2	12	10.39
Case2	0	7	0.3	15	21.82
Case3	0	9	0.4	18	37.40
Case4	15	5	0.3	18	43.05
Case5	15	7	0.4	12	73.49
Case6	15	9	0.2	15	59.07
Case7	30	5	0.4	15	60.39
Case8	30	7	0.2	18	53.76
Case9	30	9	0.3	12	89.80
K1	21.273	35.697	38.733	54.517	
K2	55.183	46.687	48.473	44.257	
K3	64.233	58.307	53.483	41.917	
R	42.960	22.610	14.750	12.600	

According to the value of range R in Table1, the air jet angle has the greatest influence on smoke control efficiency, followed by the air jet velocity and the air curtain width, and finally by the air curtain intervals. Case 9 has the best anti-smoke efficiency. Thus the optimum conditions were as below: the air jet angle of 30°, air jet velocity of 9 m/s, air curtain width of 0.3m and the air curtain interval of 12 m, and smoke prevention efficiency of 89.80%. Based on the FDS model, the ASET under fire conditions of Case 9 was estimated to be 170 s.

3.3 Supplementary evacuation simulation experiment

Based on the FDS model, the ASET under fire conditions of Case 9 is estimated to be 170 s, which is almost equal to RSET (169.3 s). The experiment with the best anti-smoke effect (case 9) has not reached the expected requirements. Therefore, we have added a comparative experiment with Case 9 (denoted as Case10) to obtain a better result. The condition of case 10 was the air jet angle of 15°, air jet velocity of 9 m/s, air curtain width of 0.3 m, and the air curtain interval of 12 m. Therefore, a group of experiments is added to determine which is better under the same conditions. Following the Formulas (1–2), the smoke prevention efficiency (η) for cas10 is calculated to be 77.49%. And the ASET of Case 10 is calculated at 225 s, much greater than RSET (169.3 s). Compared with Case 9, Case 10 has a better ASET in fire conditions.

The physical model in the project is a slanted upward staircase and there is little literature on continuous staircases. Therefore, the following experimental scenario was added based on the conditions of case 10. The additional experimental scenarios were set as follows: increasing the jet velocity (cases 11, 12, and 13), extending the distance between the fire source and the first air curtain (4 m), and also increasing the jet velocity (cases 14, 15 and 16). The simulation results are shown in Table 2.

From Table 2, when the distance between the air curtain and fire source is 1 m, the jet velocity is 15 m/s, the jet Angle is 15°, and the air curtain interval is 12 m. The air curtain has the best effect. The results of case 13 show that this parameter combination can control the flue gas temperature basically below 40°C, while the visibility in most areas is up to 30 m and the CO concentration is relatively low. The ASET under this experiment is >300s.

Because the computation will be erroneous if the fire source is too close to the air curtain. Furthermore, the air curtain will blow the upper smoke to the bottom, allowing some smoke to flow through. As a result, extending the distance between the fire source and the first air curtain helps to stabilize smoke spread and to avoid the airflow blast flue gas. When the distance between the air curtain and fire source is set at 4 m, the jet velocity is 11 m/s, and the jet Angle is 15°, the two air curtain interval is 12 m, we got the best smoke blocking effect and economy (case14). As the smoke retaining effect of case15 and case16 is basically the same, the fan with a lower jet-speed can effectively save operation costs.

Table 2. Results of evacuation simulation experiment.

Case\ Factor	The jet Angle (°)	jet velocity (m/s)	Distance between the fire source and the first air curtain (m)	Smoke prevention efficiency η (%)	ASET(s)	RSET(s)
Case9	30	9	1	89.80	170	169.3
Case10	15	9	1	77.49	225	169.3
Case11	15	11	1	94.71	225	169.3
Case12	15	13	1	111.93	>300	169.3
Case13	15	15	1	129.21	>300	169.3
Case14	15	9	4	77.49	246.3	169.3
Case15	15	11	4	94.71	251.4	169.3
Case16	15	13	4	111.90	246.3	169.3

4 CONCLUSION

All in all, a new smoke control strategy by eliminating the vertical emergency staircase and installing an air curtain over the long entrance corridor has been proposed. The effect of air curtain jet velocity, jet angle, and air curtain spacing on the smoke effect was investigated by FDS software using the three continuous staircases of Hongtudi Station of Chongqing Metro Line 10 as the prototype. By comparing the ASET with the RSET, the air curtain can be successfully used for evacuating people from a fire. When the distance between the air curtain and the fire source is 1m, the jet velocity is 15 m/s, the jet Angle is 15°, and the air curtain with an interval is 12 m, we got the best smoke retaining effect. When there is a long distance of 4 m between the air curtain and the fire source, the air curtain with a velocity of 11 m/s, a jet angle of 15° and an interval of 12 m is achieved the best smoke retaining effect and economy. However, the simulation results are only for reference, and FDS and Pathfinder simulation processes still have idealized characteristics. It is hoped that more scholars will investigate a more thorough analysis and more recommendations for smoke prevention in a subway station with long entrance and exit passageways without middle vertical evacuation stairs.

ACKNOWLEDGMENTS

This work was supported by the innovation and entrepreneurship program for college students of Chongqing University (No. S201910611413).

REFERENCES

B. Schröder, L. Arnold, A. Seyfried. (2020) A map representation of the ASET-RSET concept, *Fire Saf. J.*, 115: 103154.

J. Chi. (2013) Reconstruction of an inn fire scene using the Fire Dynamics Simulator (FDS) program, *J Forensic. Sci.*, 58: S227-234.

J. Ji, W. Lu, F. Li, X. Cui. (2022) Experimental and numerical simulation on smoke control effect and key parameters of Push-pull air curtain in tunnel fire, *Tunn. Undergr. Sp. Tech.*, 121: 104323.

L. Cheng. (2014) Study on the simulation of emergency evacuation of subway station fire based on Pathfinder, Lanzhou Jiaotong University.

M. Peng, X. Cheng, W. Cong, R. Yuen. (2021) Experimental investigation on temperature profiles at ceiling and door of subway carriage fire, *Fire Technol.*, 57: 439–459.

M. Safarzadeh, G. Heidarinejad, H. Pasdarshahri. (2021) The effect of vertical and horizontal air curtain on smoke and heat control in the multi-storey building. *J Build. Eng.*, 40:102347.

R. Huang. (2016) *Application research of air curtain smoke control in a subway station*, Chongqing University.

W. Wang, T. He, W. Huang, R. Shen, Q. Wang. (2018) Optimization of switch modes of fully enclosed platform screen doors during emergency platform fires in the underground metro station, *Tunn. Undergr. Sp. Tech.*, 81: 277–288.

X. Sun. (2018) *Research on the smoke curtain in the contact channel of subway inter-district tunnel*, Chongqing University.

Z. K.Wu, H.P. Zhang, Y.H. Sheng, Z, Chen,T. Weng, H.U. Long-Hua, X.C. Zhang. (2013) Experiment on the smoke blockage by air curtains installed at the front opening of stairs in a subway station. *Fire Sci. Technol.*, 32: 4–9.

Advances in Measurement Technology, Disaster Prevention and
Mitigation – Li & Mohd Yusof (Eds)
© 2023 The Author(s), ISBN: 978-1-032-36087-4

Stability analysis of unsaturated soil slope under rainfall infiltration

YiLiang Zhou*, Ming Li, Zhuoqun Chen & Fenghan He
School of Civil Engineering, University of South China, Hengyang, China

ABSTRACT: A slope in Hengyang Songmu Industrial Park is taken as the research object, the seepage field and the stress field of unsaturated soil slope under rainfall infiltration are analyzed by the finite element analysis software of Geo-studio, and then the stability of unsaturated soil slope is studied. Results show that the surface soil of the slope has an intense response to rainfall, and the pore water pressure is positively correlated with rainfall intensity. The reduction of effective stress in the upper slope has a negative impact on the shear strength of unsaturated soil, and results in failure and instability, which is consistent with the true state of affairs. The calculated position of the dangerous sliding surface caused by rainfall infiltration is close to the field survey result, which can provide some reference for disaster prevention and control of unsaturated soil slope.

1   INTRODUCTION

With the impacts of earthquakes and extreme weather, geological disasters occur frequently due to the diverse landform and complex geological structure in China, which poses a serious threat to people's lives, properties, and living environment. According to the "national geological disaster Bulletin (2019)" issued by the geological disaster technical guidance center of the Ministry of natural resources, 6181 geological disasters occurred in China, resulting in 211 deaths, 13 missing and 75 injured, with a direct economic loss of 2.77 billion, and 4220 landslides are the most important part of geological disasters (Ministry of Natural Resources 2018). Statistics show that 5904 geological disasters are caused by natural factors (mainly rainfall), accounting for 95.5% of the total. Rainfall is an important factor leading to slope failure in geological disasters, especially rainstorms with high intensity and long duration (Huang 2007; Li et al. 2004).

Most of the soil slope in the natural environment is generally in the unsaturated state, where the existence of matrix suction can provide part of shear strength, and increased slope stability (Liu et al. 2016; Lyu et al. 2021). As the volume of water content is increased and the mechanical strength is significantly reduced of unsaturated soil due to the impacts from rainfall infiltration, which often leads to the failure of slope. Aiming at this problem, the change of seepage field in slope caused by rainfall infiltration was analyzed, and the slope stability was studied by combining the limit equilibrium method or finite element method, while the interaction of soil and water was ignored (Lin et al. 2019; Nie et al. 2020; Shi et al. 2016). Therefore, the typical slope in Hengyang Songmu Industrial Park is taken as the research object, and based on the fluid-solid coupling theory of unsaturated soil and the finite element stress method, the numerical model of unsaturated soil slope is established by the finite element analysis software of Geo-studio. The seepage field and stress field of unsaturated soil slope under rainfall infiltration are analyzed, and then the stability of unsaturated soil slope is studied.

---

*Corresponding Author: 940387856@qq.com

DOI 10.1201/9781003330172-43

2 COMPUTING THEORY

2.1 *Coupled analysis of seepage-stress*

Based on the continuous motion equation of groundwater, the two-dimensional seepage equation is given by Darcy's law:

$$\frac{k_x}{\gamma_w}\frac{\partial^2 u_w}{\partial x^2} + \frac{k_y}{\gamma_w}\frac{\partial^2 u_w}{\partial y^2} + \frac{\partial \theta_w}{\partial t} = 0 \tag{1}$$

The coupled analysis for unsaturated soils uses incremental displacement and incremental pore-water pressure as field variables. The volumetric water content in Equation (1) is given by the following expressions (Darkshanamurthy et al. 1984):

$$\theta_w = \frac{\beta}{3}\varepsilon_v - \omega u_w \tag{2}$$

$$\beta = \frac{E}{H}\frac{1}{(1-2v)} = \frac{3K_B}{H} \tag{3}$$

$$\omega = \frac{1}{R} - \frac{3\beta}{H} \tag{4}$$

Where K_B is the bulk modulus, H is the unsaturated soil modulus for soil structure with respect to matrix suction, and R is a modulus relating the change in volumetric water content with the change in matric suction. It is assumed that the material properties remain constant within a time increment, and this equation becomes:

$$\Delta\theta_w = \beta\Delta\varepsilon_v - \omega\Delta u_w \tag{5}$$

The seepage equation can be formulated for finite element analysis by using the principle of virtual work. If virtual pore-water pressures u_w^* are applied to the seepage equation, gives:

$$\int u_w^* \left[\frac{k_x}{\gamma_w}\frac{\partial^2 u_w}{\partial x^2} + \frac{k_y}{\gamma_w}\frac{\partial^2 u_w}{\partial y^2} + \frac{\partial \theta_w}{\partial t} \right] dV = 0 \tag{6}$$

Applying integration by parts and substituting equation (5) to equation (6) gives:

$$-\int \left[\frac{k_x}{\gamma_w}\frac{\partial u_w^*}{\partial x}\frac{\partial u_w}{\partial x} + \frac{k_y}{\gamma_w}\frac{\partial u_w^*}{\partial y}\frac{\partial u_w}{\partial y} \right] dV + \int u_w^* \frac{\partial(\beta\varepsilon_v - \omega u_w)}{\partial t} dV$$
$$= \int u_w^* V_n dV \tag{7}$$

Where V_n is the boundary flux.
Using finite element approximations, the equation can be written as:

$$-\int \frac{1}{\gamma_w}[B]^T[K_w][B]\{u_w\}dV - \int \langle N \rangle^T \langle N \rangle \left\{ \frac{\partial(\omega u_w)}{\partial t} \right\} +$$
$$\int \langle N \rangle^T \{m\}^T [B]\left\{ \frac{\partial(\beta\delta)}{\partial t} \right\} dV = \int \langle N \rangle^T V_n dV \tag{8}$$

Where:

$$\int [B]^T [K_w][B]dV = [K_f]$$

$$\langle N \rangle^T \langle N \rangle = [M_N]$$

$$\int \langle N \rangle^T \{m\}^T [B]dV = [L_f]$$

$[B]$ is the gradient matrix, $[K_w]$ is the hydraulic conductivity matrix, $[K_f]$ is the element stiffness matrix, $\langle N \rangle$ is the row vector of shape functions, $[M_N]$ is the mass matrix, $[L_f]$ is the coupling matrix for flow, $\{m\}^T$ is the isotropic unit tensor, and δ is the nodal displacement.

Integrating Equation (8) from time t to time $t + \Delta t$, the backward time-stepping scheme is used, then,

$$\beta[L_f]\{\Delta\delta\} - (\tfrac{\Delta t}{\gamma_w}[K_f] + \omega[M_N])\{\Delta u_w\} =$$
$$\Delta t(\{Q\}|_{t+\Delta t} + \tfrac{1}{\gamma_w}[K_f]\{u_w\}|_t) \tag{9}$$

The finite element equilibrium equation for saturated-unsaturated soils is formulated:

$$[K]\{\Delta\delta\} + [L_d]\{\Delta u_w\} = \{F\} \tag{10}$$

Where $[K_w]$ is the stiffness matrix, and $[L_d]$ is the coupling matrix.

In summary, the coupled analysis model of seepage stress is built by the combined Equation (9) and Equation (10).

2.2 Stability analysis

In the calculation of the stability of the slope, the safety factor is calculated based on the finite element stress method. Firstly, the stress of σ_x, σ_y and τ_{xy} in each element is obtained from the coupled analysis of seepage stress. The shear stress of the soil slices is obtained by the method of Mohr's stress circle, and the effective shear strength is calculated by the substrate's normal stress. Then the sliding force is calculated by multiplying the sliding shear stress by the width of the soil slice, and the anti-sliding force is calculated by multiplying the effective shear strength by the width of the soil slice. Finally, the factor of safety is obtained by combining the total length of sliding surfaces. It is written as:

$$FS = \frac{\sum S_r}{\sum S_m} \tag{11}$$

Where S_r is the total anti-sliding force along the entire sliding surface, and S_m is the total sliding force along the entire sliding surface.

3 A CASE STUDY

3.1 Project overview

The typical slope is located in Xinli chemical plant section, Youyi Road, Hengyang Songmu Industrial Park. The original geomorphic unit belongs to the denudation residual Hill landform. The excavated height of the slope is about 14m, and it is a Grade II side slope, as shown in Figure 1. The material composition of the slope from top to bottom is mainly as follows: (1) Residual soil (silty clay) is mainly composed of clay particles and locally containing gravel. (2) Strongly weathered silty mudstone: the fissures are well developed, and the rock mass completeness is extremely broken. (3) Moderately weathered silty mudstone: the fissures are relatively developed and the rock mass completeness is complete. The soil parameters are presented in Table 1.

Figure 1. The slope is located in Xinli chemical plant section, Youyi Road, Hengyang Songmu Industrial Park.

Table 1. Soil parameters of the slope.

Material composition	volume-weight (kN/m³)	cohesive force(kPa)	internal friction angle(°)	volumetric water content(%)	seepage coefficient(m/s)
silty clay	20	15.5	16.5	40	8×10^{-6}
Strongly weathered silty mudstone	23	28	35	30	8.5×10^{-7}
Moderately weathered silty mudstone	26	160	38	25	8.5×10^{-7}

3.2 *Result analysis*

Based on site rainfall conditions and local rainfall data, the changes of slope seepage field and stress field under rainstorm (70 mm/d) are simulated. The seepage and stress coupling module in Sigma/W in Geo-Studio is used, and the calculation results of the seepage field and stress field at each time step are obtained (Figures 2 and 3). Finally, the pore-pressure and stress data are imported into the Slope/W module, and the dynamic analysis of slope stability is carried out by the finite element stress method.

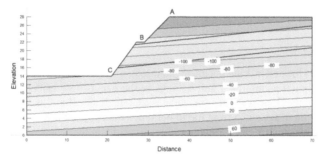

Figure 2. Initial seepage field.

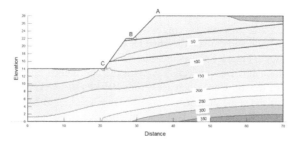

Figure 3.   Initial stress field.

The distribution of pore water pressure of the slope after 24 hours of rainfall was present in Figure 2. In the process of rainfall infiltration, the surface soil of the slope is first affected, and the dense isolines are distributed. With the continuous rainfall, the pore water pressure begins to rise, and the rainwater gradually infiltrates into the slope. However, the pore water pressure in the lower part of the slope and the seepage line have not changed significantly due to the small permeability of the soil and the limited rainfall time. Figure 5 shows the time history curve of pore water pressure at positions A, B, and C of the slope surface, respectively. The pore water pressure located at position A of the top slope will continue rising directly affected by rainfall. The pore water pressure located at position B of the slope platform increases greatly at the beginning time. Then after 0.6 days, the growth rate of pore water pressure tends to be flat and in the state of temporary saturation. The pore water pressure located at position C of the foot of the slope quickly enters the saturated state. This is because the water can be transported inside and outside the slope to the low position C, making the soil easy to adapt to the saturated state.

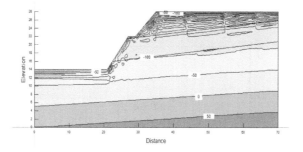

Figure 4.   Distribution of pore water pressure of the slope after 24 hours' rainfall.

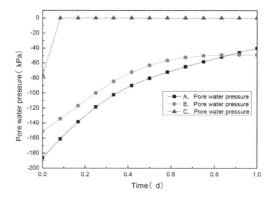

Figure 5.   The time history curve of pore water pressure at positions A, B, and C, respectively.

311

The distribution of the effective stress in the Y direction after 24 hours of rainfall was present in Figure 6. The pore water pressure varying with the changes of the seepage field destroyed its original stress balance state and caused the changes of effective stress, which also mainly occurs on the surface of the slope. Figure 7 shows the time history curve of the effective stress at positions A, B, and C of the slope surface, respectively. With the continuous rainfall, the effective stress located at position A of the top slope decreases from 195 kPa to 151 kPa, and the effective stress located at position B of the slope platform decreases from 152 kPa to 83 kPa. Based on the shear strength theory of unsaturated soil, the reduction of effective stress has a negative impact on the shear strength of unsaturated soil.

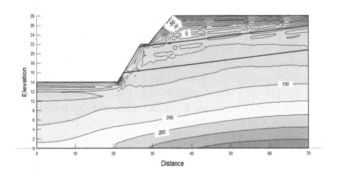

Figure 6.   Distribution of the effective stress in the Y direction after 24 hours' rainfall.

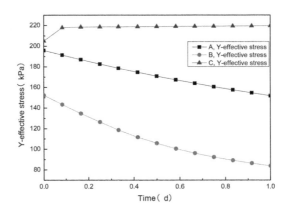

Figure 7.   Distribution of the effective stress in the Y direction after 24 hours' rainfall.

According to the analysis of the pore water pressure and stress field, the position of slope instability of the slope is mainly concentrated on the upper slope, as shown in Figure 8. A large amount of rainfall infiltration mainly occurs in the surface soil leading to the increase of pore water pressure and the volume water content, the decrease of matrix suction and effective stress, and therefore causes greater deformation of the upper slope compared with the deep soil, resulting in slope instability and failure. The safety factor of the slope decreased from 1.942 to 0.868 after 24 hours of rainfall, as shown in Figure 9.

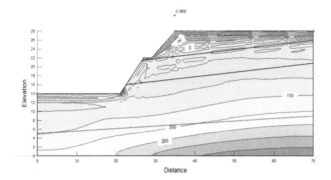

Figure 8.   Position of dangerous slip surface after 24 hours' rainfall.

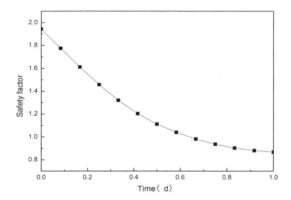

Figure 9.   Relationship between slope safety factor and time.

Based on the field survey results, it is because of the development of open cracks in the upper part of the slope caused by rainfall, which leads to the instability of the upper slope. It can be determined that the landslide type is a push-down landslide, and if it has not managed effectively, it will cause the upper part to start sliding and push the lower part to slide, resulting in serious consequences.

4   CONCLUSIONS

(1) In the process of rainfall infiltration, the seepage field and stress field of the slope change dramatically in the surface soil, the change in the seepage field will cause the change in the stress field, and vice versa. The direct coupling method used in this paper can accurately express its correlation.
(2) The finite element stress method is used to study the stability of slope under rainfall infiltration. The safety factor decreases with continuous rainfall, and the calculated position of the dangerous sliding surface is close to the true state of affairs, which can provide some reference for the deformation and stability analysis of unsaturated soil slope.

ACKNOWLEDGMENTS

The work was supported by the Scientific Research Project of the Hunan Education Department (No. 20C1620).

REFERENCES

Darkshanamurthy V, Fredlund D, Rahardjo H. (1984) Coupled three-dimensional consolidation theory of unsaturated porous media. Proceedings of the 5th international conference on expansive soils. Adelaide, *South Australia*, 99–103.

Geological Disaster Technical Guidance Center, Ministry of Natural Resources. (2018). *National geological hazard Bulletin*(2018), Beijing.

Huang R Q. (2007) Large-Scale Landslides And Their Sliding Mechanisms In China Since The 20th Century. *Chinese Journal of Rock Mechanics and Engineering*, 26(3): 433–454.

Li Y, Meng H, Dong Y, Hu S E. (2004) Main Types and characteristics of geo-hazard in China-Based on the results of the geo-hazard survey in 290 counties. *The Chinese Journal of Geological Hazard and Control*, 15(2): 29–34.

Lin G C, Xie X H, Ruan H N, Zhu Z D, Lu B, Xu C C, Lu X G. (2019) Study on slope stability weakening process along with infiltration by rainfall. *Hydro-Science and Engineering*, (3): 95–102.

Liu J, Sun S L, Liu B, Chen Y Y, Wang E X. (2016) Analysis of Seepage and stress coupling effect on the stability of unsaturated soil slope. *Journal of Hebei University of Engineering* (Natural Science Edition), 33(3):38–42+56.

Lyu Y H, Liang D X, Wang Y, Huang X. (2021) Seepage - stress coupling analysis of unsaturated soil slope under rainfall infiltration. *Journal of the Guilin University of Technology*, 41(2): 318–324.

Nie X T, Zhuang P R, Jiao Y T, Wang B. (2020) Slope stability analysis of homogeneous earth dam considering seepage stress coupling. *Journal of North China university of water resources and electric power* (natural science edition), 41(6): 54–58.

Shi Z M, Shen D Y, Peng M, Zhang L L, Zhang F W, Zheng X Z. (2016) Slope stability analysis by considering rainfall infiltration in multi-layered unsaturated soils. *Journal of Hydraulic Engineering*, 47(8): 977–985.

Advances in Measurement Technology, Disaster Prevention and
Mitigation – Li & Mohd Yusof (Eds)
© 2023 The Author(s), ISBN: 978-1-032-36087-4

The influence of the operation and management of the Cao'e River gate on the flood control and drainage of the Hangyong Canal

Chengjie Tu
Zhejiang Province Qiantang River Basin Center, China

Weijin Chen*
Shaoxing No.1 Water Conservancy Ecological Construction Co., Ltd, Shaoxing, China

Yishan Chen, Mei Chen, Xiongwei Zhang, Linjie Wang & Liya Zhu
*Key Laboratory for Technology in Rural Water Management of Zhejiang Province Zhejiang University of
Water Resources and Electric Power, Hangzhou, China*

ABSTRACT: Cao'e River sluice is located at the Cao'e River estuary. It is the largest sluice
in the strong tidal bore estuary area in China and an important pivotal project for water resource
allocation in eastern Zhejiang. Different gate opening times have different effects on the regional
water level decline and the change rate of flow velocity in the Ningshao plain river network.
Therefore, gate scheduling directly affects the flow characteristics. In order to fully study the
impact of the operation and management of Cao'e River sluice on the flood control and drainage
of Hangzhou-Ningbo Canal, this paper uses Mike21 software to calculate and analyze the relevant
data of Hangzhou-Ningbo Canal 72 hours after the sluice is opened. Finally, it is concluded that the
best time to dispatch the Cao'e River sluice and the Binhai sluice is 16 hours before the rainstorm.

1 INTRODUCTION

The production and life of human society are first restricted by the regional natural environment.
Various natural conditions and factors first standardized the type and nature of water conservancy
facilities and their functions. In all links between human use and the transformation of nature,
especially in areas with abundant water resources, it is particularly important to make full use
of water conservancy development to make the most of water. With the rapid development of the
economy and society in the basin in recent years, the ecological outlook of nature has been indirectly
changed, and floods and waterlogging disasters have occurred indirectly and frequently. Of course,
the cause of flood disaster is not only this one factor, but also meteorological factors, topographic
conditions, insufficient flood control projects, insufficient river management, insufficient drainage
capacity of plain rivers, and urban construction change the underlying surface conditions. These
factors are gradually affecting the healthy development of the city.

Cao'e River sluice is located at the mouth of Cao'e River, about 30 kilometers away from
Shaoxing City, Zhejiang Province. In December 2008, it was lowered to store water. Since then,
East Zhejiang water conservancy has formed an integrated pattern, and the Hangzhou-Ningbo
Canal has been opened. The benefits of water resources development and utilization of Cao'e
River brought by the sluice gate have begun to appear (Liu et al. 2021; Shao et al. 2015; Sun 2019;
Zhang & Wang 2017).

Based on Mike21 (Zheng 2022; Zhang et al. 2020), this paper studies the flow characteristics of
the Hangyong Canal and typical points, analyzes its impact on the surrounding streets, and avoids
unnecessary losses. Figure 1 shows the location of the Hangzhou-Ningbo Canal.

---

*Corresponding Author: chenweijin2011@126.com

DOI 10.1201/9781003330172-44

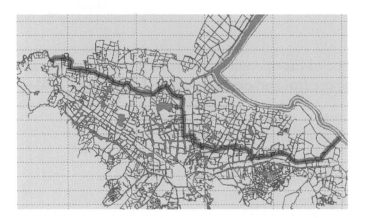

Figure 1. Location map of Hangyong Canal.

2 HYDRODYNAMIC MATHEMATICAL MODEL

The hydrodynamic module (HD) is a basic module of Mike21, which is used to simulate the water level and flow velocity in the river.

2.1 Basic equation

The principle of the two-dimensional unsteady hydrodynamic module is based on the Reynolds mean stress equation of two-dimensional incompressible fluid.

$$h = \eta + d \tag{1}$$

The two-dimensional flow continuity equation describing the plane is:

$$\frac{\partial h}{\partial t} + \frac{\partial h\bar{u}}{\partial x} + \frac{\partial h\bar{v}}{\partial y} = hS \tag{2}$$

Where: h is the total head; η is the river bottom elevation; d is still water depth; t is the time; $\bar{u}$ and $\bar{v}$ are the velocities based on the average of water depth; x and y are the Cartesian coordinates; S is the point source flow.

2.2 Definite solution condition

Initial conditions: it is needed to give the initial terrain, velocity, water depth, etc. of each calculation grid node.

Boundary conditions: the flow process at the inlet of a given model; discharge process of given water level at the outlet.

3 ANALYSIS OF THE INFLUENCE OF HANGYONG CANAL ON WATERLOGGING DRAINAGE

3.1 Calculation conditions

3.1.1 Hydrodynamic conditions of river network
The river network of Shaoxing plain extends to the East Xiaojiang River in the west, the Cao'e River in the east, the mountainous basin in the south, and the boundary of Xiaoshao plain and the

mouth of Cao'e River in the north. Among them, the water from Keqiao District and Yuecheng District is mainly supplied through the East Zhejiang Water Diversion Project, the South Cao'e River Water Diversion Project, and the Southeast Haoba Water Diversion Project, so as to increase the water fluidity of Shaoxing plain river network (Wei et al. 2022; Zhu & Liu 2005) and effectively improve the river water quality in the urban area. The Cao'e River sluice and Binhai sluice discharge water to the Qiantang River, which effectively improves the overall drainage capacity of Shaoxing plain, the plain river water distribution system, and the overall water environment of the plain.

3.1.2 Model calculation conditions

After observation and a series of preliminary simulation calculations, the time step of the proposed model is 600s and the number of steps is 432. The time is assumed to start from 0:00:00 on May 1, 2019 to 24:00:00 on May 3, 2019. The total simulation time is 72 hours. The roughness of the straight reach of the main stream is 0.025, and the roughness of the river in the upstream mountainous area is 0.04. The longer the simulation time is, the longer the computer simulation calculation time is. In fact, after a certain time, the watershed has tended to be stable.

Assuming the inflow of Hangyong Canal, the outflow of Cao'e River gate in the north, and the initial water level, the data analysis and research of the river are carried out. In the two-dimensional modeling, the boundary type of the inlet of Hangyong Canal is the inflow water area with a flow of 50m^3/s. The boundary type of the outlet of Cao'e River sluice is the outflow water area with a boundary condition of 2.5m, the initial water quality is class V water, the BOD concentration is 10 mg / L, the initial water level is 3.9 m, and the temperature is 30 °C.

3.2 Calculation results and analysis

The source of the Hangyong Canal is located in the southwest of Keqiao District. The river crosses the urban area and mainly flows through Doumen, Daoxu, Cao'e, and Dongguan streets. It can be seen from Figure 2 that when the distance is certain, the water level is decreasing with the increase of time; at a certain time, with the increase of the distance along the way, the water level is decreasing, and it continues to fluctuate from 15km to 50km, and then rapidly decreases after 50km.

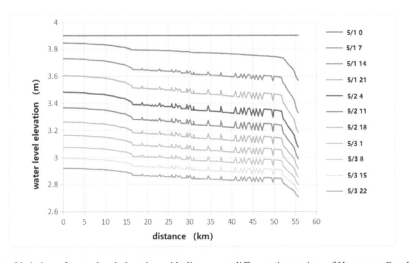

Figure 2.   Variation of water level elevation with distance at different time points of Hangyong Canal.

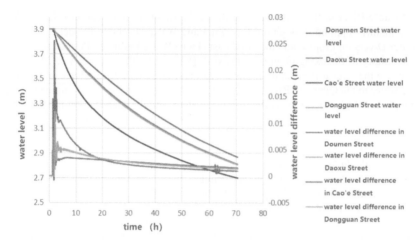

Figure 3. Variation diagram of water level and velocity with time at typical points.

Figure 3 shows that the water level of the points near Doumen, Daoxu, Cao'e, and Dongguan streets began to decrease when both gates are open. After 70 hours of drainage, the water level in the four streets dropped below 3m, and the drop in Cao'e street was the largest. After 2 hours from zero, the river velocity near the four streets reached the maximum velocity of 0.0254 m/s. After about 70 hours, the average flow velocity reached 0.0133 m/s.

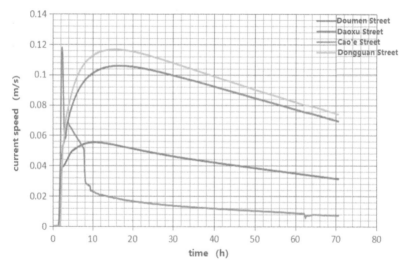

Figure 4. Four streets velocity diagram.

Figure 4 shows that the flow velocity in each street increases at first and then decreases. The flow velocity in Cao'e street changes greatly in a short time. The other three streets have similar trends.

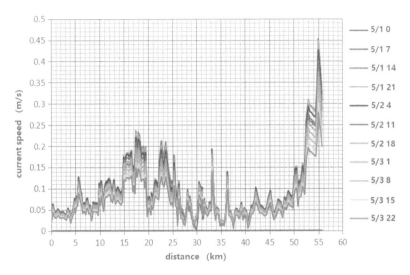

Figure 5.   Variation of velocity with distance in Hangyong Canal at different time points.

Figure 5 shows the maximum velocity at 54.970 km. The variation of velocity varies with height and bottom, that is, the velocity varies with distance at different times, and the variation trend is not as uniform as the variation trend of water surface elevation. There are many factors that cause this. According to the investigation data, the most important factor is the width of the river channel. The cross-section velocity of the wide channel is small, otherwise, it is large.

Figure 6.   Location map of Shangyu e-Game Town.

The point 54.970 km away from the source of Hangyong Canal and 0.446 m/s flow velocity—adjacent to Shangyu e-Game Town is analyzed.

Figure 7 shows that the water level drops more and more slowly with time, and the final water level is about 2.73m; at the beginning, the flow velocity rises rapidly, the maximum flow velocity is about 0.47m/s, then the flow velocity decreases linearly, and the final flow velocity is 0.275m/s, which is still large. Therefore, it is necessary to remind relevant personnel to conduct river operations carefully during this period.

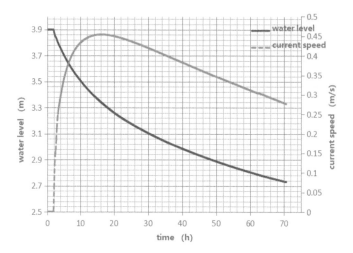

Figure 7. Flow characteristic variation diagram.

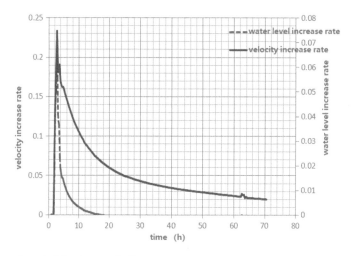

Figure 8. Flow characteristic change rate diagram.

Figure 8 shows that the increase rate of water level after the 20th hour is less than 0, which means that the water level continues to decline, which conforms to the actual opening law; in the whole drainage process, the increase rate of flow velocity is also more appropriate to the change of flow velocity. The maximum velocity change rate is about 0.075, and the minimum value is 0.0065.

4 CONCLUSION

According to the analysis, it is the most favorable time to open Caoejiang sluice and Binhai sluice 16 hours in advance. In recent years, the urban infrastructure construction has reduced the ability to regulate and store waterlogged water in the plain, and increased the pressure of waterlogging drainage in the plain. In order to ensure that the urban economic construction will not stop because of the urban flood control measures system, the gate structure can be analyzed in the future. By increasing the number of gate holes or optimizing the structure of the gate, the hydrodynamics of the river network can also be changed and the occurrence of natural disasters can be prevented.

ACKNOWLEDGMENTS

This research was supported by the Funds Key Laboratory for Technology in Rural Water Management of Zhejiang Province (ZJWEU-RWM-202101), the Joint Funds of the Zhejiang Provincial Natural Science Foundation of China (No. LZJWZ22C030001, No. LZJWZ22E090004) and the Funds of Water Resources of Science and Technology of Zhejiang Provincial Water Resources Department, China (No.RB2115, No.RC2040), the National Key Research and Development Program of China (No.2016YFC0402502), the National Natural Science Foundation of China (51979249).

REFERENCES

Liu, Y, Hua J. X, Cao D.W. (2021). *Operation and safe flood control scheme for a sluice project Harnessing the Huaihe River* (06), 93–94.

Shao, W.L., Lai, Y.N, Han, Y.F. (2015) Influence of siltation in the upper reaches of Cao'e River Sluice on waterlogging drainage in Xiaoshao Plain. *Small hydropower.*, 03: 36–37 + 65.

Sun, J.H. (2019) Textual research on the canal in eastern Zhejiang – Also on the regional water environment structure and water conservancy situation in Ningshao Plain. *Social science front.*, 12: 111–133 + 281-282.

Wei, B.W, Li, Y, K, Qi, Y.H, Yan, F, Yao S.Y.(2022) *Flood risk assessment based on hydrodynamic and fuzzy comprehensive model Water conservancy and Hydropower Technology* (Chinese and English).

Zhang, C, Zhang, L.W, Mo. Z.L, Chen, Y, Gao. W, Yan. F.(2020). *Research on the connection between urban drainage and flood prevention and watershed flood control system in China in the new era Water supply and drainage* (10), 9–13+58.

Zhang, S.Y, Wang, X.R. (2017) The formation and evolution of river network and water system in Ningshao plain and the practice of Contemporary Landscape Architecture. *Landscape architecture.*, 07: 89–99.

Zheng, Z.Z. (2022). *Study on flood control, damp proof and waterlogging control engineering system of intercepted River in Shenzhen Dakangang new city Guangdong water resources and hydropower* (02), 15–20.

Zhu, Y.M, Liu, L.J. (2005). *On the importance of water regulation and storage capacity in plain river network area from the perspective of waterlogging drainage Zhejiang water conservancy technology* (05), 26–27.

Advances in Measurement Technology, Disaster Prevention and Mitigation – Li & Mohd Yusof (Eds)
© 2023 The Author(s), ISBN: 978-1-032-36087-4

Numerical simulation of slope retrogressive thaw slumps in the permafrost region

Bao Zhou, Zhongfu Wang, Liang Wu, Yongyan Zhang & Sailajia Wei
Qinghai Institute of Geo-environment Monitoring, Xining, China

Wenfeng Zhu*
Chang'an University, School of Highway, Xi'an, China

ABSTRACT: With global warming, the retrogressive thaw slumps are widely developed in the permafrost region of Qinghai Tibet Plateau, which destroys the ecological environment and damages the infrastructure. This paper conducts the numerical simulation based on the monitor data and literature review. The results show that the freezing and thawing depth of the slope increased with the temperature increase. In the permafrost region, the significant reduction of shear strength parameters caused slope retrogressive thaw slumps. The safety factor decreased with the temperature significantly, and the corresponding horizontal displacement also increased. This paper is beneficial for engineering construction and economic construction.

1 INTRODUCTION

Slopes in permafrost regions are mainly reflected in that their instability is closely related to temperature. With the increase in global temperature, the Qinghai Tibet Plateau responds significantly to global warming (Cheng et al. 2019; Wu et al. 2009). The rise of temperature or seasonal freezing and thawing will inevitably bring a common engineering geological problem in permafrost region, that is, retrogressive thaw slumps (Luo et al. 2022), which is a kind of slope instability phenomenon. Therefore, based on the research foundation of slope stability and considering the influencing factors of temperature change, it is of great practical significance to study the instability mechanism and causes of retrogressive thaw slumps.

Permafrost is a kind of soil that is very sensitive to temperature. In other words, with the increase in temperature, the shear strength of permafrost is significantly reduced. Correspondingly, Hutchinson (Hutchinson 1974) pointed out that due to the accumulation of ice fragments on the freezing front, the water content of soil increases during thawing, and the undrained shear strength of soil decreases, resulting in slope instability. In terms of the stability evaluation method of slope in permafrost regions, Xu Jiang (Jiang et al. 2007) and others used the Monte Carlo simulation method to calculate the failure probability and reliability index, parameters C and ϕ were regarded as random variables, and the reliability of permafrost region under freezing and thawing conditions was preliminarily evaluated.

With the development of numerical simulation, a series of studies try to use the strength reduction method in the finite element and finite difference numerical simulation, which has achieved good results. Wang Lina (2008) analyzed the freeze-thaw stability of soil slope by using the limit equilibrium method and finite element method respectively, and analyzed the stability of the slope. Chen Yuchao (2008) applied the limit equilibrium method, strength reduction method, and finite difference method of water, heat, and force coupling to discuss the effects of thawing, freezing, and freezing-thawing cycles on slope stability respectively. The above research simulates the change

---

*Corresponding Author: z15374888162@163.com

DOI 10.1201/9781003330172-45

of temperature distribution, considers the change of freezing and thawing depth, displacement, and stress field, and compares it with the actual temperature situation, but it does not consider the damaging effect of temperature field change on the soil after freezing and thawing. Therefore, Chu Zhicheng and Chen Penghui (2020) introduced the freezing-thawing depth and freezing-thawing damage coefficient of soil, proposed the thermal-mechanical coupling method based on finite elements, took the stability coefficient as the system characteristic quantity, and analyzed it by orthogonal test and improved grey correlation analysis model. It is concluded that the cohesion is the most sensitive to the stability of permafrost slope; the influence of environmental factors such as freezing-thawing damage coefficient, freezing-thawing depth of seasonal active layer, internal friction angle, slope, and density on permafrost slope could not be ignored.

Permafrost slopes undergo extremely complex mechanical and thermal changes in the process of freezing-thawing cycles. It is known that the cohesion of soil deteriorates exponentially with the increase in the number of freezing-thawing cycles. Based on this, firstly, this paper determined the freezing and thawing depth of the slope using numerical simulation. Secondly, this paper considered freezing-thawing cycles and the strength reduction method to simulate the displacement, plastic zone, and safety factor of the slope in the permafrost region. Finally, this paper is beneficial for recognizing the mechanism of retrogressive thaw slumps developed in the permafrost region.

2 THEORY BACKGROUND

2.1 Temperature conduction

The built-in thermodynamic function of FLAC3D can simulate heat conduction and convection. The conduction variables involved are mainly temperature and three heat flow components. Based on the energy balance equation and Fourier heat conduction law, the transport law derived from the latter is used to establish a connection with the former to obtain the heat conduction differential equation. When the boundary conditions and initial conditions are known, the differential equation can be solved. The energy balance equation is shown as follows (Lin 2020):

$$-q_i + q_v = \rho C_v \frac{\partial T}{\partial t} \tag{1}$$

Where, q_i is the heat flux vector, W/m^2; q_v is the volumetric heat source intensity, W/m^3; ρ is the mass density, kg/m^3; C_v is the specific heat at constant volume, $J/kg°C$; T is the temperature, $°C$; t is the time, h.

$$q_i = -kT_i \tag{2}$$

Where, k is the thermal conductivity, $W/m°C$.

2.2 Strength reduction method

The basic principle of the strength reduction method is as follows: initially input the shear strength parameters cohesion c and internal friction angle, and continuously reduce the shear strength parameters by dividing by the reduction coefficient. With the increasing reduction coefficient, the reduced shear strength parameter input program is used for calculation until the slope reaches the limit equilibrium state, the unit body is a failure, and the slope is unstable. The corresponding coefficient in this state is the safety factor.

$$c' = \frac{c}{F_s} \tag{3}$$

$$\varphi' = \tan^{-1}\left(\frac{\tan \varphi}{F_s}\right) \tag{4}$$

$$\tau' = c' + \sigma \tan \varphi' \tag{5}$$

Where c is the cohesion; c' is the cohesion after reduction; φ is the internal friction angle; φ' is the internal friction angle after reduction; F_s is the safety factor.

3 ENGINEERING BACKGROUND AND NUMERICAL MODEL PARAMETERS

The research area in this paper is located in Qilian county, Qinghai province. 3500m shady slope and 3800m sunny slope are the lower boundary of permafrost. The permafrost shows weak to medium frost heaving, and the thawing settlement coefficient is close to 0.01. The frozen rock shows slight frost heaving deformation under a low-temperature environment. The frozen soil area develops collapse and landslide in some sections due to cold frost weathering and seasonal thawing.

To explore the mechanism of slope retrogressive thaw slumps, this paper divided the slope into the nonfreezing-thawing layer and the freezing-thawing layer. The freezing-thawing cycle is reflected by changing the shear strength parameters of the freezing-thawing layer. Based on the investigation and the characteristics of the local climate, it is needed to set the initial conditions of the temperature field according to the monitored temperature below the surface and influence depth. This paper studies the most dangerous month in a freezing-thawing cycle, to simulate the maximum freezing-thawing depth at the top and toe of the slope.

According to the calculated thawing depth, the slope models with an 8° belong to a high frequency of slope retrogressive thaw slumps (Luo et al. 2022). During the model creation, the height is uniformly selected as 30m, and the elastic-plastic constitutive model and Mohr-Coulomb failure criterion were adopted. And the initial stress field is selected according to the self-weight stress field. The constraint method of boundary conditions is that the displacement of the lower area is 0, the left and right sides are horizontal fixed boundaries, and the top surface is a free surface. In addition, the specific parameter is as follows:

Table 1. Physical and mechanical parameters of slope soil (Lin 2020).

Thermal conductivity$k(W/m°C)$	1.92
Specific heat $C(J/kg°C)$	800
Density ρ (kg/m^3)	2000
Bulk modulus $K(MPa)$	16
Shear modulus $G(MPa)$	5.13
Elastic modulus $E(MPa)$	20
Poisson's ratio μ	0.3
Cohesion c(kPa)	30
Internal friction angle (°)	25

4 RESULTS AND DISCUSSIONS

4.1 *Temperature filed simulation*

In the research area of this paper, the annual average temperature is below 0°C, and the extreme maximum temperature is 29.8°C; the extreme minimum temperature is −36.3°C. The temperature in the area decreases with the increase of terrain. For every 100m increase of terrain, the annual average temperature decreases by about 0.43−0.57°C.

According to the field temperature monitor data, the measuring point at the depth of 15m has begun to appear above 0 during May 2021(Figure 1). It caused the slope retrogressive thaw slumps (Figure 2). Therefore, it is necessary to simulate the temperature field of the slope in the permafrost region.

Figure 1.   Variation of soil temperature data with time.

Figure 2.   Retrogressive thaw slumps and collapse.

Figures 3 and 4 show the temperature distribution under 29.8°C and 3°C atmospheric temperatures. It can be seen that there is a great difference in the freezing-thawing depth between these two temperature conditions. The 29.8°C atmospheric temperature could thaw 13.7 m depth of permafrost religion near the slope toe, which could lead to reverse thawing and collapse of the slope much deeper than an atmospheric temperature of 3°C. In addition, the freezing-thawing depth of the slope top was deeper than that of the slope toe.

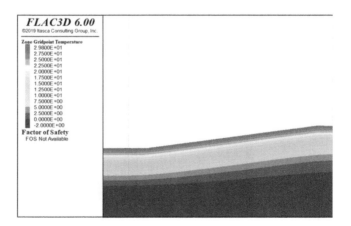

Figure 3.   Temperature distribution under 29.8°C.

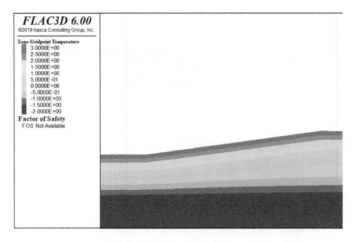

Figure 4.　Temperature distribution under 3°C.

4.2　*Retrogressive thaw slumps simulation*

Retrogressive thaw slumps are a special kind of landslide in the permafrost area of Qinghai Tibet Plateau. The scale and speed of the landslides are much smaller than those of ordinary ones.

In the permafrost region, when the angle of the slope is greater than 3°, retrogressive thaw slumps may develop. The surface of the permafrost region may thaw under a relatively high temperature. Under the condition of high ice content, the mass has low shear strength or 0 shear strength, which belongs to a plastic flow state (Jin et al. 2005). Therefore, the cohesion and internal friction angle may reduce to 1/6 (or more) of the initial value of the permafrost.

Figures 5 and 6 show the displacement of the slope under the 29.8°C and 3°C atmospheric temperatures, separately. Under the frozen condition with cohesion the initial value, the safety factor is 5.70, which is in a safety situation absolutely, mainly because the angle of the slope is much lower than that of ordinary ones. However, the retrogressive thaw slumps developed when the temperature increased. The horizontal displacement of the slope increased with the temperature increase. In addition, there is a difference between slope top and slope toe when the atmospheric temperature is 29.8°C.

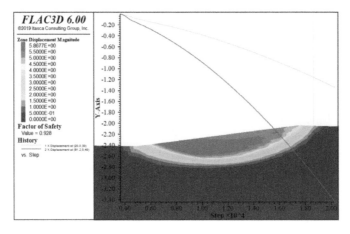

Figure 5.　Displacement distribution under 29.8°C.

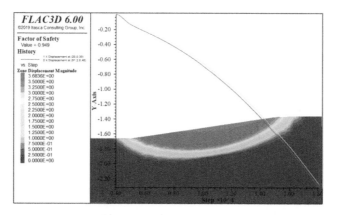

Figure 6. Displacement distribution under 3°C.

As can be seen from Figures 5 and 6 that the displacement of the slope increases with the retrogressive thaw slumps development. From Figures 3 and 4, we can obtain that the temperature decreased from the top to the bottom of the slope. Therefore, the surface mass will reach the plastic flow state, primarily. The sliding surface will not be too deep and has a parallel relationship with the slope surface. The main reason is that the thickness is limited. However, the area of retrogressive thaw slumps is large and develops rapidly, which is very easy to cause the destruction of the natural environment, continuously.

5 CONCLUSION

Firstly, this paper expounds on the basic theory of temperature field and strength reduction method; secondly, the local climatic conditions and geological conditions of the study area are introduced; thirdly, according to the literature review and climate conditions, the basic data of the model are determined; finally, according to the measured temperature data and the maximum freezing-thawing depth obtained from the monitoring, the surface temperature required for the temperature field simulation was determined, and the slope retrogressive thaw slumps is simulated. This paper draws the following conclusions:

a) Based on the measured temperature under the influence depth, the temperature field of the 8° slope was simulated and the thawing depth of slope toe and slope top was obtained. The freezing and thawing depth of the slope toe was smaller than that of the slope top.
b) With the increase in atmospheric temperature, the safety factor decreased significantly, and the corresponding horizontal displacement is also increasing. Therefore, climate warming will cause more geological disasters in the permafrost region.
c) In the permafrost region, the significant reduction of shear strength parameters is one of the important factors causing slope retrogressive thaw slumps under the freezing-thawing cycle environment. It is necessary to put forward the prevention and control measures of geological disasters in permafrost regions by using the ecological protection method.
d) Due to the limited parameters taken in this paper, the results do not fully reveal the mechanism of retrogressive thaw slumps. In future research, comprehensive research will be carried out in combination with the temperature, water content and resistivity monitored on site and the difference in mechanical properties of on-site soil before and after thawing.

ACKNOWLEDGMENT

The research was funded by the Science & Technology Platform of Qinghai Province (No.2021-ZJ-T08) and Applied Basic Research of Qinghai Province (No.2019-ZJ-7053).

REFERENCES

G.Cheng, L.Zhao, R.Li, et.al, "Characteristic, changes, and impacts of permafrost on Qinghai-Tibet Plateau" *Chinese Science Bulletin*, 2019, vol. 64 (27), pp.2783–2795.

J.Luo, F.Niu, Z. Lin, et al, "The characteristics and patterns of retrogressive thaw slumps developed in the permafrost region of the Qinghai-Tibet Plateau". *Journal of Glaciology and Geocryology*, 2022, vol. 44 (1), pp.96–105

J.Wu, Y.Sheng, Q.Wu, et al, "Processes and modes of permafrost degradation on the Qinghai-Tibet Plateau". *Sci China Ser D-Earth Sci*, 2009, vol. 39 (11), pp.1570–1578.

Jin, J.Sun, S.Fu. Discussion on landslides hazard mechanism of two kinds of low angle slope in the permafrost region of Qinghai-Tibet plateau. *Rock and Soil Mechanics*, 2005, vol. 26 (5): 774–778.

L. Wang, "*Freezing-thawing stability of slopes in the seasonally frozen region,*" Thesis, 2008, Harbin Institute of Technology.

N.Hutchinson. "Periglacial solifluxion: an approximate mechanism for clayey soils". *Geotechnique*, 1974, vol. 24 (3): 438–443.

X.Jiang, G.Yang, H.Liu."Evaluation of Permafrost Slope with Monte Carlo Simulation Method and Program Design". *Chinese Journal of Underground Space and Engineering*, 2007.

Y. Chen, "*Preliminary study on rock and soil slope stability under the freezing-thawing condition,*" Thesis, 2008, Xi'an University of Science and Technology.

Z. Lin, "Study on Freezing and Thawing Stability and Influencing Factors of Highway Slopes in Permafrost Regions of Naqu, Tibet," Thesis, 2020, Chongqing Jiaotong University.

Z.Chu, P.Chen, S.Lei, et al. Sensibility analysis of the influence factors on the stability of permafrost slope. *Journal of Henan Polytechnic University* (Natural Science), 2020, vol. 39 (5): 146–153.

Advances in Measurement Technology, Disaster Prevention and Mitigation – Li & Mohd Yusof (Eds)
© 2023 The Author(s), ISBN: 978-1-032-36087-4

Research on horizontal roadway passing through large hidden collapse column

Xianzhi Shi*
Guizhou Yuxiang Mining Group Investment Co. Ltd. Bijie, Guizhou, China

Yongjin Tang
Nuodong Coal Mine, Puding County, Southwest Guizhou, China

Yanmei Chen & Weifeng Huang
Zhangshuanglou Coal Mine of Xuzhou Mining Group, Xuzhou, Jiangsu, China

ABSTRACT: In the process of coal mining, collapse column has a great threat to mine safety production. A collapse column was exposed during the construction of the third west roadway at -500m level in Zhangshuanglou Coal Mine. The water head values of Taiyuan Formation No.4 limestone aquifer and Ordovician limestone aquifer are 4.35MPa and 4.79MPa respectively. Through the implementation of underground drilling and ground high-resolution three-dimensional seismic exploration, the collapse column is identified as a large-scale hidden collapse column with a long axis length of 220m, short axis length of 180m, and caving height of more than 346m; the rock mass in the collapse column is cemented again in the later stage, and the column has a certain strength. According to the analysis and research, the large-scale hidden collapse column is a dead collapse column that is not rich in (conducting) water, and is in the late development stage; the distances between the third West roadways and the fourth limestone and Ordovician limestone aquifers are 61m and 219m respectively, which are greater than the safe water-resisting layer thickness of 1.45m and 1.52m. The mine demonstrated and optimized the construction scheme that the roadway directly passes through the collapse column according to the current construction orientation, adjusted the roadway support mode, formulated safety technical measures to prevent water inrush, and ensured the safe excavation of the main roadway through the collapse column. The research results provide technical support and practical bases for safe and efficient passing through large-scale hidden collapse columns.

1 INTRODUCTION

The development of collapse columns in North China coal measures is common (Niu et al. 2011; Yin et al. 2005; Zhang & Tan 1998), which seriously threatens the safety production of coal mines(Shi et al. 2009; Tang et al. 2011). For the collapse columns exposed in the underground mining construction, most mines use the downhole drilling technology to explore the development scale of the collapse column (Shi et al. 2009), study the strength of the column and the water diversion (rich) situation, and adopt the excavation project to pass through the fallen column (Zhao 2015; Zhang 2011). In the measures to control the collapse column, Yin Shangxian, Wu Qiang, and Wang Shangxu [8], et al. only studied the hydrogeological conditions of the collapse column and did not propose the construction scheme for the excavation project. Zhao Shungang(2015) and Zhang Yongping (2011) only studied the supporting form of the collapse column and did not study the hydrogeological characteristics of the collapse column. Based on the study of the hydrogeological characteristics of the collapse column, this paper demonstrates that the collapse

---

*Corresponding Author: zslsxz@aliyun.com

column is a collapse column that does not be rich in (conduct) water, puts forward the feasibility of the main roadway directly passing through the collapse column, and optimizes the construction scheme and support form of the main roadway passing through the collapse column. The paper studies the prevention and control of water damage and the supporting technology of the roadway.

2 GENERAL SITUATION OF COAL MINE

2.1 Overview of the third West Lane

Zhangshuanglou Coal Mine adopts vertical shaft horizontal development, which is divided into four production levels of –500m, –750m, –1000m, and –1200m, and six mining areas including west 1, west 2, west 3, east 1, east 2 and east 3 are arranged for mining. The coal seams mined in the mine are No.7 coal and No.9 coal seams of the Permian Shanxi formation.

The xisan –500m horizontal track roadway (referred to as the third west roadways) is designed to be arranged in the sandstone layer between coal seams 7 and 9 of the Shanxi formation, with a spacing of 61m from the fourth limestone (L4) of Taiyuan Formation on the floor and 219m from the Ordovician limestone. The roadway is constructed in a 271° direction, with a total length of 3305m; the roadway is a straight wall semi-circular arch section, supported by bolt mesh and anchor cable shotcrete; the net width of the roadway is 4.6m, the net height is 4.1m, and the net cross-section is 15.5m^2.

A structural body is exposed at 2645m from the construction of the third west roadway. The stop position of the roadway has entered 15m of the structure (see Figure 1). There is no water seepage in the surrounding rock of the roadway. According to the field observation, the rocks in the structure are disorderly accumulated with complex lithology, and the main lithology is sandstone, mudstone, sandy mudstone, argillaceous sandstone, etc., with broken coal locally; the rock blocks are of different sizes and shapes, and the surface scratch is not obvious; the rock mass is argillaceous cemented. It is determined that the structural body is a collapse column (Niu et al. 2011; Shi et al. 2009).

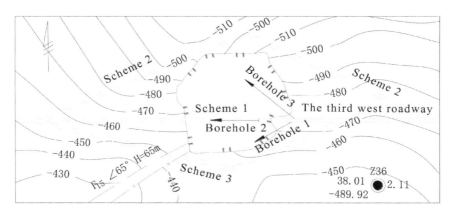

Figure 1. Borehole layout of exploration collapse column.

2.2 Coal measures stratum and structure

The main coal-bearing strata in the mine are Taiyuan Formation (C$_{3t}$), Shanxi Formation (P$_{1s}$), Lower Shihezi Formation (P$_{1x}$), and upper Shihezi Formation (P$_{2s}$). The mineable coal seams mainly occur in the upper Carboniferous Taiyuan Formation (C$_{3t}$) and the lower Permian Shanxi Formation (P$_{1s}$) strata. Now only the No.7 and No.9 coal seams of the lower Permian Shanxi Formation (P$_{1s}$) are mined.

Zhangshuanglou minefield is located in a relatively independent and completely closed geological structural unit surrounded by boundary faults in the East, West and south. The coal-bearing strata are monoclinic structures inclined to NNW, with a dip angle of 18 − 25° and an average of

22°. There are 35 large-scale faults, including 27 normal faults and 8 reverse faults. There are four large faults with a drop of 30–85m, such as F_{13}, F_{14}, F_{15}, F_{16}, etc.

A collapse column has been exposed in the West No.1 mining area at -500m level. In addition, the collapse column exposed in West No.3 mining area at −500m level has revealed two collapse columns above −500m level.

2.3   *Hydrogeological conditions of West No.3 mining area*

The aquifer of West No.3 mining area affecting the construction of West No.3 main roadway is limestone karst fissure confined aquifer of Carboniferous Taiyuan formation, with an average thickness of 34.20m and 14 limestone layers. Among them, the aquifer threatening water inrush in No.9 coal seam is the L4 limestone aquifer. The thickness of the fourth limestone aquifer is 4.80–11.69m, with an average of 8.21m. The distance from coal seam 9 is 38.8–67.5m with an average of 48.91m. According to L4 limestone pumping test data, the unit water inflow is 0.566– 0.879*l*/s.m, which belongs to a medium water yield aquifer; the water level elevation is –49.77m. Karst fissures are developed and karst caves are found locally. The water leakage rate of ground exploration boreholes exposed above –500m level is 70.4% (Shi & Sun 2001).

The borehole in the minefield reveals that the Ordovician limestone fractured karst confined aquifer is 311.86m thick, with a unit water inflow of 0.489–2.299*l*/s.m, which belongs to the medium to strong water rich aquifer; the water level elevation is –6.07m. Ordovician limestone aquifer has a hidden outcrop area of 14km² in the south of the minefield, which is rich in water content and is an indirect recharge source of mine water.

3   IMPLEMENTATION OF COLLAPSE COLUMN EXPLORATION PROJECT

3.1   *Downhole drilling*

In order to find out the spatial development scale of the collapse column in the third west roadway, three hydrogeological boreholes (see Figure 1) were designed and constructed near the front of the roadway to explore the spatial form and hydrogeological characteristics of the collapse column. The No.1 borehole is located on the south side (left side) of the roadway. The distance between the hole and the head-on is 26.5m. The construction is carried out at 241° and 4.5° elevations (see Table 1). The final hole depth is 54.18m. The No.2 borehole is located at the head of the roadway, parallel to the construction direction of the roadway, with an elevation of 4.5° and a final hole depth of 59.80m. The No.3 borehole is located on the north side of the roadway. The distance from the hole to the head is 29m. The construction direction is 310° and the elevation angle is 5° and the final hole depth is 81.61m.

Table 1.   Drilling results of exploration collapse column.

Borehole number	Drilling depth (m)	Angle (°)	Horizontal distance (m)	Angle with the third West Lane (°)	Distance between hole bottom and West Third Lane (m)
Borehole 1	54.18	4.5	54.01	30	27.01
Borehole 2	59.80	4.5	59.62	0	0
Borehole 3	81.61	5	81.30	39	40.65

3.2   *High resolution 3D seismic exploration*

After the collapse column was exposed in West No.3 mining area, in order to further identify the development scale of the collapse column and verify the exploration results of underground drilling data, high-resolution 3D seismic exploration was carried out on the ground(Yin et al. 2012). The exploration and control underground area is 550m long from east to west and 600m wide from south to north, which can effectively control the area of 250m in the south, 350m in the north, 220m in the East, and 330m in the West.

4 STUDY ON THE CHARACTERISTICS OF SUBSIDENCE

4.1 *Study on collapse column scale*

4.1.1 *Horizontal development scale*
The actual length of the collapse column exposed in the No.1 hole is 31.52m; that of the No.2 borehole is 69.80m, plus the exposed length of the third west roadway is 15.0m, and the horizontal distance of the exposed collapse column in the roadway construction direction is 74.57m. When the No. 3 hole is constructed to the hole depth of 9.5m, the collapse column is exposed, and the length of the exposed collapse column is 55.03m. None of the three boreholes revealed the edge of the other column. According to the analysis of the occurrence change law of the rock blocks exposed during the drilling process, the rock blocks in the collapse column exposed by the No.3 drilling hole (north side borehole) have the same tendency as the drilling direction at the edge of the collapse column (9.5–22.55m, length 13.05m); the rock block passing through the edge of the collapse column (22.55–34.46m, length 11.91m) has the opposite tendency to the drilling direction; after entering the deep part of the collapse column, the rock mass inclination is opposite to the drilling direction (34.46–81.61m, 47.15m in length), and the dip angle is about 5°. It is analyzed that the axial length of the collapse column in the construction direction of the No. 3 borehole is no less than 145m.

The long axis length of the collapse column is 220m and the short axis is 180m. According to the comprehensive analysis of drilling data and high-resolution 3D geological exploration data, the collapse column is a large buried collapse column with a short axis of 180m and a long axis of 220m.

4.1.2 *Development height analysis*
The lithology of the collapse column exposed in the third West Lane is gray-black, variegated, reddish brown mudstone, gray-white, grayish green, medium to coarse-grained sandstone, light flesh red quartz sandstone (oil luster, hard, coring rate of 100%). According to the lithologic analysis of the roadway and drilling, the caving rock is from the Lower Shihezi Formation (P_{2x}) boundary sandstone to the upper Shihezi Formation (P_{2s}) Kuishan sandstone section. The normal layer in front of the collapse column exposed in the third west roadway is sandstone 17m under the floor of No.7 coal seam. The caving height of the rock exposed by the roadway and drilling hole is not less than 109m (according to the data of borehole Z36 near the roadway). Therefore, the collapse column developed from Ordovician strata to Permian Lower Shihezi Formation strata, with a development height of more than 346m (see Figure 2), which is a large buried collapse column (Zhang & Tan 1998).

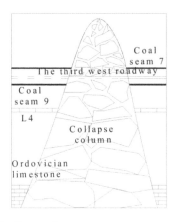

Figure 2. Development height of collapse column.

4.2 Analysis of lithology and strength of the column

The rock blocks in the collapse column are cement after hydration of mudstone, and the lithology revealed by the borehole is basically consistent with that revealed by the roadway. From the analysis of the formation process of the collapse column [8], the column body is a low-strength structural body, which is formed by the disorderly accumulation of caving rock blocks and cemented again by clay and other materials under the later stage of ground pressure. The core recovery rate of the borehole is low (0–24%) at the edge of the collapse column and 100% at the middle position of the collapse column. Therefore, the collapse column has certain strength after late cementation.

4.3 Analysis of water conductivity (enrichment) of collapse column

According to the tunnel exposure, the surrounding rock of subsided column has no water outflow phenomenon. Among the three exploration boreholes, only the No.3 borehole has 0.2m³/h water from 34.31m, and the other two boreholes have no water outflow phenomenon. Through comprehensive research and analysis, it is considered that the collapse column exposed in the third west roadway is a long axis of 220m, and the caving height is more than 346m; the rock in the column body is cemented in the later stage, with certain strength; the huge structure body without water-conducting and water-rich properties is the collapse column in the dead stage (Tang et al. 2011; Classifications and characteristics of karst collapse columns in North China coalfields[J]. Coal Geoglogy & Exploration, 2006, 34(4):53–56), and does not have water inrush conditions (Yin et al. 2005).

5 SELECTION OF SCHEME FOR PASSING THROUGH COLLAPSE COLUMN IN WEST NO.3 ROADWAY

5.1 Analysis of water inrush possibility of limestone in roadway floor

According to the observation data of aquifer level, the water level elevation of the L4 limestone aquifer is –49.77m, that of Ordovician limestone is –6.07m, and that of the third west roadway is –485m. According to the calculation formula of the safe water-resisting layer thickness of roadway floor (National Coal Mine Safety Administration, 2018)

$$t = \frac{L\left(\sqrt{\gamma^2 L^2 + 8K_{\mathrm{p}}p} - \gamma L\right)}{4K_{\mathrm{p}}} \tag{1}$$

Where: T-thickness of safe water-resisting layer, m;
L-roadway floor width, (4.6m);
γ - average density of water resisting layer of the floor, (0.025MN / m³);
K_P-The average tensile strength of the floor aquiclude, (0.365 MPa);
P-The actual head value of the floor aquiclude is (4.35MPa for four ash and 4.79MPa for Ordovician limestone).

The results show that the thickness of the safe water-resisting layer of L4 limestone on the roadway floor is 1.45M and that of Ordovician limestone is 1.52m. The distance between the roadway and L4 limestone aquifer and Ordovician limestone aquifer is 61m and 219M respectively. In the collapse column, the roadway is not threatened by the water inrush of L4 limestone water and Ordovician limestone water.

5.2 Scheme selection for passing through collapse column

Scheme 1, because there is no water inrush threat in the roadway within the collapse column, the roadway can be constructed directly through the collapse column according to the existing orientation. According to the existing 271° orientation, the roadway directly passes through the collapse column, and the total length of the roadway passing through the collapse column is 106m, as shown in Figure 1.

Scheme 2: the roadway retreats, adjusts the construction direction to the right, bypasses the collapse column, and then performs construction to the original design roadway layer position, and then adjusts the direction for construction. The roadway was retreated by 212m, and 252m was constructed according to the orientation of 297° (26° with the roadway), and then 276 m was constructed with the azimuth of 242° and then adjusted to 271° between coal seams 7 and 9. The total construction length of the roadway is 528m.

In the third case, the roadway retreats head-on, turns to the left side, bypasses the collapse column, and then constructs to the original design roadway level adjustment direction. The roadway retreated 100m, turned left, constructed at 246° direction for 149m, turned right for 295° construction, 253m to 7 and 9 coal seams, and adjusted to 271° for construction. The total construction length of the roadway is 402m.

The project quantity of Scheme 2 is 437m more than that of Scheme 1, and the construction time is 6 months longer than that of Scheme 1, which results in the delay in the formation period of the West No.3 mining area, which is not conducive to the mine replacement; although the engineering quantity of Scheme 3 is smaller than that of Scheme 2, the roadway is close to the fourth limestone, and it has to pass through F_{15} fault with 65m drop, so it is very easy to water inrush at the structural development area and has high risk, so it is not feasible.

According to the theoretical calculation results and the actual situation of the collapse column, the edge of the collapse column is the position with the worst filling, the most developed fracture, and the best water filling property. However, there is no water flow phenomenon when the third west roadway enters the subsided column 15m, and the deformation of bolt mesh supporting the roadway is not large, which indicates that the rock mass in the collapse column has a certain strength. There is no abnormal change in the water level observation hole of the fourth limestone and Ordovician limestone water level near the head-on after the collapse column is exposed by the roadway. Therefore, after comprehensive consideration of various factors, the mine decided to choose the first scheme, according to the existing roadway construction direction to continue the construction of the third West roadway.

5.3 *Roadway support design and construction*

Scheme 1 is the bolt mesh shotcreting support. The roadway is located in the collapse column, the surrounding rock is broken, and the cement between rock blocks is mainly clay material with low strength. The clay and broken surrounding rock are easy to expand when encountering water, and the moisture in the bolt construction is easy to penetrate into the surrounding rock, resulting in the increase of the loose circle of the roadway, which is not conducive to the permanent support of the roadway.

Scheme 2 is the shotcreting support. The roadway is supported by a U-shaped steel shed, full face support, shotcrete on both sides and roof, and pouring concrete on the floor.

Through comprehensive comparison, Scheme 2 has less damage to the surrounding rock of the roadway, can make full use of its own strength to support the roadway, and can effectively prevent water vapor infiltration into surrounding rock in the later stage. The roadway passing through the collapse column section is designed as a straight wall semi-circular arch section with shotcrete shed support. The section specification is consistent with that of the normal stratum block, as shown in Figure 3. The ditch is arranged in the middle of the roadway, with a section of 600 × 600mm, and the arch is built with cement.

5.4 *Safety technical measures for roadway construction*

During the period of passing through the collapse column in the third west roadway, the water prevention and control measures are adopted while exploring and excavating, with 50m exploration and 30m construction. After the initial spraying and erecting shed, the whole section must be resprayed, and the distance between U-shaped sheds shall not be greater than 700mm; the thickness of spraying layers on both sides and inside and outside the roof shall not be less than 50mm, and the thickness of spraying layer on the bottom plate shall not be less than 500mm. The tunnel ditch is built with cement mortar and followed by the harrow loader; if the mine water inflow in the western area of the collapse column is small, the drainage pipe is used to replace the ditch in the collapse column section. In order to reduce the water pressure and the threat of water inrush, drilling holes should be constructed in advance near the collapse column.

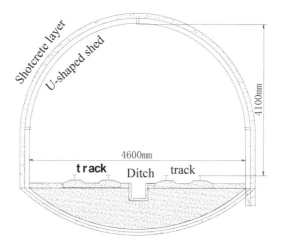

Figure 3. Section of U-shaped shed support.

6 CONCLUSION

A. Underground drilling and surface 3D high-resolution earthquakes have identified that the collapse column is a large-scale hidden collapse column with a long axis length of 220m, short axis length of 180m, and caving height of more than 346m.
B. Based on the drilling results, the upper and lower limits of the collapse column development are analyzed, and it is found that the rock mass in the collapse column has a certain strength. The collapse column is a huge structural body without water-conducting and water-rich properties, which is in the dead stage and does not have the conditions for water inrush.
C. The scheme of roadway passing through collapse column is optimized, and the safety technical measures for passing through collapse column are formulated, which ensures that the third west roadway passed through the collapse column safely, saving 437m development work amount and more than 6 months construction period for the mine.

REFERENCES

Classifications and characteristics of karst collapse columns in North China coalfields[J]. *Coal Geoglogy & Exploration*, 2006, 34(4): 53–56.
National Coal Mine Safety Administration. *Detailed rules for water prevention and control in coal mines* [M]. Beijing: Coal Industry Press, 2018.
Niu Lei, Wu Qiang, Li Bo.Classification of karst collapse columns in North China coalfields based on the generalized model of the inside structure of the columns [J]. *Coal Geoglogy & Exploration*, 2011(2): 57–60.
Shi Xianzhi,Sun Yancong.Feasibility of arranging horizontal main roadway in the fourth limestone mine of Taiyuan formation[J]. *Coal Technology*, 2001(2): 21–22.

Shi Xianzhi,Yuan Dezhu,Ge Jungang.Study on water control technology of collapse column in 21301 working face of Chensilou Coal Mine[J]. *China Coal*, 2009(4): 21–22.

Tang Junhua, Bai Haibo, Yao Banghua, et al.Theoretical analysis on water-inrush mechanism of concealed collapse pillars in the floor[J]. *Mining Science and Technology* (China), 2011(2): 57–60.

Yin Qifeng,Pan Dongming,Yu Jingcun,et a1.Research on seismic identification technique of coal mine collapse column J]. *Progress in Geophys.* 2012, 27(5): 2168–2174.

Yin Shangxian, Wu Qiang, Wang Shangxu. Water-bearing characteristics and hydrogeological model of karst collapse column in North China [J]. *Chinese Journal of Rock Mechanics and Engineering*, 2005, 24(1): 77–82.

Zhang Yonghong, Tan Zhuoying.Discussion on classification of karst collapse column in North China type coalfield [J]. *Coal Engineer*, 1998(5):19–24.

Zhang Yongping. Study on support technology of +450m levels north wing air intake roadway passing through xe7 collapse column in Gaohe mine[J]. *Coal Engineering*, 2011, (11): 49–50.

Zhao Shungang.Construction technology of driving face passing through collapse column [J]. *Coal*, 2015, 24(2): 40–41, 55.

Advances in Measurement Technology, Disaster Prevention and Mitigation – Li & Mohd Yusof (Eds)
© 2023 The Author(s), ISBN: 978-1-032-36087-4

Research on simulation pre-assembly method of bridge steel structure based on characteristic space line under safety monitoring background

Yanyi Li & Mingfang Zhu*
College of Surveying and Geo-Informatics, Tongji University, Shanghai, China

ABSTRACT: The assembly of bridge steel structures involves the body of large-scale construction. The bridge steel structure is obtained and constructed by surveying and mapping. On this basis, digital intelligent simulation assembly can effectively prevent the influence caused by assembly error in practical engineering. Therefore, this paper summarizes an automatic simulation pre-assembly method of bridge steel structure based on characteristic spatial line, which has reference significance for the data simulation assembly of this type of large steel structure. Overall, the pre-assembly method can effectively simulate a series of problems that may occur in the assembly process of large bridge structural parts, and further predict the possible assembly errors or errors in advance.

1 INTRODUCTION

The detection of the processing quality of bridge steel structure components is of great significance in bridge construction (Cantarelli et al. 2018; Park et al. 2018). Bridge construction occupies an essential position in municipal buildings and plays a significant role in social development (Wu et al. 2020; Zhang et al. 2018). In the bridge construction process, to ensure that the quality of steel structure components meet the standard, the simulated pre-assembly of steel structure components must be carried out (Li & Chen 2018; Ying et al. 2019). There are two main methods of quality inspection: manual method and model method. The manual process tests its quality through field splicing, which has some problems, such as low efficiency, high site requirements, poor quality control effect, etc. The model method is based on the point cloud component model to detect its quality, but there is a loss of modeling accuracy and a single splicing method. To test the possible problems in the installation process of bridge steel structure construction and better carry out a digital-analog pre-assembly test on the components to be assembled, a quality detection method based on the characteristic line of splice is proposed in this paper. The method proposed in this paper has high efficiency, high precision, and flexibility, which can effectively solve the problem of automatic pre-assembly and testing of ground steel structural members in bridge construction.

2 EXTRACTION AND MATCHING METHODS AND PROCESSES OF SPATIAL FEATURE LINES

2.1 *Building point cloud data*

Component point cloud refers to the component point cloud data after preprocessing. The components appear in pairs (i.e., A and B) in the component splicing. Here, we take one of them as a reference. As shown in Figure 1, there are two component point clouds to be spliced, which are placed disorderly.

---

*Corresponding Author: zhumingfang@tongji.edu.cn

Figure 1.   Component point cloud to be spliced.

2.2   *Component pose transformation*

Component pose transformation refers to the coordinate change of the component so that the element is roughly horizontal and the top surface is above the bottom surface. During data acquisition, the data are not placed in order. To facilitate the smooth progress of the following methods, some changes need to be made to the position and attitude of components to ensure the subsequent algorithm. As shown in Figure 2, it is a comparison diagram before and after the component pose transformation. The W state is placed disorderly, and its bounding box is large, while the component in the R state is roughly horizontal, the top surface is above the bottom surface, and the bounding box is small.

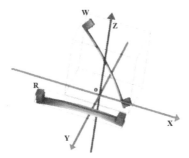

Figure 2.   Comparison before and after component pose transformation.

2.3   *Line feature extraction*

The edge line of the splice is extracted by the slice method. As shown in Figure 3, the blue part is the steel structure point cloud data, while the yellow part is the extracted edge line feature data.

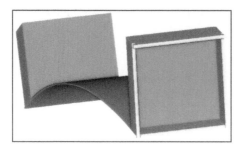

Figure 3.   Effect drawing of line feature extraction.

2.4 *Specify splicing position*

Specifying the splicing position refers to establishing a position of the component as the initial splicing position, which is similar to the coarse registration of registration, which is equivalent to a coordinate transformation.

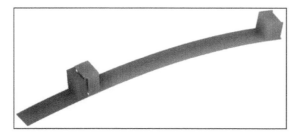

Figure 4.   Specify splice location.

2.5 *Component splicing*

The line features of two adjacent components are matched. The coordinate transformation matrix is obtained to complete the splicing of two adjacent pieces. As shown in Figure 5, it is a schematic diagram of matching by using the line features at the splicing of two adjacent steel structural members. Figure 6 is a schematic diagram of components after coordinate transformation of the component point cloud using the coordinate transformation matrix obtained after line feature ICP registration and the resulting graph of component splicing.

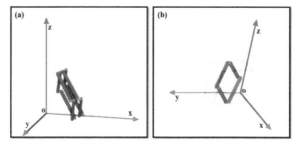

Figure 5.   Line feature matching (a) Before line feature matching; (b) after line feature matching.

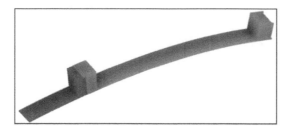

Figure 6.   Component splicing.

2.6 *Deviation analysis*

After completing component splicing, it is necessary to conduct deviation analysis and evaluate the component processing quality, including splicing deviation and overall linear deviation. The splicing error is a local error that describes the relative error between components after splicing. The general linear variation is a widespread error, representing the error between the overall piece and the expected design after joining. In general, deviation analysis evaluates the processing quality of components. However, this paper mainly studies the effectiveness of the pre-assembly method, so the evaluation process starts with the assembly accuracy.

3 EXPERIMENT AND RESULTS

3.1 *Display of experimental results*

This section uses the simulation data recovered from the design data to analyze the component splicing and deviation-based online feature matching. The accuracy of the component automatic assembly method is evaluated simultaneously.

(1) Firstly, the point cloud data is constructed by simulation. Here, the experiment is carried out with the density of 36 mm as an example.
(2) On this basis, the attitude transformation of steel structure components is carried out.
(3) The feature line is extracted to prepare for the subsequent attitude matching.
(4) Specify the splicing position and determine the matching of the splicing position according to the extracted feature line.
(5) Construction splicing is the last step of pre-assembly. Through this step, we can directly complete the complete display of the construction and installation process.

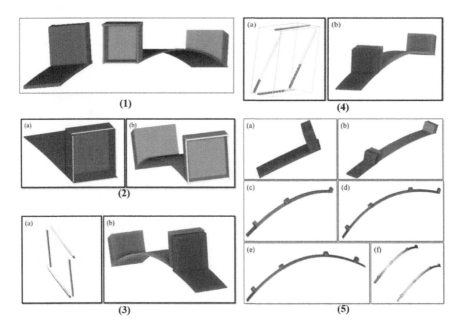

Figure 7.   Display of experimental results.

3.2 *Method evaluation and analysis*

The splicing surface deviation analysis and overall linear deviation analysis are carried out for the obtained component splicing results. Firstly, the splicing surface deviation analysis is carried out, and the required eight pairs of corresponding points are shown in Figure 9.

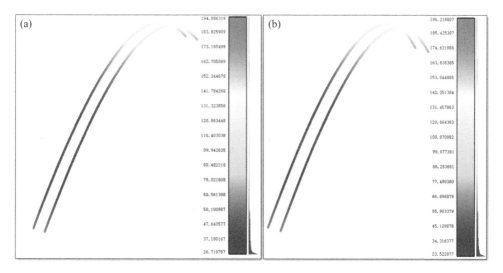

Figure 8. Analysis diagram of overall linear deviation (a. Analysis diagram of overall linear deviation of East data. b. Analysis diagram of overall linear deviation of West data.).

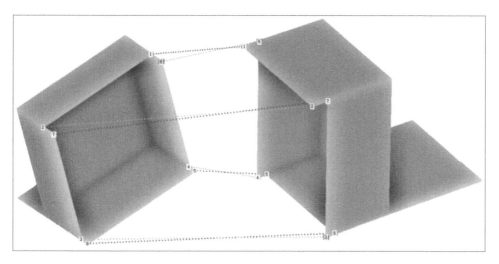

Figure 9. Corresponding points of splice surface deviation analysis.

It can be seen from Figure 8 that the color gradually brightens from left to right, that is, the error gradually increases. This is because the splicing direction is from left to right. When splicing, only the accuracy of splicing is considered, and the error caused by the overall alignment is not considered. With continuous splicing, the error will accumulate and become larger and larger.

Through the error analysis of the corresponding endpoint extracted from the line feature points, it is found that the average distance error of each splice is within 0.07 mm, and the maximum distance error is within 0.5 mm. The mean error of components in each direction is within 0.05 mm, and the maximum error of distance is within 0.4 mm, that is, the accuracy of splicing is very high.

341

4 DISCUSSION

Through the automatic simulation pre-assembly method of bridge steel structure based on a characteristic spatial line in this paper, the experimental test of pre-assembly of this type of bridge steel structure is well completed, and a complete pre-assembly process is put forward. According to this process, we can efficiently realize the rapid assembly test of the same type of construction and provide reference ideas and methods for assembling experimental equipment in the same field. At the same time, this method can test the risk of installation in the simulated environment and timely give the monitoring areas and critical points that need attention in the installation process. It has important reference significance for the intelligent analysis of measurement data.

5 CONCLUSIONS

Aiming at a series of problems arising in the process of bridge preassembly, this paper uses LiDAR to scan the steel structure of the bridge, obtains the real three-dimensional point cloud data of the components, carries out pre-simulation assembly on this basis, and puts forward a relatively complete pre assembly idea and method. The conclusions are as follows:

(1) The pre-assembly method can effectively simulate a series of problems that may occur in the assembly process of large bridge structural parts, and further predict the possible assembly errors or errors in advance;
(2) Through the scanned point cloud data, large parts are modeled, and the key features of parts are extracted through the idea of point cloud registration. The matching points of parts assembly are carried out through public features, and the error analysis before and after matching of different structural parts can objectively reflect the splicing of large parts.

ACKNOWLEDGMENT

We are very grateful to the College of Surveying and Geo-Informatics, Tongji University for providing the bridge pre-assembly test data. The source data can be obtained by contacting the corresponding author.

REFERENCES

Cantarelli, C. C., Flybjerg, B., Molin, E. J. E., & Wee, B. van. (2018). Cost Overruns in Large-Scale Transport Infrastructure Projects. *Automation in Construction*, *2*(1): 1–7.

Li, X., & Chen, W. (2018). Research on Design and Construction Optimization of Bionic Dendritic Steel Structure Based on BIM. *Journal of Physics: Conference Series*, *1087*(5): 1–6.

Park, S. I., Park, J., Kim, B. G., & Lee, S. H. (2018). Improving applicability for information model of an IFC-based steel bridge in the design phase using functional meanings of bridge components. *Applied Sciences (Switzerland)*, *8*(12): 1–10.

Wu, H. J., He, L., Wang, S. R., Li, J. Q., & Lu, P. (2020). Geometry Control Method Based on Stress-Free State Theory for Long-Span Concrete-Filled Steel Tube Arch Bridge. *Bridge Construction*, *50*(6): 1–9.

Ying, C., Zhou, Y., Han, D., Qin, G., Hu, K., Guo, J., & Guo, T. (2019). Applying BIM and 3D laser scanning technology on virtual pre-assembly for complex steel structure in construction. *IOP Conference Series: Earth and Environmental Science*, *371*(2): 1–12.

Zhang, F., He, B., Hui, J., Zhu, B., Zhang, J., Liu, Y., & Cheng, G. (2018). Smart management for steel-structure bridge industrialization construction. *ICNSC 2018 - 15th IEEE International Conference on Networking, Sensing and Control.*

Advances in Measurement Technology, Disaster Prevention and
Mitigation – Li & Mohd Yusof (Eds)
© 2023 The Author(s), ISBN: 978-1-032-36087-4

Taking the Longxian-Qishan-Mazhao fault as an example to analyze the pseudo effect of normal/inverse fault caused by strike-slip movement

Chenyi Yang, Ji Ma, Xiaoni Li*, Yifei Xu & Lina Su
Shaanxi Earthquake Agency, Xi'an, China

ABSTRACT: The Longxian-Qishan-Mazhao Fault is important boundary tectonics on the south-west edge of the Ordos Block, which has obvious sinistral strike-slip characteristics. It was found from a field survey that the Longxian-Qishan-Mazhao Fault has the phenomenon of "vertically dislocating" the Quaternary aeolian loess–paleosol. To clarify the cause of its formation, three typical geological sections of the fault were analyzed. The result shows that the "vertical dislocation" observed in the field is mainly the false appearance caused by the strike-slip movement of the fault. To further strengthen the cognition of the sections of the Longxian-Qishan-Mazhao Fault, by considering the geometric elements and constraints controlling the stratum of fault sections, the pseudo effect model of normal/reverse fault formed by strike-slip movement is supplemented and improved, which provides a basis for the study of fault activity and seismic risk in similar areas.

1 INTRODUCTION

The Longxian-Qishan-Mazhao fault is the largest and most active branch of the Longxian-Baoji Fault zone on the southwest edge of the Ordos Block, with complex tectonic background and seismic risk. There were many moderately-strong earthquakes in history, and now small earthquakes are also very active. Historical earthquakes are mainly concentrated near Longxian, Meixian, and Qishan. It may be the seismogenic structure of the M7 earthquake in 780 B.C., and there were several high-frequency and low-intensity earthquake clusters since 1970 (Shi 1996, 2011; Shi et al. 2013). It has been continuously active since the late Quaternary. The maximum potential magnitude of the fault is about 7.5 (Shi 2011; Shi et al. 2013; Wang et al. 2018). Through the observation of field geological sections, some scholars believed that the main activity mode of the fault in Quaternary is normal faulting (Li 1992; Peng et al. 1992; Shi, 2011; Wang 2011). In recent years, the fault landform, current crustal deformation monitoring data, and regional tectonic background characteristics show that the fault has had obvious characteristics of sinistral strike-slip movement since the late Pleistocene (Chen et al. 2018; Li et al. 2018; Li 2017; Liu et al. 1997; Lin et al. 2015; Zheng et al. 2016). The strike-slip movement will cause the pseudo effect of normal/reverse fault (Zhu et al. 2008). When strike-slip movement dislocates the monoclinic stratum, due to various cross-cutting and sliding relations between the stratum and fault, it often causes visual dislocation of the marker stratum in a plane or section. The dislocation marker stratum dislocation is not completely consistent with the real dislocation. For example, when the dipping fault slides along the strike of the fault plane, the section will show the pseudo effect of normal/reverse fault. It is worth noting that in addition to the monoclinic stratum formed by the structure, the typical aeolian loess-paleosol sequence sedimentary layer has the characteristics of "drape" occurrence. Its sedimentary mode is that the sediments are covered on the original landform with almost the same thickness as the snow layer, basically inheriting the original landform outline under it (Hu 1965; Liu 1965;

---

*Corresponding Author: 25413091@qq.com

Liu & Shi 2000). Due to the uneven sedimentary occurrence, this aeolian loess-paleosol sequence sedimentary layer also has the condition to form the pseudo effect of stratigraphic dislocation. Considering that the Longxian-Qishan-Mazhao Fault has obvious strike-slipping characteristics, did this movement result in the "vertical dislocation" of loess-paleosol sedimentary strata? Aiming at this common but possibly neglected problem in the study of active structures, this paper selects three typical field sections of the Longxian-Qishan Mazhao Fault to study the pseudo effect of stratigraphic dislocation caused by the strike-slip mechanism.

2 EVIDENCE OF NEW STRIKE-SLIP ACTIVITY OF LONGXIAN-QISHAN-MAZHAO FAULT

2.1 Overview Longxian-Qishan-Mazhao fault

The strike of the Longxian-Qishan-Mazhao Fault is 300°−315°, with a total length of about 130 km. Geographically, the fault starts from Caobi in the northwest, passes through Qianyang, Fengxiang, and Qishan, enters the Weihe Graben, and terminates in Mazhao in front of the Qinling Mountain. The linear characteristics of the satellite image of the fault are clear and obvious (Figure 1). From the perspective of geotectonics, the fault is located on the southwest edge of the Ordos Block. It is one of the important components of the main boundary zone between Qinghai Tibet Plateau and the North China Block in the Shaanxi-Gansu junction area. In the north, the fault relates to the Liupanshan Fault. In the south, it crosses the western section of the Weihe Graben to the east and reaches the front of Qinling mountain. The Longxian-Qishan-Mazhao Fault belongs to the largest and most active branch of the Longxian-Baoji Fault zone in the southern section of the Liupanshan Fault, with strong seismic risk (Shi 2011; Shi et al. 2013; Wang et al. 2018). The fault was normal

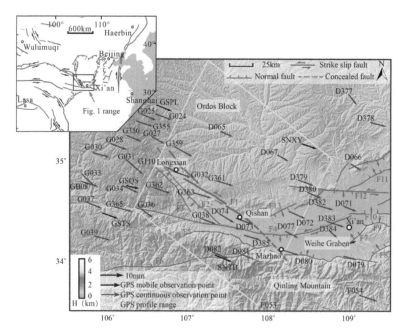

Figure 1. Map of the position of the tectonic setting of the Longxian-Qishan-Mazhao Fault and regional GPS velocity field (1998−2018). F1: Longxian-Qishan-Mazhao Fault; F2: Badu-Guozhen Fault; F3: Taoyuan-Guichuansi Fault; F4: Weihe Fault; F5: Northern margin Fault of Qinling; F6: Taochuan Fault; F7: Lintong-Chang'an Fault; F8: Western margin Fault of Huashan; F9: Lishan Fault; F10: Weinan Fault; F11: Liquan Fault; F12: Kouzhen-Guanshan Fault; F13: Fuping-Qianxian Fault.

in the early stage, but about 5.3 Ma ago, it was affected by the uplift of Qinghai Tibet Plateau and lateral extrusion of Qilian Block (Zhang et al. 2003), it absorbed the residual energy of the strike-slip movement of the Haiyuan fault in the northwest, and began the movement dominated by sinistral strike-slip (Li 2017; Chen et al. 2018).

2.2 *Geomorphic characteristics of sinistral strike-slip fault dislocation*

Many geomorphic dislocations reflect that the Longxian-Qishan-Mazhao Fault has obvious new activity characteristics dominated by sinistral strike-slip. Through the fine measurement of horizontally faulted rivers and terraces and dating results, it is inferred that the sinistral strike-slip rate of the fault since the late Quaternary is 0.5–3 mm/a (Lin et al. 2015; Li 2017; Li et al. 2018; Zheng et al. 2016), which is far greater than the vertical movement rate of the fault (< 0.1 mm/a) (Li 1992; Wang 2018). The Qianyang-Qishan section is located in the middle of the fault. In this section, the river has obvious sinistral displacement (Figures 2a–c). Among them, the sinistral terrace of the Shaao River is a typical representative (Figure 2d). The three geological sections to be discussed below are located in areas with obvious sinistral strike-slip (Figures 2b–c).

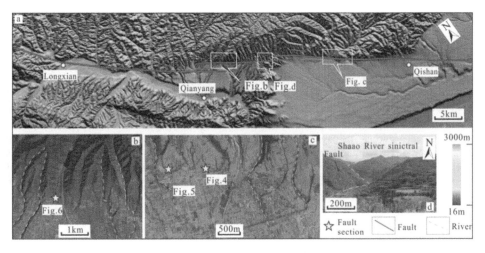

Figure 2. The terrain landscape and horizontal-offset landforms of the Longxian-Qishan-Mazhao Fault (b, c, and d).

2.3 *The present sinistral strike-slip characteristics of the fault reflected by GPS measurement*

GPS observation data have been widely used in crustal deformation and seismicity monitoring. Many achievements have been made in the analysis of the relative horizontal movement rate of the two walls of the active fault by using the regional GPS velocity field data (Hamiel et al. 2018; Holt et al. 2000; Langbein & Bock 2004; Wang et al. 2000; Zhang et al. 2003). Based on the observation data of GPS stations (1998–2018) of "China Crustal Movement Observation Network" (Yu et al. 2019), this paper selected 21 GPS stations (the relationship between section range and fault location is shown in Figure 1) and analyzed the current relative horizontal movement of the two walls of the Longxian-Qishan-Mazhao Fault. According to the GPS monitoring data, the GPS velocity profile of the Longxian-Qishan-Mazhao Fault was drawn, and the GPS velocity values of each station were projected onto the section parallel to the fault strike (Figure 3). It can be seen from Figure 5 that the south wall of the fault moves faster towards the southeast than the north wall, and the speed difference between the two walls is 1.6 ± 0.5 mm/a. Therefore, the GPS velocity field calculation shows that the Longxian-Qishan-Mazhao Fault is undergoing sinistral strike-slip (shear) movement at a rate of 1.6 ± 0.5 mm/a.

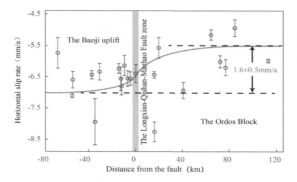

Figure 3.   GPS observations along the cross-sections of the Longxian-Qishan-Mazhao Fault (fault-parallel direction).

3   DESCRIPTION AND ANALYSIS OF FAULTED STRATUM

The above described the geomorphic characteristics of fault sinistral dislocation and calculated the current movement rate of the fault. Both reflect that the Longxian-Qishan-Mazhao Fault is dominated by sinistral strike-slip movement from the late Quaternary to the present, which is consistent with the relative tectonic movement trend between the northeast edge of Qinghai Tibet Plateau and the southwest edge of Ordos Block (Li et al. 2017; Yuan et al. 2013; Zheng et al. 2013). Along the Longxian-Qishan-Mazhao Fault, there are several natural sections of late Pleistocene strata that have been faulted. Can these "normal faults" be the pseudo effect caused by fault strike-slip movement? During this paper, the reasons for the "vertical dislocation" caused by strike-slip faults will be discussed and supplemented by the pseudo effect model of normal/reverse faults. Three typical aeolian loess-paleosol fault sections in the Qianyang-Qishan section of the fault (Figure 2 shows the section position) were selected to analyze the possibility of the "normal fault" phenomenon according to the geological section.

3.1   *Field geological sections phenomenon*

The field geological sections are described as follows:

(1) The Beishuigou Village Section is located in the abandoned brick factory in Beishuigou Village near Fengxiang (34.58° N, 107.46° E) (Figure 2c), with a section length of about 80 m and a strike of 60°. The strike of the fault is 130°, and the dip angle is 85°. The fault "vertically" dislocated the late Pleistocene paleosol layer S_1, the late Pleistocene loess layer (L_1), and the Middle Pleistocene loess layer (L_2). Among them, the apparent vertical fault throw of S_1 is about 1 m. Visually, this dislocation looks like a "normal fault" (Figure 4).

Figure 4.   Photograph (left side) and sketch geologic cross-section of the Longxian-Qishan-Mazhao Fault in Beishuigou Village near Fengxiang. (The code represents the stratigraphic unit of the loess-paleosol sequence).

(2) The Taixiangsi Section is located in Taixiangsi, Fengxiang (34.60° N, 107.44° E) (Figure 2c), with a section length of about 100 m and a strike of 70°. The strike of the fault is 310°, and the dip angle is 85°. The fault dislocated the paleosol layer S_1 at the bottom of the late Pleistocene, the late Pleistocene loess layer (L_1), and the Middle Pleistocene loess layer (L_2, L_3), and its paleosol layer (S_2). Among them, the apparent vertical fault throw of S_1 is about 1.7 m. Visually, this dislocation looks like a "normal fault" (Figure 5).

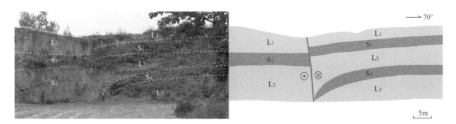

Figure 5.   Photograph (left side) and sketch geologic cross-section of the Longxian-Qishan-Mazhao Fault in Taixiangsi near Fengxiang. (The code represents the stratigraphic unit of the loess-paleosol sequence).

(3) The Zhangjiajiao Section is located in Zhangjiajianjiao, Fengxiang (34.70° N, 107.21° E) (Figure 2b), with a section length of about 50 m and a strike of 40°. The strike of the fault is 325°, and the dip angle is 85°. The fault dislocated the late Pleistocene paleosol layer S_1, late Pleistocene loess (L_1), Middle Pleistocene loess (L_2), and its paleosol layer (S_2). Among them, the apparent vertical fault throw of S_1 is about 4.0 m. Visually, this dislocation looks like a "normal fault" (Figure 6).

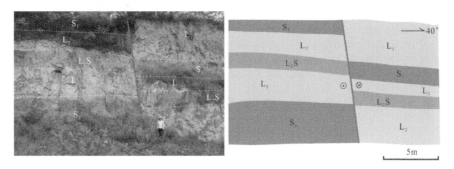

Figure 6.   Photograph (left side) and sketch geologic cross-section of the Longxian-Qishan-Mazhao Fault in Zhangjiajianjiao near Fengxiang. (The code represents the stratigraphic unit of the loess-paleosol sequence).

3.2   Principle of the pseudo effect of strike-slip fault

When the strike-slip fault dislocates the monoclinic stratum, if the stratigraphic dislocation relationship is inconsistent with the true dislocation mode of the fault, the pseudo effect of normal/reverse fault will often be formed (as shown in Figure 7) (Zhu et al. 2008). It is worth noting that in addition to the monoclinic strata formed by tectonic action, the Quaternary aeolian loess-paleosol stratigraphic sequence with "drape" sedimentation can also have similarly inclined strata. Because such deposits accumulate and cover the original terrain like snow, they will inherit the underlying terrain contour (Hu 1965; Liu 1965; Liu & Shi 2000). Quaternary aeolian loess-paleosol deposits are widely distributed in Northwest China, which has an alternating rhythmic structure of yellow loess (L) and reddish brown paleosol (S) in the vertical direction. Many chronological studies have determined the ages of different horizons of this sedimentary sequence (Heller & Liu 1982; Kukla & An 1989; Lai 2010) so that the relevant horizons of this sequence can become an important marker layer with a time scale in the study of active tectonics and seismology. However, due to weathering, leaching, or dissolution by infiltration of surface water and other reasons, clear fault scratches cannot be preserved on the section, and cannot be used to provide an analytical basis for judging the relative movement mode of the two walls of the fault in the field. The "dip-slip" or "vertical" faults observed in the field may be formed by the horizontal movement of the fault. Therefore, it is necessary to understand and identify the pseudo effect in the study of active tectonics and seismology to restore the real movement mode of the fault.

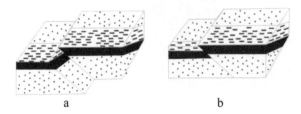

a   b

Figure 7. Pseudo effect of reverse fault caused by a strike-slip fault (a. The monoclinal stratum dislocated by the strike-slip fault. B. The profile of "reverse fault" effect.) (Zhu et al. 2008).

3.3 *The supplement and improvement of the strike-slip fault model*

The existing model of pseudo effect formed by strike-slip faulting is a schematic diagram (Zhu et al. 2008), which cannot help to be analyzed in fieldwork. In this regard, we consider the movement direction of the strike-slip fault, the tendency of faulted stratum, the angle between sections strikes and fault strike, and the observation direction. Based on the previously established model (Figure 7), we supplemented and improved the influence of the above geometric elements of the fault section on the fault pseudo effect, and established a strike-slip fault model with a pseudo effect of normal/reverse (Figure 8). Taking the sinistral strike slip fault as an example, the model in Figure 8 simulates the process and principle of the pseudo effect of normal/reverse fault formed by the horizontal movement dislocating inclined stratum. In the model, the brownish red line is

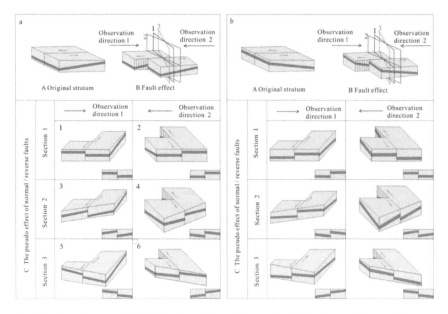

Figure 8. Supplementary model of strike-slip fault with the pseudo effect of normal/reverse fault.

the marker layer (paleosol), and profiles 1, 2, and 3 represent the relationship between fault strike and profile strike (the strike of section 1 is perpendicular to the fault strike, the strikes of section 2 and section 3 are oblique to the fault strikes). In addition, to simplify the model, the dip angle of the strike-slip fault in Figure 8 is set to 90°. However, in practical application, it is necessary to combine the model with fault tendency for analysis and judgment.

Two conditions need to be met to form the pseudo effect in the model in Figure 8. (1) The attitude of the dislocated stratum is inclined. (2) The strike of the fault and the strike of the faulted stratum cannot be parallel. When they are perpendicular to each other, the pseudo effect of the normal/reverse fault on the section is the most obvious. If they are oblique, the greater the angle of the oblique intersection, the more obvious the pseudo effect of the normal/reverse fault.

4 JUDGMENT AND ANALYSIS OF THE PSEUDO EFFECT OF NORMAL/REVERSE FAULT

4.1 *The use method and step of the supplementary model*

According to the field survey, when the fault "vertical dislocating" monoclinic stratum, if there is no other additional information (such as no scratch on the section) for reference, it can be analyzed and judged by the model in Figure 8 combined with the following steps to determine whether the "vertical dislocation" of the fault is the fault pseudo effect formed by fault strike-slip movement. The analysis methods are as follows: (1) observe the section, divide the stratum and measure the occurrence of fault and stratum; (2) judge whether the fault strike is parallel to the strike of the faulted stratum; (3) determine the observation direction; (4) determine the section strike, and the relationship between the strike of "vertically dislocated" stratum and the fault strike, and draw a 3D block diagram similar to that in Figure 8; (5) compare the 3D block diagram with the corresponding model in Figure 8 to determine whether the monoclinic stratum "vertically dislocated" is the fault pseudo effect caused by strike-slip movement.

4.2 *Analyzing the stratum dislocated by using the model above*

The three geological sections described above are located in the section with the obvious sinistral movement of the Longxian-Qishan-Mazhao Fault, and many sinistral faulted landforms were developed near these sections. The fault dislocated the Quaternary typical aeolian loess-paleosol sedimentary sequence with the original inclined attitude, which has the conditions for the formation of the pseudo effect of normal/reverse fault. To discuss the real cause of "vertical dislocation" in the section, the above model is used for analysis and judgment.

(1) We can perform the following analysis using the steps described in Section 1.3: (1) observe the strike of the fault and the dislocated stratum: the strike of the fault is 130°, and the strike of the stratum is SW, which are not parallel but have the conditions to form a pseudo effect of a normal/reverse fault; (2) observe the tendency of the dislocated stratum: the stratum tendency is NW, which is consistent with the model in Figure 8a; (3) determine the observation direction: the observation direction is the mirror direction (NW) on the left of Figure 4, which is consistent with the observation direction 2 in Figure 8a; (4) determine the relationship between the section strike and the fault strike: the measured fault strike is 130° and the section strike is 60°, and the two intersect at a large angle, which is consistent with Section 3 in observation direction 2 of Figure 8a. Based on the above analysis, it is considered that the "vertical dislocated" stratum of S_1 shown in Figure 4 is the visual pseudo effect of normal fault formed by the sinistral slip, and the actual observation phenomenon can be compared with the ideal state shown in Figure 8a–6.

(2) According to the steps of the model mentioned in Section 1.3, we carry out the following analysis: (1) observe the strike of the fault and the dislocated stratum: the strike of the fault is 310°, and the strike of the stratum is SW, which are not parallel and have the conditions to form the pseudo effect of normal/reverse fault; (2) observe the stratum tendency of the dislocated

stratum: the stratum tendency is NW, which is consistent with the model in Figure 8a; (3) determine the observation direction: the observation direction is the mirror direction (NW) on the left of Figure 5, which is consistent with the observation direction 2 in Figure 8a; (4) determine the relationship between the section strike and the fault strike: the measured fault strike is 310° and the section strike is 70°, and the two intersect at a large angle, which is consistent with Section 3 in observation direction 2 of Figure 8a. Based on the above analysis, it is considered that the "vertically dislocated " stratum of S_1 shown in Figure 5 is the visual pseudo effect of normal fault formed by the sinistral slip, and the actual observation phenomenon can be compared with the ideal state shown in Figure 8a–6.

(3) According to the steps of the model mentioned in Section 1.3, we carried out the following analysis: (1) observe the strike of the fault and the dislocated stratum: the strike of the fault is 325°, and the strike of the stratum is NE, which are not parallel and have the conditions to form the pseudo effect of normal/reverse fault; (2) observe the stratum tendency of the dislocated stratum: the stratum tendency is ES, which is consistent with the model in Figure 8b; (3) determine the observation direction: the observation direction is the mirror direction (NW) on the left of Figure 6, which is consistent with the observation direction 2 in Figure 8b; (4) determine the relationship between the section strike and the fault strike: the measured fault strike is 325° and the section strike is 40°, and the two intersect at a large angle, which is consistent with Section 3 in observation direction 2 of Figure 8b. Based on the above analysis, it is considered that the "vertical dislocated" stratum of S_1 shown in Figure 6 is the visual pseudo effect of normal fault formed by the sinistral slip, and the actual observation phenomenon can be compared with the ideal state shown in Figure 8b–6.

4.3 Using the model to analyze the dislocated stratum observed in the field

In Sections 3.1–3.3, the inclined paleosol layer S_1 is used as the fault marker layer in the fault profile. We compared the actual observation of these three sections with the derivation results of the model in Figure 8 (Table 1). The analysis shows that the "rising wall" and "falling wall" of the three actual sections are consistent with the "rising wall" and "falling wall" of the corresponding model in Figure 8, as well as the tendency of the marker layer of the actual observation section, is also consistent with the apparent tendency of the marker layer of the corresponding model, and the fault pseudo effect shown by the field observation and the model is consistent. Because the horizontal movement rate of the fault in this area is much greater than the vertical movement rate, the influence of the vertical movement of the fault on the fault displacement can be ignored. Therefore, through the model comparison results, it is considered that the "normal fault phenomenon" observed in the field is the pseudo effect caused by fault strike-slip movement.

In the model, the factors, and constraints such as the non-perpendicularity between the strike of the observed fault section and the strike of the fault are considered, and the application steps are given. However, the model is idealized. For example, the problems such as the local erosion surface of the marker layer and multi-stage activity of the fault will bring difficulties to the field analysis and judgment. Therefore, it is necessary to select a typical fault section as much as possible,

Table 1. Comparison between field sections of the Longxian-Qishan-Mazhao Fault and fault pseudo effect shown in Figure 8.

Position	ORW	ODW	OMLT	Model number	RWM	DWM	MLTM	Results
Beishuigou	RW	LW	WS	2a-6	RW	LW	WS	
Taixiangsi	RW	LW	ES	2a-6	RW	LW	ES	Consistent
Zhangjiajianjiao	LW	RW	NE	2b-6	LW	RW	NE	

(Note: ORW: observed rising wall; ODW: observed descending wall; OMLT: observed marker layer tendency; RWM: the rising wall of the model; DWM: the descending wall of the model; MLTM: marker layer tendency of the model; RW: right wall; LW: left wall)

carefully divide and measure the occurrence of fault and stratum during fieldwork, and compare it with the model, which will help to improve the accuracy and scientificity of judging the real movement of the fault.

5 CONCLUSIONS

Based on the results and discussions presented above, the conclusions are obtained as below:

(1) Based on the fault geomorphic dislocation and modern GPS observation, it is considered that the Longxian-Qishan-Mazhao Fault has obvious characteristics of the sinistral strike-slip movement. To analyze whether the strike-slip movement of the fault will form the pseudo effect, the pseudo effect model of normal/reverse fault is established based on previous studies. By comparing three typical loess-paleosol fault sections on the Longxian-Qishan-Mazhao Fault (Qianyang-Qishan segment) with the model, the result showed that the "vertical dislocation" of the strata is the pseudo effect formed by the sinistral strike-slip movement.

(2) The main function of the supplementary model is to remind and help identify that the strike-slip fault can also form the "vertically dislocated" stratum. It is a principled auxiliary analysis method and needs to be comprehensively judged by other research means. At the same time, the model is established for strike-slip movement. If the fault has oblique slip movement, the influence of the vertical movement component on formation dislocation needs to be considered. In addition, the model is also idealized, and the actual situation of the field fault sections may be more complex, which needs to be carefully analyzed and compared.

(3) The field fault section identification is one of the most direct and basic means to understand and study active faults. Its accuracy is very important to judge fault geometry and kinematics. Through the example of the Longxian-Qishan-Mazhao Fault, in this paper, the importance of identifying fault pseudo effect in the study of active faults was illustrated. It can provide further technical support for the study of fault activity properties in similar areas, and has certain reference and guiding significance for the analysis of fault activity and seismic tectonic risk in the study area.

ACKNOWLEDGMENT

This study was supported by the National Key R & D Program of China (Grant No. 2018YFC1503205) and the program of Geometric Structure and Tectonic Transformation of the Southern end of the Liupanshan Fault. We are very grateful to Xueze Wen and anonymous referees for their helpful suggestions that improved the manuscript.

REFERENCES

Aiming Lin, Gang Rao, Bing Yan. (2015) Flexural fold structures and active faults in the northern-western Weihe Graben, central Chinas[J]. *Journal of Asian Earth Scienses*, 114(1): 226–241.

Daoyang Yuan, Weipeng Ge, Zhengwei Chen, et al. (2013) The growth of northeastern Tibet and its relevance to large-scale continental geodynamics: A review of recent studies[J]. *Tectonics*, 32(5): 1358–1370.

Guangtao Hu. (1965) The mantle-overlying structure of loess layer in western Guanzhong[J]. *Quaternary Sciences*, 4(2): 85–92.

Hamiel Y, Masson F, Piatibratova O, et al. (2018) GPS measurements of crustal deformation across the southern Arava Valley section of the Dead Sea Fault and implications to regional seismic hazard assessment[J]. *Tectonophysics*, 1(16): 171–178.

Heller F, Tungsheng Liu. (1982) Magnetostratigraphical dating of loess deposits in China[J]. *Nature*, 300(5891): 431–433.

Holt W E, Chamot R N, Pichon X L, et al. (2000) The velocity field in Asia inferred from Quaternary fault slip rates and GPS observations[J]. *Journal of Geophysical Research Atmospheres*, 105(B8): 19185–19209.

Jianbing Peng, Jun Zhang, Shengrui Su, et al. (1992) *Active faults and geological hazards in Weihe Basin*[M]. Northwest University Press. Xi'an

Jiansheng Yu, Kai Tan, Caihong Zhang, et al. (2019) Present-day crustal movement of the Chinese mainland based on Global Navigation Satellite System data from 1998 to 2018[J]. *Advances in Space Research*, 63(2): 840–856.

Kukla G, Zhisheng An. (1989) Loess Stratigraphy in Central China[J]. *Palaeogeography Palaeoclimatology Palaeoecology*, 72(89): 203–225.

Langbein J, Bock Y. (2004) High-rate real-time GPS network at Parkfield: Utility for detecting fault slip and seismic displacements[J]. *Geophysical Research Letters*, 31(15): L15S20.

Mingqiu Wang. (2011) *The characteristics of active faults and analysis of seismic activity in Baoji area*[D]. Chang'an University. Xi'an

Peizhen Zhang, Min Wang, Weijun Gan, et al. (2003) Slip rates alone major active faults from GPS measurements and constrains on contemporary continental tectonics[J]. *Geoscience Frontiers*, 10(S1): 81–92.

Qi Wang, Guoyu Ding, Xuejun Qiao, et al. (2000)Research of present crustal deformation in the southern Tianshan(Jiashi), China by GPS geodesy[J]. *Acta Seismologica Sinica*, 22(03): 263–270.

Qingyu Chen, Renwei Xiong, Qinjian Tian. (2018) Segmentary characteristics of the geometrical structure of the Longxian-Qishan-Mazhao active fault[J]. *Earthquak*, 38(03): 66–80.

Shidi Wang, Yaqin Shi, Fengwen Ren. (2018) Analysis and textual research of the seismogenic structure of the Qin-Long earthquake in 600 A.D. [J]. *Journal of geomechanics*, 24(2): 158–168.

Shidi Wang. (2018) *Tectonic deformation in late cenozoic of Liupanshan-Baoji fault zone in the NE margin of Tibet Plateau*[D]. Northwest University. Xi'an

Suowang Liu, Jiasi Gan, Yunsheng Yao, et al. (1997) Strike-slip transform deformation along the northern boundary fault of western Qinling and Haiyuan fault and interaction between them and Longshan block[J]. *Crustal Deformation and Earthquake*, 17(3): 73–83.

Tungsheng Liu, Yafeng Shi. (2000) Table of Chinese quaternary stratigraphic correlation remarked with climate change[J]. *Quaternary Sciences*, 20(02): 108–128.

Tungsheng Liu. (1965) *The Chinses loess-soil sequences*[M]. Science Press. Beijing

Wei Shi. (2011) *The analysis of the development characteristics and activity about fault zone of Longxian-Baoji*[D]. Chang'an University. Xi'an

Wenhui Li, Changqiao Zou, Hongda Liang, et al. (2017) Crustal structure beneath the Liupanshan fault zone and adjacent regions[J]. *Chinese Journal of Geophysics*, 60(6): 2265–2278.

Wenjun Zheng, Daoyang Yuan, Peizhen Zhang, et al. (2016) Tectonic geometry and kinematic dissipation of the active faults in the northeastern Tibetan Plateau and their implications for understanding northeastward growth of the plateau[J]. *Quaternary Sciences*, 36(4): 775–788.

Wenjun Zheng, Huiping Zhang, Peizhen Zhang, et al. (2013) Late Quaternary slip rates of the thrust faults in western Hexi Corridor (Northern Qilian Shan, China) and their implications for northeastward growth of the Tibetan Plateau[J]. *Geosphere*, 9(2): 342–354.

Xinnan Li, Peizhen Zhang, Wenjun Zheng, et al. (2018) Kinematics of Late Quaternary Slip along the Qishan-Mazhao Fault: Implications for Tectonic Deformation on the Southwestern Ordos, China [J]. *Tectonics*, 37(9): 2983–3000.

Xinnan Li. (2017) *Deformation pattern based on geometry and kinematics of active tectonics in the southwestern Ordos Block*[D]. Institute of Geology, China Earthquake Administration. Beijing

Yaqin Shi. (1996) On the movement forms of Longxian-Qishan-Zhouzhi fault and the seismicity characters along the fault[J]. *Northwestern Seismological Journal*, 18(02): 84–86.

Yongshan Li. (1992) Research on ground fissures in Xi'an region and active faults in Weihe Basin[M]. Seismological Press. Beijing

Zhicheng Zhu, Zuoxun Zeng, Guangming Fan, et al. (2008) *Structure geology*[M]. China University of Geosciences Press. Wuhan

Zhigang Shi, Daoyang Yuan, Tingdong Li, et al. (2013) Textual research of A.D. 600 Qin-Long earthquake and discussion on its seismogenic structure[J]. *Science and Technology Review*, 31(12): 48–52.

Zhigang Shi. (2011) The recent activity features of faults and risk trend of strong earthquake in Liupanshan region[D]. Lanzhou Institute of Seismology, *China Earthquake Administration*. Lanzhou

Zhongping Lai. (2010) Chronology and the upper dating limit for loess samples from Luochuan section in the Chinese Loess Plateau using quartz OSL SAR protocol[J]. *Journal of Asian Earth Sciences*, 37(2): 176–185.

Advances in Measurement Technology, Disaster Prevention and Mitigation – Li & Mohd Yusof (Eds)
© 2023 The Author(s), ISBN: 978-1-032-36087-4

Effect of unidirectional tension on permeability of sand-geotextile system

Jiabao Hu, Xiaolei Man*, Hui Liu, Hanyue Wang & Mengyu Rong
College of Civil Engineering and Architecture, Chuzhou University, Chuzhou, China

ABSTRACT: A kind of geotextile was selected to study the influence of uniaxial tension on the permeability of the sand-geotextile system, and the permeability test of the sand-geotextile system under different tensile strains was carried out by gradient ratio permeameter. The influence of different tensile strains on permeability parameters, such as permeability, sand leakage, flow velocity, and gradient ratio, is analyzed. The test results show that under the condition of warp stretching, the permeability and anti-clogging performances of geotextiles decrease first and then increase, and the soil retention performance increases first and then decreases. When the tensile stress is 3%, the soil retention performance is the best, and the permeability and anti-clogging performance are the worst. In the weft stretching state, its soil retention performance continues to decline, and water permeability and anti-blocking performance continue to rise.

1   INTRODUCTION

Geotube dam is formed by the geotextile tube bags stack cofferdam filled with sand, in which the geotextile tube bags are prepared using geotextile with high tensile and compressive strength. Compared with the traditional river and lake construction technology, it has the advantages of a short construction period, local material selection, low cost, and green construction, and is widely used in water conservancy and water transportation, estuary, and coast engineering construction in China (Li 2003; Sun 2010; Shen 2021). In the construction process, the construction period and work efficiency are determined by the dehydration rate during the construction period and the seepage rate during the operation period. Therefore, the seepage problem of filling pipe bags has become a hot research topic at home and abroad.

Wu et al. (2020) used the LBM-DEM-DLVO coupling method to explore the relationship between the interaction force between geotextile fiber and sand particles and the resistance between soil particles FYD. The results showed that the greater the interaction force, the more easily the soil particles stay in the geotextile. Five types of non-woven sand-geotextile systems were selected to perform different hydraulic gradient tests by Zhou et al. (2018). The influence of hydraulic gradient on the filtration performance of the non-woven sand-geotextile system and the permeability coefficient of different geotextiles were different with the increase in hydraulic gradient. Non-woven, high-strength woven fabrics, common woven fabrics, and silt systems were used by Yi et al. (2015) to conduct gradient ratio tests under different hydraulic gradients to obtain the effects of different hydraulic gradients on permeability. Malik et al. (2011) studied the influence of mutual coupling of various density filling materials on the permeability of filling pipe bags by using MATLAB and the Newton methods. The results showed that different densities of filling soil have different effects on the permeability of filling pipe bags. Chang et al. (2014) carried out a bag-lifting test for filling the soil with different viscosity under the premise of applying a certain beating disturbance to the bag. The research showed that the efficiency of beating disturbance can be improved while expanding the pore size of the bag fabric. This method can greatly improve the efficiency of dehydration and consolidation of the filling bag. In practical engineering, the filling

---

*Corresponding Author: manxl@chzu.edu.cn

process of filling pipe bags was carried out under pressure, while the indoor hanging bag test can only carry out simple no-pressure dehydration and consolidation test. Wu et al. (2016) carried out a field large-scale pipe bag dewatering test. For five types of filling pipe bags, comparative tests were carried out by filling different soil materials. Through the comparative analysis of dry density, moisture content, pore water pressure, gradation, consolidation rate, and other parameters of soil in different parts of filling pipe bags, the influence of different construction methods on the dewatering and consolidation effect of filling pipe bags were studied. The research showed that the filling and drainage were adopted. The drainage method can still ensure that the soil-retaining property and permeability of the filling pipe bags were excellent, which met the actual engineering requirements.

However, during the filling process of the filling pipe bag, the geotextile will be subjected to different directions and different degrees of stress, which will lead to the deformation of the filling pipe bag, and hinder the dehydration and consolidation of the filling pipe bag and affect its overall dehydration performance. Therefore, domestic and foreign scholars have carried out a series of studies on the stress and deformation of filling pipe bags.

For the study of permeability of geotextile under uniaxial tension, Huang et al. (2012) explored the principle of existing tensile test methods of geotextile at home and abroad and put forward many tensile test methods. Ding et al. (2019) compared the influence of two calculation methods of wide strip method and the multi-ribbed method on the tensile mechanical properties of different fabrics by tensile research. The study showed that the wide strip method is more suitable for testing the tensile properties of new high-strength warp knitted composite geotextiles. Qiao et al. (2018) studied the uniaxial tension of different specifications of geotextile mat, and concluded that weft tension was more likely to play the tensile effect of fabric. Cao et al. (2014) conducted indoor tests on geotextiles using a universal testing machine and found that the deformation curve of high-strength spun geotextiles was nonlinear. For the study of permeability of sand-geotextile system under unidirectional tensile, Wu et al. (2003) explored the change of permeability characteristics of geotextile under certain strain conditions by vibrating screen test, gradient ratio test, and permeability test on geotextile after weft tension. The results showed that the equivalent pore size of the geotextile increases under unidirectional tension, which led to the increase in the permeability coefficient of the geotextile. Chen et al. (2003) tested the gradient ratio of the filter system composed of geotextiles and bad soil by using the improved gradient ratio permeation instrument. The results showed that the gradient ratio of the filter system increases when the nonwoven fabric was unidirectional tensioned. Zhou et al. (2014) used a tensile testing machine to draw five different specifications of fabrics. The results showed that the tensile properties of the same fabric were consistent, and the restriction of soil on geotextile was positively correlated with the elongation of fabric. Tang et al. (2013) studied the influence of uniaxial tension on the filtration performance of four different geotextiles by using a self-made gradient ratio permeameter. The results showed that the changes of spun geotextiles and nonwoven geotextiles with tensile strain are completely different. With increase in tensile strain, the permeability and anti-clogging performance of spun geotextiles were enhanced, and the soil retention performance is weakened.

The above scholars only considered the influence of the tensile effect on the permeability of geotextile when studying the seepage problem of filling bags. However, Man et al. (2020) believed that the different tensile directions would lead to different results of permeability characteristics. Therefore, it was necessary to study the permeability characteristics of geotextile under unidirectional force. In this paper, the commonly used geotextiles of 150 g/m^2 in engineering were stretched in warp and weft by gradient ratio permeameter, and the influence of unidirectional tensile on the permeability characteristics of 150 g/m^2 woven fabric-coated sand-geotextile system was studied and analyzed.

2 TEST METHODS

2.1 *Test device*

The permeability characteristics of the sand-geotextile system after uniaxial tension were studied by gradient ratio permeameter. As shown in Figure 1, a self-developed seepage test device that can provide unidirectional flow was set up. A fixed water tank was set up on the upper part of the test device to provide constant head hydraulic conditions. The inlet of the gradient ratio permeator was connected to the upper fixed water storage tank to ensure the constant head test. The diameter (D) of tube A was 160 mm, and the height (H) was 160 mm; the diameter (D) of tube B was 160 mm, and the height (H) was 80 mm. Cylinder A and Cylinder B were connected by a flange plate and screwed to fix the geotextile after tension. To prevent the phenomenon of water seepage and air intake of the instrument and further affect the accuracy of the test results, sponge tape was connected to the lower end of Cylinder A and the upper end of Cylinder B to increase the closeness between the geotextile and the cylinders. The soil sample height above the geotextile was 110 mm. Four pressure measuring tubes were arranged on the side wall of the middle cylinder to calculate the permeability parameters. The side wall of the bottom cylinder had an outlet, which could measure the amount of water. The lower container can collect sand particles through the fabric.

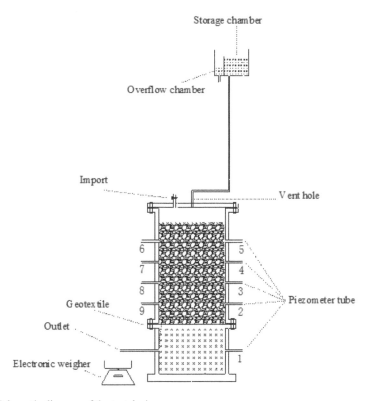

Figure 1.   Schematic diagram of the test device.

2.2   *Test materials*

The geotextile used in the test was woven fabric with 150 g/m^2 weight per unit area, with a thickness of 0.64 mm and an equivalent aperture of 0.09–0.5 mm. Due to the good gradation of soil having a good filtering characteristic, soil particles were not easy to lose, and geotextile anti-silt blocking ability was strong. Therefore, adverse soil was selected in this experiment to test the permeability of geotextile under unidirectional tension. It was composed of fine sand less than 0.075, medium sand with particle size between 0.3 and 0.6, and coarse sand with particle size between 0.6 and 1.0. The results of soil particle parameters were shown in Table 1.

Table 1. Particulate parameters of soil samples.

Fine grain Content	Characteristic particle size (mm)			Inequality Homogeneous Number Cu	Curvature Coefficient Cc
	d_{10}	d_{30}	d_{60}		
10%	0.075	0.43	0.6	8.0	4.11

2.3 Test process

The penetrator was assembled with 0%, 3%, 6%, and 9% of the tensile strain of the geotextile, and then the prepared soil was mixed and stirred evenly, and filled in the device. The filling process adopted the way of punching and pressing to ensure stability between the soil. When the soil material was filled to a certain 110 mm, the top cover related to the top cylinder by flange and the screw was tightened to improve its stability. A high-water tank was set on the A cylinder as the constant head, and the height of the head was ensured to be consistent with the height of the filled sand after compaction. The inlet of the top cover of the gradient penetrometer was connected to open the valve. When the water in the water tank slowly flows into the whole device through the inlet, the outlet valve of the lower cylinder was opened and the inlet valve of Cylinder A was opened at the same time to discharge the air in the device and improve the accuracy of the test. When the whole device was full of water, all the valves were closed, water tank was adjusted to the appropriate height, and the pressure measuring tube was connected with the instrument, so that the water flowed from top to bottom through the whole test device and then the change in height of the pressure measuring tube was observed. When the height of the pressure measuring tube was highly consistent and no longer changed, the outlet was opened, and the test data were collected. After 30 hours of the continuous test, the sand particles in the bottom cylinder were collected and dried, and weighed.

3 TEST RESULTS AND ANALYSIS

3.1 Effect of tensile strain on flow velocity

Figure 2 shows the curve of seepage flow velocity with time. From Figure 2, the change of seepage flow velocity with time under different tensile strain conditions was the same, which increased first and then decreased with the increase in time, and tended to be stable after 220 minutes. The reason was that, due to the scouring effect of unidirectional flow, fine particles were lost from the fabric pores. With time, fine particles gradually accumulated and formed a filter system, resulting in the stabilization of seepage flow rate. The stable value was shown in Figure 3.

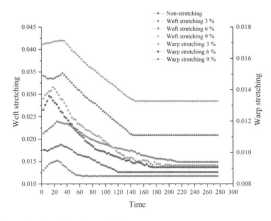

Figure 2. Curves of flow velocity changing with time.

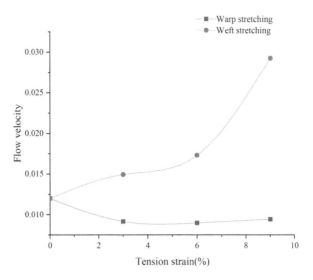

Figure 3. Curve of flow velocity changing with tensile strain.

It can be seen from Figure 3 that the relationship between seepage velocity and tensile strain of geotextile was positive under the action of weft tension, and the seepage velocity increased with the increase in tensile strain. Weft tension was the enhancement of permeability of geotextiles. Warp stretching was the weakening of the seepage velocity of the sand-geotextile system. Among them, before the weft tensile stress was 3%, the seepage velocity decreases significantly with the increase in tensile strain. It was relatively stable at 3%, and only slightly changed with the increase in tensile stress. It showed that the siltation of the sand-geotextile system was serious when the warp tension was 3%–9%.

3.2 Effect of tensile strain on soil properties

The amount of sand leakage can directly reflect the soil retention performance of the fabric, which means that the total mass of soil particles passes through the fabric within a certain time.

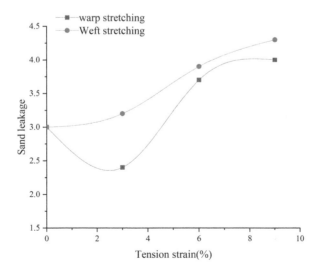

Figure 4. Curve of sand leakage changing with tensile strain.

Through multiple groups of repeated experiments to reduce the experimental error, the variation curve of sand leakage and tensile strain is shown in Figure 4. It can be seen from the figure that under the action of warp tensile, sand leakage of the sand-geotextile system decreases first and then increases with tensile strain. When the tensile strain reached about 3%, the sand leakage reached the minimum point. It indicated that when the tensile strain of geotextile reaches about 3% under the action of warp tensile, the sand leakage of the sand-geotextile system is the least, and the soil retention performance of the sand-geotextile system was the best. Under the action of weft tensile strain, the sand leakage of the sand-geotextile system showed a continuous upward trend with the tensile strain, indicating that the weft tensile strain weakened the soil retention performance of the sand-geotextile system.

3.3 Effect of tensile strain on anti-clogging performance

For the size of the gradient ratio tester in this experiment, the calculation formula of gradient ratio GR is as follows:

$$GR = \frac{H1-2}{25+\&} \Big/ \frac{H2-4}{50} \tag{1}$$

In the formula, H1-2 is the water level difference (cm) between tube 1 and tube 2; H2-4 is the water level difference (cm) between tube 2 and tube 4, which is the thickness of geotextile (cm). Figure 7 showed the variation of gradient ratio GR with time under different tensile strains.

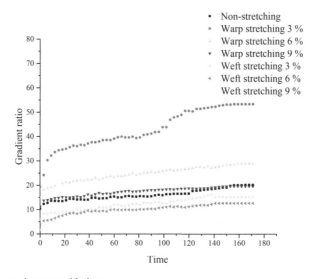

Figure 5. Gradient ratio curve with time.

It can be seen from Figure 5 that the variation of gradient ratio with time under different tensile strain conditions was the same, which increased with time and tended to be stable after 160 min. The reason was that due to the scouring effect of unidirectional flow, the fine particles moved down quickly and accumulated on the surface of geotextile, which aggravated the siltation of the sand-geotextile system and finally tended to be stable. The stable value is shown in Figure 6. It can be seen from the figure that under the action of warp tensile, the gradient ratio of the sand-geotextile system increases first and then decreases with the increase in tensile strain. When the tensile strain reached 3%, the gradient ratio reached the maximum. It indicated that the anti-silting and blocking ability of the sand-geotextile system is poor when the warp tension is 3%; under the action of weft tensile, the gradient ratio of the sand-geotextile system showed a continuous downward trend with the increase in tensile strain, indicating that the weft tensile strain had an enhanced effect on the anti-silting and plugging performance of the sand-geotextile system.

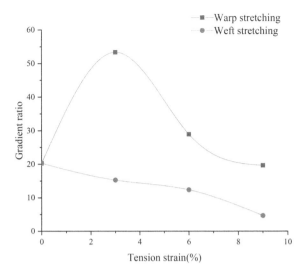

Figure 6.   Curve of gradient ratio changing with tensile strain.

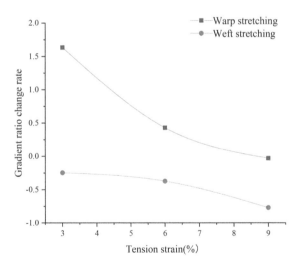

Figure 7.   Curve of gradient ratio change with tensile strain.

It can be seen from Figure 7 that the change rate of gradient ratio decreased with the increase in tensile strain under the action of warp and weft tensile. When the sand-geotextile system was subjected to warp tensile force, the change rate of gradient ratio decreased significantly, indicating that the anti-deposition and plugging ability of different sand-geotextile systems decreased first and then increased with the increase in tensile strain. When the warp tensile stress was at a stage of 0%–6%, the change was obvious. The change was relatively stable when the warp tensile stress was 6%–9%. Under the action of weft tensile stress, the gradient ratio change rate of the sand-geotextile system was relatively stable, and the silt prevention ability is improved when the weft tensile stress is between 6% and 9%.

4 CONCLUSIONS

Studying the change of permeability characteristics of the sand-geotextile system under tensile stress in warp and weft, the gradient ratio vertical permeameter was used for the permeability test. The permeability, sand leakage, flow rate, and gradient ratio of the W150 sand-geotextile system per unit time under the tensile strain of 0%, 3%, 6%, and 9% in warp and weft were studied, and the following conclusions were obtained.

(1) Unidirectional tensile has a significant effect on the soil retention performance of the sand-geotextile system. When the warp tensile strain was in the 0%–3% stage, it showed that the soil retention performance was gradually increasing. When the warp tensile strain was in the 3%–9% stage, the soil retention performance was gradually decreasing. When the warp tensile strain was 3%, the soil retention performance of the geotextile was the best.

(2) Unidirectional tension has an obvious influence on the anti-clogging performance of the sand-geotextile system. The anti-clogging performance of the geotextile was enhanced due to the warp tensile. When the warp tensile strain was at the stage of 0%–3%, the gradient ratio of the sand-geotextile system was gradually reduced, and its anti-clogging performance was gradually weakened. When the warp tensile strain was at the stage of 3%–9%, its anti-clogging performance was gradually enhanced. When the warp tensile strain was at 3%, the clogging of geotextile was the most serious, and the anti-clogging performance was the worst. The weft tension led to the improvement in the anti-clogging performance of geotextiles.

ACKNOWLEDGMENT

This study was supported by the Key Research Project of Natural Science in Colleges and Universities of Anhui Province (Grant No. KJ2021A1099) and Chuzhou University Development Experiment Project (Grant No. kfsy2138).

REFERENCES

Cao G. F., Xu B., Wang M. S., et al. Experimental study on deformation of reinforced cushion with high strength woven geotextile in enclosure engineering [J]. *Geotechnical mechanics*, 2014 (S1): 238–244.

Chang G. P., Shu Y. M., Yin J. C., etc. Efficient dehydration method for filling high viscosity mud with pipe bags [J]. *Hydroelectric energy science*, 2014 (3): 129–133.

Chen L., Tong Z. X., Effect of tensile strain on geotextile - discontinuous graded soil clogging characteristics [J]. *Journal of Hydropower*, 2003 (02): 97-102.

Ding J. H., Zhang W., Sun H. et al., a new test method for tensile properties of high-strength warp-knitted composite geotextiles [J]. *Journal of Yangtze University of Science*, 2019, 36 (10): 175–179.

Huang W. B., Ren Z. D., Cao M. J., Instruments and methods for tensile test of geotextiles [J]. *Water science, technology and economy*, 2012, 18 (012): 109–112.

Li B. Q., Research and application of geotextile filling bag in Tianjin port seawall construction [J]. *China port construction*, 2003 (5): 3.

Malik J, Sysala S. Analysis of geosynthetic tubes filled with several liquids with different densities [J]. *Geotextiles and Geomembranes*, 2011, 29(3): 249–256.

Man X. L., Wang W. S., Liu G. Y., Di Y. F., Bao Y. J., Effect of sewing methods of bag fabric on dehydration performance of pipe bags [J]. *Journal of Three Gorges University: Natural Science Edition*, 2020, 42 (6): 6.

Qiao J. G., Liu W. L., Li S. X., Comparison of tensile properties of different types of polyethylene geotextiles [J]. *Application of engineering plastics*, 2018, 046 (012): 115–119, 124.

Shen J., Case study on the performance of protective measures for geotextile bags along the Tarim River [J]. *Water resources development and management*, 2021 (7): 5.

Sun Y. K., Application of sand-filled bags of large prism geotextiles [J]. *Water transportation in China: the second half of the month*, 2010 (5): 2.

Tang L., Tang X. W., She W., etc. Experimental study on the effect of uniaxial tension on the filtration performance of geotextiles [J]. *Journal of Geotechnical Engineering*, 2013 (04): 785–788.

Wu H. M., Shu Y. M., Chang G. P., etc. Field model test of high efficiency dewatering process for filling pipe bags with high viscosity (powder) granular soil [J]. *Geotechnical Engineering Journal*, 2016, 38 (0z1): 209-215.

Wu S, Chen Y, Zhu Y, et al., Study on filtering process of filtering with LBM-DEM-DLVO coupling method [J]. *Geotextiles and Geomembranes*, 2020, 49 (1).

Wu S. C., Hong Y. S., Wang R. H. The effect of uniaxial tensile strain on the pore size and filtration characteristics of geotextiles [J]. *Geotextiles and Geomembranes*, 2008, 26 (3): 250–262.

Yi J. R., Cao M. J., Experimental study on geotextile clogging [J]. *Hydropower energy science*, 2015, 33 (4): 4.

Zhou B., Wang H. Y., Wang X. D., The effect of hydraulic gradient on the filterability of soil- nonwoven fabric system [J]. *People's Yellow River*, 2018, 40 (02): 109–112+130.

Zhouping, xuchao, Lidan, Wu Di Study on tensile test of geosynthetics in soil [J]. *Journal of Jiamusi University* (NATURAL SCIENCE EDITION), 2014, 32 (05): 675–678+682

Advances in Measurement Technology, Disaster Prevention and
Mitigation – Li & Mohd Yusof (Eds)
© 2023 The Author(s), ISBN: 978-1-032-36087-4

Seismic response of 8 MW offshore wind turbines supported on the pile-bucket composite foundation

Yanguo Sun, Chengshun Xu*, Xiuli Du, Piguang Wang, Renqiang Xi & Yilong Sun
Key Laboratory of Urban Security and Disaster Engineering, Beijing University of Technology, Beijing, China

ABSTRACT: In the construction of offshore wind farms, foundation design for offshore wind turbines is one of the urgent points as offshore wind farms extend to deep sea areas and the larger wind turbine utilizing. Pile-bucket composite foundation is a useful way to support the larger wind turbine in the deep sea. In this study, the numerical models of monopile and pile-bucket composite foundations are developed by using ABAQUS. To present the superiority of pile-bucket composite foundations, these developed numerical models are used to perform a comparative study of capacity characteristics for monopile and pile-bucket composite foundations and analyze the seismic response of offshore wind turbine structures supported by monopile and pile-bucket composite foundations. The results show that the pile-bucket composite foundation can better utilize the bearing capacity of soil around the pile than the monopile. Furthermore, the seismic performance of the pile-bucket composite foundation is better than the monopile. the seismic response of offshore wind turbine turbines supported by a pile-bucket composite foundation is less than the monopile when applying the three different seismic waves. The decreased level is 32% and 22% and 8% in the displacement of wind turbine tower top, and 6% and 2% and 26% in the largest tilt of foundation on the soil surface.

1 INTRODUCTION

As environmental pollution and the worldwide energy crisis protrude, renewable energy is actively explored all around the world to achieve the goal of "carbon neutrality". Wind energy is green renewable energy and is one of the promising ways to ensure carbon neutrality (Zou et al. 2021). Compared with onshore wind power resources, offshore wind power resources have the advantages of high wind speed, effective power generation time, and proximity to economic development zones (Esteban et al. 2011). As we all know, China has a vast coastline and abundant offshore wind power resources, and many offshore wind farms are under construction in China (Wang et al. 2009).

Recently, offshore wind farms have mainly been built on offshore sites. So far, large-diameter monopile foundations, gravity foundations, jacket foundations, tripod foundations, and suction bucket foundations have dominated offshore wind turbine foundation forms (Wang et al. 2018). At present, the super large diameter monopile of 5 m to 8 m is the main form of offshore wind turbine foundation. However, with offshore wind farms developing into the deep sea, the monopile foundation is no longer suitable for the needs of offshore wind farms (Barker 2013). As the diameter of a monopile has developed to 10 m or even larger, the cost of manufacturing and constructing a monopile foundation is extremely high. The wind turbine is gradually developing to a high-power generation capacity. The development of offshore wind turbine technology leads to higher and higher requirements for the foundations (Medina et al. 2021). Finding an appropriate foundation has become one of the major difficulties in offshore wind power construction.

---

*Corresponding Author: xuchengshun@bjut.edu.cn

                                      DOI 10.1201/9781003330172-50

In the past decade, many studies have been carried out on the new foundation of an offshore wind turbine. From many research results, the bearing performance of the new foundation has many advantages over the monopile. The new type of foundation mainly improves the bearing performance by strengthening the single pile, such as umbrella foundations (Li et al. 2017; Yang et al. 2020), pile-plate composite foundations (Wang et al. 2018), pile-turbine composite foundations (Chen et al. 2020) and others. Li et al. designed a new type of foundation for an offshore wind turbine with good bearing characteristics. The lower part of the foundation is a cylinder around the single pile, and the upper part can be placed with stones to improve bearing capacity. The bearing performance of the new type of composite foundation in four kinds of sand sites is studied through a centrifuge model test, and it is found that the ultimate bearing capacity of the new type of foundation is almost four times that of the monopile (Li et al. 2020). A composite umbrella foundation is designed for an offshore wind turbine. As it says, an umbrella structure is added to the upper part of the bucket (Li et al. 2017; Yang et al. 2020). Chen et al. studied the bearing capacity of the pile-bucket composite foundation. Its results show that the flexural capacity of the pile-turbine composite foundation is greater than the sum of the single pile and single bucket foundation (Chen et al. 2020). It can be seen from the previous research that the new composite foundation has obvious advantages in bearing performance. It is not difficult to draw such a conclusion that the new composite foundation could have a good application prospect in offshore wind farms. The pile-bucket composite foundation has been used in several experimental projects in China (Fu et al. 2021). However, some of the Chinese wind farms are built in a potential seismic region, so it is very important to analyze the seismic response of offshore wind structures (Kaynia 2019; Katsanos et al. 2016).

In conclusion, it is meaningful that the difference in bearing capacity between monopile and pile-bucket composite foundations be explored. Based on the finite element software ABAQUS, the numerical calculation models of offshore 8 MW wind turbines supported by monopile and pile-bucket composite foundations are established respectively. The lateral bearing capacity of monopile and pile-bucket composite foundations is compared by the numerical models. The seismic response of offshore wind turbine structures supported by the two foundations is then analyzed.

2 NUMERICAL MODELS OF OFFSHORE WIND TURBINES

The offshore wind turbine supported by a pile-bucket composite foundation is shown in Figure 1. The turbine blade, engine room, and hub are simplified as the concentrated mass on the top of the tower in the finite element model. The foundations above the mud surface and tower are discretized into beam elements, and the foundations below the mud surface are solid elements. Then coupling

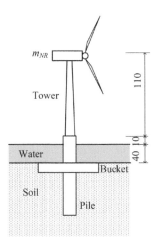

Figure 1.   The offshore wind turbine system supported on the pile-bucket composite foundation.

motions are established to connect the structures above and below the mud (Chen et al. 2020). The parameters of an 8 MW wind turbine are adopted from public data (Desmond et al. 2016), as shown in Table 1. In the study, the sea water depth is 40 m, the pile extends 10 m above sea level, and the height of the tower is 110 m. The schematic diagram of the monopile and the pile-bucket composite foundation is shown in Figure 2. The bucket and pile in the composite foundation are restrained by binding (Chen et al. 2020). Tables 2 and 3 show the dimensions of the monopile and the pile-bucket composite foundation, respectively.

Table 1. Major parameters of 8 MW wind turbine.

Offshore wind turbine	8 MW
Rotor-Nacelle-Assembly mass (t)	480
Tower height (m)	110
Rotor diameter (m)	164
Tower top diameter (m)	5
Tower bottom diameter (m)	7.7
Tower top thickness (m)	0.022
Tower bottom thickness (m)	0.036

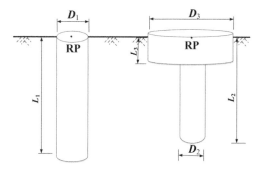

Figure 2. The monopile and the pile-bucket composite foundation.

Table 2. Specific dimensions of the monopile.

D_1 (m)	L_1 (m)	t_{P1} (m)
7.70	60	0.05

Table 3. Specific dimensions of the pile-bucket composite foundation.

D_2 (m)	L_2 (m)	t_{P2} (m)	D_3 (m)	L_3 (m)	t_B (m)
5.0	40	0.05	15	10	0.05

Referring to the actual situation of the project, the material of the pile-bucket composite foundation structure is steel. The steel density is 7,850 kg/m3. The elastic modulus is 2.1×105 MPa and the Poisson ratio is 0.3. The finite element model adopts a cuboid. To avoid the influence of the boundary effect of the model on the calculation results, a reasonable range of soil boundary values is obtained through multiple trial calculations. The bottom of the body adopts fixed constraints, and the lateral boundaries are fixed with lateral displacement (displacement in the x and y directions is 0). The soil adopts the Mohr-Coulomb elastoplastic model. The effective density is 1,000 kg/m³.

The elastic modulus, E_{sd} = 30 MPa. The Poisson ratio is 0.3 and the friction angle q is 35°. In the model, the normal contact between the foundation and the soil is "hard," and sliding friction is used in lateral contact while the lateral sliding coefficient is $\tan(3q/4)$.

Before the load is applied, the in-situ stress balance calculation is performed on the model. The overall vertical displacement of the model is less than 10–5 m after the in-situ stress is balanced, so the influence of the initial displacement of the formation on the calculation results can be ignored.

3   COMPARISON OF LATERAL BEARING CAPACITY

This section presents the comparison of lateral bearing capacity between monopile and pile-bucket composite foundations. As shown in Figure 1, a reference point RP is established at the center point of the foundation at the mud surface, which is the loading point. The response diagram of the monopile and pile-bucket composite foundation under horizontal load is shown in Figure 3. Figure 3(a) shows the relationship between displacement and load under horizontal load. It demonstrates that the horizontal bearing capacity of the monopile and the pile-bucket composite foundation designed for the 8 MW wind turbine are similar. At the same time, the pile-bucket composite foundation has a larger horizontal bearing capacity than the pile and the bucket foundation, and the pile-bucket composite foundation takes advantage of the bucket foundation and the single pile foundation at the same time. In state (a), the horizontal load is small, and the pile-bucket composite foundation has a larger initial horizontal stiffness than the single-pile foundation. At the same displacement, the pile-bucket composite foundation bears a larger horizontal load than the single-pile foundation. Figure 3(b) shows the change of the bending moment with the depth of foundation penetration when the horizontal displacement is 0.5 m. As shown in Figure 3(b), the large-diameter single-pile foundation has a larger bending moment than the pile-bucket composite structure, and the monopile is easier to enter the plastic stage. The pile-bucket composite foundation takes advantage of the bucket foundation and reduces the minimum bending moment of the structure. The research results show that the pile-bucket composite foundation can be more conducive to the synergistic effect of the foundation and the foundation soil so that the soil can share a larger load.

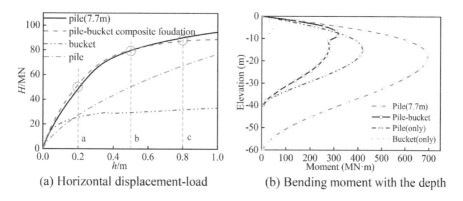

(a) Horizontal displacement-load   (b) Bending moment with the depth

Figure 3.   Comparison study on horizontal bearing capacity.

4   COMPARISON OF SEISMIC RESPONSE

Herein, the seismic response of an 8 MW offshore wind turbine under three different ground motions is studied. Figure 4 shows the acceleration time history of three seismic events.

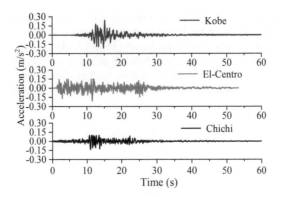

Figure 4.　Input acceleration time history.

Figure 5 shows the tower top seismic displacement response under different ground motions. The displacement of offshore wind turbines supported by the pile-bucket composite foundation is relatively smaller than that of a monopile. The seismic response of offshore wind turbines supported by a pile-bucket composite foundation is less than the monopile when applying the three different seismic waves. The decreased level is 32%, 22%, and 8% in the displacement of the wind turbine tower top, respectively. The application of pile-bucket composite foundations in offshore wind power reduces the dynamic response of offshore wind turbine structures under earthquakes.

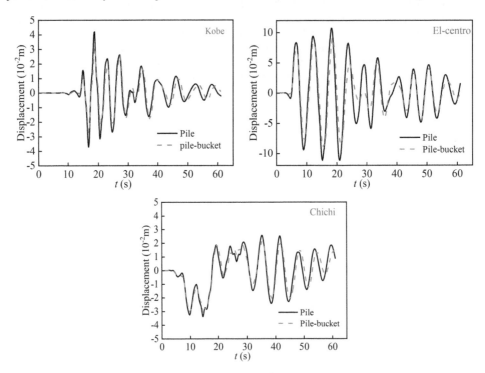

Figure 5.　Seismic response of the tower top.

To ensure the normal operation of the wind turbine, the rotation angle of the support structure of the offshore wind turbine must be limited. Figure 6 shows the angle response of foundation rotation at the mud surface under different ground motions. The results indicate that the rotation angle of the pile-bucket composite foundation at the mud surface is much smaller than that of the monopile when applying the three different seismic waves. The decreased level is 6%, 2%, and 26% in the largest tilt of foundation on the soil surface, respectively. The pile-bucket composite foundation can reduce the rotation angle of the foundation, illustrating that the seismic performance of the supporting structure has been effectively improved.

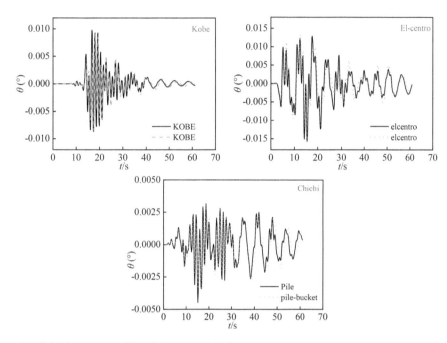

Figure 6.   Seismic response of foundation at mud surface.

5   CONCLUSIONS

A pile-bucket composite foundation is a new approach to foundation design for offshore wind turbines that can give full play to the advantages of bearing performance of both the pile and the bucket at the same time. It has great prospects in the design of offshore wind turbine support structures. Through the research on the static and seismic response of offshore wind turbines supported by the monopile and the pile-bucket composite foundation, the following conclusions are obtained in this paper:

(1) Compared with the monopile, the pile-bucket composite foundation can better transfer the upper load to the soil. And it improves the bearing performance of the foundation by reducing the bending moment of the structure.
(2) Under the seismic load, the displacement of the tower top supported by a monopile is much larger than that of an offshore wind turbine supported by a pile-bucket composite foundation. And the displacement amplification factor can be increased to 1.3, which indicates that the pile-bucket composite foundation has better seismic performance.
(3) The foundation angle deformation response at the mud surface when the offshore wind turbine is supported by the pile-bucket composite foundation is relatively smaller than that of the monopile. It shows that the pile-bucket composite foundation can better ensure the normal operation of the offshore wind turbine, which is an excellent foundation design method.

ACKNOWLEDGMENT

This study was financially supported by the National Outstanding Youth Science Fund Project of the National Natural Science Foundation of China (Grant No. 51722801).

REFERENCES

Barker, P. (2013). Deep water turbines - the next step. *Maritime Journal* (307), 39–41.

Chen, D., Gao, P., Huang, S., Li, C., & Yu, X. (2020). Static and dynamic loading behavior of a hybrid foundation for offshore wind turbines. *Marine Structures*, 71, 102727.

Desmond, C., Murphy, J., Blonk, L., & Haans, W. (2016). Description of an 8 mw reference wind turbine. *Journal of Physics Conference*, 753, 092013.

Fu, Z., Wang, G., Yu, Y., & Shi, L. (2021). Model Test Study on Bearing Capacity and Deformation Characteristics of Symmetric Pile–Bucket Foundation Subjected to Cyclic Horizontal Load. *Symmetry*, 13(9), 1647.

Katsanos, E. I., Thöns, S., & Georgakis, C. T. (2016). Wind turbines and seismic hazard: a state-of-the-art review. *Wind Energy*, 19(11), 2113–2133.

Kaynia, A. M. (2019). Seismic considerations in design of offshore wind turbines. *Soil Dynamics and Earthquake Engineering*, 124, 399–407.

Li, H., Liu, H., & Liu, S. (2017). Dynamic analysis of umbrella suction anchor foundation embedded in seabed for offshore wind turbines. *Geomechanics for Energy & the Environment*, S2352380816300491.

Li, X., Zeng, X., & Wang, X. (2020). Feasibility study of monopile-friction wheel-bucket hybrid foundation for offshore wind turbine. *Ocean Engineering*, 204, 107276.

MD Esteban, Diez, J. J., Lopez, J. S., & Negro, V. (2011). Why offshore wind energy?. *Renewable Energy*, 36(2), 444–450.

Medina, C., Álamo, G. M., & Quevedo-Reina, R. (2021). Evolution of the Seismic Response of Monopile-Supported Offshore Wind Turbines of Increasing Size from 5 to 15 MW including Dynamic Soil-Structure Interaction. *Journal of Marine Science and Engineering*, 9(11), 1285.

Wang, X., Zeng, X., Li, J., Yang, X., & Wang, H. (2018). A review on recent advancements of substructures for offshore wind turbines. *Energy Conversion and Management*, 158(FEB.), 103–119.

Wang, X., Zeng, X., Yang, X., & Li, J. (2018). Feasibility study of offshore wind turbines with hybrid monopile foundation based on centrifuge modeling. *Applied energy*, 209, 127–139.

Wang, Z., Jiang, C., Qian, A., & Wang, C. (2009). The key technology of offshore wind farm and its new development in china. *Renewable & Sustainable Energy Reviews*, 13(1), 216–222.

Yang, Q., Yu, P., Liu, Y., Liu, H., Zhang, P., & Wang, Q. (2020). Scour characteristics of an offshore umbrella suction anchor foundation under the combined actions of waves and currents. *Ocean Engineering*, 202, 106701.

Zou, C., Xiong, B., Xue, H., Zheng, D., & Wu, S. (2021). The role of new energy in carbon neutral. *Petroleum Exploration and Development*, 48(2), 480–491.

*Advances in Measurement Technology, Disaster Prevention and
Mitigation – Li & Mohd Yusof (Eds)*
© *2023 The Author(s), ISBN: 978-1-032-36087-4*

Seepage stability analysis of 3# reservoir for a ski park

Yanjing Shi
Tianjin Renai College, Tianjin, China

Wenlong Niu
China Institute of Water Resources and Hydropower Research, Beijing, China

Lingchao Meng* & Linlin Jiang
Tianjin Renai College, Tianjin, China

ABSTRACT: With the 3# reservoir of a ski park as the study case, seepage stability analysis is
carried out in this paper. Because there are many cracks in the reservoir and there is no monitoring
data, two schemes are proposed based on literature review and expert opinions to analyze the seepage
stability of the reservoir. First, it is considered that the anti-seepage function of the concrete face
slab is completely ineffective. Second, we consider the concrete face slab still has an anti-seepage
function, but it needs to be reduced. The results show that in the first case, the seepage quantity
is very large and the saturated line is very high, which is inconsistent with the actual situation.
However, the calculation results of the second case are reasonable. The present work is expected
to provide a reference for seepage stability analysis of similar projects.

1   INTRODUCTION

According to the *2020 Statistic Bulletin on China Water Activities*, the number of completed reservoirs in China reached 98,566, of which 4,872 are large or medium-sized, and the other 93,694
are small reservoirs. Due to the limitation of construction conditions, there are many problems
such as the lack of engineering data in the current safety appraisal for most of the small reservoirs.
Therefore, how to perform safety evaluation reasonably according to the available data is a major
problem in the safety appraisal of small reservoirs.

This paper focuses on how to consider the influence on the seepage stability analysis when there
are many cracks in the concrete face slab and there is no monitoring data during the safety appraisal
of the 3# reservoir for a ski park.

2   PROJECT PROFILE

There are three reservoirs for a ski park located in Wutong Avenue, Taizicheng, Chongli District,
Zhangjiakou City. From upstream to downstream, they are named 1# reservoir, 2# reservoir, and 3#
reservoir. The reservoirs were built around 2010. Their task is to supply water to make snow for the
ski park in winter and discharge flood during the flood season. The capacity of the three reservoirs
is 49,000 m^3, 67,300 m^3, and 53,300 m^3, respectively. The total storage capacity is 169,600 m^3.
The whole reservoir basin uses a concrete face slab for anti-seepage. The filling material of the
reservoirs is soil and gravel.

---

*Corresponding Author: mlch8868@163.com

DOI 10.1201/9781003330172-51

The construction process of the reservoirs is not very standardized. Furthermore, the construction quality is poor. At present, cracks, leakage, bulging, and other phenomena threatening the safety of the reservoirs have appeared. Hence, it is necessary to conduct safety evaluation tests to ensure the reservoirs' safety.

Taking 3# reservoir as an example (the two others are similar), the seepage stability is analyzed in the safety appraisal in this paper. At first, we carry out a safety inspection to make sure the factors which will affect the seepage of the reservoir. Second, based on the results of the safety inspection, experts' opinions, and previous works, two schemes of seepage stability analysis are put forward. One scheme considers the complete failure of the face slab anti-seepage system, the other one considers the reduction of the face slab anti-seepage effect. Third, finite element software is used to calculate respectively. At last, results are analyzed and suggestions are given.

3 SAFETY INSPECTION OF THE RESERVOIR

According to the national specifications (Nanjing Hydraulic Research Institute, Dam Safety Management Center of the Ministry of Water Resource 2017), safety inspection was carried out firstly in the safety appraisal to guide the seepage stability analysis of the reservoir.

The results of the safety inspection show that there are many large cracks in both the water retaining section and the slope around the 3# reservoir. The cracks with a width of more than 5 mm were measured on-site. The schematic diagram of crack distribution was drawn in Figure 1. The width, depth, and length of the cracks are shown in Table 1. The base slab of the reservoir has concrete damage whose area is about 5m×5m (Figure 2). The steel bars in the base slab are exposed, as shown in Figure 3. In addition, there are cavities on the slope of the reservoir (Figure 4).

Figure 1.   The schematic diagram of crack distribution.

Table 1. The statistics of cracks for 3# reservoir.

No.	Stake Number	Max crack width (mm)	Max crack depth (mm)	Crack length (m)
①	0+036.240-0+046.240	15.02	19.24	31.2
②	0+076.240-0+086.240	35.62	0.28	28.8
③	0+126.038-0+136.038	35.60	28.62	36.6
④	0+136.038-0+144.468	9.06	1.28	29.1
⑤	0+148.978-0+159.324	16.10	3.64	29.9
⑥	0+169.670-0+180.016	70.80	50.04	30.9
⑦	0+180.016-0+190.362	20.00	13.68	30.4
⑧	0+190.362-0+200.708	13.22	10.40	30.7
⑨	0+211.054-0+222.081	13.26	10.08	13.2
④	0+211.054-0+222.081	21.64	72.04	8.9
⑪	0+252.033-0+262.023	12.26	10.02	5.3
⑫	0+301.905-0+316.962	6.64	6.08	31.2
⑬	0+326.252-0+336.252	5.02	3.24	31.8
⑭	0+336.252-0+346.252	5.08	6.04	30.1
⑮	0+356.252-0+386.252	7.06	6.28	12.8

Figure 2. Concrete damage.   Figure 3. Exposed steel bars.   Figure 4. Cavities.

4 SEEPAGE STABILITY ANALYSIS METHOD

4.1 Schemes of seepage stability analysis

According to the results of the safety inspection, the anti-seepage concrete face slab of the reservoir has many cracks, local damage, and cavities, both of which will lead to reservoir leakage. Moreover, according to the introduction of the manager, leakage has occurred in the reservoir, but there is no specific data for seepage quantity. In summary, there are some problems in the anti-seepage function of the concrete face slab. The seepage stability analysis cannot be carried out according to the original design scheme. It is necessary to consider the influence of concrete face slab cracks on seepage reasonably. Thus, two schemes are proposed in this research for seepage stability analysis.

The safety inspection results show that there are many cracks in the concrete face slab. What's more, both the width and the length of cracks are large. According to the suggestions of experts, considering the complete failure of the anti-seepage system, the water retaining section is regarded as a homogeneous earth rock dam for seepage stability analysis in the first scheme. Based on the literature, the second scheme is proposed. In this scheme, we consider the concrete face slab still has some anti-seepage functions even though it has many cracks. We only need to reasonably reduce its anti-seepage effect which will be measured by the permeability coefficient. After that, the seepage stability can be analyzed.

In this article, the finite element method and the hydraulic structure analysis software Autobank7.7 are used to calculate the seepage stability of the two schemes.

4.2 *Model parameter*

In Scheme 1, the face slab and geomembrane lose the anti-seepage function, so the water retaining section is calculated as an earth rock dam. The dam is divided into zones according to the results of supplementary geological exploration. The permeability parameters of each zone are mainly selected through supplementary geological exploration and literature (Hua & Zheng 2018), which are shown in Table 2.

Table 2. The permeability coefficient of each zone for the dam in scheme 1.

No.	Zone	Permeability coefficient (m/d)	Remarks (m/s)
1	Filling material	70	0.81×10^{-3}
2	Crushed stone	80	0.93×10^{-3}
3	Strongly-weathered granite	0.18	2.08×10^{-6}
4	Medium-weathered granite	0.0108	1.25×10^{-7}

In scheme 2 the reduction of the anti-seepage effect caused by cracks is considered. The anti-seepage effect is measured by the permeability coefficient of the concrete face slab. Strictly speaking, the permeability coefficient used in the following calculation should be inversed according to seepage monitoring data. However, due to the small capacity, low engineering grade, and non-standard management, there are no monitoring facilities, the permeability coefficient of the damaged face slab is obtained through the literature. Shan Gao (Gao et al. 2021) pointed out in her study that when the whole face slab is a block, the equivalent homogenized permeability coefficient of the cracked face slab is 8.29×10^{-7}m/s. Therefore, in this safety appraisal, the permeability coefficient of concrete face slab is 8.29×10^{-7}m/s. The geomembrane in which the permeability coefficient is obtained through literature is still considered to work normally. On this basis, the seepage stability of the reservoir is calculated.

The permeability parameters are mainly selected by supplementary geological exploration and literature (Cen et al. 2012; Hua & Zheng 2018), which are shown in Table 3.

Table 3. The permeability coefficient of each zone for the dam in scheme 2.

No.	Zone	Permeability coefficient (m/d)	Remarks (m/s)
1	Concrete face slab	0.0716	8.29×10^{-7}
2	Mortar cushion	8.64×10^{-4}	1×10^{-8}
3	Geomembrane	8.64×10^{-6}	1×10^{-12}
4	Gravel cushion	0.3456	4×10^{-6}
5	Filling material	70	0.81×10^{-3}
6	Crushed stone	80	0.93×10^{-3}
7	Strongly-weathered granite	0.18	2.08×10^{-6}
8	Medium-weathered granite	0.0108	1.25×10^{-7}

4.3 *Calculation condition*

According to the requirements of specification (Yellow River Engineering Consulting Co., Ltd, 2021), the two working conditions as follows are selected for seepage stability analysis for both two schemes.

(1) Working condition 1: it is the normal water level (1795.00m) upstream, meanwhile there is no water downstream.
(2) Working condition 2: it is the check flood level (1795.70m) upstream, meanwhile there is no water downstream.

5 CALCULATION RESULTS AND ANALYSIS

5.1 *Calculation results*

The results of the two schemes are shown in Table 4, and the saturation lines are shown in Figures 5 and 6. Because the results under different conditions are similar, the saturation lines under Working Condition 1 of the two schemes are only shown in the paper.

Table 4. Calculation results of seepage stability.

Scheme number	Working conditions	Gradient at exit point	Allowable gradient of dam	Gradient at concrete face slab	Allowable gradient of face slab	Seepage quantity per unit width (m³/s/m)
Scheme 1	Condition 1	0.500	0.30~0.25	/	200	0.00244
	Condition 2	0.501		/		0.00299
Scheme 2	Condition 1	0.249		50.4		3.248×10^{-5}
	Condition 2	0.278		52.3		3.457×10^{-5}

Figure 5. Saturation lines under Scheme 1.

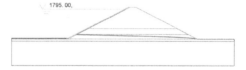

Figure 6. Saturation lines under Scheme 2.

5.2 *Result analysis*

Through the results of the two schemes under the two working conditions, it can be known:

(1) Leakage analysis
   a) The seepage quantity is larger when the water level is higher, and this law holds for both two schemes.
   b) For Scheme 1, the seepage quantity of the water remaining section is estimated to be 14651m³/d when the upstream is the normal water level. The seepage quality is very large, accounting for 27.5% of the capacity. For Scheme 2, the seepage quantity is about 179 m³/d at the same water level.
(2) Permeability stability analysis
   Table 4 lists the seepage gradient under different working conditions. The seepage gradient of Scheme 1 does not meet the allowable value, but the value of Scheme 2 can meet the requirements.
(3) Saturated line analysis
   For Scheme 1, it can be seen the saturated lines are very high under each working condition. Because Scheme 1 does not consider the anti-seepage function of the face slab and the permeability coefficient of the filling material is large. In Scheme 2, the face slab is considered to

have a certain anti-seepage function. The saturated line was significantly reduced after passing through the face slab and the geomembrane.

(4) Comparative analysis

From the above results, we can see that the seepage quantity calculated by Scheme 1 is very large. Specifically, the daily seepage quantity accounts for 27.5% of the capacity. In addition, there is no anti-seepage function of the face slab, the high saturation line may lead to the slope instability of the reservoir. For Scheme 2, the daily seepage quantity is about 179 m^3, the saturated line is low. According to the manager, although it has leakage problems, the reservoir has been running stably for more than ten years. Considering comprehensively, the daily seepage quantity of Scheme 1 is too large, the saturated line is very high and the reservoir is at risk of landslide, which is inconsistent with the actual situation. The results of the second scheme are consistent with the actual, so the subsequent structural stability evaluation and seismic safety evaluation are carried out on this basis.

6 CONCLUSIONS

In this paper, a seepage stability analysis of 3# reservoir for a ski park is carried out. Two schemes are proposed to consider the influence of face slab cracks on seepage. One is to consider the complete failure of the anti-seepage system. The other one is to consider the reduced effect of the anti-seepage system. The conclusions are obtained as below:

(1) If the permeability coefficient of the filling material for a dam is large, the seepage quantity calculated by considering the complete failure of the anti-seepage system is very large and the saturated line is very high, which is mostly inconsistent with the actual situation. Therefore, this method is not recommended in the safety appraisal.

(2) If consider the face slab still has a certain anti-seepage function, the calculation results are close to the actual situation. However, how to reasonably measure the anti-seepage effect of a cracked face slab is the key to this scheme.

(3) For reservoirs with monitoring facilities, it is recommended to inversed the permeability coefficient of the cracked face slab based on the monitoring data. For small reservoirs without monitoring facilities, it is recommended to obtain the permeability coefficient of cracked face slab by field test. For projects with neither monitoring facilities nor sufficient funds for field tests, it is impossible to obtain an accurate permeability coefficient of cracked face slab, which can only be determined by comprehensively considering the safety inspection results, similar projects, and suggestions from the manager.

REFERENCES

Jianxin Hua, Jianguo Zheng. (2018) *Geological engineering handbook.* China Architecture & Building Press, The Beijing.

Nanjing Hydraulic Research Institute, *Dam Safety Management Center of the Ministry of Water Resource.* (2017) Guidelines on dam safety evaluation. China Water & Power Press, The Beijing.

Shan Gao, Junrui Chai, Jing Cao, et al. (2021) Numerical Analysis of Seepage for Concrete Face Rockfill Dam with Cracks Based on Block Equivalent Continuum Method. In: *Arabian Journal for Science and Engineering.* pp. 10341–10354.

Weijun Cen, Meng Wang, Zhixiang Yang. (2012) Partial saturated seepage properties of (composite) geomembrane earth-rock dams. In: Advances in Science and Technology of *Water Resources*, pp. 6–9.

Yellow River Engineering Consulting Co., Ltd. (2021) *Design code for rolled earth-rock fill dams.* China Water & Power Press, The Beijing.

Advances in Measurement Technology, Disaster Prevention and Mitigation – Li & Mohd Yusof (Eds)
© 2023 The Author(s), ISBN: 978-1-032-36087-4

Scenario construction and emergency rescue analysis of high altitude mine fire

Jun Hu* & Rui Huang*

School of Resource and Safety Engineering, Central South University, Changsha, China

ABSTRACT: The occurrence of a mine fire accident will cause incalculable consequences. The emergency rescue process of mine fire accident is facing huge difficulties due to the harsh environmental conditions, such as cold climate, low pressure, and low oxygen in high-altitude areas. This paper takes a horizontal roadway at an altitude of 3720m in Yunnan Pulang Copper Mine as an example and creates a completely new accident scenario model based on the theory of scenario construction. The model is mainly composed of three different elements, which are scene elements, accident morphological elements, and emergency rescue elements. By analyzing these three elements separately, we can understand the complete process of the occurrence and development of high-altitude mine fire accidents. We can also propose corresponding solutions to terminate the accident according to the accident scenarios at different stages. This will play a huge role in improving the effectiveness of emergency rescue plans and ensuring the safety of people's lives.

1 INTRODUCTION

Mine fires are usually divided into two types by the cause: external fires and internal fires (Zhao 2016). Generally speaking, the occurrence of external fires includes sparks caused by short circuits in electrical equipment or lines. There are also high-temperature fires caused by illegal use of electrical heating, and high-temperature welding slag generated by electric welding machines. The internal fire is mainly caused by the spontaneous combustion of sulfur and phosphorus minerals in the well. At present, a large number of mining projects have been built in the western region (Yang 2018). However, the environmental conditions in high-altitude areas are different from those in plain areas, which makes the characteristics and laws of fire occurrence and development different from those in plain areas. At present, there are few studies on high-altitude mine fire accidents. Huang Rui (2020) simulated and explored the spread of mine fires at different altitudes and found that evacuation time will be increased but the evacuation distance will be shortened in the event of a mine fire accident as the altitude rises. Geng Meng (2021) studied the dynamic combustion process of trackless transportation equipment at different altitudes through FDS fire simulation software. Wang Yaning (2020) used the method of numerical simulation to conduct an optimization study on the smoke control wind speed when a fire occurred in a high-altitude tunnel in Yunnan. Zhang Ying (Zhang & Sun 2011; Zhang & Huang 2011) found through experiments that the speed of fire spread and the length of the burning area in the plateau area were both smaller than those in the plain area.

In recent years, scenario construction has gradually become a new type of emergency management method. Some scholars believe that a "scenario" is a systematic description of the whole

---

*Corresponding Authors: 2798192763@qq.com and huangrui@csu.edu.cn

process, all-around and panorama of the general law of a certain type of emergency based on the real background. It is a systematic and visual presentation of a certain type of major risk (Liu 2012; Wang 2015). Wang Yongming (Wang 2019) believes that scenario construction is not only an important starting point for risk analysis but also an important method to provide direction and goals for emergency preparedness. Zhang Yuebing (2020) used scenario construction to analyze the response and disposal of highway tunnel fire accidents. Based on the information-driven theory, Hu Renyuan (2021) analyzed the driving process of information in the four stages of the evolution of hazardous chemical fire accident scenarios and established related models. Lian Huiqing (2020) applied "scenario-response" to the field of mine flood emergency decision-making, and proposed a multi-stage mine flood emergency decision-making mechanism.

To sum up, scenario construction is an effective means to improve the level of emergency management. However, there are few studies on the construction of mine fire accident scenarios in the existing literature, and mine fire accidents involving high-altitude environments are even blank. Therefore, this paper constructs a complete fire accident scenario by simulating the process of occurrence and development of mine fires in alpine and high-altitude areas from the perspective of scenario construction. In this way, we can study the emergency capability assessment based on situation comparison, and also clarify the relevant points in the emergency rescue process to provide a certain reference for emergency rescue training guidance.

2 CONSTRUCTION OF HIGH-ALTITUDE MINE FIRE ACCIDENT SCENARIOS

2.1 *Model building*

This paper proposes a new scenario model that combines the above theories and related research contents of scenario construction. It divides the elements of the occurrence and development of emergencies into three parts, which are scene elements, accident morphological elements, and emergency rescue elements. The model is shown in Figure 1.

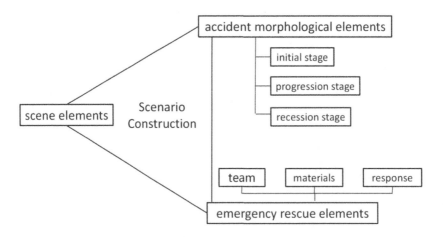

Figure 1.   Construction model of high-altitude mine fire accident scenarios.

This model is mainly constructed from three aspects, including scene elements, accident morphological elements, and emergency rescue elements.

First of all, scene elements are the basic components of scene construction, and it is the basis of the emergency scene chain. We need to make a detailed description of the constructed accident site, including the environment, initial conditions of personnel, and spatial geometry to obtain the complete accident scene elements. This will help emergency drillers better participate in the decision-making and response of the system when these elements are accurately described.

Analysis of accident morphological elements requires us to describe the whole process of accident occurrence and development in detail. We can divide mine fire accidents at high altitudes into three stages, which are the initial stage, progression stage, and recession stage.

Given the development form of accident disasters, we need to have corresponding measures to control the development conditions of accidents, to reduce or eliminate the consequences of accidents. Therefore, it is necessary to configure the corresponding measures and tools for each rescue stage according to the accident evolution form in the scene. From this, the content of emergency rescue elements is formed. The emergency rescue elements are generally described in three aspects: emergency team, emergency materials, and response and disposal.

2.2 *Analysis of scenario elements*

(1) Analysis of scene elements

According to the scenario model constructed above, we analyze different elements. In terms of scene elements, we mainly describe the spatial geometric elements of the accident to establish relevant accident simulation models. We take a horizontal roadway at an altitude of 3720m in Yunnan Pulang Copper Mine as an example to construct the model. To reduce unnecessary computations during simulation, the model can be simplified. The overall frame size of the roadway is set to be 100 m long and 4m wide, its cross-section is arc-shaped, and the top of the roadway is 3.7 m high. Since the characteristic height of the human eye is about 1.6m(Yang et al. 2005), we set up a temperature slice and a visibility slice at a height of 1.6m for recording temperature data and visibility data. At the same time, 6 temperature detectors and 6 smoke concentration detectors are set in the model to measure the temperature and smoke concentration when the fire occurs. The positions of the two are the same, and the positions along the horizontal direction are (2 m, 1.6 m, 10 m), (2 m, 1.6 m, 20 m), (2 m, 1.6 m, 40 m), (2 m, 1.6 m, 60 m), (2 m, 1.6 m, 80 m), (2 m, 1.6 m, 90 m). A 1m^2 fire source is set in the middle of the roadway, and the fuel is set to n-heptane, as shown in Figure 2.

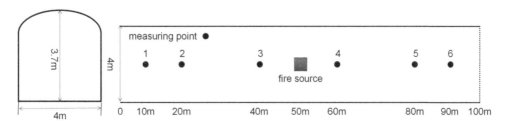

Figure 2.   Mine roadway model.

The t^2 fire model is adopted since the development of fire is an unstable process, and its mathematical expression is:

$$Q = \alpha t^2 \qquad (1)$$

In the formula: Q represents the heat release rate of the fire source (kW). α is the fire growth coefficient (kW/s2), which indicates the speed of fire development. According to the classification of the National Fire Protection Association (NFPA), the fire growth rate of the t2 model can be divided into 4 types: super fast fire, fast fire, medium fire, and slow fire (Tian & Zhang 2018). The external fire in the mine belongs to the medium speed, so we take α as 0.01127kW/s2. t represents the time (s) for the development of the fire. Based on the maximum heat release rate of a typical site fire, the maximum fire source heat release rate for a mine fire is 10MW. The longitudinal wind speed in the roadway is set as 1m/s. According to literature (Gou et al. 2012; He 2011; Wang 2009), it can be calculated that the atmospheric pressure at this altitude is 70.107kPa, the oxygen content is about 15.09%, and the initial ambient temperature is 7°C.

(2) Analysis of accident morphological elements.

1) Analysis of temperature in the roadway.

As shown in Figure 3, it is the temperature slices at different times when a fire occurs in the roadway. This slice graph shows the temperature downstream of the fire source along the wind flow direction. According to the temperature slice diagram, the temperature closer to the fire source is lower than the temperature farther away from the fire source when a fire occurs. It can be seen from the analysis that the height of the slice arrangement is 1.6m, the fire source set at the same time is fixed in the middle of the roadway, and the influence of wind speed is added so that the high temperature generated by the fire rises along the direction of wind flow. Due to the low oxygen content, low temperature, and low air density at high altitudes, it is difficult to fully burn in the event of a fire, which will also cause the peak temperature to drop.

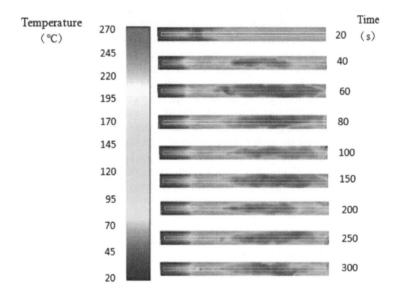

Figure 3. Temperature slice diagram.

2) Analysis of smoke and visibility.

We can observe the flow of smoke during the fire through the simulation of fire smoke as shown in Figure 4. It can be found that when a mine fire occurs in a high-altitude area, the flame is mainly spread on the floor and develops in the horizontal direction. Moreover, the black smoke will quickly fill the roadway with the action of wind flow and move to the top of the roof. At the same time, there will be a certain degree of the smoke reverse phenomenon, but it is not obvious. The main reason is that high altitudes have lower oxygen levels and fewer combustibles present, so the rate of combustion slows down. The rate at which the fire plume spreads with the air is also greatly reduced due to the lower density of the air. Shown in Figure 5 is a slice plot of visibility downstream of the fire source over time. After the fire broke out, the smoke quickly filled the roadway under the action of wind flow, and from 20 s onwards, the visibility remained unchanged with the development of time. It can also be seen that at a closer distance to the fire source, the visibility will be greater than that farther away from the fire source. At a height of 1.6m, the visible range of the human eye is only about 3m, which will cause great difficulties for people to escape.

Time (s)

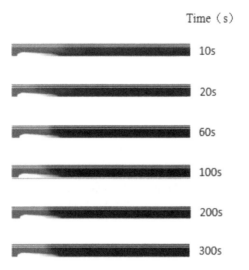

Figure 4. Smoke diffusion diagram.

Visibility (m)   Time (s)

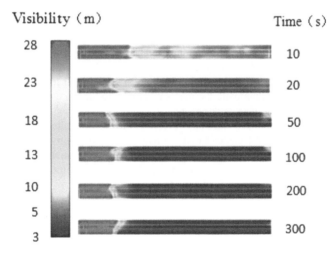

Figure 5. Visibility slice diagram.

(3) Analysis of emergency rescue elements

According to the different stages of fire development, we can propose emergency rescue measures to make the accident scene disappear. The specific description is as follows:

1) The initial stages of a fire accident

At 9:00 a.m. one day, an electrical fire accident occurred in the oil circuit system due to the high temperature of the scraper when it was in use. Personnel near the mine discovered the fire through gas detectors and rushed to the accident site immediately and started firefighting work.

In this situation, the actions that people in the mine took were as follows: 1) they immediately found a nearby fire extinguisher to put out the fire, time paid attention to the influence of the wind speed in the roadway, and stood in the upwind direction to prevent inhalation of excessive smoke; 2) they immediately cleaned up nearby combustibles to prevent the fire from expanding due to the action of wind; 3) they reported the specific situation of the fire scene to the superior and wore a self-rescuer for self-rescue.

Result: The fire was extinguished. There were no casualties and the accident scene disappeared.

2) The progression stage of a fire accident

The action of the wind caused the fire to expand rapidly and produce a lot of smoke and toxic gases. It is difficult for people to put out the fire, and at the same time, some people in the well fell into a coma due to lack of oxygen.

Actions that could be taken in this situation are as follows: 1) the underground personnel should quickly organize self-rescue and mutual rescue before the rescue and medical staff arrive. The victims should wear the self-rescuer correctly, and evacuate to the fresh air according to the disaster avoidance route in the emergency plan drill. At the same time, they need to cut off the underground power supply; 2) if people find it difficult to evacuate, they need to find the nearest refuge chamber as soon as possible to take refuge and wait for the arrival of rescuers; 3) after receiving the alarm, the relevant departments quickly start the emergency plan. According to the instructions of the superior, the rescue team was quickly organized to go down and rescue and put out the fire.

Result: People took refuge in the well and rescuers rushed to bring the fire under control. The accident scene disappeared.

When toxic fumes quickly fill the roadway, visibility is poor and rescuers cannot enter. At the same time, the high temperature generated by the fire caused the roof structure of the roadway to change and there was a danger of collapse. In this scenario, we should quickly evacuate nearby people to prevent secondary accidents. At the same time, we should apply for reinforcements and expand emergency response, and prepare to go down well for disaster relief at any time according to the development of the fire.

Results: No major casualties occurred, and the accident scene disappeared.

3) The recession phase of a fire incident

The environmental conditions of low pressure and low oxygen in high altitude areas will cause the oxygen in the roadway to be rapidly consumed, making the fire gradually reduce. Rescuers must quickly rescue the trapped people and send them to the hospital for treatment. The cleaning team will ventilate and decontaminate the roadway, eliminate hidden dangers of the accident and complete the emergency response work on site.

The enterprise should conduct a summary review of the accident after the emergency response is over. It should find out the deficiencies in the emergency response process and make corrections and should compensate and appease the disaster victims. The environmental protection department will inspect the accident scene to prevent environmental pollution. Enterprises should learn from this experience and avoid similar accidents in the future.

3   CONCLUSION

In this paper, the method of scenario construction is adopted to study the emergency rescue of mine fires in high-altitude areas. A new accident scenario model is constructed, which is mainly composed of three scenario elements. They are scene elements, accident morphological elements, and emergency rescue elements. The main conclusions can be summarized from the analysis of these three different elements as follows:

(1) The occurrence and development process of fire will be very different from that in plain areas due to the special environmental conditions in high-altitude areas. It is necessary to fully consider the influence of environmental conditions when setting the scene. For example, the selection of values such as atmospheric pressure and oxygen content.
(2) In the analysis of accident morphological elements, the simulation results pointed out that the environmental conditions of low pressure and low oxygen will have a certain impact on the diffusion of smoke when a mine fire occurs in a high-altitude area. On the one hand, the temperature gradually increases along the direction of the wind flow and the visibility gradually

decreases when a fire occurs. People should seek refuge in the nearest shelter in the direction of the wind for evacuation quickly. On the other hand, the oxygen content in the mine will be rapidly consumed when a fire occurs, which will greatly affect the physiological conditions, causing people to slow down the speed of escape.

(3) According to the analysis of the elements of emergency rescue, the accident loss will be minimal if the scenario can be quickly terminated in the initial development stage of the fire. The consequences of casualties can also be greatly reduced if the trapped people can implement correct and effective self-rescue and mutual rescue measures when the fire reaches the stage of rapid development.

In terms of future work, we will focus on constructing a complete dynamic accident scenario model. Add more possible emergencies in the model, such as derivative accidents and other emergencies, so that it can achieve a better emergency rescue training effect.

REFERENCES

Geng, M., Liu, J., Wang, D. (2021). Simulation study on the effect of altitude on the fire of trackless transport vehicles in mines. *China Safety Science and Technology*, 17: 59–64.
Gou, H.S., Li, Y.S., Luo, Z.F. (2012). Calculation of ventilation air volume and fan selection for tunnel construction in high altitude areas. *Tunnel Construction*, 32: 53–56.
He, L. (2011). Study on the numerical simulation of aerated ventilation of single tunnel in plateau mine. *Modern Mining Industry*, 27: 37–40.
Hu, R.Y., Xia, D.Y., Zhang, J. (2021). Construction of hazardous chemical fire dynamic Scenario based on information driven theory. *Fire Science and Technology*, 40: 586–589.
Huang, R., Wu, E., Wu, L. (2020). Research on the influence of altitude on the smoke spreading law of mine roadway. *Gold science and technology*, 28: 293–300.
Lian, H.Q., Yang, J.W., Han, R.G. (2020). Emergency decision-making mechanism of mine flood accident based on "scenario and response". *Coal Geology & Exploration*, 48: 120–128.
Liu, T.M. (2012). Scenario construction of emergency plan for major emergencies – one of the research on the preparation technology of emergency plan based on "situation-task-capability". *China Safety Science and Technology*, 8: 5–12.
Liu, T.M. (2012). Emergency preparedness task setting and emergency response capacity building – A study on scenario-task-capacity based emergency plan formulation technology ii. *China Safety Science and Technology*, 8: 5–13.
Tian, S.C., Zhang, C. Z. (2018). Numerical simulation research on response temperature of thermal detector in buildings with decentralized fire. *China Safety Science Journal*, 28: 56–61.
Wang, G.F. (2009). Technical research and application of construction machinery properties in plateau cold region. *Gansu Science and Technology*, 25: 44–45.
Wang, Y.M. (2015). Theoretical Framework and technical route of Major Emergency Scenario Construction. *China Emergency Management*, 2015: 53–57.
Wang, Y.M. (2019). Evolution of scenario construction theory and its enlightenment to Emergency management in China. *China Safety Science and Technology*, 15: 57–62.
Wang,Y.N., Zhang, C.L., Zhang, J.R. (2020). Research on wind speed optimization of smoke control in yunnan high-altitude tunnel fire. *China Water Transport*, 20: 206–208.
Yang, B. (2018). Thinking on key points of mine engineering design in high-cold and high-altitude area. *China Molybdenum Industry*, 42: 11–16.
Yang, L.Z., Guo, Z.F., Ji, J.W. (2005). A fire development model based on regional simulation. *Chinese Science Bulletin*, 12: 1272–1277.
Zhang, Y., Sun, J.H. (2011). Experimental study on fire spreading characteristics of timber surface in plateau environment. *Combustion Science and Technology*, 17: 274–279.
Zhang, Y., Huang, X.J. (2011). Effects of sample width on flame spread over horizontal charring solid surfaces on a plateau. *Science Bulletin*, 56: 919–924.
Zhang, Y.B., Su, H.Y. (2020). Construction of emergency scenario of highway tunnel fire accident and analysis of response and disposal. *City and Disaster Reduction*, 1: 46–49.
Zhao, G.Y. (2016). Study on the cause and prevention of mine fire. *Energy and Conservation*, 8: 113–114+122.

Advances in Measurement Technology, Disaster Prevention and
Mitigation – Li & Mohd Yusof (Eds)
© 2023 The Author(s), ISBN: 978-1-032-36087-4

Seismic performance analysis of concrete-filled steel tubular column-corrugated steel plate composite pier

Ziqi Li*, Yanyan Fan* & Wang Li
Civil Engineering College of Lanzhou Jiaotong University, China
National and Provincial Joint Engineering Laboratory of Road & Bridge Disaster Prevention and Control, China

Wenlong Zhang
Civil Engineering College of Lanzhou Jiaotong University, China

ABSTRACT: The finite element software ABAQUS was used to establish the CFST column-corrugated steel plate composite shear wall. The quasi-static simulation of the composite shear wall was carried out to analyze the seismic performance and failure mechanism of the structure. The influence of wave direction, plate thickness, wave height, and hoop coefficient on the seismic performance of the composite structure was explored. The results show that with the increase of plate thickness, the bearing capacity of the structure increases, and the ductility and initial stiffness of the structure increase, but it is unfavorable to the stiffness degradation of the structure. The energy dissipation capacity of the transverse plate is better, the ductility is 2.1 times that of the vertical plate, and the degeneration of the transverse plate is slow. The ductility of the structure increases with the increase of wave height; with the increase of hoop coefficient, the hysteresis curve of the component began to appear to "pinch" effect.

1 INTRODUCTION

CFST column-corrugated steel plate composite shear wall is composed of the embedded corrugated plate, CFST column frame, and I-beam. Guo Yanlin (2009) studied buckling-restrained steel plate shear walls, and found that the bearing capacity and energy dissipation capacity of the structure are higher than ordinary steel plate shear walls. Li Feng (2011) studied the stiffening and slitting mode of steel plate shear walls. It is considered that the cross stiffener improves the bearing capacity and ductility of the structure. Opening and slitting will lead to a decrease in the bearing capacity of the component, but the energy dissipation capacity of the component is improved. Tong Shugen (2015) studied the critical buckling stress of stiffened steel plate shear walls. It is considered that the buckling stress of the shear wall will increase with the increase of the torsional stiffness of stiffeners. Pan Yue (2019) studied the performance of stiffened repair of steel frame-steel plate shear wall after the earthquake, and found that stiffened rib repair can improve the seismic performance of the structure. The connection form of stiffened rib (welding fixation and bolt anchorage) has little effect on the energy consumption and bearing capacity of the component. Yang Zhuoqiang (2009) carried out experimental research on the seismic performance of jointed shear walls and ordinary shear walls. Wang Wei (2019) studied the shear capacity of corrugated steel plate shear walls and composite walls. The results show that corrugated steel plate shear wall has a large deformation capacity and good seismic performance.

---
*Corresponding Authors: 12062481@qq.com and 46939099@qq.com

DOI 10.1201/9781003330172-53

In this paper, the trapezoidal corrugated plate is used as the embedded steel plate, and the ABAQUS nonlinear finite element model is established to analyze the seismic performance and failure mechanism of the CFST column-corrugated steel plate composite structure. The influence of wave direction, plate thickness, wave height, and steel pipe hoop coefficient on the seismic performance of the composite structure is explored.

2 ESTABLISHMENT OF FINITE ELEMENT MODEL

2.1 Model size

At present, the corrugated plate section mainly has sine wave and trapezoidal wave, this paper uses trapezoidal wave. The shear wave and vertical wave CFST column-corrugated steel plate composite structures are established as embedded plates as shown in Figure 1. The beam of the composite structure is made of I-steel. Table 1 is the size parameters of the finite element model.

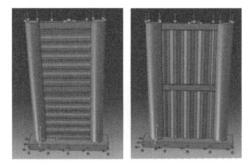

Figure 1.　Finite element model of CFST column-corrugated steel plate composite structure.

Table 1.　Dimension parameters of each model.

Test specimen number		Sectional dimension (mm×mm)	Grade of concrete	Steel type	Thickness of steel plate (mm)	Wave height (mm)	Confining factor
HW-1	column	219×6	C40	Q235	3	20	1.04
	beam	140×80×6×9	-	Q235			
HW-2	column	219×6	C40	Q235	4	20	1.04
	beam	140×80×6×9	-	Q235			
PW-3	column	219×6	C40	Q235	4	20	1.04
	beam	140×80×6×9	-	Q235			
PW-4	column	219×6	C40	Q235	4	25	1.04
	beam	140×80×6×9	-	Q235			
PW-5	column	219×7	C40	Q235	4	25	1.26
	beam	140×80×6×9	-	Q235			

2.2 Material constitutive and element selection

The concrete in the steel tube is simulated by Hanlinhai constitutive model, and the improvement of compressive strength of core concrete by steel tube is considered by introducing hoop coefficient ξ. The steel was subjected to the bilinear kinematic hardening model, and the elastic modulus E was 206000GPa in the elastic stage and 0.01 E in the strengthening stage. The corrugated plate, steel tube, and beam are all made of Q235-grade steel. The yield strength is 235MPa and the Poisson's ratio is 0.3. Steel tube, beam, and concrete are simulated by solid element C3D8R, and the corrugated plate is simulated by shell element S4R.

2.3 Boundary conditions

Restrain the freedom of each direction at the bottom of the cap and simulate the bottom consolidation. A reference point is established at the top of the composite structure, and the reference point and the top of the frame are coupled to impose reciprocating loads at the reference point. Vertical force is simulated by the surface load. Binding constraints are applied between steel tubes and concrete, steel tubes and beams, and corrugated plates and beams.

3 PARAMETER ANALYSIS OF CFST COLUMN-CORRUGATED STEEL PLATE COMPOSITE STRUCTURE

3.1 Plate thickness

The finite element models of HW-1 and HW-2 are established. The thickness of the corrugated plate in the two models is 3mm and 4mm respectively, and the other parameters are consistent. The quasi-static simulation is carried out, and the comparison between the hysteresis curve and the skeleton curve is shown in Figure 2.

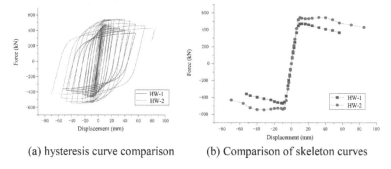

(a) hysteresis curve comparison   (b) Comparison of skeleton curves

Figure 2. Comparison of hysteretic curves and skeleton curves of CFST column-corrugated steel plate composite structures with different plate thicknesses.

Figure 2(a) shows that the thickness of the corrugated plate increases, the area of the hysteresis curve increases, and the energy dissipation capacity increases. Figure 2(b) is the skeleton curve comparison diagram of the two groups of components. It can be seen from the diagram that the bearing capacity of the component increases significantly with the increase of the thickness of the corrugated plate, but the initial stiffness is basically the same. The peak load of component HW-2 is 623.80kN, which is 13.7% higher than that of component HW-1 (548.65kN).

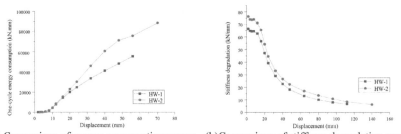

(a) Comparison of energy consumption curves   (b)Comparison of stiffness degradation curves

Figure 3. Comparison of energy dissipation curve and stiffness degradation curve of CFST column-corrugated steel plate composite structures with different plate thicknesses.

Figure 3 shows that with the increase of corrugated plate thickness, the energy dissipation capacity and cumulative energy consumption of the structure increase. In the elastic stage, the one-cycle energy dissipation capacity of the two groups of components is basically the same, but after the structure enters the plasticity, the one-cycle energy dissipation capacity of the component HW-2 is significantly greater than that of the component HW-1, and the cumulative energy dissipation and one-cycle energy dissipation are basically the same. As the thickness of the corrugated plate increases, the initial stiffness of the structure increases by 14.9%. After stiffness normalization, the stiffness degradation of the two groups of components is basically the same.

3.2 Wave direction

Wave direction is an important parameter of the corrugated plate. Two groups of vertical wave HW-2 and shear wave PW-3 components are established by ABAQUS. The quasi-static analysis of the two groups of components is carried out. The hysteresis curve and skeleton curve of the two groups of components are obtained as shown in Figure 4.

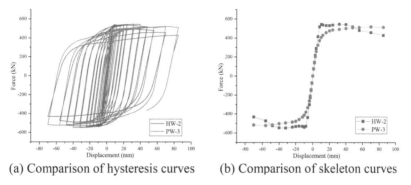

(a) Comparison of hysteresis curves   (b) Comparison of skeleton curves

Figure 4.   Comparison of hysteretic curves and skeleton curves of CFST column-corrugated steel plate composite structures with different wave directions.

It can be seen from Figure 4 (a) that the hysteresis curve of the shear wave structure is flatter than that of the vertical wave structure, and the bearing capacity of the vertical corrugated plate is greater than that of the transverse corrugated plate in the elastic stage. From the comparison of skeleton curves, it can be seen that the bearing capacity of the structure decreases slowly and the whole structure shows better ductility after the structure enters plasticity. And the shear wave structure enters the plastic after 5.61mm, which is 19.9% less than that of the vertical wave structure.

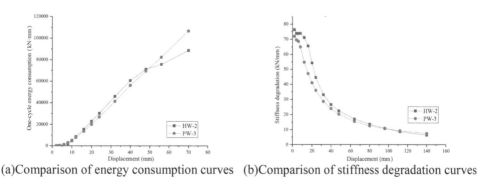

(a)Comparison of energy consumption curves   (b)Comparison of stiffness degradation curves

Figure 5.   Comparison of Energy Dissipation Curve and Stiffness Degradation Curve of CFST Column - Corrugated Steel Plate Composite Structure with Different Thickness.

385

It can be seen from Figure 5 that the one-week energy dissipation capacity of the shear wave structure is larger than that of the vertical wave structure, and the wave direction has little effect on the cumulative energy dissipation of the structure. The cumulative energy dissipation of the two groups of structures is basically the same. The initial stiffness of the vertical wave structure is greater than that of the shear wave structure, and the stiffness degradation rate of the vertical wave structure is greater than that of the shear wave structure. After normalization of the equivalent stiffness, it can be found that when the displacement angle reaches 0.01, the equivalent stiffness degradation coefficients of the two groups of structures are basically the same.

3.3 Wave height

In the Technical Specification for Steel Structures with Corrugated Webs, b/d=1.0~1.5, wave angle a=40~60°, and the ratio range of oblique straight section to wavelength is 1.15~1.4.

In order to study the influence of different wave heights on the mechanical properties of CFST column-corrugated steel plate composite structures, the finite element models of PW-3 and PW-4 components are established, and the hysteresis curves and skeleton curves of the two groups of structures under reciprocating loads are extracted as shown in Figure 6.

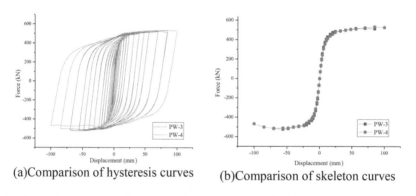

(a)Comparison of hysteresis curves   (b)Comparison of skeleton curves

Figure 6.   Comparison of hysteretic curves and skeleton curves of CFST column-corrugated steel plate composite structures with different wave heights.

Table 2.   Bearing capacities of two groups of members at different displacements.

Component number	Bearing capacity at 6 mm (kN)		Bearing capacity at 10 mm (kN)		Bearing capacity at 70 mm (kN)	
	Positive direction	Negative direction	Positive direction	Negative direction	Positive direction	Negative direction
PW-3	322.79	−333.15	405.47	−414.21	516.99	−517.02
PW-4	305.48	−315.68	387.33	−394.98	522.15	−513.98

The bearing capacity of two groups of members under displacement loading to 6mm, 10mm, and 70mm is compared, as shown in Table 2. At 6mm, each component is in the elastic stage, the bearing capacity of component PW-3 is 5.7% higher than that of component PW-4, and the negative bearing capacity is 5.5% higher. When the displacement was loaded to 10mm, the positive bearing capacity of the two groups increased by 25.6% and 26.8% respectively compared with that of 6mm. When the displacement is 70 mm, the positive bearing capacity of PW-4 is 522.15kN, and the negative bearing capacity is −513.98kN. The positive load is 0.9% larger than that of PW-3.

3.4 Hoop coefficient

The finite element models of PW-4 and PW-5 members are established to study the influence of steel tube hoop coefficient on the mechanical behavior of CFST column-corrugated steel plate composite structures. The hysteretic curves and skeleton curves of the two groups of structures under reciprocating loads are extracted as shown in Figure 7.

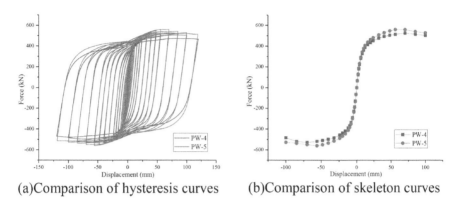

(a)Comparison of hysteresis curves   (b)Comparison of skeleton curves

Figure 7. Comparison of hysteretic curves and skeleton curves of CFST column-corrugated steel plate composite structures with different hoop coefficients.

From Figure 7, with the increase of hoop coefficient, the hysteresis curve of the component begins to appear 'pinch' effect. With the increase of steel tube hoop coefficient, the bearing capacity of the member increases. When the structure enters the plastic stage, the bearing capacity of the structure increases with the displacement loading. After reaching the peak load, the bearing capacity of the component PW-5 decreases faster than that of the component PW-4.

Table 3. Bearing capacities of two groups of members at different displacements.

Component number	Bearing capacity at 6mm (kN)	Bearing capacity at 20mm (kN)	Bearing capacity at 40mm (kN)	Bearing capacity at 70mm (kN)	Bearing capacity at 100mm (kN)
PW-4	280.78	432.89	493.18	524.70	501.66
PW-5	292.11	462.40	528.83	558.17	527.45

It can be seen from Table 3 that the bearing capacity of the structure increases with the increase of the hoop coefficient. At 6 mm displacement loading, the bearing capacity of component PW-5 is 4.0% higher than that of component PW-4, when the displacement reaches about 70mm, the structural bearing capacity reaches the peak value; at the peak displacement, the bearing capacity of the component PW-5 increased by 6.4% compared with that of PW-4, and then the bearing capacity of the structure decreased. At 100 mm, the bearing capacity of the two groups of components decreased by 4.4% and 5.5%, respectively, and the bearing capacity decline rate of the component PW-5 was greater than that of PW-4.

Table 4. Ductility coefficient of each component.

Test specimen number	Yield displacement (mm)	Limit displacement (mm)	Ductility factor
HW-1	7.9	38.1	4.8
HW-2	7.7	55.7	7.2
PW-3	5.6	84.3	15.0
PW-4	5.1	99.5	19.5
PW-5	4.9	–	–

Through analysis, Table 4 shows that as the thickness of the corrugated plate increases, the ductility of the structure increases; the ductility of the transverse corrugated plate is 2.1 times that of the vertical corrugated plate; with the increase of wave height, the ductility of the structure increases.

4 CONCLUSIONS

Using ABAQUS finite element analysis software, five groups of CFST column-corrugated steel plate composite structures were established, and the quasi-static simulation of the structure was carried out. The following conclusions were obtained:

(1) With the increase of corrugated plate thickness, the bearing capacity of the structure increases significantly and the energy dissipation capacity increases, but the initial stiffness of the structure is basically the same.
(2) The ductility of the transverse corrugated plate structure is 2.1 times that of the vertical corrugated plate structure. The energy dissipation capacity of the transverse wave structure is larger than that of the vertical wave structure, and the wave direction has little effect on the cumulative energy dissipation of the structure. The cumulative energy dissipation of the two groups of structures is basically the same.
(3) The ductility of the structure increases with the increase of wave height; with the increase of hoop coefficient, the hysteresis curve of the component began to appear 'pinch' effect.

ACKNOWLEDGMENTS

This work was supported by Gansu Natural Science Foundation (Grant Nos. 20JR10RA237).

REFERENCES

Guo Yanlin, Zhou Ming, Dong Quanli. (2009) Study on elastic-plastic shear ultimate bearing capacity and hysteretic behavior of buckling-restrained steel plate shear wall [J]. *Engineering Mechanics*, 26 (02): 108–114.
Li Feng, Wang Dong, Guo Hongchao, et al. (2011) Numerical simulation of seismic performance of different types of steel plate shear walls [J]. *World Earthquake Engineering*, 27 (04): 27–32.
Pan Yue. (2019) *Performance study on post-earthquake stiffening repair of steel frame-steel plate shear wall* [D]. Xi'an University of Architecture and Technology.
Tonggenshu, Yang Zhang, Zhang Lei. (2015) The effective bending stiffness of unilateral stiffeners in steel plate shear walls [J]. *Journal of Zhejiang University* (Engineering Edition), 49 (11): 2151–2158.
Wang Wei, Liu Gewei, Su Sanqing, Zhang Longxu, (2019) Ren Yingzi, Wang Xin. Study on the shear capacity of corrugated steel plate shear wall and composite wall [J]. *Engineering Mechanics*, 36 (07): 197–206.
Yang Zhuoqiang, Li Zhu, Liu Yuanzhen. (2009) Comparative analysis of the seismic performance of jointed shear wall and ordinary shear wall structure [J]. *Engineering Mechanics*, 26 (S2): 225–229.

Research on civil and hydraulic structure
and geological characteristics

Advances in Measurement Technology, Disaster Prevention and
Mitigation – Li & Mohd Yusof (Eds)
© 2023 The Author(s), ISBN: 978-1-032-36087-4

A comparative study on determination methods of effective area of natural smoke and heat exhaust ventilator

Hang Yin* & Longfei Tan*
Sichuan Fire Research Institute of Ministry of Emergency Management, Chengdu, China

ABSTRACT: In the engineering design of natural smoke and heat exhaust ventilation system (NSHEVS), the determination of the effective area of each NSHEV should be as accurate as possible based on the principle of reflecting the ability of smoke and heat exhaust equivalently. However, the definitions and determination methods of effective areas are quite different in different regions. In this study, the characteristics and differences of the determination methods from China, Japan, the UK, USA, Australia, and Europe are summarized and analyzed from the aspect of whether all the influence factors of smoke and heat exhaust ability are fully considered, namely geometric characteristic of the NSHEV, indoor fire environment, outdoor wind environment, and design conditions of NSHEVS. Some suggestions and future research directions are pointed out from the perspective of comprehensive assessment based on multiple influence factors.

1   INTRODUCTION

NSHEVS that utilizes thermal buoyancy to channel smoke, heat, and toxic gases into the open air is widely adopted as an essential part of any modern fire protection measures for assisting in the evacuation of buildings and reduction of fire damage and financial loss. The smoke control effect of NSHEVS, for maintaining a smoke-free layer above the floor and reducing smoke and heat build-up, mainly depends on an important design parameter, i.e., the overall effective area of the entire system. Therefore, the determination of the effective area for each independent NSHEV is also vital in design practice, according to which the type and total number of NSHEVs can be confirmed to achieve the overall effective area of NSHEVS.

On the system level, the effect of multiple factors on the smoke and heat exhaust process of NSHEVS has been studied, such as wind direction and speed, building geometry, the layout of air inlets, and location of fire outbreak (Król & Król 2017; Węgrzyński & Krajewski 2017; Węgrzyński & Lipecki 2018; Węgrzyński et al. 2018). As for the smoke exhaust effect and the effective area of a single NSHEV, most studies focused on the discharge coefficient of NSHEV and the effect of corresponding influence factors on the aerodynamic efficiency (Meessen 1993; Prahl & Emmons 1975; Steckler et al. 1984; Szikra & Gábor 2014; Węgrzyński et al. 2019). Some studies in the research area of traditional wind-driven cross-ventilation also provided use for reference to analyze the discharge coefficient and the definitions of the effective area in the research area of smoke control (Karava et al. 2004; Yi et al. 2018; 2019). However, there are relatively fewer studies regarding the differences in how to determine the effective area of NSHEV between the East and the West and the merits and shortcomings of existing determination methods. Therefore, a comparative analysis is conducted in this study to discuss the diversity and development tendency of determination methods from different regions.

---

*Corresponding Authors: bank2002yinhang@126.com, and tanlongfei@scfri.cn

DOI 10.1201/9781003330172-54

2 TECHNICAL KEYS TO DETERMINE THE EFFECTIVE AREA OF NSHEV

In order to determine the effective area of NSHEV precisely, three technical keys need to be achieved. First, the evaluation benchmark of effective area and the connotation of terminology should be unified for all kinds of NSHEVs. Up to now, the worldwide definitions of the effective area can be classified into two main categories. One is the geometric free area, which is represented by the passable area, in whole or part, of the flow path through the NSHEV. The other is the aerodynamic free area, which is defined as the product of the discharge coefficient and the opening area of NSHEV. Second, the effective area should be the area based on the effectiveness of NSHEV, which can reflect the ability of smoke and heat exhaust equivalently. However, the effectiveness of NSHEV is affected not only by the geometric characteristic of NSHEV in the traditional sense, but also by the indoor fire environment and the outdoor wind environment. Therefore, all the influence factors shall be considered when determining the effective area. Third, the determination of effective area shall be closely combined with the design conditions of NSHEVS. The effective area of the same NSHEV might be different in different design conditions.

3 METHODS TO DETERMINE THE EFFECTIVE AREA OF NSHEV

As shown in Figure 1, NSHEV can be categorized into four types according to its geometrical feature and opening way: flap type, louvre type, parallel-opening type and sliding type. The methods to determine the effective area of the above-mentioned NSHEVs can be categorized into three types: simple assessment, experimental assessment, and simple & experimental assessment.

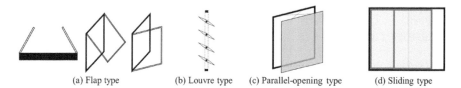

(a) Flap type   (b) Louvre type   (c) Parallel-opening type   (d) Sliding type

Figure 1.   Four typical types of NSHEV.

3.1 *Simple assessment*

Simple assessment means using the simple calculation formula to directly calculate the effective area of NSHEV without specific experimental testing. Take China as an example. According to the Chinese national standard of GB 51251, the effective areas of all kinds of NSHEV are obtained by simply calculating the geometric free area of NSHEV. For the flap type, the effective area of NSHEV is determined by calculating the projection area of the movable flap as follows:

$$A_e = \begin{cases} A_f \cdot sin\alpha, 0° < \alpha \leq 70° \\ A_f, \alpha > 70° \end{cases} \tag{1}$$

where $A_e[m^2]$, $A_f[m^2]$ and $\alpha[°]$ are the effective areas, the geometric area of the flap and the opening angle, respectively. The effective area of louvre-type NSHEV is determined by calculating the effective passable area as follows:

$$A_e = A_v \cdot \varepsilon \tag{2}$$

where $A_v[m^2]$ and $\varepsilon[-]$ are the area of the opening through the NSHEV and baffling coefficient of the louvre, respectively. The effective area of parallel-opening-type NSHEV is determined by calculating the area of the partial flow path through the NSHEV as follows:

$$A_e = \begin{cases} min\ (h \cdot L/2, A_v)\ , Roof\ mounted \\ min\ (h \cdot L/4, A_v)\ , Wall\ mounted \end{cases} \tag{3}$$

where L[m] and h[m] are the perimeter and moving distance of the movable part of the NSHEV, respectively. Finally, the effective area of sliding-type NSHEV is directly determined by calculating the opening area through the NSHEV.

The calculation system of Japan is similar to that of China, but the specific calculation methods are slightly different for some forms of NSHEVs. In the light of the handbook on design and construction of building equipment compiled by the Japan Conference of Building Administration, as shown in Figure 2, the effective areas of outward opening bottom-hung, inward opening bottom-hung, and inward opening center-hung flap-type NSHEVs and louvre-type NSHEV are calculated using the following equation:

$$A_e = \begin{cases} A_v \cdot \alpha/45°, 0° < \alpha \le 45° \\ A_v, 45° \le \alpha < 90° \end{cases} \quad (4)$$

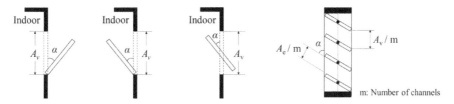

Figure 2.   Schematic diagram for the Japanese calculation methods.

Furthermore, as shown in Figure 3, in addition to directly referring to the methods from the European standard of EN 12101-2, the British standard of BS 9999 also provides some different simple methods to calculate the effective area of three kinds of flap-type NSHEVs used for the venting of fire-fighting shafts. The calculation methods corresponding to the bottom-hung NSHEV, bottom-hung NSHEV with overhang, and side-hung NSHEV are shown as follows:

$$A_e = \begin{cases} \min(A_1, A_v), Bottomhung \\ \min(A_1, A_v, A_2 + 2A_3), Bottomhung with overhang \\ \min(A_v, A_4 + A_5), Sidehung \end{cases} \quad (5)$$

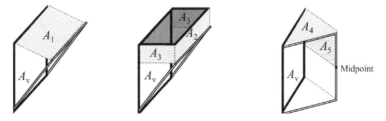

Figure 3.   Schematic diagram for the British calculation methods.

3.2 Experimental assessment

Experimental assessment means the effective area of NSHEV is obtained from the experimental testing. Generally, the aerodynamic free area is the goal achieved by testing the discharge coefficient of NSHEV. The equations to calculate the above-mentioned two parameters are shown, respectively, as follows:

$$A_e = A_v \cdot C_v \quad (6)$$

$$C_v = Q/A_v \cdot \sqrt{\rho/(2\Delta p)} \quad (7)$$

where C_v[-], Q[m³/s], ρ[kg/m³] and Δp[Pa] are the discharge coefficient, flow rate of air, air density, and pressure drop across the NSHEV.

The method specified in the Australian standard of AS 2428.5 is a typical procedure of experimental assessment. As shown in Figure 4, the test apparatus consists of four sections: air supply, flow measurement, pressure drop measurement, and specimen installation. The final value of the discharge coefficient is determined by averaging at least six sets of data corresponding to different flow rates of air. The percentages of error in the determination of the parameters on the right side of Equation (7) are also considered when calculating the discharge coefficient.

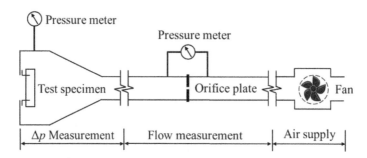

Figure 4.   Schematic diagram of the Australian test apparatus.

For certain countries, the various values of discharge coefficient can be used directly. Take the USA as an example. According to the NFPA standard of NFPA 204, the discharge coefficient of NSHEV can be obtained from the manufacturer or be taken from the data of some specific NSHEV types provided by the standard.

3.3   Simple & experimental assessment

Simple & Experimental Assessment is the procedure that combines simple assessment with experimental assessment. The most representative example is the method adopted by the European standard of EN 12101-2. In the procedure of simple assessment, for some specific roof-mounted NSHEVs, the reference values of discharge coefficient are suggested as a fixed value of 0.4; for two bottom-hung wall-mounted NSHEVs, the discharge coefficients for outward and inward opening and various opening angles are directly given in table form. In the procedure of experimental assessment, as shown in Figure 5, the test specimen is mounted onto a turntable located at the top surface of the settling chamber. The wind field acting on the specimen with different wind directions can be achieved by the side wind simulator and turntable.

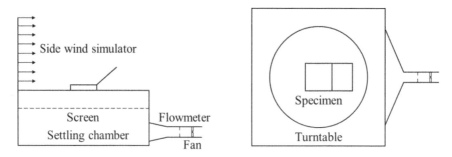

Figure 5.   Schematic diagram of the European test apparatus from the front and top view.

For the test scenarios without side wind, similarly to the Australian method, the calculated discharge coefficient should be the mean value of at least six sets of data. But the pressure drop across the NSHEV for each flow rate of the air shall be within the range of 3Pa ~ 12Pa. For the test scenarios with a side wind, at least six sets of tests shall be conducted to each wind direction under the premise of maintaining the pressure drop across the NSHEV between 0.005 ~ 0.15 times the dynamic pressure of side wind. Then the calculated discharge coefficient at a certain wind direction can be determined from the regression line as the value corresponding to the pressure drop across the NSHEV equal to 0.082 times the dynamic pressure of side wind. The final determined discharge coefficient is the minimum value for all conditions with and without side wind.

4 COMPARISON AND DISCUSSION

As shown in Table 1, the Asian countries, China and Japan, generally adopt a geometric free area to determine the effective area of NSHEV. The greatest advantage is that the process of determination is easy and convenient, but the property difference of diverse NSHEVs in local resistance cannot be fully reflected in the calculated value of the area. Moreover, the impacts of indoor fire environment and outdoor wind environment on the effectiveness of NSHEV are not considered at all. For the Western countries, the evaluating indicator, aerodynamic free area, is more scientific and reasonable because all the influence on the effectiveness of NSHEV can be reflected in the single parameter of discharge coefficient. The key point in the accurate determination of the discharge coefficient is whether or not all the influence factors and design conditions are considered sufficiently during the procedure of experimental assessment, as illustrated in Table 2.

Table 1.   The definitions and determination methods of effective area of NSHEV.

Regions	Flap type	Louvre type	Parallel-opening type	Sliding type
China		Geometric free area (Simple assessment)		
Japan		Geometric free area (Simple assessment)		
UK	Geometric free area or same as Europe		Same as Europe	
USA		Aerodynamic free area (Simple assessment)		
Australia		Aerodynamic free area (Experimental assessment)		
Europe		Aerodynamic free area (Simple & experimental assessment)		

Table 2.   The reflection degree of the impact of influence factors & combination degree with the design conditions of NSHEVS.

Regions	Impact of the geometric characteristic of NSHEV	Impact of the indoor fire environment	Impact of outdoor wind environment	Design conditions
China	Partly reflected	Not reflected	Not reflected	Uncombined
Japan	Partly reflected	Not reflected	Not reflected	Uncombined
UK	Partly reflected	Not reflected	Not reflected	Uncombined
USA	Basically reflected	Not reflected	Not reflected	Uncombined
Australia	Basically reflected	Partly reflected	Not reflected	Uncombined
Europe	Basically reflected	Partly reflected	Partly reflected	Uncombined

As for the reflection degree of the impact of the geometric characteristic of NSHEV, the specimens are mounted at the end of the horizontally arranged wind tunnel and onto the top surface of the settling chamber, respectively, for the Australian and European test apparatuses. The discrepancies in an opening way between top-open and side-open NSHEVs and between top-hung and bottom-hung flap-type NSHEVs are not reflected for both test apparatuses. All else being equal, the same calculation results of the experimental assessment will be obtained for the above-mentioned NSHEVs.

In the aspect of the reflection degree of the impact of the indoor fire environment, the gas medium adopted in the experimental assessment is the air under normal temperature driven by the mechanical power of the fan. The rising process of high-temperature smoke driven by buoyancy in the fire scene is not well simulated. It remains to be discussed whether the effectiveness of NSHEV can be accurately obtained under specific temperatures and pressure.

For the reflection degree of the impact of the outdoor wind environment, only the European test apparatus has the function of side wind simulation, but the supplied wind velocity is only a fixed value equal to or greater than 10 m/s. The complexity of the real wind field acting on the NSHEV in a mounted place is ignored. In addition, the tests under the condition with side wind are not required for the wall-mounted NSHEV during the experimental assessment.

In terms of the combination degree with the design conditions, none of the determination methods combines with the design conditions of NSHEVS. With respect to the Australian method and the European method for the test condition without side wind, the mean value of the test results under different flow rates of air is regarded as the final calculation result of the specimen. However, unlike the characteristic in the fire scene, the discharge coefficient of NSHEV is insensitive to the mass flow rates of air under normal temperature. In addition, as for the European method for the test condition with a side wind, the fitted value corresponding to the pressure drop across the NSHEV equal to 0.082 times the dynamic pressure of side wind is taken as the final determined value. Nevertheless, the ratio of pressure drop across the NSHEV to the dynamic pressure of side wind is not fixed under the interaction between indoor fire environment and outdoor wind environment for different design conditions of NSHEVS.

It can be seen from the above that no one determination method considers all the influence factors and design conditions sufficiently simultaneously. It is suggested to improve the existing methods from two aspects. The first thing is to improve the assessment system. For the test conditions, whether with side wind or without side wind, the test scenarios should combine closely with the design conditions of NSHEVS by ensuring the consistency of certain key parameters, such as the design pressure difference between the smoke layer and outdoor environment. In addition, for the test condition with a side wind, multiple sets of tests should be conducted corresponding to multiple simulated side winds less than the extreme design wind speed in the mounted place of NSHEV. The critical wind speed might be any value greater or less than 10 m/s. The next proposed thing is to develop a new test apparatus containing both the hot smoke generator and variable-speed side wind simulator. The relevant parameters required for the above-mentioned improved assessment system can be obtained by the newly constructed test apparatus with the comprehensive capability to simulate more influence factors. To sum up, the determination of the effective area of NSHEV should be a comprehensive assessment based on multi-factor coupling.

5   CONCLUSIONS

Any inaccurate or incorrect assessment of independent NSHEV could lead to the inadequate capacity of the NSHEVS, thereby threatening the fire escape. In order to achieve the above goal, the assessment should be conducted based on the same evaluation benchmark and take full account of all influence factors of natural smoke extraction and the design conditions of NSHEVS. However, according to the comparison of the definitions and determination methods of the effective area of NSHEV from China, Japan, the UK, USA, Australia, and Europe, so far, no one method could nicely make the effective area of NSHEV equivalently reflect the ability of smoke and heat exhaust. Asian countries generally adopt a geometric free area to simply determine the effective area of NSHEV, but only partly consider the impact of the geometric characteristic of NSHEV on the effectiveness of NSHEV. While European and American countries usually assess the aerodynamic free area of NSHEV by testing the discharge coefficient, the test procedure cannot simulate the influence of indoor fire and outdoor wind environments very well. A perfect determination method, tightly bound to the design conditions of NSHEVS, is expected to realize the multifactorial assessment of the effective area of NSHEV.

ACKNOWLEDGMENTS

This research was financially supported by the Research Program of the Fire and Rescue Department of the Ministry of Emergency Management (No.2018XFGG18) and the Central Public-interest Scientific Institution Basal Research Fund (NO.20218805Z).

REFERENCES

Karava P, Stathopoulos T, and Athienitis AK, "Wind-driven flow through openings – A review of discharge coefficients," *Int J Vent.*, vol. 3, pp. 255–266, 2004.

Król M and Król A, "Multi-criteria numerical analysis of factors influencing the efficiency of natural smoke venting of atria," *J Wind Eng Ind Aerod.*, vol. 170, pp. 149–161, 2017.

Meessen H, "The VdS-round-robin test. Testing-station specific influences on the aerodynamic performance of smoke ventilators," *J Wind Eng Ind Aerod.*, vol. 99, pp. 443–447, 1993.

Prahl J and Emmons HW, "Fire induced flow through an opening," *Combust Flame.*, vol. 25, pp. 369–385, 1975.

Steckler KD, Baum HR, and Quintiere JG, *"Fire induced flow through room opening – Flow coefficients,"* Proceedings of the Twentieth Symposium (International) on Combustion, Michigan, USA, pp. 1591–1600, 1984.

Szikra C, Gábor L, and Ph.D. T, *"Approach to define the aerodynamic free area for natural smoke vents in a CFD simulation environment,"* Proceedings of the Fire and Evacuation Modeling Technical Conference (FEMTC 2014), Gaithersburg, Maryland, pp. 1–6, 2014.

Węgrzyński W and Krajewski G, "Combined wind engineering, smoke flow and evacuation analysis for a design of a natural smoke and heat ventilation system," *Procedia Eng.*, vol.172, pp. 1243–1251, 2017.

Węgrzyński W and Lipecki T, "Wind and fire coupled modeling – Part I: Literature review," *Fire Technol.*, vol.54, pp. 1405–1442, 2018.

Węgrzyński W, Krajewski G, Suchy P, and Lipecki T, "The influence of roof obstacles on the performance of natural smoke ventilators in wind conditions," *J Wind Eng Ind Aerod.*, vol.189, pp. 266–275, 2019.

Węgrzyński W, Lipecki T, and Krajewski G, "Wind and fire coupled modelling – Part II: Good practice guidelines," *Fire Technol.*, vol.54, pp. 1443–1485, 2018.

Yi Q, Li H, Wang X, Zong C, and Zhang G, "Numerical investigation on the effects of building configuration on discharge coefficient for a cross-ventilated dairy building model," *Biosyst Eng.*, vol.182, pp. 107–122, 2019.

Yi Q, Zhang G, König M, Janke D, Hempel S, and Amon T, "Investigation of discharge coefficient for wind-driven naturally ventilated dairy barns," *Energ Buildings.*, vol.165, pp. 132–140, 2018.

Advances in Measurement Technology, Disaster Prevention and
Mitigation – Li & Mohd Yusof (Eds)
© 2023 The Author(s), ISBN: 978-1-032-36087-4

Research on the bearing capacity of box aluminum alloy joints using welded

Qiujun Ning & Jiawei Lu
School of Civil Engineering, Xi'an University of Architecture & Technology, Xi'an, People's Republic of China

Xiaosong Lu*
School of Civil and Architectural Engineering, Xi'an Technological University, Xi'an, People's Republic of China

ABSTRACT: The welded joint is one of the connections widely used in aluminum alloy structures. At present, there is little research on welded joints of aluminum alloy, which seriously hinders the development and application of aluminum alloy structures. Therefore, the paper studies the welded joints of aluminum alloy structures with actual site welding conditions. Four box-type aluminum alloy welded specimens are designed and tested with the static load. The force transmission components of the joints include vertical ribs and cover plates. The results show that the rib and cover plate can effectively increase the bearing capacity and ductility of the joint, and the increase of the vertical rib elongation also has a significant improvement on the capacity of the welded joint.

1 INTRODUCTION

Reinforced concrete structures and steel structures are the most common structural forms in civil engineering. However, steel surface corrosion, stress corrosion and corrosion fatigue not only cause appearance defects to the structure, but also will ultimately deteriorate the mechanical properties of the members and threaten the durability and service life of structures (Chang et al. 2018; Shi et al. 2012; Tan et al. 2021). Compared with other structural forms, the unique physical and mechanical properties of aluminum alloy (Al-Al) contribute to its success as constructional material. The low density of the material gives the advantage in weight reduction, relieves the burden on the lower supporting structure and contributes to seismic design. Corrosion resistance makes it unnecessary to protect aluminum alloy structures even in extreme environments (Mazzolani 2012; Sun et al. 2022). Thus, it can not only ensure structural bearing capacity and component strength, but also improve durability. The application of the Al-Al structure will play a vital role in the development of the civil engineering industry.

Al-Al structure belongs to the category of the metal structure, in principle the same as steel structure, but has its own characteristics. The research and application of Al-Al in Europe started early. In the course of nearly a hundred years of development, a large number of investigators have explored the constitutive relationship of materials, the force analysis of individual components, the joints on the global buckling capacity, the calculation of overall structures, and the fire protection design (Stein Hardt 1971). In 1939, Ramberg and Osgood proposed a classic constitutive relationship for Al-Al, i.e., the Ramberg-Osgood model (Ramberg & Osgood 1939). Subsequently, American scholars conducted a series of experimental studies on welded Al-Al axial compression

---

*Corresponding Author: hxsjylxs@126.com

    DOI 10.1201/9781003330172-55

components and obtained the calculation formula of the stability coefficient of welded Al-Al components (Brungraber & Clark 1962; Clark & Jom bock 1957; Gibson 1938). With the development of Al-Al materials, more scholars have invested in the research of Al-Al structure. Guo Xiaonong et al. conducted an axial compression test on 62 members of Al-Al 6061-T6. The experimental results showed that part of the instability was bending instability, and the other part was bending and torsional instability (Shen et al. 2007). Fan et al. established the FE models for single-layer latticed domes with semi-rigid joints and analyzed their buckling behavior (Fan et al. 2010).

Although Al-Al has been used as a building material in building structures, current research focuses on the strain-stress relationship of Al-Al and the bearing capacity of components. There were few results on the welding connections or joints of Al-Al. At present, the welded connection of Al-Al has been the important point which is the key technique. There is no perfect and equitable criterion for the welding processing technology. On the other side, HAZ is very evident for Al-Al. So, it is necessary to research the property of welded Al-Al connection. In this paper, the welded joints of Al-Al structures are studied with actual site welding conditions. Four box-type Al-Al welded specimens are designed and tested with a static load. The force transmission components of the joints include vertical ribs and cover plates.

2 EXPERIMENTAL PROGRAM

2.1 *Specimen description*

The cross-section size of the beam and column is 220×7mm, and the shape of the cross-section is a box. The parameters of the specimens are shown in Table1, and there are three types of specimens: basic specimen, the specimen with vertical rib and the specimen with a cover plate. The GJ-01 is the basic specimen, which is shown in Figure 1(a). GJ-02 and GJ-03 are test specimens formed by adding vertical ribs with a thickness of 7mm based on GJ-01. The outer elongation of the vertical ribs is 80mm and 120mm, respectively, as shown in Figure 1(b). One 7mm×150mm (thickness × length) cover plate is added to the upper beam flange of GJ-04 based on GJ-01, as shown in Figure 1(c). The specimens have been shaped by LGK-120T IGBT contravariant air plasma cutting machine and welded by MG-500 welding machine and SALMG-3 welding rob (Eur5356) with MIG technical.

Table 1. The parameter of specimens.

Specimens	The extend-length of vertical ribs (mm)	Cover plate length (mm)	Notes
GJ-01	—	—	basic specimen
GJ-02	80	—	with vertical rib
GJ-03	120	—	with vertical rib
GJ-04	—	150	with cover plate

2.2 *Material properties*

The property of Al-Al is 6061-T6 for testing, and test samples are made by "The rule of metallic materials–tensile testing: Method of test at room temperature" (GB/T228.1-2010) (National Standard of P.R.C. Metallic materials–tensile testing: Method of test at room temperature (GB/T228.1-2010). Beijing: China Standard Press, 2010 (in Chinese)). The material properties are obtained by the mono stretch test with DNS300 electronic universal testing machine. The plate thicknesses used for the material test are divided into 7.0 mm, and 10.0 mm, as shown in Figure 2 and the results in Table 2. The average values of yielding and tension strength are over the data in the code of China. It can be seen from Figure 3 that as the thickness of the specimen increases, the

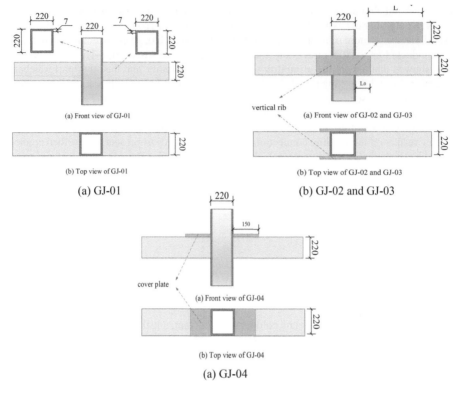

(a) GJ-01

(a) GJ-02 and GJ-03

(a) GJ-04

Figure 1. Dimensions of test specimens.

fracture of the specimen appears to be neck-down, which indicates that the 10mm thick specimen exhibits ductility. Figure 4 shows the stress-strain response of the sample.

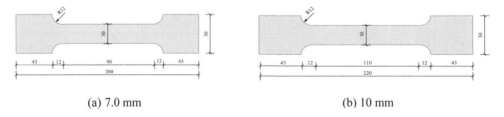

(a) 7.0 mm   (b) 10 mm

Figure 2. Schematic diagram of specimens used for the material test (unit: mm).

Table 2. Mechanics performance parameter of 6061-T6 Al-Al.

Specimen	Width (mm)	Thickness (mm)	Area (mm²)	Yield strength (MPa)	Tension strength (MPa)	Stretch ratio (%)
A-7mm	29.61	6.89	203.91	250.96	285.29	9.50
B-10mm	30.77	10.97	337.43	227.01	269.21	29.35

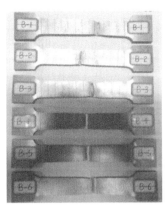

Figure 3.   Diagram of failure shape after stretching.

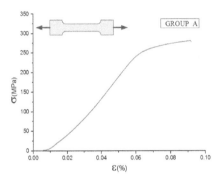
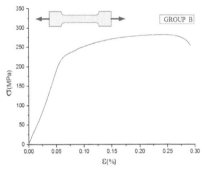

Figure 4.   Mono strain-stress relationship of 6061-T6 Al-Al.

2.3   *Test setup and loading protocol*

The concentrated force is loaded at the column top with a hydraulic jack, and the beam is pinned at each end. Simultaneously, in order to facilitate the description of the test phenomenon, the specimens are labeled as Beam-A and Beam-B. Figure 5 shows the schematic diagram of the test setup and load device.

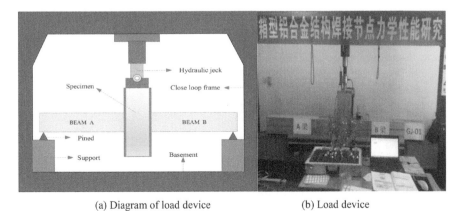

(a) Diagram of load device   (b) Load device

Figure 5.   Schematic diagram of the test setup and load device.

The preload will be on the specimen first. The preload is carried out in three stages, and each stage takes 20% of the standard load. In the formal loading, before reaching the standard load, the loading value of each level takes 20% of the standard load. When the load is added to 90% of the local buckling load, the load of each level is taken as 10% of the standard load until the specimen has a buckling failure. Table 3 shows the loading protocol applied in the experimental procedure.

Table 3. Load mechanism for the test (unit: kN).

Specimens	Load (kN)	Stage of loading					
		Preload			Formal load		
GJ-01	59.597	3-step	Step load 11.919	1~4-step	Step load 11.919	5~6-step	Step load 5.960
GJ-02	128.215	3-step	Step load 25.643	1~4-step	Step load 25.643	5~6-step	Step load 12.822
GJ-03	139.918	3-step	Step load 27.983	1~4-step	Step load 27.983	5~6-step	Step load 13.992
GJ-04	66.863	3-step	Step load 13.373	1~4-step	Step load 6.686	5~6-step	Step load 6.686

2.4 *Instrumentation*

Strain data is measured with the strain gauge and rosette at the key point of the specimen, and the displacement is measured with a displacement device. There are six displacement devices in the test (Figure 6): D1~D4 are set at the bottom of the beam surface in the zone of beam-column connection to survey the local displacement of Beam-A and Beam-B. The D5 is set at the top of the column to measure the vertical displacement at mid-span. D6 is set at the lateral side of the column to get the deformation data of the column. The strain rosette is stuck on the upper flange surface of A and B beams. The attachment position of the specimen strain gauge and the strain rosette is shown in Figure 7.

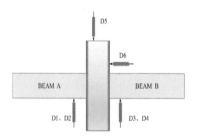

Figure 6. Displacement meter.

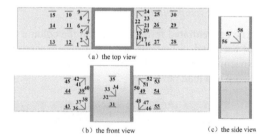

Figure 7. The strain gauge and rosette.

3 EXPERIMENTAL RESULTS AND DISCUSSIONS

3.1 *Failure mode and displacement-load curves*

Figure 12 shows the displacement-load curve of 4 beam-to-column connection specimens. During static loading tests, the experimental observations of 4 specimens, including deformation, crack, and failure mode, can be summarized as follows:

(1) The destruction processing of GJ-01

At the initial loading stage, the displacement-load curve is linear. When loading is 62.921 kN, displacement increases with no adding loading. At this point, the displacement-load curve appears to be an obvious inflection point, and the beam appears obvious buckling. As the loading continues, the displacement increases rapidly, and the load gradually decreases. Simultaneously, the welding of beam-column connection cracks at the Beam-B of A-B and B-A side (Figure 8).

 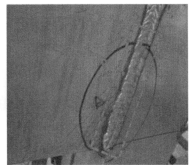

(a) Welding crack on the A-B side (b) Welding crack on the B-A side

Figure 8. Destruction process of GJ-01.

(2) The destruction processing of GJ-02

While the load reaches 124.733kN, displacement is increasing with no adding of loading. Meanwhile, deflection of the specimen is evident (Figure 9(a)). When the load continues, the vertical displacement continues to increase, but the load stabilizes at about 124kN. After the vertical displacement reached 38.58mm, the growth rate of the displacement increased, and the load gradually decreased. The weld seam cracks at the position where the vertical rib is connected to the Beam-B of the A-B side, and the crack length is 150mm (Figure 9(b)). Neck-down has occurred at the upside of the column (Figure 9(c)), and there is a bump on the downside of the column (Figure 9(d)). Meanwhile, the connection loses its capacity.

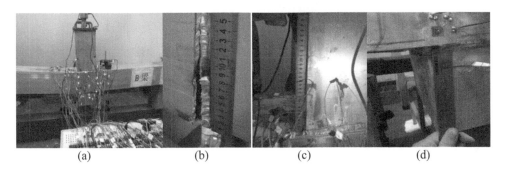

(a)   (b)   (c)   (d)

Figure 9. Destruction process of GJ-02.

(3) The destruction processing of GJ-03

Experimental observations were similar to specimen GJ-02. As illustrated in Figure 10, the failure of the specimen was a brittle mode. Compared with specimen GJ-02, the deformation and crack of the specimen were relatively decreased since the growth of the vertical ribbed plate increases the strength of the joint.

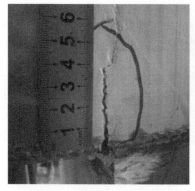 

(a) Crack of vertical rib   (b) Crack on the bottom of beam A

Figure 10. Destruction process of GJ-03.

(4) The destruction processing of GJ-04

As indicated in Figure 11, serious cracks occurred in the weld of the beam and column interface, tearing along the direction of the interface. The weld crack is at the edge where beam A of the A-B view is connected to the column crack, and the fracture length is 80mm (Figure 11(a)). Similarly, the welding crack at the edge where beam B of the A-B view is connected to the column, and the fracture length is 110mm (Figure 11(b)). The failure of the specimen was sudden and should be considered brittle.

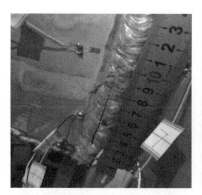 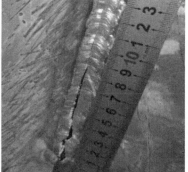

(a) Welding crack of beam A   (b) Welding crack of beam B

Figure 11. Destruction process of GJ-04.

Overall, the failure modes of the connections (GJ-01~GJ-04) were similar. In the connections, the deformation was concentrated in the weld, while no significant deformation occurred in the strengthening area. Experimental observations from the specimen GJ-02~GJ-04 demonstrated that the stress concentration of the weld was effectively avoided by using the strengthened connection. Moreover, the cracks appeared without any symptoms and developed rapidly, thereby causing the sudden failure of the specimen. Consequently, the failure mode of the connection was considered brittle.

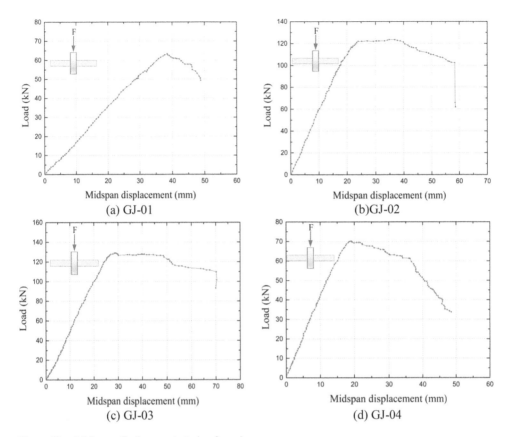

Figure 12.   Mid-span displacement-strain of specimen.

3.2   *Strain gauge readings*

In this test, by attaching resistance strain gauges to the key parts of the welding HAZ of the specimen, the strain change rule was observed to assist in the analysis of the whole process of the stress change of the specimen. Figure 13 shows the typical strain readings measured at different key locations of the specimens.

The curve in Figure 13(a) and Figure 13(b) is for the No.13 strain gauge of GJ-01 and No.28 strain gauge of GJ-02, respectively. Due to shear force, the tensile stress state of the flange of the beam is formed. However, the flanges on the far side of the joints have a relatively small strain, and the stress does not reach the yield. The curve shows that strain increases with the increase of load, the shape of the curves is almost linear. While up load to about 62 kN and 120 kN, respectively, a point of inflection appears on the curve, and the load and strain decrease sharply. The specimen is destroyed, and the load is over. In the test, the strain of the beam with ribbed joints increases, but it remains elastic.

The load-strain curve in Figure 13(c) is for the No.32 strain gauge of GJ-03, and the location of the gauge is in the horizontal direction on the A-B side vertical rib of GJ-03. It was mainly subjected to shear deformation, the stress did not reach yield, and the elastic strain recovered after unloading. The No.39 strain of GJ-04 yield and weld crack here. It can be seen from Figure 13(d) that the strain on the beam flange near the column side increases as it gets closer to the HAZ.

All specimens yield first in the welding HAZ of the joint region and then expand to all directions. All the specimens were pulled apart near the joint weld between the lower flange of the beam and the column, and all of them broke suddenly. This result is consistent with the actual earthquake damage and the test results of other scholars. The reason is that the stress of the lower flange of the beam is worse than that of the upper flange, and the weld quality is difficult to ensure. Moreover, after the occurrence of cracks, the stress concentration here is further intensified, leading to cracking soon.

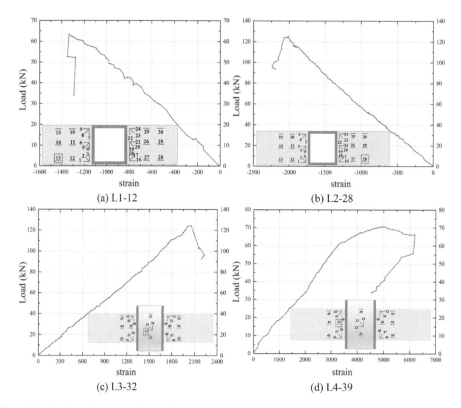

Figure 13.   Load-strain curve of a specimen.

4   CONCLUSION

With the processing and results analysis, the major conclusions are summarized as follows:

(1) The welded connection without any reinforcement will be damaged at the weld seam where the beam and column are connected. The specimens with vertical ribs were damaged on the vertical ribs at the beam-column junction, the upper column at the beam-column junction appeared to shrink, and the lower column expanded. The failure mode of the test specimen with the cover plate is similar to the basic specimen, but the width and length of the crack are more obvious.
(2) For the connection with vertical rib or cover plate, the capacity rose, and the yielding loading increased with the increased length of vertical rib. All specimens yield first in the welding HAZ of the joint region and then expand to all directions. All the specimens were pulled apart near the joint weld between the lower flange of the beam and the column and broke suddenly.
(3) The vertical or horizontal rib greatly influences the increasing capacity of connection, and the ultimate capacity of the connection is decided on the welding crack. So, the welding plate plays an important role in the capacity of the welded connection. In the future, the research can be on the size and quality of weld beads and the influence of HAZ.

REFERENCES

Brungraber R.J, Clark, J. W. Strength of welded aluminum columns. *ASCE Trans.* 1962, Part III.

Chang B, Hong X, Zeng J, et al. Fluid Flow Characteristics and Weld Formation Quality in Gas Tungsten Arc Welding of a Thick-Sheet Aluminium Alloy Structure by Varying Welding Position. *Applied Sciences*, 2018, 8(8): 1215.

Clark, J. W. and Jom bock, J.R. lateral buckling of I-beams subjected to unequal end moments, *J. eng. Mech. Div.* ASCE, EM3, 1957.

F. M. Mazzolani, 3D aluminum structures, *Thin-Walled Struct.* 61 (2012), 258–266.

Fan F, Ma HH, Cao ZZ, et al. Direct estimation of critical load for single-layer reticulated domes with semi-rigid joints. *Int J Space Struct* 2010; 25(1): 15–24.

Gibson, G. S. An approximate method for calculating the distortion of welded members, *Welded J.* 17NO. 7, July 1938.

National Standard of P.R.C. *Metallic materials–tensile testing: Method of test at room temperature* (GB/T228.1-2010). Beijing: China Standard Press,2010 (in Chinese).

Ramberg, W, Osgood, W. R. *Description of stress-strain curves by three parameters*, NACA Report 656, 1939.

Stein Hardt O. Aluminum constructions in civil engineering. *Aluminium.* 1971; 47: 131–9.

Sun J, Qu X, Gao C. Study on the Design Method of Ring Groove Rivet Joint in Aluminium Alloy Structure. *International Journal of Steel Structures*, 2022, 22(1): 294–307.

Tan Y, Zhang Y, Zhang Q, et al. Static properties and stability of super-long span aluminum alloy mega-latticed structures. *Structures*, 2021, 33(3): 3173–3187.

X. Shi, N. Xie, K. Fortune, J. Gong, Durability of steel-reinforced concrete in chloride environments: An overview, Constr. Build. *Mater.* 30 (2012) 125–138.

Z. Y. Shen, X. N. GUO, Y. Q. Li. Reliability analysis of aluminum alloy structural members. *Journal of Building Structures*, 2007, 28(06):147–152 (in Chinese).

Advances in Measurement Technology, Disaster Prevention and
Mitigation – Li & Mohd Yusof (Eds)
© 2023 The Author(s), ISBN: 978-1-032-36087-4

Comprehensive evaluation of operation reliability of IWCDSS for large water supply engineering based on cloud model

Xiaoying Liu*
Construction and Administration Bureau of South-to-North Water Diversion Middle Route Project of Henan Provence, Zhengzhou, China

Heng Zhang
Henan Water Conservancy Survey, LTD, Zhengzhou, China

Guanghui Li
North China University of Water Resources and Electric Power, Zhengzhou, China

ABSTRACT: The operation reliability of the intelligent water conservancy decision support system (IWCDSS) of large-scale water supply projects will directly affect the safety of the water supply. The operating environment and system composition of the IWCDSS will have a certain impact on its operational reliability. This article takes the reliable operation of the IWCDSS of large-scale water supply projects as the goal. It analyzes the factors affecting the operation reliability from the four aspects of the computer room environment, hardware, communication and network, and software, and establishes the operation reliability evaluation index system of the decision support system (DSS). The index system is used to construct a comprehensive evaluation model for the operation reliability of the IWCDSS of water supply engineering based on the cloud model. Taking an IWCDSS of a water supply project as an example, this paper verifies the feasibility of the comprehensive evaluation model, which provides a reference for the operation reliability evaluation of the IWCDSS of water supply projects.

1 INTRODUCTION

Water conservancy informatization and wisdom have become an important implementation path to promote the high-quality development of water conservancy in the new stage. At present, water conservancy informatization and intelligence have been applied to many aspects such as water resource scheduling, water condition monitoring and other aspects. The intelligent water conservancy decision support system (IWCDSS) of large-scale water supply projects uses modern communications, computers, big data mining, artificial intelligence and other technologies to realize functions such as water distribution, flow monitoring, water quality monitoring, remote control and emergency management. The operational reliability of IWCDSS of large-scale water supply projects will directly affect water supply security. Therefore, it is necessary to evaluate the operation reliability of the IWCDSS for water supply projects.

Scholars at home and abroad have conducted a lot of research on the reliable operation of water conservancy information systems. Liu Haiyan (Hai et al. 2014) improved the management mode of the water conservancy information management system through cloud computing and improved the operational efficiency of the system. SolheeKim (2017) et al. used the evaluation indicators of the network information system to evaluate the security of the integrated agricultural and water information system and identified the vulnerabilities in the system and obtained comprehensive evaluation results. Xu Lizhong (1998) calculated the integrity of the system by using the monitoring

---

*Corresponding Author: 1147578993@qq.com

DOI 10.1201/9781003330172-56

and evaluation model according to the operating state of the hydrological telemetry system to make a comprehensive estimation of the system operation. Chen Xing (2017) analyzed the factors that affected the safety of the automatic control system of water supply projects and proposed an optimal design for the operation of the system to improve the reliability of the system operation. Based on these studies, this article takes the reliable operation of the IWCDSS of large-scale water supply projects as the goal. It analyzes the factors affecting the operation reliability from the four aspects of the computer room environment, hardware, communication and network, and software, and establishes the operation reliability evaluation index system of DSS. The index system is used to construct a comprehensive evaluation model for the operation reliability of the IWCDSS of water supply engineering based on the cloud model. Taking an IWCDSS of a water supply project as an example, this paper verifies the feasibility of the comprehensive evaluation model and provides a reference for the operation reliability evaluation of the IWCDSS of water supply projects.

2 BUILD AN OPERATION RELIABILITY EVALUATION INDEX SYSTEM OF THE IWCDSS FOR LARGE-SCALE WATER SUPPLY PROJECTS

Cai Yang (2010) adopted a hierarchical system architecture and divided the water conservancy information system operation guarantee platform into four levels: monitoring layer, support layer, application layer and portal layer. Hou Zhaocheng (2010) divided the overall framework of the water conservancy automation system into five parts: application system, support platform, infrastructure, organizational management, and technical support according to the characteristics of project operation and management. Li Liu (2016) analyzed the risk factors affecting project operation scheduling and believed that the normal use of application software and hardware equipment and facilities is the key to the smooth operation of the automation system. According to the composition of IWCDSS of large-scale water supply projects, combined with on-site investigations, IWCDSS of large-scale water supply projects is divided into four parts: computer room environment, hardware, communication and network, and software. The impact factor of the computer room environment includes the computer room entity, generic cabling system (GCS) and air conditioning system. The impact factors of hardware include UPS system, computer system, cabinet system, early warning device and safety monitoring system. The impact factors of communication and network include optical cable and splice box, man-hand hole, network equipment, and network configuration. The impact factor of the software includes the operating system, database, middleware, and business software. These impact factors can be further divided according to the corresponding functional characteristics to form the state layer of the fuzzy evaluation model. The specific indicators are shown in Table 1.

Table 1. DSS Operational reliability evaluation index system.

Criterion layer	Factor layer	State layer
Computer room environment U_1	Computer room entity U_{11}	Dustproof in the computer room U_{111}. Esd prevention in the equipment room U_{112}. Irregular fire system U_{113}. Water or condensation in the equipment room U_{114}.
	Integrated wiring U_{12}	There is debris in the cable trench U_{121}. The wiring duct is not closed, and the threading pipe is damaged U_{122}. Irregular wiring U_{123}.
	Air conditioning system U_{13}	Temperature and humidity are not up to standard U_{131}. Air conditioning filter and air duct blocked U_{132}.
Hardware U_2	UPS system U_{21}	Battery pole climbing acid, leaking liquid U_{211}. The battery rack is unstable U_{212}. The maintenance channel is not insulated with insulation foot pads U_{213}. Pole and connection strip damage U_{214}.

(continued)

Table 1. Continued.

Criterion layer	Factor layer	State layer
	Computer system U_{22}	Aging of consumable parts U_{221}.Black screen U_{222}.Disk bad blocks or bad sectors U_{223}.The storage is invalid or full U_{224}.
	Cabinet system U_{23}	Insufficient space outside the cabinet U_{231}. The cabinet is not securely installed U_{232}. The ventilation fan is not receiving power U_{233}. Abnormal power indicator U_{234}.
	The early warning device U_{24}	Electronic fence failure U_{241}. The alarm system is manually turned off U_{242}. Monitoring equipment running abnormally U_{243}. Abnormal operation of access control equipment U_{244}.
	Safety monitoring system U_{25}	Poor battery switch contact U_{251}. Optical transceiver status is abnormal U_{252}. The MCU device is abnormal U_{253}. Abnormal display of flowmeter and liquid level gauge U_{254}.
Communication and network U_3	Cable and splice box U_{31}	Exposed fiber optic cable U_{311}. No sealing strips, door locks, and patent leather drops U_{312}. The optical cable monitoring index is not qualified U_{313}. No grounding or non-conforming grounding U_{314}.
	Man-hand hole U_{32}	There is water and foreign matter in the holeU_{321}. Brackets and trays damaged U_{322}. The protective layer of the optical cable and the splice box is corroded, damaged or deformed U_{323}.
	Network equipment U_{33}	The crystal head has a poor connection to the switch and information point interface U_{331}. Poor connection of the module interface U_{332}. Interrupted twisted-pair U_{333}. Network card failure U_{334}. Optical transceiver failure U_{335}.
	Network configuration U_{34}	The network card driver is missing U_{341}. IP address setting error U_{342}.Switch configuration failure U_{343}.Proxy server failure U_{344}.
Software U_4	Operating system U_{41}	Operating system crash U_{411}. CPU, memory exhausted U_{412}. Insufficient file system space U_{413}. External storage not recognized U_{414}.
	Database U_{42}	Listen port conflict U_{421}. Database lockout U_{422}. Bad file block U_{423}.
	Middleware U_{43}	The service process is invalid U_{431}. Application service dropped U_{432}. Configuration file error U_{433}.
	Business software U_{44}	Unable to open system platform executable U_{441}. Monitoring data is not available U_{442}. System platform data distortion U_{443}.

2.1 Multi-level fuzzy comprehensive evaluation model

The multi-level fuzzy comprehensive evaluation model is a widely used multi-objective and multi-factor decision-making technology based on fuzzy theory. It combines AHP and fuzzy evaluation to avoid the influence of subjective factors of evaluation information on the final result to the greatest extent and is suitable for the integration of evaluation subjects on multi-level and multi-index evaluation information (Zhang 2000). In view of the complexity and fuzziness of influencing factors of DSS operation reliability, this paper adopts a multi-level fuzzy comprehensive evaluation method to comprehensively evaluate the operational reliability of the system.

Let U be the set of influencing factors that affect the operational reliability of DSS with U={U1, U2,U3,U4} and U1={U11,U12,U13},..., U4 = {U41,U42,U43,U44},..., U11={U111,U112,U113, U114},..., U44 = {U441,U442,U443}.

When constructing the operation reliability evaluation index system of DSS, the operation reliability level can be used as a comment layer, which is represented by excellent V1, good V2, general V3, and poor V4, and V = {V1, V2, V3, V4}. If the state layer factor causes a very serious system failure, such as the breakdown of the chip or the motherboard due to static electricity in the automation equipment; the accumulation of water in the equipment room causes equipment damage or personal injury due to electric leakage; the sulfuric acid leakage inside the UPS battery corrodes the battery rack or causes fire; disk damage causes data loss, etc., the system operation reliability level is judged to be poor V4.

On the basis of the operation reliability evaluation index system of DSS, the weight and membership degree of each influencing factor are determined, and the final evaluation result is obtained through the transformation of the fuzzy relationship. According to the principle of maximum membership degree, we evaluate the system security level comprehensively, and the calculation is as follows:

$$S = W \otimes R \tag{1}$$

In the formula, S is the final evaluation result matrix, W is the weight vector of the factor layer, and R is the membership degree of the state layer to the factor layer.

3 IMPROVED FUZZY COMPREHENSIVE EVALUATION METHOD FOR CLOUD MODEL

The fuzzy evaluation model constructed in this paper can solve the ambiguity caused by the influencing factors of the operational reliability of DSS, but the randomness and uncertainty of the fuzzy model are ignored. The cloud model is a transformation model that realizes qualitative concepts and quantitative descriptions, and the randomness of fuzzy systems can be fully considered. Therefore, the cloud theory was introduced and improved accordingly to comprehensively evaluate the operational reliability of DSS. The specific implementation process is shown in Figure 1.

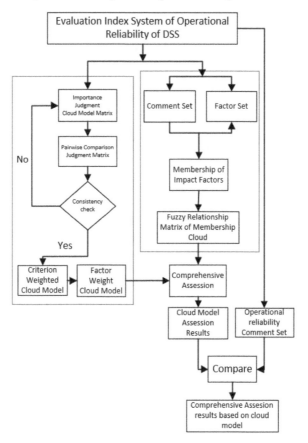

Figure 1.   Flow chart of a comprehensive evaluation for operational reliability of DSS.

3.1 Cloud model theory

Definition 1 (Li et al.1995): Suppose $U = \{x\}$ is a domain represented by exact numerical values, C is a qualitative concept of U. If a quantitative value $x \in U$, and x is a random realization of the qualitative concept, the membership degree $\mu(x) \in [0, 1]$ of x to C is a random number with stable tendency, for $\forall x \in U$ it has a corresponding $\mu(x)$. Then the distribution of x on the domain U is called the membership cloud, which is referred to as cloud for short.

The cloud model has three numerical features, which are respectively expressed as expected Ex, entropy En and super entropy He, where Ex represents the most typical sample point of the concept C homogenization; En represents the uncertainty of the qualitative concept C, reflecting the span of the cloud; He represents the uncertainty of entropy.

The cloud model can be established by cloud generators, and the more commonly used generators are forward cloud generators and reverse cloud generators. The forward cloud generator realizes the conversion from qualitative to quantitative and generates cloud droplets through the digital features of the obtained cloud model; the reverse cloud generator converts quantitative data into qualitative concepts.

3.2 Determination of comment set cloud model

Using the operational reliability rating of DSS as an evaluation layer can realize quantitative revealing of system operation reliability, so a forward cloud generator was adopted. From the above point of view, the evaluation level of system operation reliability V = (V, V2, V3, V4) can be quantified by experts based on experience to give the evaluation standard $[B_{min}, B_{max}]$. In the traditional fuzzy comprehensive evaluation method, the quantized intervals of different comments are often separated and lose intersection, thus ignoring the fuzziness generated in the quantization process (Li 2018). Therefore, this paper intersects the quantization intervals of different comments, fully considers the randomness and discreteness of the model, reduces the error caused by human factors, and improves the accuracy of the evaluation results. The calculation formula of digital features of the comment set cloud model is as follows:

$$\begin{cases} Ex = (B_{min} + B_{max})/2 \\ En = (B_{max} - B_{min})/6 \\ He = k \end{cases} \tag{2}$$

In the formula, B_{max} and B_{min} represent the maximum and minimum values of the evaluation value range, respectively. K is a constant.

3.3 Establishment of weight cloud model (cloud model scale)

In the Satty scale of classical AHP, a natural number between 1 and 9 is generally used to determine the relative importance of two factors (Deng et al. 2012). In order to avoid the deviation caused by human factors in the importance evaluation of impact factors, the cloud model can be introduced into the AHP and improved by constructing a judgment matrix based on the scale of the cloud model. The scale criteria of the cloud model are shown in Table 2.

Table 2. Scale criteria of cloud model of operation reliability impact factors for DSS.

Definition of importance		Scale cloud model C (Ex, En, He)
C_i is more important than C_j	Absolutely	$C_4(9,0.33,0.01)$
	Strong	$C_3(7,0.33,0.01)$
	Obviously	$C_2(5,0.33,0.01)$
	A little bit	$C_1(3,0.33,0.01)$
C_i and C_j are equally important.		$C(1,0,0)$

(continued)

Table 2. Continued.

Definition of importance		Scale cloud model C (Ex, En, He)
C_j is more important than C_i	Absolutely	$C_1(1/3, 0.33/9, 0.01/9)$
	Strong	$C_2(1/5, 0.33/25, 0.01/25)$
	Obviously	$C_3(1/7, 0.33/49, 0.01/49)$
	A little bit	$C_4(1/9, 0.33/81, 0.01/81)$

The judgment matrix is constructed by the cloud model scale, the root method is used to calculate the weight of influence factors of system operation reliability, and the weight cloud model is obtained. The calculation formula (Zhang et al. 2014) is as follows:

$$Ex_i^o = \frac{Ex_i}{\sum Ex_i} = \frac{\left(\prod_{j=1}^{n} Ex_{ij}\right)^{\frac{1}{n}}}{\sum_{i=1}^{n} \left(\prod_{j=1}^{n} Ex_{ij}\right)^{\frac{1}{n}}} \tag{3}$$

$$En_i^o = \frac{En_i}{\sum En_i} = \frac{\left(\prod_{j=1}^{n} Ex_{ij}\sqrt{\sum_{j=1}^{n}\left(\frac{En_{ij}}{Ex_{ij}}\right)^2}\right)^{\frac{1}{n}}}{\sum_{i=1}^{n}\left(\prod_{j=1}^{n} Ex_{ij}\sqrt{\sum_{j=1}^{n}\left(\frac{En_{ij}}{Ex_{ij}}\right)^2}\right)^{\frac{1}{n}}} \tag{4}$$

$$He_i^o = \frac{He_i}{\sum He_i} = \frac{\left(\prod_{j=1}^{n} Ex_{ij}\sqrt{\sum_{j=1}^{n}\left(\frac{He_{ij}}{Ex_{ij}}\right)^2}\right)^{\frac{1}{n}}}{\sum_{i=1}^{n}\left(\prod_{j=1}^{n} Ex_{ij}\sqrt{\sum_{j=1}^{n}\left(\frac{He_{ij}}{Ex_{ij}}\right)^2}\right)^{\frac{1}{n}}} \tag{5}$$

3.4 Establishment of the membership cloud model

The membership function of the influence factors of system operation reliability was solved by using the reverse cloud model, and the process is as follows:

(1) Calculate the expected value Ex:

$$Ex = \frac{1}{n} \cdot \sum_{i=1}^{n} x_i \tag{6}$$

(2) Calculate the entropy En:

$$En = \sqrt{\frac{\pi}{2}} \cdot \frac{1}{n} \cdot \sum_{i=1}^{n} |x_i - Ex| \tag{7}$$

(3) Calculate the excess entropy He:

$$He = \sqrt{\frac{1}{n-1} \cdot \sum_{i=1}^{n} (x_i - \bar{x})^2 - En^2} \tag{8}$$

The expectation, entropy and excess entropy of the corresponding cloud model are determined by using the status data of the influence factors of DSS operation reliability, so as to determine the corresponding membership cloud model.

4 CASE ANALYSIS

4.1 *Project overview*

A large water supply project decision support system includes a comprehensive office support system, communication transmission system, network data storage and safety management system, schedule in consultation with the physical environment, etc. The system uses the technology of modern communication, computer, information management and so on to build an advanced, practical, safe and reliable water conservancy data center, which integrates basic and global water conservancy information resources storage, management, sharing, exchange, publishing, application, service hosting, safety management, standards-setting, technical support and other functions. Thus, a standard and open water conservancy information infrastructure system and water conservancy information resource service window are formed, which can comprehensively improve the processing capacity of water dispatching and other businesses and give full play to the project benefits.

4.2 *Cloud model of operational reliability rating comment set for DSS*

Through the field investigation, experts determined the quantization interval of comment set according to experience, as the four grades of excellent [0, 4), good [2, 6), average [4, 8) and poor [6, 12). Formula (2) was used to calculate the corresponding cloud model parameters of the comment set, as shown in Table 3.

Table 3. Cloud model of operational reliability rating comment set for DSS.

Level	Excellent	Good	Average	Poor
Evaluation cloud model	(2.0,0.6667,0.1)	(4.0,0.6667,0.1)	(6.0,0.6667,0.1)	(9.0,1.0,0.1)

The forward cloud generator was used to generate the cloud map by using Python software to collect the digital features of the known comments cloud model, as shown in Figure 2.

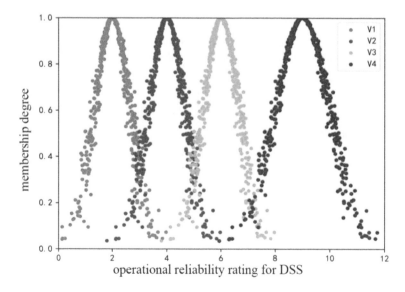

Figure 2. Cloud model of operational reliability rating comment set for DSS.

4.3 The influencing factors of the weighted cloud model of operational reliability of decision support system(IWDSS)

Experts, such as engineering and technical personnel, scholars, etc., according to the cloud model scaling criteria shown in Table 2, are hired to make a pairwise judgment on the importance of the criterion layer factors in the index system. By formulas (3)-(5), the judgment matrix is established. The above steps are repeated to determine the digital characteristics of the factor layer relative to the criterion layer. The comprehensive weight cloud model of the factor layer is derived based on the algebraic operation formula of the cloud model (Wang & Feng 2005), as shown in Table 4.

Table 4. The influencing factors of the weighted cloud model of operational reliability of decision support system (IWDSS).

Criterion	Numerical characteristics	Indicator	Numerical characteristics
Room environment U_1	(0.0725, 0.0663, 0.0663)	Physical Computer U_{11}	(0.0237, 0.0216, 0.0216)
		Comprehensive Wiring System U_{12}	(0.0272, 0.0262, 0.0262)
		Air Conditioning System U_{13}	(0.0216, 0.0185, 0.0185)
Hardware U_2	(0.4723, 0.4675, 0.4675)	UPS System U_{21}	(0.1171, 0.1138, 0.1138)
		Computer system U_{22}	(0.0620, 0.0636, 0.0636)
		Torque System U_{23}	(0.0586, 0.0592, 0.0592)
		Warning Device U_{24}	(0.0887, 0.0862, 0.0862)
		Security Monitoring System U_{25}	(0.1459, 0.1446, 0.1446)
Communications and Networking U_3	(0.1697, 0.1714, 0.1714)	Optical Cables and Transfer Boxes U_{31}	(0.0837, 0.0827, 0.0827)
		Man-hand Hole U_{32}	(0.0179, 0.0189, 0.0189)
		Network Equipment U_{33}	(0.0333, 0.0348, 0.0348)
		Network Configuration U_{34}	(0.0348, 0.0350, 0.0350)
Software U_4	(0.2854, 0.2947, 0.2947)	Operating System U_{41}	(0.1205, 0.1271, 0.1271)
		Database U_{42}	(0.0426, 0.0449, 0.0449)
		Middleware U_{43}	(0.0507, 0.0508, 0.0508)
		Service Software U_{44}	(0.0716, 0.0718, 0.0718)

4.4 The influencing factors of the membership cloud model of operational reliability of decision support system (IMDSS)

According to the specific characteristics of the status layer factors and their importance to the factor layer in Table 1, they are graded, and the status layer factors are assigned four grades: severe, heavy, general, and good, scored as 3, 2, 1 and 0 respectively. Five experts are hired to assign and score the influencing factors of the operational reliability of DSS for the water supply project. Using the reverse cloud generator to analyze and process the evaluation index data. The characteristic parameters of the cloud model of the operational reliability of the decision support system are calculated by formulas (6)–(8). The results are shown in Table 5.

415

Table 5. The influencing factors of the membership cloud model of operational reliability of decision support system (IMDSS).

Factor	Numerical characteristics	Factor	Numerical characteristics
Physical Computer U_{11}	(5.8000,0.8021,0.2379)	Optical Cables and Transfer Boxes U_{31}	(4.5000,1.0444,0.0958)
Comprehensive Wiring System U_{12}	(5.1667,1.1141,0.3543)	Man-hand Hole U_{32}	(4.6000,1.1029,0.2891)
Air Conditioning System U_{13}	(2.0000,0.7162,0.3921)	Network Equipment U_{33}	(1.8000,0.8021,0.2379)
UPS System U_{21}	(6.2000,0.8021,0.2379)	Network Configuration U_{34}	(1.7500,0.9400,0.1819)
Computer System U_{22}	(3.4000,1.1029,0.2891)	Operating System U_{41}	(1.0000,0.5013,0.4987)
Torque System U_{23}	(1.7500,0.4700,0.1706)	Database U_{42}	(0.8000,0.4011,0.1979)
Warning Device U_{24}	(2.8000,0.8021,0.2379)	Middleware U_{43}	(0.8200, 0.772, 0.2568)
Security Monitoring System U_{25}	(2.1667,0.6963,0.2861)	Service Software U_{44}	(1.4000,1.1029,0.2891)

4.5 *Comprehensive assessment*

According to the IWMSS and IMDSS and based on equation (1) and the basic operation criteria of the cloud model (Xu et al. 2014), we determine the cloud model of the comprehensive evaluation result of the operational reliability level of DSS. Its expectation, entropy, and hyperentropy are 2.8013, 0.8514, and 0.8258, respectively. Comparing the comprehensive assessment result cloud model (V) with the comment set cloud model (V1, V2, V3, V4), the results are shown in Figure 3.

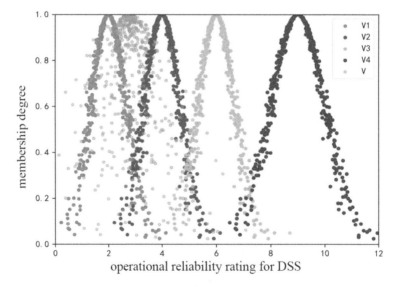

Figure 3. Cloud and cloud model scale for comprehensive evaluation of operational reliability grade of DSS.

From Figure 3, it can be seen that the comprehensive assessment result cloud model (V) is located between V1 and V2. Cloud droplets are more concentrated at the top. It shows that the deviation between the actual state of the system reliability and the estimated value of the evaluation is small, and the reliability is high.

5 CONCLUSION

This paper introduces cloud theory into the fuzzy model, establishes a judgment matrix according to cloud model scaling criteria, and then determines the weight cloud model of the impact factor. This can avoid the bias caused by excessive human subjective arbitrariness. The membership function of the cloud model is used to determine the membership degree of the influencing factors, which reduces the uncertainty and randomness. Python is used to visualize and visualize the final evaluation results. This method improves the reliability of the assessment results and provides a new idea for the comprehensive assessment of the operational reliability of the DSS.

This paper applies the proposed cloud model and multi-level fuzzy comprehensive assessment method to the operational safety and reliability assessment of intelligent water conservancy DSS for large water supply projects. The operational reliability level of DSS is calculated. The operational reliability level of DSS is between excellent and good, and cloud droplets are more concentrated at the top. It shows that the evaluation results are more reliable and more consistent with the actual situation, which verifies the feasibility of the constructed model.

REFERENCES

Cai, Y. Zhou, W.X. Zhan, Q.Z.(2010) Research and Application of Water Conservancy Information System Operation Guarantee Platform. *China Water Resources*, (03): 27–29.

Chen, X.(2017) Study on Safety and Reliability of Automatic Control System for Water Supply Project, Housing and Real Estate. *Housing and Real Estate*, (24): 285+292.

Deng, X. Li, J.M. Zeng, H.J.(2012) Research on Computation Methods of AHP Wight Vector and Its Applications. *Mathematics in Practice and Theory*, 42(07): 93–100.

Hai, Y. Liu, X.M. Liu, Z.W.(2014) Study on Water Conservancy Information System Based on Cloud Computing. *Applied Mechanics and Materials*, 3147.

Hou, Z.C. Zhai, Y.F.(2010)General Framework Design of Automatic Dispatching System for Mid-route of the South-to-north Water Diversion Project. *Water Resources Informatization*, (02): 40–45

Li, D.Y. Meng, H.J. Shi, X.M.(1995) Membership Clouds and Membership Cloud Generators. *Journal of Computer Research and Development*, (06): 15–20.

Li, L.(2016) Automatic Management of Dispatching Operation of the Middle Route Project of South-to-North Water Transfer. *Hebei Water Resources*, (06):43.

Li, S.Q.(2018) Safety Evaluation in Flight Evasion Based on Cloud-Model Fuzzy Comprehensive Assessment. *Electronics Optics & Control*, 25(12): 84–89.

Solhee, K. Chamwoo, K. Jung, C.(2017) Development of Evaluation Indicators for Web-based Agricultural Water Information System Using Mandal-Art Method. *Journal of Korean Society of Rural Planning*, 23(4), 49–59.

Wang, H.L. Feng, Y.Q.(2005) *Improved AHP Based on Judgment Matric Scaled with Cloud Model*. Proceedings of the Seventh National Member Congress and the Seventh Annual Conference of China Management Science (01):6.

Xu, L.Z.(1998) A Method and Model of Monitoring System Integrity Based on Information Fusion. *Journal of Hohai University*, (03): 33–36.

Xu, Z.J. Zhang, Y.P. Su, H.S.(2014) Application of Risk Assessment on Fuzzy Comprehensive Evaluation Method Based on the Cloud Model. *Journal of Safety and Environment*, 14(02): 69–72.

Zhang, J.J.(2000) Fuzzy Analytical Hierarchy Process. *Fuzzy Systems and Mathematics*, (02): 80–88.

Zhang, Q.W. Zhang, Y.Z. Zhong, W.(2014) A Cloud Model-Based Approach for Multi-hierarchy Fuzzy Comprehensive Evaluation of Reservoir-induced Seismic Risk. *Journal of Hydraulic Engineering*, 45(01): 87–95.

Advances in Measurement Technology, Disaster Prevention and
Mitigation – Li & Mohd Yusof (Eds)
© 2023 The Author(s), ISBN: 978-1-032-36087-4

Study on self-healing properties of reactive powder concrete with nano SiO$_2$

Tao Xu, Xiaofeng Liu & Shujie Liu
Anhui Sijian Holding Group Co., LTD. Hefei, Anhui, China

Mengjun Han
Anhui Construction Engineering Group Co. LTD. Hefei, Anhui, China

Peibao Xu*
Anhui Sijian Holding Group Co., LTD. Hefei, Anhui, China
Department of Civil Engineering, Anhui Jianzhu University, Hefei, Anhui, China

ABSTRACT: Improving the self-healing properties of concrete is of great significance for extend-ing the service life of buildings and avoiding catastrophic accidents. In this paper, concrete samples with doping amounts of 1%, 1.5% and 2% were prepared by incorporating nano-silica dioxide (SiO$_2$) reactive powder, and the influence of nano SiO$_2$ content on concrete self-healing properties were explored. Experimental results show that the nano SiO$_2$ can effectively improve the ultimate compressive strength and splitting strength of concrete. Meanwhile, the nano SiO$_2$ improves the recovery rate of the compressive and splitting strength of the preloaded concrete, and there exists the optimal nano SiO$_2$ content for the maximum recovery rate of the concrete compressive strength and the splitting strength. In addition, experiments found that the strength recovery rate of concrete decreases with the increase of the preload, which results in decreasing recovery rate of compres-sive strength and splitting strength. These test results can provide guidance for optimizing the self-healing properties of concrete.

1   INTRODUCTION

Concrete is a typical quasi-brittle material widely used in major projects such as high-rise buildings, nuclear power projects, hydraulic projects, ports and wharves [1,2]. It is prone to produce cracks or local damage under dynamic disturbance, which reduces the load-bearing capacity, durability, and waterproofness of the structure, thus shortening the service life of the building and even causing catastrophic accidents [3–6]. Currently, most structural cracking problems are solved by regular maintenance and post-repair. However, this kind of repair is mostly aimed at the macro cracks, and the repair site is generally a vulnerable part of concrete. Moreover, it is easy to produce secondary cracking due to the limitation of repair materials [7,8]. When the self-healing concrete structure is adopted, the concrete cracks formed by external forces or during use can self-heal at the early stage of cracking, and this characteristic greatly improves the service life of the structure. Therefore, research on improving the self-healing performance of concrete has attached the attention of a large number of science researchers [9–15].

Currently, self-healing concrete is mainly divided into natural self-healing concrete and engi-neering self-healing concrete [16–27]. Natural self-healing refers to the process in which concrete is damaged with unhydrated cementitious materials in the damaged part, and these cementitious materials further react to form new reaction products to bridge the cracks [16–18]. Engineering

---

*Corresponding Author: peibaoxu@ahjzu.edu.cn

                                                   DOI 10.1201/9781003330172-57

self-healing is to improve the self-healing ability of concrete by using some engineering technologies [19,20]. At present, the commonly used technologies for engineering self-healing concrete are built-in fiber, microbial or bionic technologies, respectively, where the physical and chemical properties of fibers [21–24], microbial activity and alkali resistance [25–27] have a great impact on the self-healing performance.

Recently, nanofillers have been used to prepare composite reactive powder concrete to enhance its self-healing performance [28–30]. Generally, nanofillers have nucleation and water absorption effects. Therefore, they can adsorb water to promote further hydration and act as nucleation sites for these hydration products, which accelerates the hydration process and promotes the self-healing of concrete [31]. At the same time, nanofillers significantly fill the large particle gap between cement particles, quartz powder, quartz sand, and volcanic ash particles, forming a highly filled matrix [32]. In addition, many types of nanofillers, such as nano SiO_2(NS) and nano clays, are highly volcanic ash reactive; thus, they are allowed to form additional C-S-H gels [33].

The above studies have proposed that nanofillers can enhance self-healing performance, but the current quantitative studies on the improvement of self-healing performance are insufficient. In this paper, nano active powder concretes (NRPC) with the NS content of 1%, 1.5% and 2% were prepared by adding NS into concrete. We further experimentally studied its compressive strength and splitting strength before and after damage, and mainly explored the effect of NS content on the self-healing performance of NRPC by systematically studying the strength recovery rate of NRPC. The paper is organized as follows. In Section 2, the experimental raw materials and specimen preparation are introduced, and the experimental approaches are described in detail. Then, the experimental results are presented and discussed in detail in Section 3. Finally, the whole paper is summarized.

2   EXPERIMENT

2.1   *Raw materials and specimen preparation*

The raw materials used in this paper mainly include the P.O42.5 grade ordinary Portland cement, F class I grade fly ash, sand with fineness modulus 2.7, 5–31.5mm continuous graded natural crushed stones, S95 slag powder, UEA concrete expansion agent and 20nm NS. The mix ratio of the specimens is formulated in accordance with the design regulations of JGJ 55–2011, and three contents of NS are added, as shown in Table 1. The size of the NRPC block is 150×150×150mm, as shown in Figure 1.

Table 1.   NRPC mix ratio.

Material name	Cement	Fly ash	Slag powder	Water	Fine aggregate	Coarse aggregate	Expansion agent	NS
Kg/m³	260	26	39	160	680	1200	26	4.9,6.5,8.1
Proportion	0.80	0.08	0.12	0.49	2.09	3.69	0.08	1.5%,2%,2.5%

Figure 1.   NRPC sample with NS.

2.2 Test equipment and methods

In this experiment, the strength recovery rate of NRPC is selected as an index to study the effect of NS on the self-healing properties of concrete. Following test regulations DL/T 5150-2017, the test steps are as follows. 1) We take out the test block after curing for 28 days for the strength test. We first take a group of test blocks to measure their ultimate compressive strength f_c. The remaining two groups of test blocks are preloaded to $\beta = 60\% f_c$ and $\beta = 80\% f_c$, respectively. 2) After preloaded, samples are put into the water at about 20°C for 28 days. 3) We measure the ultimate strength of preloaded concrete.

The compressive strength is

$$f_c = \frac{F}{A},$$
(1)

where F is the ultimate load and A is the bearing area of the specimen.

The splitting strength f_t is

$$f_t = \frac{2F}{\pi A}.$$
(2)

The strength recovery rate K_{iR} is defined as

$$K_{iR} = \frac{f_{iR}}{f_{i0}} \times 100\%,$$
(3)

where f_{iR} is the recovery strength of the concrete after preloaded, f_{i0} is the ultimate strength of concrete without NS, $i = c$ stands for compression and $i = t$ stands for tension.

3 RESULTS AND DISCUSSION

3.1 Compressive and splitting strength of nano concrete

Figure 2 shows the effect of NS on the compressive strength and splitting strength of concrete. It can be seen from the figure that the addition of NS significantly improves the compressive strength and splitting strength of concrete. This is because NS has a nucleation effect and water absorption effect, which promotes the hydration process and improves the strength of concrete. The different content of NS has different effects on the compressive and splitting strength of concrete. For the compressive strength, the optimum content of NS is 2%, and the corresponding strength is increased by 8.58%. For the splitting strength, the optimum content of NS is 1.5%, and the corresponding strength is increased by 27.2%.

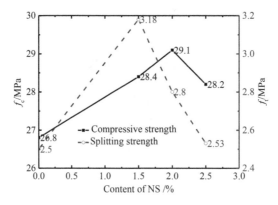

Figure 2. Influence of NS on the compressive strength and splitting strength of concrete.

Figure 3 shows the effect of NS on the compressive strength of concrete with different preloads. It can be seen from the figure that the NS promotes the recovery of the compressive strength of concrete. The compressive strengths of concretes with preload of 60% f_c and 80% f_c have been well recovered, and the compressive strength gradually decreases with the increase of preload. This is because the increase of preload leads to more serious damage to concrete, which is not conducive to strength recovery. When the content of NS is 1.5%, the compressive strength is the highest, reaching 36.4 MPa. In addition, for the fully damaged concrete samples, the strength can also be recovered to a certain extent after curing in water for 28 days. In general, when local damage occurs under the action of preload, the effect of improving its compressive strength is the best when the content of NS is 1.5%.

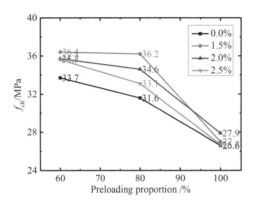

Figure 3.   Influence of NS on compressive strength of concrete with different preloads.

Figure 4 shows the effect of NS on the splitting strength of concrete with different preloads. It can be concluded that the addition of NS promotes the recovery of the splitting strength of concrete. The splitting strengths of preload 60% f_c and 80% f_c have been well recovered, and the splitting strength gradually decreases with the increase of preload. When the content of NS is 1.5%, the splitting strength is the highest, reaching 3.36 MPa. For the fully damaged concrete samples, the strength has also been recovered to a certain extent after curing in water for 28 days. Generally, when local damage occurs under the action of preload, the effect of NS on the splitting strength is optimal when the content of NS is 1.5%.

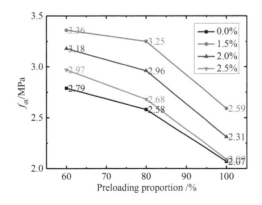

Figure 4.   Influence of NS on splitting strength of concrete with different preloads.

3.2 Strength recovery rate of nano concrete

Figure 5 shows the effect of NS on the recovery rate of concrete compressive strength. It can be seen from the figure that the addition of NS can better promote the recovery rate of compressive strength of concrete. With the increase of NS content, the compressive strength of concrete increases first and then decreases, indicating an optimal NS content. For the same preload, the strength recovery rate of concrete without NS is always less than that of concrete with NS. Among them, the optimal NS content for the largest strength recovery of concrete is 1.5%. For the concrete with preload 60% f_c, the strength recovery rate is the highest, reaching 135.8%. The strength recovery rate of concrete without NS is the lowest, which is 104.1%. This result means that the more serious the concrete damage, the lower the recovery rate of compressive strength.

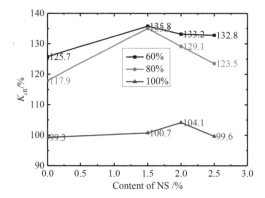

Figure 5. Influence of NS on the compressive strength recovery rate of concrete.

Figure 6 shows the effect of NS on the recovery rate of splitting strength. It can be seen from the figure that the addition of NS can better promote the recovery rate of splitting strength of concrete. With the increase of NS content, the splitting strength of concrete increases first and then decreases, which indicates that there is an optimal content of NS. For the given preload, the strength recovery rate of concrete without NS is less than that of concrete with NS. Among them, the NS content, which corresponds to the largest strength recovery rate of concrete, is 1.5%. For the concrete with preload 60% f_c, the strength recovery rate is the highest, reaching 134.4%. Meanwhile, the strength recovery rate of concrete without NS is the lowest, which is 103.6%. It can be concluded that the more serious the concrete damage, the less the recovery of splitting strength.

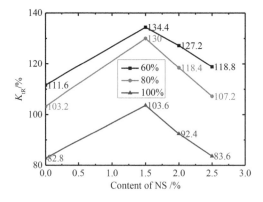

Figure 6. Influence of NS on splitting strength recovery rate of concrete.

4 CONCLUSION

In this paper, the influence of NS content on self-healing properties of concrete was mainly explored, and the recovery rates of compression strength and splitting strength of concrete with different content of NS reactive powder were experimentally studied. Experimental results show that the addition of NS improves the ultimate compressive strength and splitting strength of concrete. Meanwhile, the NS improves the recovery rate of compressive and splitting strength of preloaded concrete, and when the strength of preload is 60% f_c and 80% f_c, the recovery rate of 1.5% NS concrete compressive and splitting strength is the largest. In the case of the same content of NS, the strength recovery rate of concrete decreases with the increase of preload, which indicates that the more serious the damage to the concrete, the more unfavorable it is to the recovery of compressive strength and splitting strength. The experimental results can provide a reference for the application of nanomaterials in self-healing concrete.

ACKNOWLEDGMENTS

This work was financially supported by the Ministry of Housing and Urban-Rural Development (2019-K-060), Anhui Sijian Holding Group Co. LTD (HYB20190128) and Anhui Construction Engineering Group Co. LTD (HYB20210150).

REFERENCES

[1] Basheer, L., Kropp, J., Cleland, D.J. (2001) *Assessment of the Durability of Concrete from its Permeation Properties: A Review*. pp. 93–103.
[2] Bhaskar, S., Anwar Hossain, K.M., Lachemi, M., Wolfaardt, G., Otini Kroukamp, M. (2017) Effect of self-healing on strength and durability of zeolite-immobilized bacterial cementitious mortar composites. *Cement Concr. Compos.*, 82: 23–33.
[3] De Rooij, M., Tittelboom, K.Van., Belie, N.De., Schlangen, E. (2013) Self-healing phenomena in cement-Based materials: state-of-the-art report of RILEM technical committee 221-SHC: *self-healing phenomena in cement-based materials* (Vol. 11): Springer.
[4] Du, F., Jin, Z., She, W., Xiong, C., Feng, G., Fan, J. (2020) Chloride ions migration and induced reinforcement corrosion in concrete with cracks: a comparative study of current acceleration and natural marine exposure. *Construct. Build. Mater.*, 263: 120–099.
[5] Gardner, D., Isaacs, B., Lark, R., Joseph, C., Jefferson, A.D. (2010) Experimental investigation of adhesive-based self-healing of cementitious materials. *Mag. Concr. Res.*, 62: 831–843.
[6] Gupta, S., Pang, S.D., Kua, H.W. (2017) Autonomous healing in concrete by bio-based healing agents – a review. *Construct. Build. Mater.*, 146: 419–428.
[7] Homma, D., Mihashi, H., Nishiwaki, T. (2009) Self-healing capability of fiber-reinforced cementitious composites. *J. Adv. Concr. Technol.*, 7: 217–228.
[8] Huang, H., Ye, G., Shui, Z. (2014) Feasibility of self-healing in cementitious materials using capsules or a vascular system. *Construct. Build. Mater.*, 63: 108–118.
[9] Huang, Hao. Liang., Guang, Y.E. (2014) *A review on self-healing in reinforced concrete structures in view of serving conditions, 3rd International Conference on Service Life Design for Infrastructure*. Zhuhai. China., pp:1–14.
[10] Jiang, Z., Li, W., Yuan, Z. (2015) Influence of mineral additives and environmental conditions on the self-healing capabilities of cementitious materials. *Cement Concr. Compos.*, 57: 116–127.
[11] Jonkers, H.M., Thijssen, A., Muyzer, G., Copuroglu, O., Schlangen, E. (2010) Application of bacteria as self-healing agent for the development of sustainable concrete. *Ecol. Eng.*, 36: 230–235.
[12] Jonkers, H.M., Thijssen, A.G., Muyzer, O., Copuroglu, E., Schlangen. (2010) Application of bacteria as self-healing agent for the development of sustainable concrete. *Ecol. Eng.*, 36: 230–235.
[13] Kayondo, M., Combrinck, R., Boshoff, W.P. (2019) State-of-the-art review on plastic cracking of concrete. *Construct. Build. Mater.*, 225: 886–899.

[14] Keshavarzian, F., Saberian, M., and Li, J. (2021) Investigation on mechanical properties of steel fiber reinforced reactive powder concrete containing nano-SiO2: An experimental and analytical study. *Journal of Building Engineering.*, 44. pp.102–601.

[15] Li, K., Li, L. (2019) Crack-altered durability properties and performance of structural concretes. *Cement Concr. Res.*,124: 1–11.

[16] Li, Z., Di, S. (2017) The microstructure and wear resistance of microarc oxidation composite coatings containing nano-hexagonal boron nitride (HBN) particles. *J Mater Eng Perform.*, 26(4): 1–11.

[17] Liu, H., Huang, H., Wu, X., Peng, H., Li, Z., Hu, J., Yu, Q. (2019) Effects of external multi-ions and wet-dry cycles in a marine environment on autogenous self-healing of cracks in cement paste. *Cement Concr. Res.*, 120: 198–206.

[18] Mayhoub, O.A., Nasr, E.S.A., Ali, Y.A. and Kohail, M. (2021) The influence of ingredients on the properties of reactive powder concrete: A review. *Ain Shams Engineering Journal.*,12(1), pp.145–158.

[19] Mihashi, H., Nishiwaki, T. (2012) Development of engineered self-healing and self-repairing concrete-state-of-the-art report. *J. Adv. Concr. Technol.*, 10: 170–184.

[20] Nishiwaki, T., Kwon, S., Homma, D., Yamada, M., Mihashi, H. (2014) Self-healing capability of fiber-reinforced cementitious composites for recovery of water tightness and mechanical properties. *Materials.*, 7: 2141–2154.

[21] Palin, D., Wiktor, V., Jonkers, H.M. (2015) Autogenous healing of marine exposed concrete: characterization and quantification through visual crack closure. *Cement Concr. Res.*, 73: 17–24.

[22] Qian, S.Z., Zhou, J., Schlangen, E. (2010) Cement & concrete composites influence of curing condition and pre-cracking time on the self-healing behavior of engineered cementitious composites. *Cem. Concr. Compos.*, 32 (9): 686–693.

[23] Qiu, J., Tan, H.S., Yang, E.H. (2016) Coupled effects of crack width, slag content, and conditioning alkalinity on autogenous healing of engineered cementitious composites. *Cement Concr. Compos.*, 73: 203–212.

[24] Reinhardt, H.W., Jooss, M. (2003) Permeability and self-healing of cracked concrete as a function of temperature and crack width. *Cement Concr. Res.*, 33: 981–985.

[25] Ruan, S., Qiu, J., Weng, Y., Yang, Y., Yang, E.H., Chu, J., Unluer, C. (2019) The use of microbial-induced carbonate precipitation in healing cracks within reactive magnesia cement-based blends. *Cement Concr. Res.*, 115: 176–188.

[26] Ruan, Y., Han, B., Yu, X., et al. (2018) Mechanical Behaviors of Nano-Zirconia Filled Reactive Powder Concrete Under Compression and Flexure. *Construction and Building Materials.*, 162: 663–673.

[27] Wang, J.Y., Soens, H., Verstraete, W., De Belie, N. (2014) Self-healing concrete by use of microencapsulated bacterial spores. *Cem. Concr. Res.*, 56: 139–152.

[28] Wiktor, V., Jonkers, H.M. (2011) Quantification of crack-healing in novel bacteria-based self-healing concrete. *Cement Concr. Compos.*, 33: 763–770.

[29] Wiktor, V., Jonkers, H.M. (2016) Bacteria-based concrete: from concept to market. *Smart Mater. Struct.*, 25: 084006.

[30] Xu, J., Deng, H., Shen, X. (2014) Safety of moxibustion: a systematic review of case reports. *Evidence-based Complementary and Alternative Medicine.*, 1: 783–704.

[31] Zhang, L., Ma, N., Wang, Y., Han, B., Cui, X., Yu, X., Ou, J. (2016) Study on the reinforcing mechanisms of nano-silica to cement-based materials with theoretical calculation and experimental evidence. *Journal of Composite Materials.* 50(29)., pp.4135–4146.

[32] Zhang, W., Han, B., Yu, X., Ruan, Y., Ou, J. (2018) Nano boron nitride modified reactive powder concrete. *Constr Build Mater.*, 179: 186–197

[33] Zhu, H., Hu, Y., Li, Q., Ma, R. (2020) Restrained cracking failure behavior of concrete due to temperature and shrinkage. *Construct. Build. Mater.*, 244: 118–318.

Advances in Measurement Technology, Disaster Prevention and Mitigation – Li & Mohd Yusof (Eds)
© 2023 The Author(s), ISBN: 978-1-032-36087-4

In-place evaluation of resilient modulus of roadbed with iron tailings materials using portable falling weight deflectometer

Zhibin Wang
Hehei Xiongan Jingde Expressway Limited Liability Company, Baoding, China

Ting Li & Guangqing Yang
School of Civil Engineering, Shijiazhuang Tiedao University, Shijiazhuang, Hebei, China

Yunfei Zhao*
School of Urban Geology and Engineering, Hebei GEO University, Shijiazhuang, China

ABSTRACT: As an inspection method for roadbed compaction quality, the portable falling weight deflectometer (PFWD) has many advantages, such as easy operation and time-saving. This paper investigates the resilient modulus of iron tailings roadbed using PFWD and discusses the relationship between resilient modulus and differential settlement. The results show that the resilient modulus of roadbed with iron tailings materials is significantly increased with compaction numbers. The resilient modulus of the roadbed with iron tailings materials needs to be higher than 107.4 MPa in order to meet the compaction requirement in practice.

1 INTRODUCTION

A large number of iron tailings materials are piled up due to the overexploitation of mineral resources in the past, leading to the waste of land resources and the imbalance of the ecological environment (Jia et al. 2015; Ju et al. 2018). With the rapid development of highway engineering in China, filling materials for roadbeds are becoming increasingly scarce, while iron tailings can be used as filler instead. Nowadays, more and more iron tailings materials are used for roadbed filler because they have the potential to undertake significant loads (Wang 2017).

Some scholars have investigated the characteristics of iron tailings materials. For instance, Guo et al. (2020) investigated the grading distribution and strength characteristics of iron tailings by conducting a field test. Wang et al. (2021) studied the applicability of iron tailings as filling materials for roadbeds of highways in Zhangjiakou in China. Yi et al. (2014) carried out the test on the mix ratio of iron tailings as roadbed fillers, and it showed that when the amount of iron tailings reaches 70%, the compressive strength of roadbed with iron tailings materials will meet the requirement for first-class highways. Therefore, iron tailings have been widely used as roadbed fillers. The compaction quality of the roadbed with iron tailings materials will significantly influence the service life of the roadbed. Qin et al. (2008) and Gash et al. (1992) analyzed the relationship between construction parameters and compaction quality of iron tailings roadbed. During compaction, the surface settlement method is typically used to control the compaction quality, which is quantified by the differential settlement tested by the Level. However, the tested results usually are affected by the standardization of the operation of humans (Qiao et al. 2013). As a method for inspection of roadbed compaction quality, the portable falling weight deflectometer (PFWD) has the advantages of easy operation and time-saving (Li et al. 2018). It can obtain the resilient modulus and then evaluate the compaction quality of the roadbed. However, there is limited research on evaluating the resilient modulus of roadbeds with iron tailings materials. Further studies are increasingly needed.

*Corresponding Author: 89zyf@163.com

DOI 10.1201/9781003330172-58

This paper investigates the resilient modulus of the roadbed with iron tailings materials using PFWD. The relationship between resilient modulus and differential settlement is also discussed. This study will provide a better understanding of the compaction quality of roadbeds with iron tailings materials during the construction of highways.

2 TEST DEVICES

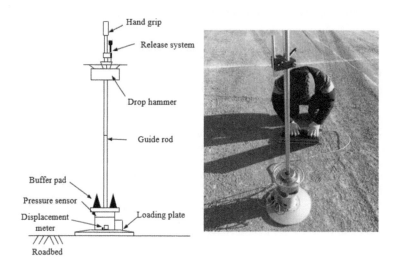

Figure 1. Portable falling weight deflectometer (PFWD) test.

The PFWD comprises the loading system, sensors, transmission and data analysis device. The handgrip, release system, drop hammer, guide rod, buffer pad and loading plate are the loading system, as shown in Figure 1. The sensors are composed of pressure sensors and displacement meters. The weight of the drop hammer is 10 kg. The radius of the loading plate is 30 cm. And the drop distance is 60 cm.

The roadbed will undertake an impact load when the hammer is dropped into the buffer pad. Since the time usually is less than 25 ms, the deformation of the roadbed under impact load is typically linear and elastic. Therefore, the resilient modulus of the roadbed can be calculated as the following equation:

$$E_p = \frac{5\pi p \delta (1 - \mu^2)}{L} \tag{1}$$

where E_p is the resilient modulus of the roadbed; δ is the radius of the loading plate; p is the dynamic stress of the roadbed under the loading plate; μ is the Poisson's ratio; L is the maximum deformation of the roadbed.

3 EXPERIMENTAL STUDY

3.1 *Test site*

The field compaction quality test on the roadbed with iron tailings materials is carried out at Yanchong Highway in China. This highway is from Beijing to Zhangjiakou during Winter Olympics in 2022. The distance of this highway is 113.684 km and the construction period is relatively short. Therefore, the PFWD will play an essential role in evaluating the compaction quality of roadbeds with iron tailings materials since its easy and quick detection characteristics.

3.2 *Test points*

The length of the test section on compaction quality of iron tailings roadbed is 300 m, as shown in Figure 2. The bulldozers and road rollers are used before the test to eliminate the influence of the foundation settlement on the compaction test. The test points are distributed every 40 m in the longitudinal direction. The test points on the left roadbed are marked as Ai, while the test points on the right roadbed are marked as Bi (i=1,...,6). There are 12 test points in this section, as shown in Figure 2.

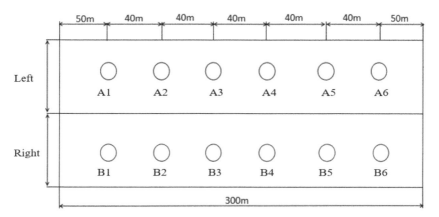

Figure 2.   Distribution of the test points.

The loose paving thickness of roadbed with iron tailings materials is set as 41 cm and 33 cm, respectively. The compaction process is carried out five times. The first compaction is the static press, and the rest four compactions are dynamic vibrations to compact the roadbed. The resilient modulus of the roadbed with every compaction under the two loose paving thicknesses is tested. The resilient moduli of the roadbed are selected after the testing data becomes stable.

4   RESILIENT MODULI OF THE ROADBED WITH IRON TAILINGS MATERIALS

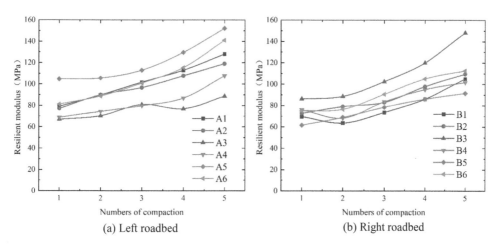

Figure 3.   Resilient moduli of the iron tailings roadbed with the loose paving thickness of 41 cm.

When the loose paving thickness of the roadbed with iron tailings materials is 41 cm, the resilient moduli on the left and right roadbed are tested using PFWD, and the results are shown in Figure 3. The resilient modulus of the iron tailings roadbed is increased with the compaction numbers. When the compaction number is one, the resilient moduli are changed from 66.78 MPa to 104.71 MPa at the left roadbed. When the compaction number is five, the resilient moduli are increased from 76.79 MPa to 129.52 MPa. The resilient moduli at the right roadbed exhibit similar characteristics. In addition, the resilient moduli vary in the longitudinal direction due to different compaction quality.

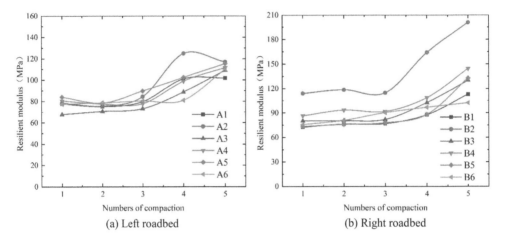

(a) Left roadbed (b) Right roadbed

Figure 4.   Resilient moduli of the iron tailings roadbed with the loose paving thickness of 33 cm.

When the loose paving thickness of the roadbed with iron tailings materials is 33 cm, the resilient moduli on the left and right roadbed are tested using PFWD, and the results are shown in Figure 4. Similar to the characteristics in Figure 3, the resilient moduli are slightly increased with the compaction numbers as well. When the compaction number is five, the resilient modulus at A4 exhibits a pretty high value. The possible reason is that the compaction quality at this time becomes perfect for this location. The compaction quality at B2 is also better than those of other locations.

5   RELATIONSHIP BETWEEN RESILIENT MODULI AND DIFFERENTIAL SETTLEMENTS

The compaction quality of the iron tailings roadbed can also be quantified by the differential settlement tested by the Level. The differential settlement is tested at the 12 test points with two loose paving thicknesses.

When the loose paving thickness is 33 cm, the relationship between the resilient modulus and differential settlement is shown in Figure 5. The natural logarithm equations can describe the correlation curve between resilient modulus and differential settlement, as shown in Figure 5. It is noted that the equation at the right roadbed has a quite high correlation coefficient (R2 = 0.848). Therefore, the following equation is selected to describe the relationship between resilient modulus and differential settlement of iron tailings roadbed:

$$E_P = -25.03 \ln \Delta H + 124.83 \qquad (2)$$

Where ΔH is the differential settlement of the roadbed with iron tailings materials.

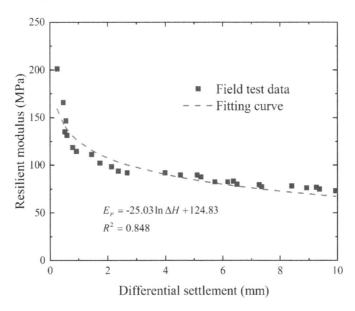

Figure 5.   Relationship between resilient modulus and differential settlement.

Table 1.   Required resilient modulus.

Differential settlement (mm)	Resilient modulus (MPa)
6	80.0
5	84.5
4	90.1
3	97.3
2	107.4
1	124.8

When the compaction of the roadbed is completed, the differential settlement is generally distributed in 2–6 mm. The resilient modulus can be calculated from equation (2). The required resilient modulus of roadbed with iron tailings materials to reflect compaction quality is shown in Table 1. Therefore, if the differential settlement of the iron tailings roadbed is required to be less than 2 mm during compaction, the corresponding resilient modulus of the iron tailings roadbed needs to be higher than 107.4 MPa to make sure the compaction quality meets the requirement in practice.

6   CONCLUSIONS

This paper evaluates the compaction quality of roadbed with iron tailings materials using a portable falling weight deflectometer. The relationship between the resilient modulus and differential settlement is also proposed. The following conclusions can be drawn:

(a) The resilient modulus of roadbed with iron tailings materials is significantly increased with compaction numbers. It also varies in the longitudinal direction due to different compaction qualities.

(b) Natural logarithm equations can describe the relationship between resilient modulus and differential settlement. The resilient modulus of the roadbed with iron tailings materials needs to be higher than 107.4 MPa to meet the compaction requirement in practice.

This study would provide a reference for the compaction quality of roadbeds with iron tailings materials in highways and show that the waste materials could be useful.

REFERENCES

Gash B, Volz R, Potter G, et al. (1992) The effects of cleat orientation and confining pressure on cleat porosity permeability and relative permeability in coal. *The Log Anal*, 33(2): 176–177.

Guo X, Chen Z, Shao L, et al. (2020) Experimental study on sedimentary behavior and basic physical mechanical properties of fine iron tailings. *Chinese Journal of Geotechnical Engineering*, 42(07): 1220–1227.

Jia Q, Liu B, Yu F, et al. (2015) Statistics analysis and management suggestion for the emergent environment accident of tailing pond in China. *Safety and Environment Engineering*, 22(02): 92–96.

Ju J, Huang X, Xue Y, et al. (2018) Thoughts of mineral resources conservation and comprehensive utilization in China in the new era. *China Mining Magazine*, 27(01):1–5.

Li S, Li H, Han X, et al. (2018) Field compaction tests on rockfill dams based on surface settlement method. *Chinese Journal of Geotechnical Engineering*, 040(0z2): 127–131.

Qiao L, Pang L, Sui Z, et al. (2013) Field test study of filling and penetration of highly plastic clay located between the dam core wall and cushion concrete. *Rock and Soil Mechanics*, 34(S1): 97–102.

Qin S, Chen S, Song H. (2008) Research on filling test of high embankment with over coarse-grained soil. *Chinese Journal of Rock Mechanics and Engineering*, 27(10): 2101–2107.

Wang G. (2017) *The stability analysis of iron tailings sand subgrade slope based on the strength reduction method*. Shangxi Science and Technology of Communications, 06: 40–43.

Wang Z. (2021) Comparative analysis on measures for advanced support of shallow-buried tunnel. *Northern Traffic*, 08: 73–74.

Yi S, Wang L, Wang Z. (2014) Application research of Jinshadian's iron tailings as subgrade materials. *Mining R&D*, 34(04): 88–91.

Advances in Measurement Technology, Disaster Prevention and Mitigation – Li & Mohd Yusof (Eds)
© 2023 The Author(s), ISBN: 978-1-032-36087-4

Research on application technology of in-situ loading test of cast-in-place concrete slab

Jianpeng Zhang*
Shandong Academy of Building Research Co. Ltd, Shandong Testing Center for Quality of Construction Engineering Co. Ltd, Jinan, China

Jianyang Zhang*
Jinan First Construction Group Co. Ltd, Jinan, China

ABSTRACT: Taking a six-storey frame structure as an example, due to the large deviation of the protective layer of negative moment reinforcement in excess of the allowable deviation in a floor on the fourth floor, in order to ensure the safe use of structural engineering and reflect the most unfavorable loading factors, this paper uses the hierarchical load method, static load in the floor in situ safety inspection, measurement of the load, deflection and residual strain of concrete site casting such as indicators, and theory of contrast. The experimental results and theoretical results are discussed and analyzed to provide a reference for similar projects in the future.

1 INTRODUCTION

In recent years, with the development of the economy, the construction industry also has a qualitative leap, at the same time, the quality of construction is also accompanied by a series of problems. For example, the steel protection layer does not meet the design requirements, and the concrete protective layer refers to the thickness of the concrete from the outer edge of the stressed steel bar to the outer surface of the concrete, also known as the thickness of the concrete protective layer. Its effect on the structure is mainly reflected in three aspects: 1) Bond anchorage between reinforcement and concrete; 2) Protect steel reinforcement from rust; 3) Influence on the effective height of force on components. The location of the reinforcement during construction and the quality of the concrete construction are critical to ensuring the thickness of the concrete protective layer. Because the process of structural design has a certain rich degree, some engineering problems can be further confirmed by structural calculation or testing whether meet the requirement of structure safety and use, and it can not only guarantee the safety of the structure, and avoids unnecessary structure reinforcement structure, causing unnecessary economic loss and construction delays (Chen et al. 2013). In this paper, through an engineering example, the thickness of the protective layer of the negative moment reinforcement of a four-storey floor is too large, the support model of the floor is established, and the load, deflection, residual strain, and other indexes of the concrete cast-in-situ slab are measured through graded in-situ loading, which is compared with the theory, and the test results and theoretical results are discussed and analyzed (Zhang 2015), so as to provide a reference for subsequent similar projects.

2 PROJECT OVERVIEW

An apartment building is a six-storey frame structure, and the roof of the building adopts reinforced concrete cast-in-situ slab. The structural design safety level is grade II, the seismic fortification

*Corresponding Authors: 764285929@qq.com and zjysdwf@163.com

DOI 10.1201/9781003330172-59

category of the building is standard fortification category, the seismic fortification intensity is grade 7 (0.10g), the design earthquake is divided into group III, the construction site category is class II, and the seismic fortification level of the frame is grade III. The elevation of the project is shown in Figure 1. According to the inspection of relevant departments, the maximum allowable deviation of the reinforcement protective layer of the negative bending moment reinforcement along the 1 / 4 axis of the four-story 1 / 4 ~ 5 / A~ C axis floor is 38mm. In order to ensure the safe use of the structural engineering, considering the most unfavorable factors of the load, the safety static load inspection is carried out in the original position of the four-storey 1 / 4 ~ 5 / A~ C axis floor along the 1 / 4 axis floor.

Figure 1.   Project elevation.

3   FIELD TEST

3.1   *Test preparation and loading*

Considering the most unfavorable factors of load, when conducting safety static load inspection for 4 ~ 1 / 4 / A ~ C-axis floor slab and 1 / 4 ~ 5 / A ~ C-axis floor slab on the fourth floor of the apartment building, 3 ~ 1 / 3 / A ~ C-axis floor slab, 4 ~ 1 / 4 / A ~ C-axis floor slab, 1 / 4 ~ 5 / A ~ C-axis floor slab and 1 / 5 ~ 6 / A ~ C-axis floor slab shall be loaded at the same time. The specific loading layout is shown in Figure 2 (the beam support in the shadow in the figure is 1 / 4 axis support, and the inclined line plate is the plate that needs to be loaded at the same time).

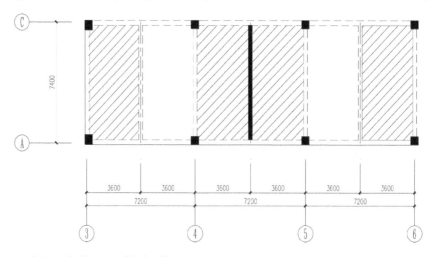

Figure 2.   Schematic diagram of the loading position.

According to the design drawings and atlas, the floor dead load is counted. The apartment floor live load is taken as 2.0kN/m^2, the balcony floor live load is taken as 2.5kN/m^2, and the toilet floor live load is taken as 2.5kN/m². According to the *Technical standard for On-site inspection of concrete structures (GB/T 50784-2013)* and *Load code for building structures (GB 50009-2012)*, the inspection load of safety static load inspection shall adopt the design value of load effect combination in the limit state of bearing capacity(Ministry of Housing and Urban-Rural Development, PRC 2012). After calculation, the ground of the apartment is 9.22kN/m², the ground of the balcony is 11.12kN/m², and the ground of the toilet is 17kN/m². In this test, water is used as the load, and a dial indicator is set at the short-span support and mid-span on the bottom of the floor slab. The loading is divided into 10 levels, of which the first level is the self-weight state of the slab, and the load holding time of each level is in strict accordance with the provisions of the *Standard Test Method for Concrete Structures (GB/T 50152-2012)*. The loading site is shown in Figures 3 ~ 14.

Figure 3. Second load.   Figure 4. Fourth load.   Figure 5. Sixth load.

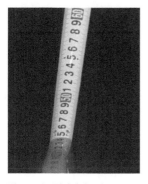  

Figure 6. Eighth load.   Figure 7. Tenth load.   Figure 8. Deflection measurement.

3.2 *Test result*

Under a load of component safety inspection, cracks were found on the floor bottom of the 4~1/4/A~C inter-axle floor when loaded to grade 6. The cracks were all located at the bottom box of the floor. The maximum crack width was 0.10mm. When loaded to grade 10, the maximum crack width was 0.15mm and the maximum crack width was 0.12mm after unloading. 1/4~5/A~C floor slab between axles is cracked at the bottom of the slab when loaded to the 5th stage. The cracks are all located at the bottom boxes of the slab. The maximum crack width is 0.11mm. When loaded to the 10th stage, the maximum crack width is 0.17mm and the maximum crack width is 0.13MM after unloading. The measured deflection values are less than the corresponding theoretical calculation values. The concrete results are shown in Table 4 (all deflections in the table are short-term deflections caused by external loads, excluding initial deflections). The measured deflection basically keeps a linear relationship with the load. The curves of the measured values for each load and deflection are shown in Figure 9 (No. 1 plate in the figure is 4~1/4/A~C inter-axle floor, No. 2 plate in the figure is 1/4~5/A~C inter-axle floor). Residual deflection of components measured by 4~1/4/A~C axle-to-axle floor slab bottom short-span directional dial gauge is 18.6%, and it is measured to be 18.9% by 1/4~5/A~C axle-to-floor slab bottom short-span directional dial gauge, which is not more than 20% of the maximum deflection. Actual deflection measurements and floor cracking under various loads are shown in Tables 1~2. The floor panel inspected meets the safety requirements.

Table 1. 4~1/4/A~C Inspection result of Inter-axle floor under various loads.

Load series	Apartment load value (kN/m^2)	Loading value of balcony (kN/m^2)	Loading value for toilet (kN/m^2)	Deflection (mm)	Cracking condition	Crack width (mm)	Remarks
1	0	0	0		Uncracked	/	Plate dead weight
2	0.8	1.2	2.4	0.18	Uncracked	/	/
3	0.8	1.2	2.4	0.33	Uncracked	/	/
4	0.5	0.6	2.4	0.42	Uncracked	/	/
5	0.5	0.6	2.4	0.51	Uncracked	/	/
6	0.6	0.7	0.6	0.63	Uncracked	0.10	/
7	0.6	0.7	0.6	0.73	Uncracked	0.11	/
8	0.6	0.7	1.2	0.89	Uncracked	0.12	/
9	0.6	0.7	1.2	1.04	Uncracked	0.13	/
10	0.6	0.7	1.2	1.18	Uncracked	0.15	/
Uninstall				0.22	Uncracked	0.12	Uninstall

Table 2. 1/4~5/A~C Inspection result of Inter-axle floor under various loads.

Load series	Apartment Load Value (kN/m^2)	Loading value of balcony (kN/m^2)	Loading value for toilet (kN/m^2)	Deflection (mm)	Cracking condition	Crack width (mm)	Remarks
1	0	0	0		Uncracked	/	Plate dead weight
2	0.8	1.2	2.4	0.20	Uncracked	/	/
3	0.8	1.2	2.4	0.42	Uncracked	/	/
4	0.5	0.6	2.4	0.55	Uncracked	/	/
5	0.5	0.6	2.4	0.67	Cracked	0.11	/
6	0.6	0.7	0.6	0.81	Cracked	0.12	/
7	0.6	0.7	0.6	0.96	Cracked	0.14	/
8	0.6	0.7	1.2	1.14	Cracked	0.14	/
9	0.6	0.7	1.2	1.34	Cracked	0.15	/
10	0.6	0.7	1.2	1.53	Cracked	0.17	/
Uninstall				0.29	Cracked	0.13	Uninstall

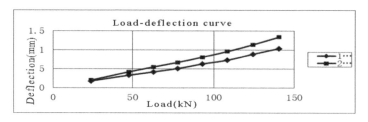

Figure 9. Load deflection curve.

This test is carried out in situ. The surrounding supports are frame beams. During the test, the settlement of the support needs to be considered. Therefore, the influence of the support needs to be considered in the deflection data processing(Ministry of Housing and Urban-Rural Development, PRC 2012). After eliminating the influence of bearing settlement, the measured maximum deflection in the middle of the span and the corrected maximum deflection in the middle of the span considering self-weight are calculated according to Formula (1) ∼ (3) (Ministry of Housing and Urban-Rural Development, PRC 2012), and the specific calculation results are shown in Table 3.

$$\alpha_q^0 = \mu_m^0 - \frac{\mu_l^0 + \mu_r^0}{2} \tag{1}$$

$$\alpha_s^0 = (\alpha_q^0 + \alpha_g^0)\psi \tag{2}$$

$$\alpha_g^c = \frac{M_g}{M_b}\alpha_b^0 \tag{3}$$

In the formula: α_q^0 is to eliminate the influence of bearing settlement, the measured maximum deflection at mid span, (Unit: mm); μ_l^0 is the measured settlement displacement of the left end support, (Unit: mm); μ_r^0 is the measured settlement displacement of the left end support, (Unit: mm); μ_m^0 is the measured value of mid span deflection including bearing settlement, (Unit: mm); α_s^0 is the maximum deflection in the middle of the span after considering the self-weight and other corrections, (Unit: mm); α_g^c is the mid span deflection value generated by the self-weight of the component and the weight of the loading equipment, (Unit: mm); ψ is the correction factor when the concentrated load replaces the uniformly distributed load; M_g is the mid span bending moment value generated by the self-weight of components and the weight of loading equipment, (Unit: KN. M); M_b and α_b^0 are the mid-span bending moment value (KN. m) and the measured mid-span deflection value (mm) generated by the previous load from the beginning of the applied load to the inflection point of the bending moment deflection curve.

Table 3. Comparison between theoretical deflection and measured deflection of floor slab.

Detection site		Deflection (mm)		Conclusion
Four layers 4∼1/4/A∼C floor slab between axles	Short span direction	Theoretical calculation value	4.62	Meet specification requirements
		Measured value	1.18	
Four layers 1/4∼5/A∼C floor slab between axles	Short span direction	Theoretical calculation value	4.62	Meet specification requirements
		Measured value	1.53	

4 EXPERIMENTAL ANALYSIS

The four-storey 4∼1/4/A∼C axis floor and 1/4∼5/A∼C axis floor of the project have no obvious damage. The measured deflection value is less than the corresponding theoretical calculation value. The measured deflection basically maintains a linear relationship with the load. The residual deflection of components measured by the dial indicator in the direction of short span and long span at the bottom of the plate is not more than 20% of the maximum deflection, according to article 12.2.14-2 of the *Technical Standard for On-site Inspection of Concrete Structures(GB/T 50784-2013)*, the structural evaluation of static load inspection: under the action of the component safety inspection load, when the tested component has no obvious signs of damage and the measured deflection value meets one of the following conditions, the safety of the tested component can be evaluated to meet the requirements. (1) The measured deflection value is less than the corresponding theoretical calculation value. (2) The measured deflection basically maintains a linear relationship with the load. (3) The residual deflection of the member shall not be greater than 20% of the maximum deflection. The safety of the four floors between4~1/4/A~C axes and between 1/4~5/A~C axes inspected in the project meets the specification requirements.

5 EPILOGUE

In this paper, combined with the actual project, according to the relevant specifications and considering the most unfavorable factors of the load, the step-by-step loading method is adopted to load the cast-in-situ slab. Under the action of the load effect combination design value in the limit state of bearing capacity, the test floor does not produce obvious damage, the measured deflection value is less than the corresponding theoretical calculation value, the measured deflection basically maintains a linear relationship with the load, and the residual deflection of the member is not greater than 20% of the maximum deflection. According to article 12.2.14-2 of *The Technical Standard for On-site Inspection of Concrete Structures GB/T 50784-2013*, the safety of floor slabs between 4~1/4/A~C axes and between 1/4~5/A~C axes of the fourth floor inspected in this project meets the requirements of the code and can be used safely according to the original design function. The experimental results are in good agreement with the theory, which can provide a reference for similar projects.

6 RESEARCH PROSPECT

In this paper, although the research on the thickness of the negative reinforcement protective layer has achieved preliminary results, there is still a lot of further research work to be carried out, briefly discussed as follows:

(1) The design and construction personnel need to further understand the local or the specific environmental conditions of the project, and further refine the design value of the thickness of the protective layer according to the concrete design specifications. In addition, the construction quality of the thickness of the protective layer in recent years can be counted, and the results can be applied to the design of the protective layer to guide the design value so that the design value can better meet the requirements of the service life of the structure.
(2) Further research and development of steel bar pads and horse bench forming machines, and conduct pad forming process tests. Develop horse bench machines and pad machines with unique performance. The key technology of the spacer block with holes and special grooves is the processing and manufacturing of molding equipment.
(3) Further develop specific instruction manuals, and verify the manuals in practice. After achieving good results, they will be popularized and used to guide construction and completely improve the current problems that have no basis for protective layer control.

REFERENCES

J. W. Chen, Q. Pan and B. Han. (2013) Experimental study on the structural performance of a prestressed concrete double T slab. *J. Shanxi Architecture*, 39(9): 41–42
Ministry of Housing and Urban-Rural Development, PRC. (2012) *Load code for building structures: GB 50009-2012*. China Construction Industry Press, Beijing.
Ministry of Housing and Urban-Rural Development, PRC. (2012) *Standard for test methods of concrete structures: GB/T 50152-2012*. China Construction Industry Press, Beijing.
Ministry of Housing and Urban-Rural Development, PRC. (2012) *Technical standard for on-site inspection of concrete structures: GB/T 50784-2013*. China Construction Industry Press, Beijing.
S. M. Zhang. (2015) Test of limit state of the normal use of prestressed concrete double T slab in grain storage. *J. Construction quality*, 33(5): 93–95

Advances in Measurement Technology, Disaster Prevention and Mitigation – Li & Mohd Yusof (Eds)
© 2023 The Author(s), ISBN: 978-1-032-36087-4

Study on vibration response characteristics of subgrade filled with granite residual soil

Junlong Hu*
Key Laboratory of Environment and Safety Technology of Transportation Infrastructure Engineering, CCCC, Guangzhou, China
CCCC Fourth Harbor Engineering Institute Co., Ltd., Guangzhou, China

Yongjia Nong
CCCC Guanglian Expressway Investment Development Co., Ltd., Qingyuan, China

Yao Xie, Jing Wang & Deyong Wang
Key Laboratory of Environment and Safety Technology of Transportation Infrastructure Engineering, CCCC, Guangzhou, China
CCCC Fourth Harbor Engineering Institute Co., Ltd., Guangzhou, China

ABSTRACT: The continuous compaction control technology based on the coupling effect of soil and the vibratory drum has been gradually applied to subgrade engineering in recent years to realize the comprehensive monitoring of the compaction quality of the working face. Granite residual soil is widely distributed in South China and is widely used as subgrade filler. In order to study the vibration response characteristics of the vibratory drum in the process of vibratory compaction for granite residual soil filler, field tests are carried out. The amplitude spectrum of the acceleration signal collected from the vibratory drum is obtained by FFT, and the influence laws of the number of rolling passes and dry density of filler on the vibration response characteristics are studied. The selection of continuous compaction measuring value and the applicability of the compaction degree evaluation model for granite residual soil filler is further discussed. The research results can provide technical references for the compaction state evaluation of granite residual soil filler.

1 INTRODUCTION

With the development of the economy and society, tens of thousands of kilometers of highway mileage are added every year in China, which brings a large number of subgrade filling projects. The compaction state evaluation of subgrade is one of the core contents of subgrade engineering, which is closely related to the stability and settlement of subgrade. The traditional method is to sample and detect the compactness of the working face. Due to the low sampling frequency and randomness, it is unable to comprehensively evaluate the compaction quality of the working face, which is easy to cause the phenomenon of lack or excess compaction. In recent years, with the improvement of China's infrastructure construction standards, digital and intelligent construction technology is increasingly being introduced into engineering construction. Under this background, a continuous compaction control technique is proposed. Its basic working principle is to indirectly reflect the compaction condition of the soil layer by measuring the vibration response of the vibratory drum. In the process of vibratory compacting, the compaction state of the filler is constantly changed by the excitation of the vibratory drum, and the vibration state of the vibratory drum is also changed by the action of the soil, that is, the vibration response of the vibratory drum is related to the compaction state of the filler. The continuous compaction control technique is to reflect the compaction state of filler through the vibration response of the vibratory drum and calculates the corresponding

---

*Corresponding Author: hjunlong@cccc4.com

DOI 10.1201/9781003330172-60

continuous compaction measuring value (Xu et al. 2015; Zhang et al. 2015). Cooperating with the global positioning system (Wang et al. 2020), the compaction state distribution map of the whole compaction face is obtained, and then the compaction parameters can be adjusted in real time.

Granite residual soil is widely used as subgrade filler in highway engineering in South China. However, the continuous compaction control technique is rarely used in highway engineering. Thus the vibration response characteristics of vibratory drum rolling granite residual soil filler have not been revealed, and the corresponding continuous compaction measuring value and its correlation model with traditional compaction measuring value have not been reported.

This paper analyzes the vibration response characteristics of the vibratory drum in the process of compacting granite residual soil filler through field tests, puts forward the continuous compaction measuring value of granite residual soil, and establishes its correlation equation with the traditional compaction measuring value.

2 FILLER PROPERTIES AND TEST SCHEME

2.1 *Physical and mechanical properties of the filler*

Granite residual soil is sampled in the field and geotechnical tests are carried out. The particle size distribution curve is shown in Figure 1. The gravel group accounts for 37.5%, the sand group accounts for 37.6%, and the fine-grained group accounts for 24.9%. The filler is silty sand, with a coefficient of uniformity of 162.5 and a coefficient of curvature of 2.22. It is well graded. The optimum water content is 8.67% and the maximum dry density is $1.82g/cm^3$.

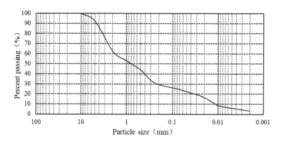

Figure 1. Particle size distribution curve.

2.2 *Equipment*

A XCMG XS263J single drum vibratory roller with an overall dimension of 6530×2470×3260mm is used for vibratory compaction of filler and the working quality is 26t.

A High Target V30 continuous compaction system is adopted for position information and acceleration data acquisition, as shown in Figure 2. The system is mainly composed of the base station, mobile station, data acquisition box, acceleration sensor, and onboard plate. The base station and mobile station are used to solve the differential signal to obtain the real-time position of the roller, and the mobile station is fixed on the top of the cockpit of the roller.

The acceleration sensor which is rigidly connected to the drum is used to collect the vertical acceleration of the drum and its sampling frequency is 1000Hz. The high-frequency vibration acceleration data is collected and recorded by a portable computer and corresponding software.

2.3 *Test site*

The test site is 25m long and divided into three strips horizontally, as shown in Figure 3. The width of each strip is a rolling drum track, i.e., 2170mm. From right to left, they are 1#, 2#, and 3# test strips respectively. The total rolling passes of 1#, 2#, and 3# strips are 2, 4, and 6 respectively. A conventional compaction measurement point is arranged at an interval of 5m along the longitudinal direction of each strip, 1# strip corresponds to P1 ~ P5 measurement points, 2# strip corresponds to P6 ~ P10 measurement points, 3# strip corresponds to P11 ~ p15 measurement points, and each conventional compaction measurement point is arranged in the middle of the strip.

Figure 2.   Site layout.

2.4   *Test procedure*

The field test process is as follows:

(1) For filler paving, it needs to draw a square grid within the test area, calculate the filler quality required for each grid according to the requirements of 30cm loose paving thickness, and use dump trucks for unloading and paving.

(2) The site shall be roughly leveled. After the filler is paved, the 26t roller shall be used for statically rolling for one time, and then the scraping grader shall be used for rough levelling.

(3) Site layout: it is needed to draw a range of 25m × 6.5m on the roughly leveled site, and mark the starting and ending lines of rolling and the positions of conventional compaction measurement points.

(4) It is required to check the instruments and equipment, check the working status of the vibratory roller, continuous compaction system, and RTK, and check the positioning error between RTK and continuous compaction system. The next test can be carried out only when all instruments and equipment are in normal working status.

(5) Vibratory compaction: the vibratory roller drives to the rear of the starting line, starts weak vibration, and starts rolling the test 1#, 2#, 3# strips after reaching the stable state. The vibratory compaction is carried out in the order of 1# round-trip compaction for 2 times, 2# round-trip compaction for 4 times, and 3# round-trip compaction for 6 times. During the rolling process, the stability of the traveling speed and direction of the roller is ensured. In this test, the traveling speed of the vibratory roller is controlled at 3km/h.

(6) Data acquisition and recording: during the driving process of the vibratory roller, the position data of the roller and the acceleration data of the vibratory drum are continuously collected and recorded through the acquisition software.

(7) Conventional compaction measurement: for 1#, 2#, and 3# strips, the sand filling method is used to detect the dry density and water content of rolling every 5m from the starting line. Before sand filling, RTK is used to record the coordinates of each detection point, which corresponds to the position data and acceleration data in the continuous compaction system.

3   TEST RESULTS AND DISCUSSION

3.1   *Characteristics of acceleration spectrum*

Since it is difficult to obtain effective acceleration characteristics through the acceleration time history curve, a fast Fourier transform (FFT) along 1024 points is performed on the acceleration time history of each measurement point to obtain the amplitude spectrum of each measurement point under different rolling passes, as shown in Figure 3.

It can be seen from Figure 4 that in the process of rolling granite residual soil filler, the fundamental wave component of 30Hz is the most obvious in the acceleration response spectrum, which is the excitation frequency of the drum, with an amplitude of 30.12~49.24 m/s^2 and an average value of 43.15m/s^2.

At the same time, the amplitude spectrum of each measurement point also contains harmonic and interharmonic components. The significant harmonics are 60Hz and 270Hz. In addition, there are interharmonic components near 285hz, 420hz and 450hz. The second harmonic (60Hz) is the most representative among those harmonics, whose amplitude is between 0.09~2.28m/s^2 and the average value is 0.91m/s^2.

The generation of harmonics is due to the nonlinearity of the soil-drum system, and the generation of interharmonics may be due to the additional excitation caused by the local unevenness of the site.

The above analysis of the acceleration amplitude spectrum of the vibratory drum shows that the characteristics of the fundamental wave (30Hz) and the second harmonic (60Hz) are obvious. Therefore, these two parameters are mainly selected for subsequent analysis in this paper.

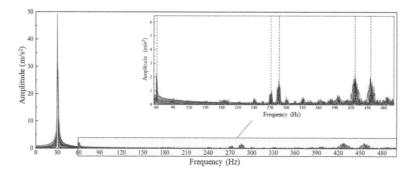

Figure 3.   Summary of the amplitude spectrum of each detection point.

3.2   Influence of rolling passes on characteristics of acceleration spectrum

The amplitudes of the fundamental wave and second harmonic corresponding to different rolling passes are shown in Figure 4~Figure 5. With the increase of rolling passes, both the fundamental wave and the second harmonic show a fluctuating upward trend. After rolling 6 times, the amplitude of the fundamental wave is 10.3% higher than that of the first rolling, and the amplitude of the second harmonic is 55.2% higher than that of the first rolling.

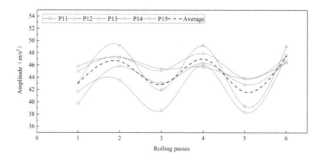

Figure 4.   Relation curve between rolling passes and fundamental wave amplitude.

At the same time, it can also be noted that the amplitude of the fundamental wave and the second harmonic of the vibratory drum are related to the driving direction of the roller, which shows that there is a large difference in the response amplitude between odd and even passes. In the fundamental wave response, the amplitude difference between the latter pass and the previous pass is up to 24.7% and the average value is 10.5%. In the second harmonic response, this effect is more significant. The maximum amplitude difference between the latter pass and the previous pass is 588.7%, and the average value is 77.5%.

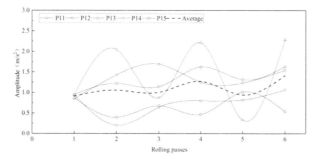

Figure 5.   Relation curve between rolling passes and second harmonic amplitude.

In addition, from the fluctuation trend of the average value, it can be seen that the above effect will become more and more significant with the increase of rolling passes, that is, with the increase of rolling passes, the oscillation amplitude of the fundamental wave and second harmonic amplitude tends to expand. The oscillation amplitude increases by 13% and 136% on average for the fundamental wave, and the second harmonic for each additional rolling passes, respectively.

The above analysis results show that although the acceleration response of the vibratory drum generally shows an upward trend with the increase of rolling passes. The driving direction of the roller has a nonnegligible influence on the acceleration response. In the actual continuous compaction measurement, the continuous compaction measuring value shall be obtained in the same rolling direction as far as possible for the same working face.

3.3 Influence of dry density on characteristics of acceleration spectrum

The relationship between dry density measured at 1#~3# strip and amplitude of the fundamental wave and the second harmonic is shown in Figure 6. It can be seen that the amplitude of the fundamental wave is approximately positively correlated with the dry density, that is, the amplitude of the fundamental wave increases with the increase of the dry density, and the two are generally linearly positively correlated. For every increase of the dry density by 0.1g/cm^3, the amplitude of the fundamental wave increases by about 2.9m/s^2. Within the range of dry density measured in this test, the amplitude of the fundamental wave increases by 13.9% with the increase of dry density.

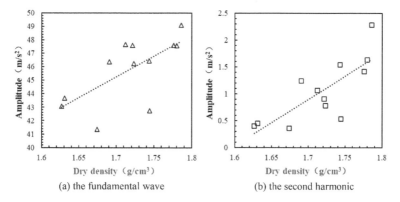

Figure 6.   Relationship between dry density and amplitude.

From the relationship between the second harmonic amplitude and dry density, the second harmonic is also positively correlated with dry density, that is, with the increase of dry density, the second harmonic amplitude is also increasing. Within the range of dry density measured in this test, the maximum amplitude of the second harmonic increases by 535.6% with the increase of dry density.

3.4 Discussion on the selection of continuous compaction measuring value

The above analysis results show that there is a good positive correlation between the amplitude of the fundamental wave and the secondary harmonic wave and dry density. Their applicability as measured values can be considered. In order to facilitate comparison, min-max normalization is carried out for the amplitude of the fundamental wave and the second harmonic, which is linearly transformed to [0,1] interval to obtain the normalized amplitude of fundamental wave A_Ω and normalized amplitude of the second harmonic $A_{2\Omega}$. At the same time, the commonly used CMV index (Nie et al. 2017)is introduced for comparison. The calculation of CMV is shown in Formula (1):

$$CMV = \frac{A_{2\Omega}}{A_\Omega} \tag{1}$$

The Pearson's correlation coefficients between A_Ω, $A_{2\Omega}$, CMV, and degree of compactness are calculated, and the correlation coefficients are 0.68, 0.78, and 0.78 respectively. The results show that the correlation coefficient for CMV and $A_{2\Omega}$ is higher, and meets the requirement that the correlation coefficient between the vibration compaction value output by the measuring equipment and the conventional measuring value should not be less than 0.7 in the *Technical Requirements for Continuous Compaction Control System of Fill Engineering of Subgrade for Highway (JT/T 1127-2017)*, indicating that they have a good linear correlation with the degree of compactness. The correlation coefficient between A_Ω and degree of compactness is only 0.68, indicating a weak linear correlation between them. Therefore, CMV and $A_{2\Omega}$ are more suitable for the establishment of the compaction discrimination equation.

Linear regression is furtherly carried out to obtain the linear regression equation between normalized CMV, $A_{2\Omega}$, and degree of compactness, as shown in Formula (2) and Figure 7. The established linear regression equation is:

$$\begin{cases} K = 7.4646A_{2\Omega} + 91.6760, R^2 = 0.61 \\ K = 7.3480CMV + 91.6820, R^2 = 0.61 \end{cases} \tag{2}$$

Therefore, in practical engineering, for granite residual soil, the amplitude spectrum is obtained by Fourier transform of the collected vibratory drum acceleration, the second harmonic amplitude is extracted or the CMV index is calculated, and then the conventional compaction measuring value, that is, the degree of compactness of this point, can be calculated by substituting into Formula (2).

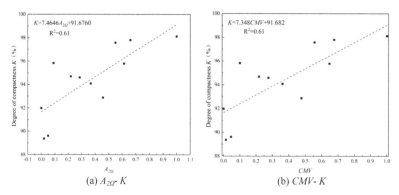

(a) $A_{2\Omega}$- K (b) CMV- K

Figure 7.   Correlation between $A_{2\Omega}$, CMV, and K.

4 CONCLUSION

Vibration response characteristics of vibratory drum and influencing factors are analyzed through field tests in this paper, and the discriminant equation of degree of compactness is established. The main conclusions are as follows:

(1) The acceleration amplitude spectrum of the vibratory drum during the rolling process of granite residual soil includes fundamental wave, harmonic and interharmonic components, in which the fundamental wave (30Hz) and the second harmonic (60Hz) are more significant.
(2) With the increase of rolling passes, the amplitudes of the fundamental wave and the second harmonic generally fluctuate and rise, but they are related to the traveling direction of the roller. In the actual project, the continuous compaction measuring values in different traveling directions should be avoided.
(3) The dry density has a good positive correlation with the amplitude of the fundamental wave and the second harmonic. With the increase of dry density, the amplitude of the fundamental wave and the second harmonic increase.
(4) There is a high correlation coefficient between the second harmonic amplitude, *CMV*, and degree of compactness, while the correlation coefficient of fundamental wave amplitude is low. The second harmonic amplitude or *CMV* can be used as the continuous compaction measuring value to characterize the compaction state of granite residual soil filler.

Due to the wide range of particle size distribution and water content of natural granite residual soil, its corresponding physical and mechanical properties vary greatly. Since the vibration response of the vibrating drum under drum-soil coupling conditions is closely related to the physical and mechanical properties of the compacted soil, the above response characteristic analysis and compactness discriminant model may be only applicable to the specific site. Carrying out a large number of field tests and introducing artificial intelligence algorithms into compactness discriminants may be a method to establish a more universal compaction discrimination model in further research.

ACKNOWLEDGMENTS

This work was financially supported by the Science and Technology Project of Guangzhou-Lianzhou Expressway.

REFERENCES

Nie Zhihong, Xie Yang, Jiao Tan.(2017) Analysis of spatial distribution characteristics of continuous measuring parameters of subgrade compaction quality. *China Railway Science*, 38(2):6–10.
Wang Jingwei, Jiang Hua, Zhou Yuefeng.(2020) Design and application of intelligent compaction management and control system based on GPS/LoRa. *Highway Engineering*, 45(5):117–122.
Xu Guanghui, Luo Zehua, Tian Bo. (2015) Summary of the development of continuous compaction control technology. *Road Machinery & Construction Mechanization*, 32(08):34–38.
Zhang Jialing, Xu Guanghui, Cai Ying.(2015) An investigation on quality inspection and control for continuously compacting subgrade. *Rock and Soil Mechanics*, 36(4):1141–1146.

Advances in Measurement Technology, Disaster Prevention and Mitigation – Li & Mohd Yusof (Eds)
© 2023 The Author(s), ISBN: 978-1-032-36087-4

Study on vertical deformation law of existing pipeline caused by tunnel construction

ShouJia Chen*, HeXin Wei, YunPeng Fan, XiuJun Ai, ChaoLong Zhao & XinZhen Luan
China Construction Sixth Bureau Installation Engineering Co., Ltd., Xi'an, Shaanxi Province, China

ABSTRACT: In this paper, the Pasternak foundation model is used to study the deformation law of the existing pipeline caused by shield construction. The two-stage method is used to establish the balance differential equation of the existing pipeline vertical deformation, and the finite difference method is used to derive the analytical solution of the pipeline deformation. The rationality and validity of the theoretical calculation are verified through the comparative analysis of theoretical calculation and field monitoring. The sensitivity analysis of parameters to the vertical deformation of the pipeline is carried out. Through parameter analysis, it is found that the buried depth and diameter of the pipeline have little influence on the vertical deformation of the pipeline. The formation loss rate has a significant effect on the maximum vertical deformation of the pipeline. As the formation loss rate increases, the vertical deformation and tangent slope of the pipeline increase significantly.

1   INTRODUCTION

Tunnel shield construction will cause deformation of the surrounding soil, resulting in the settlement of adjacent buildings and settlement of existing pipelines. Excessive soil deformation will cause deformation or even damage to adjacent buildings and pipelines. It is urgent to study the influence law of shield tunnel excavation on existing pipelines.

At present, there are many studies on the vertical deformation of existing pipelines caused by shield tunnel construction (Cai 2015; Chambon & Core 1994; Long 1982; Lin & Huang 2019; Wei et al. 2019; Zhang 2004; Gao & Cai 2015; Chambon & Core 1994; Wu 2013). In terms of theoretical calculation, Rongzhu Liang et al. (2016) simplified the existing tunnel into a continuous Euler-Bernoulli with equal stiffness. In the two-parameter model, the deformation law of the new tunnel to the existing tunnel is obtained by the two-stage stress method, and finally, the parameter sensitivity analysis is carried out for different influencing factors. Gu Shuancheng et al. (2015) deduced the influence of tunnel excavation on adjacent pipelines and the deflection expression based on the Winkler theory. Li Haili et al. (2018) proposed an analysis method for pipeline response under tunnel excavation. Huang Xiaokang et al. (2017) derived the theoretical solution of the interaction between adjacent pile foundations and tunnel construction based on the Pasternak foundation model. Zhang Zhiguo et al. (2016) proposed that the Pasternak two-parameter model has advantages over the Winkler foundation model. Because the Winkler model does not consider the continuity of foundation deformation, there is a big difference between theoretical calculation and practical engineering.

In this paper, the Pasternak two-parameter model considering the shear effect between soils is used to solve the problem of soil layer continuity. And a two-stage analysis method is introduced to analyze the settlement and stress of the existing pipeline caused by the tunnel construction. That

---

*Corresponding Author: 1140741773@qq.om

DOI 10.1201/9781003330172-61

is, in the first stage, the vertical displacement of the soil along the axis of the existing pipeline caused by the tunnel construction can be obtained, and in the second stage, the displacement of the soil displacement field in the first stage is applied to the pipeline, and the force balance differential equation of the existing pipeline is established. The finite difference method is used to solve it, the matrix form solution is obtained, and the parameter sensitivity analysis is carried out. Finally, the theoretical calculation results are verified by combining them with engineering examples.

2 VERTICAL DISPLACEMENT OF SOIL MASS CAUSED BY EXCAVATION

2.1 Mechanical model and basic assumptions

The research in this paper is based on the basic assumptions proposed by the Pasternak model method:

(1) The pipeline and the soil layer are a uniform continuum, and the pipeline is regarded as an infinite beam. (2) The shear layer only produces shear deformation (3) The pipeline and the soil layer are always in close contact without relative sliding. (4) The friction between the pipeline and the surrounding soil layer is ignored. It is assumed that R is the radius of the tunnel, D is the diameter of the pipeline, Z_0 is the buried depth of the existing pipeline axis center, and h is the buried depth of the tunnel axis. The simplified diagram of the model is as follows:

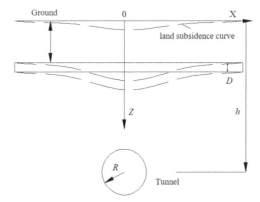

Figure 1. Simplified calculation model of pipeline deformation caused by tunnel shield construction.

2.2 Vertical displacement of free soil

As the tunnel excavation causes the stratum to be disturbed and the stratum is lost, according to Loganathan et al., the settlement $s(x)$ of the soil mass induced by the tunnel excavation is proposed as:

$$s(x) = \varepsilon_0 R^2 \left(-\frac{z_0 - h}{x^2 + (z_0 - h)} + (3 - 4\mu) \cdot \frac{z_0 + h}{x_2 + (z_0 + h)^2} - \frac{2z_0[x^2 - (z_0 + h)^2]}{[x^2 + (z_0 + h)^2]} \right)$$
$$\cdot \exp -\left[\frac{1.38x^2}{(h + R)^2} + \frac{0.69z_0^2}{h^2} \right] \tag{1}$$

In the above formula: R- Tunnel radius. ε_0- average formation loss ratio. h - depth of tunnel axis. Z_0- depth of pipeline axis. μ- Poisson's ratio of soil. x- Horizontal distance from the tunnel centerline.

$$q(x) = -G\frac{d^2 s(x)}{dx^2} + k_w s(x) \tag{2}$$

In the formula: $q(x)$- additional stress acting on the pipeline. EI- pipeline stiffness. k_w- spring stiffness coefficient. G- substratum soil shear stiffness.

3 ANALYTICAL SOLUTION FOR PIPE DEFLECTION

3.1 *Differential equation of pipeline deflection*

In order to establish the differential equation of the deflection W of the existing pipeline caused by the tunnel shield construction, the force analysis is carried out on the micro-element of dx at any position of the pipeline.

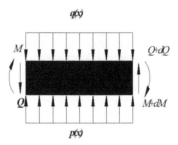

Figure 2.   Force analysis of pipeline elements.

Figure 3 is a schematic diagram of the distribution of $q(x)$ in the pipeline cross-section. It is assumed that it acts on the axis of the pipeline and is evenly distributed along the y direction.

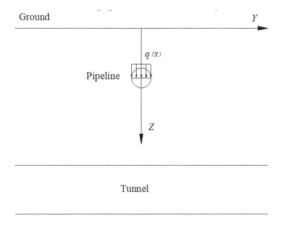

Figure 3.   Schematic diagram of the distribution of q(x) along the y direction.

For the force analysis of the micro-element body, it can be obtained from the static balance equation:

$$\sum Y = 0, Q - (Q + dQ) + Dp(x)dx - Dq(x)dx = 0 \tag{3}$$

$$\sum M = 0, M - (M + dM) + (Q + dQ)dx + q(x)\frac{(dx)^2}{2} - p(x)\frac{(dx)^2}{2} = 0 \tag{4}$$

In the above formula, dx is the width of the element (m); Q is the shear force of the cross-section (N); M is the bending moment of the cross-section (K/m); dQ is the increment of the shear force of the section along the length of the pipeline dx (N); dM is the increment of the bending moment of the inner section along the length of the pipe dx (K/m); q(x) is the foundation reaction force acting on the pipe (N/m²); D is the diameter of the pipe (m), which is the resulting line deflection from the construction effect of the shield n.

$$p(x) = -G\frac{d^2\omega(x)}{dx^2} + k_w w(x) \tag{5}$$

Assuming that $w(x)$ is known, the curvature of the pipeline can be obtained at any point:

$$\frac{1}{\rho} = \frac{w''(x)}{(1+w'^2(x))^{3/2}} = -\frac{M}{EI} \tag{6}$$

where: X is the curvature of the pipeline deflection, and W is the equivalent bending stiffness of the underground pipe. Since the deflection curve of the pipeline is a flat curve, the comparison with 1 in Equation (6) can be omitted, and the differential equation of the deflection curve is obtained as follows:

$$EIw''(x) = M \tag{7}$$

Arranging Formulas (2), (5), (6), and (8), the following can be obtained:

$$EI\frac{d^4 w(x)}{dx^4} + k_w D\omega(x) - GD\frac{d^2 w(x)}{dx^2} = Dq(x) \tag{8}$$

3.2 Finite difference method

Using the finite difference method, as shown in Figure 3, the total length of the pipeline is L, and it is divided into n sections of micro-elements, then the length of each section of the micro-element is -1, n+1, and n+2.

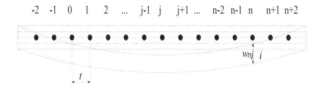

Figure 4. Pipeline differential node division.

The Taylor series expansions of $w(x)$ with two points $x_i - t$ and $x_i + t$ on both sides of any point are:

$$w(x_i - t) = w_i - w_i' t + \frac{w_i'' t^2}{2} - \frac{w_i''' t^3}{3} + \cdots + o(t^n) \tag{9}$$

$$w(x_i + t) = w_i + w_i' t + \frac{w_i'' t^2}{2} + \frac{w_i''' t^3}{3} + \cdots + o(t^n) \tag{10}$$

Simultaneously solving Equations (9) and (10) by adding and subtracting respectively and ignoring higher powers, the difference formula of the fourth derivative is obtained as:

$$\left(\frac{d^4 w}{dx^4}\right)_i = \frac{d^2}{dx}\left(\frac{d^2 w}{dx^2}\right) = \frac{1}{t^2}\left[\left(\frac{d^2 w}{dx^2}\right)_{i+1} - 2\left(\frac{d^2 w}{dx^2}\right)_i + \left(\frac{d^2 w}{dx^2}\right)_{i-1}\right] = \frac{1}{t^4}[(w_{i+2} - 2w_{i+1} + w_i) -$$
$$2(w_{i+1} - 2w_i + w_{i-1}) + (w_i - 2w_{i-1} + w_{i-2})] = \frac{1}{t^4}(w_{i+2} - 4w_{i+1} + 6w_i - 4w_{i-1} + w_{i-2}) \tag{11}$$

From Equations (8) and (11), the differential form of the vertical displacement of the pipeline is obtained:

$$\frac{EI}{t^4}w_{i-2} + \left(-\frac{4EI}{t^4} - \frac{GD}{t^2}\right)w_{i-1} + \left(\frac{6EI}{t^4} + 2\frac{GD}{t^2} + kD\right)w_i$$

$$+ \left(-\frac{4EI}{t^4} - \frac{GD}{t^2}\right)w_{i+1} + \frac{EI}{t^4}w_{i+2} = Dq(x)_i \tag{12}$$

where i is between 2 and n-2, now that:

$$\alpha w_{i-2} + \beta w_{i-1} + \gamma w_i + \beta w_{i+1} + \alpha w_{i+2} = Q(x)_i \tag{13}$$

where α, β and γ can be expressed as:

$$\begin{bmatrix} \alpha \\ \beta \\ \gamma \end{bmatrix} = \begin{bmatrix} 1 & 0 & 0 \\ -4 & -1 & 0 \\ 6 & 2 & 1 \end{bmatrix} \cdot \left[\frac{1}{t^4} \ \frac{GD}{EI^2} \ \frac{kD}{EI}\right]^T \tag{14}$$

$Q(x)_i = Dq(x)_i$, $q(x)_i$ q is the additional stress caused by tunnel excavation on the i-th section of the pipeline:

$$q(x)_i = -G\frac{d^2 S(x)_i}{dx^2} + kS(x)_i \tag{15}$$

Solving the deformation, the deflection equation of the pipeline can be obtained:

$$[W]_{(n+1)\times 1} = [K]^{-1} \cdot [q]_{(n+1)\times 1} \tag{16}$$

$$\theta_i = \left(\frac{dw}{dx}\right)_i = \frac{1}{2h}(w_{i-1} - w_{i+1}) \tag{17}$$

$$Q_i = \frac{dM}{dx} = \frac{EI}{2h^3}(w_{i-2} - 2w_{i-1} + 2w_{i+1} - w_{i+2}) \tag{18}$$

$$M_i = EI\frac{d\theta}{dx} = \frac{EI}{h^2}(w_{i-1} - 2w_i + w_{i+1}) \tag{19}$$

Since both ends of the pipeline are far away from the tunnel, the boundary conditions can be regarded as fixed at both ends, and the displacement and rotation angle of the pipeline are 0, so that:

$$w_{-2} = w_{-1} = w_{n+1} = w_{n+2} = 0 \tag{20}$$

When x = 0, it can be simplified into (19) as:

$$\left(\frac{6}{h^4} + 2\frac{GD}{EIh^2} + \frac{kD}{EI}\right)w_0 + \left(-\frac{1}{h^4} - \frac{GD}{EIh^2}\right)w_1 + \frac{1}{h^4}w_2 = \frac{q(x)_0 D}{EI} \tag{21}$$

$$\text{Get: } (\alpha + \gamma)w_0 + \beta w_1 + \alpha w_2 = Q(x)_0 \tag{22}$$

In the same way, the boundary conditions will be brought in, and when $i = 1, n-1, n$, (19) will be brought into:

$$\left(-\frac{4}{h^4} - \frac{GD}{EIh^2}\right)w_0 + \left(\frac{6}{h^4} + 2\frac{GD}{EIh^2} + \frac{kD}{EI}\right)w_1 + \left(-\frac{4}{h^4} - \frac{GD}{EIh^2}\right)w_2 + \frac{1}{h^4}w_3 = \frac{q(x)_1 D}{EI} \tag{23}$$

$$\frac{1}{h^4}w_{n-3} + \left(-\frac{4}{h^4} - \frac{GD}{EIh^2}\right)w_{n-2} + \left(\frac{6}{h^4} + 2\frac{GD}{EIh^2} + \frac{kD}{EI}\right)w_{n-1} + \left(-\frac{4}{h^4} - \frac{GD}{EIh^2}\right)w_n = \frac{q(x)_{n-1} D}{EI} \tag{24}$$

$$\frac{1}{h^4} w_{n-2} + \left(-\frac{4}{h^4} - \frac{GD}{EIh^2}\right) w_{n-1} + \left(\frac{6}{h^4} + \frac{1}{h^4} + 2\frac{GD}{EIh^2} + \frac{kD}{EI}\right) w_n = \frac{q(x)_n D}{EI} \qquad (25)$$

Finishing Formulas (23)~(25), the following can be obtained:

$$\beta w_0 + \gamma w_1 + \beta w_2 + \alpha w_3 = Q(x)_1 \qquad (26)$$

$$\alpha w_{n-3} + \beta w_{n-2} + \gamma w_{n-1} + \beta w_n = Q(x)_{n-1} \qquad (27)$$

$$\alpha w_{n-2} + \beta w_{n-1} + (\alpha + \gamma) w_n = Q(x)_n \qquad (28)$$

When $i = 2, 3 \cdots n - 3, n - 2$, the differential form of the pipeline deflection curve is shown in formula (12). Combined with Equations (12), (22), (26), (27), and (28), the stiffness matrix K in the vertical deflection equation (23) generated by the pipeline under tunnel shield construction is:

$$[K] = \begin{bmatrix} \alpha + \gamma & \beta & \alpha & & & & \\ \beta & \gamma & \beta & \alpha & & & \\ \alpha & \beta & \gamma & \beta & \cdot & & \\ & \alpha & \beta & \cdot & \cdot & \cdot & \\ & & \cdot & \cdot & \cdot & \cdot & \alpha \\ & & & \cdot & \cdot & \gamma & \beta & \alpha \\ & & & & \alpha & \beta & \gamma & \beta \\ & & & & & \alpha & \beta & \alpha + \gamma \end{bmatrix}_{(n+1)\times(n+1)} \qquad (29)$$

For the determination of the parameters of the Pasternak model, Tanahashi gives the formula for calculating the shear stiffness G of the foundation.

3.3 Determination of model parameters

For the parameter determination of the Pasternak model, Tanahashi (Tanha 2004) is selected to give the calculation formula of foundation shear stiffness G.

$$G = \frac{E_s H_t}{6(1 + v_s)} \qquad (30)$$

Where, E_s is the elastic modulus of the foundation soil (MPA), v_s is the Poisson's ratio of the foundation soil (dimensionless), and H_t is the thickness of the shear layer of the foundation soil (m). According to literature (Xu 2005), the thickness of the shear layer under the pipeline is taken as 8D.

4 SENSITIVITY ANALYSIS OF MODEL PARAMETERS

In this paper, the effects of pipeline burial depth, pipeline diameter, and formation loss rate on pipeline deformation are changed. This paper only considers the pipeline located in the general influence area, its length can be 40m, and the number of divisions is n, then the length of the unit segment is $\Delta t = 40/n$. In this paper, only the influence of single-line tunnel excavation is considered under the shield tunneling condition, and the calculation parameters of soil and pipeline are as follows: $E_s = 13.5 MPa$, $v_s = 0.35$, $k = 40 MPa/m$, $2Rh = 15m$, $z_0 = 4m$, D=0.8m.

4.1 Influence of buried depth of pipeline on vertical deformation

The buried depth of pipelines for different purposes varies greatly. When the tunnel depth is constant, the deformation law of pipelines with different depths can be reflected by the distance between the pipeline and the tunnel axis.

The vertical displacement of the pipeline under the same buried depth is shown in Figure 5. When the buried depth of the pipeline changes, the corresponding vertical deformation of the pipeline follows the Gaussian normal distribution as a whole. The main settlement range of the pipeline is within 2R on both sides of the tunnel central line. With the increase of the buried depth of the pipeline, the maximum vertical displacement of the pipeline and the maximum tangent slope of the displacement curve increase. The maximum slopes of the pipeline displacement under the five buried depths are 0.63, 0.8, 0.93, 1.02, and 1.33mm/m respectively. When the buried depth of the pipeline decreases, the maximum displacement decreases slowly.

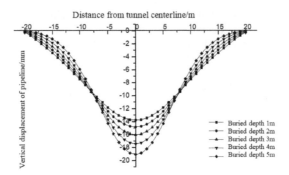

Figure 5.   Vertical displacement of pipeline under different buried depths.

4.2   Influence of pipeline diameter on settlement of underground pipeline

The diameters of underground pipelines vary greatly. The vertical displacement of pipelines under different diameters caused by shield tunnel construction is shown in Figure 6.

The vertical displacement of the same diameter pipeline follows the Gauss normal distribution as a whole, and when the diameter of the pipeline is 1.0 m, the vertical displacement of the pipeline is the largest. When the diameter is far from D = 1m, the maximum displacement of the pipeline decreases gradually. There is a linear relationship between the additional stress of the pipeline and its diameter, but the bending stiffness of the pipeline section is proportional to the 4th power of the pipeline diameter. When the diameter of the pipeline is small, the additional stress of the pipeline is small, and its bending rigidity is also small. The additional stress contributes little to the displacement of the pipeline, so the displacement is small. When the pipeline diameter is larger than a certain threshold value, the maximum vertical displacement of the pipeline decreases with the increase of the pipeline diameter.

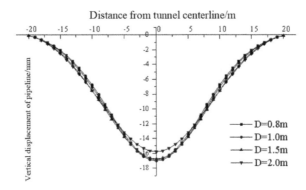

Figure 6.   Vertical displacement of pipeline under different diameters.

4.3 Influence of stratum loss rate change on underground pipeline settlement

Even for the same area, the change of soil layer, the adjustment of construction parameters, and human and environmental factors may lead to great changes in the formation loss rate. Therefore, it is necessary to study the impact of the change of formation loss on the linearity of the pipeline. Figure 7 shows the vertical displacement of existing pipelines under different ground loss rates caused by shield tunnel construction.

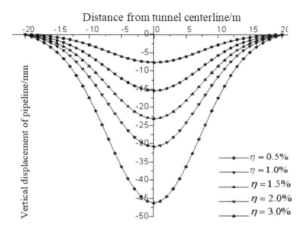

Figure 7.   vertical loss rate of pipeline under different stratum displacement.

The variation law of vertical displacement of pipeline under different formation loss rates obeys Gaussian normal distribution as a whole. As the formation loss rate increases, the vertical displacement and tangent slope of the pipeline increase significantly. The formation loss rate increased from 0.5 to 2 and 3, and the maximum settlement value of the pipeline increased by 4 times and 6.5 times respectively. It shows that the formation loss rate has a great impact on the settlement of soil and pipelines. According to the analysis, the excavation volume should be strictly controlled, the supporting force of the excavation face should be close to the static earth pressure, and the reasonable parameters of grouting behind the wall should be selected in time to reduce the influence of construction on adjacent pipelines.

5   EXAMPLE VERIFICATION

Taking a tunnel section in Guangxi as the engineering background, the measured soil and pipeline parameters are $E_s = 17.8MPa$, $v_s = 0.31$, $k = 50.1MPa/m$, $2R = 6m$, $h = 15m, z_0 = 5m$, and $D = 1.0m$. Through the comparative analysis between the theoretically calculated value and the field measured value, the comparison diagram is shown in Figure 8.

It shows that the settlement curve obtained from the theoretical calculation value of pipeline settlement is generally consistent with the field measured value. The theoretical method in this paper can accurately predict the vertical displacement of the pipeline. Within 10m from the left side of the road centerline, the theoretical calculation value is in good agreement with the actual monitoring value. The maximum displacement of actual monitoring is less than that of theoretical calculation, which shows that the theory in this paper is slightly conservative. Within 10m from the right side of the road centerline, the maximum displacement actually monitored is greater than the theoretical displacement. This is because the subway tunnel is a double-track tunnel, and when the subsequent tunnel is excavated, it will disturb the soil layer above the previous tunnel, increase the original stratum loss rate, and then increase the vertical settlement of the pipeline.

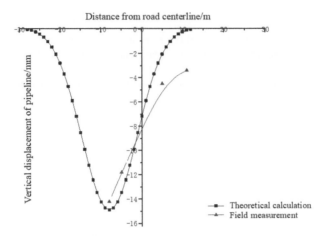

Figure 8.   Comparison between theoretical calculation and field monitoring.

6   CONCLUSION

Based on the Pasternak two-parameter model, combined with the boundary conditions, using the finite difference method, this paper deduces the theoretical solution of the shield construction to the vertical displacement of the existing pipeline, compiles the relevant program with MATLAB, and analyzes the parameter sensitivity of the influencing factors of the vertical settlement of the pipeline. The following conclusions are drawn:

(1) In this paper, the Pasternak model is applied by a two-stage displacement method. The model can reflect the shear stiffness of soil to some extent. Based on the model, a vertical deformation differential equation is created. Based on the boundary condition and finite difference method, the settlement law of the existing pipeline during tunnel construction is deduced.
(2) The maximum vertical displacement of the pipeline in tunnel construction and the maximum tangent slope of the displacement curve increase with the increase of the buried depth of the pipeline. The maximum vertical displacement of the pipeline decreases with the increase of pipeline diameter. With the increase in formation loss rate, the vertical displacement and tangent slope of the pipeline increase significantly.
(3) By comparing it with the on-site monitoring, it is proved that the theoretical calculation results are in good agreement with the existing pipeline settlement curve, and can better describe the existing pipeline deformation law in practical engineering.

REFERENCES

Chambon P, Core J. (1994) Shallow Tunnels in Cohesionless Soil: Stability of Tunnel Face. *Journal of Geotechnical Engineering*, 120(7):1148–1165.
GAO Bing- li, CAI Zhi-yun. (2015) Study on influence law of deformation for adjacent pipelines perpendicular to the metro line due to construction of the metro tunnel. *Journal of Safety Science and Technology*, 11(12):59–64.
Gu S, He H, Ru G. (2015) Analysis of Underground Pipeline Stress Caused by Metro Tunneling. *Urban Mass Transit*, 18(05):14–18+23.
Huang X, Lu K. Zhu D. (2017) Simulation test study of deformations of pipelines located at different geometric positions arising from shield tunneling. *Rock and Soil Mechanics*, 38(S1):123–130.
Li H, Zhang C, Lu K. (2018) Nonlinear analysis of the response of buried pipelines induced by tunneling. *Rock and Soil Mechanics*, 39(S1): 289-296.

Liang R, Xia T, Hong Y, et al. (2016). Effects of above-crossing tunnelling on the existing shield tunnels. *Tunnelling and Underground Space Technology*, 5(58): 159–176

Lin C, Huang Mao-song. (2019) Deflections of discontinuous buried pipelines induced by shield tunnelling based on Pasternak foundation. *Chinese Journal of Geotechnical Engineering*, 41(07): 1200–1207.

Long Y. (1982) Calculation of beam on elastic foundation. People's education press, Beijing

Tanha Shi H. (2004) Formulas for an infinitely long Bernoulli-Euler beam on the Pasternak model. *Journal of the Japanese Geotechnical Society*, 44(5): 109–118.

Wei G, Wang C, Cai S., et al. (2019) Model tests on influences of quasi-rectangular shield construction on underground pipelines. *Chinese Journal of Geotechnical Engineering*, 41(08): 1489–1495.

Wu W. (2013). Settlement Calculation Method of Rigid Connection Underground Pipeline Affected by Tunnel Construction. *China Water & Wastewater.*, 29(10):105–108.

Xu L. (2005) Longitudinal Settlement of Shield Tunnel in Soft Soil. Tongji University, Shanghai:

Zhang Fengxiang. *Shield tunnel*. Beijing: People's Communications Press, 2004: 867–8693

Zhang Z, Xu G. (2016) Jian-fei. Influence of tunneling on the deflection of adjacent piles considering shearing deformation of foundation and 3D effects of lateral soils beside piles. *Chinese Journal of Geotechnical Engineering*, 38(05):846–856.

Advances in Measurement Technology, Disaster Prevention and Mitigation – Li & Mohd Yusof (Eds)
© 2023 The Author(s), ISBN: 978-1-032-36087-4

Acoustic-to-seismic coupling characteristics of a specific poroelastic site

Liangyong Zhang
College of Meteorology and Oceanography, National University of Defense Technology, Changsha, Hunan, China
Northwest Institute of Nuclear Technology, Xi'an, Shaanxi, China

Xin Li, Weiguo Xiao, Xiaolin Hu, Shuai Cheng, Kaikai Li & Dezhi Zhang*
Northwest Institute of Nuclear Technology, Xi'an, Shaanxi, China

ABSTRACT: Air-coupled seismic waves contain a wealth of acoustic information and provide an alternative way to invert explosion source parameters in addition to the precursor seismic waves, which is of great significance. Based on the measured data, this paper analyzed the acoustic-to-seismic coupling characteristics of chemical explosions at a poroelastic site with a sandy soil surface sparsely covered by short vegetation and discussed the change law and quantitative relationships of the arrival time, amplitude, and acoustic-to-seismic coupling coefficient of air-coupled seismic waves. The results show that the sound velocities obtained through theoretical calculations and the arrival time of the air-coupled seismic waves were almost the same as that obtained from the arrival time of acoustic waves, and the particle velocity amplitude of the air-coupled seismic waves had a significant linear relationship with the distance on the logarithmic scale. Additionally, the coupling coefficient of peak acoustic pressure was related to the difference between the attenuation coefficients of acoustic pressure and particle velocity of the air-coupled seismic waves and remained unchanged with the variation of distance as a whole. Through data analysis, it was found that for the coupling coefficients of acoustic pressure peak, Zhalf and ENZhalf were the least discrete on logarithmic and equidistant scales respectively, with corresponding values of 0.57 Pa/(um/s) and 0.34 Pa/(um/s), while for the acoustic-to-seismic coupling coefficients of positive acoustic impulse peak, Z2 and ENZ2 0&1 are the least discrete on logarithmic and equidistant scales respectively, with corresponding values of 0.0081 Pa•s/(um/s) and 0.0047 Pa•s/(um/s).

1 INTRODUCTION

The energy produced by a near-surface chemical explosion will couple into the air and ground medium to produce acoustic and seismic waves, respectively (Arrowsmith et al. 2010; Bonner et al. 2013; Ford et al. 2014; Pasyanos & Kim 2018). The seismic measurement point will receive the seismic waves coupled to the ground medium nearby the explosion source (precursor seismic waves) and the seismic waves generated by the acoustic waves coupled to the ground in the air near the measurement point (air-coupled seismic waves). Compared to the precursor seismic waves, the air-coupled seismic waves contain a wealth of acoustic information. If the transfer function of the acoustic and air-coupled seismic waves nearby the seismic measurement point is known, the acoustic information coupled to the air from the explosion can be obtained by the air-coupled seismic waves measured by a seismometer. The acoustic information obtained by a seismometer provides another way of inverting the explosion source parameters in addition to ways of the

---

*Corresponding Author: zhangdezhi@nint.ac.cn

 DOI 10.1201/9781003330172-62

precursor seismic waves, playing a prominent role in the explosion accident monitoring (Jiang et al. 2019), damage assessment(Albert & Taherzadeh 2013; Driels 2012), and other aspects.

Since air-coupled seismic waves are closely related to the ground structure in the vicinity of the seismic measurement point, the ground surface structure is complex, and geological parameters are hard to obtain directly, the analysis of air-coupled seismic waves has been a difficult issue (Albert 1993; Albert & Taherzadeh 2013; Kaynia et al. 2011). Flohr et al. (1979) proposed the acoustic coupling theory of a non-poroelastic ground for the analysis of air-coupled seismic waves in a purely elastic medium. After that, Biot (1956a; 1956b; 1962a; 1962b) proposed the acoustic coupling theory of the poroelastic medium, which was further developed by Stoll (1970; 1980; 1981) and applied to the acoustic propagation process in underwater sediments, forming the Biot-Stoll theory (Sabatier et al. 1986; Sabatier & Hickey 2000) which well explains the acoustic propagation process in a poroelastic medium. Subsequently, the theory was applied to the single-porous-layer ground(Albert 1993a, 1993b; Yamamoto 1983), which solved the problem of air-coupled seismic waves of single-porous-layer ground. After then, the theory was further applied to a two-layer ground structure composed of porous media and nonporous substrate, which addressed the air-coupled seismic wave problem of a nonporous surface covered by snow (Bass & Bolen 1984; Sabatier et al. 1986; Sabatier & Hickey 2000). Kaynia et al. (2011) modeled a multilayer ground surface structure of a porous medium, proposing a general approach to modeling air-coupled seismic waves and analyzing the characteristics of ground impedance and acoustic attenuation. Based on measured data, Albert et al. (2013) analyzed the variation of the acoustic-to-seismic coupling characteristic parameter ratio with distance under various geological conditions and predicted the waveforms of air-coupled seismic waves at a poroelastic site based on the above theory. In conclusion, the theoretical analysis method of air-coupled seismic waves has been established, but more in-depth studies need to be conducted on acoustic-to-seismic coupling laws of the various ground medium, especially on the quantitative analysis of acoustic-to-seismic coupling. In this paper, an experimental analysis of the acoustic-to-seismic coupling characteristics of explosions at a poroelastic site with a sandy soil surface sparsely covered by short vegetation was carried out to discuss the change law and quantitative relationships of the arrival time, amplitude, and acoustic-to-seismic coupling coefficient of air-coupled seismic waves.

2 EXPERIMENTAL INFORMATION

A total of three experiments with yields of 100 kg and 8 kg were conducted, and the explosion sources and their meteorological conditions are shown in Table 1. Seven acoustic and seismic measurement points are set and co-located, and the layout of the measurement points is not the same in the three experiments with the distribution of the measurement points shown in Figure 1. The ground is covered by sandy soil and sparsely distributed short vegetation as seen in Figure 2.

Table 1. Explosion sources and meteorological conditions.

Explosion source number	Yield/kg	Height of source/m	Meteorological condition		
			Air pressure/hpa	Temperature/°	Wind speed
EX01	100	10	860	−0.7	Light winds
EX02	8	0.5	864	−1.45	Light winds
EX03	100	0.5	861	−2.65	Light winds

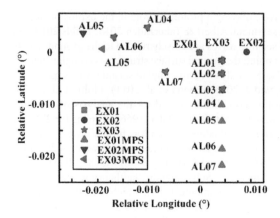

Figure 1.   Distribution of measurement points. MPS means measurement points.

Figure 2.   Ground conditions.

3   ACOUSTIC AND SEISMIC DATA

3.1   *Sound propagation characteristics*

3.1.1 *Waveform characteristics.* Typical waveforms of the measured acoustic data are shown in Figure 3, with the relative time in the horizontal coordinate. As can be seen from the figure, the acoustic waveforms have high signal-to-noise ratios. As the distance increases, the high-frequency quantity decays rapidly, the positive pulse width becomes larger, the rising edge of the waveforms gradually slows down from a steep slope, and the characteristics of the negative part of the waveforms also change significantly.

3.1.2 *Arrival time and wave velocity.* Figure 4 shows the relationship between the arrival time of acoustic waves and the distance, from which the wave velocity is fitted, as shown in Table 2. Theoretically, the sound velocity is temperature-dependent and can be calculated as follows (Ma & Shen 2004).

$$c = 331.45\sqrt{1 + \frac{T}{273.16}} \qquad (1)$$

Where T is the ambient Celsius temperature (°C). The theoretical sound velocity can be obtained by substituting the ambient temperature into the equation, as shown in Table 2. As can be seen from the table, the measured and theoretical values are almost the same, with a relative error of no more than 0.8%, indicating that the on-site sound velocity estimated by means of the theoretical method has high accuracy.

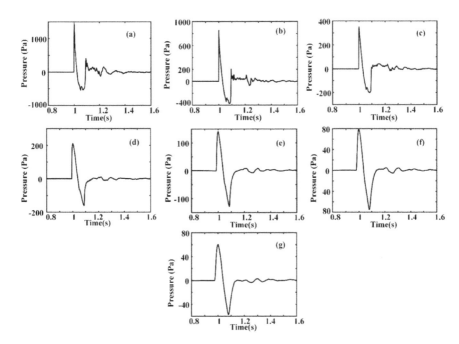

Figure 3.   Typical waveforms of acoustic pressure (EX01). (a)-(g) correspond to AL01-AL07.

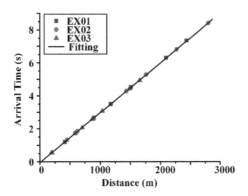

Figure 4.   Arrival time of acoustic waves.

Table 2.   Velocities of acoustic waves.

Explosion source number	Fitted sound velocity (m/s)	Theoretical sound velocity (m/s)	Relative error (%)
EX01	330.03	331.03	0.30
EX02	332.23	330.57	0.50
EX03	332.23	329.84	0.72
Comprehensive result	330.7	–	–
Comprehensive standard deviation	3.75	–	–

3.1.3 *Amplitude attenuation law.* Since this paper focuses on the acoustic-to-seismic coupling characteristics, which means that both acoustic data and seismic wave data need to be considered, and the amplitude of the acoustic measurement points co-located with abnormal seismic measurement points is not taken into consideration. The acoustic attenuation data of EX01-EX03 is shown in Figure 5, where the solid line represents the fitted curve on the $\log_{10}$ scale. The fitting formula is given in Equation (2), and the coefficients to be determined are given in Table 3. As can be seen from the figure, the acoustic pressure in the logarithmic coordinates has a good linear relationship with distance, and the values of the coefficients to be determined show that the acoustic pressure attenuation coefficients of the three explosion sources have a difference, which may be incurred by different locations of the explosion sources, atmospheric environments, and ground conditions.

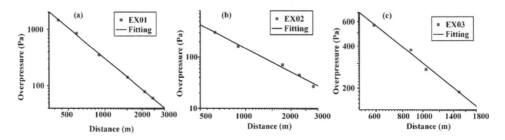

Figure 5.   The attenuation of acoustic pressure amplitude with distance. (a)–(b) correspond to EX01-EX03.

$$\log_{10}(p) = a_1 + b_1 \log_{10}(r) \tag{2}$$

where p is the acoustic pressure amplitude (Pa), r is the distance (m), and a_1 and b_1 are the coefficients to be determined.

Table 3.   Values of coefficients to be determined.

Explosion Source Number	Coefficients to be Determined	
EX01	a_1	7.94
	b_1	−1.82
	SD	0.02
EX02	a_1	6.76
	b_1	−1.53
	SD	0.05
EX03	a_1	6.11
	b_1	−1.21
	SD	0.03

*SD represents the standard variation in the $\log_{10}$ scale

3.2   Propagation characteristics of air-coupled seismic waves

3.2.1 *Waveform characteristics.* The seismic waves generated by an explosion can be divided into two types, that is, precursor seismic waves and air-coupled seismic waves. The precursor seismic waves are coupled from the vicinity of the explosion source to the underground and then propagate to the seismic measurement point, while the air-coupled seismic waves are coupled from the wave in the air nearby the seismic measurement points. Figure 6 shows two typical air-coupled seismic waves measured. In Figures 6(a) and (b), obvious superimposition exists between the air-coupled seismic waves and the precursor seismic waves, while in Figures (c) and (d), the air-coupled seismic waves and the precursor seismic waves are clearly separated so that purely air-coupled seismic waves can be obtained. To study the acoustic-to-seismic coupling characteristics, the air-coupled seismic waves with obvious superimposition of the precursor seismic waves will not be considered. As shown in Figure 6, the air-coupled seismic waveforms have steeper rising or falling edges, narrower pulse widths, and relatively larger amplitudes than the precursor seismic waveforms.

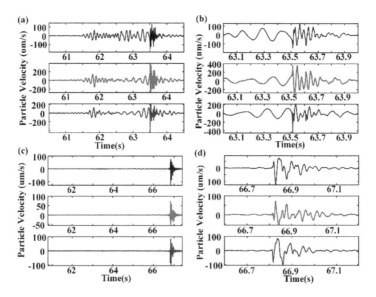

Figure 6. Typical waveforms of air-coupled seismic waves. The waveform graphs from top to bottom in each figure show the east (E) component (black line), the north (N) component(red line), and the vertical (Z) component (blue line). (a) and (c) Overall waveforms of EX01's AL04 and EX03's AL06 respectively, (b) and (d) Air-coupled seismic waveforms of EX01's AL04 and EX03's AL06 respectively.

3.2.2 *Arrival time and wave velocity.* Figure 7 shows the arrival times of the air-coupled seismic waves, from which the wave velocities are fitted, as shown in Table 4. The fitted wave velocities of the air-coupled seismic waves are close to those of the acoustic data, with a relative error of no more than 1%, indicating that the on-site sound velocity is estimated by means of the air-coupled seismic waves has high accuracy.

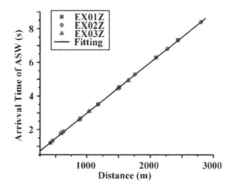

Figure 7. Arrival times of air-coupled seismic waves. ASW means air-coupled seismic wave.

Table 4.   Air-coupled seismic wave velocities.

Explosion source number	Fitted wave velocity of air-coupled seismic wave (m/s)	Fitted wave velocity of acoustic data (m/s)	Relative error (%)
Ex01	331.13	330.03	0.33
Ex02	332.23	332.23	0
Ex03	335.57	332.23	1.0
Comprehensive result	331.92	330.7	0.37
Comprehensive standard deviation	6.19	3.75	-

3.2.3 *Amplitude attenuation law.* Figure 8 shows the typical waveforms of air-coupled seismic waves in the E, N, and Z directions (from top to bottom in the figure). The E and N components show three typical peaks and valleys from the start of the acoustic coupling, both of which are labelled 0, 1, and 2. The Z component shows two typical peaks and valleys from the start of the acoustic coupling, both of which are labeled 1 and 2, in order from left to right. The amplitude characteristics of air-coupled seismic waves are characterized by a variety of characteristic parameters. This paper adopts valleys in the Z direction (Z1), peaks in the Z direction (Z2), and half of the peak-to-peak value in the Z direction (Zhalf), RMS values of peaks or valleys or half of the peak-to-peak value in N, E and Z directions (ENZ1, ENZ2, and ENZhalf). 0&1 represents 0 and 1 peak or valley in E and N directions, while 1&2 represents the 1 peak and 2 peaks or valley in E and N directions. ENZhalf1 represents the average of ENZ1 and ENZ2, while ENZhalf2 represents the RMS values of half of the peak-to-peak values in the E, N, and Z directions.

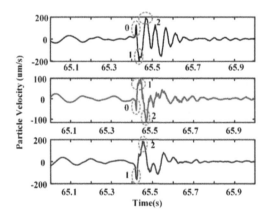

Figure 8.   Typical waveform characteristics of air-coupled seismic waves. Waveform graphs from top to bottom show the east (E) component (black line), north (N) component(red line), and vertical (Z) component (blue line).

The attenuation characteristics of the air-coupled seismic waves in EX01-EX03 are shown in Figure 9, where the solid lines are the fitted curves of Zhalf and ENZhalf1 on the $\log_{10}$ scale. The fitted equation is given in Equation (3), and the coefficients to be determined are given in Table 5. As can be seen from the figure, the particle velocities of the air-coupled seismic waves in the logarithmic coordinates have a good linear relationship with distance, and the values of the coefficients to be determined show that the particle velocity attenuation coefficients of the three explosion sources are different. This is due to differences in explosion source locations, atmospheric conditions, and surface conditions. For the EX01 source, Zhalf shows the best linearity among the Z-direction particle velocities, while ENZ1 shows the best linearity among the ENZ 1&2 particle velocities, followed by ENZhalf, and ENZhalf shows the best linearity among ENZ 0&1 particle velocities. For the EX02 source, Z2 shows the best linearity among the Z-direction particle velocities, followed by Zhalf, while ENZhalf shows the best linearity among the ENZ 1&2 and ENZ 0&1 particle velocities. For the EX03 source, Z1 shows the best linearity among the Z-direction particle velocities, followed by Zhalf, while ENZ2 shows the best linearity among the ENZ 1&2 particle velocities, followed by ENZhalf, and ENZhalf shows the best linearity among the ENZ 0&1 particle velocities. To sum up, Zhalf, ENZhalf 0&1, and ENZhalf 1&2 have good overall linearity (in logarithmic coordinates).

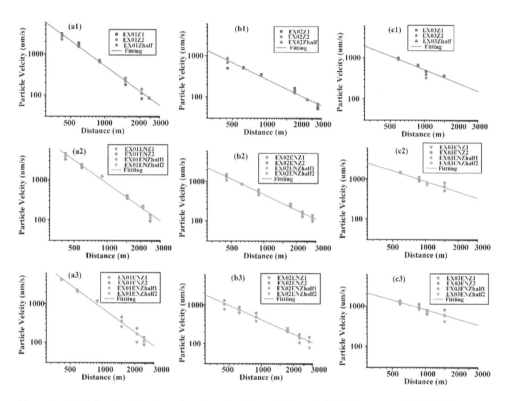

Figure 9. Particle velocity attenuation characteristics of air-coupled seismic waves. (a1)–(c1) Z components of EX01-EX03 respectively, (a2)–(c2) ENZ 1&2 of EX01–EX03 respectively, (a3)–(c3) ENZ 0&1 of EX01–EX03 respectively.

As can be seen from Table 5, ENZhalf 0&1 has the smallest standard deviation in two of the three trials and one in the middle, with relatively small dispersion and good linearity on the $\log_{10}$ scale.

$$\log_{10}(v) = a_2 + b_2 \log_{10}(r) \tag{3}$$

where v is the particle velocity (um/s), r is the distance (m), and a_2 and b_2 are the coefficients to be determined.

461

Table 5. Coefficients to be determined.

Explosion source number	Coefficients to be determined	Zhalf of air-coupled seismic waves	1&2 ENZhalf of air-coupled seismic waves	0&1 ENZhalf of air-coupled seismic waves
EX01	a	8.78	8.77	8.97
	b	−2.02	−1.96	−2.03
	SD	0.04	0.07	0.06
EX02	a	6.46	6.60	6.32
	b	−1.34	−1.31	−1.24
	SD	0.05	0.03	0.03
EX03	a	6.00	5.58	5.31
	b	−1.10	−0.88	−0.80
	SD	0.08	0.05	0.04

* SD represents the standard variation on a log10 scale.

4   ANALYSIS OF ACOUSTIC-TO-SEISMIC COUPLING CHARACTERISTICS

The ratio of the acoustic-to-seismic waveform characteristic parameter is defined as the acoustic-to-seismic coupling coefficient, e.g., the ratio of the peak acoustic pressure to the particle velocity amplitude, as shown in Equation (4), and its variation with distance is related to the difference of the coefficients to be determined for the acoustic pressure and the particle velocity.

$$A_{cp} = \frac{P}{v} = 10^{a_1-a_2} r^{b_1-b_2} = 10^{d_a} r^{d_b} \tag{4}$$

where d_a and d_b are the coefficients to be determined.

Figure 10 shows the variation of the coupling coefficient of peak acoustic pressure with distance for EX01-EX03. As can be seen from the figure, the acoustic-to-seismic coupling coefficient of EX01 has an overall increasing trend with the increase of distance, while the acoustic-to-seismic coupling coefficient of EX02-EX03 has an overall decreasing trend with the increase of distance, which is caused by the differences in the ground surface in different directions (measurement points of EX01 and EX02 and EX03 are distributed in different directions). As can be seen from the figure, the results of ENZhalf1 and ENZhalf2 are consistent, with the discrete points overlapping exactly, which will not be distinguished in subsequent analysis. The solid lines in the figure show the fitted curves of Zhalf and ENZhalf1 obtained from Equation (4) on the $\log_{10}$ scale, and the coefficients to be determined are shown in Table 6. It can be seen from the table that the overall slope of the acoustic-to-seismic coupling coefficient Zhalf is relatively small, followed by ENZhalf 1&2.

Table 6. Coefficients to be determined.

Explosion source number	Coefficients to be determined	Zhalf of air-coupled seismic waves	ENZhalf 0&1 of air-coupled seismic waves	ENZhalf 1&2 of air-coupled seismic waves
EX01	d_a	−0.84	−1.02	−0.83
	d_b	0.20	0.21	0.14
EX02	d_a	0.31	0.44	0.16
	d_b	−0.19	−0.29	−0.22
EX03	d_a	0.11	0.81	0.53
	d_b	−0.11	−0.41	−0.33

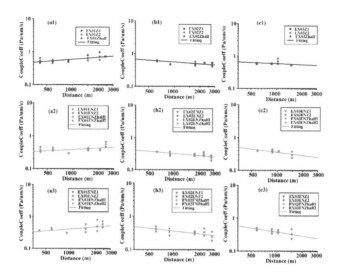

Figure 10.   Characteristics of air-coupled seismic waves. (a1)–(c1) Z components of EX01–EX03 respectively, (a2)–(c2) ENZ 1&2 of EX01–EX03 respectively, (a3)–(c3) ENZ 0&1 of EX01–EX03 respectively.

Figure 11 shows the variation of each characteristic parameter coupling coefficient of acoustic pressure peak coupling coefficient with distance, and each characteristic parameter data contains the data of the three explosion sources. As can be seen from the figure, the acoustic-to-seismic coupling coefficient is not obviously correlated with distance and oscillates around a certain value with distance changing. The dashed lines in the figure show the fitted curves of the acoustic-to-seismic coupling coefficients obtained through Equation (4) on the $\log_{10}$ scale, and the solid lines

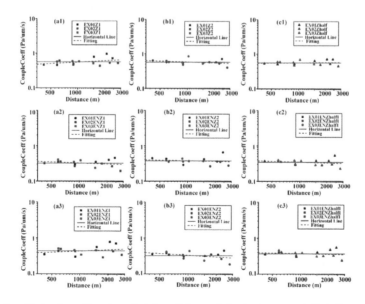

Figure 11.   Amplitude attenuation characteristics of air-coupled seismic waves. (a1)–(c1) correspond to Z1, Z2 and Zhalf respectively, (a2)–(c2) correspond to ENZ1 1&2, ENZ2 1&2 and ENZhalf1 1&2 respectively, (a3)–(c3) correspond to ENZ1 0&1, ENZ2 0&1 and ENZhalf1 0&1 respectively.

show the fitted results when the slope is 0, and the coefficients to be determined are shown in Tables 7–8.

Table 7.   Coefficients determined in the Z direction.

Characteristic parameters	Coefficients to be determined	Acoustic-to-seismic coupling coefficient as depicted by the dashed lines in the above figure	Fitted value of acoustic-to-seismic coupling coefficient when the slope is 0 as depicted by the solid lines in the above figure
	d_a	-0.67	-0.25
Z1	d_b	0.14	0
	SD	0.09	0.097
	d_a	0.021	-0.24
Z2	d_b	-0.086	0
	SD	0.07	0.073
	d_a	-0.32	-0.25
Zhalf	d_b	0.023	0
	SD	0.07	0.064

Table 8.   ENZ coefficients to be determined.

Characteristic parameters	Coefficients to be determined	Acoustic-to-seismic coupling coefficient of 1&2 as depicted by the dashed lines in Figure 11	Fitted value of acoustic-to-seismic coupling coefficient of 1&2 when the slope is 0 as depicted by the solid lines in Figure 11	Acoustic-to-seismic coupling coefficient of 0&1 as depicted by the dashed lines in Figure 11	Fitted value of acoustic-to-seismic coupling coefficient of 0&1 when the slope is 0 as depicted by the solid lines in Figure 11
	d_a	-0.20	-0.50	-0.68	-0.36
ENZ1	d_b	-0.097	0	0.10	0
	SD	0.099	0.10	0.125	0.12
	d_a	-0.27	-0.43	0.023	-0.49
ENZ2	d_b	-0.052	0	-0.17	0
	SD	0.099	0.10	0.1	0.11
	d_a	-0.22	-0.46	-0.26	-0.43
ENZhalf1	d_b	-0.078	0	-0.057	0
	SD	0.089	0.09	0.094	0.09

*SD represents the standard variation on a $\log_{10}$ scale.

As can be known from the fitted results depicted by the dashed lines in Tables 7–8, the fitted slope of each characteristic parameter does not exceed 0.15, and the distribution of fitted slopes focuses on 0.1, indicating that the acoustic-to-seismic coupling coefficient remains a constant value with the variation of distance. From the fitted results depicted by the solid lines in Tables 7–8, it can be seen that the fitted results of the acoustic-to-seismic coupling coefficients of the characteristic parameters in the Z direction are basically the same, with the value of around –0.25 on the $\log_{10}$ scale, which can be translated into an acoustic-to-seismic coupling coefficient of 0.57 Pa/(um/s) (see Table 9), where

Zhalf has the smallest dispersion (smallest standard deviation) among the characteristic parameters in the Z direction. The difference between the acoustic-to-seismic coupling coefficients of 1&2 is relatively small among the characteristic parameters of ENZ, while the difference between the acoustic-to-seismic coupling coefficients of 0&1 is significant. ENZhalf1 1&2 has the smallest dispersion (smallest standard deviation) with an acoustic-to-seismic coupling coefficient of 0.34 Pa/(um/s) (see Table 9). As a whole, ENZhalf1 1&2 outperforms ENZhalf1 0&1. Compared with the Z-direction and ENZ coefficients to be determined, the Z-direction characteristic parameters are generally less discrete than ENZ on the logarithmic scale (see Tables 7–8). The acoustic-to-seismic coupling coefficient of Zhalf is the least discrete. In contrast, ENZ 1&2 and 0&1 are generally less discrete than the Z-direction characteristic parameters on an equally spaced scale (see Table 9). The acoustic-to-seismic coupling coefficient of ENZhalf1 1&2 is the least discrete.

Table 9. Acoustic-to-seismic coupling coefficient (fitting slope on a logarithmic scale is 0).

Parameters	Z1	Z2	Zhalf	ENZ1 1&2	ENZ2 1&2	ENZhalf1 1&2	ENZ1 0&1	ENZ2 0&1	ENZhalf1 0&1
Coupled coeffcient									
Pa/(um/s)	0.57	0.57	0.56	0.32	0.37	0.34	0.44	0.32	0.37
SD	0.13	0.10	0.08	0.07	0.08	0.07	0.13	0.08	0.08

*SD denotes the standard deviation on the equally spaced scale, calculated based on the uncertainty propagation theory.

The fitted values of the acoustic-to-seismic coupling coefficients (with a slope of 0) of the positive acoustic impulse peak are obtained using the same method as above, as shown in Table 10. It can be seen from the table that the Z2 acoustic-to-seismic coupling coefficient is the least discrete on the logarithmic scale with a value of 0.0081 Pa•s/(um/s), and the acoustic-to-seismic coupling coefficient of ENZ2 0&1 is the least discrete on the equally spaced scale with a value of 0.0047 Pa•s/(um/s).

Table 10. Acoustic-to-seismic coupling coefficients of positive acoustic impulse peak (fitted slope on the logarithmic scale is 0).

Characteristic parameter	Coeffici-ents	Z1	Z2	Zhalf	ENZ1 1&2	ENZ2 1&2	ENZhalf1 1&2	ENZ1 0&1	ENZ2 0&1	ENZhalf1 0&1
Acoustic-to-seismic coupling coefficients on the $\log_{10}$ scale	d_a	-2.09	-2.09	-2.09	-2.34	-2.27	-2.31	-2.20	-2.33	-2.28
Acoustic-to-seismic coupling coefficients on the equally spaced scale	SD Values Pa·s/(um/s)	0.32 $\times10^{-3}$	0.28 $\times10^{-3}$	0.29 $\times10^{-3}$	0.30 $\times10^{-3}$	0.30 $\times10^{-3}$	0.29 $\times10^{-3}$	0.33 $\times10^{-3}$	0.29 $\times10^{-3}$	0.30 $\times10^{-3}$
		8.13	8.18	8.10	4.58	5.35	4.91	6.33	4.66	5.30
	SD	6.01 $\times10^{-3}$	5.19 $\times10^{-3}$	5.44 $\times10^{-3}$	3.14 $\times10^{-3}$	3.68 $\times10^{-3}$	3.34 $\times10^{-3}$	4.81 $\times10^{-3}$	3.09 $\times10^{-3}$	3.64 $\times10^{-3}$

5 CONCLUSIONS

Based on the measured data, this paper analyzed the acoustic-to-seismic coupling characteristics of chemical explosions at a poroelastic site with a sandy soil surface sparsely covered by short vegetation and discussed the change law and quantitative relationships of the arrival time, amplitude, and acoustic-to-seismic coupling coefficients of air-coupled seismic waves. The following conclusions were drawn.

1) The sound velocities obtained from theoretical calculations and the arrival time of air-coupled seismic waves are almost the same as that obtained from the arrival time of acoustic waves, and the relative error does not exceed 1%, indicating high estimation accuracy of sound velocity.
2) The characteristic parameter of particle velocity amplitude of air-coupled seismic waves has an obvious linear relationship with distance in the logarithmic scale and is related to the acoustic attenuation characteristics. The coupling coefficient of peak acoustic pressure is related to the difference between the attenuation coefficients of the acoustic pressure and the particle velocity of the air-coupled seismic waves, remaining unchanged with the variation of distance as a whole.
3) The coupling coefficients of acoustic pressure peak, Zhalf and ENZhalf were the least discrete on logarithmic and equidistant scales respectively, with corresponding values of 0.57 Pa/(um/s) and 0.34 Pa/(um/s), while for the acoustic-to-seismic coupling coefficients of positive acoustic impulse peak, Z2 and ENZ2 0&1 are the least discrete on logarithmic and equidistant scales respectively, with corresponding values of 0.0081 Pa•s/(um/s) and 0.0047 Pa•s/(um/s).

REFERENCES

Albert, D.G. (1993a) A comparison between wave propagation in water-saturated and air-saturated porous materials. *Journal of Applied Physics*, 73(1): 28–36.

Albert, D.G. (1993b) *Attenuation of outdoor sound propagation levels by a snow cover*. US Army Cold Regions Research and Engineering Laboratory, CRREL Report 93–20.

Albert, D.G., Taherzadeh, S. (2013) Attenborough K, et al. Ground vibrations produced by surface and near-surface explosions. *Applied Acoustics*, 74: 1279–1296.

Arrowsmith, S.J., Johnson, J.B., Drob, D.P., et al. (2010) The seismoacoustic wavefield: a new paradigm in studying geophysical phenomena. *Reviews of Geophysics*, 48(RG4003): 1–23.

Bass, H.E., Bolen, L.N. (1984) *Coupling of airborne sound into the earth*. Physical Acoustics Research Group, University of Mississippi, AD-A146231.

Biot, M.A. (1956a) Theory of propagation of elastic waves in a fluid-saturated porous solid. I. Low-frequency range. *J. Acoust. Soc. Am.*, 28(2): 168–178.

Biot, M.A. (1956b) Theory of propagation of elastic waves in a fluid-saturated porous solid. II. Higher frequency range. *J. Acoust. Soc. Am.*, 28: 179–191.

Biot, M.A. (1962c) Generalized theory of acoustic propagation in porous dissipative media. *J. Acoust. Soc. Am.*, 34(9): 1254–1264.

Biot, M.A. (1962d) Mechanics of deformation and acoustic propagation in porous media. *Journal of Applied Physics*, 33(3): 1482–1498.

Bonner, J., Waxler, R., Gitterman, Y., et al. (2013) Seismo-acoustic energy partitioning at near-source and local distances from the 2011 Sayarim explosions in the Negev Desert, Israel. *Bulletin of the Seismological Society of America*, 103(2A): 741–758.

Driels, M.R. (2012) *Weaponeering: Conventional weapon system effectiveness*. 2 ed. Virginia: American Institute of Aeronautics Astronautics, Inc.

Flohr, M.D., Cress D.H. (1979) *Acoustic to seismic coupling properties and applications to seismic sensors*. Vicksburg: U. S. Army Engineer Waterways Experiments Station Environmental Laboratory.

Ford, S.R., Rodgers, A.J., Xu, H., et al. (2014) Partitioning of seismoacoustic energy and estimation of yield and height-of-burst/depth-of-burial for near-surface explosions. *Bulletin of the Seismological Society of America*, 104(2): 608–623.

Jiang, W.B., Chen, Y., Peng, F. (2019) The yield estimation of the explosion at the Xiangshui, Jiangsu chemical plant in March 2019. *Chinese J. Geophys.* (in Chinese), 63(2): 541–550.

Kaynia, A.M., Løvholt, F., Madshus, C. (2011) Effects of a multi-layered poroelastic ground on attenuation of acoustic waves and ground vibration. *Journal of Sound and Vibration.*, 330: 1403–1418.

Ma, D.Y., Shen H. (2004) *Handbook of acoustics.* Science Press, Beijing.

Pasyanos, M.E., Kim, K. (2018) Seismoacoustic analysis of chemical explosions at the Nevada national security site. *Journal of Geophysical Research: Solid Earth*, 124: 908–924.

Sabatier, J.M., Bass, H.E., Bolen, L.N. (1986) The interaction of airborne sound with the porous ground: the theoretical formulation. *J. Acoust. Soc. Am.*, 79(5): 1345–1352.

Sabatier, J.M., Hickey, C.J. (2000) The acoustic to seismic transfer function at the surface of a layered outdoor ground: James. *Proceedings of SPIE*, 4038: 633–644.

Stoll, R.D. (1980) Theoretical aspects of sound transmission in sediments. *J. Acoust. Soc. Am.*, 68(5): 1341–1350.

Stoll, R.D., Bryan, G.M. (1970) Wave attenuation in saturated sediments. *J. Acoust. Soc. Am.*, 47(5): 1440.

Stoll, R.D., Kan, T.K. (1981) Reflection of acoustic waves at a water-sediment. *J. Acoust. Soc. Am.*, 70(1): 149–156.

Yamamoto, T. (1983) Propagation matrix for continuously layered porous seabeds. *Bulletin of the Seismological Society of America*, 73(6): 1599–1620.

Advances in Measurement Technology, Disaster Prevention and Mitigation – Li & Mohd Yusof (Eds)
© 2023 The Author(s), ISBN: 978-1-032-36087-4

The coupling effect of fluid and solid in the red rock slope of Yunnan Province

Jiachen Xu & Qingwen Zhang*
School of Civil Engineering, Southwest Forestry University, Kunming, China

ABSTRACT: To study the stability of the red bed slope in Central Yunnan under rainfall conditions. According to the current situation of red bed soft rock in Central Yunnan, the geological characteristics of red bed in Central Yunnan, rock mass structure of excavated slope, types of excavated slope disasters, and prevention and treatment measures are systematically analyzed. The seepage coupling analysis of slope instability under rainfall is carried out by using geo studio and can provide a reference for disaster treatment.

1 INTRODUCTION

The inclination of national policies has enabled the southwest region to enter the ranks of rapid development, especially the rapid development of highway construction. The topographic relief in Central Yunnan, especially the existence of the special geological condition of the "red layer in Central Yunnan", makes the highway tunnel and other projects form a typical slippery stratum due to the special engineering nature of the red layer in the construction process and later maintenance process (Li et al. 2007). In engineering construction, it is very easy to induce various accidents, among which the slope disaster is the most likely to occur (Xu 2014). Therefore, we should study the slope disaster control technology in the red bed area of central Yunnan, and analyze it by using geo studio finite element simulation, thereby providing a reference for the slope disaster problems involved in future projects.

2 RED LAYER

2.1 *Distribution and scope of red beds in Central Yunnan*

"Red bed in Central Yunnan" includes lower Triassic, Jurassic, Cretaceous, and Upper Tertiary strata, which are mainly composed of purple red mudstone, shale, and purple gray sandstone mixed with silty mudstone, siltstone, and other clastic rocks formed by continental sedimentation. A small number of coal seams are distributed in lower Triassic and upper Tertiary strata, while grayish green peat rocks are exposed in Jurassic mudstone. It is a special rock formation with half soft and half hard. The red beds in Central Yunnan are distributed in the south of Jinsha River, the east of Ailao Mountain, and the west of Lufeng Donghu village. Some small basins are more common in mountainous and canyon areas (Gao 2010).

2.2 *Geological characteristics of red bed area in Central Yunnan*

The clay minerals in the red bedrock have small particles, strong hydrophilicity, and a low permeability coefficient. When water seeps into the pores of the rock, the adsorbed water film of the

---

*Corresponding Author: 904829146@qq.com

   DOI 10.1201/9781003330172-63

small particles will thicken and collect a large amount of free water and bound water. The water content will increase, and the uneven expansion of the rock mass will produce uneven stress in the rock mass, and the binding of the cement will deteriorate, resulting in the reduction of the strength of the rock mass, even easily disintegrated (Yang 2018).

As a sedimentary rock, the bearing capacity of the rock mass itself is not enough. The force of the rock mass is non-molecular, and it is also an anisotropic material, especially in expansion and contraction. In case of water loss, the rock will produce new fractures on the original basis and form a more serious fracture network, which may eventually disintegrate into small blocks in repeated cycles (Wang et al. 2016; Zhao et al. 2020). In the subtropical humid plateau monsoon climate area, due to the uneven distribution of precipitation time, there are soluble salts in some interlayers. When rainwater enters, it will cause some dissolution phenomena and accelerate disintegration. When the rock stratum collapses, the cohesiveness of the original cement will also be greatly reduced, and the integrity is very poor and irreversible (Zhang et al. 2008).

Viscous minerals are easily saturated by water absorption. After saturation, they are not easy to evaporate under strong light or wind. Moreover, after disintegration, the rock stratum broken into small pieces is more likely to be weathered and form soil (Zhu 2009).The red bed soft rock in Central Yunnan has disintegration and expansibility. For the same kind of rock, the more complete the rock is, the better the disintegration resistance is, and vice versa (Zhu 2013).

3 SLOPE ROCK MASS STRUCTURE IN RED BED AREA OF CENTRAL YUNNAN

The sloping structure of red bed soft rock is generally divided according to the inclination angle of the slope rock stratum: near horizontal, inclined, and upper covering accumulation.

3.1 *Near horizontal slope rock mass*

The rock mass stability of the near-horizontal slope is mainly determined by the lithology and combination relationship of the rock stratum, while the weak interlayer between the rock stratum and the rock stratum does not play a main control role because of the near-horizontal reason. Therefore, we can distinguish a subclass according to the lithology of rock stratum: the slope dominated by soft mudstone, the slope interbedded by soft and hard rock stratum, and the slope dominated by hard sandstone.

3.2 *Inclined slope rock mass*

The inclined slope rock mass is divided into a subclass according to the inclined direction of rock stratum and the inclined direction of slope: bedding slope, anti-inclined slope, and oblique slope structure. Bed slope structure refers to the inclination of the rock stratum in accordance with that of the excavated slope, and the angle of the two is close to or even parallel.

The structure of an anti-inclined slope can be divided into the structure dominated by thick sandstone with developed joints along the slope and the interbedded structure of sand and mudstone with developed layers. The structural joint structural plane of the first structure is in the same direction as the excavated slope. The super thick sandstone is divided by the structural joint plane and becomes a super thick sandstone layer, and the composition direction of the rock layer is nearly perpendicular to the slope direction. Therefore, the joint plane plays a major role in controlling the stability of the slope. The second structure is that the development of interlayer dislocation is obvious, but they are inclined toward the mountains, which generally does not affect the slope's stability, and the development of joint surface is very small.

The bedding oblique slope structure refers to that the interlaminar dislocation development is obvious, there will be more weak interlayers between the layers, and there is an included angle between the slope excavation surface and the layers. The weak interlayer between the layers and the structural plane inclined to the free surface are the main factors affecting the stability of the slope.

The other is the anti-inclined intersection structure slope. The interlayer dislocation surface and soft release layer of this structure are also relatively developed, but it is opposite to the direction of the excavation slope or even close to vertical, so it does not play a control role. On the contrary, the structural joint surface of this structure may be close to the direction of the excavation slope, so the structural joint surface is the most important factor affecting this slope.

3.3 *Slope rock mass accumulated by upper overburden*

For the red bed slope rock mass with upper overburden, the stability of this slope mainly depends on the nature of the upper overburden, and because most of the upper accumulation is looser, the slope disaster generally occurs in the upper accumulation.

Moreover, the different shapes of the upper cover will also have different effects on the stability of the slope. If the cover is in the inverted V shape, the rainwater can be discharged in time to reduce the impact of rainwater on the stability of the slope. It is also in the m or even double M shape, then the retained rainwater will infiltrate and cause greater harm to the stability of the slope.

4 TYPES OF RED BED SLOPE DISASTERS

4.1 *Near horizontal slope disaster*

The excavation disasters of near-horizontal red bedrock slope mainly include weathering erosion, landslide, rock mass cracking, and collapse. According to the combination of different lithologic layers, different disasters will also be caused.

The geological hazard of soft mudstone slope dominated by mudstone and siltstone in red bed slope is mainly weathering erosion. If the protection is not timely after excavation, it will be more obvious. After a rainy season, weathered debris will be accumulated at the foot of the slope.

Sand mud interbedded slope with sandstone and mudstone as the main material is also the most common red bedrock structure. Some geological disasters such as horizontal landslides and tension cracks will occur. Before the slope is unstable, there are usually cracks. Under the action of self-weight or excavation load, the cracks expand, rainwater seeps in, and the fissure water is not discharged in time to form a horizontal thrust. Under the combined action of the uplift pressure of the seepage water along with the layer, a horizontal push landslide or tension crack is caused.

If the upper layer is sandstone and the lower layer is mudstone during slope excavation, the difference in weathering time will be caused due to their weak weathering resistance, resulting in mudstone cavity under sandstone and collapse.

The lithology of this kind of slope mainly composed of sandstone is relatively hard. In the slope disaster, it is mainly collapsed, which can be divided into dumping type, sliding type, staggered type, and tensile fracture type. When excavating, excavate according to the design slope ratio and protect in time, which is easier to deal with such disasters.

5 MODEL AND RESULT ANALYSIS

5.1 *Calculation model and initial boundary conditions*

As shown in Figure 1, section 1–2 of the model is 40 m long; section 1–6 is 20 m long; section 6–5 of the slope top is 19.2 m long; the horizontal distance of section 5–4 of the slope surface is 5.8 m; section 4–3 of the slope bottom is 15 m long; the slope height is 10 m; the slope angle is 60°. Three monitoring sections are set at the slope shoulder, slope center, and slope toe to observe the change of pore pressure stress in the slope. The initial seepage boundary water heads on the left and right sides of the slope are 10 m and 5 m respectively. Section 1–2 at the bottom of the slope is an impervious boundary, and the slope top, slope surface and slope bottom are simulated rainfall boundary conditions. The initial displacement boundaries of sections 1–6 and 3–2 on the left and right sides restrict horizontal displacement, and sections 1–2 at the bottom of the slope restrict both horizontal and vertical displacement, as shown in Figure 1.

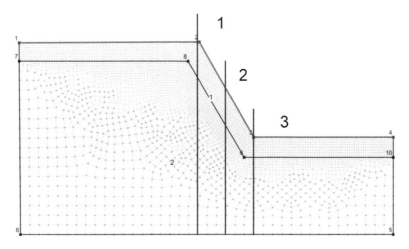

Figure 1.　Slope model.

As shown in Figure 2, the overall change trend of average effective stress and pore water pressure at the center and shoulder of the slope is similar. However, due to the easier loss of rainwater at the center and shoulder of the slope than at the foot of the slope, there is little change in effective stress and pore water pressure at the deeper section at the center of the slope. The fluctuation is mainly concentrated within 2 m of the slope surface. Compared with the foot of the slope, the change of stress and seepage from the center of the break and shoulder is more intense. When it is the most severe, the fluctuation values of average effective stress and pore water pressure exceed 200 kPa and 1,000 kPa. The displacement of slope soil is also more obvious. In contrast, the displacement change at the center of the slope is the largest. When the rainfall stops, it tends to be gentle faster. However, different from the displacement at the foot of the slope, the displacement at the center of the slope and slope shoulder does not increase after the rainfall stops but slightly recovers.

5.2　Unsaturated two-dimensional seepage control equation

Based on Darcy's law, a basic unit is taken out from the unsaturated soil, and the two-dimensional seepage equation when the soil is saturated is deduced through mass conservation law and Darcy's Law:

$$\frac{\partial}{\partial x}[k_{sx}\frac{\partial h}{\partial x}] + \frac{\partial}{\partial y}[k_{sy}\frac{\partial h}{\partial y}] = 0 \tag{1}$$

The two-dimensional seepage control equation with unsaturated soil is expressed as:

$$\frac{\partial}{\partial x}[k_{wx}\frac{\partial h}{\partial x}] + \frac{\partial}{\partial y}[k_{wy}\frac{\partial h}{\partial y}] + Q = \rho_w g m_2^w \frac{\partial h}{\partial t} \tag{2}$$

In the above two equations, k_{sx} and k_{sy} are the permeability coefficient of saturated soil in the X direction and Y direction; h is the total head; k_{wx} and k_{wy} are the permeability coefficient in the X direction and Y direction; Q is the applied flow boundary; ρ_w is the fluid density; m_2^w is the slope of the soil water characteristic curve, which is related to matrix forces.

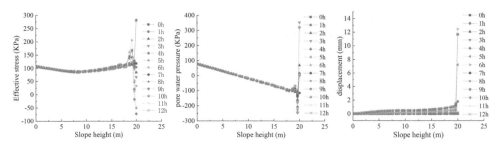

Figure 2.   Time history variation of seepage stress in soft rock slope.

6   CONCLUSIONS

(1) In the red bed area of central Yunnan, the rock strata are hydrophilic, easy to expand and hydrate, easy to lose water and disintegrate, low cohesiveness, easy to saturate, and difficult to evaporate, and readily weathered and eroded. It belongs to soft rock.

(2) The rock mass structure of red bed slope can be divided into near horizontal, inclined, and overlying accumulation according to the inclination angle, and can be divided into eight subclasses according to the factors that play a major role in controlling the slope stability.

(3) With continuous rainfall, the slope top and shoulder of soft rock slopes are more prone to instability. With the decrease of effective stress and pore water pressure, the slope displacement shows a slowly increasing trend, and finally increases suddenly when it is unstable.

REFERENCES

Gao Fangfang *Analysis of red bed distribution and slope engineering diseases in Yunnan* [D] Southwest Jiaotong University, 2010.

Li Chuanbao, Cheng Qiangong, Xu Caifeng Study on slope deformation and failure mode in red bed area of central Yunnan [J] *Subgrade Engineering*, 2007 (05): 102–104.

Wang Fei, Cao Chenfei, Chen Jiwen Experimental study on Microstructure and hydraulic properties of red bed soft rock in Central Yunnan [J] *Geology and exploration*, 2016,52 (06): 1152-1158 DOI:10.13712/j.cnki.dzykt. 2016.06.016.

Xu Guangzheng Geological characteristics analysis and engineering application of red beds in Central Yunnan [J] *Highway transportation technology* (Applied Technology Edition), 2014,10 (12): 143–148.

Yang Yi *Study on deformation characteristics and reasonable support system of red bed soft rock tunnel in Central Yunnan* [D] Southwest Jiaotong University, 2018.

Zhang Yong'an, Li Feng, Chen Jun Study on water rock interaction characteristics of red mudstone [J] *Journal of engineering geology*, 2008 (01): 22–26.

Zhao Likui, Li Xiaoli, Feng Lihong, Zhang Rui Study on support measures of deep buried soft rock tunnel in red bed in Central Yunnan [J] *Journal of underground space and engineering*, 2020,16 (S2): 737–743 + 761

Zhu Chunlin, Xing Zhihui, Rao Chunfu, Hu Guohua Hydrogeological characteristics of salt bearing red beds in Central Yunnan [J] *Yunnan geographic environment research*, 2009,21 (06): 1–7.

Zhu Junjie *Study on mechanical properties of red bed soft rock of central Yunnan water diversion project and classification of tunnel surrounding rock* [D] Chengdu University of technology, 2013.

Advances in Measurement Technology, Disaster Prevention and Mitigation – Li & Mohd Yusof (Eds)
© 2023 The Author(s), ISBN: 978-1-032-36087-4

Analyzing the characteristics of the flow field and laws of tunnels at different inclinations using CFD

Jie Xu & Qingwen Zhang*

School of Civil Engineering, Southwest Forestry University, Kunming, China

ABSTRACT: With the continuous development of underground space technology, the underground excavation space for transportation, hydropower, underground railway, and other projects is becoming larger. In underground tunnel construction, it is increasingly important to protect the health of construction personnel; ventilation is one of the main measures in tunnel construction. Tunnel construction in a high-altitude area differs from tunnel construction in a low-altitude area, including the application of construction ventilation technologies. Under the environment of low pressure and hypoxia, long excavation length, and poor construction conditions, construction ventilation technology has become a technical problem in high-altitude highway tunnel construction. In this paper, Fluent, a fluid calculation software, is used to carry out numerical simulation calculations of jet fan layout under different inclination angles of the roadway. Finally, the optimal layout angle of the jet fan and corresponding tunnel ventilation characteristics are obtained by analyzing the flow field under four working conditions. The calculation results show that the inclination angle of the jet fan has great influence on the wall layout at the branch tunnel intersection. When the jet fan is arranged at the intersection of the main and branch wind tunnels, its wind speed streamline is stable and electric conductivity is good. When the inclination angle of the jet fan is 0°, the flow diversion effect is the best.

1   INTRODUCTION

In the process of tunnel construction, tunnel ventilation design is very important. In the construction of many expressways, especially in mountainous areas and hilly areas, tunnel schemes are often used to solve the problems to overcome terrain and elevation obstacles, shorten mileage, increase speed, reduce time, and improve alignment. China's highway construction continues to extend to the central and western regions; there will be increased highway tunnel design and construction. Under the condition of mechanized operation, ventilation can provide fresh air for the construction site in the cave, eliminate toxic and harmful gases and all kinds of dust to create a good working environment, and ensure the health and safety of construction personnel. Ventilation can provide fresh air for the tunnel construction site, and eliminate toxic and harmful gases and various dust, thus creating a good working environment, to ensure the health and safety of construction personnel. When pressure ventilation is used, it is difficult to discharge harmful gas at the junction of the tunnel with the slope support tunnel because of its length limitation. Moreover, when the branch tunnel is long, under the influence of its slope, the discharge of harmful gases and dust pollutants in the tunnel must overcome a certain elevation to do work, which requires optimization of the ventilation system under technical and economic conditions, that is, the arrangement of jet fan ventilation.

---

*Corresponding Author: 904829146@qq.com

DOI 10.1201/9781003330172-64

For the study of tunnel ventilation, Fang Yong (Fang et al. 2009) used CFD software to simulate the jet field of jet fans in a three-lane highway tunnel and determined the reasonable installation position of jet fans in a highway tunnel. Through numerical simulation of a hydraulic tunnel, Chang Xiaoke (2017) concluded that there would be obvious eddy currents when air flows through the branch tunnel. Xing Yao (2018) corrected the ventilation design parameters of high-altitude tunnel construction. For the Xueshanliang tunnel project, the uncorrected air volume is only 30% of the corrected air volume, and the uncorrected air volume of the parallel guide tunnel is only 36% of the corrected air volume, and the corresponding fan selection is obtained. Nie Qingwen (2021) studied the effect of position and angle of jet fan arrangement on pressure boost and diversion and obtained that the diversion effect of an arrangement at the back of the intersection between the main tunnel and the cross passage was better than that of an arrangement at the cross passage.

Analysis of the existing numerical simulation research results of tunnel ventilation during the construction period indicated that the tunnel ventilation research is focused on a single tunnel, the tunnel with branch tunnel, especially the slope branch tunnel research is very few, and the intersection jet fan layout research is not clear. In this paper, Solidworks is used to establish a tunnel ventilation model, numerical simulation software Fluent is used to analyze the influence of jet fans with different heights and angles on tunnel ventilation, and a reasonable installation position of jet fans is proposed for Banbanshan tunnel.

2 PROJECT OVERVIEW

The No. 1 branch tunnel of Bandengshan is 372 m long. CO in the stratum exceeds the prescribed limit by 2.6 times, and H2S exceeds the prescribed limit by 27 times. The problem of toxic and harmful gases is prominent, which has a great influence on construction safety. Construction area of Bungshan Branch tunnel No. 1: After the branch tunnel is connected, two working faces are carried out to assist the upstream and downstream construction of the main tunnel, followed by downstream construction and upstream construction of the main tunnel. The upstream direction of construction is about 2,164 m, and the downstream direction of construction is about 3,413 m. Two SDF (C)-4-No.14 axial flow fans are used for pressure-in ventilation at branch openings. Seamless steel tubes are used for air ducts, and pipe joints relate to flanges and are laid along the tunnel side walls. To eliminate or reduce the influence of this phenomenon on the ventilation effect, a fan is installed in the intersection area.

3 CALCULATION MODEL

3.1 *Mathematical model*

The models of turbulence equations can be divided into the Direct Numerical Simulation (DNS), Reynolds Average Method (RANS), and Large Eddy Simulation (LES). RANS model includes the stress model and vortex-viscosity model. The stress models are divided into the Reynolds Stress Equation Model (RSM) and Algebraic Stress Equation Model (ASM). Vortex-viscose model can be divided into the first-order equation model and the bivariate equation model. The most widely used models are the standard K-ε model, the Realizable K-ε model, and RNG K-ε model.

As the air pressure and density in the tunnel do not change much, although the wind speed at the exit of the jet fan is large, its Mach number is lower than 0.3, thus the gas can be considered at an incompressible flow. The airflow in the tunnel is usually turbulent. In the numerical simulation, the airflow field in the tunnel is defined as a three-dimensional viscous turbulent field. Its numerical model includes continuity equation, momentum equation, turbulent kinetic energy equation, and turbulent kinetic energy dissipation rate equation. The fan layout is shown in Figure 1.

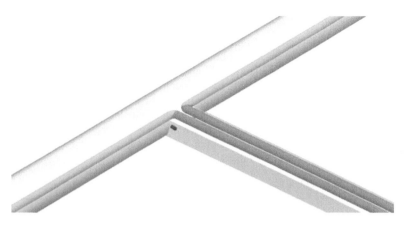

Figure 1.   Tunnel model diagram.

3.2   *Grid model division*

For the jet fan layout problem at the intersection of the branch tunnel and main tunnel in Bandenshan Tunnel, SolidWorks was used to establish a THREE-DIMENSIONAL model. To simplify the calculation, the 100 m upstream and downstream of the main tunnel and the 200 m branch tunnel were taken as the research object. ICEM CFD was selected for grid division, and Tetrahedrons grid was adopted for partition. The wall size was 0.5 m. As the airflow velocity and pressure gradient at the inlet and outlet of the jet fan and the air outlet of the ventilation pipe are very large, it is necessary to carry out grid encryption on the inlet and outlet and push ventilation to ensure the accuracy of calculation and better convergence of residual error. The size of the outlet of the push ventilation is 0.05 m and that of the jet fan is 0.02 m.

3.3   *Boundary conditions and solution settings*

In tunnel construction ventilation, pipe wall and tunnel wall boundary are specified as rotational velocity and tangential velocity components in Cartesian coordinates, including shear rate under slip conditions. The setting of inlet wall boundary conditions for turbulence parameters are: Fan characteristic curve function; the boundary conditions of the axis and the periodic boundary of movement in the flow field; and the setting of periodic boundary conditions for streamlines in the flow field.

The standard two-path turbulence model is adopted, and the ε equation includes a term that cannot be calculated on the wall surface, so the wall function must be used. The tunnel wall surface adopts the non-slip solid wall boundary. Considering the effect of the roughness of the wall function, the roughness constant of the tunnel wall is 0.5, and the reflux turbulence intensity is 5%. According to Detailed Rules for Ventilation Design of Highway Tunnel, the air volume required by excavation face, the breathing of staff in the tunnel, smoke dilution, exhaust gas dilution of the mechanical internal combustion engine, and consumption of windpipe for 100 m can be calculated, and the air volume required by tunnel is 968.09 m³/min, and the wind speed at the outlet of the forced ventilation pipe is V = 11.6 m/s.

A jet fan, model SDS-6.3-2P-4-18°, is selected based on the field measured data and energy-saving requirements. The length of the fan is 1.27 m, the diameter is 0.74 m, and the wind speed at the fan outlet is 26 m/s. The air duct outlet was set as the speed outlet, the wind speed of the upstream and downstream air duct outlet was 11.6 m/s, and the jet fan inlet and the jet fan outlet were set as the speed inlet boundary conditions. According to the fan performance parameters, the fan inlet speed was set as -26 m/s, and the air outlet speed was set as 26 m/s. The tunnel outlet is the pressure outlet, and the relative atmospheric pressure is 0. According to the tunnel ventilation system and design requirements, the convergence judgment condition is that the quality difference between inlet and outlet is less than 0.5%. After 1,800 iterations, the convergence solution is obtained. The quality difference between the inlet and outlet is 0.0021, and the calculation results converge.

3.4 *Calculated operating condition*

According to the Rules for Highway Tunnel Ventilation Design, the arrangement of jet fans on the tunnel cross section should not invade the building limit, and the net distance between the edge of jet fans and the tunnel building limit should not be less than 15 cm. Considering the existing pressure ventilation pipe layout, crosswalk on the right side of the branch tunnel, fan noise, and other factors, the tunnel jet fan is located at the intersection of the main tunnel and the branch tunnel, it is arranged on the left side of the branch tunnel, and placed horizontally and parallel to the slope of the branch tunnel. The distance between the center and the left side wall of the branch tunnel is 0.55 m. To study the influence of jet fan installation height on the ventilation effect, the layout form of the jet fan is shown in Table 1.

Table 1.   The layout of the jet fan.

Working condition	Jet fan arrangement
Working condition 1	The height is 3.0 m and the inclination is 0°
Working condition 2	The height is 3.5 m and the inclination is 0°
Working condition 3	The height is 3.0 m and the inclination is 5.52°
Working condition 4	The height is 3.5 m and the inclination is 5.52°

4   ANALYSIS OF NUMERICAL SIMULATION RESULTS

To study the variation of velocity streamline of the branch tunnel, in the branch tunnel, a surface was determined at $z = 0$ on the xy plane, denoted as section Z_1. The wind velocity flow charts of four different working conditions were shown in Figure 2.

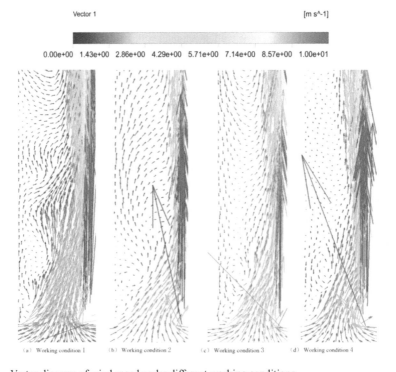

(a) Working condition 1   (b) Working condition 2   (c) Working condition 3   (d) Working condition 4

Figure 2.   Vector diagram of wind speed under different working conditions.

Figure 2 shows that when the jet fan starts to work, the wind speed in the middle of the branch tunnel is smaller than its inner wall, and the jet fan layout height is 3 m. When the inclination angle is 0°, the jet will impact the tunnel floor, and low-speed areas are observed near the tunnel top. These low-speed areas are vortex backflow areas, which account for 42% of section Z_1. After the jet develops 80 m, the velocity streamlines of the flow field at the bottom of the branch tunnel are gradually uniform, and the wind speed is stable. The tilt angle is 5.52°. The Koen effect will cause a portion of the jet to stick inside the tunnel wall and roof, and as a result, the jet over a short distance will enter a fully developed flow region, and the central hole under the jet fan injection branch can form more eddy zones, which cause a greater amount of flow disorder, leading to higher wind speed fluctuations.

When the layout height of the jet fan is 3.5 m and when the inclination angle is 0°, the eddy current area accounts for 34% of section Z_1. The flow line of the bottom wind speed is smooth without the disorder, and the overall wind speed is fast, indicating that the resistance of the gas in the bottom flow process is less, and the gas flow in the working area at the height of 0–3.2 m is laminar flow. When the inclination angle is 5.52°, the axial velocity of the jet fan decreases due to the entrainment effect, and the cross section of the jet expands gradually. As the height increases, the air jet begins to move away from the lower part of the tunnel, and the number of vortex zones generated under the air jet decreases. It can be found that when the layout height of the jet fan is 3.5 m, the inclination angle is 0°, the energy loss of gas in the flow process is less, and the stability is faster. The pressure boost effect of the jet fan is the best, and the harmful gas and dust particles can be carried out more effectively to achieve the best ventilation effect.

When the layout height is 3 m and 3.5 m, it can be seen from Figure 2 that when the jet fan layout height is 3 m, the upstream and downstream wind speed of the tunnel and the streamline of the intersection are stable, and harmful gases and suspended particles can be effectively discharged from the intersection. However, a large vortex zone will be generated in the middle of the branch tunnel, and the low gas velocity will lead to the slow discharge of dust and harmful gases in the tunnel. When the height of the jet fan is 3.5 m, the low-speed fluid near the intersection is sucked into the jet, and the eddy current area of the branch tunnel is also small. After 100 m operation of the branch tunnel, the water flow gradually stabilizes. The wind velocity streamlines at the intersection of the integrated branch wind tunnel and the three-dimensional wind velocity streamlines at the intersection of the main wind tunnel are obtained. When the jet fan layout is low, the ventilation efficiency of the branch tunnel is low; when the jet fan layout is high, the ventilation effect of the branch tunnel is good.

According to the above analysis, when the height of the inner wall of the jet fan is different, its inclination angle has a great influence on the jet development. At a small inclination angle, the air jet cannot escape from the tunnel floor effectively, and a larger inclination angle will produce a larger vortex zone in the upper part of the tunnel. Therefore, there is an optimal height and angle to achieve the most efficient ventilation in the tunnel.

5 CONCLUSIONS

(1) When the inner wall of the jet fan is arranged at the branch tunnel intersection, its jet flow is affected by the layout height and inclination angle. At a small inclination angle, the air jet cannot escape from the tunnel floor effectively, and a larger inclination angle will produce a larger vortex zone in the middle of the tunnel.

(2) During the construction of a tunnel made with a side hole and jet fan model, a computational fluid dynamics analysis was performed for a different arrangement of height, slope angle, jet fan flow field analysis comparison, and the bench mountain tunnel pressure is drawn into the ventilation conditions, and the main cross-section of the jet fan hole and hole layout height of 3.5 m and angle of 0° to achieve the most effective construction ventilation effect.

(3) The layout of the jet fan is affected by the longitudinal slope of the tunnel branch, and there are backflow and vortex phenomena near the confluence section. In the specific layout, it must be determined according to the actual tunnel geometry and ventilation parameters.

REFERENCES

Chang Xiaoke. *Numerical Simulation research on Construction Ventilation of Hydraulic Tunnel with support Tunnel* [D]. Xi'an University of Technolgy,2017.

Fang Yong, Lan Yu, Zeng Yanhua, et al. Research on setting position of jet fan in three-lane highway tunnel [J]. *Modern tunnel technology*, 2009, 46(02):90–93+98.

Nie Qingwen. Study on Optimization of Jet Fan Layout at Construction Cross Passage Intersection [J]. *Railway Construction Technology*, 2021(10):35–39.

Xing Yao. *Research on Key Technologies of Ventilation in Construction of Highway Tunnel with High altitude and long Length* [D]. Chongqing Jiaotong University, 2018.

*Advances in Measurement Technology, Disaster Prevention and
Mitigation – Li & Mohd Yusof (Eds)*
© 2023 The Author(s), ISBN: 978-1-032-36087-4

Review on the development of bridge health monitoring system

Beilin Han, Yansong Huang, Genghao Song, Lixuan Luo, Zhongheng Wu & Siyue Zhu*
School of Civil Engineering and Architecture, Wuhan Polytechnic University, Wuhan, Hubei, China

ABSTRACT: The bridge health monitoring system collects bridge information through sensors and transmits it to the data center, and realizes real-time monitoring and early warning of bridges through data processing and analysis. To improve the bridge health monitoring system, this paper summarizes and studies the traditional bridge health monitoring system, and introduces the method of selecting and optimizing the layout of bridge health monitoring sensors in detail. By comparing different transmission systems of bridge health monitoring, the advantages and disadvantages of each system are obtained. Finally, a 5G-based bridge health monitoring system is proposed, which can provide a reference for improving the problems of high delay caused by insufficient data transmission bandwidth in the monitoring system.

1 INTRODUCTION

The bridge health monitoring system extends the life of the structure, protects the normal operation of the structure, and provides reference and decision-making for the management and maintenance of the bridge by monitoring and evaluating the condition of the bridge structure. Therefore, bridge health monitoring is crucial for transportation safety. The bridge health monitoring system includes a sensor system, data acquisition, and transmission system, data management system, and structural health early warning and evaluation system (Zhou et al. 2020). The bridge health monitoring system generates a large amount of signal data every day. To ensure the real-time monitoring of the bridge health monitoring system, the collection and transmission of data have become a hot research direction at present. This paper summarizes and analyzes the status quo of bridge health monitoring systems, focuses on the sensor system and data acquisition and transmission system, and discusses 5G and augmented reality (AR). The application scenarios of artificial intelligence in the future bridge health monitoring system have also been prospected.

2 BRIDGE HEALTH MONITORING SYSTEM

At present, the bridge automatic monitoring system has been widely used in bridges at home and abroad. According to incomplete statistics, more than 400 large bridges in the country are equipped with health monitoring and management systems, and the development speed has reached the world's leading level. In recent years, with the vigorous development of a new generation of information technology and the continuous improvement of the national strategic system, the comprehensive application of the concept and technology of "5G + Internet + mass data + cloud computing + monitoring factor." We implement full-process, all-weather, full-coverage online monitoring of bridge safety, and then establish a more advanced and scientific bridge health monitoring

---

*Corresponding Author: zhusiyue@whpu.edu.cn

and evaluation system. To achieve more accurate data analysis, evaluation, forecasting, and decision support have become the focus of attention.

3 BRIDGE HEALTH MONITORING WAYS

Visual inspection is the most primitive method for bridge health monitoring, but it is often carried out through auxiliary equipment, which is not only time-consuming and laborious, but also affects the normal traffic order, and has great security risks. In recent years, with the help of modern technologies, such as drones, robots, and high-definition photography, the traditional manual detection of local locations, such as the bottom, sides, and piers of the bridge body can be upgraded to precise robot monitoring to achieve multi-dimensional and integrated monitoring. In terms of data interpretation, automatic disease image recognition technology replaces manual interpretation, which greatly improves the monitoring efficiency and monitoring accuracy.

With the development of sensor technology and communication technology, the way of bridge health monitoring has also ushered in vigorous development. For example, the principle of optical fiber strain sensor has the advantages of no temperature increase, time drift, and very little electromagnetic interference in the process of information transmission, excellent accuracy, and stability, real-time monitoring of multi-point measurement, etc. The principle of the grating sensor is based on the wavelength modulation type, and the optical fiber Bragg is detected by external physical parameters. Wavelength modulation to obtain sensing information. Web geographic information system (WebGIS) based on Map Internet geographic information system server (GISIGS) 3D controls, realize 3D rendering of bridge models, and complete 2D and 3D interactive data sharing control configuration. BIM is an intelligent work processing mode in the 3D modeling of bridges (Dang & Shim 2018; Davis et al. 1997). With the convenient operation characteristics of the mobile terminal, a personal digital assistant (PDA) can realize static and dynamic co-processing and real-time system evaluation and monitoring. In SGS-THOMSON Microelectronics 8 (STM8), the microprocessor is the control core of the whole system, and the integrated real-time monitoring, display, and acquisition modules form a framework intelligent plate. Drone output close-up high-definition pictures of visualized parts, and video of dynamic structures (Alexander & Sigrid 1998; Culshaw et al. 1996; Shahawy & Arockiasamy 1996).

At present, there are still a few bridge monitoring devices in mechanized operation and have not carried out intelligent reform on a large scale and comprehensively. With the advent of the 5G era, the widespread application of the Industrial Internet is accompanied by the advancement of the intelligence of many monitoring sensor technologies. The application scenario of 5G in bridge health monitoring should be the main research direction of bridge monitoring in the future.

4 CLASSIFICATION AND SELECTION OF BRIDGE HEALTH MONITORING SENSORS

The sensor has good static and dynamic characteristics. In the bridge health monitoring activities, not only is the data collected by the sensor accurate enough but also the data converted by the sensor is close to linearity. Therefore, the static and dynamic characteristics of the sensor should be considered in the selection of the sensor. Sensors used for bridge health monitoring generally have small errors, fast feedback, high reliability, good compatibility, and strong real-time performance (Lin et al. 2021; Miao et al. 2011). Feature sensors commonly used in monitoring are mainly classified according to the measurement method of the sensor, the information conversion method, and the production material. In the process of bridge health monitoring, most of the sensors choose appropriate sensors according to different monitoring items, as shown in Table 1.

In real monitoring activities, sensors often interfered with the surrounding environment. To obtain accurate and real-time data obtained by sensors in complex environments, the designer should select the appropriate sensor type and optimize it according to the structural characteristics of the monitoring project and the characteristics of the surrounding environment. layout.

Table 1. Common sensor types in bridge health monitoring system.

Monitoring program	Sensor	The layout of the monitoring spot
Stress monitoring	Steel bar meter	Bridge pier, tower, girder, etc.
Strain monitoring	Surface strain gauge	Bridge pier, tower, girder, etc.
Deflection monitoring	Differential pressure deformation measurement sensor	Bridge pier, tower, girder, etc.
Tilt monitoring	Box fixed inclinometer	Bridge pier, tower, girder, etc.
Tension monitoring	Magnetic flux sensor	Main tower rod, main span hangers, etc.
Vibration monitoring	Magnetic-electric transducer	Bridge pier bridge tower, etc.
Crack monitoring	Crack meter	Maximum seam width of the bridge body
Habitat monitoring	Wet temperature sensor	Main tower of bridge deck
Wind speed and wind direction monitoring	Wind instrument	Main tower deck, etc.
Main tower displacement monitoring	Displacement transducer	Bridge body, etc.
Vehicle flow monitoring	Load monitor	Bridge body, etc.
Velocity monitoring	Image displacement measuring camera	Bridge body, etc.
Settlement monitoring	Differential pressure settler	Bridge pier, etc.

5 OPTIMAL LAYOUT OF BRIDGE HEALTH MONITORING SENSORS

Traditional sensor placement methods mainly include the modal kinetic energy method (MKE) and effective independence method (EI) (Kammer 1991; 2005; Papadopoulos & Garcia 1998). Since MKE and EI optimize the sensor layout according to the principle of maximum data information, the amount of calculation is large. To simplify the calculation and improve the calculation efficiency, researchers at home and abroad have carried out a series of optimization on the algorithm, such as orthogonal triangular decomposition (QR) (Li et al. 2008), effective independent driving point residual method (EFI-DPR) invasive weeds optimization algorithm (IWO) (Zubir & Yahaya 2021), firefly algorithm, hierarchical immune firefly algorithm, universal gravitation search algorithm (GSA), and effective independent drive point residue method (Wan et al 2021; Zhang et al. 2021). The above optimization algorithm has the advantages of small design parameters and simplified calculation, and EFI-DPR has good noise resistance. However, in the optimal layout of actual projects, the above methods are prone to slow convergence and local optimization due to small parameters and excessive pursuit of optimal solutions.

To improve the optimization ability and convergence of the algorithm and improve the problem of the local optimal solution in the process of the optimal layout. Common methods include double structure coding genetic algorithm, double coding adaptive gravity algorithm, energy coefficient effective independence method, etc. The above methods introduce other coding methods into the traditional optimization algorithm to improve the algorithm. In addition, by increasing the search space and the number of various factors of particles in the quantum particle swarm optimization algorithms and the artificial fish swarm algorithm (AFSA). In the research of sensor optimal layout, we not only need to improve the algorithm's searching ability and convergence performance but also to discuss the universality of its searching ability. Bridge health monitoring is the comprehensive monitoring of the whole bridge from underground to air components, so the optimal arrangement of sensors should not only consider the optimal arrangement of local components but the bridge itself and its environment. From the application of AFSA, the optimal arrangement of sensors should be integrated with a variety of disciplines. For example, artificial intelligence (AI) should be combined with the algorithm of optimal arrangement to explore a set of arrangements through AI's continuous learning and simulation, which is the focus of our future research.

6 DATA ACQUISITION AND TRANSMISSION OF BRIDGE HEALTH MONITORING SYSTEM

6.1 *Collection of bridge health monitoring data*

The data acquisition system of bridge health monitoring first collects and then converts the signals sent by sensors into identifiable signals. There are many design methods for the data acquisition system, including centralized and distributed data acquisition systems (Li et al. 2014). A centralized data acquisition system mainly collects dynamic sensor signals, while distributed data acquisition system mainly collects static sensor signals. The data acquisition system is supplemented by static and dynamic systems to give full play to their advantages.

The software for the data acquisition system is the key to the data acquisition system. In large bridges, the number of sensors is about 400. Aiming at such a large data acquisition project, a developer proposed a data acquisition method using virtual instrument technology by using LabVIEW and LabWindows/CVI software development platform, and a data acquisition system suitable for bridge health monitoring is developed. At present, the main research direction is on how to realize system integration, namely with data acquisition, analysis, integration, and storage of data acquisition systems, such as the combination of the Internet of Things (IoT) and Geography Information System (GIS) technology to bridge health monitoring. The bridge health monitoring data acquisition system based on Zigbee wireless sensor network technology is more intelligent than the traditional data acquisition system.

6.2 *Transmission of bridge health monitoring data*

Data transmission also plays a key role in bridge monitoring. At present, wired and wireless data transmission methods are widely used in bridge monitoring, as shown in Figure 1. The method of data collection determines the method of data transmission. In most monitoring systems, a variety of communication methods are used for data transmission. Nowadays, as cable transmission mode gradually becomes incompatible with the environment, it becomes necessary to perform regular maintenance on the line, increasing the costs and risks of bridge health monitoring systems. With hundreds or even thousands of large bridges in the process of data transmission, the traditional transmission technology is weak, network latency makes it difficult to ensure the real-time performance of bridge health monitoring systems. With the characteristics of high bandwidth, low latency, and interconnection of everything in 5G networks, the health monitoring, and management of bridges can be realized more accurately.

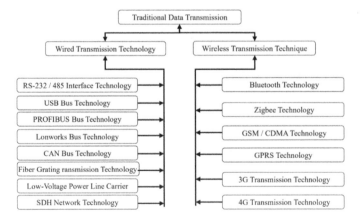

Figure 1. Traditional data transmission.

Contemporary is the era of vigorous development of information. All kinds of new concepts are put forward to help drive the bridge health monitoring system. "5G+AI+ big data + Internet of things" and the integration of the bridge health monitoring system will continue to deepen. Safe, efficient, and sustainable development of "smart road bridge" is bound to become a new trend in the development of bridge safety monitoring systems. 5G in bridge health monitoring RS-232/485 interface technology, USB bus technology, process field bus (PROFIBUS) technology, LonWorks bus technology, controller area network (CAN), bus technology, fiber Bragg grating transmission technology, low-voltage power line carrier, SDH network technology, wired transmission technology, traditional data transmission mode, wireless transmission technology, Bluetooth technology, Zigbee technology, GSM/CDMA technology, GPRS technology, 3G transmission technology, 4G transmission technology, and the applications of early warning are worth paying attention to. 5G communication technology has unique advantages with its security, speed advantages, and deep integration with the IoT. Combining it with the bridge monitoring system, which needs a large amount of data storage and transmission, can not only improve the data transmission rate of the bridge monitoring system but also analyze the bridge data with big data and AI to give early warning of the possible risks of the bridge.

The bridge health inspection system based on 5G industrial Internet can make full use of sensor technology, 5G communication technology, and industrial IoT data analysis technology to realize the new digital operation, and the intelligent transformation of bridge infrastructure. Combined with 5G and the depth of the sensor technology integration for 24 hours bridge to implement real-time monitoring, the deformation and stress distribution of bridge and surrounding environment conditions for online real-time monitoring, through 5G high-speed network of sensor collected data conversion and connected with a traffic management system, coupled with large of data analysis based on a set of 5G high bandwidth delay, at the end of the bridge health monitoring system, as shown in Figure 2. In addition, deep learning and reasonable calculation are carried out on the data of possible safety risks of bridges through big data and AI, and multi-tiered early warning and emergency plan processing are carried out on the safety of bridges. On this basis,

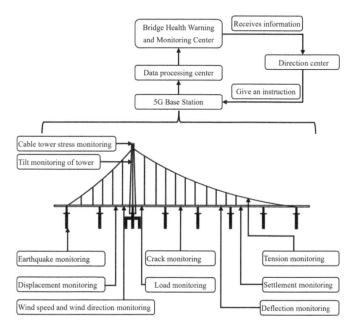

Figure 2. Bridge health monitoring system based on 5G.

the calculation model of the structure is modified, the service safety status of the structure is analyzed and quantitatively assessed, the remaining life of the structure is predicted, the safety and life status of the structure is grasped and early warning, and the management maintenance and repair reinforcement of the safety life of the bridge structure is scientifically supported. The bridge health detection system based on 5G can also be combined with BIM to generate digital models through data processing so that managers can intuitively understand the stress and health status of each structure of the bridge. At the same time, AR technology can also be combined to enable customers to monitor the real state of the bridge more intuitively. When the system monitors abnormal data signals, the alarm system will be triggered. The system will automatically generate processing methods by comparing the data in the database and take timely measures such as personnel intervention and traffic control to eliminate security risks. The bridge health monitoring system based on 5G can make full use of the advantages of 5G to realize the interconnection of all things to build intelligent bridges and realize the intelligent management of bridge monitoring, which is the development direction of the bridge health monitoring system in the future.

8 CONCLUSION AND PROSPECT

As an important auxiliary tool for the management and maintenance of bridge structures in the later period, the bridge health monitoring system is favored by bridge management units at home and abroad. It can provide a lot of reliable data information both in real-time monitoring and regular inspection of the bridge. Great progress has been made in bridge health monitoring, but the data transmission and processing technology of bridge health monitoring are not mature enough, resulting in insufficient bandwidth, high energy consumption, and high delay. The high-speed data transmission of 5G communication technology ensures low delay and high reliability of data, and low power consumption will save more monitoring costs for customers. Therefore, the combination of 5G communication technology and bridge health monitoring system can improve the problem of high data transmission delay in bridge health monitoring system; consequently, accelerating the application of 5G in bridge health monitoring is worth for further research.

REFERENCES

Alexander K, Sigrid B. (1998) 3D-Gis for Urban Purposes. J. Geo Informatica, 2(1):79–103.
Culshaw B, Michie C, Gardiner P, et al. (1996) Smart structures and applications in civil engineering. *J. Proceedings of the IEEE*, **84**(1):78–86.
Dang N S, Shim CS. (2018) BIM authoring for an image-based bridge maintenance system of existing cable-supported bridges. *J. Earth and Environmental Science*, **143**(1): 012032.
Davis M, Bellemore D, Kersey A. (1997) Distributed fiber Bragg grating strain sensing in reinforced concrete structural components. *J. Cement and Concrete Composites*, **19**(1):45–57.
Kammer D C. (1991) Sensor placement for on-orbit modal identification and correlation of large space structures. J. Journal of Guidance, Control, and Dynamics, 14(2):251–259.
Kammer D C. (2005) Sensor set expansion for modal vibration testing. *J.Mechanical Systems and Signal Processing*, **19**:700–713.
Li D S, Li H N, Fritzen C. (2008) A note on fast computation of effective independence through QR downdating for sensor placement. *J. Mechanical Systems and Signal Processing*, **23**(4):1160–1168.
Li P, Fang X, Chen Z P, et al. (2014) Wireless Distributed Data Acquisition System for EIT. *J. Applied Mechanics and Materials*, **2958**:1298–1301.
Lin Y B, Li F Z, Chang K Z, et al. (2021) The Artificial Intelligence of Things Sensing System of Real-Time Bridge Scour Monitoring for Early Warning during Floods. *J. Sensors* (Basel, Switzerland), **21**(14):12–15.
Miao C Q, Chen L, Feng Z X. (2011) Study on effects of environmental temperature on dynamic characteristics of Taizhou Yangtze River Bridge. *J.Engineering Sciences*,**9**(02):78–82+92.
Papadopoulos M, Garcia E. (1998) Sensor placement methodologies for dynamic testing. *J.AIAA Journal*, **36**(2):256–263.

Shahawy A, Arockiasamy M. (1996) Analytical and Measured Strains in Sunshine Skyway Bridge. II. *J. Journal of Bridge Engineering*, **1**(2):87–97.

Wan P, He H L, Guo L, et al. (2021) InfoGAN-MSF: a data augmentation approach for correlative bridge monitoring factors. J. Measurement Science and Technology, **32**(11):7–44.

Zhang L Y, Fu M, Fei T, et al. (2021) The Artificial Fish Swarm Algorithm Optimized by RNA Computing. *J. Automatic Control and Computer Sciences*, **55**(4) : 346–357.

Zhou C Y, Wu Y T, Cui G J, et al. (2020) Comprehensive measurement techniques and multi-index correlative evaluation approach for structural health monitoring of highway bridges. *J. Measurement*, **152**(C):107360.

Zubir M A, Yahaya F M. (2021) Experimental Investigation into Flexural Performance of Reinforced Profiled Steel Decking. *J. Key Engineering Materials*, **6206**:243–253.

Advances in Measurement Technology, Disaster Prevention and
Mitigation – Li & Mohd Yusof (Eds)
© 2023 The Author(s), ISBN: 978-1-032-36087-4

Analysis of reservoir damage and optimization on completion fluid system of adjusting wells of Weizhou 12-1 oil field

Cheng Jian
CNOOC (China) Limited-Zhanjiang, Zhanjiang, Guangdong, China

Xingjin Xiang*
Hubei HANC New-Technology Co. Ltd., Jingzhou, Hubei, China

ABSTRACT: In this paper, we systematically analyzed and evaluated the damage factors of the target reservoir of adjusting wells of the Weizhou 12-1 oil field. Moreover, we put forward the suggestions for the completion of fluid system optimization and the measures on the release of production capacity and optimized the major treating agent. Furthermore, we evaluated the performance of the optimized completion fluid system. The results showed that the main damage factors in drilling and completion were low-permeability water lock and reservoir water sensitivity. The optimized completion fluid system has the characteristics of good dissolution, strong anti-swelling, low corrosion, and good anti-water lock. The optimized single fluid has good reservoir protection and reservoir reconstruction performance. The suggestions on the drilling and completion fluid system for adjusting wells of Weizhou 12-1 oil field are conducive to fully releasing production capacity.

1 INTRODUCTION

After years of development, the drilling and completion fluid technologies used in the oilfields of the Western South China Sea have resulted in several recognized systems, including oil-based and water-based systems, solid-phase, and non-solid-phase systems. They have been used widely with mature on-site maintenance measures and satisfactory effects. However, compared with the oilfields in other seas, the oilfields of the Western South China Sea feature relatively complicated reservoirs. The physical properties and sensitivity characteristics of reservoirs in different blocks or different reservoir groups but the same block vary significantly. Especially in recent years, many complicated reservoirs have been drilled in the sea area of the Western South China Sea. They are characterized by poor physical properties and high reservoir protection requirements. Under such circumstances, it is necessary to optimize the drilling and completion fluid system and protect the reservoirs with all efforts. In addition, the optimization of completion fluid is beneficial to the full release of reservoir production capacity.

The target reservoirs for the 12 adjusting wells of Weizhou 12-1 oil field are located in L_1 Member, which has never been subject to the research of reservoir sensitivity and drilling and completion fluid system. In addition, these reservoirs feature poor physical properties, with mainly low permeability and super-low permeability, so it is very difficult to protect them. Therefore, we evaluated the adaptability of the existing drilling and completion fluid systems based on analyzing the damage factors of the target reservoirs of the adjusting wells and proposed measures on reservoir protection and release of reservoir production capacity to optimize the completion fluid system and provide technical guarantee for safe and efficient development, good reservoir protection effect, and full release of production capacity of the adjusting wells of Weizhou 12-1 oil field.

*Corresponding Author: kyxmtb66@163.com

    DOI 10.1201/9781003330172-66

2 RESERVOIR DAMAGE TYPES AND ANALYSIS OF POTENTIAL DAMAGE FACTORS

2.1 Analysis and evaluation of damages resulting from porosity and permeability characteristics of reservoirs

The target reservoirs for the designed 12 adjusting wells are L_1I_{lower} of Wells C1 and C2H in the WZ12-1-2 Well Area, L_1I, L_1II_{upper}, and L_1II_{lower} of Wells C3H, C4H, C5H and C6H in WZ11-2E-4d Well Area, L_1I of Wells C7H and C8H in WZ12-1-8 Well Area, L_1I_{upper} of Wells C17H and C18H in WZ12-1-B12H1 Well Area, and L_1II_{upper} of Wells C19H and C20H in WZ12-1-2dSa Well Area. As the porosity is 13.40–15.90% and the permeability is 1.9–22.0 mD, they are middle-low-porosity, low-permeability, and super-low-permeability reservoirs, and water lock damages are of high possibility during drilling and completion process.

The improved H-Water Lock prediction software based on the water lock index APT_i model proposed by Canadian scholar D. B. Bennion was used to predict water lock damage. The results are that the water lock indexes APT_i in WZ12-1-2, WZ11-2E-4d, WZ12-1-8, WZ12-1-B12H1, and WZ12-1-2dSa well areas are smaller than 0.2, indicating super-high water lock damage characteristics.

Filtered seawater is used as the water-based fluid of completion fluids without waterproof measures to prevent water sensitivity damage. Artificial cores are used for water lock damage evaluation.

Experimental steps are as follows: (1) measure the gas permeability of artificial core Ka, use CBH-II core oil-water saturation test device to vacuum the saturated formation water, and soak it at the reservoir temperature for more than 20 hrs for standby application; (2) use forward permeability test to measure the permeability of kerosene K_0 on the core flow test device; (3) use the 5 PV filtered seawater for reverse pollution, place it for 1h without any operation, and simulate that the operating fluids enter the reservoir; (4) use forward permeability test to measure the permeability of kerosene K_1; (5) calculate the damaging rate of permeability. See Table 1 for the calculation results.

Table 1. Evaluation results of water lock damage.

Core number	Length cm	Diameter cm	Porosity %	Ka mD	K_0 mD	K_1 mD	Damage ratio %
226	7.05	2.50	29.5	43.62	21.810	10.845	50.28
235	7.0	2.51	36.6	18.61	10.242	3.554	65.29
237	6.98	2.52	32.5	2.16	0.964	0.225	76.66
239	7.05	2.52	30.2	0.85	0.394	0.073	81.60

Based on Table 1, it can be found that the lower the core permeability is, the greater the water lock damage is. Besides, the initial permeability, pore structure, and initial water saturation of the reservoirs are important factors influencing the water phase invasion damages to the reservoirs. As the invading water phase results in a higher water saturation of a reservoir, the potential effective permeability damages will be greater, namely, the water lock damages are severer. Therefore, if water-based drilling and completion fluids are used for adjusting wells subsequently, proper considerations are recommended for the water lock issue to minimize the surface tension and interfacial tension of the fluids pumped in wells.

2.2 Analysis and evaluation of damages resulting from lithologic characteristics of reservoirs

Rock debris is sampled from representative target reservoirs for X-ray diffraction analysis of relative contents of whole-rock minerals and clay minerals. See Table 2 for the results.

Table 2. Contents of whole rock mineral and clay mineral in reservoir cuttings.

Well number	Well depth (m)	Horizon	Content of whole-rock minerals/%					Clay mineral content/%				
			Quartz	Potassium feldspar	Anorthosite	Dolomite	Clay minerals	Illite/ montmorillonite interlayers (I/S)	Illite (I)	Kaolinite (K)	Chlorite (C)	Interlayer ratio S (%)
WZ12-1-2	2825	L_1l_{lower}	47.3	2.8	–	1.0	48.9	37	21	23	19	30
WZ12-1S-2	3182.5-3183.49	L_1V_{upper}	82.0	4.0	3.8	–	10.2	59	39	–	2	30

Based on Table 2, it can be found that the content is 47.3%–82.0% for non-clay minerals (mainly quartz) and 10.2%–48.9% for clay minerals (mainly illite/montmorillonite interlayers). The results imply that the reservoirs are vulnerable to potential water sensitivity, velocity sensitivity, and acid sensitivity damages.

2.3 Analysis and evaluation of damages resulting from sensitivity characteristics of reservoirs

A reservoir sensitivity evaluation test is carried out to analyze the potential damage factors for reservoirs and further provide scientific bases for future development and design. This is of great realistic significance for reservoir protection and improving the integrated economic development benefits of oil and gas fields (Li & Li 1998; Zhang et al. 2012). The natural cores from target reservoirs are subject to velocity sensitivity, water sensitivity, and acid sensitivity analysis and evaluation tests according to *Formation Damage Evaluation by Flow Test* (Standard ID: SY/T5358-2010). See Table 3 for the results.

Table 3. Evaluation results of sensitivity of target reservoirs.

Sensitive type	Well number	Core number	Well depth (m)	Ka (mD)	Porosity (%)	Damage ratio (%)	Sensitivity
Velocity sensitivity	WZ12-1-A7	6	3173.00-3175.00	2.41	13.3	12.50	Weak
Water sensitivity	WZ12-1S-2	4	3182.50-3183.99	56.9	15.7	54.59	Medium to strong
Hydrochloric acid sensitivity	WZ12-1-A7	6-1	3173.00-3175.00	2.43	13.3	−7.41	Weak
Soil acid sensitivity	WZ12-1S-2	3	3182.50-3183.99	16.4	15.5	14.4	Weak

Based on Table 3, it can be found that target reservoirs are subject to a damage rate of velocity sensitivity of 12.50%, with weak velocity sensitivity but without critical velocity, showing that there is no or minor fines migration, and no particular attention may be paid to velocity sensitivity damage; the damage rate of water sensitivity is 54.59%, with middle-to-high damage. In drilling and completion operations, the mineralization degree of the operating fluids should be adjusted with its inhibition enhanced to prevent water sensitivity damage. The target reservoirs suffer weak hydrochloric acid and mud acid sensitivity and can be acidized for the release of production capacity.

2.4 Analysis and evaluation of damages resulting from fluid characteristics of reservoirs

The target reservoirs contain crude oil with good properties, which are conventional light crude, featuring "6-low and 1-high": low density, low viscosity, low pour point (21–27°C), low sulfur content (about 0.15%), low colloid content (1.45%–5.00%), low asphaltene content (1.48%–5.53%), and high wax content (12.50%–17.33%). The formation water is $NaHCO_3$ type, so it should be prevented from contacting scale-forming cations Ca^{2+} and Mg^{2+} and forming inorganic scales to damage the reservoirs in the drilling and completion process.

3 ADAPTABILITY EVALUATION ON EXISTING DRILLING AND COMPLETION FLUID SYSTEMS

3.1 Use of drilling and completion fluid systems

In recent years, oil-based drilling fluid systems have been generally used in drilling Weizhou formations of Weizhou oilfield groups. Such operations are smooth in general and some wells are subject to the PLUS/KCl system with frequent difficulties in POOH of the drill string and holding torque. Solid-free organic salt drilling fluid systems are used in horizontal well intervals in target formations. The supporting completion fluids are generally implicit acid completion fluids. JPC breaker fluids are added to the water-based completion fluids for horizontal wells. White oil is seldom used in oil-based drilling fluids. PF-HDM of 1% content is added into the implicit acid completion fluids of some wells to prevent emulsification. According to the use of existing drilling and completion fluid systems (Table 4), the initial production (%; in production allocation) for middle-low-permeability wells with poor physical properties in L_1 Member is lower than 60%. Corresponding measures for the release of production capacity are in urgent need.

Table 4. Application of existing drilling and completion fluid systems.

Oilfield	Well number	Production horizon	Permeability (mD)	Initial production (%)	Drilling and completion of fluid system
Weizhou 11-4N	B17	$L_1 I_{upper}$, $L_1 I_{lower}$, $L_1 V$	28.6-71.4	90	PDF-MOM, implicit acid completion fluid + 1%PF-HDM
	B9	$L_1 I_{upper}$, $L_1 I_{lower}$, $L_1 V$	5.3-19.2	60	PDF-MOM, implicit acid completion fluid + 1%PF-HDM
Weizhou 12-2	A20	$L_2 II$	25.3	0	PDF-MOM, implicit acid completion fluid + 1%PF-HDM
	B44	$L_2 II$	4.3-21.6	50	PDF-MOM, implicit acid completion fluid + 1%PF-HDM
	B43	$L_2 II$	4.3-21.6	49	PDF-MOM, implicit acid completion fluid + 1%PF-HDM
Weizhou 12-1	B29S2	W4I, W4II	10-150	18	PDF-MOM, implicit acid completion fluid + 1%PF-HDM
Weizhou 11-1N	A7H	$L_1 II$	34.81-1240.9	49	PRD, JPC gel breaker, implicit acid completion fluid
	A17H	$L_1 IV_{upper}$	3.2-174	72	PRD, JPC gel breaker, implicit acid completion fluid
Weizhou 12-1W	A7H	$W3X_{lower}$	30.4	159	Solid-free organic salt drilling fluid, JPC gel breaker, implicit acid completion fluid
	A8H	$W3X_{lower}$	30.4	100	PDF-MOM, white oil, implicit acid completion fluid

3.2 Reservoir protection performance of drilling and completion fluid systems

Samples are taken on site with oil-based drilling fluids and implicit acid completion fluids well prepared. The existing drilling and completion fluid systems are subject to reservoir protection performance evaluation according to the *Lab Testing Method of Drilling and Completion Fluids Damaging Oil Formation* (Standard Code: SY/T 6540-2002).

The testing steps are as follows: (1) vacuum the core with saturated formation water and measure the porosity; (2) use the forward permeability test to measure the initial gas permeability at room temperature, and use the core displacement device to measure oil-phase permeability; (3) use drilling fluids for reverse core pollution for 2 hrs on the dynamic pollution instrument at 3.5 MPa and 130°C; (4) remove the external mud cake after the pollution with the completion fluid, squeeze 2 PV implicit acid completion fluids reversely, and remain well shut-in for 6 hrs; (5) take out the core, remove the core face mud cake, and measure the oil-phase permeability; (6) cut the polluted core by 0.5 cm and then measure the oil-phase permeability; (7) squeeze 2 PV acidizing productivity release fluid reversely to test the productivity release fluid, remain well shut-in for 6 hrs, and then measure the oil-phase permeability; (8) calculate the core permeability recovery values according to the oil-phase permeability before and after pollution in each stage. See Table 5 for the results.

Table 5. Reservoir protection performance of existing drilling and completion fluid systems.

Core number	2#	5#
Well number	WZ12-1-B33	WZ12-1-A7
Well depth (m)	3681.25-3683.25	3173-3175
Length (cm)	6.77	5.19
Diameter (cm)	2.48	2.48
Gas permeability K_a(mD)	10.52	2.51
Porosity ϕ (%)	13.8	13.7
Oil-phase permeability K_1(mD)	6.76	0.86
Oil-based drilling fluid pollution	3.5 MPa, 130°C	3.5 MPa, 130°C
	Reverse pollution for 2 hrs	Reverse pollution for 2 hrs
Implicit acid completion	–	Squeeze 2 PV reversely and shut in
fluid pollution	–	for 6 hrs
Oil-phase permeability K_2(mD)	4.98	0.61
Permeability recovery value Rd1 (%)	73.6	71.1
Release of production capacity	–	Squeeze 2 PV acidizing productivity release fluid reversely and shut in for 6 hrs
Oil-phase permeability K_3(mD)	5.98 (0.5 cm)	0.95
Permeability recovery value Rd2 (%)	88.5	110.9

Based on Table 5, it can be found that after the 2# core is polluted by the oil-based drilling fluid, the permeability recovery value after the removal of the face mud cake is 73.6%; while after the core is cut by 0.5 cm, the permeability recovery value reaches 88.5%, indicating a good reservoir protection performance and showing that the existing oil-based drilling fluids are suitable for the target reservoirs; after 5# core is polluted by the oil-based drilling fluid and the implicit acid completion fluid, its permeability recovery value is only 71.1%, but the value reaches 110.9% after it is further polluted by the acidizing productivity release fluids. The possible reason may be the weak dissolution performance of the implicit acid completion fluids to the cores from the target reservoirs and the lack of preventive water lock measures.

The reservoir protection performance and productivity release performance of completion fluid require further optimization. See Table 6 for detailed measures and suggestions.

4 OPTIMIZATION AND EVALUATION OF COMPLETION FLUID SYSTEM

4.1 Optimization of completion fluid system

Major treating agents for corresponding reservoir protection and reconstruction measures should be carefully selected based on the damage factors for the target reservoirs (Guo & Huang 2017;

Table 6. Suggestions on optimization measures of reservoir protection and productivity release performances of existing completion fluid systems.

Well type	Completion fluid system	Reservoir protection performance	Productivity release performance
Directional well	Implicit acid completion fluid + 1%PF-HDM	Anti-water lock is not considered	Measures to increase productivity release after perforation can be considered
Horizontal well	Implicit acid completion fluid + 1%PF-HDM	Oil-based drilling fluid mud cake removal and anti-water lock are not considered	Mud cake remover or organic plugging remover can be pumped to soak in the open hole section
Horizontal well	JPC gel breaker + implicit acid completion fluid	Anti-water lock is not considered	Replacing the JPC gel breaker with acidizing productivity release fluid can be considered

Qiu 2017; Yu et al. 2017; Zhang & Wang 2009; Hang et al. 2011; 2018). First, it is needed to select the water lock proof agent RF-STWL that can reduce the oil-fluid interface tension to below 0.1 mN/m according to *The Method for Measurement of Surface Tension & Interfacial Tension* (Standard Code: SY/T 5370-1999) with the recommended dosage of 2%–4%. Second, the direct gel-breaking method and core displacement method are used to compare and evaluate different types of breaker fluids and select the dual effect solid acid gel breaker RF-HWNA of the complex organic acid type. The gel breaker can increase the viscosity reduction rate of gel to more than 95%. After the plugged core is soaked and de-plugged, the filter loss greatly exceeds the blank loss before the core is polluted. The dual effect gel breaker can decompose the polymer at the core plugging layer and dissolve partial minerals in the core throat, significantly increasing the permeability. Therefore, its reservoir reconstruction effect is prominent and it is beneficial to the release of production capacity. Third, for horizontal wells drilled by oil-based drilling fluids, the mud cake remover PF-HCF for oil-based drilling fluids should be selected carefully to realize a mud cake weight loss ratio of above 95%. It is very useful in cleaning polluted slotted pipes.

4.2 *Evaluation of completion fluid system*

4.2.1 *Evaluation of dissolution performance*

The selective dissolution effect of the acids selected for the acid reconstruction of sandstone reservoirs is crucial to the effective dredging of pores and throats and the dredging effect of the target reservoirs. An excessively high dissolution rate may cause the collapse of the reservoir framework, but a small dissolution rate is not suitable for reservoir reconstruction. According to the acid selection and evaluation methods in *Formation Damage Evaluation by Flow Test* (Standard Code: SY/T 5358-2010), it is needed to evaluate the dissolution performance of reservoir rock debris by the acidizing productivity release fluid and the on-site implicit acid completion fluid. Then, a PHS-3C acidity meter is used to monitor the pH values before and after the dissolution. See Table 7 for the results.

Based on Table 7, it can be found that the on-site implicit acid completion fluid features a small dissolution rate of rock debris of the target reservoirs, which is only 0.12%–1.41%; while the dissolution rate of the acidizing productivity release fluid is 5–19 times of that of the on-site implicit acid completion fluid, indicating that the productivity release fluid is more useful in reservoir reconstruction; in addition, 15% hydrochloric acid equals the acidizing productivity release fluid in the dissolution rate of rock debris of the target reservoirs but features a higher pH

Table 7. Dissolution performance of optimized completion fluid system.

No.	Acid fluid	Cuttings well number	Cuttings horizon (m)	Reaction time (h)	Dissolution rate (%)	pH value before dissolution	pH value after dissolution
1	Acidizing productivity release fluid	WZ12-1-2	2825.00	16	7.51	0.50	0.59
2	Acidizing productivity release fluid	WZ12-1-2	2836.00	16	8.65	0.50	0.61
3	Acidizing productivity release fluid	WZ12-1S-2	3182.5-3183.99	16	2.25	0.50	0.53
4	On-site implicit acid completion fluid	WZ12-1-2	2825.00	16	1.26	1.23	2.27
5	On-site implicit acid completion fluid	WZ12-1-2	2836.00	16	1.41	1.23	3.33
6	On-site implicit acid completion fluid	WZ12-1S-2	3182.5-3183.99	16	0.12	1.23	1.57
7	15% HCl	WZ12-1-2	2825.00	16	7.51	-1.26	0.62
8	15% HCl	WZ12-1-2	2836.00	16	9.19	-1.26	0.63
9	15% HCl	WZ12-1S-2	3182.5-3183.99	16	2.31	-1.26	0.49

value change before and after the dissolution. This means that the acidizing productivity release fluid is more suitable for a deeper reconstruction effect.

4.2.2 Evaluation of anti-swelling performance

According to the *Performance Evaluation Method for Clay Stabilizer for Water Injection* (Standard Code: SY/T 5971-1994), a high-speed centrifuge and other equipment can be used to evaluate the anti-swelling performance of the optimized completion fluid system. See Table 8 for the results.

Based on Table 8, it can be found that compared with seawater, the optimized implicit acid completion fluid system features a good anti-swelling performance for reservoir rock debris with an anti-swelling rate higher than 93%. This is associated with the clay stabilizer added into the completion fluid system.

4.2.3 Evaluation of corrosion performance

In combination with the anti-corrosion scheme (13Cr for the remaining 11 wells except for N80 for Well C18H) and according to *Performance Test Method and Evaluation Indexes of Corrosion Inhibitors for Acidizing Fluids* (Standard Code: SY/T 5405-1996) and *Water Quality Standard and Practice for Analysis of Oilfield Injecting Waters in Clastic Reservoir s* (Standard Code: SY/T 5329-2012), the coupon corrosion weight-loss method is used to evaluate the corrosion performance of the completion fluid system, and a HTHP corrosion instrument is used to measure the corrosion rate at 130°C (Table 8). The findings are that the average corrosion rates of the implicit acid completion fluid system for 13Cr and N80 pipes are lower than 0.076 mm/a before and after the optimization at 130°C× 72 hrs; while the average corrosion rates of the oil-based drilling fluid mud cake remover and the acidizing productivity release fluid are less than 2 g/m².h at 130°C × 4 hrs. All of them comply with the requirements in the petroleum industry standards.

4.2.4 Evaluation of anti-water lock performance

Compared with seawater and on-site implicit acid completion fluid, the optimized implicit acid completion fluid and acidizing productivity release fluid with the anti-water lock agent feature

Table 8. Anti-swelling performance, corrosion performance, anti-water lock performance of optimized completion fluid system.

System	Anti-swelling rate	Corrosion temperature	Corrosion time	Average corrosion rate		Gas-fluid surface tension	Oil-fluid interfacial tension
	%	°C	h	13Cr	N80	mN/m	mN/m
Seawater	60.2	130	72	0.0598 mm/a		64.81	16.2361
On-site implicit acid completion fluid	93.3	130	72	0.0649 mm/a		62.39	15.3621
Optimized implicit acid completion fluid	93.2	130	72	0.0657 mm/a	0.0698 mm/a	21.50	0.0992
Oil-based drilling fluid mud cake removal fluid	93.6	130	4	0.9265 g/m^2.h			
Acidizing productivity release fluid	97.6	130	4	1.2365 g/m^2.h	1.5698 g/m^2.h	23.80	0.0861

less gas-fluid surface tension (21.50–23.80 mN/m) and oil-fluid interfacial tension (<0.1 mN/m) (Table 8). The anti-water lock performance is significantly improved, which is more beneficial to target reservoir protection and release of production capacity.

4.2.5 Evaluation of compatibility

The optimized implicit acid completion fluid and optimized oil-based drilling fluid filtrate are mixed in different proportions and heated at 130°C for 12 hrs. Then, it is needed to measure the oil and water volumes and observe the interface. See Table 9 for the results.

Table 9. Compatibility evaluation of optimized implicit acid completion fluid and optimized oil-based drilling fluid filtrate.

V (implicit acid completion fluid): V (oil-based drilling fluid filtrate)	Volume after heating at 130°C for 12 h/ml		Interface condition
	Oil	Water	
1:9	13.5	1.5	Clear interface
5:5	7.5	7.5	Clear interface
9:1	1.5	13.5	Clear interface

Based on Table 9, it can be found that after the optimized implicit acid completion fluid and oil-based drilling fluid filtrate are mixed in different proportions, the oil-water interface is clear without obvious turbidity, emulsification, etc. It shows that the optimized implicit acid completion fluid and oil-based drilling fluid filtrate feature satisfactory compatibility.

Next, the optimized implicit acid completion fluid or acidizing productivity release fluid is mixed with the optimized solid-free organic salt drilling fluid filtrate or formation water in different proportions, and the turbidity method is used to evaluate the compatibility. See Table 10 for the results.

Table 10. Compatibility evaluation of optimized implicit acid completion fluid and optimized solid-free organic salt drilling fluid filtrate or formation water.

| Item | Volume ratio | Turbidity value/NTU | | Phenomenon |
		Before heating	After heating at 130°C for 12 hrs	
V (implicit acid completion	1:9	6.1	10.9	Colorless and transparent
fluid): V (solid-free organic	5:5	2.8	9.2	
salt drilling fluid filtrate)	9:1	0.3	1.2	
V (acidizing productivity	1:9	6.9	3.9	Light yellow and transparent
release fluid): V (solid-free	5:5	5.8	2.5	
organic salt drilling fluid filtrate)	9:1	2.6	1.4	
V (implicit acid completion	1:9	0.2	0.5	Clear and transparent
fluid): V (formation water)	5:5	0.3	0.6	
	9:1	0.5	0.9	

Based on Table 10, it can be found that after the optimized implicit acid completion fluid or acidizing productivity release fluid is mixed with the optimized solid-free organic salt drilling fluid filtrate or formation water in different proportions, no incompatible phenomena occur, e.g., turbidity. The low turbidity value means that these fluids have good compatibility and are beneficial to reservoir protection and reconstruction.

The viscosity method is used to evaluate the compatibility of the optimized implicit acid completion fluid and reservoir oil. The reservoir oil in Well WZ12-1W-A4H is mixed with the optimized implicit acid completion fluid or acidizing productivity release fluid by 100:0, 80:20, 60:40, 50:50, 40:60, 20:80, and 0:100 respectively, which is then stirred evenly in a thermostat water bath at 50°C. Then a DV-II type Brookfield Viscometer is used to measure the viscosity at 50°C. See Figure 1 for the results.

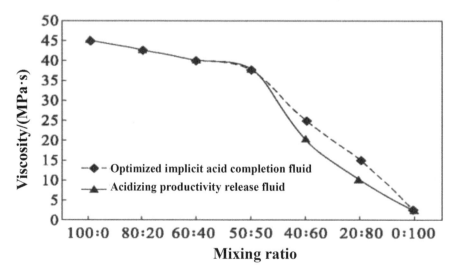

Figure 1. Compatibility of optimized implicit acid completion fluid or acidizing productivity release fluid and reservoir oil in Well WZ12-1W-A4H.

Based on Figure 1, it can be found that after the reservoir oil in Well WZ12-1W-A4H is mixed with the optimized implicit acid completion fluid or acidizing productivity release fluid, no emulsifying and thickening occurs, indicating satisfactory compatibility.

4.2.6 Evaluation of reservoir protection performance

The reservoir protection and reconstruction performances of the optimized completion fluid system are evaluated according to the *Lab Testing Method of Drilling and Completion Fluids Damaging Oil Formation* (Standard Code: SY/T 6540-2002). See Table 11 for the results.

Table 11. Evaluation of reservoir protection performance of optimized completion fluid system.

	Core number	Well number	Contaminated medium	Length cm	Diameter cm	Gas permeability mD	Permeability recovery value %	Measures for productivity release	Permeability recovery value after measures (%)
Single fluid reservoir protection performance	3#-2	WZ12-1S-2	Seawater	5.14	2.48	15.9	40.2		
	6#-3	WZ12-1-A7	Optimized implicit acid completion fluid	4.42	2.48	2.66	99.6		
	4#	WZ12-1S-2	PF-HCF	5.68	2.48	52.7	112.6		
	6#-4	WZ12-1-A7	Acidizing productivity release fluid	4.42	2.48	2.66	158.9		
Productivity release effect of reservoir reconstruction	4#-2	WZ12-1S-2	Oil-based drilling fluid, PF-HCF, optimized implicit acid completion fluid	5.68	2.48	54.2		Replace PF-HCF with an open hole section	115.6
	4#-3	WZ12-1S-2	Oil-based drilling fluid, PF-HCF, optimized implicit acid completion fluid	5.68	2.48	55.3		PF-HCF plugging removal and stimulation	116.2
	3#-5	WZ12-1S-2	Optimized solid-free organic salt drilling fluid, acidizing productivity release fluid, and optimized implicit acid completion fluid	5.14	2.48	15.8		Replace JPC gel breaker with acidizing productivity release fluid	126.9
	5#-4	WZ12-1-A7	Optimized solid-free organic salt drilling fluid, JPC gel breaker, optimized implicit acid completion fluid, and acidizing productivity release fluid	5.19	2.48	2.56		Plugging removal with acidizing productivity releases fluid	121.3

Based on Table 11, it can be found that, from the perspective of the single fluid reservoir protection performance evaluation effect, after the low-permeability reservoir is polluted by seawater, the permeability recovery value is only 40.2%; while the optimized implicit acid completion fluid brings enhanced anti-water lock performance and the permeability recovery value increases significantly, reaching 99.6% and indicating satisfactory reservoir protection performance; after the productivity release and fluid pollution with the oil-based drilling fluid and solid-free organic salt drilling fluid, the permeability recovery value exceeds 112%, showing good reservoir reconstruction performance. For the oil-based drilling fluid and solid-free organic salt drilling fluid, they increase the permeability recovery values significantly, which are higher than 115% based on the evaluation effect with regular operations and additional measures for productivity release. The results show a good reservoir reconstruction performance and are beneficial to the full release of production capacity.

5 SUGGESTIONS ON OPTIMIZING THE DRILLING AND COMPLETION FLUID SYSTEMS OF VARIOUS WELLS

To fully release the production capacity, the drilling and completion fluid systems for various wells at the target reservoirs are recommended as follows:

1) Directional Well C1: oil-based drilling fluid, PF-HCF, and optimized implicit acid completion fluid.
2) Upper perforation intervals of horizontal Well C3H: oil-based drilling fluid, PF-HCF, and optimized implicit acid completion fluid.
3) Horizontal Wells C2H-C8H and C17H-C20H: optimized solid-free organic salt drilling fluid, acidizing productivity release fluid, optimized implicit acid completion fluid; or optimized solid-free organic salt drilling fluid, JPC gel-breaking fluid, optimized implicit acid completion fluid, with acidizing productivity release fluid as the remedial measure for release of production capacity.

6 CONCLUSIONS

The damage factors of target reservoirs are systematically analyzed and evaluated and the major damage factors in the drilling and completion process are determined to be low-permeability water lock and reservoir water sensitivity. Combined with the early drilling and completion fluid systems, output, and reservoir protection performance of drilling and completion fluids, the measures and recommendations for completion fluid system optimization and productivity release are proposed. Furthermore, the major treating agents, the anti-water lock agent RF-STWL, dual-effect solid acid gel breaker RF-HWNA and mud cake remover PF-HCF are carefully selected, and the optimized completion fluid system formula is determined. Optimized completion fluid systems exhibit good dissolution, strong anti-swelling properties, low corrosion, good water lock performance, and good reservoir protection. The recommended drilling and completion fluid systems for various wells in the target reservoirs are proposed in combination with the well type and the selected drilling fluid system, which is beneficial to the full release of production capacity.

REFERENCES

Guo, Y.F., Huang, J.Q. (2017) *Application of reservoir acidizing and deplugging technology in low-permeability oilfields* [J]. Liaoning Chemical Industry, 46(11): 1132–1134.

Li, Z.Z., Li, X.H. (1998) *Study on damage of reservoir water sensitivity and its influencing factors* [J]. 8 (2): 45–50.

Qiu, Y.P. (2017) *Study on drilling and completion fluid technology for reservoir protection in Hafaya Oilfield, Iraq* [D]. Qingdao: China University of Petroleum.

Yu, X.Y., Su, Y., Sun, L. (2017) *Acidizing enhancement technology and its application in offshore low-permeability reservoir* [J], Inner Mongolia Petrochemical Industry. 43(8): 62–66.

Zhang, B., Wang, H.B. (2009) *Discussion on acidizing reconstruction technology of low-porosity, low-permeability and high-temperature reservoir in offshore exploration wells* [J]. Tianjin Science and technology, 36 (5): 50–52.

Zhang, C., Ren G.L., Jin, S.K. et al. (2018) Study on high-temperature completion productivity release fluid for offshore low-porosity and low-permeability gas field [J]. *Drilling Fluid and Completion Fluid*. 35(6): 126–130.

Zhang, L., Jin, Q.H., Liang, W., et al. (2011) Application of organic acid in acidizing offshore oilfield [J]. *Petrochemical Technology*. 40 (7): 770–774.

Zhang, R., Zang S.B., Ren, X.J. (2012) *Analysis on water sensitivity damage factors of low-permeability reservoir rocks* [J]. Inner Mongolia Petrochemical Industry. 25(12): 15–17.

Advances in Measurement Technology, Disaster Prevention and
Mitigation – Li & Mohd Yusof (Eds)
© 2023 The Author(s), ISBN: 978-1-032-36087-4

Effect of cement ratio on work and bonding properties of styrene acrylic emulsion-based cement composites

Yipeng Ning*, Biao Ren*, Zhihang Wang, Ao Yao, He Huang & Tengjiao Wang
Airport Construction Engineering Department, Air Force Engineering University, Xi'an, China

ABSTRACT: Aiming at the joint disease problem of cement concrete pavement of airport, to
further improve the basic performance of styrene acrylic emulsion cement composite pavement
joint sealant, the effects of cement ratio on the performance and bonding properties of styrene acrylic
emulsion cement composite pavement joint were studied using leveling test, pouring consistency
test, elongation bonding test, scanning electron microscope test and mercury injection test. The
leveling property, pouring consistency and elastic recovery rate were tested. Combined with a
scanning electron microscope experiment and a mercury injection hole test, the micro modification
mechanism was revealed. The results show that with the increase in cement ratio, the elastic recovery
rate of the joint filler decreases, and the pouring consistency of the joint filler increases gradually,
but the overall increase is small, and the pouring consistency values of each group of joint filler are
less than 10 s. A proper increase in the cement ratio is conducive to reducing the number and size
of pores in the joint filler and improving the pore size distribution. This will improve the density of
the sealant and increase the working and setting bonding properties of the styrene-acrylic emulsion
cement composite.

1 INTRODUCTION

Airport cement concrete pavement joint disease is an important factor affecting the normal use of
military airport pavement and military fighter combat training, and the failure of airport pavement
joint filler is the main cause of airport joint disease. Styrene acrylic emulsion-based cement composite material is a high-performance polymer cement composite material, which uses styrene acrylate
copolymer emulsion as matrix material, cement, and inorganic filler as reinforcement material, and
uses inorganic cement cementitious material and organic polymer material effectively compounded.
These composites have the advantages of high bonding strength, good flexibility, excellent durability, economy, and environmental protection. Ariffin et al. (Ariffin et al. 2015) prepared epoxy
resin modified cement mortar with different dosages (5%, 10%, 15%, and 20%), and studied the
working performance, setting time, and mechanical properties of epoxy modified mortar under
different curing conditions. The results show that the setting time of modified mortar is prolonged
with the increase of epoxy resin dosage, The working performance is reduced. The modified mortar
is wet cured first and then dry cured. When the content of the epoxy resin is 10%, its compressive,
flexural, and splitting tensile strength reaches the maximum at the same time, and the epoxy resin
can react with OH- ions generated by cement hydration. Yang Qianrong et al. (2016) compared and
studied the effects of different polymers (re-dispersible latex powder, cellulose ether, and starch
ether) on the construction performance and sagging resistance of cement mortar by extrusion time
test and sagging test. The results show that the mixing of three polymers improves the workability
and sagging resistance of cement mortar. Almeida et al. (2007) studied the mechanical properties

---

*Corresponding Authors: ningyipeng1998@163.com and 1124038494@qq.com

DOI 10.1201/9781003330172-67

and microstructure of styrene acrylic emulsion modified cement mortar and analyzed the applicability of styrene acrylic emulsion modified cement mortar as a ceramic tile bonding material. The results show that after adding styrene acrylate emulsion, the pore size of cement mortar is refined, the average pore size decreases, the density increases, and the interfacial bond strength increases. Zhu Mingsheng (2011) studied the influence of the amount of styrene acrylic emulsion on the performance and mechanical properties of cement repaired mortar. The results show that the compressive strength of cement mortar is significantly decreased, the flexural strength is slightly lower than that of the reference group, and the workability of the cement mortar is significantly improved.

2 EXPERIMENT

2.1 *Experimental materials*

The raw materials of styrene acrylic emulsion-based cement composites include styrene acrylic emulsion, ordinary Portland cement, water, inorganic filler, dispersant, film builder, and defoamer. This paper selects the Acronal S400F AP styrene acrylic emulsion produced by BASF Company as raw material, and its main technical indicators are shown in Table 1.

Table 1. Main technical indexes of Acronal S400F AP styrene acrylic emulsion.

Solid content (%)	Viscosity (mPa·s)	pH	T_g (°C)	MFFT (°C)	Average particle size (μm)
56 ± 1	400–1800	7.0–8.5	−7	0	0.1

The cement adopts Shaanxi Lantian Yaobai brand 42.5 ordinary Portland cement. The inorganic filler adopts the ultra-fine talc powder produced by Liaoning Dashiqiao Wantong powder factory for basic mix proportion design, the defoamer adopts NOPCO NXZ produced by Japan SAN NOPCO (Shanghai) Co., Ltd., the defoamer adopts NOPCO NXZ defoamer produced by Japan SAN NOPCO (Shanghai) Co., Ltd., and the dispersant adopts Japan SAN NOPCO (Shanghai) Sn-dispersant 5040 dispersant and film-forming additive produced by Jiangsu Tianyin Chemical Co., Ltd. are alcohol ester-12 (DN-12).

2.2 *Specimen preparation*

The above main materials are prepared according to the mix proportion shown in Table 3. The specific preparation process is as follows:

(1) Polymer emulsion and dispersant are mixed and stirred for 3 minutes to disperse the polymer emulsion evenly. After adding the defoamer and film assistant, the mixture was mixed for six minutes to ensure an equal distribution of the emulsion.
(2) In preparing polymer cement flexible composite materials, Portland cement, and talcum powder are mixed until they are fully mixed and mixed evenly, then the above-prepared emulsion is added, stirred at low speed for five minutes, stirred at high speed for ten minutes, and finally stirred for ten minutes artificially.
(3) Polymer-based cement flexible composites are poured into molds and cured for 28 days before being demolded. The specimen and its specific dimensions are shown in Figures 1 and 2.

Figure 1.   Standard joint filler test piece.   Figure 2.   Standard joint filler specimen size.

Table 2.   Cement ratio single factor test mix proportion.

	Basic parameters			Consumption of raw materials for one test (unit: g)					
Specimen number	Powder liquid ratio	Cement ratio	Filler type	Styrene acrylic emulsion	Cement	Filler	Dispersant	Defoamer	Coalescent
SNB-1	0.40	20%	Talc	100	8	32	1.12	0.7	5
SNB-2	0.40	25%	Talc	100	10	30	1.12	0.7	5
SNB-3	0.40	30%	Talc	100	12	28	1.12	0.7	5
SNB-4	0.40	35%	Talc	100	14	26	1.12	0.7	5
SNB-5	0.40	40%	Talc	100	16	24	1.12	0.7	5
SNB-6	0.40	45%	Talc	100	18	22	1.12	0.7	5

2.3   *Experimental method*

In this paper, the basic performance test of styrene acrylic emulsion cement composite pavement joint sealant is carried out based on the *Technical Specification for the Construction of Cement Concrete Pavement Joint Materials at Military Airports* (GJB 6951 - 2010) and the *Building Sealing Material Test Method* (GB/T 13477 - 2002).

3   TEST RESULTS AND DISCUSSIONS

3.1   *Working performance*

According to the leveling test results of the joint filler, the joint filler with different cement ratios has good leveling performance and can be self-leveled during the joint filling process. After the joint filing, the surface is smooth and flat. Further combined with the influence law of cement ratio on the pouring consistency of joint filler (as shown in Figure 3), the pouring consistency of joint

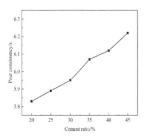

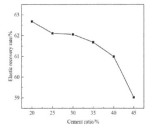

Figure 3.   Variation of pouring consistency of joint filler with cement ratio.

Figure 4.   Change of elastic recovery rate of joint filler with cement ratio.

filler gradually increases with the increase of cement ratio, but the overall increase is small, and the pouring consistency values of each group of joint filler are less than 10 s, so it is not enough to affect the leveling of joint filler.

3.2 *Constant elongation bonding performance*

The bond failure modes of joint fillers with different cement ratios are shown in Figure 5. After the constant extension bonding performance test of joint fillers with different cement ratios, except SNB-1 with a cement ratio of 20% has slight bonding failure, all other groups of joint fillers have no cohesion or bonding failure, and have good constant extension bonding performance. Further combined with the influence of cement ratio on the elastic recovery rate of sealants (as shown in Figure 4), it is known that with the increase of cement ratio, the resilience recovery rate of the sealant decreases continuously, and when the cement ratio increases to 45%, the elastic recovery rate of the sealant will lower than 60%, which does not meet the elastic recovery requirement of styrene acrylic emulsion cement composite pavement sealant. This indicates that the excessive cement ratio can lead to a significant decrease in the resilience recovery ability of styrene acrylic emulsion cement composite pavement sealants (Dong & Zhang 2008).

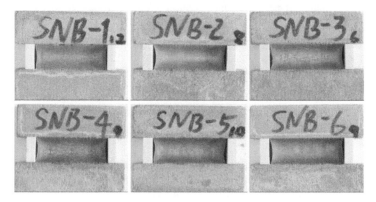

Figure 5. Effects of cement ratio on the bond failure mode of joint filler at constant elongation.

4 MICRO MECHANISM ANALYSIS

The micromorphology of joint fillers with different cement ratios under the SEM test is shown in Figure 6, and the magnification is 100 times. It can be seen from the diagram that with the increase in cement ratio, the number of internal pores in the composite pavement sealant of styrene acrylic emulsion cement decreased significantly, and the pore size decreased significantly. When the cement ratio was 20%, the number of pores in the joint was larger and the pore size was larger. Thereafter, with the increase in cement ratio, the number of pores gradually decreased and the pore size decreased. When the cement ratio increases to 40% and 45%, The micromorphology, and structure of the sealant are becoming denser. Only a few pores are observed and the pore size is relatively small. This indicates that the reasonable increase of cement proportion in the composite pavement sealant of styrene acrylic emulsion (Dong 2002; Joint Sealing Material of Cement Concrete Pavement for Highway (JT/T 203-2014); Xu et al. 1998; Zhang et al. 2004) can significantly reduce the number of internal pores in the sealant, effectively reduce the pore size and optimize the microstructure.

It can be seen from the diagram that with the increase in cement ratio (SNB-1 to JC to SNB-4 to SNB-6), the pore volume and the characteristic pore size of the sealant decrease continuously, and the pore size distribution shifts in a decreasing direction. The percentage of macropores decreases

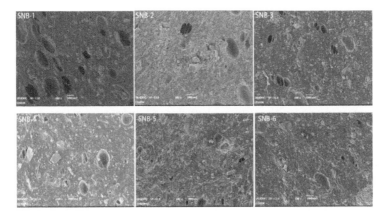

Figure 6.   Effects of cement ratio on the microstructure of joint filler.

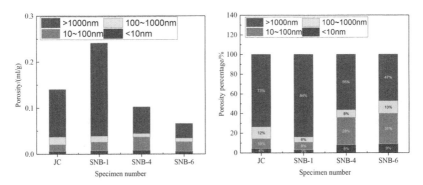

Figure 7.   Effect of different cement ratios on pore volume and pore percentage of joint filler.

significantly, and the percentage of gel pores, transition pores, and pores increases. Properly increasing the cement ratio is conducive to reducing the number of pores in the joint filler and reducing the pore size. At the same time, it can improve the pore size distribution and make the pore size develop in the decreasing direction, to improve the compactness of the joint filler and improve the working and constant elongation bonding performance of the joint filler. The main reason for this phenomenon is that the increase in cement ratio leads to the continuous increase of cement hydration products in the joint filler. On the one hand, the cement hydration products are gradually filled in the pore structure (Xu et al. 1998), resulting in the reduction of the number of pores and pore size in the joint filler. On the other hand, more cement hydration products have a cross-linking reaction with the polymer membrane, so the compactness of the joint filler is further improved.

5   CONCLUSIONS

(1) The pouring consistency of the joint filler gradually increases with the increase of the cement ratio, but the overall increase is small, and the pouring consistency values of each group of joint filler are less than 10 s, so it is not enough to affect the leveling of the joint filler.

(2) The high cement ratio can lead to a significant decrease in the resilience recovery ability of the composite pavement sealant. A proper increase in the cement ratio is conducive to reducing the number of pores in the joint filler and reducing the pore size. Meanwhile, the joint filler can improve the pore size distribution and the pore size will develop in a decreasing direction,

thus improving its compactness as well as its constant elongation bonding performance. The reasonable value range of the water mud ratio is 30% to 35%.

(3) The research ideas, methods, and relevant conclusions of this paper can provide a certain scientific basis for the application of polymer-based cement composites in airport pavement joint filling engineering and related fields.

REFERENCES

Almeida A E F D S, Sichieri E P. Experimental study on polymer-modified mortars with silica fume applied to fix porcelain tile[J]. *Building & Environment*, 2007, 42(7):2645–2650.

Ariffin N F, Hussin M W, Sam A R M, et al. Strength properties and molecular composition of epoxy-modified mortars[J]. *Construction & Building Materials*, 2015, 94:315–322.

Dong S, Zhang Z Q. Study of the microstructure of polymer-cement compounded waterproof coating[J]. *Green Building*, 2008, 24 (4):35–38.

Dong S. *Research on the preparation of polymer-modified cement-based waterproof coating and its coating properties and microstructure*[D]. Chongqing: Chongqing University, 2002.

Joint Sealing Material of Cement Concrete Pavement for Highway (JT/T 203–2014).

Technical Regulations on Jointing Material Construction of Cement Concrete Pavement in Military Airport (GJB 6951-2010) [S].

Xu Y J, Li X C, Weng T Y, et al. Study on blend system of styrene-acrylic ester copolymeric emulsion and cement mortar (I) modified mechanism and microstructure forms of the blend system[J]. *Journal of Beijing University of Chemical Technology* (Natural Science Edition), 1998 (4):28–10.

Yang Q R, Jiang C D. Effect of polymer on properties of cement-based building sealing mortar [J]. *Journal of Tongji University* (NATURAL SCIENCE EDITION), 2016, 44 (1): 107–112

Zhang Z Q, Dong S, Ding H. The effect of the components of VAE latex modified cementitious compound waterproofing coating on its properties[J]. *Journal of Chongqing Jianzhu University*, 2004, 26 (4):83–87.

Zhu M S. Preparation and properties of styrene acrylic emulsion modified cement repair mortar [J]. *cement technology*, 2011 (3): 31–33.

Advances in Measurement Technology, Disaster Prevention and Mitigation – Li & Mohd Yusof (Eds)
© 2023 The Author(s), ISBN: 978-1-032-36087-4

The application status of red mud in building materials

Guilin Huang*

School of Materials Science and Engineering, Wuhan University of Technology, Wuhan, China

ABSTRACT: Red mud is a kind of strong alkaline solid waste that is inevitably generated in the aluminum production process of alumina enterprises due to the characteristics of the production process. According to the assessment, the total amount of red mud in the world has exceeded 3.5 billion tons, and a large amount of red mud cannot be processed in time, which will cause huge harm to soil, air, water resources, and the human body. For a long time, many researchers have developed several red mud treatment and utilization methods, which are applied in different fields. This paper will summarize and summarize the application status of red mud in the field of construction in my country from six specific preparation applications: preparation of cement materials, preparation of concrete materials, preparation of brick materials, preparation of roadbed materials, preparation of glass-ceramic materials, and preparation of ceramic materials. And analyze the existing problems, and provide the basis for the application of red mud in building materials in the future.

1 INTRODUCTION

Red mud is a solid waste produced by the components of bauxite after a series of physical and chemical reactions to aluminum in the industrial aluminum production process, often containing many iron oxide components showing a red appearance, so it is called red mud. The red mud contains Na_2O, Fe_2O_3, Al_2O_3, SiO_2, and a small amount of TiO_2 MnO, Cr_2O_3, and metal elements such as iron, aluminum, and titanium (Cao et al. 2007), which have great recycling value. According to the differences in the aluminum production process, red mud can be divided into three categories, namely Bayer red mud, sintered red mud, and combined red mud (Nan et al. 2009). China is a big producer of alumina, and the production of a large amount of alumina at the same time, inevitably will produce a large number of red mud, according to the assessment, the production of 1 ton of alumina will produce about 1-2 tons of red mud, of which more than 90% for Bayer red mud (Zhu & Lan 2008). Red mud components vary greatly, with high alkali content and high-water content. Most red mud cannot be directly processed and consumed. They can only be stacked treatment. The common storage methods are dry storage and wet storage. Red mud stored and treated will cause harm to the ecological environment and the human body, such as soil pollution, groundwater pollution, air pollution, and leaching of heavy metal ions (Xue et al. 2017). How to comprehensively recycle red mud without secondary pollution has become a major research hotspot. Among the many ways of utilizing red mud, the red mud applied to building materials consumes the most, and is the most likely way to achieve large-scale comprehensive utilization of red mud.

---

*Corresponding Author: 1377096084@qq.com

DOI 10.1201/9781003330172-68

2 APPLICATION STATUS

China is a big country in the production and demand of building materials. If we can realize the industrialization of low-cost preparation, simple process, then no secondary pollution of red mud building materials will achieve a major leap in the comprehensive utilization of red mud. Because the red mud contains a large number of silicate and aluminosilicate substances, small particle size, porous structure, large water content, and other properties, with a certain degree of water hardness, while also stimulating the potential activity of some mineral admixtures (Jing et al. 2001), for its application in the field of building materials to provide the basic conditions, so it can be applied to the field of building materials. At present, the specific application direction of the comprehensive utilization of red mud in this field in China can be divided into six specific application directions: preparation of cement materials, preparation of concrete materials, preparation of brick materials, preparation of subgrade materials, preparation of glass-ceramic materials, preparation of ceramic materials.

2.1 *Preparation of cement materials*

Cheng et al. (2021) added red mud to the cement of the oil well and carried out relevant performance tests. The results show that when the optimal amount of red mud is mixed at 5%, the degree of hydration, structural density, and strength of the product increase, while the content is too high, and the effect of reducing the relevant properties will be reduced. In the study of Li et al. (2018), red mud-based cementitious material (RCM) with excellent strength and low leaching rate of heavy metal ions was prepared using related raw materials such as red mud and arsenic sludge under the appropriate ratios and maintenance conditions. Figure 1 shows the schematic diagram of the preparation process of the RCM.

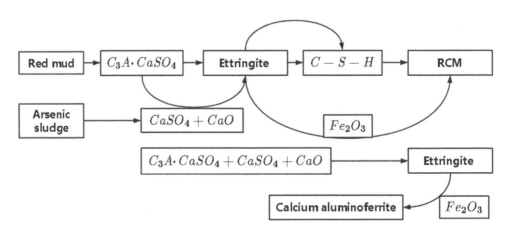

Figure 1. Schematic diagram of the preparation of RCM from red mud and arsenic sludge as raw materials.

2.2 *Preparation of concrete materials*

Li et al. (2019) prepared using municipal solid waste incineration fly ash (MSWIFA) and red mud under conditions such as sodium silicate with a mass ratio of 3:7 and a mass fraction of 14% as a binder, which produced a high strength and good durability of the red clay base polymer material (RGM). Hou et al. (2021) studied the development of ultra-high performance concrete (UHPC), and the results showed that UHPC was replaced by red mud after replacing part of the cement. The early strength of the UHPC has been improved, while also reducing its condensation time, which provides new ideas for UHPC applications in rapid repair engineering.

2.3 Preparation of bricks

In the research of Yin et al. (2020), .red mud, gangue, mud, and other materials were used as raw materials to prepare sintered porous bricks. The results showed that when the number of alternative materials was mixed up to about 75%, porous sintered bricks with strength and durability indicators that meet the corresponding national standards can be prepared. Arroyo et al. (2020) used red mud instead of 80% clay as raw material to manufacture sintered bricks, and the results showed that sintering at a temperature higher than that of red mud, when the content of red mud increased, the final product had higher compressive strength and density, and lower water absorption. And the natural radioactive element content after density correction can meet the relevant standards. Agrawal et al. (2022) found that by adding a certain percentage by weight of BaSO4 and binder (kaolin or sodium hexametaphosphate), the red mud through the route of producing ceramics into diagnostic X-ray shielding bricks. The results show that the strength of the prepared shielded brick is suitable for wall applications, and no heavy metal elements have been found to leach out of the sintered brick. Because of its special shielding and protection characteristics, it has a good application prospect. The preparation process is as shown in Figure 2.

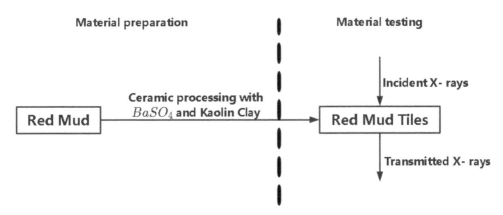

Figure 2.   Schematic diagram of the preparation process of X-ray shielding bricks.

2.4 Preparation of subgrade materials

Zhang et al. (2020) prepared using Bayer red mud, superabsorbent resin, and aluminate sulfate cement as raw materials when the amount of red mud was 70% and the amount of superabsorbent resin was 0.4%. The 28-d strength is 5.2 MPa, and the shrinkage rate of only 1.8% is in line with the needs of subgrade filling engineering high amount of red mud subgrade filling material. Chen et al. (2019) added red mud to the loess subgrade project for partial substitution of raw materials, and the results showed that due to its strong alkalinity and hydrolysis, red mud can promote the formation of more cemented hydrates, and can improve the pore mechanism of the material and make it denser to obtain good durability. Among them, the optimal dosage is proposed to be 15–20%.

2.5 Preparation of glass-ceramic materials

Qu et al. (2019) selected raw materials such as red mud and SiO_2 and tested their properties. The results showed that the addition of red mud contributes to the nucleation process and the prepared glass-ceramic has excellent bending strength and has high application value. Li et al. (2021) used red mud and chromite instead of clay as raw materials to prepare glass-ceramic. The results showed that in the sintering and solidification of glass-ceramic and preparation process, red mud not only plays the role of flux and nucleating agent as raw material, and the product indicators are quite excellent and good at the same time. Liu et al. (2019) used red mud and electrical insulation waste as raw materials to prepare a glass ceramic material with excellent performance, and the results showed that the material performs satisfactorily in terms of bulk density, porosity, compressive strength, and acid corrosion resistance. Among them, when the mass ratio of red mud to electrical insulator waste is 2:8, the prepared sample can show the best comprehensive performance.

2.6 *Preparation of ceramic materials*

Wei et al. (2019) took red mud, combustible solid particulate matter, glass powder, potassium feldspar powder, and other materials as related raw materials, and prepared target ceramics by drying and calcining after air maintenance. The results show that under the optimal conditions after optimization, thermal insulation ceramic products with excellent compressive and thermal insulation properties can be prepared. Raghubanshi et al. (2021) developed heavy aggregates (HWA) using red mud, barium sulfate, and activated carbon powder as reducing agents, and prepared a maximum density of up to 4.1 g/cm^3. High-density ceramics, compared with natural aggregates, has increased in crushing rate and crushing index. The preparation process is shown in Figure 3.

Raw Materials

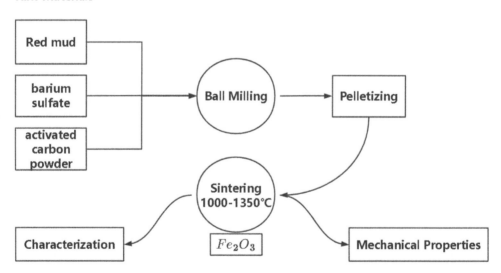

Figure 3.   Preparation of high-density ceramics from related raw materials such as red mud.

3   CONCLUSIONS

Red mud is a valuable secondary resource. If it can achieve large-scale industrial applications, it will have great ecological and environmental protection benefits. At present, the application of red mud in the field of building materials accounts for the main part of the comprehensive utilization of red mud. In recent years, scholars at home and abroad have also carried out a lot of research. China's red mud comprehensive utilization in this field of advantages is shown in Table 1.

Table 1. Analysis of the advantages and disadvantages of the specific application of red mud in the field of building materials.

Apply the orientation	Advantage	Inferior position
Preparation of cement	The raw materials are simple, inexpensive, and easy to obtain, and the preparation process is simple	The utilization rate of red mud is low, the cost of red mud de alkali removal is high, and there is a problem with pan alkali
Preparation of concrete materials	The preparation process is simple, and the utilization rate of red mud is high	The cost of red mud desalination treatment is high, and there is a problem with pan-alkali
Preparation of bricks	The preparation process is simple, the utilization rate of red mud is high, the raw materials are cheap, and there is no secondary pollution	The relevant reaction mechanism during the preparation process is not yet clear
Preparation of subgrade materials	Red mud utilization rate is high, no secondary pollution, and simple raw materials	The working nature of subgrade materials varies greatly, and it is difficult to match the construction period
Preparation of glass-ceramic materials	The process is simple and there is no secondary pollution	The price of raw materials is higher, and the utilization rate of red mud is lower
Preparation of ceramic materials	The process is simple and there is no secondary pollution	The relevant reaction mechanism during the preparation process has not been clarified, and the utilization rate of red mud is low

However, there are still some problems that need further research and solutions, which can be summarized as follows:

(1) Current research often only focuses on the physical and chemical properties of materials prepared from red mud, but few studies focus on the environmental evaluation of the materials
(2) When red mud bricks are made and sintered products are prepared, the relevant complete reaction mechanism and the occurrence state of red mud components are relatively scarce in the literature.
(3) Environmental problems such as pan-alkali problem, heavy metal ion leaching problem, and radioactive concentration limit industrial production and need to be paid attention to. How to achieve high dosage, low comprehensive cost, and no secondary pollution are three important factors that determine the industrialization of red mud in the field of building materials. These three factors have led to the following recommendations concerning the application of red mud as a building material:
    (1) To realize the application of high content of red mud, the preparation of bricks from red mud should be the key development direction, which can well solve the secondary pollution problems such as efflorescence and radioactive pollution, and the overall cost is low.
    (2) When preparing roadbed materials, it is important to consider the influence of red mud content on the workability of roadbed materials, as well as to coordinate well with the construction period.
    (3) Local aluminum production enterprises should select bauxite ore and optimize aluminum production parameters according to their production conditions to optimize the properties of red mud from the source and reduce the processing cost of red mud.
    (4) Enterprises should make full use of the advantages of the resources of the surrounding industries, optimize the freight, water, electricity, and related raw material costs, and establish the environmental advantages of the enterprise to reduce the cost of red mud utilization.

REFERENCES

Agrawal V, Paulose R, Arya R, et al. (2022) Green conversion of hazardous red mud into diagnostic X-ray shielding tiles. *Journal of Hazardous Materials*, 424: 127–507.

Arroyo F, Luna-Galiano Y, Leiva C, et al. (2020) Environmental risks and mechanical evaluation of recycling red mud in bricks. *Environmental Research*, 186: 109–537.

Cao Y, Li W D, Liu Y A. (2007) Properties of Red Mud and Current Situation of Its Utilization. *Bulletin of Silicate*, (01): 143–5.

Chen R F, Cai G J, Dong X Q, et al. (2019) Mechanical properties and micro-mechanism of loess roadbed filling using by-product red mud as a partial alternative. *Construction and Building Materials*, 216: 188–201.

Cheng X W, Yang X Z, Zhang C, et al. (2021) Effect of red mud addition on oil well cement at high temperatures. *Advances in Cement Research*, 33(1): 28–38.

Hou D S, Wu D, Wang X P, et al. (2021) Sustainable use of red mud in ultra-high performance concrete (UHPC): Design and performance evaluation. *Cement and Concrete Composites*, 115: 103862.

Jing Y R, Jing Y Q, Yang Q. (2001) *Basic properties and engineering properties of red mud. Light Metals*, (04): 20–3.

Li B, Chen C Z, Zhang Y W, et al. (2021) Preparation of glass-ceramics from chromite-containing tailings solidified with Red Mud. *Surfaces and Interfaces*, 25: 101210.

Li Y C, Min X B, Ke Y, et al. (2018) Utilization of red mud and Pb/Zn smelter waste for the synthesis of a red mud-based cementitious material. *Journal of hazardous materials*, 344: 343–349.

Li Y C, Min X B, Ke Y, et al. (2019) Preparation of red mud-based geopolymer materials from MSWI fly ash and red mud by mechanical activation. *Waste Management*, 83: 202–208.

Liu T Y, Zhang J S, Wu J Q, et al. (2019) The utilization of electrical insulators waste and red mud for fabrication of partially vitrified ceramic materials with high porosity and high strength. *Journal of cleaner production*, 223: 790–800.

Nan X L, Zhang T A, Liu Y, et al. (2009) Main Categories of Red Mud and Its Environmental Impacts. *Chinese Journal of Process Engineering*, 9(S1): 459–64.

Qu Z M, Zhang S, Zhang Y L. (2019) Preparation of CaO-SiO2-Fe2O3-Al2O3 system glass-ceramics from red mud with high iron. *Nonferrous Metals Science and Engineering*, 10(04): 34–8+71.

Raghubanshi A S, Mudgal M, Kumar A, et al. (2021) Development of heavyweight aggregate via in-situ growth of high density ceramics using red mud. *Construction and Building Materials*, 313: 125376.

Wei H S, Ma X E, Guan X M, et al. (2019) Preparation of Bayer Red Mud Light-weight Thermal Insulation Ceramics. *Bulletin of Silicate*, 38(03): 749–51+61.

Xue S G, Li Y B, Guo Y. (2017) Environmental impact of bauxite residue:a comprehensive review. *Journal of the University of Chinese Academy of Sciences*, 34(04): 401–12.

Yin Q Y, Lou G H, Li F, et al. (2020) Research on technical properties of fired perforated bricks from coal gangue and red mud. *New Building Materials*, 47(04): 73–6.

Zhang N, Gao Y F, Li Z F, et al. (2020) Experimental Research on Subgrade Filling Materials with High Mixed Red Mud. *China Comprehensive Utilization of Resources*, 38(11): 10–3.

Zhu J, Lan J K. (2008) Comprehensive Recovery and Utilization of Red Mud. *Mineral Protection and Utilization*, (02): 52–4.

Advances in Measurement Technology, Disaster Prevention and Mitigation – Li & Mohd Yusof (Eds)
© 2023 The Author(s), ISBN: 978-1-032-36087-4

Research on optimum design of reinforced concrete frame structure

Man Chen*

Department of Civil Engineering, Xi'an Traffic Engineering Institute, Xi'an, China

ABSTRACT: By achieving the established construction goals, a design method for engineering structure optimization is proposed, combined with various advanced technologies to make the design more perfect, and finally achieve the optimal design of the building structure. It not only satisfies the building structure but also satisfies the basic functions, improves the space utilization rate, improves the material utilization rate, and meets the livability requirements. In the process of building structure optimization, in addition to improving the structural design, it can also be optimized from the selection and preparation of raw materials. This paper provides ideas for the optimal design of reinforced concrete frame structures from multiple perspectives.

1   INTRODUCTION

With the growth of the population, the number of residential constructions has increased significantly, and people's requirements for architectural design are also getting higher and higher to achieve a better living environment. In recent years, the design of concrete structures has often encountered some problems due to the lack of optimization and rationality, making the structure less stable and difficult to meet specific design standards. Reducing the cost and considering the quality of the project is the most basic goal in the implementation process of the project, and to achieve this goal, it is necessary to design the building structure reasonably. The research goal of this paper is to optimize concrete from the perspective of structure and material, to make the building structure more perfect. Under the premise of satisfying the basic functions, the building can make full use of materials and advanced technology to improve the design level to win the fierce competition in the construction market.

2   MAIN CONTENT OF THE STUDY

To improve the quality and standard of building design, it is required to design from the perspective of building structure design and raw materials (Shao 2020). It adapts to the requirements of human society for the utilization performance, safety, and appearance of houses, and optimizes the structural design from multiple angles to reduce energy consumption and construction costs, thereby improving enterprise benefits. Finally, it can improve the competitiveness of the company in the market. To efficiently perform all aspects of the optimization design task and achieve the goal of managing safety and stability, the principles of safety must be followed in the work. The area around the building is densely populated, and once a disaster occurs, it will cause huge losses. Therefore, design work must follow safety points to minimize the impact of disasters (Qian 2020).

Optimizing the prestressed concrete frame structure can improve the overall structural characteristics of the house on the one hand, and increase the economic benefits of the building on the

*Corresponding Author: manchen1992@126.com

DOI 10.1201/9781003330172-69

other hand. It is an effective means to reduce engineering costs and improve competitive advantages. When designing the engineering structure, the optimization principle of the original design scheme can be used to optimize the building structure, which mainly includes the optimization of the structure type, the advantages of space layout, careful design, and precise calculation. To achieve the expected work goals, it is necessary to follow scientific principles in actual work, formulate corresponding optimal design schemes and schemes, and continuously strengthen the overall mechanism of optimal design work. Therefore, the structural design of the building needs to be reasonably optimized from the perspective of the design scheme to ensure its feasibility, stability, and usability.

3 KEY TECHNOLOGIES FOR RESEARCH

The optimization of concrete structures must strictly implement industry standards. The safety factor is the most important consideration in structural design. Therefore, when optimizing the structure, it is necessary to not only ensure the safety of the building but also ensure that the strength and stiffness of the building structure meet the requirements of relevant legislation and technical requirements (Song 2020).

3.1 *Pre-planning control of building structure*

In the pre-planning stage of residential projects, the government should control the quality of the plan, because the development direction of the plan has determined the structural design standards, which has an impact on the final construction of the physical structure. In the study of space structure, the comprehensive design of various functions is studied in combination with the overall planning and design, and different overall design methods are studied. method, and choose the overall design method of a reasonable structure. In addition, in optimizing the overall design method, it is necessary to continuously improve the engineering design details to increase the overall feasibility of the design scheme (Yang 2020).

3.2 *Strengthen detail processing*

To enhance the stability of the building space structure, it is necessary to strengthen the management of the detailed structure in the process of architectural design. The staff must not only rely on the traditional building structure experience but also rely on more detailed analysis to make design decisions (Yang et al. 2020). The size and position of the structural details are all complete to avoid small omissions and major mistakes, change the traditional architectural design concept, further improve the quality and level of architectural design, and enhance the rationalization and safety of the architectural structural design, thus realizing the green energy saving of architectural design effect.

3.3 *Model design of data*

After the basic structural design of the house is completed, a large amount of construction engineering data will appear. Because of the characteristics of the building structure, the different external environmental conditions and load requirements have interlocking factors on the building structure. At this time, people use a virtual three-dimensional system formed by a computer to optimize the structural design scheme (Xie et al. 2019). The structural simulation results and relevant numerical information of the building can be injected into the architectural design software system, and the feasibility of building architecture design can be studied by simulating the changes of external factors, such as the hazards of building structures, the hazards of fire to building structures, the hazards of land subsidence to building structures, etc.

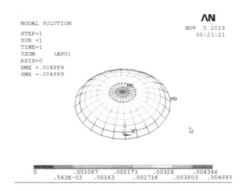

Figure 1. Model of the space reticulated shell.

3.4 *Optimizing the design scheme*

When optimizing the architectural design of the building, special attention should be paid to optimizing the fitting degree of the overall design scheme. Because the structural design of the building is a multi-level 3D engineering design task, it affects the integrity and credibility of the structural design of the building. Therefore, when designing building components such as supporting beams, main beams, secondary beams, floor slabs, and load-bearing walls, it is necessary to analyze the relationship between various components in detail to determine the closeness of the connection between building components (Sun 2017).

3.5 *Computer simulation technology*

With the increasing development of building technology, many computer technologies have been widely used. Among them, simulation technology is to use special application software to establish the simulation result model of the building to simulate the architectural design effect of the building. An actual building can be tested by performance. For example, computer simulation technology can be used to inject the optimized building components into the system, and the actual stress value of the components can be analyzed and calculated, and then the actual strength of the components can be directly deduced. value. It can also check the engineering design results in various fields, thereby reducing structural accidents, enhancing the rationality of structural design, avoiding subsequent design changes, and playing an important role in optimizing the building structure. In addition, the technology can also be used to calculate the engineering quantity and cost, making the results more accurate, the data more reliable, and the design scheme selection more efficient (Zhou 2016).

4 THE MAIN METHOD OF RESEARCH

4.1 *Optimization of structural design scheme*

Excellent building architecture design can not only improve the seismic and dynamic characteristics of buildings but also save land resources and increase investment benefits. Thereby producing great practical significance and economic value. The project is based on an original construction plan that has been rationally optimized considering current construction requirements and economic costs. Taking the main structure and the enclosure structure as the content, the structure optimization scheme design mode is formed, and then the structural optimization scheme design finally obtained by scientific calculation is used (Dou 2016). The building structure optimization scheme design may include the main structure optimization scheme design, the basic structure optimization scheme design, the building detail structure optimization scheme design, and the building interior envelope structure optimization scheme design. The optimal design of the building structure should comprehensively consider the comfort and function issues, implement cost management based on economy, and reduce the cost of the building structure as much as possible.

4.2 *Optimization of raw material selection and preparation*

By promoting China's urbanization projects and building infrastructure projects, more adequate development space will be created for the construction and building materials industries. It is the most used building material in construction, and cement is the most used building material. Huge pollution and its damage to ecosystems have seriously damaged development and the development on which human beings depend. Therefore, low-carbon green concrete is an inevitable choice for the optimal design of reinforced concrete frame structures (Wang 2021).

4.2.1 *Low carbon cement*

Figure 2. Generation of cement.

In the production process of ordinary concrete, some raw materials (such as siliceous and calcareous raw materials) must be fired with high heat, which will consume a large amount of fuel and generate a large amount of carbon dioxide, sulfur dioxide, nitrogen oxides, and other important pollutants. By using clinker-free concrete or non-combustible mortar to produce concrete, energy can be reduced and pollution can be reduced, thus achieving the effect of saving and environmental protection. In the manufacturing process, it is possible to replace traditional fuels with low-fossil fuels and limestone with non-carbonated raw materials. For another example, in concrete production, waste tires, waste plastics, sewage sludge, and other combustible wastes can be used to replace traditional fuels. Or by using raw materials such as slag, gypsum waste, various tailings slag, or properly disposed of industrial waste to replace cement clinker, and then replace limestone. This can effectively reduce the direct emission of carbon dioxide in the cement manufacturing process, and there is also huge room for development in this regard (Liu et al. 2020).

4.2.2 *Regenerated aggregate*

Recycled aggregate is the recycling of waste cement blocks, waste cement, and waste blocks after a series of treatments. Recycled aggregate can also be used to make building materials such as cement, concrete block bricks, cobblestone bricks, and low-strength tiles. Alternatively, after screening, sorting, and cleaning, it can be used as recycled stubby aggregate or as natural cement sand (Zhou 2020). The use of recycled aggregate can not only reduce the environmental pollution caused by construction waste to the natural environment, but also save natural aggregate resources, reduce the contradiction between supply and demand of aggregate, and reduce natural resources. Fundamentally addressing emissions patterns has significant energy consumption and social implications. To meet the needs of sustainable social development, it is an important channel for optimizing concrete production and has a significant impact on economic and environmental effects.

Figure 3.   Waste concrete.   Figure 4.   Regenerated aggregate.

4.2.3 *Admixtures*

Various materials, such as pulverized blast furnace slag cement, pulverized zeolite powder, silica fume, and pulverized coal, which can form active gel particles, are mixed with cement and fillers to form concrete. The addition of mineral fillers can not only reduce the amount of concrete but also improve the workability of cement. It can also change the microstructure of cement, thereby reducing the hydration and exothermic stage of concrete, and increasing the dimensional stability of cement. It plays a role in increasing the strength and durability of cement and is an important method for concrete optimization (Tian 2019).

Figure 5.   Silica fume.   Figure 6.   Mineral powder.

The traditional architectural design concept is based on high-strength design. With the rapid development of the economy, the architectural design concept has also changed, and green concrete materials have been introduced as the architectural design concept. The green design aims to reduce environmental pollution, improve people's quality of life, and put people in the first place. The specific ecological design should start from the comprehensive consideration of safety, applicability, durability, economy, ecology, and so on. Concrete as a heavy building structure should emphasize material safety and green design. Green concrete is not only excellent in performance, but also economical. The design of green concrete mainly focuses on the raw materials and preparation of concrete materials, and specific factors should be considered.

5   CONCLUSION

Residential buildings are the primary places in which humans live, and the existence of quality issues could greatly impact their health. Therefore, the design of all building structures must go through repeated discussions, demonstrations, and examinations, and finally determine the possibility and safety of the design. As an important part of the design optimization, the design unit, the construction unit, and the engineering supervision department can refer to suggestions to correct the deficiencies and absurdities of the original design. Moreover, to improve the quality of the optimization scheme design, it is necessary to establish a professional research and development organization for design and optimization technology, conduct a multidisciplinary and expert-level analysis and discussion on the overall design scheme of the engineering structure, and propose appropriate solutions.

Not all optimal methods are the best, and it is always necessary to compare and study the optimal method and the scheme, analyze its possibility, and choose a more reasonable design method. In engineering practice, relevant designers should fully understand the specific needs of customers, proceed from the real object, improve the professional technical level through continuous practice and accumulation, and effectively optimize the design of concrete structures. Doing so can make the structure fully benefit from the comprehensive optimization, further improve the safety and stiffness of the overall structure, and improve the design efficiency.

After the optimization of the house structure design, many complex construction processes can be simplified. At the same time, it is convenient to use a variety of new energy-saving and environmentally friendly building materials, thereby effectively reducing engineering costs, and increasing living comfort and aesthetic ecological value. Structural design optimization can improve the overall efficiency of home architecture.

REFERENCES

Dou P.(2016) On the Optimization Design Method of Steel Concrete Frame Structure. *Building Materials and Decoration*, 2,82–83.
Liu H.B, Yin F.T, Yang Y.P, Shao Z.J, Bian X, Hu S.W.(2020) Development and Application of Green Concrete. *Science and Technology Innovation and Application*, 27,176–177.
Qian W. (2020) Analysis on the optimization design of concrete structures [J]. *Anhui Architecture*, 27/4, 94–95+99.
Shao C.Q. (2020) Optimization measures for construction technology of mass concrete structures. *Housing and Real Estate*, 21,185.
Song L.Y. (2020) Optimal Design of Concrete Structures in Structural Engineering. *Building Materials and Decoration*, 11,104.
Sun Z.P.(2017) *Research on optimization design of reinforced concrete frame structure*. Shandong University of Science and Technology.
Tian C.H.(2019) *Preparation and performance of green concrete*. Shijiazhuang Railway University.
Wang W.J. (2021) *The scheme and implementation method of green concrete prepared by recycled sand powder from construction waste*. Construction Technology Development, 48/20,149–150.
Xie J, Wang Q, Ni Y.J, Bao S.X, etc.(2019) Design method of quasi-full internal force for optimization of reinforced concrete frame structure . *Science, Technology and Engineering*, 19/21,15–20.
Yang D.X. (2020) Research on optimization design of building concrete structure. *Smart City*, 6/3, 161–162.
Yang Z.T, Liu Y, Yin G.S, Shi M.H, Wei P.F. (2020) Optimal design analysis of multi-row ceramsite concrete composite block structure. *Bulletin of Silicate*, 39/01,113–123.
Zhou J.F. (2016) *Multi-objective optimization analysis of a reinforced concrete frame structure*. Hebei University of Engineering.
Zhou S.Y.(2020) Research status and application prospects of green concrete. *Sichuan Building Materials*, 46/5,5+14.

Advances in Measurement Technology, Disaster Prevention and Mitigation – Li & Mohd Yusof (Eds)
© 2023 The Author(s), ISBN: 978-1-032-36087-4

Comparison of VVCM and marshall method for ATB-30 asphalt mixture design

Mei Song*
Weinan Highway Administration Bureau, Weinan, Shaanxi, China

Sheng Li* & Yu-hao Bao
Key Laboratory for Special Area Highway Engineering of Ministry of Education, Chang'an University, Xi'an, Shaanxi, China

ABSTRACT: The reliability of vertical vibration compaction method (VVCM) specimens was determined by indoor and outdoor tests, and the influence of the VVCM method and Marshall method on the design of the ATB-30 asphalt mixture was studied comparatively. The results indicated that the volumetric index of the VVCM specimens of ATB-30 asphalt mixture was closer to the pavement core samples, and the correlation between the mechanical strength of VVCM specimens and pavement core samples was as high as 94%, while the correlation of Marshall specimens was less than 61%; compared with the Marshall method for designing ATB-30 asphalt mixture, the VVCM design of ATB-30 asphalt mixture saved about 8% asphalt, increased the density by 1.6%, and the Marshall stability, dynamic stability, shear strength, and tensile strength increased by 54%, 39%, 55%, and 54%, respectively. Moreover, the water stability was improved.

1 INTRODUCTION

At present, the representative international asphalt mixture design methods are the Marshall design method, Hveem design method, GTM design method, and Superpave design method (Jiang et al. 2018). One of the main differences between these design methods is the different forming methods used: the Marshall method uses the compaction forming method; the Hveem method uses the kneading compaction forming method; the GTM and Superpave methods use the rotary compaction forming method. China has been using the Marshall method since the 1970s and has incorporated it into its specifications. However, with the development of modern traffic and construction technology and the depth of engineering practice, it was found that the Marshall method has lagged the production reality, mainly because the Marshall compaction standard lags the traffic status quo, and the Marshall specimen has poor correlation with the field performance, and the Marshall design standard cannot guarantee its design mix has good long-term performance (Chai & Li 2020). The more widely used methods are foreign rubbing compaction method and rotary compaction method, although the correlation between the formed specimens and field performance is good, its equipment is more expensive, and it is difficult to promote the application in China (Yu et al. 2005). In recent years, China and foreign countries have carried out research on asphalt mixture vibration molding method, and found some results. These foreign countries include France, Switzerland, and Yugoslavia. and China and these countries have developed vibration molding equipment, but foreign asphalt mixture vibration compaction on its performance and other aspects of the report are not found. Studies conducted at China's Harbin Institute of Technology, South China University of Technology, and Chang'an University, as well as other studies on the vibration compaction

---

*Corresponding Authors: songm_weinan@163.com and 2021121194@chd.edu.cn

DOI 10.1201/9781003330172-70

and crushing of coarse aggregates have less impact, and can make the asphalt mixture structure to achieve a skeletal dense structure. Large particle size gravel asphalt mixture is suitable for vibration molding method (Asi 2007; Sha et al. 2008), but did not confirm the effect of vibration compaction specimens and the actual effect of the field match, and did not involve the corresponding design methods (Jiang et al. 2014). To this end, we conducted relevant studies to address the selection criteria of vibratory compaction apparatus, verify the correlation between vertical vibratory compaction molding specimens and field performance, and propose corresponding design methods. In the paper, the effect of the VVCM method and Marshall method on the design of ATB-30 asphalt mixes was studied by comparing them with indoor tests.

2 MATERIALS AND TEST METHODS

2.1 *Materials*

Asphalt used is Xinjiang Karamay A grade 70# asphalt. Its technical indicators meet the requirements of JTG E20 (Chinese Standards 2011). In this paper, crushed limestone, machine-made sand, and limestone powder were used as the coarse aggregate, fine aggregate, and mineral powder, respectively. All of them were procured from Liulin County, Shanxi Province, China. The technical properties of aggregate were measured according to the JTG E42 (Chinese Standards 2005).

2.2 *Mineral grading*

The mineral gradation used for ATB-30 asphalt mixture is shown in Table 1.

Table 1. Gradation of ATB-30 asphalt mixture.

Sieve size (mm)	31.5	26.5	19	16	13.2	9.5	4.75	2.36	1.18	0.6	0.3	0.15
Passing rate (%)	100	89.4	69.3	62.1	55.7	44.1	28.9	19.5	13.7	10.4	6.5	5.2

2.3 *Test method*

The mechanical properties of the mixture include Marshall stability, compressive strength, splitting strength, shear strength, and tensile strength tests. Marshall stability (MS), compressive strength (R_C), and splitting strength (R_T) following the test methods in the JTG E20 (Chinese Standards 2011).

2.3.1 *Shear strength test*

The uniaxial penetration test was used to test the shear strength (τ_d) characterization of the high-temperature properties (Zak et al. 2017), and the uniaxial penetration test model is shown in Figure 1. The uniaxial penetration test is similar to the CBR test, which is applied to the specimen by a steel indenter, and then the shear resistance of the mixture is derived by the derivation of the mechanical equation. The test was performed with an electronic universal testing machine, using a 42 mm diameter indenter, and the test temperature was 60°C under the same unfavorable conditions as the rutting test, with a loading rate of 1 mm/min.

The calculation formula of uniaxial penetration strength is shown in Equations (1) and (2).

$$Rg = \frac{4P}{\pi \phi^2} \tag{1}$$

$$\tau d = 0.339 \times Rg \tag{2}$$

where τ_d is the shear strength; R_g is the penetration strength of the specimen; P is the maximum load when the specimen is damaged.

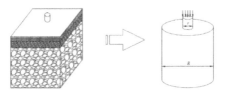

Figure 1.   Uniaxial penetration test model.

2.3.2   Tensile strength test

The low-temperature performance was characterized by the semi-circular bending test (SCB) to test the tensile strength (σ) (Lü et al. 2008). The principle of the SCB test is shown in Figure 2. Tests of semi-circular bending are conducted by applying a concentrated load until the specimen fractures and breaks in the span of a simply supported semi-circular specimen with specified dimensions. The tensile strength of the material is determined by the maximum load of damage. The test was conducted using a 1.2 cm diameter round bar as the support and compression bar. The test temperatures were −20°C, 15°C, and 25°C, and the corresponding loading rates were 5, 50, and 50 mm/min, respectively.

The tensile strength is calculated in Equation (3):

$$\sigma = \frac{12Pa}{t\varphi^2} \tag{3}$$

where P is the ultimate load; a is the support spacing; generally, $a = 0.4\ \Phi$; t is the thickness of the specimen, which is generally chosen as 50 mm.

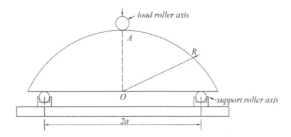

Figure 2.   Schematic diagram of SCB test principle.

3   VVCM DESIGN METHODOLOGY

3.1   Vertical Vibration Compaction Method (VVCM) and evaluation

To simulate the effect of vibration rolling in the field, Sha (2018) [6] proposed the vertical vibration compaction method. The vertical vibration compaction instrument is used to simulate the working principle of the directional vibration roller. The centrifugal force in the horizontal direction is zero when the vibration compaction instrument works, and the sinusoidal excitation force is generated in the vertical direction. Vibrator parameters are set as follows: working frequency 37 Hz, nominal amplitude 1.2 mm, working weight 3.0 kN (1.2 kN for the on-board system and 1.8 kN for the off-board system). When forming the specimen, first load the ATB-30 asphalt mixture into the test mold, and then use the vibration instrument vibration compaction 100 s. The specimen size is Φ150 mm $\times$ h 95.3 mm.

The comparison results of volume index and mechanical properties of ATB-30 mixture Marshall specimens, VVCM specimens, and pavement core samples under the same conditions of material and composition are shown in Tables 2 and 3.

Table 2. Comparison of volume indicators of core samples and samples produced by the Marshall method and VVCM.

Volume index (V)	Pavement core sample (V_x)	Marshall specimens (V_M)	VVCM specimens (V_V)	V_M/V_x (%)	V_V/V_x(%)
ρ_f (g·cm^{-3})	2.497	2.433	2.483	97.4	99.4
VV (%)	2.5	5.4	3.5	216	140
VMA (%)	10.6	12.6	10.8	119	102
VFA (%)	76.7	57.1	68.0	74.4	88.7

Table 3. Comparison of mechanical properties of core samples and samples produced by the Marshall method and VVCM.

Mechanical index (S)	Test temperature (°C)	Pavement core sample (S_x)	Marshall specimens (S_M)	VVCM specimens (S_V)	S_M/S_x (%)	S_V/S_x (%)
MS (kN)	60	38.44	22.72	34.86	59.1	90.7
	20	9.34	6.27	9.04	67.1	96.8
R_C (MPa)	40	4.97	3.43	4.79	69.0	96.4
	60	3.87	2.44	3.52	63.0	91.0
	−20	4.02	2.58	3.82	64.2	95.0
R_T (MPa)	15	2.47	1.57	2.26	63.6	91.5
	20	1.78	1.07	1.69	60.1	94.9
τ_d (MPa)	60	1.98	1.1	1.91	55.6	96.5
	−20	9.14	5.68	8.93	62.1	97.7
σ (MPa)	15	7.53	4.74	6.99	62.9	92.8
	25	4.71	2.73	4.43	58.0	94.1

From the data in Table 2, the correlation between VVCM specimens ρ_f and VFA and pavement core samples are 99.4%, 88.7%, 1.4 times, and 1.02 times of VV and VMA, respectively; the correlation between Marshall specimens ρ_f and VFA and pavement core samples are 97.4%, 74.4%, 2.16 times and 1.19 times of VV and VMA, respectively. This indicates that the volume index of the VVCM specimen of ATB-30 asphalt mixture is closer to that of the pavement core sample, i.e., the VVCM specimen is closer to the actual pavement mixture structure, and the design of ATB-30 asphalt mixture by the VVCM method can make the pavement more stable.

From the data in Table 3, the correlation between the mechanical strength of ATB-30 asphalt mixture VVCM specimens and pavement core samples is 94.1% on average, while the correlation between the mechanical strength of Marshall specimens and pavement core samples is 60.9% on average. This indicates that VVCM can better simulate the actual rolling effect and accurately predict the pavement performance, and it is more practical to use VVCM to study the composition and performance of the ATB-30 asphalt mixture.

3.2 *ATB-30 VVCM design criteria for asphalt mixture*

The design criteria for ATB-30 asphalt mixtures VVCM are shown in Table 4.

Table 4. ATB-30 VVCM design criteria for asphalt mixtures.

Design Method	Design requirements for the following indicators					
	VV (%)	VMA (%)	VFA (%)	MS (kN)	τ_d (MPa)	σ (MPa)
VVCM	2.8~4.0	≥ 10.5	64~74	2	≥ 1.25	≥ 4.0
Marshall	3~6	≥ 12.5	55~70	≥ 15	–	–

4 VVCM METHOD AND MARSHALL METHOD DESIGN COMPARISON

4.1 Comparison of physical and mechanical properties

The physical and mechanical test results of VVCM specimens and Marshall specimens of ATB-30 asphalt mixture with different asphalt-aggregate ratios are shown in Figures 3–10.

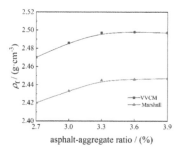

Figure 3. Relationship between ρ_f and asphalt-aggregate ratio.

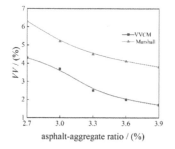

Figure 4. Relationship between VV and asphalt-aggregate ratio.

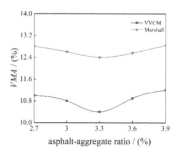

Figure 5. Relationship between VMA and asphalt-aggregate ratio.

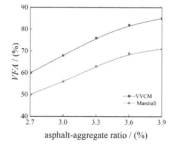

Figure 6. Relationship between VFA and asphalt-aggregate ratio.

From Figures 3 to 10, the compaction work in VVCM specimens is greater than that in Marshall specimens, and the density of VVCM specimens is about 1.02 times greater than that of Marshall specimens; therefore, compared with Marshall specimens, the VMA in VVCM specimens decreases, the VFA increases, and the VV decreases. Second, VVCM specimens are stronger than Marshall specimens as their density is higher and their particle arrangement is more reasonable thanks to the vibration method. The shear strength of VVCM specimens is 1.23-1.57 times that of Marshall specimens, and the average is 1.37 times; the tensile strength of VVCM specimens is 1.47–1.59 times greater than that of Marshall specimens, and the average is 1.51 times.

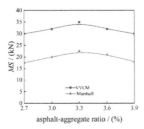

Figure 7.   Relationship between MS and asphalt-aggregate ratio.

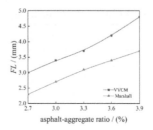

Figure 8.   Relationship between FL and asphalt-aggregate ratio.

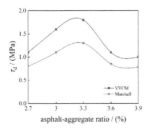

Figure 9.   Relationship between τd and asphalt-aggregate ratio.

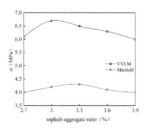

Figure 10.   Relationship between σ and asphalt-aggregate ratio.

4.2   Design results and performance verification

According to the test results, VVCM determined the optimal asphalt-aggregate ratio of 3.07% and the Marshall method determined the optimal asphalt-aggregate ratio of 3.34%, and the volume parameters corresponding to the optimal asphalt-aggregate ratio are shown in Table 5.

Table 5.   Volume parameters corresponding to an optimum asphalt-aggregate ratio.

Design method	OCA (%)	ρ_f (g·cm^{-3})	VV (%)	VFA (%)	VMA (%)
VVCM	3.07	2.486	3.3	70.2	10.7
Marshall	3.34	2.447	4.3	65.4	12.5

According to the results of the mixing proportion in Table 5, molded VVCM specimens and Marshall specimens were verified for road performance, and the comparison results are shown in Table 6.

Table 6.   Comparison of pavement performance of ATB-30 asphalt mixture based on different design methods in the optimum asphalt-aggregate ratio.

	Road performance (S)	VVCM specimens (S_V)	Marshall specimens (S_M)	S_V/S_M
	MS (kN)	35.24	22.83	1.54
Water stability	Residual stability (%)	90.5	86.5	1.05
	Freeze-thaw splitting strength ratio (%)	89.6	78.7	1.14
Rut resistance	Dynamic stability (times/mm)	5000	3595	1.39
	τ_d (MPa)	1.92	1.24	1.55
Low-temperature crack resistance	σ (MPa)	6.78	4.41	1.54

Using the ATB-30 asphalt mixture designed by VVCM, the Marshall stability, dynamic stability, shear strength, and tensile strength were increased by 54.4%, 39.1%, 54.8%, and 53.7%, respectively. Water stability was improved as well.

5 CONCLUSIONS

Based on the results and discussions presented above, the conclusions are obtained as below:

(1) The correlations of VVCM specimens ρf and VFA with pavement core samples were 99.4% and 88.7%, and VV and VMA were 1.4 times and 1.02 times of pavement core samples, respectively; the correlations of Marshall specimens ρf and VFA with pavement core samples were 97.4% and 74.4%, and VV and VMA were 2.16 times and 1.19 times of pavement core samples, respectively.
(2) The correlation between the mechanical strength of ATB-30 asphalts mixture VVCM specimens and the pavement core samples was 94.1% on average, while the correlation between the mechanical strength of Marshall specimens and the pavement core samples was 60.9% on average.
(3) With the Marshall method design ATB-30 asphalt mixture, VVCM design ATB-30 asphalt mixture saves about 8% of asphalt, increases density by 1.6%, reduces VV and VMA by 23.3% and 14.4%, respectively, and increases VFA by 7.3%.
(4) Compared with the design results of the Marshall method, the Marshall stability, dynamic stability, shear strength, and tensile strength of ATB-30 asphalt mixture designed by VVCM increased by 54.4%, 39.1%, 54.8%, and 53.7%, respectively, and the water stability was also improved.

ACKNOWLEDGMENT

This work was financially supported by the Shaanxi Provincial Transportation Technology Project (Grant No. 20-02K).

REFERENCES

Asi I M. Performance evaluation of SUPERPAVE and Marshall asphalt mix designs to suite Jordan climatic and traffic conditions[J]. *Construction and Building Materials*, 2007 21(8): 1732–1740.
Chai Jin-Ling, Li Wei. GTM-based Experimental Study on the Asphalt Mixtures Design[J]. *Materials Reports*, 2020, 34(S2): 1283– 1287.
Jiang Y, Deng C, Xue J, et al. Investigation into the performance of asphalt mixture designed using different methods[J]. *Construction and Building Materials*, 2018, 177: 378–387.
Jiang Ying-Jun, Chen Zhe-Jiang, Xu Xiao-He, et al. Evaluation on Vertical Vibration Compaction Method of ATB-30 Asphalt Mixture[J]. *Journal of Building Materials*, 2014, 17(04): 638–643.
Lü Guang-Yin, Hao Pei-Wen, Pang Li-Guo, et al. Mechanical Simulation of Semi-circular Bending Test in Asphalt Mixtures [J]. *Journal of Wuhan University of Technology*, 2008, 30 (3): 50–52.
Sha ai-min, Wang Ling-Juan, Geng Chao. Vibrating Compaction Method of Large Stone Asphalt Mixture [J]. *Journal of Chang'an University: Natural Science Edition*, 2008, 28 (2): 1–4.
Yu Hong-Xing, Wei Lian-Yu, Zhao Ke, et al. Research on Asphalt Mixture Dynamic Stability Design with GTM [J]. *Journal of Highway and Transportation Research and Development*, 2005, 25 (9): 23–26.
Zak J, Monismith C L, Coleri E, et al. Uniaxial Shear Tester-new test method to determine shear properties of asphalt mixtures[J]. *Road Materials and Pavement Design*, 2017, 18(sup1): 87–103.

Advances in Measurement Technology, Disaster Prevention and
Mitigation – Li & Mohd Yusof (Eds)
© 2023 The Author(s), ISBN: 978-1-032-36087-4

Establishment and validation of finite element model for GFRP confined basalt fiber reinforced concrete long columns under compression

Jingshan Jiang
School of Civil Engineering and Architecture, Nanjing Institute of Technology, Nanjing, China
Nanjing Geo Underground Space Technology Co., Ltd., Nanjing, China

Xin Huang*
Key Laboratory of Ministry of Education for Geomechanics and Embankment Engineering, Hohai
University, Nanjing, China

Zhihua Wang
Nanjing Geo Underground Space Technology Co., Ltd., Nanjing, China

Chao Zhang & Youxin Wei
School of Civil Engineering and Architecture, Nanjing Institute of Technology, Nanjing, China

ABSTRACT: The eccentric compressive numerical analysis model of glass fiber reinforced polymer (GFRP) confined basalt fiber reinforced concrete long column was established using the finite
element software ABAQUS. The random distribution of basalt fiber in the core concrete was implemented with the secondary development function of ABAQUS. The accuracy of the model was
verified by the test data of thirteen specimens in other literature. The results show that the model can
effectively reflect the stress state of the specimen under eccentric compression. The ultimate bearing capacity, load-axial strain curve, and load-midspan deflection curve of the specimen simulated
by the finite element model are in good agreement with the test results.

1 INTRODUCTION

In recent years, glass fiber reinforces polymer (GFRP) restrained concrete structures have received
a lot of attention from scholars due to their high pressure-bearing capacity, high ductility, and good
corrosion resistance, which are widely used in high-rise structures and offshore buildings (Zeng et
al. 2014). Based on the previous research, many new types of GFRP confined structures have been
proposed and studied by scholars (Bazli et al. 2020; Jin et al. 2020; Zhang et al. 2019). It should be
noted that the influence of eccentric compression was not considered in most of the above studies.
Almost all the building components are under an eccentric compression state and are long columns.
Therefore, it is necessary to study the eccentric compression performance of long columns.

Basalt fiber has high strength and high modulus of elasticity, which can be added to concrete
to improve the strength of concrete and effectively inhibit the development of cracks (Zhou et al.
2020). The current researches on basalt fiber concrete are mostly indoor experimental research,
while the investigations on GFRP confined basalt fiber reinforced concrete components are still
relatively few.

Most studies on the mechanical properties of GFRP confined concrete columns are studied
through tests. However, in the laboratory test, the sample preparation has the disadvantages of

---

*Corresponding Author: 2540588261@qq.com

          DOI 10.1201/9781003330172-71

long time-consuming, high cost, and poor repeatability, so it is necessary to use the numerical simulation method for analysis. To the best of our knowledge, few works consider the influence of basalt fiber on mechanical properties of GFRP confined concrete long columns, the influence of basalt fiber and eccentric compression on mechanical properties of GFRP confined concrete long columns are even less considered. Therefore, the finite element analysis software ABAQUS was used in this paper to establish the eccentric compression model of GFRP confined basalt fiber reinforced concrete long column. The numerical analysis results are compared with the test results of the relevant literature to validate the accuracy of the model.

2 ESTABLISHMENT OF FINITE ELEMENT MODEL

2.1 Constitutive model

2.1.1 Concrete
The concrete damaged plasticity (CDP) model embedded in ABAQUS was used to describe the compressive and tensile behavior of concrete. The model defines the evolution of the concrete damage surface as controlled by the compressive equivalent plastic strain ε_c^{pl} and the tensile equivalent plastic strain ε_t^{pl} (Cao & Li 2017). When GFRP-confined reinforced concrete columns are under compression, the inner core concrete is in a three-dimensional compression state, and the ordinary concrete constitutive model is not applicable.

The core concrete was defined using the compression and tensile constitutive models for restrained concrete proposed by Han Linhai (Han, 2007), where the compression constitutive model is:

$$y_1 = \begin{cases} 2x_1 - x_1^2 & (x_1 \leq 1) \\ \dfrac{x_1}{\beta(x_1-1)^2+x_1} & (x_1 > 1) \end{cases} \tag{1}$$

where
y_1=normalized value of compressive stress, $y_1= \sigma_c/\sigma_0$;
x_1=normalized value of compressive strain, $x_1= \varepsilon_c/\varepsilon$;
σ_c=compressive stress;
σ_0=compression peak stress, $\sigma_0 =[1 + (-0.054\times\xi^2+0.4\times\xi)\times(24/f_c)]\times f_c$;
ε_c=compressive strain;
ε_0=compression peak strain, $\varepsilon_0=1300\times10^{-6}+12.5f_c\times10^{-6}+[1400+800\times(f_c/24-1)]\times\xi^{0.2}\times10^{-6}$;

$$\beta = (2.36 \times 10^{-5})^{[0.25+(\xi-0.5)^7]} f_c^2 \times 5.51 \times 10^{-4};$$

f_c=axial compression strength of concrete, f_c= 51.1 MPa;
ξ=constraint effect coefficient, $\xi= A_{gfrp}f_{gfrp}/(A_c f_c)$;
A_{gfrp}=surface area of GFRP tube, mm^2;
f_{gfrp}=axial tensile strength of GFRP tube, f_{gfrp}= 680 MPa, A_c=core concrete area (mm^2).
The tensile constitutive model is:

$$y_2 = \begin{cases} 1.2x_2 - 0.2x_2^2 & (x_2 \leq 1) \\ \dfrac{x_2}{\alpha(x_2-1)^{1.7}+x_2} & (x_2 > 1) \end{cases} \tag{2}$$

where
y_2= normalized value of tensile stress ($y_2=\sigma_t/\sigma_1$);
x_2= normalized value of tensile strain ($x_2=\varepsilon_t/\varepsilon_1$);
σ_1= axial tensile peak stress (σ_1=3.1 MPa);
ε_t= tensile strain;
ε_1= axial tensile peak strain (ε_1=43.1$\times\sigma_1\times10^{-6}$);
α= model parameter (α=0.31$\times\sigma_1^2$).

Compression and tensile damage factors are used to describe the stiffness degradation of concrete after damage. The expression of compression damage factor d_c is:

$$d_c = \frac{(1 - \beta_c) \, \varepsilon_c^{in} E_0}{\sigma_c + (1 - \beta_c) \, \varepsilon_c^{in} E_0} \tag{3}$$

where

ε_c^{in}= compressive inelastic strain ($\varepsilon_c^{in}=\varepsilon_c$-$\sigma_c/E_c$);
ε_c= compressive strain;
σ_c= compressive stress;
E_c= compressive elastic modulus of concrete;
β_c= ratio of compression plastic strain to inelastic strain ($\beta_c = 0.70$) (Liu et al. 2014).

The expression of tension damage factor d_t is:

$$d_t = \frac{(1 - \beta_t) \, \varepsilon_t^{ck} E_0}{\sigma_t + (1 - \beta_t) \, \varepsilon_t^{ck} E_0} \tag{4}$$

where

ε_t^{ck}= tensile cracking strain ($\varepsilon_t^{ck}=\varepsilon_t$-$\sigma_t/E_t$);
ε_t= tensile strain;
σ_t= tensile stress;
E_t= tensile elastic modulus of concrete;
β_t= ratio of tensile plastic strain to cracking strain ($\beta_t=0.95$) (Liu et al. 2014).

Many other parameters need to be defined when using the CDP model. The expansion angle φ was 30°, the plastic potential offset m was 0.1, the ratio of biaxial to uniaxial ultimate compressive strength f_{b0}/f_{c0} was 1.16, the ratio K of the second stress invariant on the meridian plane of tension and compression was 0.6667, and the viscosity coefficient μ was 0.0005.

2.1.2 GFRP

GFRP is a special anisotropic elastic brittle material, and its constitutive model is defined by Lamina linear elastic model. The fiber direction is defined as Axis 1, the direction perpendicular to the fiber direction on the GFRP surface is defined as Axis 2, and the direction perpendicular to Axis 1 and Axis 2 is defined as Axis 3. After trial calculation, the elastic modulus E_1, E_2 of Axis 1 and Axis 2, the main Poisson's ratio v_{12}, the shear modulus G_{12} in the face of Axis 1 and Axis 2, the shear modulus G_{13} in the face of Axis 1 and Axis 3 and the shear modulus G_{23} in the face of Axis 2 and Axis 3 were defined. The GFRP tube was designed by using Composite Layup. E_1, E_2, v_{12}, G_{12}, G_{13}, G_{23}, the thickness of each layer, and the winding angle of the fiber were input according to the measured value (Liu et al. 2021; Wang et al. 2010; Wang et al. 2015).

2.1.3 Steel bars

The ideal elastic-plastic bilinear model was used to define the constitutive model of steel bars (longitudinal reinforcements and stirrups), with the slope of the reinforcement section taken as one percent of the elastic modulus.

2.2 Realization of random distribution of basalt fiber

Theoretically, basalt fibers present a completely random distribution in concrete with random direction, random length, and random position (Wang 2020). The reasonable establishment of the random distribution model of basalt fiber in ABAQUS is the key to the numerical analysis of mechanical properties of GFRP confined basalt fiber reinforced concrete long columns. A Python script for cyclic random generation of basalt fibers was written for modeling with the help of the secondary development platform of ABAQUS. The specific method is as follows: use the line unit to simulate the fiber, randomly generate 3D coordinates of one end of the fiber (x, y, and z), and then input the length of the fiber to randomly generate the relative displacement of the fiber in three directions (Δx, Δy, and Δz) to determine the 3D coordinates of the other end of the fiber ($x + \Delta x$, $y + \Delta y$, and $z + \Delta z$) to complete the creation of a fiber. The overall number of fibers generated can be determined by the volume rate and the diameter of the monofilament, and the fibers generated by running the script are shown in Figure 1.

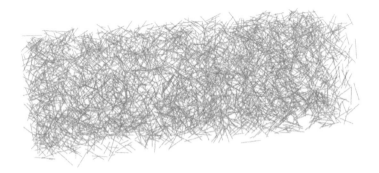

Figure 1.   Randomly generated basalt fibers.

2.3   *Grid division*

High-quality grid division is the key to 3-D model research. After trial calculations, a grid size with better accuracy and calculation costs was determined. Here, we set 30 mm for the core concrete, 35 mm for the GFRP tube, and 10 mm for the reinforcement. The reinforcement, core concrete, and GFRP tube were set as Truss units, Solid units C3D8R, and Shell units S4R, respectively.

2.4   *Interaction and boundary conditions*

The contact between the GFRP tube and the core concrete was described using a finite slip equation and a Coulomb friction model with a friction coefficient of 0.6 (Zhou & Li 2013), and the normal behavior was set to "hard contact". The reinforcement was embedded in the concrete. The load and boundary conditions for GFRP confined reinforced concrete long columns are defined at reference points *Load* and *Fix*. Coupling constraints are applied to the upper and lower surfaces of the long column using a kinematic coupling, and the control points are *Load* and *Fix*, respectively. As a result, the load can be applied at different eccentric distances by changing the coordinates of *Load* and *Fix*.

To simulate the eccentric compression condition of the long column hinged at both ends, the U1 (x-axis direction), U2 (y-axis direction), UR2 (rotation around the y−axis), and UR3 (rotation around the z-axis) of the *Load* were constrained, and the displacement load was applied in the U3 (z-axis direction), and the U1, U2, U3, UR2 and UR3 of the *Fix* were constrained, the specific load and boundary conditions are shown in Figure 2.

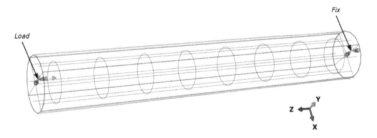

Figure 2.   Load and boundary conditions for GFRP confined reinforced concrete long columns.

3   VERIFICATION OF FINITE ELEMENT MODEL

The finite element model for relevant literature tests (Liu et al. 2021; Wang et al. 2010; 2015) is established using the above modeling method, and the simulation results were compared with the test results to verify the accuracy of the finite element model analysis results. Table 1 gives the test-related parameters, the test measured ultimate bearing capacity, and the calculation results of the ABAQUS numerical analysis of Wang et al. (Liu et al. 2021; Wang et al. 2010; 2015), where GFRP fiber winding angles are $\pm57.5°$. According to Wang et al. (2010) and Wang et al. (2015), the longitudinal reinforcements of the specimens are 6B12 and the stirrups are A6.5@180; the specimens that are specified in Liu et al. (2021) do not set longitudinal reinforcements and stirrups.

Table 1.   Comparison between experimental parameters and numerical simulation results obtained by other researchers (Liu et al. 2021; Wang et al. 2010; 2015).

Source	Number	L (mm)	D (mm)	T (mm)	V_b (%)	λ	E (mm)	f_{cc} (MPa)	f_{y1} (MPa)	f_{y2} (MPa)	N_{ue} (kN)	N_{uc} (kN)	E (%)
Wang et al. (2010)	L4	1600	200	5		32		54.6	385	290	2740	2620	−4.38
	L5	1800	200	5		36		54.6	385	290	2596	2609	0.50
	L6	2000	200	5		40		54.6	385	290	2568	2556	−0.47
	L7	2200	200	5		44		54.6	385	290	2479	2507	1.13
	L8	2400	200	5		48		54.6	385	290	2390	2229	−6.74
Wang et al. (2015)	Z5-H-7-40-U	1400	200	5		28	40	46.7	345	241	877	980	11.74
	Z7-H-7-40	1400	200	5		28	40	46.7	345	241	1087	1080	−0.64
	Z8-H-7-60	1400	200	5		28	60	46.7	345	241	756	737	−2.51
	Z9-H-7-100	1400	200	5		28	100	46.7	345	241	414	461	11.35
	Z10-H-7-160	1400	200	5		28	160	46.7	345	241	237	232	−2.11
	Z11-H-9-40	1800	200	5		36	40	46.7	345	241	1005	1054	4.88
Liu et al. (2021)	0L-0-C	200	100	1			8	22.6	/	/	175	183	4.57
	0L-0.1B-C	200	100	1	0.1		8	22.6	/	/	209	227	8.61

Notes: L= length of the column; D= diameter of core concrete; t= thickness of GFRP tube; V_b= volume rate of basalt fiber; λ= length-diameter ratio of the column, $\lambda = 4L/D$; f_{cc}= axial compressive strength of core concrete; f_{y1}= yield strength of longitudinal reinforcements; f_{y2}= yield strength of stirrups; N_{ue}= test measured value of ultimate bearing capacity; N_{uc}= finite element simulation value of ultimate bearing capacity; E= error between finite element simulation value of ultimate bearing capacity and test measured value of ultimate bearing capacity, $E= (N_{uc} - N_{ue}) \times 100/N_{ue}$ (%).

As can be seen from Table 1, the errors between the ultimate bearing capacity test value and the finite element simulation value of most specimens are within 10%, and most of them are within 5%. It means that the ultimate bearing capacity of the specimen simulated by finite element has a good consistency with the test ultimate bearing capacity.

Figure 3 shows the comparison between the load–axial strain curve of finite element analysis and the measured load–axial strain curve obtained by Wang et al. (Liu et al. 2021; Wang et al. 2010), and the comparison between the load–midspan deflection curve of finite element analysis and the measured load–axial strain curve obtained by Wang et al. (2015). The curves obtained from the finite element analysis of each specimen can well match the measured curve of the test. The development trend of the curve obtained from the finite element analysis is consistent with the test curve. Although the curves of individual specimens have a slight error, they are still within the engineering allowed range. In summary, the finite element model established in this paper can be used to simulate the compression behavior of GFRP confined basalt fiber reinforced concrete long columns, and can accurately simulate the real stress and deformation process and ultimate state of the column under axial compression and eccentric compression.

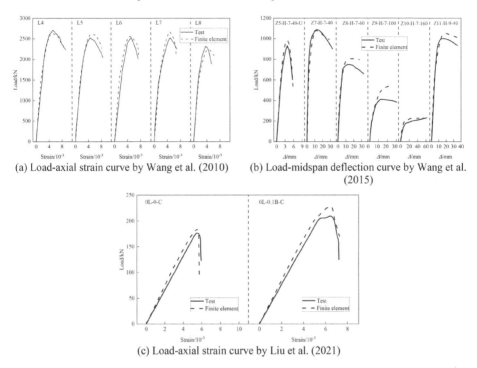

(a) Load-axial strain curve by Wang et al. (2010)

(b) Load-midspan deflection curve by Wang et al. (2015)

(c) Load-axial strain curve by Liu et al. (2021)

Figure 3.   Comparison of measured results of load-axial strain and load-median deflection curves with finite element analysis results obtained by other researchers (Liu et al. 2021; Wang et al. 2010; 2015).

4   CONCLUSION

The secondary development function of ABAQUS was used to realize the random distribution of basalt fiber in concrete, and the finite element model of GFRP confined basalt fiber reinforced concrete long column is established. By comparing with test data from other literature, the model can not only effectively simulate the influence of random distribution for basalt fiber on the eccentric compression bearing capacity of the specimen but also can effectively reflect the real force situation during the eccentric compression of the specimen. The ultimate bearing capacity, load-axial strain curve, and load-midspan deflection curve of the model under eccentric compression are consistent with the test results, indicating the good accuracy of the finite element model.

ACKNOWLEDGMENT

This study was financially supported by Jiangsu Industry-University-Research Cooperation Project (Grant No. BY2021044).

REFERENCES

Bazli, M., Zhao, X. L., Raman, R. K. S., et al (2020) Bond performance between FRP tubes and seawater seas and concrete after exposure to seawater condition. *Constr Build Mater*, 265:120342.

Cao, S. T., Li, Z. S. (2017) An elastoplastic damage constitutive model for confined concrete under uniaxial load. *Eng Mech*, 34: 116–125.

Han, L. H. (2007) *Concrete filled steel tubular structures from theory to practice*. Science Press, Beijing.

Jin, L., Chen, H., Wang, Z., et al (2020) Size Effect on axial compressive failure of CFRP-wrapped square concrete columns: tests and simulations. *Compos Struct*, 254:112843.

Liu, W., Xu, M., Chen, Z. F. (2014) Parameters calibration and verification of concrete damage plasticity model of ABAQUS. *Ind. Constr*, 44:167–171,213.

Liu, X. F., Zhang, Y., Yan, S. T., et al (2021) Mechanical properties of basalt-fiber-reinforced concrete confined with BFRP. *Ind. Constr*, 51:187–193.

Wang, B. L., Xin, G. H., Wang, Q. X. (2015) Experimental study of long reinforced concrete-filled GFRP tubes columns subjected to eccentric load. *Compo Sci Engg*, 21–27.

Wang, J., Bai, X. S., Xue, H. (2020) Experimental study on basalt fiber modifying HSCC columns under eccentric compression. *J. Civ Envir. Engg*, 42:90–97.

Wang, Q. X., Guan, H. B., Cui, W. T. (2010) Experimental study on mechanical properties of concrete filled GFRP tubular long columns under axial loading. *Bldg Struct*, 40:80–83.

Zeng, L., Li, L.J., Chen, G.M., et al (2014) Experimental study on mechanical behavior of GFRP-recycled concrete-steel tubular columns under axial compression. Chin. *Civ. Engrg*. J, 47: 21–27.

Zhang, B., Wei, W., Feng, G. S., et al (2019) Influences of fiber angles on axial compressive behavior of GFRP-confined concrete stub column. *J. Bldg Struct*, 40: 192–199.

Zhou, C. D., Li, C. L. (2013) Strain evolution of concrete square columns confined by GFRP. *J. Bas Scie Engg*, 21: 137–146.

Zhou, H., Jia, B., Huang, H., et al (2020) Experimental Study on Basic Mechanical Properties of Basalt Fiber Reinforced Concret. *MTLS*, 13:1362.

Advances in Measurement Technology, Disaster Prevention and Mitigation – Li & Mohd Yusof (Eds)
© 2023 The Author(s), ISBN: 978-1-032-36087-4

Research and practice on gas control technology of long-distance adjacent strata in large inclined coal mining face

Zhanjin Lu & Zunyu Xu*
China Coal Technology Engineering Group Chongqing Research Institute, Gas Prevention, and Control Branch, Chongqing, China
State Key Laboratory of the Gas Disaster Detecting, Preventing and Emergency Controlling, Chongqing, China

ABSTRACT: In this paper, combined with the actual gas geological situation of 211112 large dip angle coal face in Xinji No. 2 mine, the gas source of large dip angle coal face is analyzed and judged. The proposed measures are based on the analysis of the impact of distant adjacent strata on the mining of large inclined coal seams, while not affecting the normal mining of the working face. They are used to adjust the number of boreholes in the roof strike of the working face, increase the diameter of the directional boreholes in the working face, and increase the drainage system capacity. A series of plans and measures, such as an advance shed at the end of the working face and strict management at the site, is implemented to prevent the gas overrun in the working face effectively and ensure safe and smooth mining of the working face.

1 INTRODUCTION

It is easy to exceed the limit of the gas in the airflow during the movement of the working face. At the same time, due to the large inclination of the coal seam, the "triangular coal" at the tail section of the machine is likely to be flaked due to the movement of the frame. The working face gas exceeds the limit. The tail section of the machine is affected by the change in the coal seam inclination angle and the layout of the roadway. In some areas, the gas concentration in the upper corner wall is relatively small, and the gas concentration in the rear pillar and between the supports of the tail section is higher, which leads to drainage of large amounts of gas to the upper corner of the working surface. Unfavorable factors affect the drainage effect of the upper corner. In addition, high-concentration gas accumulates for a long time along the roof of the tail section of the working face, the head of the frame, and between the frames, which poses a considerable threat to safety during the moving and lowering of the working face.

This article takes the 211112 large-incline coal mining face of Xinji No. 2 mine as the research object to analyze and judge the gas source of the large-incline coal face. By expounding the impact of gas emission from adjacent layers at a relatively long distance on the mining of high-incline coal seams, based on the established gas treatment plan of the working face, the number of drilling holes in the roof direction of the working face is adjusted, the number of large-diameter directional drilling holes in the working face is increased, and the number of drilling holes, in general, is increased. Several plans and measures, including the capacity of the working face extraction system, the advancement of the tail section of the working face, and the strict monitoring of the site, were implemented to prevent the working face from exceeding the gas limit effectively and ensure safe and smooth mining.

---

*Corresponding Author: 347933098@qq.com

2   WORKING FACE PROFILE

The 211112 working face is located in the east wing of the central mining area at the second level of Xinji No. 2 mine, which is at a distance of 10.6–43.3 m from the 11-1 coal, with an average of 28.0 m. The upper distance to the 13-1 coal seam is 56.1–95.3 m, with an average of 72.5 m. The 13-1 coal seam has a gas content of 2.7–9.2 m^3/t, a gas pressure of 0.25–1.39 MPa, and a firmness coefficient of 0.37–0.54. It is an outburst coal seam. The central mining areas (-600 m below and -750 m above) are the protruding danger zone.

    The coal seam is black, powdery, and fragmented. Dark coal is the main form, followed by bright coal. The fissures are relatively developed and the structure of the coal seam is complex. The thickness of the coal is between 0 (0) 0 (0) 0.8 m and 1.7 (4.5) 2.0 (0.5) 2.2 m, with an average of 1.0 (0.9) 1.5 (0.4) 1.2 m. The direct roof of the coal seam is dominated by sandy mudstone or siltstone, with an average thickness of 3.65 m.

3   GAS CONTROL MEASURES AT THE WORKING FACE

3.1   *Drainage method*

The 211112 working face adopts the long-wall retreat-type comprehensive mechanized coal mining method of full height at one time, all caving and forced roof management. The gas emissions arise from the adjacent layers of the coal seam. According to coal seam occurrence conditions, gas emission composition, and roadway layout, the comprehensive drainage method of roof strike drilling and upper corner buried pipe is adopted.

3.2   *Gas control measures during the mining face*

3.2.1   *Drilling holes in the direction of the roof*
In the 2111112 wind tunnel, drilling sites are constructed every 70 m (15 drill sites in total), and each drill site is responsible for drilling six to eight holes. The final hole of the drill hole is located within 15–35 m of the roof of the coal seam. Boreholes are adjacent to this hole stubble below 30 m. Strike drilling construction must advance the progress of the working face by at least one group and at most two groups.

3.2.2   *Drainage of buried pipe in the upper corner*
During the mining of the working face, a DN200 mm drainage branch pipeline was pre-buried in the upper corner to drain the gas in the goaf. According to the advancement of the working face, the construction period of the upper corner retaining wall, and the upper corner gas emission, adjust the upper corner buried pipe drainage flow in a timely manner to ensure the stability of the upper corner gas.

4   PRODUCTION STATUS OF WORKING FACE

At the early stage, the working face is equipped with an air volume of 1,400 m^3/min, and the gas emission from the working face is small (the gas concentration of the return air flow is about 0.1%). After the first pressure is applied to the working face, the gas concentration of the top-slab drilling drainage pipeline (DN250 mm) is between 5% and 10%, and the gas concentration of the upper corner buried pipe drainage pipeline (DN200 mm) is about 1.0%. On October 30, the working face advanced by about 130 m, the roof direction drilling hole, the upper corner buried pipe drainage concentration began to rise. When the working face advanced to about 150 m, the roof direction drilling (DN250 mm) drainage pipeline gas concentration was reached. At the same time, the gas concentration in the upper corner retaining wall of the working surface increased slowly from about 1.5% to about 8%. The gas concentration on the top and between the racks of the 20 racks

(78#-97#) in the tail section increased rapidly, as well as the gas concentration in the partial racks. The maximum concentration of top gas is about 40%. The gas concentration of the return airflow at the working face slowly increases from about 0.1% to about 0.4%. The gas concentration in the airflow of the working face fluctuates considerably during the shifting and lowering of the working face, and the maximum gas concentration of the return air flow reaches 0.81%, and there is a great safety hazard in the working face. At the same time, the absolute gas emission from the working face increased slowly from 5 m^3/min (3.5 m^3/min for drainage, 1.5 m^3/min for air drainage) in the early stage of mining to 35 m^3/min (30 m^3/min for drainage, 5 m^3/min for air drainage).

In view of the changes in gas, the amount of gas emission from the working face was re-predicted. At the same time, the "triangular coal" of the tail section of the working face was moved and dropped during the period of flaking. After the coal flaked, high-concentration gas quickly poured into the working face. In the wind and current, according to the characteristics of the gas emission of the working face, supplementary plans and measures are adopted for the gas control project of the working face.

5   SUPPLEMENTARY MEASURES TAKEN AT THE LATER STAGE OF MINING FACE

5.1   *Increase the air supply of the working face*

According to the gas emission, the air volume at the working face should not be less than 1,500 m^3/t. By adjusting the air volume distribution in the mining area, the return air capacity of the 2111 mining area is increased. The actual air volume distributed during the later stage of the working face is 2,000 m^3/t.

5.2   *Adjust the drilling design of the roof direction*

Construct a roadside drilling field every 70 m in the wind tunnel of the working face. In the drilling field, 6–8 drilling holes shall be constructed in the early stage of the working face recovery, 32 drilling holes shall be adjusted for each drilling field, and the final hole height of the drilling holes shall be increased at the same time. The final drilling point in the early stage of the mining is within 15–35 m of the coal roof, adjusted to the range of 15–50 m from the coal roof. The drainage range is 40 m downwards, and the adjacent drilling yard bores 50 m of stubble to intercept the upper part. With regard to coal seam pressure relief gas effect, advance the progress of the working face not less than 1 drilling yard and not more than 2 drilling yards. The roof direction drilling and drainage adopt a ϕ250 mm main pipe pipeline, and the ϕ200 mm metal drainage hose is connected to the extractor for drilling connection.

5.3   *Add directional drilling design*

Due to the existence of drainage inefficiency zones in the roof direction drilling holes between adjacent drilling sites, the drainage volume of the working face decreases and the amount of air exhaust gas increases when the working face passes through the drilling field. To ensure the safe mining of the working face, supplementary construction of large-diameter directional drilling. There are 5 directional holes in the 211112 working face, with a total engineering volume of 2,500 m. The boreholes control the lower edge of the wind tunnel at 6–30 m and are 25 m away from the roof of the coal seam. To ensure the drainage volume of the directional drilling, the drilling uses a ϕ300 mm pipeline to connect to the extractor through two ϕ200 mm metal extraction hoses, and the drilling uses a ϕ159 mm metal extraction hose to connect with the extractor.

5.4   *Adjust the drainage volume of the buried pipe in the upper corner*

We increase the drainage flow of the upper corner pipeline according to the situation of closed gas in the upper corner of the working surface. The upper corner drainage adopts the method of centralized drainage of the main line and multi-channel hose decentralized drainage to change the upper corner of the gas. The flow field distribution reduces the gas concentration in the closed chamber. The ϕ200 mm pipeline is used for centralized extraction at the upper corners of the working surface, and the depth of entry into the retaining wall is not less than 2.0 m. According to the gas distribution in the exit and upper corner of the working surface, the ϕ76 mm extraction hose and the metal cannula are used to decentralize the 20 (78#-97#) section of the tail section of the machine to reduce the frame head and the frame.

5.5 Add two drainage pipelines

In the early stage of mining at 211112 working faces, the working face adopts two-way gas pipeline drainage. The ϕ200 mm pipeline is used for drainage at the upper corner, and the ϕ250 mm pipeline is responsible for the drainage through the roof strike. According to the gas emission of the working face, the pressure relief gas of the 13-1 coal seam on the working face is poured into the working face. At the same time, combining with the type and quantity of gas control drilling holes in the working face, it is decided to increase the capacity of the working face drainage system. Based on the original two drainage pipelines in the working face, two additional drainage pipelines (ϕ250 mm, ϕ300 mm) are added to increase the roof direction drilling and upper corner drainage capabilities, and at the same time, add the top of the tailstock of the intubation drainage and directional drilling on the working face. The extraction capacity of the working face has been increased from the 65 m³/min extraction mixing volume of DN200 mm (25 m³/min) and DN250 mm (40 m³/min) in the early stage of mining to 160 m³/min extraction mixing volume.

6 GOVERNANCE EFFECT

By adopting a series of practical and effective measures, the gas-control work face achieved remarkable results. The overall gas emission of the working face is maintained at 30–35 m³/min, but with the increase in the gas drainage of the working face, the gas concentration in the airflow of the working face is between 0.1 and 0.15, and the air exhaust gas emission is 2–3 m³/min. The upper corner retaining wall, tail section frame head, and the local gas concentration of 1.0%–1.5% on the top of the frame can fully meet the needs of safe mining, ensure the smooth and orderly mining of the working face, and completely solve the problem of gas over-limit and local point high at the working face. Concentration of gas accumulation and other safety hazards. See Figures 1 and 2 for

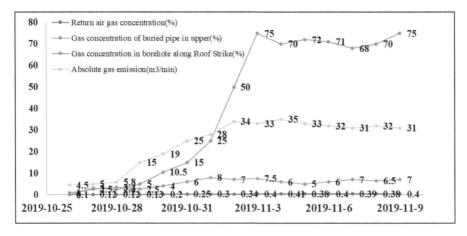

Figure 1. The gas change diagram of the working face that is about 130–150 m in advance of the working face.

details of the gas changes in the working face when the working face is advanced from 130 m to 150 m, and the gas changes after the working face are supplemented with gas control measures.

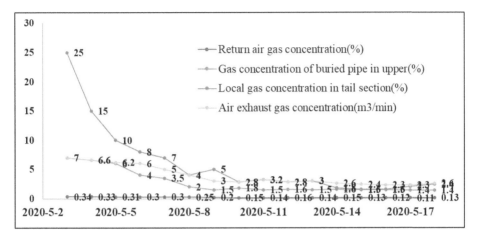

Figure 2.   Gas change diagram after supplementing gas control measures at working face.

7   CONCLUSIONS

(1) The gas emission from the 211112 working face of Xinji No. 2 mine increased rapidly and significantly. Affected by the large inclination of the coal seam, the 13-1 coal overlying long distance coal seam (spacing 56.1–95.3 m; average 72.5 m) pressure relief gas. The working face has a greater impact.

(2) In the early stage of the working face, the roof caving was insufficient, and the "three belts" of the working face roof were not fully formed. When it is advanced to 130 m, the roof is about to form a square (the working face has an oblique length of 135 m), and the roof caving is sufficient. Affected by the "three zones" of the roof, the pressure relief gas in the overlying 13-1 coal seam poured into the working face and the goaf, and the concentration of drainage at the working face, the support head, the local points of the supporting roof, and the gas concentration in the airflow increased rapidly.

(3) By adopting safety measures such as increasing the roof direction of the working face, large-diameter drilling, increasing the mining capacity of the working face, and strict on-site gas management of the working face, no gas exceeded in its limit in the later mining of the working face. No high-concentration gas accumulation occurs between the roof and the frame, which eliminates the hidden gas safety hazards during the working face recovery period, and the working face can be recovered safely and efficiently.

(4) We comprehensively analyze the impact of pressure relief gas in the long-distance outburst coal seam on the large-incline coal mining face, study and formulate many plans for gas control, and manage technical measures to provide help for gas control in the large-incline mining face under similar conditions.

ACKNOWLEDGMENT

This work was supported by the key projects of the Joint Fund for Regional Innovation and Development (Grant No. U21A20110).

REFERENCES

Cheng Yuanping. *Coal Mine Gas Prevention Theory and Engineering Research*, Xuzhou: China University of Mining and Technology Press, 2010.

Xie Ping. *Study on the migration and accumulation of pressure relief gas in the cracked space caused by overburden mining*[D]. Hunan University of Science and Technology. 2013.

Yuan Liang et al. *Technical Manual of Chief Coal Mine Engineer*[M]. Beijing: Coal Industry Press, 2013.

Yuan Liang. Theory of pressure relief mining and gas extraction and technology system of coal and gas co-excavation[J]. *Journal of China Coal Society*, 2019.01, 34(1).

Zhang Guoshu, etc. *Ventilation Safety*[M]. Xuzhou: China University of Mining and Technology Press, 2007.

Advances in Measurement Technology, Disaster Prevention and Mitigation – Li & Mohd Yusof (Eds)
© 2023 The Author(s), ISBN: 978-1-032-36087-4

Study of the deformation characteristics of overlying surrounding rock in floor roadway of Qujiang coal mine

Zhonghua Wang*

National Key Laboratory of Gas Disaster Detecting, Preventing and Emergency Controlling, Chongqing, China
China Coal Technology Engineering Group Chongqing Research Institute, Chongqing, China

ABSTRACT: To study the deformation characteristics of the overlying surrounding rock in the floor roadway of Qujiang Coal Mine, the displacement of the surrounding rock at different depths of the fixed roadway was determined. The top is the largest, the top and bottom are 7.5 m, and the top and bottom are the smallest at 15 m. Different depths of the surrounding rock display a pattern of alternating crests and troughs from the outside to inside. The crests with a larger displacement of the surrounding rock are the damaged areas, and the troughs with smaller displacements of the surrounding rocks are non-destructive areas.

1 INTRODUCTION

The level of in situ stress is relatively low in shallow coal mines. The surrounding rock of the roadway is generally divided into the areas of reduced crushing stress, the area of increased plastic concentrated stress, and the area of original rock stress elasticity (Jing 2020). Under the high-stress environment in the deep coal mine, the coal rock mass is in the post-peak characteristic state, and there may be complex stress areas in the surrounding rock of the roadway. Under certain conditions, the surrounding rock of the roadway alternates between expansion zones and compression zones, resulting in regional ruptures (Wang 2019). The peak stress of the surrounding rock of the roadway weakens, and the width of the pressure relief zone increases significantly (Chen 2017). Due to the large difference in situ stress between shallow and deep mining, there is a difference in the width of the pressure relief zone near the surrounding rock of the roadway (Yuan 2014). The in situ stress level is high in deep mining zonal rupture may occur near the surrounding rock of the roadway, and the width of the pressure relief zone is significantly larger than that in the shallow part (Chen 2014). Therefore, it is of great significance to study the deformation characteristics of the surrounding rock overlying the roadway of the coal mine floor (Li 2008).

2 MINE OVERVIEW

2.1 The basic situation of the mine

The horizontal layout (–850 m) of Qujiang Company has six mining areas, namely, the east one that goes up the mountain, the west one that goes up the mountain, the east one that goes down the mountain, the west one goes down the mountain, two on the west that go up the mountain, and the three in the east that go up the mountain. Currently, there are 603 fully mechanized mining faces. There are 9 and 3 excavation working faces for rock roadway and coal roadway, respectively, and 12 excavation working faces in total. The long-wall mining method is adopted, and the back-propelling method, blasting, and falling coal, manual coal loading, and mechanical transportation are adopted.

---

*Corresponding Author: boaidajia2007@126.com

DOI 10.1201/9781003330172-73

2.2 Mine development and mining

The mine adopts the vertical shaft single-level downhill mining method. The production level elevation is -850 m, and the total return airway elevation is −660 m. There are four shafts in the mine industry plaza: main shaft, auxiliary shaft, central air shaft, and Dongli air shaft, with net sections of 19.6 m^2, 33.2 m^2, 38.5 m^2, and 36.5 m^2, respectively. The wellhead elevations of the main, sub, central, and Dongli air wells are +66.85 m, +67.23 m, +45.95 m, and +43.5 m, respectively, and the bottom elevation is −850.0 m. The depths are 1,000 m, 988 m, 895.95 m, and 873.5 m, respectively.

At present, there are four mining areas arranged horizontally at −850 m, namely, the east-uphill, the west-uphill, the east-downhill, and the west-downhill mining areas, which are alternately produced by jumping up and down. There are 602 fully mechanized mining faces, 303 shot mining faces, 2 coal mining faces, and 215 fully mechanized mining faces to spare. There are 8 and 4 excavation working faces for rock roadway and coal roadway, respectively, and 12 excavation working faces in total.

The long-wall mining method is adopted, and the back-propelling method, blasting, and falling coal, manual coal loading, and mechanical transportation are adopted.

2.3 Mine ventilation gas

The mine was put into operation from 2003 to 2012, and the gas grade identification was carried out, and the identification results were all coal and gas outburst mines. According to the mine gas grade identification data from 2003 to 2012, the graph of the changing trend of the gas gushing volume was drawn; 16.53~44.49 m^3/min, with an average of 33.07 m^3/min. In 2012, the absolute gas gushing volume of the mine reached 58.85 m^3/min, and the relative gas gushing volume of the mine was 32.29 m^3/t.

The gas gushing of Qujiang Company has the characteristics of zonal distribution, according to the gas grade identification results of the whole well from 2008 to 2011. Since the gas source is mainly concentrated in the east, the gas gushing volume is high in the east and low in the west, and prominent in the east and west. Judging from the 2012 full-well gas grade appraisal results, since the mine production is mainly concentrated in the west wing, showing a trend of high in the west and low in the east, the gas output in the west wing will increase. Therefore, it is necessary to strengthen anti-outburst measures, increase gas drainage efforts, and ensure safe production in mines.

A total of 6 outbursts have occurred since the well was built, and the outburst types are large and medium outbursts, including 4 large outbursts and 2 medium outbursts. There is an increasing trend with the increase of mining depth.

The mine is centrally juxtaposed and diagonally mixed with exhaust ventilation, the main shaft and the auxiliary shaft allow entry of air, and the central air shaft and the Dongli air shaft discharge air. The central air shaft fan has two BDK-8-No. 32 type counter-cyclone fans, one for use and the other for standby. The rated power of the fans is 2×630 kW. The blade installation angle is 35° for the first stage and 30° for the second stage. The two stages operate simultaneously. The negative pressure is 315 mm. Dongfengjing fans are two FBCDZ-8-No. 26 type counter-cyclone fans, one for use and one for standby. The rated power of the fans is 2×355 kW, and the blade installation angle is 18.5° for the first stage and 13.3° for the second stage, and the two stages run at the same time. The total inlet air volume of the mine is 11,643 m^3/min, and the total return air volume is 11,983 m^3/min. Among them, the return air of the central air shaft is 9,892 m^3/min, and the return air of the Dongli air shaft is 2,091 m^3/min. The mine ventilation grade hole is 4.3 m^2. Each excavation face is an independent ventilation system. Qujiang Company's -850 m horizontal B4 coal seam has a natural tendency for coal dust explosions and coal spontaneous combustion. The ground temperature in this area is the normal ground temperature, but due to the large mining depth, the ground temperature is relatively high. At present, the production level temperature of the mine is generally about 38°~39°, and the minefield has one or two high-temperature zones.

3 INVESTIGATION PLAN OF SURROUNDING ROCK DEFORMATION OF FLOOR ROCK ROADWAY

Table 1 shows the investigation plan for the surrounding rock deformation of the floor rock roadway of Qujiang Company. Each inspection content is arranged with 5 inspection holes, that is, 5 inspection positions, which are 15 m on the lower side, 7.5 m on the lower side, directly above, 7.5 m on the upper side, and 15 m on the upper side. The DW-6 multi-point displacement meter was used to measure the displacement of the surrounding rock of the roadway at different depths.

Table 1. Investigation plan for pressure relief in floor rock roadway area of Qujiang Company.

Roadway	Inspection position (from the head of the excavation: m)			Drilling diameter (mm)
	Group 1	Group 2	Group 3	
702 Floor Lane	180	100	60	32
213 Floor Lane	248	212	120	32

4 DEFORMATION OF SURROUNDING ROCK OF ROCK ROADWAY WITH PRESSURE RELIEF FLOOR

After drilling, use the guide rod to send the base point anchor (pawl type) to the specified depth, install a base point (6 base points in total) from the inside to the outside, record the readings every day after installation, and count the surrounding rock displacement at different depths. The survey results of the 15 m of surrounding rock on both sides of the roof of the 702 and 213-floor lanes are shown in Tables 2 and 3, respectively.

Table 2. Displacement and deformation of surrounding rock in the monitoring section of 702-floor roadway along trough (Groups 1 to 3).

Position	Orifice distance (m)	Bottom 15 m	Bottom 7.5 m	Right above	Top 7.5 m	Top 15 m
Head-to-head 180 m (The first group)	10	0.5	1.6	1	1.3	1.8
	8	0.2	0.2	1	1.5	1
	6	0.5	3.7	0.5	0.1	0.2
	5	0.2	0	3	1	1
	4	0.4	1.5	0.5	0.5	0.2
	2	0.7	0.5	1	1.4	1.5
Head-to-head 100 m (The second group)	10	/	0.1	0.3	0.2	0.3
	8	/	0.1	1.3	0.2	0.5
	6	/	0.4	1.5	0.7	0.8
	5	/	0.2	2.3	1	0.1
	4	/	0.2	1.5	0.4	0.4
	2		0.3	1.1	0.2	0.3
Head-to-head 60 m (The third group)	10	0.4	0.5	1	0.7	0.1
	8	0.2	0.4	1.1	0.8	0.1
	6	0.5	0.5	1.2	0.7	0.6
	5	0.1	0.3	0.8	0.4	0.2
	4	0.1	1.1	1.9	1	0.8
	2	0.2	0.1	0.7	0.6	0.1

537

Table 3. Displacement and deformation of surrounding rock at monitoring section of 213-floor roadway along the trough (Groups 1 to 3).

Position	Orifice distance (m)	Bottom 15 m	Bottom 7.5 m	Right above	Top 7.5 m	Top 15 m
	6	0.3	0.2	0.8	0.3	0.2
Head-to-head	5	0.5	0.3	0.1	0.5	0.3
248 m	4	0.3	0.6	1.1	0.3	0.5
(The first group)	3	0.9	1.1	1.7	1.2	1.4
	2	0.3	0.4	0.8	0.9	0.2
	1	1	0.6	2.2	1.2	0.3
	6	0	0.2	0.3	0.1	0.4
Head-to-head	5	1	2.1	2.6	3.6	1.2
212 m	4	0.2	0.8	0.5	1	1
(The second group)	3	0.8	1	2	1.4	1.1
	2	0.1	0.6	0.4	0.2	0.3
	1	1.6	2.2	2.6	3	1.5
	6	0.4	0.3	0.2	0.6	0.7
Head-to-head	5	0.7	1.2	1.9	0.4	0.8
120 m	4	0.4	0.6	0.6	0.4	0.3
(The third group)	3	0.6	0.8	1.8	1	1
	2	0.3	0.4	1.8	0.9	0.9
	1	1.4	1	2	2.1	1.2

It can be seen from Tables 2 and 3 that the displacement of the surrounding rock within 15 m of the upper and lower sides of the floor roadway basically shows the characteristics of being the largest directly above, and gradually decreasing to 7.5 m and 15 m on both sides; different base points of the same borehole appear as wave crests from the outside to the inside. The characteristics of the wave trough change, that is, the displacement and deformation at 2–3 places are small, and the displacement and deformation at 2–3 places are large. The large and gradually decreasing monotonic variation law preliminarily judges that there is partition rupture in the surrounding rock after the excavation of the roadway. The crest with a large displacement is the damaged area of the surrounding rock, and the trough with a small displacement is the non-destructive area of the surrounding rock. Figures 1 and 2 show the representative data.

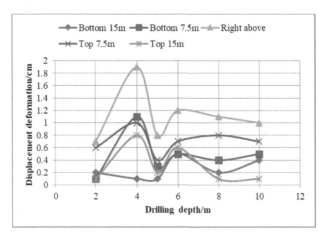

Figure 1. Displacement and deformation of surrounding rock in the monitoring section of 702-floor roadway along trough (Group 1).

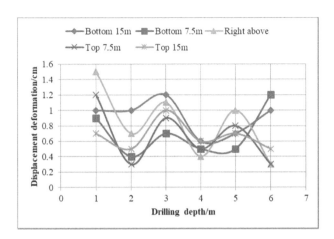

Figure 2. Displacement and deformation of surrounding rock at monitoring section of 213-floor roadway along the trough (Group 2).

5 CONCLUSION

The displacement and deformation of the surrounding rock within the range of 15 m from the upper and lower sides of the rock roadway of the pressure relief floor shows that the deformation of the surrounding rock directly above is the largest, followed by the upper and lower sides of 7.5 m, and the smallest of the upper and lower sides of 15 m.

The maximum displacement deformation directly above the floor lane is 3.7 cm; the maximum displacement deformation within the range of 7.5 m above and below the floor lane is 3 cm; the maximum displacement deformation is 1.8 cm.

Different base points of the same borehole show the characteristics of alternating peaks and troughs from the outside to the inside. This variation rule is different from the monotonic variation rule of the displacement of the surrounding rock of the shallowly buried cavern and the gradual decrease with the increase of the distance from the cave wall. The larger wave crest is the damage zone of the surrounding rock, and the trough position with the smaller displacement is the non-destruction zone of the surrounding rock.

ACKNOWLEDGMENT

This study was financially supported by the Technological Innovation and Entrepreneurship Fund Special Project of Tiandi Technology Co., Ltd. (Grant No. 2021-2-TD-ZD008), the General project of Chongqing Research Institute Co., Ltd. (Grant No. 2019YBXM31), and the General project of Chongqing Research Institute Co., Ltd. (Grant No.2020YBXM22).

REFERENCES

Chen Haoxiang, Qi Chengzhi, Li Kairui, Xu Chen, Liu Tiantian. Study on Nonlinear Continuous Phase Change Model of Subregional Rupture of Surrounding Rock in Deep Roadway (J). *Geotechnical Mechanics*, 2017, 38(04):1032–1040.
Chen Xuguang, Zhang Qiangyong, Yang Wendong, Li Shucai, Liu Dejun, Wang Hanpeng. Comparative analysis of test and field monitoring of surrounding rock partition failure in deep roadway (J). *Chinese Journal of Geotechnical Engineering*, 2011, 33(01):70–76.

Jing wei, Guo Rui, Yang Renshu, Jing Laiwang, Xue Weipei. Theoretical analysis of surrounding rock deformation in deep roadway considering rock rheology and strain softening (J). *Chinese Journal of Mining and Safety Engineering*, 2021, 38(03): 538–546.

Li Shucai, Wang Hanpeng, Qian Qihu, Li Shuchen, Fan Qingzhong, Yuan Liang, Xue Junhua, Zhang Qingsong. On-site monitoring of the fracture phenomenon of surrounding rocks in deep roadways (J). *Chinese Journal of Rock Mechanics and Engineering*, 2008(08):1545–1553.

Wang Xuebin, Bai Xueyuan, Ma Bing, Zhang Zhihui, Lv Jinguo. Influence of the heterogeneity of surrounding rock of roadway on its zonal fracture (J). *Journal of China University of Mining and Technology*, 2019, 48(01): 78–86.

Yuan Chao, Wang Weijun, Zhao Yanlin, Yu Weijian, Peng Gang, Wu Hai, Tang Hai. Theoretical analysis on the deformation of roadway surrounding rock considering the plastic hardening and softening characteristics of rock mass (J). *Coal Journal*, 2015, 40(S2): 311–319.

Advances in Measurement Technology, Disaster Prevention and
Mitigation – Li & Mohd Yusof (Eds)
© 2023 The Author(s), ISBN: 978-1-032-36087-4

Research on the law of pressure relief and permeability increase in overlying coal seam of floor roadway

Zhonghua Wang*

National Key Laboratory of Gas Disaster Detecting, Preventing and Emergency Controlling, Chongqing, China
China Coal Technology Engineering Group Chongqing Research Institute, Chongqing, China

ABSTRACT: To study the law of pressure relief and permeability enhancement in the overlying coal seams of different rock pillar floor roadways, the air permeability coefficients at different positions of the overlying coal seams of 710-floor roadways and 505-floor roadways were measured on site. Through investigation, it is found that the air permeability coefficient of the coal seam directly above the pressure relief floor rock roadway is the largest, followed by the upper and lower banks at 7.5 m, and the upper and lower banks at 15 m. Comprehensive analysis shows that with the increase in excavation time, the permeability of the coal seam corresponding to the drill hole gradually increases, and the pressure relief effect gradually increases. The smaller the distance between the floor roadway and the coal seam, the more significant the anti-reflection effect.

1 INTRODUCTION

Coal seam permeability represents the difficulty of coal seam gas flow (Cao 2019). It is an important indicator to measure the difficulty of coal seam gas pre-extraction (Han 2021). At present, the method of the Chinese Academy of Mines is the most widely used determination method in China. It is established on the basis that the state of coal seam gas flowing to the borehole is radially unstable. The coal seam permeability coefficient is calculated by measuring the radially unstable flow of coal seam gas (Men 2016). This method is also called the borehole radial flow method. Gas drainage needs to test the change law of the permeability coefficient of the coal seam after the pressure relief of the overlying coal seam in the floor roadway (Wang 2016).

2 OVERVIEW OF THE TEST AREA

2.1 710-Floor Lane

The bottom extraction roadway of 702 wind tunnel is located at the fourth level (–650 m), belonging to the east III uphill mining area. The design length of this roadway is 536 m, and the service life is more than 2 years. The surface elevation is from +34.6 m to +38.2 m, and the underground elevation is from –527.6 m to –575.3 m. The design section of this lane is a semi-circular arch, with a clear width of 4,200 mm, a clear height of 3,300 mm, and a design tunneling area (S) of 11.92 m². The lane is designed to be 21.7 m (horizontal distance) before the new midnight point in the east III transportation uphill and dig down 400 m (horizontal distance) according to the true azimuth of 226°30′, with a slope of –4°30′. To the east is east III transportation uphill, south to the 369 mined-out area, west to the 501 mined-out area, and north to the unmined area. The roadway is in the rock layer 26 m below the B4 coal seam floor and gradually enters the rock layer 12 m below the B4 coal seam floor. Its lithology is light gray fine sandstone and dark gray fine siltstone

---

*Corresponding Author: boaidajia2007@126.com

DOI 10.1201/9781003330172-74

interbedded with thin layered mud. The quality siltstone contains quartz fine-grained sandstone with a layer thickness of about 15 m. The upward is light gray thick layered medium-grained sandstone with obvious horizontal bedding. The thickness is about 3.5 m~4.5 m, and upwards is light gray to dark gray, thin-layered argillaceous fine siltstone, with occasional siderite nodules. The geological structure of this roadway is relatively simple, revealing a group of oblique normal faults with east-west trending, with a drop of H = 0.7 m, and a dip angle of ∠68°.

2.2 505-Floor Lane

The 505 roadways along the bottom of the trough are located at the fourth level (−650 m), belonging to the east three downhill mining area; the surface elevation is +38.3 m~+40.1 m, and the underground elevation is −651.7 m~−697.3 m. The roadway is first dug 150 m in the true direction of 266.5° at point 1 of the east III transportation downhill measure, and then dug down 300 m (horizontal distance), with a slope of −6°. The lane is west of the east III transportation downhill and south of the 503-working face. The southeast belongs to the unmined area. The road is excavated on the floor about 15~27 m away from the B4 coal seam. Its lithology is light gray fine sandstone and dark gray fine siltstone interbedded with thin layered argillaceous siltstone containing quartz fine-grained sandstone, and the layer thickness is 15 m from left to right. The upwards are light gray fine siltstone with obvious horizontal bedding, with a thickness of 3.5~4.5 m. The upwards are light gray to dark gray thin layered argillaceous fine siltstone, occasionally siderite nodules, with a thickness of 5.0 ~6.0 m. The development of the B4 coal seam is relatively stable. The coal thickness is about 2.8~3.8 m, and there are gangues of about 0.4~0.6 m in the middle. The inclination of the coal seam is about 10° ~ 13°, the strike is N70~ 85° E, and the trend is SW. The absolute gas emission rate is 6.01 m^3/min, the relative emission rate is 14.46 m^3/t. Coal dust is explosive, with an explosion index of 17.65%. Coal seams have spontaneous combustion, and the spontaneous combustion period is 4–6 months. The geological structure of this roadway is relatively complicated. During the tunneling process, 6 faults were exposed, with a drop of more than 2 m and a drop of 0.5–2 m.

3 INVESTIGATION PLAN FOR COAL SEAM ANTI-REFLECTION EFFECT

The simple drainage method is used to test the gas flow rate in the borehole, and then the coal seam permeability is calculated (Zhang 2021). The hole diameter is Φ75 mm. The inspection sites are the 710-floor roadway and the 505-floor roadway. The thickness of the two roadways and the rock pillars of the overlying coal seam are 15 m and 19 m, respectively. After the drilling construction, 5 inspection holes are arranged for each inspection content, that is, 5 inspection locations, which are 15 m on the lower side, 7.5 m on the lower side, directly above, 7.5 m on the upper side, and 15 m on the upper side. Each roadway is tested in three groups, and the specific inspection locations are shown in Table 1.

Table 1. Investigation of the location of coal seam permeability coefficient.

Roadway	Inspection location (from the head of the tunneling: m)		
	The first group	The second group	The third group
710-Floor Lane	139	95	55
505-Floor Lane	204	168	100

4 CHANGE LAW OF COAL SEAM PERMEABILITY COEFFICIENT

Using the borehole radial flow method, the permeability coefficients of coal seam B4 within 15 m from both sides of the roof of 710 and 505-floor roadways of Shangzhuang coal mine are calculated as shown in Tables 2 and 3, Figures 1 and 2, respectively.

Table 2. Measurement results of permeability coefficient of coal seam B4 in 710 coal road.

Number of groups	Lower 15 m	Lower 7.5 m	Right above	Upper 7.5 m	Upper 15 m
The first group	0.0200	0.0277	0.0793	0.0437	0.0211
The second group	0.0178	0.0263	0.0639	0.0355	0.0197
The third group	0.0151	0.0220	0.0458	0.0285	0.0167

Table 3. Measurement results of permeability coefficient of the B4 coal seam in 505 coal road.

Number of groups	Lower 15 m	Lower 7.5 m	Right above	Upper 7.5 m	Upper 15 m
The first group	0.0229	0.0493	0.1444	0.0869	0.0253
The second group	0.0237	0.0464	0.1228	0.0869	0.0313
The third group	0.0229	0.0449	0.0969	0.0657	0.0253

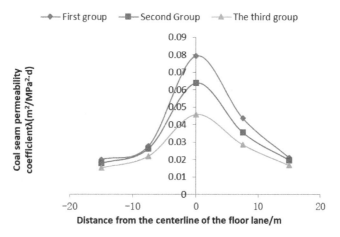

Figure 1. Measurement results of permeability coefficient of coal seam B4 in 710 coal road.

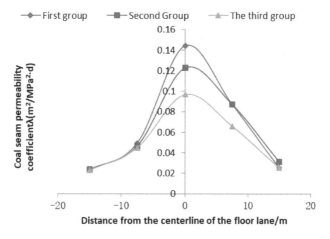

Figure 2. Measurement results of permeability coefficient of the B4 coal seam in 505 coal road.

It can be seen from Tables 2 and 3 that the air permeability coefficient of the coal seam directly above the floor roadways 710 and 505 is the largest, followed by 7.5 m on both sides, and the smallest at 15 m on both sides; the maximum air permeability coefficient of the coal seam directly above the floor roadway 710 is 0.0793, 0.0437, 0.0211 $m^2/MPa^2 \cdot d$, respectively. The maximum permeability coefficient of the coal seam directly above the 505-floor roadway is 0.1444, 0.0869, 0.0313 $m^2/MPa^2 \cdot d$; the permeability coefficient of the coal seam at the same position is from the first group to the third group.

5 CONCLUSION

Through investigation, it is found that the air permeability coefficient of the coal seam directly above the pressure relief floor rock roadway is the largest, with a maximum value of 0.1444 $m^2/MPa^2 \cdot d$, followed by the upper and lower sides at 7.5 m, and the maximum value is 0.0869 $m^2/MPa^2 \cdot d$. At 15 m from the upper and lower sides, the maximum value is 0.0303 $m^2/MPa^2 \cdot d$.

Comprehensive analysis shows that with the increase in excavation time, the permeability of the coal seam corresponding to the drill hole gradually increases, and the pressure relief effect gradually increases. The smaller the distance between the floor roadway and the coal seam, the more significant the anti-reflection effect, which means that the gas flow in the same relative position of floor roadway 710 is greater than that of floor roadway 505.

ACKNOWLEDGMENT

This study was financially supported by the Technological Innovation and Entrepreneurship Fund Special Project of Tiandi Technology Co., Ltd. (Grant No. 2021-2-TD-ZD008), the General project of Chongqing Research Institute Co., Ltd. (Grant No.2019YBXM31), and the General project of Chongqing Research Institute Co., Ltd. (Grant No.2020YBXM22).

REFERENCES

Cao Jianjun, Wang Zhonghua, Yang Huiming. Research on pressure relief and outburst prevention technology for strip pre-excavation floor rock roadway in deep coal roadway(J). *Mining Safety and Environmental Protection*, 2019, 46(06): 14–19.

Han Ying, Dong Bowen, Zhang Feiyan, Lu Shuai, Li Weidong. Research progress of hydraulic punching pressure relief and anti-reflection technology in my country(J). China Mining Industry, 2021, 30(02): 95–100.

Meng Xianzheng, Wang Zhonghua, Chen Guohong, Zhang Yongjiang. Pressure-relief and outburst prevention technology for deep coal roadway driving in a single severely outburst seam(J). Coal Science and Technology, 2016, 44(12): 75–80.

Wang Zhonghua. Pressure relief and permeability enhancement method for pre-digging floor rock roadway(J). China Coal, 2016, 42(06): 101–104.

Zhang Xiaogang, Jiang Wenzhong, Du Feng. Development status and prospects of antireflection technology for high gas and low permeability coal seams(J). Coal Mine Safety, 2021, 52(02): 169–176.

Advances in Measurement Technology, Disaster Prevention and Mitigation – Li & Mohd Yusof (Eds)
© 2023 The Author(s), ISBN: 978-1-032-36087-4

Analysis of hydration heat and pipe cooling of the mass concrete cap of the main tower of a self-anchored suspension bridge

Zhaobing Wang, Fajiang Luo, Kui Huang, Peng Li, Hualong Li, Kun Guo*, Jianning Li, Zizhou Dai, Qinxiong Zhang, Leping Ren, Litao He & Jincheng Fan
China Construction Third Engineering Bureau Group Co., Ltd., Xi'an, China

ABSTRACT: The Yuanshuo Bridge is a typical self-anchored suspension bridge structure. The construction of the main tower cap is carried out in a combination of layering pouring and pipe cooling to ensure that the temperature is controlled within the range of specifications and design requirements. Using large-scale finite element analysis software, this paper describes the fine modeling of temperature diffusion area, the layout of each layer tube cold, and their locations and layouts with precision based on the construction of actual simulation of the temperature diffusion conditions, and by controlling the cooling pipe diameter, flow parameters such as temperature, water flow and time. The layout of pipe cooling is optimized and analyzed, and the analysis results show that the temperature control effect is good.

1   INTRODUCTION

With the continuous development of bridge engineering, especially in the foundation engineering of some large bridges, the application of mass concrete has been very common. The American Concrete Institute (ACI) defines "mass concrete" as concrete structures that must be poured on site when the minimum section size of the structure is sufficiently large that measures must be taken to address the heat of hydration and the volume deformation of the structure caused by the heat of hydration. In China, concrete with a large volume mixing size ranging from 1~3 m and a temperature difference greater than 25°C caused by hydration heat must be avoided by taking measures (Zhu 1999). In the process of hydration and hardening of cement in concrete, considerable hydration heat will occur, which makes the temperature of concrete rise significantly and then gradually dissipate. Concrete is a poor conductor of temperature, and the temperature difference between the inside and outside of mass concrete will cause cracks in the concrete due to the internal and external constraints of concrete (Wang 1987).

In the western part of the country, where the environment is harsh, the concrete has a high content of cement, resulting in greater hydration heat within the structure. The low-temperature environment will make the concrete surface temperature drop very fast; at the same time, the internal temperature is maintained at a higher level, and it is difficult to reduce it in a short time. The huge temperature difference between inside and outside will cause great temperature stress, and when it is greater than the tensile strength of the concrete surface, temperature cracks will occur. Therefore, controlling the temperature difference between the inside and outside of mass concrete in the alpine area can effectively prevent temperature cracks and improve the durability

---

*Corresponding Author: gk_changan@163.com

DOI 10.1201/9781003330172-75

of concrete structure (Technical code for the construction of highway bridges and culverts of the people's Republic of China 2011; Hao 2012).

2 ENGINEERING SURVEY

Yuanshuo Bridge has a total length of 1054.18 m. The whole bridge hole span layout is 4×30 m + 4×30 m + (50+116+300+116+50) + 3×30 m + 3×30 m. The spindle-shaped bridge tower is in the central separation zone of the bridge. It is a single column with a steel box section. The height of the cap is 5 m, the plane size is 35.25 m×29.00 m, and the concrete is made of C40 micro-expansion concrete. To prevent the influence of cement hydration heat from damaging the normal use of the structure, it is necessary to analyze the concrete temperature field of the cap. The overall structure of the bridge is shown in Figures 1–2.

Figure 1. Side view of the main bridge of Yuanshuo Bridge.

Figure 2. Top view of the main bridge of Yuanshuo Bridge.

3 HEAT CONDUCTION EQUATION AND BOUNDARY CONDITIONS

3.1 *Theory of hydration heat of cement*

At present, there are mainly three kinds of the exponential formula: hyperbolic formula, and compound exponential formula. Different formulas will produce greatly different amounts of calculation and results, and the exponential formula is the most used. The calculation results of the exponential formula are closer to the actual engineering value (Li & Huang 2009; Lu 2007; Vratislav et al. 2014; Yang et al. 2011). Therefore, the empirical exponential formula is adopted in the finite element calculation in this study, and the formula is as follows:

$$Q(t) = Q_0 \left(1 - e^{-mt}\right)$$

Where $Q(t)$ is the cumulative heat of hydration of cement per unit mass. t is concrete age, Q_0 is the heat loss per unit mass of cement, and m is the correlation coefficient between cement varieties and hydration heat.

3.2 *The equivalent heat conduction equation of concrete with tube cooling*

In the process of concrete pouring and cooling and hardening, if water pipes are used for cooling and cooling, the internal temperature field will be unstable with time. In engineering, the equivalent negative heat source method proposed by Zhu Bofang is mostly adopted, and the equivalent heat conduction equation (Tang et al. 2012; Wang et al. 2007) is:

$$\frac{\partial T}{\partial \tau} = \alpha \left(\frac{\partial^2 T}{\partial x^2} + \frac{\partial^2 T}{\partial y^2} + \frac{\partial^2 T}{\partial z^2} \right) + \frac{\partial \theta}{\partial \tau} \quad \frac{\partial \theta}{\partial \tau} = (T_0 - T_w) + \frac{\partial \phi}{\partial \tau} + \theta_0 \frac{\partial \psi}{\partial \tau}$$

$$\phi = e^{-b\tau} \quad b = ka/D^2 \quad k = 2.09 - 1.35\eta + 0.32\eta^2 \quad \eta = \lambda L/c_w \rho_w q_w$$

Where a is the temperature conductance, T_0 is started to cool the temperature of the concrete, T_w is the cooling water inlet temperature, ϕ is considering the cooling water of hydration heat cut after hydration heat change coefficient, $\frac{\partial \phi}{\partial \tau}$ is changing rate of equivalent heat of hydration, ψ is to assume that no heat source condition, cooling water temperature and the cooling coefficient of concrete formation temperature is not balanced, θ is a concrete adiabatic temperature rise, t is the age of concrete, and $\frac{\partial \psi}{\partial \tau}$ is the cooling rate without the heat source term.

4 STRUCTURAL MODELING AND ANALYSIS

The whole process of hydration heat of the cap is analyzed and calculated by using the tube cooling module of the Midas Civil finite element program, and the foundation is built into components with specific heat and thermal conductivity characteristics. It is divided into 21,660 units and 26,064 nodes, as shown in Figure 3.

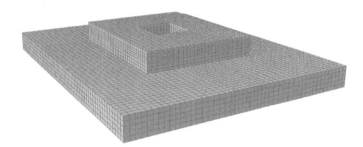

Figure 3.   Diagram of the finite element calculation model.

During the numerical simulation analysis, the following assumptions are made:

a. Assume that the mass concrete is a uniform body, the initial temperature is the same, and the heat ratio of each node is the same.
b. The influence of reinforcement and other materials in the mass concrete is ignored (Chu Inyeop et al. 2013; El-Hadj & Rger 2009; Kadri & Duval 2009; Seungjae et al. 2013; Vratislav Tydlitát et al. 2014).

4.1 Original construction plan

The cooling water pipes are arranged in a snake shape, with a total of 4 layers. The cooling pipes are standard steel pipes with a nominal diameter of 32 mm. The flow rate of a single cooling water pipe is 1.2 m³/h. The layout of each layer is shown in Figures 4–5:

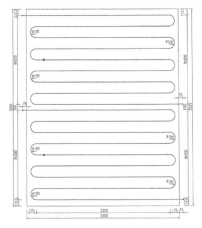

Figure 4. Cold pipe layout of the first floor.

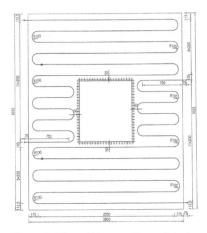

Figure 5. The cold layout of layer 2-4 pipe.

4.2 *Parameter analysis*

a. The change in pipe diameter will cause the change in cooling effect, but it will also bring a significant change in economic effect. We use the standard steel pipe with a nominal diameter of 51 mm, which is 19 mm larger than the original diameter.
b. Climate characteristics have a positive effect on heat dissipation. Water circulation times that are short will not have a cooling effect. We change the water supply time to 150 hrs after the completion of top pouring, which is 50 hrs longer than the original plan.
c. The change of cooling water flow means the change of pump power, which has a great influence on cooling and the economy. We increase the water flow to 3 m³/h, which is 1.8 m³/h more than the original plan.

4.3 *Data summary and analysis*

The control optimization calculation was carried out from four aspects of pipe cold pipe diameter, water flow time, and flow change respectively. The bottom surface of the first layer, the top surface of the first layer, the top surface of the second layer, the top surface of the third layer, the top surface of the fourth layer, and the outside of the cap were taken for the calculation. In the results chart, dark blue circles represent no temperature control measures; the red circle represents the original plan; the green circle represents flow increase; the purple circle represents the increase of pipe diameter. Light blue circles represent increased water flow time. Figures 6–11 show the calculation results.

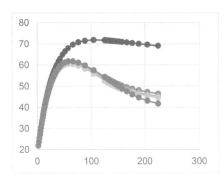

Figure 6. The temperature trend at the junction of the first and second floors.

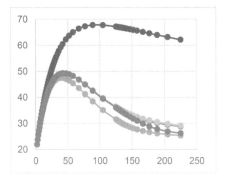

Figure 7. The temperature trend of the top surface of the first floor.

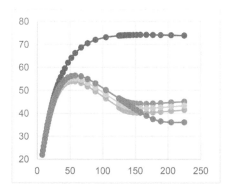

Figure 8. The temperature trend at the junction of the second and third layers.

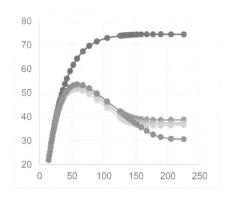

Figure 9. The temperature trend at the junction of the third and fourth layers.

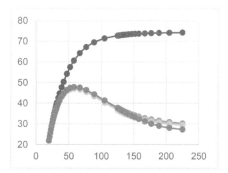

Figure 10. Fourth temperature trend.

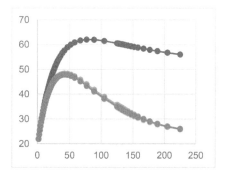

Figure 11. Outside temperature trend.

It can be seen from the analysis that the variation of each parameter has an obvious change in cooling amplitude, and the continuous cooling effect is different over time:

The temperature control effect of tube cooling is very significant. The maximum core temperature is reduced from 76°C to 62°C, which is 18%. With the increase in pipe diameter, the maximum core temperature further decreases by 2~3°C. However, since the total water inflow did not increase, the single pipe diameter had little effect on the cooling effect. The effect of increasing the water circulation time on internal temperature control is significantly increased, avoiding the re-rise of temperature in the later stage, and the control range reaches more than 50%. With the increase in flow rate, the maximum core temperature is reduced by about 5~8°C, which is about 12% higher than the original scheme.

5 CONCLUSIONS

Based on the above analysis, the following conclusions can be drawn:

a. The traditional pipe cooling mode has good applicability in the alpine region and high accuracy of finite element simulation analysis.
b. The core factor of pipe diameter is water flow, and the pipe diameter has little influence on the overall temperature control effect under the condition of constant flow.
c. The water supply time should be determined based on the local temperature. The bridge is in a cold area, and the climatic and environmental conditions are favorable for cooling down.

Therefore, the water supply time that is too long will lead to an increase in construction costs, thereby resulting in a large internal cooling rate, which is not conducive to the structure.

d. In the case of changes in thermal insulation effect caused by atmospheric temperature, actual flow and flow rate of cooling water, convection coefficient of cooling water affected by flow rate, wind speed, and rainwater, the model should be corrected in real time according to measured values to ensure accuracy.

REFERENCES

Chu Inyeop, Lee Yun Amin, Muhammad Nasir, et al. Application of a thermal stress devicefor the prediction of stresses due to hydrationheat in mass concretestructure. *Construction and Building Materials*, 2013, 45: 192–198.

El-Hadj K, Rger D. Hydration heat kineticsof concrete with silica fume. *Construction and Building Materials*, 2009, 23(11): 3388–3392.

Hao C. *Study on quality control of concrete construction in winter in cold regions*. Xi'an University of science and technology. 2012.

Kadri E H, Duval R, Aggoun Setal. Silicafume effect on hydration heat and compressive strength of high-performance concrete. *ACI Materials Journal*, 2009, 106(2): 107–13.

Li M, Huang Y. Humidity control in winter construction of mass concrete for pile caps in bridge engineering. *Sinotrans*, 2009, 02:194–195+200

Lu E. *Analysis and control of hydration heat temperature of mass concrete pile cap*. Changsha: Hunan University, 2007

Seungjae L, Geumsuk L, Sudeok S. G strain sensors for creak measurement due to hydration heat in massive concrete. *Advanced Materials Research*, 2013, 689: 329–32.

Tang S, Tang C, Liang Z Z, et al. Study on meso characteristics and thermal cracking process of concrete heat conduction and thermal stress, *Chinese Journal of civil engineering*, 2012, 45(2): 11–19.

Technical code for construction of highway bridges and culverts of the people's Republic of China (JTG / T F50-2011). Ministry of transport of the people's Republic of China.

Vratislav T, Tomá M, Robert C. Effect of w/c and temperature on the earlystage hydration heat development in Portland limestone cement. *Construction and Building Materials*, 2014, 50(15): 140–147.

Vratislav Tydlitát, Tomás Matas, RobertC erny . Effect of w/c and temperature on the earlystage hydration heat development in Portland limestone cement. *Construction and Building Materials*, 2014, 50(15): 140–147.

Wang J C, Yan P, Yu H. Analysis and prediction of early age cracking of concrete structures. *Journal of Shenyang Jianzhu University: Natural Science Edition*, 2007, 27 (1): 85–87.

Wang T M. *Crack control of buildings*. Shanghai Science and Technology Press, 1987.

Yang J K, Lee Y, Kim J K. Heat transfer coefficient in flow convection of pipe-cooling system in massive concrete[J]. *Journal of Advanced Concrete Technology*, 2011, 9(1): 103–114.

Zhu B F. *Temperature stress and temperature control of mass concrete*. China Electric Power Press, 1999.

Advances in Measurement Technology, Disaster Prevention and Mitigation – Li & Mohd Yusof (Eds)
© 2023 The Author(s), ISBN: 978-1-032-36087-4

The practice of optimal and fast drilling technology in complex formation environment of tight reservoir

Wen Li, Dawei Zhao & Fu Tao
No.5 Drilling Engineering Company BHDC, CNPC, Hejian, China

Ning Li
Downhole Services Company of CNPC Bohai Drilling Engineering Company Limited, Renqiu, China

Zhuolin Lv
Huabei Oil Communication Co., Ltd., Renqiu, China

Xianlong Zhao
No.5 Drilling Engineering Company BHDC, CNPC, Hejian, China

Hanlie Cheng
Faculty of Contemporary Sciences and Technologies, South East European University, Tetovo, Macedonia

Shiela Kitchen
College of Arts and Sciences, University of New England, Armidale, NSW, Australia

Graciela Daniels*
Central Arizona College, Coolidge, AZ, USA

ABSTRACT: Tight sandstone reservoir core microfractures are relatively developed and brittle, resulting in poor drill ability, a long drilling and completion cycle, and easy collapse of the wall of the well. Improving drilling efficiency is important to accelerating oilfield development. In this paper, principal component analysis and field statistical analysis are combined to optimize the used bit in the target area, determine the type of bit with good use effect, and analyze the optimized bit type. The PEM drilling fluid was evaluated by using high temperature and high-pressure dynamic water loss instruments under simulated drilling conditions, and the PEM system was optimized. The experiment shows that the drilling fluid system can effectively improve the borehole stability and reduce the borehole wall collapse in early development. At the same time, underbalanced drilling is used, and the fluid in the negative pressure formation can enter the hole more smoothly during drilling. According to the monitoring of wellhead returned fluid, we can timely find the underground oil and gas layer and make a judgment, which improves the exploration effect significantly and saves the drilling cost.

1   INTRODUCTION

From the perspective of reservoir genetic evolution, the formation of a low permeability reservoir is closely related to sedimentation, diagenesis, and tectonics. The proportion of low permeability reservoirs in different strata of different ages and lithology varies considerably in China, mainly in the Paleozoic, Triassic, Cretaceous, and Neogene. Lithology is mainly sandstone and siltstone. The reserves of low permeability oil and gas fields in China are rich and wide, accounting for about

---

*Corresponding Author: graciela_daniels@stu.centralaz.edu

DOI 10.1201/9781003330172-76

551

30% of the total oil and gas resources in China. Tight reservoirs have relatively simple geological structures, high reserves, poor porosity and permeability, and their oil-bearing range is mainly controlled by lithology and reservoir physical properties.

For efficient development of low permeability oil and gas fields, major foreign countries increase the discharge area of drilling technology, increase the rate of reservoir drilling in technology, improve the efficiency of drilling technology, improve the effect of reservoir protection technology, improve the horizontal well drilling well completion technology, improve the degree of technology, such as a key to improve the recovery of low permeability oil and gas fields, a higher economic efficiency. In foreign countries, complex-structure wells, especially branch wells, are drilled to increase discharge area, increase oil and gas production, and enhance oil and gas recovery. In foreign countries, advanced measurement, and monitoring instruments such as anti-high temperature and high-pressure MWD measurement systems, multi-geological parameter MWD measurement systems and electromagnetic wave MWD measurement systems are used to improve the drilling rate of reservoirs. However, conventional MWD and even electronic single-multi-point measurement and monitoring instruments are widely used in China. Underbalanced drilling technology originated in the 1930s and developed rapidly in the world petroleum industry. It performs well in protecting reservoirs and improving drilling efficiency. According to statistics, underbalanced drilling technology has been applied to more than 4,000 wells in more than 20 countries and regions. Underbalanced drilling plays an important role in solving engineering geological problems, discovering oil and gas layers in time, and improving geological understanding.

Shielding temporary plugging technology has been widely used to protect such reservoirs when the reservoir is drilled. This technique, using the pressure difference between the drilling fluid and reservoir, forces the drilling fluids that intentionally added various types and sizes of solid particles into the pore throat or fracture stenosis in a very short period. This leads to the formation of a shielding of temporary plugging ring near the wellbore and permeability close to zero. Thus, the shielding of temporary plugging ring effectively prevents the drilling fluid and subsequent construction damage to the reservoir continually. Finally, the use of perforation or other methods unblocks the shielding temporary plugging ring to recover the reservoir permeability. Then, the seepage capacity of the reservoir can be restored using perforation and acidification to protect the reservoir. To meet the requirements of drilling efficiency in low permeability reservoirs, the under-balanced drilling and completion technology and new polycrystalline diamond compact bits are adopted to improve the drilling speed, shorten the drilling cycle, and effectively reduce the drilling cost of low permeability reservoirs.

2 BIT OPTIMIZATION

To realize optimal drilling, the most important thing is to correctly understand the rock's mechanical resistance to the drilling of the drilled strata. On this basis, the reasonable selection of the bit is an important way to improve the drilling rate and reduce the drilling cost. Rock drill ability refers to the relative resistance of rock-to-rock breaking by bit or the ability of the rock to resist drilling and rock breaking by bit during drilling. It is a measure of the difficulty of drilling through rock. Therefore, drill ability distribution directly affects the selection of drill bit. If the bit type and formation drill ability are not provided, the rock breaking performance of the bit will be severely affected and the drilling speed will be limited.

The principal component analysis is an effective tool for dimension reduction. It converts a given set of variables into a new set of variables. In the process of solving practical engineering problems, to comprehensively analyze problems, many variables (or factors) are often put forward, because each variable reflects some information about the problem to varying degrees. However, when using the statistical analysis method to study this kind of variable problem, many variables will increase the complexity of the problem research, and people naturally hope that the fewer the number of variables, the more information. From a mathematical point of view, it is required to establish Q new variables ($P \leq Q$) for all the previously proposed variables (set as P). On the one hand, it is required that the Q new variables are unrelated in pairs; on the other hand, it is required that the Q new variables should reflect the information of the problem as far as possible under the original information principle, so that P < Q.

The essence of drilling bit optimization is to compare the advantages and disadvantages of the drilling effect, so the application effect is used as the selection index. Key bit parameters should also be selected for the best overall evaluation possible. This paper selected many indicators, such as footage, etc. In addition, to make the results more comparable, data must be taken from the same formation, the same lithology, and the same bit size.

The formation drill ability level is positively correlated with the depth of the well. With the increase of depth, the drilling ability decreases. At the same time, the trend reflected by the characteristic parameters is consistent with the formation properties that the increase in drill ability indicates that the formation is becoming difficult to drill.

Table 1. Drill bit optimization results.

Formation	Bit size (mm)	Bit type
Quaternary-Neogene system	444.5	MP2G
Paleogene system	311.15	FR1934S
The carboniferous	215.9	SD6443, X516

3 UNDERBALANCED DRILLING

Underbalanced drilling is the core of the drilling wherein fluid column pressure is lower than the formation pore pressure because there is no one way to predict the formation pore pressure. This requires that the density of the underbalanced drilling fluid system in a certain range is adjustable, and can realize the lowest density. In case the actual formation pressure coefficient is lower than the prediction of formation pressure coefficient, the density of drilling fluid can be reduced to achieve an underbalanced status. Underbalanced drilling fluid technology is one of the key technologies to realize underbalanced drilling, which is the main content of underbalanced drilling. Underbalanced drilling using natural conditions does not need to add additives to reduce the density of the drilling fluid and does not need the necessary auxiliary equipment and peripheral auxiliary equipment, so the drilling cost is low. There are two ways to achieve the underbalanced condition by artificial means: the first is to directly use low-density air, atomization, foam, and other drilling fluids; the other is to inject one or more non-condensable gases into drilling fluid base fluid to reduce drilling fluid density and achieve the purpose of underbalanced drilling. This method requires the addition of abatement to the drilling fluid and the use of abatement preparation and filling equipment, as well as peripheral equipment for the treatment of the returned fluid, which is costly to drill.

Underbalanced pressure wells and the difference between conventional well killing, and regular kill weight (circulation) after heavy drilling fluid may continue to be used as a comprehensive drilling fluid, drilling fluid, and underbalanced pressure well. They cannot use comprehensive drilling of well-killing fluid, but use light to continue drilling. Therefore, heavy drilling fluid and underbalanced drilling fluid replacement fast circulation system must be developed. In the process of underbalanced drilling technology research, according to exploratory well in the east of well depth range of Changyuan, we calculate the range of heavy drilling fluid reserves, to determine the scope of the heavy volume of drilling fluid tank; at the same time, considering the pressure of well liquid and underbalanced drilling fluid replacement, we study to determine the connection between the reserve tank of well-killing fluid and the connection between the circulating tank, etc. Although we have the theory of underbalanced kill, but never in practice. Under the condition that the annulus is filled with gas, the positive cycle of the kill mud pump must be used. Therefore, the study determined that two Halliburton pumps were used for reverse circulation kill. And another process conversion, such as drilling up to change bits and so on, should use positive circulation kill.

The problems of borehole enlargement, serious borehole collapse, and deep mud invasion occurred in the drilling engineering of exploratory evaluation wells in the study area, indicating that drilling fluid should be further strengthened in the protection of reservoir and borehole wall stability, and the influence of adverse factors should be minimized. Therefore, the focus of drilling fluid optimization research is to prevent clay swelling and migration, strictly control fluid loss, and maintain wellbore stability and good reservoir protection and anti-pollution ability. According to the development plan and the technical requirements of horizontal well and directional well development, the drilling fluid and the completion fluid of exploratory well are optimized to make the drilling fluid and completion fluid have excellent performance and meet the requirements of reservoir protection and safe and fast drilling.

In the laboratory study, the drilling fluid and completion fluid system of exploratory evaluation wells were evaluated comprehensively, and then the corresponding drilling fluid and completion fluid were optimized. The drilling fluid has good rheological properties, water loss, wall building, anti-collapse inhibition, good reservoir protection ability, good sand carrying, and lubrication ability. Completion fluid has good anti-swelling and anti-corrosion abilities. Drilling and completion fluids should be compatible with formation fluids. In the process of drilling, the drilling fluid will inevitably be mixed with inorganic salts, and drilling cuttings, which will pollute the drilling fluid, and the performance of the drilling fluid will change. If the change is large, it will affect the smooth operation of drilling. Therefore, the anti-pollution ability of the drilling fluid system is evaluated indoors.

According to the experimental data in Table 2, the apparent viscosity, shear force, and filtration loss of the PEM drilling fluid system did not change significantly with the increase in NaCl dosage. The experimental results show that the drilling fluid system has good anti-NaCl pollution performance.

Table 2. Evaluation of anti-NaCl ability of drilling fluid system.

NaCl Add the amount (%)	AV (MPa·s)	PV	YP	FL(API) (cm^3)
0	18.5	17.9	4.8	2.8
4	22.3	18.1	4.9	2.9
8	22.9	18.1	5.0	3.0
12	26.8	19.5	6.8	3.2

The rolling recovery method was used to evaluate the dispersion inhibition performance of the drilling fluid system. The experimental method is to take 50 g of a 6–10 mesh rock sample, add it into the system, roll it at 140°C for 12 hours, and measure the rolling recovery rate of 40 mesh. The experimental results are shown in Table 3. According to the experimental data on dispersion inhibition, the water recovery rate of the rock sample is low, and the rolling recovery rate of the PEM system is as high as 93.8%. Therefore, the PEM system has excellent dispersion inhibition performance.

Table 3. Experimental results of dispersion inhibition of drilling fluid system.

System	Well depth (m)	Sample (g)	Rolling recovery sample (g)	Recovery (%)
Distilled water	2865.5	50.0	22.6	45.2
PEM		50.0	46.9	93.8

The experimental results show that the rolling recovery rate of the PEM drilling fluid system is as high as 93.8%, which performs well in inhibiting dispersion. The system has excellent salt and calcium resistance and good cutting resistance. The system has good thermal stability and is suitable for high-temperature oil wells with high formation temperatures. The mud cake of the system lacks quality and is a thick cake before high temperature, indicating poor film formation before high temperature. The water loss of the system before the high temperature is large, resulting in high transient water loss, which is detrimental to shaft wall stability. The solid polyol PF-GJC was used to give full play to the inhibition ability of polyol in drilling fluid, improve the quality of mud cake, improve the anti-sloughing ability to drill fluid, and increase lubricity.

Problems such as borehole enlargement, borehole wall collapse, and drilling fluid invasion depth occur in exploration well-drilling engineering in the study area. From the perspective of drilling fluid, these problems indicate that drilling fluid should be further strengthened in reservoir protection, borehole wall stability, and other aspects to minimize the impact of adverse factors. Therefore, the focus of drilling fluid optimization research is to prevent clay swelling and migration, strictly control the fluid filtration loss, maintain wellbore stability, and improve plugging ability, so that the drilling fluid has good reservoir protection ability.

Low permeability reservoir damage is serious, of which liquid phase and solid phase particle invasion harm should be given priority. To reduce the damage of clay particles, people developed the drilling fluid system that has no clay, but so far, no clay mud is difficult to achieve. It generally uses a polymer tackifier at the same time. This kind of tackifier adsorption in the reservoir pore throat size for sealing the pores and throats is difficult to remove late during the completion, which can cause more serious reservoir damage. Saltwater is commonly used to clean the well during completion. This method is inefficient, resulting in the poor effectiveness of clay-free drilling fluids in low-permeability reservoirs.

In clayless drilling fluid, the addition of a waterproof locking agent can reduce the damage caused by the water phase trap. The addition can reduce the hydration expansion of clay minerals in the reservoir, and reduce the damage caused by water sensitivity. There is no insoluble solid phase, and the solid phase pollution is eliminated. According to the requirements of clay-free drilling fluid, the following treatment agents are selected in this paper: waterproof locking agent surfactant B, cuttings remover KPAM, anti-high temperature viscosities TV-2, a conventional viscosifier XC, density reducer hollow glass microbeads, and inhibitor HCOOK. Then, the orthogonal experiment was carried out. Through orthogonal experimental analysis, the appropriate drilling fluid formulation can be obtained, as shown in Table 4.

Table 4. Basic formula optimization results of clay-free drilling fluid.

The serial number	Component name/code	Add the amount (kg/m^3)
1	TV-2	3 ~ 45
2	NaOH	2 ~ 3
3	XC	3 ~ 6
4	KC00H	2 ~ 35
5	KPAM	2 ~ 5
6	Waterproof lock agent	3 ~ 60
7	Hollow glass sphere	5 ~ 100

4 CONCLUSIONS

(1) The collapse of a large section of the stratum is caused by a combination of various reasons. On the one hand, the main factor is that the density of drilling fluid is not high enough to support the shaft wall under the condition of a high-stress field to ensure that the shaft wall instability collapse occurs. On the other hand, there are microcracks and pores with relatively good connectivity in the stratified rocks. This provides a path for drilling fluid to enter the borehole formation. When fluid enters the formation, the interaction between the drilling fluid and the formation exacerbates borehole instability. The PEM drilling fluid system has been evaluated

in the laboratory and the overall design is reasonable, thus improving the performance of losing water and building walls, enhancing the stability of shaft walls, and reducing the shaft wall collapse in early development.

(2) Because underbalanced drilling can bring high comprehensive benefits, underbalanced drilling technology has been developed and applied at home and abroad. However, there are still many problems in the development and application of underbalanced drilling technology, which restrict its wide application. With the further deepening of people's understanding of underbalanced drilling technology and the deepening of exploration and development of various oil fields, the number of underbalanced wells will increase, and the corresponding problems will become increasingly important.

REFERENCES

Feng, Z. Hongming, T. Yingfeng, M. Gao, L. Xijin, X. (2009). Damage evaluation for water-based underbalanced drilling in low-permeability and tight sandstone gas reservoirs. *Petroleum Exploration and Development*, 36(1), 113–119.

Liu, S. Chen, G. Lou, Y., Zhu, L. Ge, D. (2020). A novel productivity evaluation approach based on the morphological analysis and fuzzy mathematics: insights from the tight sandstone gas reservoir in the Ordos Basin, China. *Journal of Petroleum Exploration and Production Technology*, 10(4), 1263–1275.

Wang, X. Qiao, X. Mi, N. Wang, R. (2019). Technologies for the benefit development of low-permeability tight sandstone gas reservoirs in the Yan'an Gas Field, Ordos Basin. *Natural Gas Industry B*, 6(3), 272–281.

Xuelong, W. A. N. G. Xuanpeng, H. E. Xianfeng, L. I. U. Tianhui, C. H. E. N. G. Ruiliang, L. I. Qiang, F. U. (2020). Key drilling technologies for complex ultra-deep wells in the Tarim Keshen 9 Gas Field. *Oil drilling technology*, 48(1), 15–20.

Zhang, D. Kang, Y. You, L. Li, J. (2019). Investigation of formation damage induced during drill-in process of ultradeep fractured tight sandstone gas reservoirs. *Journal of Energy Resources Technology*, 141(7).

Advances in Measurement Technology, Disaster Prevention and Mitigation – Li & Mohd Yusof (Eds)
© 2023 The Author(s), ISBN: 978-1-032-36087-4

Research on Jiangxiyan irrigation system

Bo Zhou*, Xiaoming Jiang, Yunpeng Li & Jun Deng
Institute of Water Conservancy History of China Institute of Water Resources and Hydropower Research, Beijing, China
Research Center on Flood Control, Drought Control and Disaster Mitigation of the Ministry of Water Resources, Beijing, China
Key Scientific Research Base of Water Heritage Protection and Research, State Administration of Cultural Heritage, Beijing, China

ABSTRACT: Located in Longyou County, Zhejiang Province, Jiangxiyan Irrigation System is an important ancient mountain stream diversion irrigation project on the Lingshangang River. Built between 1330 and 1333, the project has been in use for over 680 years. In 2018, Jiangxiyan Irrigation System was included in the World List of ICID Heritage Irrigation Structures. Yet academic research on the project is still very scarce. This paper attempts to analyze Jiangxiyan Irrigation System in terms of historical development, engineering system, management system and value evaluation through field research, literature survey and qualitative analysis; and concludes that the Jiangxiyan Irrigation System is a model of ancient ecological water conservancy projects, and its architectural form, engineering techniques and management system were innovative in the region at that time, which can provide an ideological reference for the construction of contemporary water conservancy projects, and a research basis for better conservation and utilization in the future.

1 INTRODUCTION

In 2018, Jiangxiyan Irrigation System in Longyou County, Zhejiang Province, was included in The World List of ICID Heritage Irrigation Structures. Built between 1330 and 1333, the project has been in use for over 680 years. The project maintains the shape and structure when it was first built 680 years ago, still irrigates 35,000 mu of farmland, and is a model of ancient mountain stream diversion irrigation projects. Compared with the famous Dujiangyan Irrigation System and Lingqu Canal, Jiangxiyan Irrigation System is not as well known, and academic research on the project is very sparse. In order to present this small but exquisite water conservancy heritage to the world and disseminate to a wider audience the technical core and unique value of the project, this paper attempts to study and analyze Jiangxiyan Irrigation System in terms of historical development, engineering system, management system, and value evaluation to pass on the technical essence and value core of this ancient water conservancy project and provide a reference for contemporary water conservancy construction and water control.

2 JIANGXIYAN IRRIGATION SYSTEM

2.1 *Historic evolution*

During AD 1330–1333, the then Longyou Daluhuachi Mongolian Tsar Koma attached importance to local agricultural development and water conservancy construction and built the Jiang-Xi Weir (Compilation Group of Hydraulic Engineering Annals of Longyou County Hydroelectric Board 1990).

---

*Corresponding Author: 369129958@qq.com

DOI 10.1201/9781003330172-77

In the Ming Dynasty, the Jiang-Xi Weir Project had a relatively complete system consisting of a barrage (the Jiang Weir), an overflow weir (the Xi Weir), and an inlet. In the 4th year of the Jiajing reign (1525), the Jiang-Xi Weir was damaged by a flood, and Zheng Dao, a local official, built a 500-meter dike to secure the site. In the 22nd year of the Jiajing reign (1543), the dike was breached due to flooding. In the 24th year of the Jiajing reign (1545), Qian Shi, the county magistrate, raised money to rebuild the Jiang Weir and the Xi Weir. He built a 500-meter stone dike, and dredged the tributaries of the weir canal, restoring the irrigation function of the Jiang-Xi Weir. During the Longqing and Wanli reigns (1571 and 1574), Tu Jie, the county magistrate, rebuilt the Jiang-Xi Weir twice. At this time, a stable system of annual repairs was formed and the weir was closed on the first day of June every year. In the 13th year of the Chongzhen reign (1640), Huang Dapeng, the county magistrate, built a new weir tunnel to divert water (Figure 1).

Figure 1.   Ruins of weir cave.

In the Qing Dynasty, the Jiang-Xi Weir fulfilled a variety of functions such as irrigation, shipping and supplying water for urban production and living, and the management system was improved. In the 19th year of the Kangxi reign (1641), the Jiang-Xi Weir was damaged and blocked by a flood, and was repaired by Lu Can, the county magistrate. In the 25th year of the Kangxi reign (1647), the Jiang-Xi Weir was destroyed by a flood. In the 1st year of the Qianlong reign (1736), Xu Qiyan, the county magistrate, rebuilt the Jiang Weir and the Xi Weir, diverting water to the city, flowing around the schoolhouse, converging in pools, and flowing to the North Gate Water Gate via the Bailian Bridge within the county, then interflowing with the weir water outside the city and injecting into the Huxi Stream. In the 12th year of the Guangxu reign (1886), Gao Ying, the county magistrate, organized the squires to raise funds for the construction of the Jiang-Xi Weir, set up a management body–the Weir Works Bureau, developed standardized systems for weir repair, financing, accountability, etc., and issued the "the Income and Expenditure Register for Rebuilding the Jiang-Xi Weir Works in Longyou".

From 1927 to 1948, the Management Committee for the Jiang-Xi Weir was established; the Management Regulations for the Jiang-Xi Weir were formulated and published, and the Jiang-Xi Weir Farmland Irrigation and Utilization Cooperative was established. During this period, there were still overhead expenses incurred by public properties of the Jiang-Xi Weir, including the Jiang-Xi Weir Public House, the Weir God Gathering and the Weir God Temple, as well as the Management Committee for the Jiang-Xi Weir and the Cooperative.

After 1950, the canal head hub of the Jiang-Xi Weir was continuously repaired, the irrigation canal system was comprehensively transformed, and the supporting facilities were further improved (Editorial Board of Longyou County Annals 1991). On January 7, 2011, the Jiang-Xi Weir was declared a "Cultural Relics Protection Unit of Zhejiang Province" by the People's Government of Zhejiang Province. In 2018, the Jiang-Xi Weir was listed in the "World Heritage Irrigation Structures" for its good preservation of the form and technique at the time of initial construction.

2.2 *Irrigation project system*

The irrigation project system of the Jiang-Xi Weir consists of the canal head diversion hub, irrigation, and drainage canal system, as well as control works. These features are described in Figure 2 and Table 1.

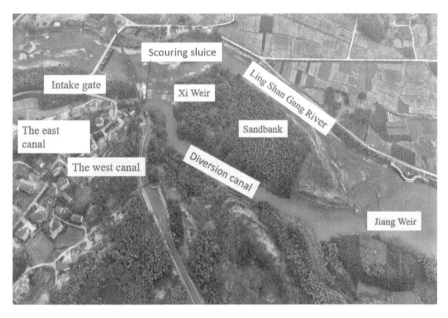

Figure 2.   Arrangement Plan of Jiangxiyan irrigation system pivot.

The canal head hub consists of the Jiang Weir, the Xi Weir, 2 inlet sluices and 1 scouring sluice (Figure 2). Located upstream, the Jiang Weir runs east-west across the Lingshan Port, with a length of 100 m, a width of 32 m at the bottom, and an elevation of 63.2 m at the top. The water level falls by about 2.3 m, and the river is intercepted by the Jiang Weir and flows to the western diversion channel. The Xi Weir is located about 360m downstream of the Jiang Weir and is arc-shaped, tangential to the direction of the water flow, with a length of 50 m, a width of 30 m at the bottom, and an elevation of 63.1 m at the top. The water level falls by about 3.5 m, and the water is blocked by the Xi Weir and diverted into the main canal on the left bank. The sandbar is connected to the Jiang Weir upstream and the Xi Weir downstream and is 450 m long from east to west, and 150 m wide from north to south.

The central sandbar contributes to reinforcing the weir. To the south of the central sandbar is the main channel of Lingshan Port, while to the north is the diversion channel, which is a tributary of Lingshan Port. The sandbar is linked to the Jiang Weir and the Xi Weir in tandem to form a 570-meter-long angular ruler-shaped barrage. Located on the west side of the Jiang Weir, the diversion channel is 25 m wide and 300 m long and is a curved bow-shaped channel naturally formed by the intersection of Snake Hill and the sandbar.

There are 2 inlet sluices. One is located on the upstream right bank of the Jiang Weir, diverting water for irrigation to the villages of Guancun and Linjia on the right bank; the other is located at the left dam head of the Xi Weir, diverting water for irrigation to the villages of Shantouwai and Houtianpu on the left bank through the General Main Canal and the Eastern Main Canal.

The scouring sluice was constructed in 1986 at the eastern end of the Xi Weir and was installed with 2 cast-iron openers, which are 3.3 m high, 7.0 m long and 2.0 m wide, with 2.3 m wide holes to prevent silting at the inlets.

The ancient irrigation canal system of the Jiang-Xi Weir consists of 2 main canals, i.e., Eastern and Western Main Canal. It was historically recorded that the maximum irrigation area reached more than 50,000 mu during the Kangxi reign in the Qing Dynasty. After 1973, the canal system of the Jiang-Xi Weir was integrated and optimized. Currently, the irrigation canal system consists of the General Main Canal and 4 main canals, i.e., Eastern, Central, Western and Guancun Main Canal, with a total length of 18.8 km. There are also 15 branch canals under the main canals, with a total length of 30.87 km, irrigating 35,000 mu of land in 21 local administrative villages.

There are 24 sluice gates of various sizes distributed along the Jiang-Xi Weir canals, fulfilling such functions as water diversion, water regulation and flood control. As the irrigation area is located in a hilly area, the water level needs to be raised in some locations through sub-weirs to regulate the water diversion for irrigation. It is documented that the Jiang-Xi Weir had 72 sub-weirs in the 17th century. Hydraulic tools such as water-powered trip-hammers were also equipped along the canals to facilitate the processing and lifting of water for irrigation by farmers.

2.3 *Cultural heritage*

To commemorate the main builders, the local people built a temple of about 300 square meters near the site of Jiangxiyan. The weir temple was a place for holding ceremonies, such as opening and closing the weir, praying for rain, etc. and it is also a place for ancient officials to talk with villagers. It is a link between officials and the people. Now, a plaque remained, which was donated by the Zhejiang provincial government to the irrigation area in 1927. It was inscribed with four Chinese characters–"Benefiting Our Farmers" and once hung from the central scroll of the temple. An ancient camphor tree beside the temple has been more than 360 years old and still flourishing. The local people called it the weir tree, which offered incense to the people throughout the year to commemorate the people who built the weir and pray for good weather.

Jiangxiyan still has 12 stele inscriptions, recording the history of construction, management system, water conservancy disputes, and so on. Among them, most of the inscriptions were from the Qing Dynasty. There are many inscriptions, art articles, constitutions, receipts, and protocol documents that were retained in the period of Gaoying repairing the weir in the Qing Dynasty, which are important historical materials for the study of Jiangxiyan.

Table 1. Inventory list of Jiangxiyan irrigation system.

Category	Type	Description
Irrigation System	Headwork	Jiang Weir Xi Weir Two headwork intake gates One headwork scouring sluice
	Irrigation Canal Network	General main canal and four sub-main canals: east main canal, middle main canal, west main canal, and Guancun main canal, with an overall length of 18.8 km. 15 branch canals with an overall length of 30.87 km.
	Control Work	One water diversion gate 24 control gates 19 sub-weirs

(continued)

Table 1. Continued.

Category	Type	Description
Relevant Heritages	Water-Power Utilization Facilities	Four water-powered trip-hammers and scoop waterwheels
	Architecture for Sacrifice	Ruins of Weir God Temple
	Cultural Relics	Weir God Tree
		Plaque inscribed with "Benefiting Our Farmers"
		Report of contributions to weir construction by the Weir Administration of the Qing Dynasty
		File related to the management of Jiangxiyan Irrigation System of the Republic of China Period, etc.
	Document Literature	Annals of Longyou County, Annals of Quzhou Prefecture, etc.
	Tablets	12 relevant tablets with inscriptions

3 VALUE ASSESSMENT OF JIANGXIYAN

The Jiang-Xi Weir is a model of ancient mountain stream-based river diversion irrigation projects, which has been in use for more than 680 years. It has boosted local agricultural development, as well as the economic development of the whole region, with outstanding historical, cultural, technological, and landscape value.

3.1 *Historical value*

Longyou in Zhejiang Province has a long history. Hehua Mountain, Qingdui, and other cultural sites in the area have been in existence for as long as 9,400 years. In the 14th century AD, the population in South China increased dramatically, which made the development of agriculture imperative. A large number of important regional irrigation projects were built at that time. Jiangxiyan was constructed in AD 1330–1333.

Mongolian Chaerkema, then Dalu Huachi of Longyou, made good use of governance of Han nationality, attached great importance to agriculture and started the construction of water conservancy projects. Jiangxiyan was one of the weirs presided over by him. During the presence of Jiangxiyan from the three generations of Yuan, Ming, and Qing dynasties to date, local officials of each dynasty have been placing importance on the maintenance and irrigation benefits of Jiangxiyan. As a result, a wealth of experience has been accumulated in the maintenance and management of Jiangxiyan over hundreds of years, still with a high reference value in terms of management mechanisms till today.

The completion of Jiangxiyan enables the 2,333 ha of grain fields to be irrigated by gravity, and secures harvests regardless of floods and droughts, making the irrigation area the most famous granary in Longyou County and even Zhejiang Province. For more than 600 years, a profusion of food has provided the most important material guarantee for the economic development of Longyou County. As the Longyou Region is well-known for developing agriculture with weirs and boosting commerce with agriculture, Jiangxiyan contributes to the rise of Longyou Business Group in the following aspects. First, the grain produced in the irrigation area, serves as an important trading material for Longyou Business Group. Second, stable, and high-yield grain has provided a rich material basis and steady social order for the rise of the Longyou Business Group over the past hundreds of years. Third, the Weir provides convenient shipping conditions for the transport of bamboo, hand-made paper, and firewood, which are produced in large amounts in the mountains of Nanxiang, Longyou.

In the long-term development, Jiangxiyan witnesses both the evolution of irrigation agriculture in Longyou and the profound history of social, economic, and cultural development of the region, with important historical value.

3.2 Scientific and technological value

The Lingshan River is a typical mountain stream-based river. Located at the throat of the Lingshan River where it transitions from the valley to the plain, the Jiang-Xi Weir can make full use of the waterfall to ensure gravity irrigation of the maximum farmland area in the downstream plain.

The entire canal head hub of the Jiang-Xi Weir is ingeniously laid out, with a sandbar in the river channel as the link between the upper Jiang Weir and the lower Xi Weir, forming a 600-meter-long barrage. The main function of the Jiang Weir is to intercept and bank up the water, diverting the stream water to flow northwards, while the Xi Weir intercepts the river water and diverts it into the canals. The sandbar replaces the conjoined wall between the two weirs, and the water is diverted by using the sandbar to divide the two weirs, making it much less difficult to complete the project than building a large weir over the river. This bold idea of using a sandbar on the river channel to form an integral part of a weir is rare in the history of water control and is of great research value in the scientific construction of weirs.

The Jiang-Xi Weir was constructed in a scientific and reasonable form, facilitating water diversion, underwater operations as well as drainage construction, while contributing to flooding diversion, and reducing the pressure of excessive river flow scouring. The Jiang Weir is located on a rocky outcrop, contributing to the safety of the Weir and saving investment; the weir site has a wide river channel, which is conducive to flooding. The diversion canal is adapted to the local context, making full use of the natural channel formed between the Snake Hill rock formations and the sandbar. The Jiang Weir, intersecting the main river channel from south to north, has a straight-line shape, with the weir axis sloping upstream at an angle of 123° to the main river channel. It also has a gentle slope downstream with a weir surface-slope ratio of 1:8, providing good energy dissipation effects. The Xi Weir, built in an east-west direction between the sandbar and Snake Hilltop, is circular, which not only widens the inlet, but also extends the overflow length, reducing the unit width flow of flood water across the weir (Ye 2014).

The weir belly of the Jiang-Xi Weir is filled with river pebbles of varying sizes. The weir penetrates deep into the intervening bed, and an apron was laid dry with large river pebbles at the foot of the downstream slope. The Xi Weir adopted the foundation with a pebble-filled pine frame, which has enhanced the foundation integrity and erosion resistance, thus effectively protecting the weir. This form of construction is still in use today.

The weir was built with local materials, mainly river pebbles, clay, pine wood and sand. The masonries of the upstream and downstream slopes were made of long river pebbles, with the large end facing downwards and buried vertically on the weir slopes. As a result, the masonries are stable and compact, and the weir is solid, providing friction and force dissipation effects; the pore space between the masonry is triangular, ensuring that the masonry is subject to a balanced transfer of water impact. The weir's foundation with pebble-filled pine frame, commonly known locally as the "oxbow barn" structure, was linked by mortise and tenon joints, and the pine material used is so bulky that it has not yet been corroded. The material used is tightly bound to the pebbles and has the same shape as concrete mortar, making it structurally stable. The Jiang-Xi Weir has survived many floods over the centuries and has been restored several times, but the foundation and skeleton of the weir have not been significantly damaged, showing the extraordinary skills of the ancient weir builders.

3.3 Cultural value

A rice cultivation culture has long existed in Longyou, Zhejiang Province. The Qingdui Site, with a history of more than 9,400 years, is the origin of local rice cultivation culture and lays the foundation for the agricultural civilization of the Qiantang River Basin.

Over the 680 years since its founding in the Yuan Dynasty, Jiangxiyan has been managed in the fashion of governmental supervision and private operation, which has effectively promoted the sustainable running of the weir and the canal. In this long-term management system, a rich and unique barrage culture was born. The Yanshen Tree, which grows next to Jiangxiyan's general main canal through the village beyond the mountain, is a century-old camphor tree in front of Yanshen Temple. It is 15 meters high, with the canopy covering an area of more than 100 square meters. Despite being over 360 years old, it is still lush. Therefore, local people call it the God of Camphor and always enshrine and worship it all year round in memory of the weir constructor and to pray for a good time, well-being, and peace. As was recorded in the Quzhou Annals, Yanshen Temple once held a private school whose operating fund was borne by Jinshan Temple Fair. For this reason, the school was called "Jinshan Elementary School." Besides, Yanshen Temple was also an occasion where Jiangxi societies and associations gathered for meetings and discussions of major issues. At the same time, many large ancestors, worship, and performance activities were also put on in Yanshen Temple. In the old days, the Yanshen Tree was an important presence; communications used to be held under the tree between the official and the ordinary and among villages.

Meanwhile, records were made each time weir controllers conducted maintenance and governance of the Weir in every dynasty, including history, systems, water controversies, customs, and other aspects of the Weir. These historical documents have been passed on to date, which is of vital significance to research into the regional history and the history of water conservancy development and manifests the development of local culture in Longyou from another perspective.

3.4 *Landscape value*

The idea of weir construction embodies the Taoist philosophy of conforming to nature. Located in the sub-range of Xianxia Mountain, Jiangxiyan Weir displays the picturesque scenery of the South Yangtze River, with water and mountains blended, providing extraordinary landscape value.

The southern and northern Jiangxiyan Weirs connect Gui Mountain and She Mountain, respectively. The weir, which is more than 20 meters high and stands on the bank of the Lingshan River, turns river water from the north to the northwest due to water strikes on the foot of the mountain, protecting dozens of mu of farmland behind Gui Mountain. The mountain is tree-lined, and clean trickles of water flow beneath it. The mountain, which lies to the north of the Weir, is more than 30 meters high and strikes from north to south. Shaping like a snake, its head faces south, while its meandering tail extends northward and connects the mountain with Yingpan Mountain. The intake sluice of Jiangxiyan Weir is built just under the head of the snake. She Mountain and Mountain Gui Mountain face each other across the mountain; river water that flows over the Weir strikes the masonry, forming tiny waves like fish scales with a buzzing sound, and occasionally, groupers swim up against the water, all of which demonstrates the harmony between water conservancy projects and nature.

Jiangxiyan Central Island, which is also called Green Sandbank Island, is formed by sand and gravel that slow down and deposit due to the washing of river water; it is surrounded by water, covering an area of 80 mu. The south part of Central Island is the main river channel of the Lingshangang River, with Jiangyan connecting Central Island and the south bank of the Lingshangang River. The plants on the Central Island are all towering big Moso bamboo and evergreen broad-leaved trees; the river surface over the Weir turns into a water storage pool of dozens of mus, forming a micro-climate green island that blends water with mountain, and is an ideal habitat for wildlife. On the island there are rabbits, squirrels, and other animals and birds like egrets and owls, rife with birds' twitter and the fragrance of flowers throughout the year, making visitors reluctant to leave (Toponymy Annals of Longyou County Annals of Zhejiang Province 1990).

Mountains and rivers mix in Jiangxiyan with the arrival of each harvest season; in the irrigation area, melons and fruits are ripe; rice pervades the field with an abundant fragrance; farmers are busy around; rivers are teeming with waves, with fish jumping within. What comes into your sight is a profusion of beautiful harvests.

4 CONCLUSIONS

Based on the results and discussions presented above, the conclusions are obtained as below:

(1) Jiangxiyan Irrigation System is an example of an ancient ecological water conservation project. The Lingshangang River Irrigation System makes full use of the natural conditions of the Lingshangang River flowing from the mountainous area to the plains. Based on the sandbank in the riverway, the weirs were built rationally and scientifically, featuring dynamic integration of the upper weir and the lower weir. It has multiple functions: water diversion irrigation, preventing or controlling floods, and regulating water yield. Moreover, it takes full advantage of the water resources of the natural rivers to irrigate the farmland and improve people's survivability. In this sense, it is a model for respecting nature and attaching importance to sustainable development.

(2) The architectural form of the Jiangxiyan Irrigation System was innovative in its construction period. The Jiangxiyan Irrigation System makes use of the sandbank in the riverway and features an upper weir and a lower weir. The upper weir cuts across the main channel of the Lingshangang River and backs up water into the canal, while the lower weir, as a lateral overflow weir, controls water flow, fulfilling a wide range of functions such as water diversion, flood drainage, and sand removal with minimal engineering facilities. The Weir's construction techniques were advanced and innovative in their construction period.

(3) The Jiangxiyan Irrigation System is an outstanding example of sustainable operation and management. Though over 680 years have passed since the completion of the Jiangxiyan Irrigation System, it still plays its role, which owes a great deal to the effective management and operation system. Since its construction in the 14th century, the Jiangxiyan Irrigation System has been supervised by the local government and managed by the public. Traditionally, county-level officials granted the management power and assigned specific tasks to prestigious local gentry, who would later distribute the work of maintenance to farmers benefiting from the irrigation system. The annual repair system was established no later than the end of the 16th century, and the post of weir official was also established. The weir administration was established in the Qing Dynasty (1644–1911). Under the supervision of the prefectural and county governments, the local gentry were specifically responsible for repairing weirs and canals, managing relevant funds, and formulating relevant rules and regulations. This approach of integrating government supervision and private management is still in use today, which guarantees the sustainable operation of the Jiangxiyan Irrigation System.

The science and technology, architectural concepts, and management systems of the Jiangxiyan Irrigation System are representative and exemplary in history. Understanding these features and values will provide an ideological reference for the construction of contemporary water conservancy projects, provide a research basis, and pave the way for better conservation and utilization in the future.

ACKNOWLEDGMENT

This work was supported by the Key Scientific Research Base Project of the State Administration of Cultural Heritage (Grant No. 2020ZCK207), and IWHR Scientific Research Projects (Grant Nos. JZ1208B022021, JZ0199B212019, and JZ1003A112020).

REFERENCES

Compilation Group of Hydraulic Engineering Annals of Longyou County Hydroelectric Board. (1990) *Water Conservancy Annals of Longyou County*. Unity Press, Beijing.

Editorial Board of Longyou County Annals. (1991) *Annals of Longyou County*. Zhonghua Book Company, Beijing.

Toponymy Annals of Longyou County Annals of Zhejiang Province. (1990) *Toponymy Annals of Longyou County*. Zhonghua Book Company. Beijing.

Ye, Z. K. (2014) *Tracing the Jiangxiyan Irrigation System*. Zhonghua Book Company, Beijing.

Advances in Measurement Technology, Disaster Prevention and
Mitigation – Li & Mohd Yusof (Eds)
© 2023 The Author(s), ISBN: 978-1-032-36087-4

Study on internal relative humidity and hydration characteristics of early-age concrete

Xiaofei Zhang*, Wenwei Zhang & Peipei Wei
School of Water Resources and Hydro-electric Engineering, Xi'an University of Technology, Xi'an, Shaanxi, China

ABSTRACT: To study the variation and distribution characteristics of relative humidity of early-age concrete and establish the relationship between temperature and relative humidity of concrete, an internal humidity monitoring test and adiabatic temperature rise test was carried out for three groups of concrete with different strengths. According to the test results, in the time domain, the concrete humidity field is divided into the vapor saturated stage and the humidity decline period. In the spatial domain, the humidity is distributed in a gradient along the height direction. The initial moisture distribution with an increasing trend from bottom to top makes the crash time of concrete in a sealed state decrease from top to bottom along the depth direction, while external drying will advance the crash time. The self-drying of concrete decreases with the increase of the water-binder ratio. The influence of external drying on the internal relative humidity of concrete decreases with the increase of the distance from the dry surface and does not affect the range beyond 8 cm from the dry surface. When the relative humidity of C30, C50 and C60 concrete is lower than 97%, 93% and 88%, the rate of hydration is close to 0.

1   INTRODUCTION

With the further development of water conservancy project construction in China, it can be predicted that concrete will still be an indispensable and important engineering material in water conservancy project construction in the next few decades (Schutter 2020). Due to the low tensile strengths of concrete materials, the occurrence of cracks in concrete buildings is often very common in the process of construction and use, and there is even a saying that there is no concrete without cracks and no dam without cracks (Jiang & Yuan 2009). According to the engineering experience, at the initial stage of concrete pouring, due to the hydration reaction and external drying, a humidity gradient will be generated inside the concrete, resulting in moisture deformation (Jiang & Yuan 2009). In this period, the cracking phenomenon is more common due to the low strength of concrete.

Many scholars have studied the development process and distribution law of the humidity field of early-age concrete to evaluate and predict the influence of drying shrinkage stress on concrete cracks and to provide a basis for the selection of anti-cracking measures in concrete construction. Huang Yaoying (2020) measured the changing course of relative humidity of concrete structures exposed to air and found that the humidity drop caused by self-drying is very small compared with external drying. Wang Jiahe (2018) monitored the relative humidity at several locations inside ordinary concrete and internally cured concrete and found that the vapor saturation stage of internally cured concrete is long and the humidity decreases relatively slowly. Zhang Jun (2012) systematically analyzed the effects of admixtures, curing conditions, and other factors on concrete humidity. The current experimental research on the concrete relative humidity field has some problems (Wang et al. 2020; Zhang et al. 2018; 2010), such as the number of observation points is small, the

---

*Corresponding Author: Zhangxiaofei_lgd@163.com

                        DOI 10.1201/9781003330172-78

vapor saturated stage and humidity decline period are not separated, and the relationship between relative humidity and temperature is not clear. As a result, the current simulation of the concrete humidity field mostly ignores the coupling relationship between relative humidity and temperature. It is difficult to obtain accurate results, which provide a basis for the selection of anti-cracking measures.

Therefore, taking three groups of concrete with different strengths as the research object, this paper studies the variation and distribution characteristics of concrete humidity fields in the sealed state and dry state from the aspects of the space domain, time domain, and initial moisture distribution, and establishes the relationship between concrete temperature and relative humidity from the perspective of hydration degree.

2 EXPERIMENTAL DESIGN

2.1 *Materials and mix proportion*

To study the influence of different water-binder ratios on the relative humidity of concrete, combined with the concrete mix ratio of different parts of the foundation constraint area and non-foundation constraint area of a typical high-concrete dam project, three groups of concrete with a water-binder ratio of 0.33 (C60), 0.43 (C50), and 0.57 (C30) are designed, representing ordinary concrete, middle-strength concrete, and high-strength concrete, respectively. And the content of fly ash is uniformly controlled at 15%. See Table 1 for concrete mix proportions.

Table 1. Concrete mix proportion.

Concrete	Water-binder ratio	Cement (kg/m^3)	Water (kg/m^3)	Fly ash (kg/m^3)	Sand (kg/m^3)	Gravel (kg/m^3)
C60	0.33	387	150	68	595	1155
C50	0.43	346	175	61	685	1090
C30	0.57	277	186	49	740	1140

2.2 *Test scheme*

2.2.1 *Internal humidity monitoring test*

Each group of water-binder ratios is formed into 8 pieces with the size of 150 mm × 150 mm × 150 mm concrete test blocks, including 4 sealed and 4 dry specimens. They are used to monitor the relative humidity of concrete at different depths. The monitoring depths are 12 cm, 8 cm, 5 cm, and 2 cm respectively. For the sealed specimen, six surfaces of the concrete are covered with tin foil paper to block the water exchange between the concrete and the external environment, so that the decrease in relative humidity is completely caused by self-drying. The five surfaces of the dry specimen are covered with tin foil paper. The upper surface is not treated so that it is in a one-dimensional drying state in the depth direction. The internal humidity begins to decrease under the combined action of self-drying and external drying. Nine temperature and humidity sensors are laid out in the test. Sensors from No.1 to No. 4 are used to monitor the dry concrete specimens, and the implantation depths are 12 cm, 8 cm, 5 cm, and 2 cm respectively. When sensors from No. 5 to No. 8 are used to monitor the sealed concrete specimens, their implantation depth is the same as that of the dry specimens. The No. 9 sensor is used to monitor the changes in ambient temperature and humidity.

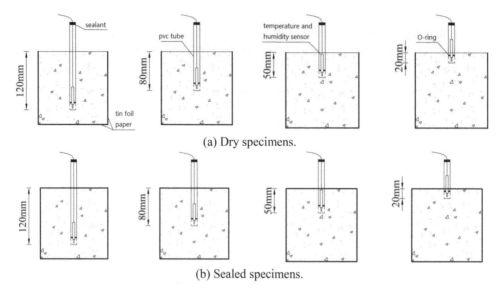

(a) Dry specimens.

(b) Sealed specimens.

Figure 1.   Temperature and humidity sensor layout schematic diagram.

The specific test process is as follows: The inner surface of the concrete mold is lined with tin foil paper. After the concrete pouring is completed, the aluminum rod is inserted into the PVC pipe. The position of the PVC pipe can be prevented from shifting since the volume weight of the aluminum rod is similar to that of the concrete. Then, the PVC pipe and the aluminum rod are extended into the designated position inside the concrete. After the concrete pouring is completed for a while (about 3–5 hours), pull out the aluminum rod, absorb the residual liquid at the bottom of the pipe with a sponge, put an O-ring on the upper end of the sensor probe, extend into the PVC pipe, and seal the top of the PVC pipe with sealant. The temperature and humidity sensors start monitoring, and the monitoring data is collected through computer software. For the sealed specimens, it is also necessary to seal the upper surface of the concrete, so that the six surfaces do not exchange water with the external environment.

2.2.2   Adiabatic temperature rise test

The adiabatic temperature rise test is carried out for the concrete of three groups of mix proportions in Table 1. The specific test process is as follows: After the concrete is poured and vibrated in the adiabatic test barrel, the copper pipe prepared in advance is implanted into the center of the concrete barrel. Then the concrete barrel is put into the adiabatic temperature rise box. The two temperature sensors in the adiabatic temperature rise box are put into the copper pipe in the barrel and outside the adiabatic temperature rise test barrel, respectively, to measure the temperature change process of concrete under adiabatic conditions, and automatically collect data through computer software. The data acquisition time interval is 5 minutes.

3   TEST RESULTS OF CONCRETE INTERNAL HUMIDITY MONITORING

3.1   Crash time analysis of concrete

The internal humidity monitoring test of concrete measured the humidity changing the course of three groups of concrete with a water-binder ratio of 0.33, 0.43, and 0.53 respectively, under sealing conditions and one-dimensional drying conditions. The test results are shown in Figure 2.

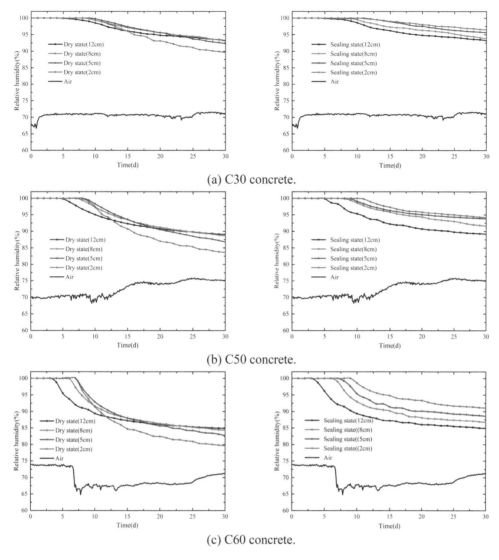

(a) C30 concrete.

(b) C50 concrete.

(c) C60 concrete.

Figure 2.   The curve of relative humidity inside concrete changes with time.

It can be seen from the figures that the variation law of relative humidity with the age of each group of concrete under different drying conditions shows two-stage characteristics. In the beginning, the relative humidity of concrete is maintained at 100%, which is the vapor saturated stage. With the continuous self-drying and external drying, the relative humidity of concrete begins to decrease and enters the humidity decline period. The demarcation point between the vapor saturated stage and the humidity decline period can be expressed by crash time. Figure 3 shows the crash time distribution of concrete.

As can be seen from the figure, under the sealing state, the crash time of each group of concrete at different depths is different. The crash time of C30 concrete at the depth of 2 cm, 5 cm, 8 cm, and 12 cm is 262 h, 239 h, 189 h, and 121 h, respectively, and the crash time of C50 concrete at the depth of 2 cm, 5 cm, 8 cm, and 12 cm is 255 h, 218 h, 180 h, and 108 h, respectively, and the crash time of C60 concrete at the depth of 2 cm, 5 cm, 8 cm, and 12 cm is 222 h, 189 h, 156 h, and 74 h, respectively. With the increase in depth, the crash time is constantly moving ahead. The size of the

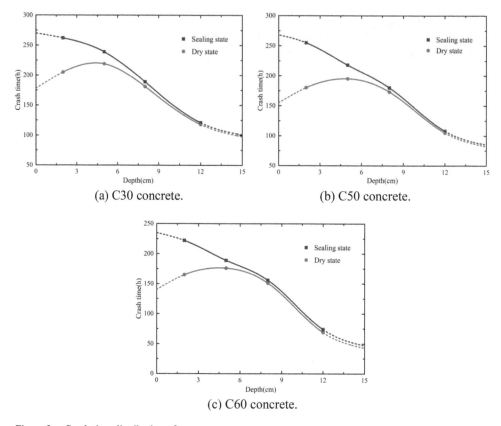

(a) C30 concrete.

(b) C50 concrete.

(c) C60 concrete.

Figure 3.   Crash time distribution of concrete.

concrete specimen in the test is small, so the hydration degree at each depth has little difference. It can be assumed that the water loss caused by self-drying at each point is the same. Therefore, the reason for the reduction of the crash time along the depth direction is that there is a settlement and bleeding at the initial stage of concrete pouring, resulting in more water content in the upper part of the concrete and less water content in the lower part. At the same time, compared with the crash time at the same depth with different water-binder ratios, the crash time is advanced with the decrease in water cement ratio. This is because the concrete with a small water-binder ratio has less moisture content at the initial time, and because the hydration reaction rate is faster, the free water in the pores will be consumed in a shorter time, so the vapor is saturated stage is shorter.

Under a dry state, the crash time of C30 concrete at the depth of 2 cm and 5 cm is reduced from 262 h and 239 h under a sealing state to 205 h and 219 h, and the crash time of C50 concrete at the depth of 2 cm and 5 cm is reduced from 255 h and 218 h under sealing conditions to 180 h and 195 h, and the crash time of C60 concrete at the depth of 2 cm and 5 cm is reduced from 222 h and 189 h under sealing conditions to 165 h and 176 h. The crash time of each group of concrete at a depth of 8 cm and 12 cm under the two states has little difference. The external drying makes the crash time of concrete advance by about 8 cm close to the concrete drying surface. For the range more than 8 cm away from the drying surface, the crash time is hardly affected by the external drying.

3.2   Influence analysis of self-drying and external drying

Analyze and calculate the total relative humidity reduction value θ, relative humidity reduction caused by self-drying θ_d and its proportion in the total reduction of relative humidity θ_d/θ, relative humidity reduction caused by external drying θ_s and its proportion in the total reduction of relative humidity θ_s/θ The calculation results are shown in Table 2.

Table 2. Calculation results of concrete relative humidity reduction.

Concrete strengths	Depth	θ (%)	θ_d(%)	θ_s(%)	θ_d/θ (%)	θ_s/θ (%)
C30	2	10.4	3.2	7.2	30.7	69.3
	5	7.5	4.0	3.5	53.3	46.7
	8	7.0	5.7	1.3	81.4	18.6
	12	6.2	6.2	0.0	100.0	0.0
C50	2	16.4	5.8	10.6	35.4	64.6
	5	13.2	6.2	7.0	46.9	53.1
	8	11.4	8.4	3.0	73.7	26.3
	12	10.9	10.8	0.1	99.1	0.9
C60	2	20.4	8.9	11.5	43.6	56.4
	5	17.3	11.4	5.9	65.9	34.1
	8	15.7	13.3	2.4	84.7	15.3
	12	15.1	15.1	0.0	100.0	0.0

It can be seen from the table that the relative humidity reduction value of each group of concrete under dry conditions decreases with the increase of depth. At 2, 5, and 8 cm, it is affected by both self-drying and external drying. And at 12 cm, it is only affected by self-drying. With the increase of depth, the proportion of the relative humidity reduction value caused by external drying decreases gradually, which indicates that the closer to the surface, the weaker the external drying is. It can be considered that the influence of external drying on the internal relative humidity of concrete only exists in the range of 8 cm from the dry surface.

In summary, the distribution and development of relative humidity of early-age concrete are affected by time, space, and initial moisture distribution at the same time. In the time domain, the relative humidity change process can be divided into the vapor saturated stage and the humidity decline period, and the relative humidity decreases with the increase of time. In space, due to self-drying and external drying, the relative humidity is distributed in a gradient along the height direction. The initial moisture distribution with an increasing trend from bottom to top makes the crash time of concrete in the sealed state decrease from top to bottom. The external drying effect will advance the crash time, but the range within 8 cm of the dry surface is affected.

4 EFFECT OF HUMIDITY REDUCTION ON CONCRETE HYDRATION DEGREE

4.1 Hydration degree of concrete based on adiabatic temperature rise test

In the practical project, since the cementitious materials in the concrete cannot all participate in the hydration reaction, the final hydration degree of the concrete is always less than 1. The three groups of concrete in the adiabatic temperature rise test are cement-fly ash cementitious material systems, and the final hydration degree can be calculated according to Formula (1) combined with the concrete mix proportion. The final hydration degrees of C30, C50, and C60 concrete are 0.8692, 0.8105, and 0.7261 respectively.

$$\alpha_\mu = \frac{1.031w/(c+FA)}{0.194 + w/(c+FA)} + 0.50p_{FA} \tag{1}$$

where α_μ is the final hydration degree, FA is the mass of fly ash in the mix proportion (kg); P_{FA} is the proportion of fly ash in cementitious materials (%).

The hydration degree of concrete can be obtained according to the ratio of adiabatic temperature rise and the final adiabatic temperature rise of concrete at different ages and combined with the final hydration degree. Then, the age corresponding to different temperatures is converted into the equivalent age when the reference temperature is 20°C, and the relationship between concrete hydration degree and equivalent age is obtained, which is fitted by Formula (2).

$$\alpha\,(t_e) = \alpha_u \cdot \exp\left(-\left(\frac{m}{t_e}\right)^r\right) \tag{2}$$

where t_e is equivalent age, h; m, r is fitting parameters.

The fitting parameters of each strength concrete are shown in Table 3. The fitting curve is shown in Figure 4.

Table 3.   Fitting parameters of concrete hydration degree with equivalent age.

Concrete strengths	α_u	m	r
C30	0.8692	25.2136	1.2192
C50	0.8105	21.7009	1.1715
C60	0.7261	17.1707	1.2135

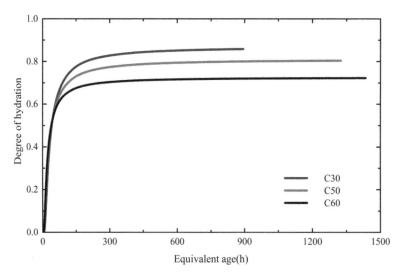

Figure 4.   The fitting curve of the relationship between hydration degree of concrete and equivalent age.

4.2   *Correction of hydration degree under drying conditions*

According to reference [7], the diffusion rate of water molecules through the Hadley layer is directly related to hydration degree and relative humidity. See Formula (3).

$$\frac{d\alpha}{dt_e} = k\,(RH)^n + p \tag{3}$$

where RH is the relative humidity (%); K is the hydration rate constant; n is the reaction order; p is the correction parameter.

Formula (2) establishes the relationship between hydration degree and equivalent age, and Formula (4) is obtained by deriving t_e on both sides at the same time.

$$\frac{d\alpha}{dt_e} = \alpha \cdot \frac{nm}{t_e^2} \left(\frac{m}{t_e}\right)^{r-1} \tag{4}$$

The relative humidity of concrete will not begin to decrease immediately after pouring but will begin to decrease at a certain time. At this time, the critical hydration degree α_c can be used as a sign to distinguish the vapor saturated stage and the humidity decline period, the critical hydration degree refers to the hydration degree when the relative humidity of concrete begins to decrease. When the hydration degree of concrete is equal to the critical hydration degree, the relative humidity RH is 100%. By substituting $\alpha = \alpha_c$, $RH = 100\%$ into Formula (3) and Formula (4), the relationship between hydration reaction rate and relative humidity of concrete can be found in Formula (5).

$$\frac{d\alpha}{dt_e} = \left\{\alpha_c \cdot \frac{r}{m}\left[\ln\left(\frac{\alpha_u}{\alpha_c}\right)\right]^{\frac{r+1}{r}} - p\right\}(RH)^n + p \tag{5}$$

The above formula establishes the hydration reaction rate calculation formula considering the critical hydration degree, final hydration degree, and the change of temperature and relative humidity when the concrete is in the late stage of hydration reaction. By integrating the equivalent age, the hydration degree of concrete under any temperature and humidity history can be obtained, which can provide a theoretical basis for the temperature and humidity coupling simulation of concrete.

The relationship between the relative humidity at the depth of 12 cm under the sealing conditions of concrete and equivalent age, as well as the development process of hydration degree, are shown in Figure 5. Formula (5) is fitted according to the test data. The fitting curve is shown in Figure 6. The parameters of the fitting curve are shown in Table 4.

Table 4. Fitting parameters of hydration reaction rate with relative humidity.

Concrete strengths	α_c	n	p
C30	0.6795	160.35	0.0000088
C50	0.6210	70.24	0.0000101
C60	0.5668	30.45	0.0000055

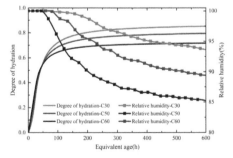

Figure 5. Relationship between hydration degree and relative humidity and equivalent age.

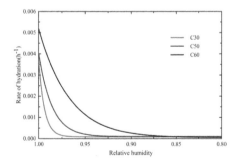

Figure 6. The fitting curve of hydration reaction rate with relative humidity.

It can be seen from the figure that when the relative humidity begins to decline, it is close to the later stage of the hydration reaction. For different strengths of concrete, when the relative humidity is the same, the hydration reaction rate of C60 is the largest, followed by C50 and C30. With the decrease in relative humidity, the hydration reaction rate decreases rapidly. When the relative humidity of C30 concrete is lower than 97%, the hydration reaction rate is close to 0. When the relative humidity of C50 concrete is lower than 93%, the hydration reaction rate is close to 0. When the relative humidity of C60 concrete is lower than 88%, the hydration reaction rate is close to 0.

5 CONCLUSIONS

(1) The self-drying of concrete decreases with the increase of the water-binder. The influence of external drying on the internal relative humidity of concrete decreases with the increase in the distance from the dry surface and does not affect the range beyond 8 cm from the dry surface.
(2) The distribution and development of internal relative humidity of early-aged concrete are affected by time, space, and initial moisture distribution at the same time. Under the sealing condition, the crash time of concrete decreases with the increase in the distance from the dry surface. The external drying will advance the crash time of the dry surface and increase the reduction of the concrete relative.
(3) For different strengths of concrete, when the relative humidity is the same, the hydration reaction rate of C60 is the largest, followed by C50 and C30. With the decrease in relative humidity, the hydration reaction rate decreases significantly. When the relative humidity of C30, C50 and C60 concrete is lower than 97%, 93% and 88%, the hydration reaction rate is close to 0.
(4) A concrete hydration degree model considering the critical hydration degree, final hydration degree, and temperature and relative humidity when the concrete relative humidity begins to decline is established, which can provide a theoretical basis for the temperature and humidity coupling simulation of concrete.

ACKNOWLEDGMENT

This study was supported by the National Natural Science Foundation of China (Grant No. 51479168).

REFERENCES

Huang Yaoying, Liu Yu, Gao Jun. (2020) Experimental Study on Relative Humidity of Hydraulic Concrete at Early Age Under Real Environmental Conditions. *Journal of Basic Science and Engineering*, 27:744–752.
Jiang Jianhua, Yuan Yingshu. (2009) Relationship of moisture content with temperature and relative humidity in concrete. *Cement and Concrete Composites*, 31:699–704.
Schutter G D. (2020) Fundamental study of early age concert behaviour as a basic for durable concert structures. *Materials and Structures*, 4920:181–188.
Wang Jiahe, Zhang Jun, Ding Xiaoping. (2018) Effect of cementitious permanent formwork on moisture field of internal-cured concrete under drying. *Mechanics of Time-Dependent Materials*, 22:95–127.
Wang Yongbao, Jia Yi, Zhao Renda. (2020) Calculation of internal humidity field of concrete based on ANSYS. *Journal of southwest jiaotong university*, 42:108–115.
Zhang Jun, Hang Yu, Qi Kun. (2012) Interior relative humidity of normal and high strength concrete at early age. *Journal of materials in civil engineering*, 24:615–622.
Zhang Jun, Hou Dongwei, Sun Wei. (2010) Experimental study on the relationship between shrinkage and interior humidity of concrete at early age. *Magazine of Concrete Research*, 62:191–199.
Zhang Jun, Wang Jiahe, Ding Xiaoping. (2018) Test and simulation on moisture flow in early age concrete under drying. *Drying Technology*, 36:221–233.

Advances in Measurement Technology, Disaster Prevention and Mitigation – Li & Mohd Yusof (Eds)

Analysis of the causes of abnormal seepage in Hualiangting reservoir

Dewei Yang
Nanjing Hydraulic Research Institute, Nanjing, China
Dam Safety Management Center of the Ministry of Water Resources, Nanjing, China

Chengjun Xu
Jiangsu Kexing Project Management Co., Ltd., Nanjing, China

Huiwen Wang*
Nanjing Hydraulic Research Institute, Nanjing, China
Dam Safety Management Center of the Ministry of Water Resources, Nanjing, China
Hohai University College of Water Conservancy and Hydropower Engineering, Jiangsu, Nanjing, China

ABSTRACT: Abnormal seepage seriously affects the safety of dams and is one of the major causes of earth-rock dam failure, so it is essential to analyze the seepage flow situation. This paper takes the Hualiangting Reservoir as an example and analyses its seepage volume, seepage pressure head and other observations. The results indicate that its abnormal seepage is mainly caused by severe seepage around the left shoulder of the dam, and there may be seepage channels at the foot of the dam.

1 INTRODUCTION

The earth-rock dam is a kind of granular structure. Under the water pressure, the reservoir water will leak downstream through the pores in the dam body, the dam foundation, the abutment soil, or the rock mass. Generally, the seepage flow of normal water seepage is small, the water quality is clear, and it does not contain soil particles. When the following phenomena appear: the seepage flow is too large and the water quality is turbid; the seepage water contains many soil particles; the water or sand turns over in the back of the dam, it indicates that the dam body has abnormal leakage (Ding & Cai 2008; Ma 2019). Abnormal leakage seriously affects the safety of the dam and should be given high priority. According to the statistics, of the 241 large earth-rock dams in China, infiltration damage accounted for 31.9% of the total number of 1,000 accidents; of the 2,291 reservoir dam accidents, infiltration damage accounted for 29% of the total (Jie & Sun 2009; Liu & Xie 2011). According to statistics from other countries in the world, the rate of earth-rock dams caused by infiltration damage is 39% in the United States (206 accidents in the dam according to the statistics), 44% in Japan, 40% in Sweden (119 accidents in the dam according to the statistics), and 40% in Spain (117 accidents in the dam according to the statistics) (Ru & Niu 2001).

For the large-scale project at Hualiangting Reservoir, which is in Anqing City, Anhui Province, the concentrated seepage, and sand surges appeared on the dam downstream slope at W65 point and 0+464 section. To ensure the operational safety of the dam, it is necessary to analyze the causes of concentrated leakage at the downstream toe of the left dam end and to assess its current seepage safety.

---

*Corresponding Author: 453590381@qq.com

DOI 10.1201/9781003330172-79

2 PROJECT DESCRIPTION

2.1 *Project overview*

Hualiangting Reservoir is located on the Changhe River, a tributary of the Wanhe River in the Taihu River Basin in Anhui Province. It is a large water conservancy project, mainly for flood control and irrigation. The reservoir has multi-year regulation performance, with a normal storage level of 88.00 m, a flood limit level of 85.50 m, a dead water level of 74.00 m, a design flood level of 95.21 m for the one-thousand-year event, and a calibration flood level of 97.30 m for the ten-thousand-year event, with a total storage capacity of 2.368 billion m3. The project consists of dams, spillways, flood discharge tunnels, diversion tunnels, and powerhouses. The dam is a clay core wall sand shell dam with a crest elevation of 99.25 m, a length of 566.00 m, a width of 6.70 m, a maximum dam height of 58.00 m, and a wave wall crest elevation of 100.25 m. The dam body is divided into the anti-seepage body (i.e., the core wall and horizontal paving), the dam shell, the filter layer, the drainage prism, and slope protection.

The base layer of the dam is the Quaternary alluvial deposit (alQ4) and the underlying Archean strata of the Dabie Mountains group (Arq).

The bedrock section on the left bank includes the dam section from pile number 0+360 to 0+565 and the left bank hillside section. The geological conditions are complex and the permeability is generally moderately permeable or weakly permeable. The dam section from pile 0+360 to 0+565 was slightly cut during the construction of the dam, and the weathered rock was not completely removed.

The station number of the river bed section is 0+098~0+360. The surface of the dam foundation is alluvial, which is mainly medium-coarse sand, gravel, and mud with a high mud content, adding a small amount of silty loam and silt layer. A bedding layer fault is developed in the dam foundation rock joint, which is the structure of the breccia and fault mud.

The bedrock section on the right bank, pile numbers 0+000 to 0+098, is located in the right dam section and the right shoulder, and is a bedrock dam foundation, with the dam body in direct contact with the bedrock. The weathering of the rock is relatively light; the rock is intact and the fissures are not developed. The permeability of the rock is poor and it is weakly permeable.

2.2 *Project construction and risk reinforcement*

The reservoir started construction in August 1958 and began to store water in June 1960. It was suspended in 1962 and resumed construction in 1970. By the end of 1976, except for the installation of the generator set, the rest of the project was completed according to the original design scale.

The first clay grouting of the core wall was carried out from 1966 to 1968, with a total of 57 holes filled with 2,124 m^3 of clay slurry. There were serious cracks in the core wall of the clay in August 1968. In 1971, the second clay grouting was carried out mainly on the section of the clay core wall from 0+100 to 0+260, with a total of 74 holes and 1,640 m^3 of grout. From 1980 to 1981, the third grouting was carried out on the clay core wall, with a total of 7 holes and 680 m^3 of grout.

On September 30th, 2009, the dam reinforcement and spillway, flood discharge tunnel, and diversion tunnel reinforcement projects started at the same time. The concrete cut-off wall was completed on April 2nd, 2010, and the curtain grouting was completed on May 14th, 2010. In early December 2010, the risk identification and water storage safety appraisal were completed. On December 14th, 2010, the reinforcement project passed the acceptance inspection of the main project.

2.3 *Engineering seepage problem*

Before reinforcement, on September 2nd, 1997, a concentrated seepage point was found at the foot of the dam slope at the left end of the gully. The water has been flowing all year round since then, with a flow rate of 0.03-0.05 L/s. The water flow was sometimes clear and sometimes turbid, causing the water at the left end of the seepage collection ditch to be yellow and muddy. The drainage prism at the water seepage point had an overhead phenomenon, and the sand was taken out to form a sand table with a range of 4.00 m〜3.00 m, where the composition was mainly medium fine sand, the same as the sand content of the dam shell.

According to the underwater topography measurement, there was a long strip of low-lying land along the original channel of the river between 192 and 270 m upstream of the dam axis. The length in the direction from the upstream to the downstream (EW direction) was 86.00 m and the width in the direction from the left to right (SN direction) was 44.00 m. The bottom area was about 760 m2 and the top area was about 2,800 m2. The bottom of the depression was 3.00~4.00 m lower than the top. Here, the reservoir cover was seriously damaged.

After the completion and acceptance of the reinforcement project in 2013, the actual leakage of the dam was still large. According to the observation data, the leakage amount reached a maximum of 328.00 L/s in June 2013. In August 2014, the leakage amount reached a maximum of 381.33 L/s while the reservoir water level was not high, at 80.05 m. There were many obvious water seepages at the foot of the drainage prism. The initial stage was clear, and after one or two days, it became a turbid rust color.

On July 6th, 2016, during the flood control inspection, three concentrated seepage points were found at 48.70 m from the station at 0+464 elevation.

On August 10, 2016, the reservoir water level was 80.33 m, the downstream water level was 43.668 m, and the seepage flow was 51.18 L/s. The water in the collecting ditch on the outside of the drainage prism was orange in color, and there was no seepage in the original sediment leakage at the bottom of the drainage prism at pile 0+337, with traces of sand surges. The surrounding sand is saturated and sinks when stepping on it. It is fine and easy to liquefy. The three concentrated seepage points, which were found at the base of the drainage prism at elevation 48.70 m on the left bank at pile 0+464, had no concentrated seepage at this point.

To ensure the safe operation of the project, an analysis of the causes of excessive leakage from the dam and concentrated leakage at the downstream toe of the left dam end is required, as well as an assessment of the current seepage safety of the dam.

3 ANALYSIS OF THE SEEPAGE PROBLEMS

3.1 *Instrumentation arrangement*

(1) Measuring weir

The measuring weir is located on the downstream side of 0+285 and is a trapezoidal weir. Before reinforcement, when the power station was generating electricity, the tailwater would flood the weir plate and the amount of seepage could not be measured. To solve this problem, weir plates were designed so that when the water level downstream is high, the first layer of the weir plate can be closed. Then the seepage level can be raised behind the weir plate, and the second trapezoidal weir will be used to continue to observe the amount of leakage. In the past, the water level on the weir was automatically measured using a tracking water level meter. At the end of 2015, the water level measurement facility on the weir was replaced with an IF-05 digital water level sensor.

(2) Piezometric tube

After reinforcement, there were 41 existing and newly buried pressure measurement tubes. The automatic collection of the water level in the pressure tubes was achieved by lifting a seepage gauge in the pressure tubes. The osmometer is a 4500S type from the American Keikang Company, with a range of 350 kPa and an accuracy of 0.1%.

The left dam end of the piezometric tube includes R42, R32, R34, R35, R36, and 0+540 sections of R42, R44, and R45 of 0+455 stations, for a total of eight. The right bank end of the piezometric tube includes R12, R14, R15, R22, R24, and R25, six pressure measuring tubes. The dam foundation piezometric tube includes UP10, UP12, UP14, UP15, UP16, UP17 six piezometric tubes of 0+080 section, UP20, UP21, UP22, UP23, UP24, UP26, UP27 seven piezometric tubes of

0+160 section, UP30, UP31, UP32, UP33, UP35, UP36, UP37 seven piezometric tubes of 0+255 section, UP40, UP41, UP42, UP43, UP44, UP45, UP46 seven piezometric tubes of 0+335 section.

3.2 *Analysis of the seepage observation data*

3.2.1 *Seepage flow*

The seepage process line is not measured much, and the regularity is poor. There are several anomalies in spikes and even 'extra-large' anomalies. It is understood that the 'extra-large' abnormal data is calculated from the 'extra-large' seepage volume, caused by a rise in water level at the weir due to the closure of the first stage of the weir plate. The remaining spike abnormalities are mostly the problems of the measuring instrument. The process line for leakage after removing obvious erroneous data is shown in Figure 1.

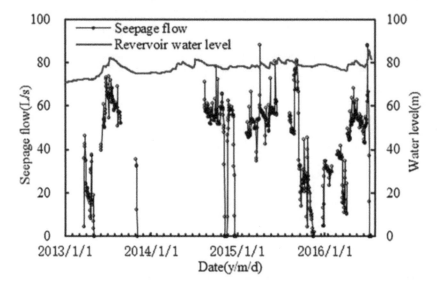

Figure 1.   Leakage process line diagram.

As can be seen from the graph, in general, leakage is smaller when the reservoir level is low and larger when it is high, especially in the observation data of 2016. The range of the seepage flow varies from 0 to 90.00 L/s. The median seepage volume is around 50.00 L/s, with most concentrated in the range of 30.00 L/s to 70.00 L/s. As a rule of thumb, it is normal for earth-rock dams of the size of the Hualiangting Reservoir to have an observed long-term seepage of between 30.00 and 70.00 L/s of open flow after reinforcement.

3.2.2 *Osmotic head*

(1) Analysis of piezometric tube data at the dam foundation

The concentrated seepage point W65 is located at the exit of seepage at section 0+335, so the following analysis focuses on its pressure tube information of it. Figure 2 shows the process line of the piezometer section since 2013.

As can be seen from Figure 2, the UP40 pipe is buried in the sand crust upstream and its pipe level almost coincides with the reservoir level process line, which is reasonable. UP41 and UP42 are buried in the medium-fine sand base on the upper and downstream side of the impermeable wall respectively, which has a good correlation with the reservoir water level, but the difference in water level between the two tubes is only 2 to 3 m, which is difficult to explain. If there is no problem with the burial and measurement, it can be explained that the horizontal paving cover and

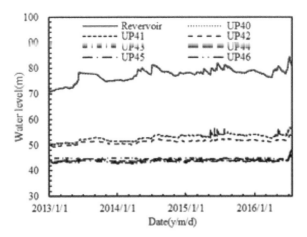

Figure 2. The 0+335 section piezometric tube water level process line.

the clay core wall here have a very good impermeable effect, eliminating 25 m of water head, while the impermeable wall eliminates 2 to 3 m of water head again, which is the very section where the horizontal the cover is 1.5 m thick loam and the cover has been damaged by a large collapse pit in the outer cover of the dam, so this should be further identified. The three pressures measuring pipes UP44, UP45, and UP46 are all located behind the horizontal cover of the dam and the water level of each pipe is about the same. Here, the water level of the UP46 pipe located downstream is higher than that of the UP45 upstream, which needs to be further identified. If it is not caused by burial, measurement errors, or the accuracy of the seepage pressure gauge, it is influenced by seepage around the dam, the latter being more likely.

(2) Outlet penetration slope

The downstream seepage outlet at section 0+335 (at W65) had problems with pipe surges and the precipitation of fine and medium sands before reinforcement. After that, a considerable amount of medium-fine sand had been precipitated in the original location, scattered between the stone joints at the bottom of the drainage channel behind the dam.

Figure 3 is a graph of the average seepage slope for 2016 at the exit of the 0+335 section (W65) calculated from observations.

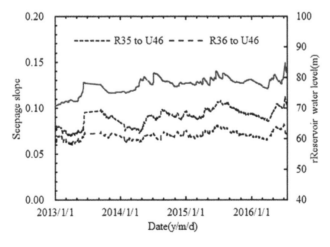

Figure 3. Osmosis slope process line for section 0+335 exit (W65) for 2016.

From Figure 3, we can obtain the following conclusions:

(1) The average infiltration slope drops between R36 and UP46 (72 m) and between R35 and UP46 (89 m), which is much greater than that between UP45 and UP46, at least 8 to 10 times. This indicates that the seepage flow from the left dam head to the downstream outlet is the main cause of infiltration damage on the downstream side of UP46 (W65).

(2) The average infiltration slope drops between R36 and UP46 and between R35 and UP46 correlates well with the reservoir level, and the average slope drop also reaches a maximum when the reservoir level is at its highest. This indicates that if the reservoir level continues to rise, the infiltration slope drop at the outlet will also continue to rise.

(3) The average infiltration slope drops between R35 and UP46 had exceeded the allowable slope drop in the fine sand of 0.10. This phenomenon indicates that the anti-filtration protection layer of the anti-filtration body here did not play a protective role.

3.3 *Causes of the seepage problem*

(1) Causes of the concentrated seepage in W65 point

This point is located at the seepage outlet of the original channel and the combined section of the dam, which is a weak part of the dam's seepage. The section of the impermeable cover was inherently inadequate, and the survey before reinforcement revealed that there was a seepage channel between the upstream and downstream directions of the section, the left dam section, and this point. After the reinforcement, the concrete impermeable wall blocked the seepage channel at the original channel, making up for the inherent deficiency in this area.

The geological conditions on the left shoulder are complex, and it is difficult to guarantee the construction quality of the impermeable treatment. If the upstream bread seal on the left bank is not in place, there is a dead end around the seepage, resulting in serious seepage flow around the dam, as confirmed by the analysis of seepage pressure observation data. The severe seepage around the dam flows to the low water level (outlet) near point W65 at the foot of the dam, raising the infiltration gradient of the seepage there, which is the main cause of the infiltration damage at W65.

(2) Causes of the concentrated seepage and sedimentation at the point of the seepage at the dam foot of the 0+464 section (three points)

This is located at the outlet of the seepage around the dam at the left end. Exploration indicates that a seepage channel already existed between the left shoulder of the dam and point W65, which is relatively close to point W65 and may have a seepage channel.

The filter layer of the drainage prism at this location fails to provide effective anti-filter protection for the medium-fine sand dam shell.

4 CONCLUSIONS

This paper assesses the analysis of the dam's seepage causes and seepage safety based on information from the pre-project safety appraisal and the reinforcement project's design, construction, acceptance, and operation management.

It was found that the infiltration slope at the seepage outlet of the main channel section of the dam was less than the permissible one of 0.10, and the infiltration stability met the requirements. However, the infiltration slope drop at the seepage outlet of the dam ends was larger than the permitted value of 0.10 due to the influence of seepage around the dam, with the left end being more serious.

The main causes of the concentrated seepage damage at point W65 and the concentrated seepage of sand downstream of section 0+464 during operation are the serious seepages around the dam (including mountain fissure water) and the failure of the drainage prismatic back filter to provide effective protection for the medium and fine sand dam shell. Also, the seepage around the right end of the dam should be taken seriously.

To ensure the safety of seepage around the dam, it is recommended that the reservoir should be operated at a controlled level, and the maximum operating level should not be higher than 83.00 m. The dam ends should be treated for impermeability, and the grouting range should be appropriately extended to prevent seepage around the dam. The downstream dam foot should be renovated to ensure that it is protected against infiltration damage in the future.

ACKNOWLEDGMENT

This study was supported by the National Natural Science Foundation of China (Grant No. 51909174).

REFERENCES

Jiabi Jie, Dongya Sun. Dam failure statistics and analysis of causes of dam failure in China[J]. *Water Resources and Hydropower Technology*, 2009, 40(12):124–128.

Jie Liu, Dingsong Xie. Current development of seepage control theory for earth-rock dams in China[J]. *Journal of Geotechnical Engineering*, 2011, 33(05):714–718.

Jing Ma. *Research on key technologies of earth-rock dam disease risk identification and dam failure risk analysis*[D]. Xi'an University of Technology, 2019.

Naihua Ru, Yunguang Niu. Accident statistics and analysis of earth-rock dams[J]. *Dams and Safety*, 2001, (01):31–37.

Shuyun Ding, Zhengyin Cai. *A review of seepage studies of earth-rock dams*[J]. People's Changjiang, 2008, (02):33–36+108.

Advances in Measurement Technology, Disaster Prevention and Mitigation – Li & Mohd Yusof (Eds)
© 2023 The Author(s), ISBN: 978-1-032-36087-4

Laboratory experimental study on leakage of dams considering low-temperature water leakage and composite geomembrane defects

Ke Li

State Key Laboratory of Eco-Hydraulics in Northwest Arid Region, Xi'an University of Technology, Xi'an, Shaanxi, China

Jie Ren*

State Key Laboratory of Eco-Hydraulics in Northwest Arid Region, Xi'an University of Technology, Xi'an, Shaanxi, China
Dike Safety and Disease Prevention Engineering Technology Research Center of Ministry of Water Resources, Zhengzhou, Henan, China

Lei Zhang

Yellow River Institute of Hydraulic Research, YRCC, Zhengzhou, Henan, China

ABSTRACT: The composite geomembrane defects can cause safety hazards to the dam's seepage control. The seepage law of the dike under the action of different seepage heads and composite geomembrane defects was studied, and the leakage temperature field, seepage line, and leakage rate of the dike were analyzed. The results show that different seepage heads have a significant impact on the temperature field of the dam body around the composite geomembrane defect. The temperature change around the defect is significantly faster and closer to the seepage water temperature as the seepage head increases; the average rise and fall rate of the seepage line of the dam increases rapidly with time. The higher the seepage head is, the earlier the leakage of the dam will occur. When the leakage is stable, the leakage rate of a single-width dam with a high head is always greater than that of a dam with a low head.

1 INTRODUCTION

Due to its good anti-seepage performance, low cost, fast construction speed, and ease of use, composite geomembrane is widely used in seepage prevention projects, such as dams, reservoir bottoms, and reservoirs. However, due to human and environmental influences, the composite geomembrane will be damaged during the construction process, and the water will pass through the defects to form concentrated leakage, and the seepage mechanism will change (Cheng et al. 2019; Gu 2009; Shu et al. 2016). Wu Dazhi et al. explored the leakage of dams with geomembrane defect rate and defect pore size through laboratory tests (Wu & Yu 2016). Sun Dan et al. analyzed the seepage characteristics of the composite geomembrane defect sand-gravel dam under the action of different pressure heads through numerical simulation (Sun et al. 2013). Cartaud F et al. studied the effect of a different composite geomembrane on the flow rate through small-scale experiments (Cartaud et al. 2005). Saidi F et al. conducted a leakage study on the interacting square composite geomembrane defects through numerical simulation (Saidi et al. 2008). Most of the current research is through numerical simulation or theoretical analysis methods, and qualitative research on the effect of composite geomembrane defect dam body leakage is rarely involved. In this paper, the seepage law of embankments under the action of different seepage heads and

---

*Corresponding Author: renjie@xaut.edu.cn

DOI 10.1201/9781003330172-80

composite geomembrane defects is investigated using an indoor sand tank embankment test, which provides theoretical support for the evaluation of dam safety as much as possible while reflecting the actual engineering situation.

2 MATERIALS AND METHODS

2.1 *Test device*

The flume experiments were conducted in a flume that was 2.3 m long, 0.5 m wide, and 1 m deep with transparent walls. Water inlets with heights of 0.2 m, 0.3 m, and 0.4 m are set on the left side of the glass sand tank, and the water head can be connected to a constant temperature water tank according to the demand. Both sides of the glass sand tank are provided with plexiglass orifice plates and permeable gauze to stabilize the water seepage head. The plexiglass orifice plate divides the glass sand tank into three sections. The left side is the upstream water inlet tank section, with a length of 0.15 m. The 2 m test embankment section, on the right side, is the downstream water outlet tank section with a length of 0.15 m; a water outlet is set at a height of 0.05 m in the downstream water outlet tank, which can be used to collect leakage and drainage. The layout of the test site is shown in Figure 1. In the dam test, the center of the composite geomembrane defect was selected to select the middle cross section of the dam for experimental research. The height of the dam is 0.46 m, the length is 0.5 m, the dam slope ratio is 1:2, the width of the crest is 0.16 m, and the length of the dam bottom is 2 m.

Figure 1.   The layout of test site testing.

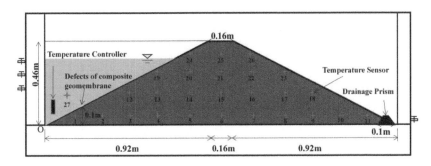

Figure 2.   Variation diagram of the layout of dam cross section and temperature sensor.

The temperature sensor acquisition device used in this test is the OMB-DAQ-54/55/56, as shown in Figure 3. The signal is transmitted to the computer in real time using the OMEGA data transmitter.

Figure 3.   Temperature sensor acquisition instrument.

In the experiment, 26 temperature sensors were arranged to monitor the temperature change of the dam section, and a temperature sensor is arranged in front of the dam to monitor the change in the seepage water temperature. A total of 27 temperature sensors are used in this test, with O in front of the dam as the origin. All the coordinates of the temperature sensor are shown in Table 1.

Table 1.   Temperature sensor measuring point location.

Measuring point	T1	T2	T3	T4	T5	T6	T7	T8	T9
x (m)	0.2	0.35	0.5	0.65	0.825	1	1.175	1.35	1.5
y (m)	0.05	0.05	0.05	0.05	0.05	0.05	0.05	0.05	0.05

Measuring point	T10	T11	T12	T13	T14	T15	T16	T17	T18
x (m)	1.65	1.8	0.5	0.65	0.825	1	1.175	1.35	1.5
y (m)	0.05	0.05	0.17	0.17	0.17	0.17	0.17	0.17	0.17

Measuring point	T19	T20	T21	T22	T23	T24	T25	T26
x (m)	0.65	0.825	1	1.175	1.35	0.825	1	1.175
y (m)	0.29	0.29	0.29	0.29	0.29	0.41	0.41	0.41

Twelve glass piezometer tubes are set on the outside of the glass water tank to monitor the change in the total pressure head of the dam. Piezometer tubes are numbered 1 to 10 based on the O in front of the dam as the origin. The horizontal positions and labels of each piezometer tube are shown in Table 2.

Table 2.   The horizontal position and label of the pressure measuring tube.

Pressure measuring tube designation	1	2	3	4	5	6	7	8	9	10
Horizontal distance from point O (m)	0.1	0.3	0.5	0.7	0.9	1.1	1.3	1.5	1.7	1.9

2.2   Test plan

The experimental protocol is summarized in Table 3. The test is divided into three working conditions, and a total of three groups of tests need to be done. First, we connect the required sensors to the computer and record the initial temperature of the dam body before the test. Secondly, we adjust the required seepage water temperature in the constant temperature water tank. The water inlet is opened to the required level and connected to the constant temperature water tank, then a small water pump is connected to the sand tank water intake. To ensure the stability of the test water head, we circulate the water supply to the water inlet tank through a small water pump. To determine the temperature of the seepage water, we placed a temperature sensor in the water next to the dam. After the seepage head required for the test is stable, the piezometric head is measured and the data is recorded. The measurement interval is 30 minutes. After running the test for a while and observing water seepage from the downstream water outlet tank, the water outlet on the right side of the sand tank was opened, and a measuring cylinder was placed to measure the dam's leakage, with the measurement data being recorded and the measurement interval being 5 minutes.

Table 3. Test conditions.

Working condition	Seepage head (m)	Seepage water temperature (°C)	Composite geomembrane defect size (m)	Composite geomembrane defect shape	The vertical distance from the damage center of the composite geomembrane to the dam floor (m)
1	0.2	5	0.03	Square	0.1
2	0.3	5	0.03	Square	0.1
3	0.4	5	0.03	Square	0.1

3 TEST RESULTS AND ANALYSIS

3.1 Analysis of temperature field of the dam under the action of seepage head

A leakage test was performed on an indoor sand tank dam with a composite geomembrane defect. Through the temperature data measured by the temperature sensor and the screening and analysis data, the composite geomembrane defect as the center is drawn. The temperature cloud map of the dam section is shown in Figure 4. It can be seen from the overall figure that under the action of different infiltration heads, with the increase of seepage time, the temperature of the dam body around the composite geomembrane defect gradually diffuses into the dam body as the seepage time increases and makes the defect mouth. The temperature of the dam body reaches 5°C, which is close to the seepage water temperature. When the test is carried out for 30 minutes, the temperature

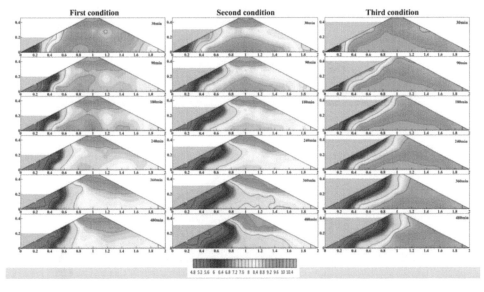

Figure 4. Cloud image of temperature field variation of dyke at different moments with composite geomembrane defect under the action of different seepage heads.

585

field upstream of the dam body and the center of the dam body has changed. It shows that leakage has occurred in the dam around the composite geomembrane defect. When the seepage head is 0.2 m, the temperature in the dam body gradually transitions from around the defect to the center of the dam body, but the temperature spreads slowly to the center of the dam body. When the seepage head is 0.4 m, the high-water head makes the dam body suffer. The pressure is high, and the location of the composite geomembrane defect is relatively low. The temperature field of the entire dam body changes in a gradient, and the leakage of the downstream dam body under the action of the infiltration head of 0.2 m and 0.3 m is stable earlier. When it is 0.3 m, the temperature field of the seepage water spreads from around the defect of the composite geomembrane to the temperature field around the middle of the entire dam body.

3.2 Analysis of wetting line of dam body under the action of seepage head

In this test, according to the data measured by the piezometric tube under three working conditions, the wetting line was drawn for the selected dam body section, as shown in Figures 5, 6, and 7. It can be seen from the figure that with the increase of seepage time under the action of different seepage heads, the seepage line of the dam body rises rapidly from 30 min to 90 min, and rises slowly from 90 min to 360 min. The average rise-fall rate of the wetting line of the dam body over time under the action of different seepage heads is calculated, as shown in Figure 8. The average rise-fall rate of

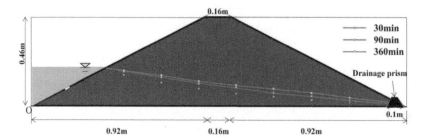

Figure 5. The wetting line diagram of the dam body at different times when the seepage water is 0.2 m.

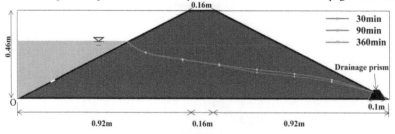

Figure 6. The wetting line diagram of the dam body at different times when the seepage water is 0.3 m.

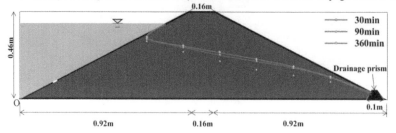

Figure 7. The wetting line diagram of the dam body at different times when the seepage water is 0.4 m.

the wetting line of the dam body also increases significantly, but from 90 minutes to 360 minutes, the average rise-fall rate of the wetting line of the dam body increases slowly, especially for the high-water head. The average rate of ascent is slower.

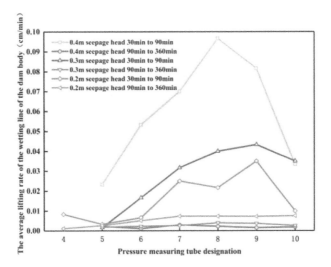

Figure 8.   Average rise-fall rate of the wetting line of the dam body under the action of different seepage water of heads.

3.3   *Analysis of single-width seepage rate of the dam under the action of seepage head*

According to the leakage data measured by the test, the curve of the single-width leakage rate of the dam body under the action of different seepage water heads with time when there is a composite geomembrane defect at the seepage water temperature of 5°C is drawn (Figure 9). Under the three working conditions, the leakage of the dam with a seepage head of 0.4 m occurs earlier than that of the dams with a seepage head of 0.2 m and 0.3 m, and the leakage begins to occur in about

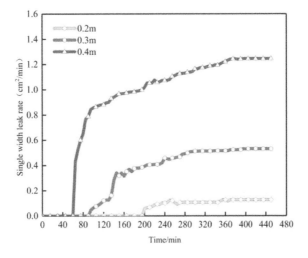

Figure 9.   Variation curve of the single-width leakage rate of the dam under different seepage head.

60 minutes, while the seepage head of 0.2 m starts to seepage at about 190 min. The single-width leakage rate of the dam body in the three working conditions reaches a stable state, indicating that the seepage of the dam body reaches a stable state currently.

4 CONCLUSIONS

(1) When the seepage water of the head increases, the temperature change of the dam body around the composite geomembrane defect is significantly accelerated and is closer to the temperature of the seepage water.
(2) The average rise-fall rate of the wetting line of the dam body changes with time, which first rises rapidly and then falls slowly to a steady state.
(3) When the seepage head is higher, the leakage of the dam body will occur earlier, and the existence of composite geomembrane defects will cause safety hazards to the dam body.

ACKNOWLEDGMENTS

This study was supported by the Shaanxi Provincial Innovative Research Team (Grant No. 2022TD-01); the Research Center on Embankment Safety and Disaster Prevention Engineering Technology of the Ministry of Water Resources of China (Grant No. DFZX202007); the Fund of National Dam Safety Research Center of China (Grant No. CX2020B03); and the National Natural Science Foundation of China (Grant No. U1965110).

REFERENCES

Cartaud F, Touze-Foltz N, Duval Y. (2005) Experimental investigation of the influence of a geotextile beneath the geomembrane in a composite liner on the leakage through a hole in the geomembrane. *J. Geotextiles & Geomembranes*, 23(2): 117–143.

Cheng W C, Ni J C, Huang H W, et al. (2019) The use of tunnelling parameters and spoil characteristics to assess soil types: a case study from alluvial deposits at a pipejacking project site. *J. Bulletin of Engineering Geology and the Environment*, 78(4): 2933–2942.

Gu J C. (2009) Experience of applying geomembrane in reservoir seepage control. *J. Advances in Science and Technology of Water Resources*, 163: 51–59.

Saidi F, Touze-Foltz N, Goblet P. (2008) Numerical modelling of advective flow through composite liners in case of two interacting adjacent square defects in the geomembrane. *J. Geotextiles and Geomembranes*, 26(2): 196–204.

Shu Y M, Wu H M,Jiang X Z. (2016)The development of anti-seepage technology with geomembrane on reservoirs and dams in China. *J. Chinese Journal of Geotechnical Engineering*. 38(S1):1–9.

Sun D,Shen Z Z, Cui J J. (2013)Seepage Numerical Simulation of Geomembrane Gravel Dam Caused by Geomembrane Defect. *J. Water Resources and Power*. 31(04):69–73.

Wu D Z, Yu L. (2016) Experimental Study on the Influence of Defects in Geomembrane on the Seepage of a Dam. *J. Journal of Zhejiang Sci-Tech University* (Natural Science Edition). 35(03): 474–478.

Advances in Measurement Technology, Disaster Prevention and Mitigation – Li & Mohd Yusof (Eds)
© 2023 The Author(s), ISBN: 978-1-032-36087-4

An experimental study on the effect of unidirectional tension on permeability of geotextile sand covering systems

Chengbin Lu, Xiaolei Man*, Hao Xiao, Ying Ge & Yingjie Wu
College of Civil and Architecture Engineering, Chuzhou University, Chuzhou, China

ABSTRACT: To explore the influence of uniaxial tension in different directions on the permeability characteristics of geotextile-coated sand systems, two kinds of woven geotextile are selected. Penetration tests of two geotextile coated sand systems under non-stretching, warp stretching, and weft stretching conditions are conducted by a gradient ratio permeameter. The effects of longitude and latitude stretching on permeability parameters, such as permeability, anti-silting and plugging performance, and soil conservation performance, are analyzed. The experiments show that when the geotextile is subjected to warp tension, its permeability and anti-clogging performance are weakened, and its soil retention performance is enhanced; when the geotextile is subjected to latitudinal stretching, its permeability and anti-silting performance are enhanced, and its soil retention performance is weakened. The thickness of the woven fabric also has a great influence on its permeability, soil retention, and anti-fouling performance.

1   INTRODUCTION

Compared with the rolling and throwing construction technology of earth-rock dams, the filling pipe bag embankment construction technology is more widely used, and it developed rapidly due to its advantages of low carbon and energy conservation, simple technology, reliable technology, cost efficiency, and controllable construction period. Among them, the overall stability of the pipe-bag dam is affected by the permeability characteristics of the pipe bag. The filling pipe bag is a wrapping system composed of filled sand and geotextile. In recent years, to study the theoretical basis of filling pipe bag permeability characteristics, the construction technology of pipe-bag dams, the dewatering and the retaining performance of filling soil, and the stability of filling pipe-bag are deeply explored by many scholars across the globe.

   At present, more in-depth research on the permeability characteristics of filling pipe bags has been carried out by scholars globally. Su et al. (2014) finally obtained the empirical formula of the permeability coefficient of different graded soil through a constant head permeability test of various graded soil materials. Tang et al. (2021) studied the influence law of geotextile placement position and number of layers on permeability characteristics and deduced the corresponding relationship between geotextile buried depth and soil permeability coefficient. Wei et al. (2021) summarized the variation laws of gradient ratio, permeability coefficient, and other parameters by conducting permeability and siltation tests of non-woven fabrics with different equivalent pore sizes. Wu et al. (2018) summarized the high-efficiency drainage consolidation construction technology of soil filling with high powder particles through an indoor bag hanging test. Ding et al. (2019) analyzed the influence of various graded coarse-grained soil on the permeability coefficient and concluded that the influence degree is different under different particle sizes. Zhan et al. (2018) used the indoor large-scale stress seepage coupling model to study the influence of stress in different layouts on the permeability characteristics of pipe bag accumulation. Man et al. (2021) obtained that

---

*Corresponding Author: manxl@chzu.edu.cn

DOI 10.1201/9781003330172-81

geotextile can inhibit the loss of fine particles in the sand by comparing the permeability coefficient without sand covering. Pang et al. (2019) analyzed the silting of geotextiles through the independently modified gradient-specific permeability instrument. Li et al. (2021) used the independently improved gradient-specific permeability instrument to conduct permeability tests on two kinds of geotextiles and obtained the permeability characteristic conditions under different needs. Wu et al. (2017) studied the difference between the permeability coefficient of geotextile under the condition of covering soil and that of pure soil and obtained that the permeability coefficient under the condition of covering silt is slightly greater than that of pure silt.

However, in the actual construction, the permeability of the filling pipe bag is inevitably affected by different flow and extrusion deformation. It is necessary to study the permeability characteristics under these two conditions. Liang et al. (Ding et al. 2019) conducted permeability tests on ultra-fine clay by using a constant head test device with variable head position and obtained the influence relationship of hydraulic gradient on permeability coefficient. Wang et al. (2016) studied the soil retention performance of biaxial tensile geotextile under reciprocating water flow through a gradient ratio infiltration device and concluded that the soil retention of geotextile decreased with the increase of strain under this condition. Zhuang et al. (2008) studied the influence principle of multiple frequency water flow on a "geotextile soil" system by using the self-developed reciprocating water flow penetration test instrument. Chen et al. (Cheng 2015) studied and analyzed the process of the filling pipe bag with soil material, and concluded that the pipe bag was stressed and deformed during the filling process. Man et al. (2020) used geotextiles with different sewing directions to carry out a bag hanging test, and studied that when the weft stress is greater than the warping stress, the bag has better dehydration performance. Lei et al. (2016) studied the permeability characteristics of geotextiles under tension and load by using the self-developed multifunctional permeability device. She et al. (Yu 2011) obtained the aperture fitting formula of geotextile under tension by measuring the fabric aperture under different tensile states. Liu (2012) measured the water permeability of geotextile under tension by using a self-developed device. Bai et al. (2015) studied the relationship between equivalent pore size and tensile strain of geotextile by using a self-developed biaxial tensile device and dry screening method. Tang et al. (2013) laid a foundation for practical construction by studying the permeability characteristics of two types of Geotextiles in different tensile states. Wu et al. (2007) conducted permeability experiments on the unidirectional tensile geotextile and soil composition system and obtained the influence law of the equivalent pore diameter of excavated geotextile on the permeability velocity.

The preparation method of woven geotextiles makes the silk lines in different directions of the fabric deformed differently under tension. The deformation of the warp is greater than that of the weft. In the construction process, When the load is the same, the deformation of the warp and weft are different, and the permeability of the filling pipe bag changes. Man et al. (2019) found in the indoor pipe bag filling test that when the compression deformation degree of warp and weft is different, the change results of permeability characteristics are different. At present, only uniaxial tension and biaxial tension are two types of domestic and foreign filling pipe bag deformation tests for research. It was not noticed that the permeability of the "soil-geotextile" system was affected by the deformation of warp and weft to different degrees. Therefore, it is necessary to further study the influence of unidirectional tensile of geotextile warp and weft yarns on the permeability of the system. To explore the influence of uniaxial tension in warp and weft on the permeability characteristics, the gradient ratio permeameter was used to conduct the permeability tests under the non-stretched coated sand state, the non-stretched coated sand state, and the tension in warp and weft of 3% coated sand state.

2 EXPERIMENTAL SETUP

2.1 *Experimental materials and apparatus*

Figure 1 shows the experimental apparatus.

One device used for the constant head permeability test is the gradient ratio permeability tester. A water storage chamber is above the fixed device at a fixed height to ensure the supply of a constant head. The main part of the gradient-specific permeability tester is divided into cylinders A (plexiglass cylinder with a 200 mm inner diameter, 200 mm height and 8mm thick has a flange at the top and bottom for connection) and B (plexiglass cylinder with a 200 mm inner diameter, 100 mm height and 8mm thick has a flange at the top and bottom for connection). A layer of foam tape is attached to the flange of A and B to ensure that the compressed foam tape and the geotextile are fully integrated during the test, effectively preventing water outflow and air entry, and making the test data more accurate. 200mm soil (Table 1 shows the soil gradation) material is filled in the gradient-specific permeability tester and the hydraulic gradient is set to $i = 6$. Four pressure measuring tubes with a spacing of 25mm are set on the upper cylinder of the gradient-specific permeability tester from the lower end. While collecting the soil particles passing through the fabric, the water outlet for measuring the water output is also arranged in the lower cylinder.

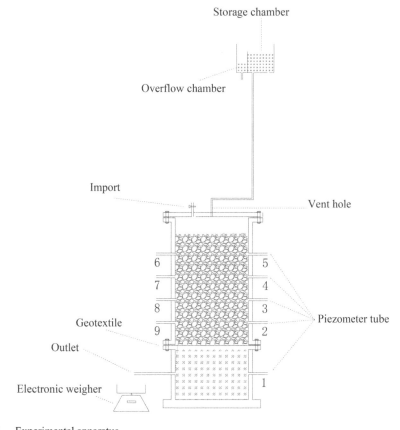

Figure 1.   Experimental apparatus.

Table 1.   Particulate parameters of soil samples.

Fine soil content	Characteristic grain size /mm			Coefficient of nonuniformity	Coefficient of curvature
	d_{10}	d_{30}	d_{60}		
10%	0.075	0.43	0.6	8.0	4.11

2.2 Testing procedure

The self-developed gradient-specific permeability instrument is used in the permeability test. The testing procedure is as follows:

(1) The geotextile cut into a rectangle is placed on the stretching machine for stretching according to a fixed rate and strain. After the tensile test, the excess geotextile is cut from the installed permeameter geotextile system.
(2) The soil prepared according to the fixed gradation is evenly filled in the removed gradient permeability meter.
(3) The water supply device is opened, and the water enters from the water outlet. By controlling the flow rate, the water can slowly soak the filled soil.
(4) When the water completely saturates the whole soil material, gradually increase the water head height and finally fix it in a fixed position. Soak for 12 hours before the test to exhaust the bubbles in the soil and reduce the adverse effects.
(5) After soaking, connect the water supply device with the water inlet, open the water outlet and start data collection. The data is tested every 15 minutes until the data is stable.
(6) Sand leakage is the data measured by drying and weighing the particles through the geotextile taken out after the test.

2.3 Testing scheme

Table 2 shows the specific experimental conditions of the tested samples.

The geotextile stretched along the meridional and latitudinal directions is placed in the infiltration device, clamped with a flange, and then the permeability test under the condition of not covered with sand and the permeability test under the condition of covered with sand is carried out respectively.

Table 2. Experimental conditions of the tested samples.

Sample No.	Test No.	Geotextile specification	Penetration test environment	unidirectional tension direction	Tensile strain
	1			unstrengthed	0%
I	2	$100g/m^2$	Without sand coating	warp	3%
	3			weft	3%
	4			unstrengthed	0%
II	5	$100g/m^2$	Sand coating	warp	3%
	6			weft	3%
	7			unstrengthed	0%
III	8	$120g/m^2$	Without sand coating	warp	3%
	9			weft	3%
	10			unstrengthed	0%
IV	11	$120g/m^2$	Sand coating	warp	3%
	12			weft	3%

3 DISCUSSION

Geotextile samples I, II, III, and IV are uniaxially stretched to 0% and 3% strain at the rate of 10 mm/min respectively. The constant head permeability tests of the stretched geotextile under sand-covered and no sand-covered conditions are carried out respectively.

3.1 Effect of unidirectional tension on permeability

Figures 2 and 3 show the seepage velocity of geotextile with or without sand coating.

Under the conditions without sand coating, the warp tension significantly reduces the seepage velocity of geotextile. Among them, Test 2 decreased by 67.6% and testing 8 decreased by 34.4%. On the contrary, the weft tension slightly increased the seepage velocity of the geotextile, and the penetration rates of Test 3 and Test 9 increased by 4.2% and 3.4% respectively. As described above, when the geotextile is pressed along the warp direction, the pore size of the geotextile became smaller, leading to the permeability rates decreased obviously. The gap of geotextile subjected to zonal force increases slightly, resulting in a slight increase in its penetration rate, which is consistent with the conclusions obtained by Man Xiao lei and others on the influence of geotextile weaving mode on the dehydration rate of filled pipe bag; When the geotextile is stretched along the warp direction to the same extent, the smaller the weight per unit area, the more severe the pore shrinkage of the geotextile, so the reduction of the penetration rate is greater. When the geotextile is stretched along the latitudinal direction, its pores will become larger, which will lead to a slight increase in the permeability rate of the stretched geotextile in the state without sand coating.

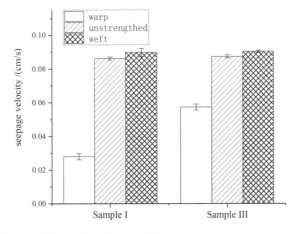

Figure 2.   Permeability rate of geotextile without sand coating.

Under the condition of sand coating, the warp tension reduces the seepage velocity of geotextile. Among them, Test 5 decreased by 32.0%, and Test 11 decreased by 27.1%. On the contrary, the weft tension increases the seepage velocity of the geotextile, and the penetration rates of Test 5 and Test 12 increased by 18.6% and 101.6%, respectively. When the geotextile is in the sand-covered

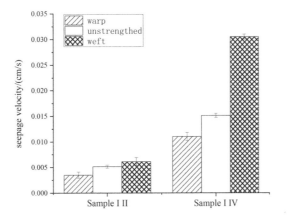

Figure 3.   Permeability rate of geotextile with sand coating.

state, the meridional tension will still reduce the porosity of the fabric, resulting in a decrease in its permeability rate. However, in the sand-covered state, the permeability rate of the geotextile with greater weight per unit area decreases more obviously after meridional tension; The tensile along the zonal direction makes the pores of the geotextile larger, which will lead to the obvious increase of permeability rate under the state of sand covering, and the greater the weight per unit area, the more obvious the increase.

Through comprehensive analysis of Figures 2 and 3, it can be obtained that the penetration rate of geotextiles of the same specification under the condition of not covered with sand and covered with sand is significantly different when they are subjected to unidirectional tension in the same direction and to the same degree. It is mainly reflected in the following three aspects:

(1) The penetration rate is reduced more obviously after the geotextile is stretched along the meridional direction, compared with the geotextile not stretched in the state of not covered with sand, it indicates that the flow rate of water will be reduced in the device by the sand body.
(2) Geotextiles are of these two specifications. After being stretched to the same extent along the meridional direction, whether in the state of uncoated sand or coated sand, the unit mass is increased and the permeability is enhanced; In the state of covered sand, the blocking effect of seepage is affected by the soil.
(3) The increasing proportion of penetration rate is much greater than that in the non-stretched state of the geotextile after the geotextile is stretched along the zonal direction. In the uncoated sand state, it shows that the influence of penetration rate can be approximately ignored. For the geotextile stretched along the zonal direction and not stretched in the uncoated sand state, it can not be ignored in the covered sand state.

3.2 Effect of unidirectional tension on soil conservation performance

Figure 4 shows the sand leakage of geotextile during penetration tests under the sand-covered state.

The soil retention performance of geotextile to filled sand sample can be directly reflected. Based on Figure 4, when Sample II and III are tensioned in the warp direction, the sand leakage decreases and increases in the weft direction. It can be seen that for the horizontal comparison in Figure 4, whether the geotextile is stretched or not, the sand leakage per unit mass is increased, the soil retention performance is reduced, and the most obvious performance is meridional stretching.

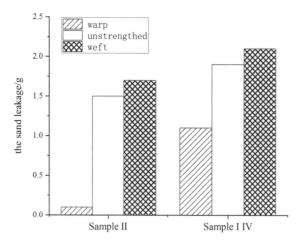

Figure 4.   Sand leakage in unidirectional tension under the sand condition.

3.3 Effect of unidirectional tension on anti-silting performance

The calculation formula of gradient ratio GR selected for this test is as follows:

$$GR = \frac{H_{1-2}}{25+\delta} \Big/ \frac{H_{2-3}}{25} \qquad (1)$$

where H_{1-2} is the water level difference between pressure measuring Pipe 1 and Pipe 2 (cm); H_{2-3} is the water level difference between piezometric Pipe 2 and Pipe 3; δ is the thickness of the geotextile (cm).

Figure 5 shows the variation law of gradient ratio with time in each group of tests in Condition 2 and Condition 4. It can be seen that each group increased with the passage of test time, and tended to be stable after 16 h.

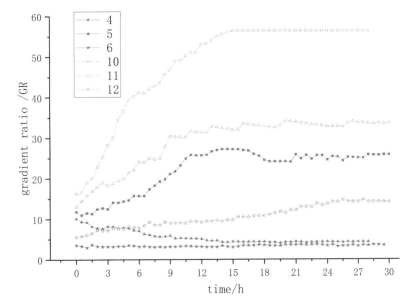

Figure 5.   Curve of gradient ratio with time.

Figure 6 shows the gradient ratio of Geotextile in the sand-covered state after seepage stability. It can be seen that the gradient ratio of geotextile stretched along with the warp increases, and the smaller the geotextile specification, the more obvious the increase is; the tensile along the zonal direction will reduce the gradient ratio of geotextile, and the greater the unit mass, the more obvious the reduction.

This shows that warp tension weakens the silt resistance of geotextiles, and the silt resistance of geotextiles is strengthened by weft tension. The effect of warp tension on the silt resistance of geotextiles is greater than that of weft tension. This is due to the manufacturing process. Warp tension makes the fabric more compact, reduces the effective aperture, and is more likely to cause deposition. The influence of geotextile thickness on its anti-deposition performance is that the thicker the geotextile is, the greater the gradient ratio is regardless of whether it is stretched or not.

The initial judgment is that geotextiles are seriously clogging. When the tensile rate is the same, the larger the geotextile specifications, the higher the gradient ratio. This can be compared through horizontal comparison in Figure 6.

To more objectively and comprehensively evaluate the siltation of the system composed of geotextile and discontinuous graded soil, the permeability coefficient of the system composed of geotextile and discontinuous graded soil is calculated and analyzed. Figure 7 shows the permeability coefficient of the geotextile sand covering system under uniaxial tension. The warp stretching of geotextile will reduce the permeability coefficient of the geotextile sand covering system, and the smaller the specification of geotextile, the more obvious the decrease. The weft stretching of geotextile increases the permeability coefficient of the geotextile sand covering system, and the greater the unit mass of geotextile, the more obvious the increase. The permeability coefficient of the system composed of geotextile and discontinuous graded soil in Condition 2 is greater than that in Condition 4 under non-tension and meridional tension, and the permeability coefficient under zonal tension is closer to that in Condition 4. Combined with the above data analysis, during the test, the small particles in the middle and upper layer of the soil gradually move down, resulting in the siltation of the geotextile. Under Condition 4, the equivalent pore diameter of the geotextile is

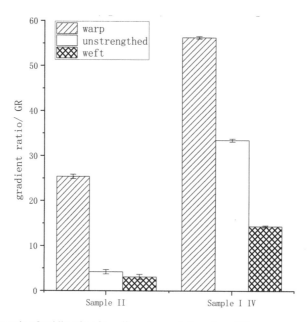

Figure 6.   Gradient ratio of unidirectional tensile under coated sand condition.

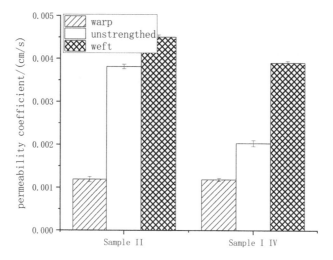

Figure 7.   Permeability coefficient of unidirectional tensile under the coated sand condition.

large, the amount of sand leakage is large, and a large amount of fine sand loss causes the soil to collapse and become dense. Therefore, the permeability coefficient of Condition 4 is smaller than that of Condition 2.

When the tensile strain of geotextile reaches 3%, the equivalent pore diameter of geotextile becomes smaller; When the zonal tensile strain reaches 3%, the equivalent pore size of geotextile increases, which is indirectly proved by the above four parameters.

4 CONCLUSION

Based on the results and discussions presented above, the conclusions are obtained as below:

(1) Different unidirectional tensions have different effects on all aspects of geotextiles. The parameters of permeability after unidirectional tension should be comprehensively analyzed. In practical application, reasonable measures should be taken according to the actual needs to achieve the optimal choice.
(2) The influence of geotextile specification on its water permeability, soil retention, and anti-silting performance can not be ignored. The warp tension weakens the water permeability and anti-silting performance of the textile fabric, and enhances the soil conservation performance; The weft tension enhances the water permeability and anti-silting performance of the woven fabric and weakens the soil retention performance.
(3) Due to the manufacturing process, the warp of woven geotextile is bent, and the warp tension leads to the compactness of the whole plane and the reduction of the equivalent aperture.

The above four parameters show that the equivalent pore size of spun geotextile decreases under meridional tension; The equivalent aperture becomes larger under zonal stretching.

ACKNOWLEDGMENTS

This research was funked by Chuzhou University Development Experiment Project (kfsy2138).

REFERENCES

Bai B., Effect of biaxial tensile strain on pore size change of nonwoven geotextile [J]. *Geotechnical mechanics*, 2015, 36 (06): 1615–1621 + 1626.

Cheng L., Stress and deformation analysis of geotextile bags in the process of mud filling [J]. *Journal of Hehai University* (Natural Science Edition), 2015,43 (01): 22–27.

Cho-Sen Wu The influence of uniaxial tensile strain on the pore size and filtration characteristics of geotextiles[J]. *Geotextiles and Geomembranes*, 2007, 26 (3): 250–262.

Ding J. H., Zhang W., Sun H. et al., a new test method for tensile properties of high-strength warp-knitted composite geotextiles [J]. *Journal of Yangtze University of Science*, 2019, 36(10): 175–179.

Ding Y., Effect of particle gradation and void ratio on permeability coefficient of coarse-grained soil [J]. *Hydrogeological engineering geology*, 2019, 46 (03): 108–116.

Lei G. D., Geotextile biaxial stretchable multifunctional permeability test device [J]. *Journal of Geotechnical Engineering*, 2016, 38 (S1): 119–124.

Li G. D., Experimental Study on Geotextile Filter Layer in Tailing Dam [J]. *Hydroelectric Energy Science*, 2021, 39 (03): 45–49.

Liu W. C., *Study on filling characteristics and calculation theory of geotextile bags* [D]. Zhejiang University, 2012.

Man X. L., *Effect of geotextile sewing method on dewatering performance of geotubes: an experimental study.* 2019, 7 (4): 205–213.

Man X. L., Effect of Sewn Methods of Bag Fabrics on Dehydration Performance of Tubular Bags [J]. *Journal of Three Gorges University* (Natural Science Edition), 2020, 42 (06): 6–11.

Man X. L., Experimental study on permeability of geotextile coated sand [J] *Journal of Heilongjiang Institute of Engineering*, 2021, 35 (05): 16–19.

Pang X. Z., Research on improvement of geotextile clogging test device and method [J]. *Journal of Yangtze Academy of Sciences*, 2019, 36 (03): 68–73.

Su L. J., Experimental study on permeability characteristics of sand with different particle sizes [J]. *geotechnics*, 2014, 35(05):1289–1294.

Tang G. H., Experimental study on the influence of geotextile on permeability of sandy soil [J]. *Journal of Beijing Normal University* (Natural Science Edition), 2021, 57(04): 563–570.

Tang L., Experimental study on the effect of uniaxial tension on the filtration performance of geotextiles [J]. *Journal of Geotechnical Engineering*, 2013, 35 (04): 785–788.

Wang S., Effect of biaxial stretching and flow conditions on the performance of filter system [J]. *Journal of Wuhan University* (Engineering Edition), 2016, 49 (02): 264–268.

Wei S M., Experimental study on infiltration and clogging of nonwoven geotextiles [J]. *Journal of Hefei University of Technology* (Natural Science Edition), 2021, 44(03): 383–388.

Wu G. Experimental study on permeability of woven geotextiles under soil condition [J]. Water science, technology and economy, *Journal of Geotechnical Engineering*, 2017, 39 (S1): 161–165.

Wu H M, High-efficiency dehydration technology of high-viscosity (powder) granular soil filling bag [J]. *Advances in water conservancy and hydropower technology.*, 2018, 38(01):19–27+35.

Yu W., Research on pore size change of nonwovens under uniaxial tension [J]. *Journal of Civil Engineering*, 2011, 44 (S2) : 9–12.

Zhang M. L., Stress - Seepage Coupling Model Test of Fill Pipe Bag [J]. *Advances in Water Conservancy and Hydropower Technology*, 2018, 38(03):48–53+65.

Zhuang Y. F., Study on the mechanism of reciprocating flow on the filtration system [J]. *Geotechnical mechanics*, 2008 (07): 1773–1777.

Advances in Measurement Technology, Disaster Prevention and Mitigation – Li & Mohd Yusof (Eds)
© 2023 The Author(s), ISBN: 978-1-032-36087-4

Three dimensional finite element analysis of a soft foundation sluice

Xiaona Li* & Yuchen Li*
School of Water Conservancy and Hydropower Engineering, Xi'an University of Technology, Xi'an, China

ABSTRACT: Sluice is widely used in hydraulic engineering. It is a low-head hydraulic structure using the gate to retain and discharge water. The working conditions of the sluice are similar to the overflow gravity dam on the rock foundation, but the water head is low, the flow is large, and it is mostly built on the soft-soil foundation in the plain area, so it has its characteristics in stability and settlement. Therefore, the design and calculation of sluice structures have been intensively studied globally. In this paper, three-dimensional finite element analysis is carried out on the whole structure of a soft foundation sluice, and the displacement and location of each structure are obtained by analyzing and calculating each sluice structure. Then the influence of different working conditions on the bending moment of the bottom plate is studied by finite element analysis. The single spring coupling element method is used to simulate the pile foundation of the interaction of piles and soil, the pile axial force, and the bending moment value of reinforcement to provide important reference data.

1 INTRODUCTION

In eastern China, sluices are usually built on soft foundations. The proposed sluice structure is usually located on the newly deposited unconsolidated soil with a deep silt layer. To ensure safety and reduce construction costs, it is necessary to explore how to control the decrease in the bending moment. In this paper, through three-dimensional finite element analysis (Fu 2003), the displacement of each part of the structure and its location are obtained. The influence of different working conditions on the bending moment value of the bottom plate is studied by the finite element method (Li & Wen 2002). Pile soil interaction problems at the same time, especially when the pile under horizontal load are more significant because the elastic modulus of soil is relatively soft, under the joint action of vertical load and horizontal load. Slipping phenomena may occur at this point, and adopting the form of contact elements to simulate the interaction of pile and soil problems is necessary. However, for some projects with a small horizontal load, the head difference of maximum value between upstream and downstream is small, so the main function of pile foundation is to improve the vertical bearing capacity of the soft soil foundation. The single spring connection element method (Yan et al. 2011; Zhao et al. 2008) is applied to the pile-soil interaction problem where vertical bearing capacity is the main force. The single spring connection element method is used to simulate the pile-soil interaction. The simulated results provide a reference for the reinforcement of pile foundation (Li et al. 2002).

2 ENGINEERING EXAMPLES

2.1 *Project overview*

A sluice for sea water control, which belongs to the first-line seawall sluice, sets tide, waterlogging, restricted navigation, water resources, and other functions in one control sluice (Zheng 2014).

---

*Corresponding Authors: lixjnj@163.com and 351866501@qq.com

2.2 Calculation model and working conditions

The finite element model of a soft foundation sluice takes the whole sluice structure, traffic bridge, and maintenance bridge as the calculation object, considers the interaction between the structure of the block base and foundation, and the roof structure above is transferred to the hoisting floor as load. The whole concrete structure with elevation from –4 m to 21 m and the foundation within a certain range is regarded as a whole structure, which is separated into eight-node hexahedral isoparametric elements. The bottom elevation of the foundation is –50 m, the elevation of the inner riverside fill is 4.5m, the elevation of the outer riverside fill is 5.5m, the width of the inner riverside is 24.5m, and the width of the outer riverside is 23m. The width of the levee on both sides of the empty container quay wall is simulated 20 m along the vertical flow direction. The overall finite element model of the sluice is shown in Figure 1, and the finite element model of the pile foundation is shown in Figure 2.

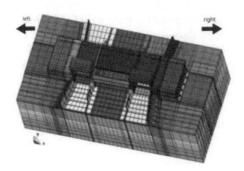

Figure 1.　Structural model of sluice.

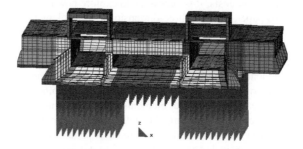

Figure 2.　Finite element model grid of pile foundation model.

For the calculation and analysis of the above models, the calculation conditions are the same, which are the completion of pile foundation construction conditions. The loads considered in the calculation process are structural dead weight, lateral earth pressure, side load, and seepage pressure.

3 ANALYSIS OF CALCULATION RESULTS

3.1 Analysis of displacement

In the finite element calculation, the displacement, strain, and stress of each node can be obtained according to the equilibrium equation and constitutive relation. According to the above calculation model and parameters, the static finite element calculation is carried out under the pile foundation construction condition, and the static load displacement of the sluice structure under the pile foundation construction condition is obtained. The maximum displacements of each specified position are shown in Table 1, where, and are the maximum displacements of vertical flow direction X, downstream flow direction Y, and vertical direction Z at the specified position respectively.

Table 1. Maximum displacement of each part of sluice structure (cm).

| location | direction | Pile foundation construction condition | |
		numerical	location
gate floor	U_x	−0.21	Close to the bottom of the right gate floor on the inland side
	U_y	−0.24	Near the outer riverside Middle top of gate bottom plate
	U_z	−8.32	Close to the bottom of the right gate floor on the outer riverside
Stilling tank wing wall	U_x	0.53	on the left side of the Inland river at the top of the Wing wall
	U_y	−0.99	Outer riverside right Top of stilling tank wing wall
	U_z	−7.19	Outer riverside right Bottom of stilling tank wing wall
Shore chamber wall	U_x	−4.85	Near the top of left bank wall on the outer riverside
	U_y	−0.89	To the right near the outer river at the top of the shore wall
	U_z	−12.50	Near the bottom of the left bank wall on the outer riverside
gate pier	U_x	−1.76	Near the outer river to the left Bottom of side pier
	U_y	−0.68	On the right side near the river Side pier pier wall top
	U_z	−6.27	Near the outer riverside Bottom of right pier
breast wall	U_x	0.47	Near the outer riverside On the left side of the bottom
	U_y	0.63	Close to the inland side On the right side of the top
	U_z	−1.09	Close to the inland side On the left side of the border
Traffic bridge	U_x	−1.40	Top of right gate bottom plate on outer riverside
	U_y	0.75	Around the seam of the right and middle sluice floor near the riverside
	U_z	−1.10	Inland riverside Traffic bridge middle bottom

According to the above calculation results, it can be seen that the maximum x-direction displacement is −4.85 cm, which occurs at the top of the left quay wall. The maximum Y displacement is −0.99 cm, which occurs at the bottom of the wing wall of the right damping pool on the outer riverside. The maximum z-direction displacement is −12.50 cm, which occurs when the bottom plate of the empty container shore wall is close to the causeway side.

3.2 Internal force analysis of gate floor structure

Finite element calculation is the displacement and stress on the node of the structure, and the design of structural reinforcement needs the internal force value of the structure. Therefore, in this paper, the sluicing structure cut off several sections, using the finite element method to solve the internal force, to find each section of the node constraint internal force value, because the need is the section of internal force rather than the node internal force, so the cross-section of the node constraint internal force synthesis of the axial force, shear and bending moment.

In X normal direction, the horizontal direction of the vertical river, three sections were selected, namely X5, X24, and X39 sections. The X5 section is located in the middle of the sluice gate floor and the parapet wall, which is used to analyze the internal force distribution of the middle sluice gate floor and the middle of the parapet wall. X24 section is located on the left side of the sluice gate floor section, and X39 is located on the right side of the sluice gate floor section. Select section Y94 for Y normal direction, along the river direction. Section Y94 is located in the section of traffic bridge, bank wall, gate pier, and gate bottom plate near the inland river, which is used to analyze the distribution of internal forces in these parts.

The internal force calculation is carried out under three working conditions: pile foundation construction, forward water-retaining, and reverse water retaining. The bending moment values of the section under various working conditions are shown in Table 2.

Table 2. Bending moment values at four sections under various working conditions (kN·m).

	X24			X5			X39			Y94		
	Left sluice gate floor	Middle sluice gate floor	Right sluice gate floor	Left sluice gate floor	Middle sluice gate floor	Right sluice gate floor	Left sluice gate floor	Middle sluice gate floor	Right sluice gate floor	Left sluice gate floor	Middle sluice gate floor	Right sluice gate floor
Pile foundation construction condition	−1192.3	—	—	—	−1175.6	—	—	—	−1243.1	−2831.7	−1746	−2925
Positive water blocking condition	−1149.2	—	—	—	−727.41	—	—	—	−1128.1	−2725.9	−1314.6	−2838.2
Reverse water retaining condition	−2558	—	—	—	−684.3	—	—	—	−921.08	−2302.1	−2144.4	−2340.7

It can be seen from Table 2 that the bending moment value of the Y94 section of the left sluice gate floor under the pile foundation construction condition is the largest, reaching 2831.7 kN·m. The bending moment value of the Y94 section of the middle sluice gate floor under reverse water retaining conditions is the largest, reaching 2144.4 kN·m. The bending moment value of the Y94 section of the right sluice gate floor is the maximum of 2925 kN·m under the pile foundation construction condition. According to the above results, it can be concluded that the bending moment of the bottom plate near the inland riverside is large under the construction condition of the pile foundation and the reverse water retaining condition, so the reinforcement near the inland riverside should be reinforced.

3.3 Pile foundation analysis

The overall deformation of the pile is the overall settlement. The piles at the left and right ends bend left and right due to the pressure from the upper part. The overall deformation is shown in Figure 3.

Figure 3.   Overall pile deformation diagram of pile foundation after construction.

To understand the internal forces on piles more accurately, the values of axial force and bending distance on typical piles under construction conditions are obtained by simulating pile-soil interaction with the single spring connection element method, as shown in Table 3.

Table 3.   Values of axial force and bending distance on typical piles under construction conditions.

Working condition	Pile no.	Pile axial force 10^6N	Maximum bending moment on pile kN.m (Mx)	Maximum bending moment on pile kN.m (My)	Pile no.	Pile axial force 10^6N	Maximum bending moment on pile kN.m (Mx)	Maximum bending moment on pile kN.m (My)
	1#	−2.28	−12.78	399.26	23#	−1.29	−15.94	−231.90
	2#	−1.30	−12.07	241.66	24#	−2.27	−16.833	−382.19
	3#	−0.96	9.30	228.92	25#	−0.18	10.16	31.46
	4#	−1.14	14.79	185.85	26#	−0.30	11.52	19.42
	5#	−0.99	20.10	148.79	27#	−0.27	12.78	9.37
	6#	−0.95	25.25	127.30	28#	−0.36	13.21	1.07
	7#	−0.93	29.62	108.55	29#	−0.44	13.00	−5.84
	8#	−0.95	25.25	127.30	30#	−0.42	12.21	−12.53
	9#	−0.76	33.94	75.61	31#	−0.34	10.72	−20.45
	10#	−0.78	35.96	61.61	32#	−0.29	9.29	−31.03
Pile foundation	11#	−0.74	38.02	51.39	33#	−2.11	84.52	−64.87
construction	12#	−2.50	43.16	47.61	34#	−0.71	22.12	−47.03
condition	13#	−2.27	43.67	−48.22	35#	−0.94	6.03	−50.12
	14#	−0.87	39.44	−50.48	36#	−0.72	−5.29	−46.44
	15#	−0.81	36.89	−59.88	37#	−0.85	−14.98	−49.34
	16#	−0.79	34.19	−72.86	38#	−0.53	−21.54	−45.44
	17#	−0.92	31.96	−86.78	39#	−0.77	−31.46	−50.00
	18#	−0.93	28.82	−101.27	40#	−0.66	−40.40	−48.67
	19#	−0.95	24.12	−117.06	41#	−0.69	−55.73	−49.17
	20#	−1.00	18.60	−136.97	42#	−0.70	−73.88	−49.06
	21#	−1.13	12.97	−172.98	43#	−0.90	−99.52	−50.69
	22#	−0.97	8.20	−214.87	44#	−2.33	−171.60	−56.42

The calculation results show that the displacement of the pile under static load is mainly caused by settlement. Under the construction condition of the pile foundation, the maximum axial force occurs on the boundary of the internal and external river gate bottom plate and the lower pile of the side gate bottom plate 1#, 12#, 13#, 24#, 33#, and 44#, and the maximum bending moment also occurs at this part. For the pile reinforcement of the above part, due to its large bending moment value, it should be properly considered in the process of reinforcement.

4 CONCLUSIONS

By establishing the calculation model of a sluice and considering all kinds of loads that may be generated during the completion of pile foundation construction, three-dimensional finite element analysis is carried out on the main structure of the sluice. At the same time, the date indicates that the bending moment of the bottom plate near the inland riverside is high, so the reinforcement should be encrypted. By simulating pile-soil interaction with the single spring connection element method, it is found that the axial force and bending moment on the boundary of the side sluice floor and the lower pile 1#, 12#, 13#, 24#, 33# and 44# of side sluice floor at the junction between middle sluice floor and side sluice floor are higher, which is not conducive to reinforcement. The maximum axial force and bending moment values and their locations provide data support for reinforcement.

REFERENCES

Fu, Y.H. (2003) *Fundamentals of finite element analysis*. Wuhan University Press, Wuhan

Li, C.C., Wang, X.M., Liu, X. (2002) Design method of reinforcement for pile foundation cap of high-rise building according to stress pattern. *Journal of Civil Engineering*, (02): 86–91.

Li, T.C., Wen, Z.W. (2002) finite element analysis method in stress analysis of arch dam *Journal of hydropower*, (04): 18–24

Yan, T.Y., Li, T.C., Zhao, L.H. (2011) improved algorithm of single spring connection element method *Journal of Three Gorges University* (NATURAL SCIENCE EDITION), 33 (06): (23–26 + 32)

Zhao, L.H., Li, T.C., Niu, Z.W. (2008) Single spring connection element method for bond slip problem of reinforced concrete. *Journal of Huazhong University of Science and Technology* (Urban Science Edition), 25(4):147–149.

Zheng, K.H. (2014) Influence of pile foundation layout on internal force of pump station gate floor. *Water conservancy and hydropower technology*, 45(07):30–32.

Author index

For Product Safety Concerns and Information please contact our
EU representative GPSR@taylorandfrancis.com Taylor & Francis
Verlag GmbH, Kaufingerstraße 24, 80331 München, Germany